ICME-13 Monographs

Series editor

Gabriele Kaiser, Faculty of Education, Didactics of Mathematics, Universität Hamburg, Hamburg, Germany

More information about this series at http://www.springer.com/series/15585

Gabriele Kaiser · Helen Forgasz
Mellony Graven · Alain Kuzniak
Elaine Simmt · Binyan Xu
Editors

Invited Lectures from the 13th International Congress on Mathematical Education

Springer Open

Editors
Gabriele Kaiser
University Hamburg
Hamburg
Germany

Alain Kuzniak
Université Paris Diderot
Paris
France

Helen Forgasz
Monash University
Clayton, VIC
Australia

Elaine Simmt
University of Alberta
Sherwood Park, AB
Canada

Mellony Graven
Rhodes University
Grahamstown
South Africa

Binyan Xu
East China Normal University
Shanghai
China

ISSN 2520-8322   ISSN 2520-8330   (electronic)
ICME-13 Monographs
ISBN 978-3-319-72169-9   ISBN 978-3-319-72170-5   (eBook)
https://doi.org/10.1007/978-3-319-72170-5

Library of Congress Control Number: 2017960201

Printed on acid-free paper

This Springer imprint is published by Springer Nature
The registered company is Springer International Publishing AG
The registered company address is: Gewerbestrasse 11, 6330 Cham, Switzerland

Preface

This book is an outcome of the 13th International Congress on Mathematical Education (ICME-13) that was held in Hamburg, Germany, from 24th to 31st July 2016. ICME-13 was hosted by the Gesellschaft für Didaktik der Mathematik (Society of Didactics of Mathematics), under the auspices of the International Commission on Mathematical Instruction (ICMI).

There were 3,486 participants at ICME-13, with 360 accompanying persons, making ICME-13 the largest ICME to date. Congress participants came from 105 countries, that is, more than half of the countries in the world were represented. Two hundred and fifty teachers attended additional activities during ICME-13.

The invited lectures (formerly known as regular lectures) are an important feature of the programme of the four-yearly ICME congress. These lectures are delivered by prominent researchers in mathematics education from different parts of the world. The International Programme Committee of ICME-13 issued the invitations to present, and the 64 invited lectures at ICME-13 covered a wide spectrum of topics, themes and issues.

Included in this volume are 44 of the 64 invited lectures from ICME-13. Not all presenters submitted papers for publication and all submissions were subjected to a strict peer-review process to insure high quality. The editors of this volume thank all reviewers for their work and Springer for providing language editing for selected contributions.

ICME-13 supported more than 223 scholars from less-affluent countries to enable them to participate in ICME-13. Consequently, this book is made available on open access to allow broad access to all mathematics education scholars across the developed and developing countries of the world.

We hope that this book will receive broad attention in the mathematics education community and that its contents will enrich international discussions on the issues raised.

Hamburg, Germany

Gabriele Kaiser
On behalf of the editors Helen Forgasz
Mellony Graven
Alain Kuzniak
Elaine Simmt
Binyan Xu

Contents

Chapter 1
Practice-Based Initial Teacher Education: Developing Inquiring Professionals

Glenda Anthony

Abstract Practice-based initial teacher education reforms are typically organised around a set of core teaching practices, a set of normative principles to guide teachers' judgement, and the knowledge needed to teach mathematics. Developing more than understandings, practices, and visions, practice-based pedagogies also need to support prospective teachers' emergent dispositions for teaching. Based on the premise that an inquiry stance is a key attribute of adaptive expertise and teacher professionalism this paper examines the function and value of inquiry within practice-based learning. Findings from the Learning the Work of Ambitious Mathematics Teaching project are used to illustrate how opportunities to engage in critical and collaborative reflective practices can contribute to prospective teachers' development of an inquiry-oriented stance. Exemplars of prospective teachers' inquiry processes in action—both within rehearsal activities and a classroom inquiry—highlight the potential value of practice-based opportunities to learn the work of teaching.

Keywords Teacher education · Practice-based · Rehearsals · Inquiry stance
Professionalism

1.1 Introduction

Initial teacher education (ITE) curricula and pedagogies reflect prevailing notions of classroom instruction at different moments in history within specific culturally ascribed educational systems. Current calls for reforms, designed to shift away from a perceived disconnect between university-based course work and practical experiences in the classroom, reflect the need to prepare teachers for the complex demands of teaching in 21st century schools. In some countries (e.g., Australia, New Zealand, United Kingdom, and United States) these reforms call for a

G. Anthony (✉)
Massey University, Palmerston North, New Zealand
e-mail: g.j.anthony@massey.ac.nz

© The Author(s) 2018
G. Kaiser et al. (eds.), *Invited Lectures from the 13th International Congress on Mathematical Education*, ICME-13 Monographs,
https://doi.org/10.1007/978-3-319-72170-5_1

reconfiguration of how teacher education is distributed between university and school sites. However, reforms are not without their critics. Researchers urge that we need to be careful that changes represent more than a pseudo-approach involving teacher candidates spending more time in clinical field placements (Zeichner 2012). Brown et al. (2015) argue that new partnerships require ITE programs to support prospective teachers in becoming more independent research-active teachers. However, in critiquing the move to school-based reforms in the UK, Meierdirk (2016) warns of the consequence concerning the "knowledge base that is needed for fruitful reflection is missing" (p. 376).

In New Zealand, the Ministry of Education has recently prioritised funding masters-level ITE programs that involve close collaboration between partner schools and universities *and* demonstrate a commitment to a teaching as inquiry stance (Aitken et al. 2013; Sinnema et al. 2017). In this paper, I draw on findings from a 3-year design-based study *Learning the Work of Ambitious Mathematics Teaching* (Anthony et al. 2015c) to argue that practice-based ITE reforms can support the development of an inquiry disposition:

> a way of knowing and being in the world of educational practice that carries across educational contexts and various points in one's professional career and that link individuals to larger groups and social movements intended to challenge the inequities perpetuated by the educational status quo. (Cochran-Smith and Lytle 2009, p. viii)

However, whilst an inquiry stance is increasingly advocated as a key attribute of professionalism associated with teacher adaptive expertise and continuous learning, little is currently known about ways to support its development within ITE settings (Parker et al. 2016). The intent of this paper is to argue for the potential of practice-based learning to afford opportunities for prospective teachers (PTs) to develop an inquiry stance. My discussion begins with an introduction to theoretical framings concerning inquiry, followed by an overview of practice-based pedagogies utilised in the Learning the Work of Ambitious Mathematics Teaching design phases. Vignettes from university in-class rehearsals, involving PTs practising core routines associated with ambitious mathematics teaching, serve to illustrate concurrent opportunities to model, practise, and engage in inquiry practices. Moving from the university to the school setting, I discuss PTs' experience of teaching instructional activities associated with rehearsals. PTs' perceptions of the challenges and their progress within the school setting serve to further illustrate how the use of inquiry practices can facilitate the development of an inquiry stance.

1.2 Inquiring Professionals

To be effective in preparing teachers for the complex demands of 21st century classrooms, PTs need opportunities to learn not only knowledge of content and students, and specific techniques and routines to manage that work, but also a vision of practice that can guide decision making, and dispositions that support student and teacher learning (Ghousseini and Herbst 2016). As Sinnema et al. (2017) note, "to teach well, and to improve their teaching, teachers need, in our view, to demonstrate their ability to inquire into that uncertainty in ways that address the particular complexities, conditions, and challenges they face" (p. 9). Informing the recommended ITE changes incorporating an inquiry stance in New Zealand, Sinnema et al. propose the adoption of six inquiry-oriented standards for teaching: inquiry in learning, teaching strategies, enactment of teaching strategies, impact of teaching, professional learning, and education systems. Each standard emphasises "high-quality teacher inquiry closely connected to learners' experience that draws on education's body of knowledge, competencies, dispositions, ethical principles, and commitment to social justice" (p. 12). For example, their proposed *Learning Priority Inquiry Standard* requires that teachers identify learning priorities for each student and be able to defend their decisions. Mediated by beliefs and commitments to social justice, defensible decisions must necessarily draw on a complex array of knowledge resources including knowledge about the learner, the discipline, and the community.

It is evident, that these inquiry-based standards pose significant challenges of judgements for the professional teacher. Positioned as agentic, the inquiring professional must decide on the learning priorities, decide on the teaching strategies, enact these strategies, and examine their impact in tandem with assessment of the relative merits of competing alternatives. In this sense, it is clear to see that being an inquiring professional is also an attribute associated with adaptive expertise (Aitken et al. 2013; Athanases et al. 2015)—a "gold standard for becoming a professional" (Hammerness et al. 2005, p. 360). Timperley (2013) described the adaptive teacher as one who is driven by a "moral imperative to promote the engagement, learning and well-being of each of their students" and who engages in "ongoing inquiry with the aim of building the knowledge that is the core of professionalism" (p. 5). As Lampert (2010) puts it, adaptive expertise enables teachers to "innovate when necessary, rethinking key ideas, practices, and values in order to respond to non-routine inputs" (p. 24). Focused on better learning for themselves and their students, adaptive teachers pursue the knowledge of why and under which conditions certain approaches have to be used or new approaches have to be devised.

Despite advocacy for adaptive expertise, little is currently known about beginning teachers' adaptive expertise capabilities and their associated development of an inquiry stance within ITE contexts (Anthony et al. 2015b; Athanases et al. 2015; Meierdirk 2016; Soslau 2012). Research on the nature and impact of PTs' reflective practice typically concerns field-based experiences (Körkkö et al. 2016), and more recently portfolio assessments (Toom et al. 2015).

Critiquing reflective practices in ITE, Ord and Nuttall (2016) argue that reflection should be accompanied by "close attention to the *embodied sensation* of learning … as a legitimate part of the content of learning to teach" (p. 361). Likewise, Thompson and Pascal (2012) argued that reflective learning needs to involve "more sociologically informed *critically* reflective practices" (p. 322) that take greater account of collaborative and emotional dimensions. They proposed that Schön's (1983) seminal constructs of reflection-in-action and reflection-on-action be expanded to include reflection-for-action: "the process of planning and thinking ahead about what is to come, so that we can draw on our experiences (and the professional knowledge base implicit within it) in order to make the best use of the time resources available" (p. 317). In this regard, Bronkhorst et al. (2011) argued that for meaning-orientated learning anticipatory reflection should "go beyond the planning of teaching and focus on why teaching should be done in a certain way" (p. 1128).

Despite these suggestions there remains considerable evidence that the potential of inquiry for professional learning is difficult to realise (Horn and Little 2010). Researching in New Zealand classrooms, Benade (2015) noted that the 'teaching as inquiry' model (Ministry of Education 2007) is frequently reinterpreted as an "instrumental formula for teachers to follow, with no requirement they examine their fundamental beliefs and assumptions" (p. 116). Moreover, the commonly reported practice of treating inquiry as a linear process with a fixed solution to a finite task constrains engagement in systematic and analytical examination of the tensions and problems teachers encounter. According to Lawton-Stickor and Bodamer (2016), genuine inquiry involves a "balance between constantly reflecting on and problematizing current structures and practices, and carrying out inquiry practices that seek to develop, and systematically explore questions that arise from reflection" (p. 395).

1.3   Inquiry Within Practice-Based Initial Teacher Education

In looking to support PTs learn how to *do* the complex practices of teaching as they relate to unpredictability and improvisation, teacher education researchers are increasingly exploring ways to avoid the dualism of theory and practice (Sinnema et al. 2017). In particular, ITE has witnessed a turn towards practice-based approaches that "view teaching not only as a resource for learning to teach but as a central element of learning to teach" (McDonald et al. 2014, p. 500). Grossman et al. (2009) proposed a framework for practice-based instruction that draws on three pedagogical approaches: representation of teaching (e.g., modelling, examining video or written case exemplars); decomposition of practice (e.g., focus on core/high–leverage practices); and approximation of practice (e.g., rehearsals). In combination, these approaches are used to occasion shifts in PTs' professional vision about teaching and support the development of productive dispositions, while simultaneously providing opportunities to learn the practices of ambitious teaching practices; practices that "position students' thinking and strategies as central means to drive learning forward" (Singer-Gabella et al. 2016, p. 412).

In mathematics education, research associated with the *Learning in, from, and for Teaching Practice* project (Lampert et al. 2013) provides us with what is arguably the most sustained study of practice-based ITE. This project is structured around Cycles of Enactment and Investigation involving PTs planning and teaching purposefully designed instructional activities that serve as containers of core practices, pedagogical tools, and principles of high-quality teaching. Teaching within rehearsals involves constructing experiences "around the critical tasks and problems that permeate teachers' daily work" (Ghousseini and Herbst 2016, p. 80). Within each rehearsal "the variations of the practice as it relates to particular students and mathematical goals" (Lampert et al. 2013, p. 238) highlight the complex relational and situated nature of teaching.

The pedagogy of rehearsals, involving modelling of practice, in-the-moment coaching and shared consideration of teaching moves and aspects of the rehearsal activity, supports collaborative inquiry in multiple ways. The cycles of enactment and investigation of deliberate practice provide a space for PTs to "open up their instructional decisions to one another and their instructor" (Kazemi et al. 2016, p. 20). For example, Lampert et al. (2013) analysis of 90 rehearsals across three ITE sites categorised teacher educator interactions as either involving directive or evaluative feedback, scaffolding enactment, or facilitating a reflective discussion of instructional decisions. The researchers noted that "discussions often entailed much work on the development of novices' judgement in adapting to the uncertainties of practice" (p. 234). In particular, feedback interactions within rehearsals that prompted PTs to reconsider and/or retry specific teaching moves enabled direct links to student outcomes related to learning a mathematical concept, offering an explanation, or developing feelings of competency. Developing an inquiry stance was also fostered through individual and collective accountability within the rehearsal process. For example, using a framework of Accountable Talk (Greeno 2002), Lampert et al. (2015) argued that the process of PTs making and defending assertions and interpretations of what they are observing and what they are doing within a rehearsal, provides an opportunity for teacher educators to actively position PTs as "authors and agents in developing knowledge of teaching" (p. 353).

1.4 Developing an Inquiry Stance Within Rehearsals

In this section, vignettes—in the form of sequences of exchanges within rehearsal scenarios from our 3-year design study *Learning the Work of Ambitious Mathematics Teaching* (Anthony et al. 2015c)—are used to illustrate the way that practice-based pedagogies can support the development of PTs' inquiry stance. Building on the work of Lampert et al. (2013), the project utilised pedagogies of practice associated with cycles of investigation and enactment of instructional activities in the form of rehearsal activities in the university and group teaching in classroom settings. The purpose of these activities was to provide opportunities for PTs to learn the work of ambitious mathematics pedagogy (Lampert 2010) through enactment of high-leverage practices. Practices identified as key to the principles and vision of ambitious mathematics teaching were those that placed students' mathematical thinking and reasoning at the centre of instruction, and supported equitable engagement of diverse learners in rich mathematical activity.

As part of the cycle of enactment and investigation, the teaching of instructional activities was rehearsed in the mathematics methods courses, and then with groups of students in school-based settings. In a rehearsal, the PT was responsible for teaching an instructional activity (e.g., Choral Count, Number String, Launching a Problem) to a group of peers acting as students, with the teacher educator acting as coach. These approximations of practice scenarios provided PTs with teaching and observational opportunities that involved controlled complexity and feedback from peers and teacher educators. Coaching, in the form of in-the-moment pauses by the teacher educator, was used to scaffold the learning of practice. This was achieved in multiple ways: stepping in and modelling aspects of practice; suggesting alternative moves to retry; prompting teacher or peer group reflection related to students' thinking, learning, and participation; asking for teacher explanation of teacher moves in order to highlight effective practice; or inputting a student response that the teacher has to address.

In the project, rehearsals conducted in the early stages of each course occasioned opportunities for PTs to attend to presentation and managerial skills (e.g., writing on the board and establishing pair-share routines). However, the focus quickly progressed to high-leverage routines associated with eliciting and responding to students' thinking. In learning to notice students' thinking, rehearsals facilitated a trajectory of practising to elicit students' thinking towards a consideration of how to elicit students' thinking in ways that enabled explanations to act as reflective tools for the learners. To illustrate, I zoom in on a rehearsal in which the eliciting process used by the teacher is extended from having peers engage with a particular response, towards using the response as a building block to further the discussion. We enter the rehearsal of a choral count, which involved counting in fives beginning from one (see Fig. 1.1), immediately after the rehearsing teacher (RT) records Robert's suggested pattern of "55 being added to each number" (pointing to diagonal numbers pairs):

RT: That's good. Does anyone have another pattern?
Coach: Pause. That's quite a complex idea and it might be one which you want to throw back to them and say does everyone agree? Like, "Let's look at what Robert said; he said that they increase by 55. Do you agree, why or why not"?

Fig. 1.1 Choral count pattern

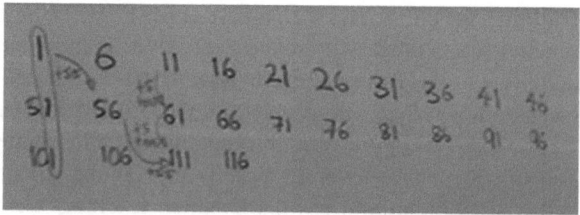

RT: Right, I would like you all to have a think about what Robert just shared with us because that is quite a complex idea, and think about what Cath said at the start about how she adds five, and somebody else said that when we are going down we are adding five tens, so think about that, adding five [pause]. *Oh I am giving it away aren't I?* Have a chat to your neighbour about how that works.

After the rehearsing students had talked for a few minutes, the rehearsing teacher asked them to share their ideas:

Megan: If you go across it is plus 5 and then going down is five tens so 5 times 10 is 50 so the 5 plus the 50 is 55 [RT notates the explanation].
RT: So that way is the same as those two? Is that what you are saying [notating the explanation with arrows]?
Megan: Yes you can add them together.
RT: Great.
Coach: Pause. You know you said *I am kind of giving it away* but what I think RT did was you really structured it so they could work out why that pattern was. If you had just said just look at it, with Year Fours they may not have seen it. You didn't say what you need to do is…, but you said look at that idea, and look at that idea, and that gave a foundation for them to then see that and use that, so that was a good thing to do.

In this vignette we see how the coach's suggested teacher move enabled the rehearsing teacher to trial a way to support students to engage with their peers' reasoning. Notably, the coach's feedback made reference to impact in terms of the how the learner was scaffolded to engage with the structural nature of the pattern. In this way, it served to draw attention to the importance of linking the teacher move to the opportunity to learn. This explicit shift from teaching to learning enabled PTs to access essential processes in their practice and become students of their students and learners of their own practice. This shift represents an important component of inquiry. As Hadar and Broady (2016) note, "when teachers explore their students' learning they adopt a different stance, placing themselves in the role of learners" (p. 102). This change in focus from self to student is also a signifier of developing adaptive expertise (Timperley 2013).

With experience of more rehearsals, the norms associated with engagement in sharing mathematical thinking shifted. The rehearsal students, placing themselves in the role of learners, became more willing to take risks, and in doing so they offered partial solutions, conjectures, or simulated student errors involving complex or incomplete explanations. This provided an opportunity for PTs to notice and learn how to use errors as an important resource. For example, in the following String activity involving a linked set of multiplication calculations the rehearsing teacher asked the students to solve 35×5:

RT: Would anyone like to share their answer?
Dan: One hundred and fifty-five.

RT: So Dan you think it is 155?

At this point, the rehearsing teacher, noticing the student error, paused indecisively, and the coach intervened:

Coach: Pause. This is a really good moment to say agree, disagree, not sure. Don't indicate what the answer is.
RT: So does everyone agree, disagree, or are you unsure about the answer?
Coach: And now you need to say remember if you agree or disagree you have to have a mathematical reason, but Dan may first want to say whether he agrees or disagrees with a mathematical reason.

Here the coach deliberately introduced an alternative to the 'agree/disagree' talk move that had not surfaced in earlier discussion—that of allowing the contributor to disagree with their own response, to change their mind and reconstruct their reasoning. As the rehearsal proceeds, Dan takes up this option as part of his role play:

RT: So Dan do you agree or disagree?
Dan: Yes, I disagree with my answer now.
RT: Do you have a new answer or would you like more time to think about it?
Coach: Well done.
Dan: One hundred and seventy five.
RT: And how did you get that answer?
Dan: For some reason what I originally did was that I knew that 30 times 5 was 150 and I don't know why but I just added 5.
RT: Because you saw another five there?
Dan: Yeah because I saw another five there and then when everyone disagreed I was wondering why. But then it clicked, so it is 5 times 5 and that is 25. So I know that 30 times five is 150 and I know that 5 times 5 is 25 because we did that before, so I just added 150 and 25 together to make 175.

In this vignette, we again see how the participants were able to experience the effects of a teacher move that provided additional thinking space for the student. The teacher's response meant that the student's erroneous thinking became a learning tool that supported reconstruction and justification of the reasoning, using mathematics as the authority. Learning to value students' erroneous thinking offers a direct challenge to many PTs' epistemological beliefs about the nature of mathematics and mathematics learning. PTs' willingness to question personal assumptions and beliefs is another example of an inquiry stance (Le Fevre et al. 2014).

In attending to students' thinking, a teacher also needs to be able to steer the discussion towards the important mathematical idea (Leatham et al. 2015). The following episode from a Choral Count rehearsal (see Fig. 1.2) illustrates how the coach explicitly surfaced the need to connect students' mathematical thinking to a mathematics point.

We enter the rehearsal with the rehearsing teacher eliciting different patterns, supported by revoicing, and press for elaboration of the solution strategies. Responding to a request to justify the claim that the pattern increased by eight, Mai noted:

Fig. 1.2 Choral count pattern

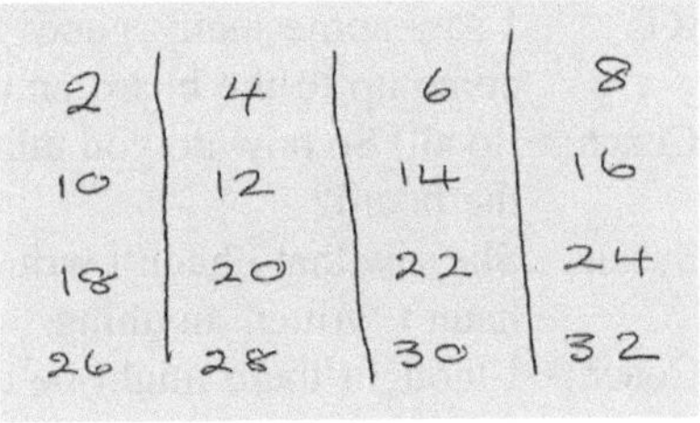

2	4	6	8
10	12	14	16
18	20	22	24
26	28	30	32

Mai: It was ten take away two.

RT: Okay, so you say ten take away two and that's eight [recording the calculation in the first column of the choral count].

Coach: Pause. Try to think at this point about getting other students to agree or disagree. You are getting some interesting patterns here.

RT: Okay does anyone disagree with Mai's observation there? What do you think Ben?

Ben: I can see the same thing.

RT: You can see the same thing, so you agree with Mai.

RT: What do you think Tui?

Tui: Yes, and the second row seems to be the same, like 28–20 is 8.

RT: So you see it in the second row as well [recording the calculation on the choral count].

C: Pause. So thinking about your questioning here, rather than just "do you agree or disagree", try a more structured approach. For example, taking what Mai said, you could have said, "Ben can you have a look at what Mai said and see if that works in the fourth column?"

Here we see the coach prompting the PTs to reflect on what might be the bigger picture in getting students to disagree or agree. Noting that the rehearsing teacher's immediate response was to attend only to Mai's single instance, the coach pressed the PTs to consider how they could use this opportunity to link the rehearsing student's thinking to the generalisation of the pattern across the rows. In effect, the coach engaged PTs in practice and reflection on how they could use talk moves to support students to "articulate a mathematical idea that is closely related to the student mathematics of the instance" (Leatham et al. 2015, p. 92).

These previous examples relate well to specific routines associated with professional noticing of students' thinking (see Anthony et al. 2015a), but could rehearsals also involve the development of an inquiry stance around issues of social justice? In supporting PTs to learn how to establish communities of mathematical inquiry (Alton-Lee et al. 2011) we wanted PTs to experience and experiment with ways to position students as competent and valued. In the next vignette we see how the coach's prompt to explain a teacher move surfaces a discussion on ways that teachers' formative assessment practices can be used to position students as 'achieving' within a class plenary session:

RT: I saw some really good work. Susan or Troy, please could one of you come up to the board and show us your thinking for the next two lines?

Coach: (to all) So how do you think RT made a decision about who to invite up to the board?

Susan: She saw that I hadn't written any of the work. I had contributed ideas but I hadn't written anything.

Coach: I thought there might be a strategic mathematical reason?

Troy: She recognised that we knew the strategy. She doesn't want us coming up if we are going to get it all wrong.

RT: That's part of it; with my Year 9 class I would have picked the weakest overall pair who got it right—they are the ones not used to being good at maths, so that was why. You were right, I had seen you got it right, but I gave you the choice of Susan or Troy.

Importantly, the ensuing discussion positioned the PTs within the activity as having valid opinions that are worth sharing—as authors and agents in developing knowledge of teaching (Lampert et al. 2015). But also the coach's response in pressing for alternative meanings modelled the expectation that PTs engage in practices that enable reflection as both a process and an outcome (Toom et al. 2015).

Within the New Zealand context, the drive towards realising the vision of Indigenous Māori students enjoying and achieving education success as Māori, demands the development of cultural competencies (Ministry of Education 2011) be central to an inquiry stance. While the instructional activities used in the research phase of the project did not incorporate explicit contextual contexts, Averill et al. (2015) makes the case that the enactment of the rehearsal activity, in itself, modelled culturally responsive pedagogy. In particular, the use of wānanga—participating with learners and communities in robust dialogue for the benefit of [Māori] learners—was evident in the co-construction of mathematical ideas through mathematical talk within the rehearsal and in the co-construction of knowledge for teaching within the PT/coach interactions around practice. For example, in the following rehearsal episode we see how wānanga was experienced through expectations for PTs to share, respect, and attend to multiple contributions from the PTs' learning community:

Coach: Is there a way to increase the proportion of learner talk? Talk in pairs about how to adapt what Michael has done to increase the amount of learner talk.

Student1: Asking others for similar ideas.

Student2: Pairs, then giving specific maths terms and asking them to discuss again in pairs using the terms.

Student3: Other ideas, like students making up their own example for everyone to do next.

Other cultural competencies such as whanaungatanga—engaging in respectful working relationships; manaakitanga—showing integrity, sincerity and respect towards Māori beliefs, language and culture; and ako—taking responsibility for their own learning and that of Māori learners, were embedded in the social norms associated with the rehearsal. The integration of these values within the community of learners meant that opportunities to take intellectual and social risks were readily adopted as a way of learning. As a PT noted in a post-rehearsal interview: "It was useful to see others at work, for one thing it was comforting to see others make mistakes, and to see we are all learners, even the lecturers".

1.5   Developing an Inquiry Stance in Classroom-Based Rehearsals

This section provides further exemplars of how practice-based pedagogies—this time associated with PTs' enactment and investigation cycle within a school—can support the development of an inquiry stance and associated adaptive expertise. Working in groups of four, the PTs were required to plan, teach, and review their teaching of a group of students aged 9–11 years over a six lesson sequence. Teaching a range of instructional activities afforded PTs opportunities to experience the relational demands associated with launching a problem, eliciting and responding to students' mathematical thinking, utilising a range of representations, connecting the big ideas in mathematics (Stein et al. 2008), and positioning students as competent (Boaler 2008).

Opportunities to engage in a more complex form of approximation of practice within a collaborative teaching inquiry supported the development of adaptive expertise—at least in an emergent sense (see Anthony et al. 2015b). In the process of working collaboratively to seek feedback to improve performance, PTs were afforded opportunities to develop metacognitive awareness about the value of an inquiry stance. For example, awareness of the collaborative aspect of learning through inquiry was evident in Chris' post school-lesson comment attributing learning as a function of their teamwork: "I think we have to think a lot about how we talk to children to get them to think, and that's definitely something that I need to work on—we actually did much better in the second visit."

Learning to work and learn within a group was challenging. However, many PTs expressed that, despite perceptions of intellectual and social risks, there were benefits. For example, Pip remarked early on in the teaching inquiry:

Even though it's a group and you're teaching and you're learning, you are getting videoed. So I feel that you are on show; that you're going to be critiqued. But as I've done one or two of the lessons you just get in and you just forget about that. My thoughts are that if you make mistakes, that's good. I'm here to learn, we're here to learn. [PI#1]

Moreover, Pip noted the value of evidence-based feedback from team members:

You don't know you do stuff, you think you are being an effective teacher, an equitable teacher but sometimes you're not. [PI#2]

Group and whole-class reviews of weekly teaching sessions helped PTs investigate thorny questions and "figure out what they do and do not yet understand about how their students are performing and what to do about it" (Hammerness et al. 2005, p. 377). These reviews surfaced many dilemmas of practice, especially in the early stages. As Chris noted, "probably the biggest thing was just the fact that a teacher is really a multi-tasker—there is just so much going on". Maximising the "public declaration of knowledge and information, and intrinsic goal setting" (Benade 2015, p. 111) supported discussions around anticipatory reflections. For example, in reviewing their video of the teaching episode Sandra noted:

> In our group we had one little girl who did it completely differently, like she was just adding on, like just counting all of them, so I think next time I would get her to repeat how one of the boys had done it, like 8 times 3, to start her thinking about other ways to do it. Like she explained her thing, but I didn't get her to repeat any other ways to get her thinking about it. [SJ#1]

To develop teacher agency and dampen the effects of enculturation into existing teaching modes, PTs were challenged to build theories of practice that bridged formal and everyday knowledge (Lampert 2010). Given repeated opportunities to experiment with teaching the instructional activities to the same group of students, PTs were pressed to evaluate what they were doing in relation to aspects of practice, the underlying principles of ambitious mathematics, and through explicit attention to student learning outcomes—a feature of developing expertise (Anthony et al. 2015b). For example, in gathering evidence of the interactions with and between students when working with groups, Troy remarked:

> Lots of kids come in with their ideas and lots of groups working well. I think they can take those ideas and use them. It's giving everyone a bit of expression; hopefully they can see themselves as more of a mathematician than they would have otherwise. [TJ#1]

However, through sharing and interpreting evidence, PTs also came to realise how their inquiry lenses were mediated by their personal histories, beliefs, and everyday practice theories (Fairbanks et al. 2010). For example, Pip, a PT who had struggled as a mathematics learner, was keenly focused on the impact of her teaching for diverse learners in terms of participatory practices. In attempting to resolve tensions between the research-based literature and her everyday knowledge of ability grouping structures, Pip was able to incorporate new evidence from her teaching inquiry:

> I can see that thinking about your groupings, not just letting the students randomly choose is a big part. I can see it being another way to change the perception that maths is only for those people with a maths brain ... and making this fun for everyone, it's not just for the bright and clever, it's for everybody. [PI#2]

Overall, there was a sense that these practice-based learning opportunities enabled PTs to appreciate that learning to manage uncertainty and develop confidence in one's improvisational capability is something that develops over time—not just with repeated practice, but with sustained professional inquiry into that practice. However, like others (e.g., Campbell and Elliot 2015; Kazemi and Wæge

2015), PTs in our study exhibited differing levels of commitment to, and confidence with, inquiry based practices. For some, willingness to take an agentic position towards improving practice appeared to be moderated somewhat by the authority of the status quo. For example, Chris near the end of his ITE, when asked whether he would like to continue to use rich group tasks responded:

> I think coming out as a new teacher it would be something that I would implement slowly … now that I have experienced this, I don't know if I would be confident to go into the classroom on the first day and go right so this is how we are doing maths. Maybe when I am comfortable in the teacher role it would definitely be something I would look at implementing one day a week to start with, then maybe two days a week. So just giving those problems out, and doing much like we done in the inquiry, creating that environment where the children are willing to discuss their thoughts and ideas. [I#2]

Rayna, in contrast, draws on her practice-based teaching experiences to argue that ambitious teaching is "doable":

> …it's not just something that people have researched and decided it works. It works, and it has benefits for everybody, like it's not just picking the mainstream and teaching to them or trying to extend them or help them, it actually works for everybody and I've seen the benefits myself so I can stand there on my own two feet and say "I've done it and it works". I think that is the biggest thing for me is that I can stand in a staffroom and say "well I've done it and it works".

1.6 Supporting Teaching Inquiry-Orientated Standards

It seems that these practice-based opportunities, designed to learn the complex work of teaching, can also be structured to develop PTs' disposition to inquire into their practice. In reviewing the preceding exemplars, it is evident that the practice-based opportunities within rehearsal cycles involving enactment of investigation can usefully contribute to the six teaching inquiry-orientated standards proposed by Sinnema et al. (2017): Learning priorities; Teaching strategies; Enactment of teaching strategies; Impact inquiry; Professional learning inquiry; and Education system.

Rehearsals were designed using instructional activities that afforded opportunities to inquire into the effects of particular instructional moves, that is, to "get deep enough into authentic interactions with specific learners to practice inventing educative responses" (Lampert et al. 2010, p. 135). I have provided examples of how, as part of this experimentation process, PTs were required to make defensible decisions on *learning priorities* for each of their learners and for those teaching strategies most likely to be successful for prioritised learning.

In *selecting and enacting teaching strategies*, PTs were expected to draw on education's body of knowledge, both theoretical and informal. The process of collaborative planning and public explication of theories of practice within reflection sessions also supported PTs to develop skills at anticipating the reactions and questions that students bring to a given topic, as well as how particular instructional strategies are likely to work. Moreover, opportunities to repeat rehearsal activities with different peer groups and different problems, including practice in how to adjust instructional activities for student learning needs, supported PTs' developing awareness of the situated nature of practice.

Central to the classroom inquiry was a focus on what happened and whether this made enough of a difference for learners. In examining the *impact of teaching* on each of their students, PTs were, in the first instance, able to draw on their experience as learners in the university-based rehearsal process. In particular, these early experiences of being a learner challenged PTs' expectations for providing explanations, sharing their thinking, and listening and learning from others. Moreover, discussion of these experiences surfaced issues of social justice related to socio-political positioning and participatory practices that framed explorations of impact for each of the students in the school-based settings.

Sinnema et al. (2017) describe the *Professional learning* inquiry as one that requires teachers to be metacognitive and self-regulated learners, as evidenced by "teachers increasingly becoming their own teachers and demonstrating the skills to learn from practice and also to learn for practice" (p. 10). Engagement in the classroom inquiry required that PTs identify their own learning needs as teachers in relation to impact. For example, Troy's journal entry noted the importance of team planning for individual student outcomes and anticipated next steps in their enactment of teaching strategies as follows:

> E [a student] is a very reluctant participant. We aim to encourage her participation by devising simpler problems and highlighting how her strategies/solutions relate to other more complex problems. C's [another student] change, in contrast, will be providing clear, accessible explanation of his strategies to his peers. [TJ#2]

Moreover, participatory norms that affirmed the entitlement and obligation for PTs to challenge information presented by the teacher educators fostered an attitude of open-mindedness. Being "open to alternative possibilities", being "willing to acknowledge that one's beliefs could be incomplete or misinformed" and engaging in "critical examination of evidence" (Le Fevre et al. 2014, p. 2) are key inquiry processes.

Sinnema et al. (2017) final inquiry standard—*Education system* inquiry—references the broader context of school, teaching, and learning. The standard emphasises the need for teachers to "participate in moving education-related debates forward and to contribute to system-wide improvements" (p. 10). As noted above, teacher educator efforts to model culturally inclusive pedagogies, combined with practice-based opportunities involving mathematical inquiry communities, went some way to challenge the hegemonic participatory practices associated with ability-based groupings in our schools (Anthony and Hunter 2017). Moreover, learning experienced as social and dialogical inquiry within communities of practice acknowledged that learning is integrally connected to worldly experiences and emotions. As Pip explained towards the end of her course:

I wasn't good at maths and knowing about the research about how teachers who are confident and have good attitudes about maths pass that on to their students, but doing maths how we've done it this way I feel more confident that I can go into the classroom. It's changed my attitude about how I feel about myself. Being able to facilitate discussion and bringing children's thinking out has been a really important part for my learning. [PI#2]

In grappling with the inherently situated, relational, and practical nature of teaching, it appeared as if PTs' practice-based experiences of teaching—of coming to know about teaching—existed "in relation to themselves, others, and contexts of time, space, and resources" (Ord and Nuttall 2016, p. 359). Potentially, these experiences of learning to construct and analyse practice with peers could lay the foundation for participation in collegial teacher inquiry as an ongoing part of professional and career development.

1.7 Challenges and Implications Going Forward

Designing and enacting practice-based activities are based on the belief that learning the work of teaching cannot be separated from its enactment; that is, teachers do not learn new things and then learn how to implement them. Exploring the function and value of inquiry in practice-based teaching, I argue that inquiry must be regarded both as a process and product. That is, in supporting PTs' development *of an* inquiry stance, it is imperative that PTs *engage* in critical and collaborative reflective practices, including reflections on, in, and for practice.

Exemplars from the *Learning the Work of Ambitious Teaching* project have shown how practice-based activities can occasion PTs learning of attributes of professionalism associated with inquiry, collective responsibility, and knowledge co-construction—attributes that signify adaptive expertise. Going forward, such expertise is crucial for mathematics teachers to "do teaching that is more socially and intellectually ambitious than the current norm" (Lampert et al. 2013 p. 241). However, in shaping this proficiency, I argue that it is imperative that teacher educators explicitly attend to the development of inquiry stance. For, without explicit attention to the development of an inquiry stance we run the risk of PTs learning a toolbox of core practices that are 'nice to know' but difficult to implement in the 'real' classroom setting. Moreover, in claiming that teacher inquiry in practice-based settings supports continuous learning and improvement, we need to be wary of pseudo-practice-based reforms that do little more than increase the amount of time spent in schools. In particular, we need to ensure that PTs have access to the full resource set of: education's body of knowledge; cultural, technical and relational competencies; dispositions; ethical principles; and commitment to social justice (Sinnema et al. 2017). Without appropriate access to this resource set the enactment of reflective practice would surely be in a technical sense rather than a critical sense (Meierdirk 2016).

These conjectures are based on my own and colleagues' emergent experiences of practice-based ITE. The challenge of how successfully we have supported PTs to examine in a critical way their fundamental beliefs and assumptions and develop an inquiry stance remains real. To develop courageous teachers who are willing to share their reflective thoughts with colleagues, invite feedback, question their own practice, and commit to change, requires that we all commit to the collaborative community of practice. Without such commitment, the preparation of teachers who can survive outside of the previously privatised practice that 21st century learning is focused on eradicating is less certain. This work will undoubtedly require ongoing theorisation of the concept of inquiry, and its relationship to adaptive expertise, particularly as it applies within practice-based teacher education.

Acknowledgements This work was supported by the Teaching and Learning Research Initiative fund administered by the New Zealand Council of Educational Research. I am indebted to my colleagues in the *Learning the Work of Ambitious Mathematics Teaching* project at Massey University (Roberta Hunter, Jodie Hunter, and Peter Rawlins) and Victoria University of Wellington (Robin Averill, Dayle Anderson, and Michael Drake). Our collaborative partnership has contributed to many of the ideas described in this paper.

References

Aitken, G., Sinnema, C., & Meyer, F. (2013). *Initial teacher education outcomes: Standards for graduating teachers*. Auckland: University of Auckland.

Alton-Lee, A., Hunter, R., Sinnema, C., & Pulegatoa-Diggins, C. (2011). *BES Exemplar1: Developing communities of mathematical inquiry*. Retrieved from http://www.educationcounts. govt.nz/goto/BES.

Anthony, G., & Hunter, R. (2017). Grouping practices in New Zealand mathematics classrooms: Where are we at and where should we be? *New Zealand Journal of Educational Studies*, 1–20. https://doi.org/10.1007/s40841-016-0054-z.

Anthony, G., Hunter, J., & Hunter, R. (2015a). Learning to professionally notice students' mathematical thinking through rehearsal activities. *Mathematics Teacher Education & Development, 17*(2), 7–24.

Anthony, G., Hunter, J., & Hunter, R. (2015b). Prospective teachers' development of adaptive expertise. *Teaching and Teacher Education, 49*, 108–117.

Anthony, G., Hunter, R., Anderson, D., Averill, R., Drake, M., Hunter, J., & Rawlins, P. (2015). *Learning the work of ambitious mathematics teaching: TLRI report*. Wellington: New Zealand Council of Educational Research.

Athanases, S. Z., Bennett, L. H., & Wahleithner, J. M. (2015). Adaptive teaching for English language arts: Following the pathway of classroom data in preservice teacher inquiry. *Journal of Literacy Research, 47*(1), 83–114.

Averill, R., Anderson, D., & Drake, M. (2015). Developing culturally responsive teaching through professional noticing within teacher educator modelling. *Mathematics Teacher Education and Development, 17*(2), 64–83.

Benade, L. (2015). Teaching as inquiry: Well intentioned, but fundamentally flawed. *New Zealand Journal of Educational Studies, 50*(1), 107–120.

Boaler, J. (2008). Promoting 'relational equity' and high mathematics achievement through an innovative mixed-ability approach. *British Educational Research Journal, 34*(2), 167–194.

Bronkhorst, L., Meijer, P., Koster, B., & Vermunt, J. (2011). Fostering meaning-oriented learning and deliberate practice in teacher education. *Teaching and Teacher Education, 27*, 1120–1130.

Brown, T., Rowley, H., & Smith, K. (2015). *The beginnings of school led teacher training: New challenges for university teacher education*. Manchester: Manchester Metropolitan University.

Campbell, M. P., & Elliott, R. (2015). Designing approximations of practice and conceptualising responsive and practice-focused secondary mathematics teacher education. *Mathematics Teacher Education and Development, 17*(2), 146–164.

Cochran-Smith, M., & Lytle, S. L. (2009). *Inquiry as stance: Practitioner research for the next generation*. New York: Teachers College Press.

Fairbanks, C. M., Duffy, G. G., Faircloth, B. S., He, Y., Levin, B., Rohr, J., & Stein, C. (2010). Beyond knowledge: Exploring why some teachers are more thoughtfully adaptive than others. *Journal of Teacher Education, 61*(1–2), 161–171.

Ghousseini, H., & Herbst, P. (2016). Pedagogies of practice and opportunities to learn about classroom mathematics discussions. *Journal of Mathematics Teacher Education, 19*(1), 79–103.

Greeno, J. (2002). *Students with competence, authority, and accountability: Affording intellective identities in classrooms*. New York: The College Board.

Grossman, P., Hammerness, K., & McDonald, M. (2009). Redefining teaching, re-imagining teacher education. *Teachers and Teaching: Theory and Practice, 15*(2), 273–289.

Hadar, L. L., & Brody, D. L. (2016). Talk about student learning: Promoting professional growth among teacher educators. *Teaching and Teacher Education, 59*, 101–114.

Hammerness, K., Darling-Hammond, L., & Bransford, J. (2005). How teachers learn and develop. In L. Darling-Hammond & J. Bransford (Eds.), *Preparing teachers for a changing world* (pp. 358–389). San Francisco: Jossey-Bass.

Horn, I., & Little, J., W. (2010). Attending to problems of practice: Routines and resources for professional learning in teachers' workplace interactions. *American Educational Research Journal, 47*(1), 181–217.

Kazemi, E., Ghousseini, H., Cunard, A., & Turrou, A. C. (2016). Getting inside rehearsals: Insights from teacher educators to support work on complex practice. *Journal of Teacher Education, 67*(1), 18–31.

Kazemi, E., & Wæge, K. (2015). Learning to teach within practice-based methods courses. *Mathematics Teacher Education and Development, 17*(2), 125–145.

Körkkö, M., Kyrö-Ämmälä, O., & Turunen, T. (2016). Professional development through reflection in teacher education. *Teaching and Teacher Education, 55*, 198–206.

Lampert, M. (2010). Learning teaching in, from, and for practice: What do we mean? *Journal of Teacher Education, 61*(1–2), 21–34.

Lampert, M., Beasley, H., Ghousseini, H., Kazemi, E., & Franke, M. (2010). Using designed instructional activities to enable novices to manage ambitious mathematics teaching. In M. K. Stein & L. Kucan (Eds.), *Instructional explanations in the disciplines* (pp. 129–141). New York: Springer.

Lampert, M., Franke, M. L., Kazemi, E., Ghousseini, H., Turrou, A. C., Beasley, H., … Crowe, K. (2013). Keeping it complex: Using rehearsals to support novice teacher learning of ambitious teaching. *Journal of Teacher Education, 64*(3), 226–243.

Lampert, M., Ghousseini, H., & Beasley, H. (2015). Positioning novice teachers as agents in learning teaching. In L. Resnick, C. Asterhan, & S. Clarke (Eds.), *Socializing intelligence through academic talk and dialogue* (pp. 363–374). Washington, DC: American Educational Research Association.

Lawton-Sticklor, N., & Bodamer, S. F. (2016). Learning to take an inquiry stance in teacher research: An exploration of unstructured thought-partner spaces. *The Educational Forum, 80* (4), 394–406.

Leatham, K., Peterson, B., Stockero, S., & Zoest, L. (2015). Conceptualizing mathematically significant pedagogical opportunities to build on student thinking. *Journal for Research in Mathematics Education, 46*(1), 88–124.

Le Fevre, D. M., Robinson, V. M., & Sinnema, C. (2014). Genuine inquiry: Widely espoused yet rarely enacted. *Educational Management Administration & Leadership, 43*(6), 883–899.

McDonald, M., Kazemi, E., Kelley-Petersen, M., Mikolasy, K., Thompson, J., Valencia, S. W., & Windschitl, M. (2014). Practice makes practice: Learning to teach in teacher education. *Peabody Journal of Education, 89*(4), 500–515.

Meierdirk, C. (2016). Is reflective practice an essential component of becoming a professional teacher? *Reflective Practice, 17*(3), 369–378.

Ministry of Education. (2007). *The New Zealand curriculum*. Wellington: Learning Media.

Ministry of Education. (2011). *Tātaiako: Cultural competencies for teachers of Māori learners*. Wellington, New Zealand: Ministry of Education.

Ord, K., & Nuttall, J. (2016). Bodies of knowledge: The concept of embodiment as an alternative to theory/practice debates in the preparation of teachers. *Teaching and Teacher Education, 60*, 355–362.

Parker, A., Bush, A., & Yendol-Hoppey, D. (2016). Understanding teacher candidates' engagement with inquiry-based professional development: A continuum of responses and needs. *The New Educator, 12*(3), 221–242.

Schön, D. A. (1983). *The reflective practitioner: How professionals think in action*. London: Temple Smith.

Singer-Gabella, M., Stengel, B., Shahan, E., & Kim, M.-J. (2016). Learning to leverage student thinking: What novice approximations teach us about ambitious practice. *The Elementary School Journal, 116*(3), 411–436.

Sinnema, C., Meyer, F., & Aitken, G. (2017). Capturing the complex, situated, and active nature of teaching through inquiry-oriented standards for teaching. *Journal of Teacher Education, 68*(1), 9–27.

Soslau, E. (2012). Opportunities to develop adaptive teaching expertise during supervisory conferences. *Teaching and Teacher Education, 28*(5), 768–779.

Stein, M. K., Engle, R. A., Smith, M. S., & Hughes, E. K. (2008). Orchestrating productive mathematical discussions: Five practices for helping teachers move beyond show and tell. *Mathematical Thinking & Learning, 10*(4), 313–340.

Thompson, N., & Pascal, J. (2012). Developing critically reflective practice. *Reflective Practice, 13*(2), 311–325.

Timperley, H. (2013). *Learning to practise: A paper for discussion*. www.educationcounts.govt.nz/publications/schooling/2511/learning-to-practise.

Toom, A., Husu, J., & Patrikainen, S. (2015). Student teachers' patterns of reflection in the context of teaching practice. *European Journal of Teacher Education, 38*(3), 320–340.

Zeichner, K. (2012). The turn once again toward practice-based teacher education. *Journal of Teacher Education, 63*(5), 376–382.

Chapter 2
Mathematical Experiments—An Ideal First Step into Mathematics

Albrecht Beutelspacher

Abstract Since the foundation of the Mathematikum, Germany, in 2002 and Il Giardino di Archimede, Florence, Italy, in 2004 there have been many activities around the world to present mathematical experiments in exhibitions and museums. Although these activities are all very successful with respect to their number of visitors, the question arises what is their impact for "learning" mathematics in a broad sense. This question is discussed in the paper. We present a few experiments from the Mathematikum and shall then discuss the questions, as to whether these are experiments and whether they show mathematics. The conclusion will be that experiments provide an optimal first step into mathematics. This means in particular that they do not offer the whole depth of mathematical reasoning, but let the visitors experience real mathematics, insofar as they provide insight by thinking.

Keywords Mathematical experiments · Science centers · Learning by experience
Mathematics in leisure time

In the last years, quite a few mathematical exhibitions have been developed and mathematical museums ("science centers") have been opened. In these, mathematics is typically not presented in the traditional way using the mathematical language. On the contrary: visitors find "exhibits", in which they may see or explore mathematics. In other words, visitors are challenged to perform "mathematical experiments". In addition, also several books with easy-to-perform experiments have been published, which aim at teachers, students or the general public.

In this article we look at mathematical experiments, and investigate their potential for formal and informal learning of mathematics. The basic reason for the success of science centers in general is expressed in the slogan "hands-on, minds-on, hearts-on". In other words, in performing the experiments, visitors get

A. Beutelspacher (✉)
Giessen, Germany
e-mail: albrecht.beutelspacher@mathematikum.de

experience. This experience leads to understanding, and understanding gives pleasure.

2.1   Mathematical Experiments and Science Centers

Probably the first man-made experiments are due to the time of Galilei (for instance experiments with pendula). In mathematics, models and instruments became important in the 19th century. The book of Dyck (1892) shows an impressive collection of mathematical models, apparatuses and instruments.

Some mathematical experiments have been known for a long time, mostly under the name of "mathematical games". Famous games are for instance Hamiltons's Icosian Game (1857), the "Tower of Hanoi" (Lucas 1883), and the Soma cube (Hein 1934).

The first initiative to collect and develop mathematical experiments as such was undertaken by the Italian professors Franco Conti and Enrico Giusti, who very successfully developed and organized the exhibition "Oltre iI compasso—the mathematics of curves", which was first shown in 1992. Since 2004 it has been enlarged to form the "Giardino di Archimede" in Florence. Nearly at the same time, the first step towards the Mathematikum was taken: in 1994 the first German exhibition under the name "hands-on mathematics" ("Mathematik zum Anfassen") was shown in Giessen, Germany. This exhibition was a work of a group of students, who organized this exhibition as a follow-up of a mathematical seminar. Mathematikum, the world's first mathematical science center, was opened in 2002. Since then, quite a few institutions of different size followed these ideas, for instance "Adventure Land Mathematics" in Dresden, "MoMath" in New York, and "Maison des Maths" in Mons, Belgium.

The idea of all these institutions is basically that the combination "interactive exhibits and visitors" works. It is fascinating to observe that in all science centers visitors start working, without a guide, without a teacher, even without reading the label, and have lots of fun. In most science centers, certainly in all mathematical science centers, the responsible people take science serious. "Fun" should not arise from strange colors, noise, fog and so on, but from insight into the phenomena. Looking at the visitors, we see experience, understanding and pleasure. In the science center-terminology: hands-on, minds-on, hearts-on.

2.2   Mathematikum Giessen

The Mathematikum in Giessen, Germany (near Frankfurt) is a mathematical science centre founded in 2002. It aims to make mathematics accessible to as many people as possible, in particular to young people. On its 1200 m^2 exhibition area it shows about 180 interactive exhibits. From the very beginning, it was a great success.

Between 120,000 and 150,000 people visit the Mathematikum each year. About 40% are group visitors, mainly school classes, 60% are private visitors, mainly families.

Visitors like the Mathematikum. In particular they like the way mathematics is presented. They are entertained by performing the experiments and trying to understand what they have experienced. The Mathematikum is a house full of communication. When one listens to what people are talking about, one notices that it is always about the exhibits.

The permanent exhibition of Mathematikum is complemented by several other formats, which address different target groups.

- *Temporary exhibitions* on special topics, such as randomness, calculating devices, mathematics in everyday life, mathematical games, etc.
- *Popular lectures* on special topics such as cryptography, astronomy, etc.
- *Lectures for children* on topics as, for instance, mathematics and—the bicycle, the bees, the heaven, the kitchen, the Christmas tree, and so on.

2.3   Some Experiments

The experiments in Mathematikum cover many mathematical disciplines, such as geometry (shapes and patterns), arithmetic (numbers and calculating), calculus (functions), probability (randomness and statistics), algorithms, and history of mathematics. No mathematical discipline is generally excluded.

We shorty describe some exhibits; more can be found in Beutelspacher (2015).

Figure 2.1 shows an invention of John H. Conway. It is a puzzle consisting of three small cubes of side length 1 and six $2 \times 2 \times 1$-cuboids, which should be

Fig. 2.1 Conway's cube

Fig. 2.2 Lights on!

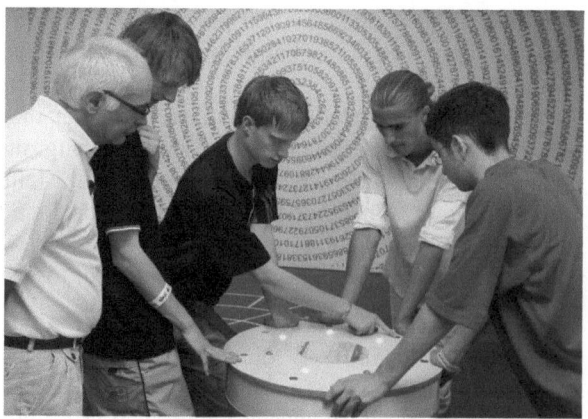

assembled to form a cube. One first calculates how big the cube will be. Even with this knowledge, most people struggle—until they get the idea where to locate the small cubes in the big cube.

In Fig. 2.2 we see seven lamps in a circle. To each lamp a switch is attached. When trying the switches, one notes that each switch affects three lamps, precisely the lamp is attached to the switch and the lamps on the right hand side and on the left hand side of the switch. When activating a switch, the status of these three lamps changes: those which have been off, are on now, and those which have been on, are off now.

The task, which is already included in the title of the experiment, is to put all lamps on.

Fig. 2.3 Tetrahedron in the cube

Many people start by randomly pressing the switches. Also in this way, we arrive at situations which are promising. For instance, if four lamps in a row are on, then it is easy to switch on the remaining lamps. Also, if only one lamp is on, one has a promising situation. By pushing one switch one gets four enlightened lamps in a row and one can proceed to finish as above.

The experiment shown in Fig. 2.3 consists of two parts.This experiment consists of two parts. One part is a cube made of glass with its upper face removed. The other part is a rather big tetrahedron which is supposed to be put inside of the cube. Most likely, first attempts will fail. Describing failed attempts, one gets an idea of how to succeed. If one holds the tetrahedron so that one vertex points downwards, it won't work. Also, if one vertex points upwards (and its face downwards), it will not work. Now, one could think of trying to let an edge point downwards. In fact, putting edge on a diagonal of the cube's upper square the tetrahedron automatically slides inside.

The experiment shown in Fig. 2.4 provides a challenging task. There is a poster showing a pattern of equilateral triangles. Following the task, one has to hold a framed irregular triangle in-between the lamp and the poster. Of course, we see a shadow. Moreover, the shadow is an irregular triangle. The task is now to put the triangle in a position so that its shadow perfectly fits onto one of the smaller equilateral triangles. For this, one has to move the triangle; back and forth, rotating in all possible ways. Eventually, the perfect shadow is found.

Fig. 2.4 All triangles are equal

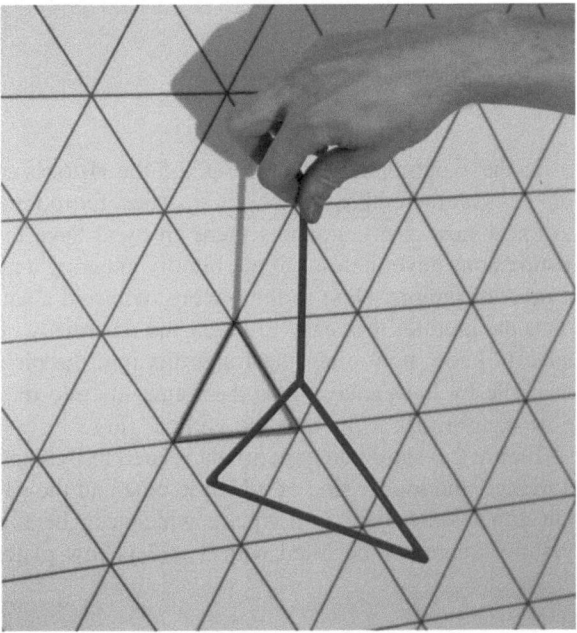

Fig. 2.5 The smarties

Fig. 2.6 Pythagoras puzzle

In the experiment shown in Fig. 2.5 the vistor is confronted with a poster, where one sees an incredible number of smarties, far too many to be counted. If you want to know how many smarties there are, you have to rely on estimation strategies. Estimations nevertheless are not blindly guessing a number but using the method of a random sample. Next to the picture, you find a square frame. Holding the frame onto the picture, it is easy to count the smarties within the frame. Now, you only have to know how often the frame fits into the picture. You find this number for example by how many times the frame fits into the upper side of the picture and how many times it fits into the vertical side.

Figure 2.6 shows an experiment related to Pythagoras' theorem. In front, there is a triangle the longer side of which is blue and the shorter sides are red and yellow. On either side, there is a square which can be filled with coloured plates. The yellow square can be filled with 3×3 yellow plates, the red square can be filled

Fig. 2.7 Two in a row

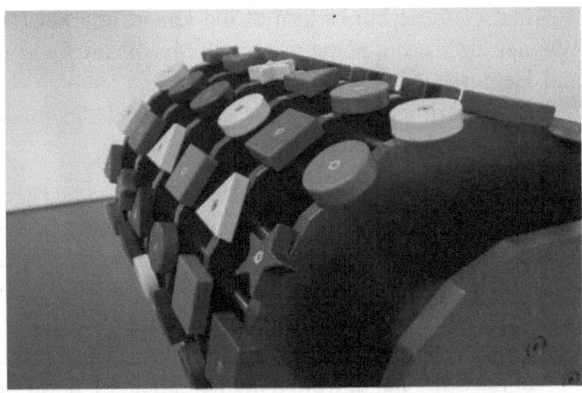

with 4 × 4 red plates so that all plates are used. By turning these 9 + 16 plates, the 25 blue plates perfectly fit into the 5 × 5 square above the triangle's blue side.

This experiment illustrates the Pythagorean theorem which states that in a rectangular triangle, the size of the legs' squares ($a^2 + b^2$) equals the size of the hypotenuse's square (c^2). In short, $a^2 + b^2 = c^2$.

In the experiment shown in Fig. 2.7 six wheels invite us to turn them. Each wheel has colorful pieces on it, which vary in form and color. Four different shapes (triangle, square, star, and circle) occur in four different colors, so that we have in total 16 symbols. Each wheel is adorned with these 16 symbols in some random order.

Now we let the wheels rotate. The wheels come to a standstill at random positions. The question is, whether "by accident" two equal symbols (shape and color identical) are at the same line.

Naively, we would conjecture that this will be a rare event, since it is no problem at all to put the six wheels in a position where no equal symbols are at the line. But when performing the experiment, we often see the bewildering situation that two equal symbols are in the same row.

2.4 Books and Easy-to-Built Experiments

In recent years, quite a few books have appeared which contain experiments with cheap material. Many of them are based on paper folding, assembling objects with sticks, and so on. Classical books on this subject are van Delft and Botermans (1978), and, on a higher level, Cundy and Rollett (1952), and Wenninger (1974).

Most of these books aim at the leisure market (for instance Beutelspacher and Wagner 2008), but some are explicitly meant for teachers (e.g. Schmitt-Hartmann and Herget 2013).

The experiments in science centers and the models which can be built using these books share several properties.

- Everybody can perform it. The experiments are deliberately simply to perform. In a strong way it is "mathematics for everybody".
- People like it. One reason why people like the experiments is their success. Each experiment has the possibility of a positive ending, and "all's well that ends well". What is more: the success is undoubtable. When I have composed the pyramid, it stands there and nobody can question it.
- On the other hand, from a mathematician's point of view, people often stop at an early stage and are satisfied with a superficial effect.

2.5   Two Critical Questions

2.5.1   Are These Experiments at All?

One of the main features of mathematics is that the truth of an assertion is obtained by a proof, that is by purely logical arguments, and not, for instance, by experiments. This distinguishes mathematics from sciences such as physics or chemistry, where experiments are used to verify a theory or to falsify a wrong hypothesis.

Also, mathematical experiments are not used to simply illustrate a definition or a theorem.

The role of a mathematical experiment is quite different. Its basic property is to stimulate thinking. In science centers, experiments do not come second (after a theory), but experiments come first. They provide a strong impulse. Basically, a person working with a mathematical experiment is challenged by a mathematical problem. As in research, one has to get the right conception, the right idea of what's going on. And sometimes, after a while of thinking, and sometimes with luck, one finds the solution.

A big advantage of such experiments is the fact that the solution is beyond any doubt, because it is materialized: the cube is there, the bridge is stable, the pattern is correct.

To put it short, a mathematical experiment works "bottom-up": starting from experience, leading to insight. It is an impulse. If the experiment is good, this impulse is so strong that it enables the visitor to ask the right questions, to get the right conceptions and, finally to get by an "Aha-moment" the right insight.

2.5.2 Is This at All Mathematics?

Certainly, it does not look like mathematics, in particular not like school mathematics. In fact, in Mathematikum we explicitly stated at the beginning that we want to make a place that doesn't look like school. Mathematical experiments do not show the mathematical language: no point is called "P", no variable is called x, in fact, there are no formulas. Also, no definitions, no theorems, no proofs.

On the other hand, an important part of mathematical activity is clearly present, namely problems. And, if visitors solve the problems, they activate mathematics-related competences, such as arguing, and communicating.

Mathematical experiments have two main target groups. (a) School classes, (b) private visitors.

When a school class visits Mathematikum, the students may deal with experiments closely related to the topics in math education. For instance, they may look at experiments dealing with the theorem of Pythagoras, or with number systems or with randomness. The teacher then can talk with the students the next day in school about their experiences and insights.

Private visitors, in particular families, behave quite differently. First of all, they have no idea, whether an exhibit represents important mathematics or mathematics at all. They do not care whether the formal mathematics behind it is difficult or easy.

For all visitors it is true that when they start to deal with an experiment, they have a chance to perform a first step into mathematics. The most important aspect is that they think. In fact, they automatically start thinking, for instance asking questions and making conjectures. They try out ideas to solve the problem and eventually they experience the Aha!-moment, in which the whole situation becomes clear.

In addition, when trying to solve a mathematical experiment, the visitors concentrate on important mathematical notions such as edge, angle, it fits, etc. and also they get acquainted with important mathematical concepts, such as patterns, correspondence, infinity, etc.

Finally, they meet not only mathematics taught in school but many aspects which go far beyond school, for instance the travelling salesman problem, minimal surfaces, prime numbers, conic sections, etc.

To sum up, working with mathematical experiments is *a first step into mathematics*. This statement has two sides.

Firstly, it is a step into mathematics. In fact, the problems posed by the experiments can only be solved by thinking, by carefully observing, by looking for the right idea.

On the other hand, dealing with experiments provides only a first step into mathematics. Many more steps could follow. In particular, in this context, there is no formal description of the mathematical phenomena.

In other words, mathematical experiments offer extremely good possibilities to "do" mathematics, but have also clear limitations: they stimulate enthusiasm and true motivation, but also they neither give formal arguments nor can replace a course in a mathematical subject.

Working with mathematical experiments goes far beyond "learning mathematics". It *empowers* people: When visitors see that they have achieved something by thinking by themselves, they become more self-confident.

2.6   Effects and Impact on the Visitors

The main effect of all science centers is *experience*. Visitors experience real phenomena. This is also what visitors like. It is not a virtual experience, which we have by working with computer programs. When we feel real physical objects and work with them, it is clear that we cannot be cheated.

Mathematical experiments stimulate *thinking*. One has to consider several possibilities, one has to develop the right idea for a solution and one verifies whether a solution is correct.

The unquestionable experience of many years of Mathematikum is that dealing with mathematical experiments makes the visitors happy. They become happy because they have *understood* something, which is very satisfying (see also Beutelspacher 2016).

The fact that experiments activate people's brain can be seen—or heard—by the noise in the exhibition. Sometimes it is really loud. But in fact, it is *communication*. People talk to each other, ask questions, give advice—and enjoy the common solution.

A final point: if mathematics is interesting, then it is also interesting outside school. In mathematical science centers as the Mathematikum, mathematics is part of the visitor's leisure time. Adult people and whole families spend hours to experience the power of mathematics. Thus, mathematical experiments and mathematical science centers have a great impact on a *mathematical education of the general public*.

References

Beutelspacher, A. (2015). *Wie man in eine Seifenblase steigt. Die Welt der Mathematik in 100 Experimenten*. Munich: C.H. Beck.

Beutelspacher, A. (2016). What is the impact of interactive mathematical experiments? In W. König (Ed.), *Mathematics and society* (pp. 27–35). Zurich: European Mathematical Society.

Beutelspacher, A., & Wagner, M. (2008). *Wie man durch eine Postkarte steigt*. Freiburg: Herder.

Cundy, H. M., & Rollett, A. P. (1952). *Mathematical models*. Oxford: Clarendon Press.

Dyck, W. (1892). *Katalog mathematischer und mathematisch-physikalischer Modelle, Apparate und Instrumente*. München: Wolf.

Schmitt-Hartmann, R., & Herget, W. (2013). *Moderner Unterricht: Papierfalten im Mathematikunterricht.* Stuttgart: Klett.

van Delft, P., & Botermans, J. (1978). *Creative puzzles of the world.* New York: H.N. Abrams.

Wenninger, M. J. (1974). *Polyhedron models.* Cambridge: Cambridge University Press.

Chapter 3
Intersections of Culture, Language, and Mathematics Education: Looking Back and Looking Ahead

Marta Civil

Abstract This paper draws from a research agenda focused on the interplay of culture, language and mathematics teaching and learning, particularly in working-class Mexican-American communities in the United States. Drawing on data collected over several years, I emphasize the need for a coordinated effort to the mathematics education of non-dominant students, an effort that involves teachers and other school personnel, the students' families, and the students themselves. Through the voices of parents, teachers, and students, I illustrate the resources that non-dominant students bring to school but often go untapped, and the tensions that this may carry. Following a socio-cultural approach grounded on the concept of funds of knowledge, I argue for the need to develop stronger communication among the interested parties to develop learning experiences in mathematics that build on the knowledge, the language and cultural resources, and the forms of participation in the students' communities.

Keywords Culture · Language of learning and teaching · Immigrant students
In-school and out-of-school mathematics · Parental engagement in mathematics

That's in mom's home. Let's do it the way that we do it in the school. [Dina]

Dina was a fifth-grade teacher (students are ten years old) in a Teacher Study Group focused on issues around mathematics, language, and culture. She was teaching in a school in a working-class community with a large number of students of Mexican origin, some of whom were classified as English Learners (ELs). In the excerpt below, Dina is reflecting on some of the challenges she thinks students face in regard to mathematics learning.

M. Civil (✉)
The University of Arizona, Tucson, USA
e-mail: civil@math.arizona.edu

© The Author(s) 2018
G. Kaiser et al. (eds.), *Invited Lectures from the 13th International Congress on Mathematical Education*, ICME-13 Monographs,
https://doi.org/10.1007/978-3-319-72170-5_3

Dina: One of the problems that I saw here was that we were teaching multiplication skills to the children, and I thought, 'Oh, that is something easy that the parents at home can help the children.' Well we ended up with totally different answers, and the children came back with the homework, and the answer was not even close to what the answer was, and they start to say to us, 'But that's an approximation.' … Then we asked some parents to come to the school and teach us what they were doing at home, and they were doing something that we never understood, but it was close to the answer. They didn't understand that math is a precise subject. You cannot change the right answer unless it's a … what is it…what is the one that is close?

Interviewer: An estimation.

Dina: An estimation, but if you do multiplication you have to give me exactly the answer, and for a while, division and multiplication became a problem because we couldn't get the children, you know, they learned it at home, how to do things, and they came, and they do it on the board, and I didn't know what were they doing. … I don't know exactly what they were trying to teach the children, but now the children are doing the math the way we asked them to do it. They are still making mistakes, but they are getting better. At least they understand.

Interviewer: Do the kids ever complain to the moms, that 'that's not the way my teacher does it'?

Dina: No, they did it the other way; 'Well, that's not the way my daddy does it at home. This is not the way my mom does it. It's the other way.' Then we ask them, 'Come to the board; we want to learn.' But we never, at least I never understood how they got the answers…. We talked to the parents, and we explained to them that we need to be precise, and we need the correct answer, and we explain how we teach it here, how we do it. And the question was, 'Well, what if he doesn't know?' 'Well you tell them to come early in the morning, and we will help them here.' We didn't want to say don't do it but… And now every Wednesday we are teaching division and multiplication, and the children are doing it the way we ask. This Wednesday when we did it, Eliseo said, 'Oh no, my mom did it different.' And he went to the board and did it that way, and I said 'Yes, but that's in mom's home. Let's do it the way that we do it in the school.' And it was again very close, but not the answer. It was an approximation. It was an approximation.

3.1 School Versus Home

The opening vignette illustrates the main theme that I will address in this paper, namely a tension between the school way and the home way. Differences between parents' ways of doing mathematics and the ways that their children may be learning at school are quite typical and can be attributed to generational changes in approaches to teaching and learning. However, as I have argued elsewhere (e.g., Civil and Planas 2010), when this is placed in the context of non-dominant communities whose knowledge is often not recognized or valued, the implications need to be considered. In the context of the work I report here, the non-dominant communities are working-class and of immigrant origin, largely from Mexico. In several cases, the parents were schooled in Mexico while their children have been mostly schooled in the United States. Suárez-Orozco and Suárez-Orozco (2001) underscore the difficulties that immigrants often face as they try to navigate the culture from their country of origin and that of their new country. In particular, "Children of immigrants become acutely aware of nuances of behaviors that although 'normal' at home, will set them apart as 'strange' and 'foreign' in public" (pp. 88–89).

Some of the differences in the ways of doing mathematics may be attributed to cultural aspects, for example, different algorithms being used in different countries. But underlying the home-school tension captured in the vignette presented above is the concept of valorization of knowledge (Abreu 1995). In her study of how children experienced the relationship between home and school mathematics, de Abreu presents an interview with 14-year-old Severina, whose father is a sugarcane worker in rural Brazil:

Interviewer: Can you tell me what you think about the way your father did the sums, is it the same or different from the way you learned in school?
Severina: It is a different way, he does it in his head, mine is with a pen.
Interviewer: Which do you think is the proper way?
Severina: School.
Interviewer: Which do you think gives a correct result?
Severina: My father.
Interviewer: Why?
Severina: Because I just think so (p. 137).

I often use this exchange between Severina and the interviewer to illustrate the concept of valorization of knowledge. While Severina knows that her father can do computation in his head and get the right answer, she still believes that the school way with pencil and paper is the proper way. What are the implication for children like Severina or the ones in the opening classroom vignette when their parents' ways are different from the school's ways? What message is Dina (the teacher in the vignette) sending when she says, "Yes, but that's in mom's home. Let's do it the way that we do it in the school"?

3.2   Some Context

This paper draws on over 20 years of a research agenda focused on the interplay of culture, language and mathematics teaching and learning, particularly in working-class Mexican-American communities in Southern Arizona, in the United States. It is important to understand that there are wide differences in these communities, with some families having been there in Arizona for generations. The area was part of Mexico until the Gadsden purchase treaty in 1853 (Sheridan 1995). A popular saying is that "we didn't cross the border, the border crossed us." Other families are recent immigrants. In some households the primary language is Spanish, while in others it is English; many families are bilingual (or multilingual). In many homes, children may speak English among themselves and use Spanish to speak with their parents and older relatives. Overall these students attend de facto segregated schools in that a majority of the students in their school is of Mexican origin and working-class. Students in these schools tend to do less well by traditional testing measures than those in schools in middle to upper class neighborhoods with fewer numbers of students of non-dominant backgrounds.

My approach to both research and outreach in schools in working-class, immigrant origin communities rejects a deficit view of these communities that tends to blame children and their families for their "lack of success" in school. Instead, my work is grounded on the theory behind the Funds of Knowledge for Teaching project (González et al. 2005). The main premise is that in all communities and households there is knowledge, resources, experiences that allow families to get ahead. Moll et al. (2005) define funds of knowledge as "these historically accumulated and culturally developed bodies of knowledge and skills essential for household or individual functioning and wellbeing" (p. 72). I argue that our obligation as educators, teachers, researchers is to learn about and from these funds of knowledge and build on them for the advancement of students in school.

Hodge and Cobb (2016) describe this approach as an example of the "Cultural Alignment Orientation", which they say "has become the default theoretical framework for research on issues of equity in mathematics education" (p. 2). In this orientation, the authors say that the focus is on "aligning the practices established in the mathematics classroom with the out-of-school practices in which students participate. Given this framing, it becomes critical to learn about and leverage students' out-of-school practices as resources to address inequities in learning opportunities" (p. 2). The authors argue for a "Classroom Participation Orientation," which they consider broadens our approach to developing equitable approaches to teaching and learning. In their view, "the Classroom Participation Orientation is grounded in the view of culture as a network of local hybrid practices that people jointly constitute as they negotiate their places in specific settings such as the mathematics classroom" (p. 4). While the two orientations are different, the authors themselves note that the Funds of Knowledge project, which is at the basis of much of my work, has moved from the notion of alignment to also incorporate that of participation, particularly building on the concept of hybridity. Following a

Funds of Knowledge orientation means that the concepts of resources, participation, and valorization are central to the data analysis (Civil 2002b, 2016a).

An underlying question in my work has been how to develop mathematical learning experiences that are culturally responsive in the sense that they reflect and build on the learners' everyday/out-of-school experiences, but are also responsive to the mathematical agenda that needs to be met (Civil 2002b, 2007). As we started bringing in the voices of parents through their participation in mathematics workshops where there is an exchange of experiences, our approach has somewhat moved back and forth between the notion of alignment, in that we have a deep interest in out-of-school mathematical experiences and the notion of participation, particularly as we see students and families navigate multiple spaces (Díez-Palomar et al. 2011; Menéndez et al. 2009). Underlying this work is a need for a two-way dialogue between home and school about mathematics (Civil 2002a; Civil and Andrade 2003; Civil and Planas 2010). In the next section I look at some avenues towards this two-way dialogue.

3.3 Towards a Two-Way Dialogue Home—School

In the opening vignette it seems that some parents had ways of multiplying that were different from what the teachers were teaching. Dina tried to learn them, when she says "Then we asked some parents to come to the school and teach us what they were doing at home, and they were doing something that we never understood, but it was close to the answer" or later on when she says that she invited students to come to the board and show how their parents had taught them, "'Come to the board; we want to learn.' But we never, at least I never understood how they got the answers." Throughout the vignette, Dina provides evidence that she tried to understand the methods but did not succeed. Dina did not feel comfortable in her understanding of mathematics and in fact in that same interview she acknowledged that if it were up to her she would not be teaching mathematics. Dina is not alone. Over the years, I have shown teachers and preservice teachers different algorithms that students may bring to class from home and I have noticed their trepidation (Civil 2016b). Dina opted for making sure that the children learned the school method and that they used that one while in school. She wanted parents to understand that there is a certain way that children are being taught, that is different from how they do it at home, and as she says "we didn't want to say don't do it but…" and does not finish the sentence:

> We talked to the parents, and we explained to them that we need to be precise, and we need the correct answer, and we explain how we teach it here, how we do it. And the question was, 'Well, what if he doesn't know?' 'Well you tell them to come early in the morning, and we will help them here.' We didn't want to say don't do it but…

This situation leads to several unanswered questions such as, were the parents' methods incomplete and indeed producing only an approximation? Even if that was the case, was there some mathematics in them worth exploring? How could the teachers have turned this source of knowledge from the parents into a learning opportunity? Would the teachers have had the support from the school administration? What role does valorization of knowledge play? Closely related to issues of valorization of knowledge and whose knowledge is valued, are issues of power:

> The border between knowledge and power—can be crossed only when educational institutions no longer reify culture, when lived experiences become validated as a source of knowledge, and when the process of how knowledge is constructed and translated between groups located within nonsymmetrical relations of power is questioned. (González 2005, p. 42)

How can we address these issues of power and valorization of knowledge when working with teachers, parents, and students? This has become a central question in my work. One activity I have been using is to have parents and teachers read quotes that other parents and teachers have said in relation to issues on the teaching and learning of mathematics. The quotes are posted around the room and the participants are to stand by the one that speaks to them the most (either because they agree with it, or they do not, or any other reason). One of the quotes I have used was related to the opening vignette and read as follows:

> We are teaching division and multiplication, and the children are doing it the way we ask. This Wednesday when we did it, Eliseo said, 'Oh no, my mom did it different.' And he went to the board and did it that way, and I said 'Yes, but that's in mom's home. Let's do it the way that we do it in the school.'

One mother and one teacher stood by that quote and this is what each said:

Mother: I identified with this quote because I did that with my child and it seems that I confused him and so the part on "let's do it the way we do it in school", we need to get involved and learn the way they teach it at school so that we can continue [the support].

Teacher: I do have students who say my mom/dad taught me this way, and for me, I do have to teach them certain ways but I encourage them if mom and dad want to teach them a different way, then my student has the strategy from school and the one from mom and dad and they can check and make sure that both answers match up, so they can check twice...

These are different reactions to the same quote. The mother underscores the importance of her learning the way they teach it at school so that she can support her child. This is most likely why she joined the project. As discussed elsewhere (Civil et al. 2005), originally parents joined a mathematics project to be able to help their children. However, in reading this mother's comment, I wonder, does she recognize the value of her own methods? Does she appreciate the potential richness of multiple methods? Is she in a way according more value to the school method than to her own method? On the other hand, the teacher, while acknowledging the institutional pressures she feels to teach specific methods, is open and in fact encourages children to bring other ways from home. Her approach is what I characterize as resource-based. This teacher views home knowledge as an asset towards her students' education. How do we capitalize on a resource-based view that encourages the use of home knowledge without pushing away parents who feel that their role is to learn the school ways? But also, how do we strike a balance between recognizing and building on home knowledge while recognizing that "merely glorifying popular knowledge does not contribute to the process of social change" (Knijnik 1993, p. 25).

Bringing quotes such as the one in the vignette for discussion with parents and teachers is an effective way to promote a dialogue around teaching and learning mathematics. Other approaches have included teams of parents and teachers presenting mathematics workshops to other parents in the community (Civil and Bernier 2006), parents and teachers participating in mathematics workshops together, parents visiting a mathematics classroom and then debriefing the visit (Civil and Quintos 2009), and teachers visiting the homes of some of their students to learn about the family's funds of knowledge. In what follows, I briefly present some key findings from these different avenues to promoting a dialogue between home and school around mathematics teaching and learning.

3.4   Cultural Aspects

Many immigrant parents (like everybody else) bring deeply rooted views of what mathematics teaching and learning should look like, for example expressing surprise at physical arrangements of the classroom where students are sitting in groups, some with their backs to the board or to the teacher's desk; or showing concern when students in 3rd grade and higher do not know their multiplication facts yet. But at the same time, as parents engage as learners of mathematics themselves, some develop an appreciation for a focus on conceptual understanding and not only memorization of facts; or an appreciation for joint problem-solving. Yet, as the example of the mother's reaction to the quote indicates, there are some parents who seem ready to give into the school ways. This points to the complexity of the situation where parents, and in particular immigrant parents, are interpreting what they see through their own cultural experiences but also are listening to their children and trying to make sense of their schooling experience. Over the years, we have collected evidence from mothers mentioning the tension they feel when they try to teach their children how they were taught (e.g., Civil and Planas 2010; Civil et al. 2005).

Parents and children are likely to have other ways of doing mathematics, or interpret problems in ways that are based on their everyday experiences and may be different from what the teacher expected. What can teachers do? Some express concern as to how to work with these different approaches. Their concern may be based on their own understanding of mathematics like Dina's case in the vignette, where she tried to understand but did not succeed. Other teachers seem comfortable encouraging students to use strategies that their mom or dad may have taught them, like the teacher earlier whom I described as having a resource-based approach.

The excerpt that follows captures some of the complexity that teachers have expressed on the issue of home versus school mathematics. Caroline was a 6th grade teacher who was reflecting on the challenges and advantages that children in general, and Latina/o children in particular faced in regard to mathematics learning.

Caroline: The Latino children, if their parents came from Mexico, then they probably did it a different way than what they did here, and even the algorithms maybe look a little different. So, I think that causes part of the problem. I think maybe part of it may be language and to translate, some of them, you know especially the students whose first language is Spanish …

Interviewer: So what do you think are the advantages that these same children bring to the classroom?

Caroline: I think it's, like when you're making the connection, if you are doing it orally they may see it in a little different way and if, if you're discussing it, the students build off of each other … so I think that's one; and then also even while you're discussing, even if you're looking at algorithms or something, they're going to be like "Oh well, my dad does it this way" or "My mom does it this way." And so, then you're bringing in another way, so that they're seeing maybe even a third or a fourth or a fifth way to attack a problem.

This excerpt represents what teachers in our work have often said. They are aware that students may have other ways to do mathematics particularly in the case of children of immigrant origin but they do not all see them as an advantage or a challenge. Caroline first refers to this as a possible problem, "So, I think that causes part of the problem" but then at the end of the excerpt these different algorithms are seen as an advantage, "And so, then you're bringing in another way, so that they're seeing maybe even a third or a fourth or a fifth way to attack a problem." Woven throughout Caroline's excerpt is the issue of language, which she sees as a potential problem when children have to go back and forth between English and Spanish. But she did not discourage the use of both languages and in fact she emphasized students discussing mathematics and sharing ideas, which is how she would learn about knowledge from home, "Oh well, my dad does it this way" or "My mom does it this way." In the next section I focus on language issues, which are particularly relevant when working with students and families whose home language is different from that of the school.

3.5   Language Aspects

It is important to briefly describe the language policy in Arizona to provide some context. In 2000, Proposition 203 was passed, which had the effect of limiting access to bilingual education for students who had been classified as English language learners (ELLs). Instead ELLs were to be taught with structured English immersion. Furthermore, in 2006 additional legislature established what is known as the 4-h English requirement, which means that ELLs were to receive 4 h of English language instruction daily. Considering that a school day has about 7 h of instruction, what does this mean for the learning opportunities of ELLs? The idea behind this approach was that students would be proficient in one year and thus be able to move out of the 4-h block. Research has documented that it takes longer than one year for students to become proficient in the command of the language needed for schooling purposes (Cummins 2000). As Gándara and Orfield (2012) point out, the case of Arizona is particularly important because, "having spawned a series of anti-immigrant and highly restrictionist language policies, Arizona stands as the embodiment of this struggle, and pending legal decisions in that state have the potential to reinforce these hegemonic practices and shape the way that English learners are educated across the U.S. for some time to come" (p. 9). But this, I argue, is relevant not only for the U.S., but for any country that is faced with educating children whose home language(s) is different from the language of instruction. As it has been pointed out, discussion on language policy in schools are closely related to issues around immigration and the education of immigrant students (e.g., Alrø et al. 2009; Barwell et al. 2016; Civil 2012; Gándara and Orfield 2012; Wright 2005).

What are the implications on different language policies on students' mathematics education? This is a broad question that cannot be fully addressed in this chapter, but I can certainly offer some snippets based on my research with parents, teachers, and students. A main finding is the impact of language policy on parental engagement in their children's mathematics education. In Acosta-Iriqui et al. (2011), we present a contrasting case of two states, Arizona and New Mexico, with radically different language policies. Through parents' voices we hear how their engagement and feeling of confidence at being able to help their children were drastically different in the two contexts. Many of the parents in the research studies in Arizona had experienced bilingual education with some of their children and then saw the switch to structured English immersion for their other children. The following quote captures the frustration of Verónica whose child was in bilingual classrooms for kinder and first grade but was moved to an English only classroom in second grade:

> I liked it while they were in a bilingual program, I could be involved …. When he was in kindergarten … I even brought work home to take for the teacher the next day. In first grade, it was the same thing, I went with him and because the teacher spoke Spanish, she gave me things to grade and other jobs like that. My son saw me there, I could listen to him, I watched him. By being there watching, I realized many things. And then when he went to second grade into English only and with a teacher that only spoke English, then I didn't go, I didn't go.

On a different occasion, Verónica also shared that she felt comfortable with her knowledge of mathematics and that she could probably help her son (by then in middle school) but that he did not come to her for help because he did not feel comfortable translating from English into Spanish so that she could understand the problems. Hence, while he could communicate in Spanish with her, he did not have the level of Spanish that would allow him to converse about mathematics. Would this be different, had he stayed in bilingual classes?

The question of how to teach mathematics in multilingual contexts has been widely studied (e.g., Barwell et al. 2016) and is certainly a complex situation tied to social and political considerations. In my specific context where it is essentially a Spanish/English situation and given the geographical location so close to Mexico, it seems that providing access to a solid bilingual education for all students would be a benefit to all. As one of the teachers said on reflecting on the advantages that his students brought to the classroom: "The ability to be bilingual, biliterate, I think is a huge advantage. I think that where we are in our country ..., so close to the border, the ability to speak two languages is not a hindrance, it's an advantage. I think you have more opportunities available to you."

Up to this point the focus of this paper has been on looking back at some of my work with parents and teachers around the general theme of seeking ways to connect home and school in relation to the teaching and learning of mathematics. The children are of course at the center, since they are often caught in the middle, trying to navigate both worlds, home and school. In what follows, I present the case of a student, Larissa, to further illustrate the intersection of language, culture, and mathematics education.

3.6   The Case of Larissa

Larissa arrived to the United States three months into the school year, at the age of 13. In Mexico, she had attended a bilingual (Spanish/English) school for most of her schooling, but based on the English placement test she was placed in classes with other students classified as ELLs, but that I will describe as bilingual learners.[1] This meant that for all their classes but an elective and maybe one more class, the bilingual learners were segregated from the students who were considered proficient in English. This case is an example of what Valdés (1998) also found in her study where ELLs were segregated from non-ELLs, thus resulting in two schools within one school. Furthermore, many schools in my local context are already de facto segregated by ethnicity and social class. For example, the school that Larissa attended was at that time 95% Latino/a, 25% ELLs and 85% eligible for the free or reduced-price lunch program.

I am focusing on Larissa as representative of the case of many adolescent students who have to adapt to different cultural and language norms while keeping up with their academic learning and going through typical adolescence growing pains. Suárez-Orozco et al. (2009) noted that in general, recent immigrant youth

---

[1]While students whose home language is not English are often labeled as ELLs, a label that emphasizes what they "lack" (i.e., English), I prefer the term "bilingual learners" as it emphasizes that they know two languages (and in some cases more than two), even if it is with different levels of proficiency.

have positive attitudes towards learning and show optimism for their future. However, as years go by, things can change, "despite their initial academic advantage, for nearly all immigrant groups, length of residence in the United States appears to be associated with declines in academic achievement and aspirations." (p. 714).

Larissa seemed to fit the profile of the academically ready learner upon arrival, as portrayed by Suárez-Orozco et al. (2009). Her knowledge of mathematics was sound and she soon emerged as a leader in mathematical discussions in the class. She was not happy about being in the segregated section for bilingual learners because, as she said in an interview three months after her arrival: "I don't really like the classes with everyone speaking Spanish. I wouldn't like to forget all the English I learned in elementary school." In that same interview, she went onto saying that she preferred English to Spanish and that if she could, she would speak mostly in English. She made friends who were English dominant and she made a clear effort to work on her English. She also shared that she was happy to be in this school because it offered more electives than her prior school in Mexico. An interview with her mother also confirmed that Larissa was happy at school: "She doesn't want to leave school. She likes it here. Larissa expresses very little of what she feels. But, what she has always told me, 'Mom, I don't want to go. I want to be here at school. I've liked it very much'" (Civil and Menéndez 2011, p. 55).

Elsewhere I have discussed the effect on students' ability to engage in mathematical discussions when access to their home language is encouraged (despite the language policy in place) (Civil 2011; Civil and Hunter 2015). Here I just give a brief illustration of the role of language in communicating a mathematical idea. Larissa was asked to explain a probability game they had just been working on to the next class, a group of 6th graders (so, a year younger), who had just entered the classroom:

Larissa: We played a game that's called the multiplication game and the rules are that, two players that are A and B, take turns rolling two number cubes, and when, the, if the product of the numbers rolled is an odd, is an odd number, player A wins a point, and if the product of the numbers rolled is even, player B wins a point. ¿Lo decimos en español? [*Do we say it in Spanish?*] …

Larissa: Es un juego que se llama *multiplication game*; entonces, dos jugadores, que son el A y el B, toman turnos tirando dados. Entonces, cuando tiras dos dados, ese número lo vas a multiplicar por el otro número del otro dado, y si el número es impar, el *player* A gana un punto; si el número es par, multiplicándolo, el *player* B tiene un punto. ¿Ya me entendieron? [*It's a game called multiplication game; so, two players, which are A and B, take turns rolling dice. So, when you roll two dice, you are going to multiply that number by the number on the other die, and if the number is odd, player A gets a point; if the number is even, multiplying it, player B has a point. Did you all get it?*]

As soon as she switches to Spanish, Larissa appears more relaxed and engaged with the audience. While she was speaking in English she was mostly looking at the handout where the game was explained. When she turns to Spanish, while she still looks at the handout, as soon as she moves away from a literal translation of "if the product of the numbers rolled is…" and says "*So, when you roll two dice, you are going to multiply that number by the number on the other die*," she is looking at the students and using her hands to gesture in the air the two numbers that are being multiplied. Granted, her expression is not precise when she says "*and if the number is odd, player A gets a point; if the number is even, multiplying it, player B has a point*", as it is not clear which "number" she is referring to. It seems implicit that she means the product but she does not express it with this level of precision. Earlier in the discussion of the problem she had found the word "product" confusing, something that is not uncommon for students (note that the word in Spanish is "*producto*", which is not that different but it is one of these terms that have a specific meaning in mathematics, different from its meaning in everyday life). I argue that though her explanation may have lacked some precision, she probably reached the 6th graders (also bilingual learners) better than through her reading of the problem in English.

When Larissa first arrived, she would take notes in class, including copying the problems from the book into her notebook. No other student in the classroom was doing that. When I asked her about this, she said that it helped her study. The notebook is part of Mexican schooling. Students have a notebook for each of their subjects and they take notes from what is on the board and do their work there. It becomes a record of what they are doing. In contrast, in the many schools I have been in for my research, I have seen very limited use of notebooks, and in fact little emphasis on students taking notes and writing in the mathematics classroom. As part of the current standards in place in mathematics education, there is an expectation that students communicate about mathematics, explain and justify their reasoning. Would developing the habit of keeping a notebook help towards this goal? Are teachers aware that students who have been schooled in Mexico already bring this habit of study? It is important to note that by April of that first year, Larissa was no longer using a notebook. As the teacher and I were trying to develop norms that involved students writing explanations in mathematics using a blank sheet of paper (instead of trying to squeeze in their writing in the handouts they were working on), I pointed out to Larissa's group that she knew how to do this:

Marta: Larissa knows this because she usually does it, though lately she seems to have picked up other habits…
Larissa: Oh, but now I'm here. Over there [Mexico], it's another story.

That the academic expectations between school in Mexico and school in the United States were perceived as quite different became evident in the interviews with parents. For example, Larissa's mom commented that the school should expect more since what she saw her daughter doing was very easy for her. This is something that other parents and students have noted in different interviews, where they mention that what they are doing in mathematics is something that they had already seen in Mexico. Larissa indeed confirmed this in her first interview (three months after her arrival). Larissa's mother also brought up that while she had seen her bring some homework at the beginning, she was not seeing that anymore. Another mother who was part of the same interview said the same thing about her son. They both adamantly said that if they were in Mexico, they would have much more homework (Civil and Menéndez 2011). In a different group interview with immigrant parents including Larissa's mother, the group brought up several aspects that they all agreed that were stricter in schools in Mexico. They mentioned that the school should have more homework, demand more from the students, expect them to bring tools (pens, notebooks) to the classroom, and ask for higher quality products from the students. As one mother said, "here they put any scribble on the notebook, or on the sheet they bring from school and that's it." Below is an excerpt for that interview:

Marcos:	[They should] give them homework so that they bring it home, so that we can see what kinds of mathematics they are doing; because there are many children who don't bring any; my daughter doesn't bring any, they don't give her any
Iliana:	Mine neither, I always ask her
Mila (Larissa's mother):	Besides giving them homework, they need to demand more [from the children], because what I see, with my daughter and school here and school in Mexico, they should demand more, because for her it's very easy here and then she just kind of glosses over… and in addition to more demanding, the school should be stricter with them… they are too lax.

I move now to Larissa the following year. She was placed in the algebra class which is the highest level of mathematics at the middle school. She was also no longer in the segregated section for bilingual learners, hence taking all her classes with students who spoke English (though with different levels of proficiency, but had met the requirements to be in those classes). Furthermore, she was taking some electives available only to students in gifted education. So, by these indicators she seemed to be doing quite well academically. Towards the end of that year I had one more interview with her. I asked her to reflect on her almost two years at that school and what her impressions were:

Larissa:	What I like about the school?
Marta:	Yes.
Larissa:	Nothing … So, nothing is really all that interesting; I mean I think that for someone to like something, the teacher has to be more sociable and make it more engaging so that we do the work with more interest.

It could be that Larissa was acting like many young people her age. It could also be that she was not being challenged enough since the only class she said she liked was the one that was for gifted students. What was interesting to me in this second interview is that she did not seem as enthusiastic about the school as in the first interview. Even more interesting were her comments about her language preference. Recall that in the first interview she had said that she preferred English over Spanish. In this second interview, we talked about how she was no longer in the section for bilingual learners and how there was much more English being used around her. While she spoke English to some of her peers, she spoke Spanish to others:

Larissa: I don't like to speak English much.
Marta: How come?
Larissa: I don't know.
Marta: It's interesting because last year you told me that you wanted to speak in English.
Larissa: No, I just wanted to practice it.

...

Larissa: I hardly like it [English].
Marta: What is it that you don't like about English?
Larissa: That I haven't learned it well... that is, there are times where I need to stay quiet because I feel embarrassed if I don't say something well. So, that's the reason.

This exchange points to the difficulty that immigrant students may face when trying to fit in an environment when they do not feel comfortable with the dominant language yet. Larissa's story traces the journey from her perhaps initial optimism when she first arrived since she already knew some English to the realization that it takes longer than a year to have a good command of another language. It is worth noting what Matilde, the mathematics teacher of the bilingual learners, said when reflecting on what happens when they move to the "regular" classes.

Matilde: I work only with ELL students ... Our kids feel afraid to be in the regular classroom because they feel the other kids have the power. So, even if I have a very brilliant a kid, he goes to a regular classroom, and he is going to be student X [meaning anonymous]. Because he is not going to be that brilliant because they're going to ask them questions in English so they don't know how to explain themselves and they're going to be quiet. So they're going to be, relegated to the back of the class. So they are afraid to go to a regular class.

The case of Larissa highlights some of the issues that immigrant students face in a different school setting. Larissa brought multiple resources with her, such as study and work habits (e.g., note taking); good mathematics background; bilingualism. But she also had to learn new cultural norms of what it means to go to school and improve her knowledge of English. All of this while going through adolescence. In the next and final section I look at the main points presented and use them to address the looking ahead part of this paper.

3.7 Looking Ahead

Throughout this paper my focus has been on the need to take a resource-based view towards the mathematics education of non-dominant students. While in the context of my work, many of these students and their families navigate varying cultural and linguistic terrains that at times may make their learning of mathematics hard, they also bring multiple resources, such as knowledge of mathematics and study habits that may go untapped; knowledge of more than one language, which can be seen as an advantage as they can provide access to more representations; experiences with mathematics at home that can be used to strengthen the connection between home and school. In Civil (2016a) I argue for the need to get a better understanding of the nature of engagement in mathematically-rich everyday practices particularly in non-dominant communities. How do parents and students participate in mathematical practices in their everyday life; how do they relate (or could) to the practices in school mathematics? As I look ahead, I wonder about the potential of the notion of culturally sustaining pedagogy (Paris 2012) for future work in mathematics education:

> The term *culturally sustaining* requires that our pedagogies be more than responsive of or relevant to the cultural experiences and practices of young people—it requires that they support young people in sustaining the cultural and linguistic competence of their communities while simultaneously offering access to dominant cultural competence. (p. 95)

In this paper, I have argued for the need to develop an integrated model that connects mathematics teaching and learning to the cultural, social, linguistic and political contexts of non-dominant students. In particular, the case of Dina calls for teacher education initiatives that provide opportunities for teachers to engage in using their students' funds of knowledge as resources for teaching. The case of Larissa underscores the potential loss of learning opportunities when students' funds of knowledge (e.g., home language(s); different ways to do mathematics and study habits) are not developed in a culturally sustaining way. To this end, as I look ahead, I suggest that we need to work on developing stronger and meaningful communication between home and school; challenge the different valorizations given to different forms of mathematics; probe the effects of language policies on students' mathematics education; engage with teachers in conversations about the mathematics education of non-dominant students; and share narratives of non-dominant students' successful participation in mathematical discussions.

Acknowledgements Some of the work in this paper was funded by the National Science Foundation under grants ESI 9901275 and ESI 0424983 and by the Heising Simons Foundation, Grant #2016-065. The views expressed here are those of the author and do not necessarily reflect the views of the funding agencies.

References

Abreu, G. de (1995). Understanding how children experience the relationship between home and school mathematics. *Mind, Culture, and Activity, 2*, 119–142.

Acosta-Iriqui, J., Civil, M., Díez-Palomar, J. Marshall, M., & Quintos-Alonso, B. (2011). Conversations around mathematics education with Latino parents in two Borderland communities: The influence of two contrasting language policies. In K. Téllez, J. Moschkovich, & M. Civil (Eds.), *Latinos/as and mathematics education: Research on learning and teaching in classrooms and communities* (pp. 125–147). Charlotte, NC: Information Age Publishing.

Alrø, H., Skovsmose, O., & Valero, P. (2009). Inter-viewing foregrounds: Students' motives for learning in a multicultural setting. In M. César & K. Kumpulainen (Eds.), *Social interactions in multicultural settings* (pp. 13–17). Rotterdam, The Netherlands: Sense Publishers.

Barwell, R., Clarkson, P., Halai, A., Kazima, M., Moschkovich, J., Planas, N., et al. (Eds.) (2016). *Mathematics education and language diversity: The 21st ICMI study*. New York: Springer.

Civil, M. (2002a). Culture and mathematics: A community approach. *Journal of Intercultural Studies, 23*(2), 133–148.

Civil, M. (2002b). Everyday mathematics, mathematicians' mathematics, and school mathematics: Can we bring them together? In M. Brenner & J. Moschkovich (Eds.), *Everyday and academic mathematics in the classroom. Journal for Research in Mathematics Education Monograph #11*, pp. 40–62. Reston, VA: NCTM.

Civil, M. (2007). Building on community knowledge: An avenue to equity in mathematics education. In N. Nasir & P. Cobb (Eds.), *Improving access to mathematics: Diversity and equity in the classroom* (pp. 105–117). New York: Teachers College Press.

Civil, M. (2011). Mathematics education, language policy, and English language learners. In W. F. Tate, K. D. King, & C. Rousseau Anderson (Eds.), *Disrupting tradition: Research and practice pathways in mathematics education* (pp. 77–91). Reston, VA: NCTM.

Civil, M. (2012). Mathematics teaching and learning of immigrant students: An overview of the research field across multiple settings. In O. Skovsmose & B. Greer (Eds.), *Opening the cage: Critique and politics of mathematics education* (pp. 127–142). Rotterdam, The Netherlands: Sense Publishers.

Civil, M. (2016a). STEM learning research through a funds of knowledge lens. *Cultural Studies of Science Education, 11*(1), 41–59. https://doi.org/10.1007/s11422-014-9648-2.

Civil, M. (2016b). "This is nice but they need to learn to do things the U.S. way": Reaction to different algorithms. In D. Y. White, S. Crespo, & M. Civil (Eds.), *Cases for mathematics teacher educators: Facilitating conversations about inequities in mathematics classrooms* (pp. 219–225). Charlotte, NC: Information Age Publishing.

Civil, M., & Andrade, R. (2003) Collaborative practice with parents: The role of researcher as mediator. In A. Peter-Koop, A. Begg, C. Breen, & V. Santos-Wagner (Eds.), *Collaboration in teacher education: Working towards a common goal* (pp. 153–168). Boston, MA: Kluwer.

Civil, M., & Bernier, E. (2006). Exploring images of parental participation in mathematics education: Challenges and possibilities. *Mathematical Thinking and Learning, 8*(3), 309–330.

Civil, M., Bratton, J., & Quintos, B. (2005). Parents and mathematics education in a Latino community: Redefining parental participation. *Multicultural Education, 13*(2), 60–64.

Civil, M., & Hunter, R. (2015). Participation of non-dominant students in argumentation in the mathematics classroom. *Intercultural Education, 26*, 296–312. https://doi.org/10.1080/14675986.2015.1071755.

Civil, M., & Menéndez, J. M. (2011). Impressions of Mexican immigrant families on their early experiences with school mathematics in Arizona. In R. Kitchen & M. Civil (Eds.), *Transnational and borderland studies in mathematics education* (pp. 47–68). New York, NY: Routledge.

Civil, M., & Planas, N. (2010). Latino/a immigrant parents' voices in mathematics education. In E. Grigorenko & R. Takanishi (Eds.), *Immigration, diversity, and education* (pp. 130–150). New York, NY: Routledge.

Civil, M., Planas, N., & Quintos, B. (2005). Immigrant parents' perspectives on their children's mathematics. *Zentralblatt für Didaktik der Mathematik, 37*(2), 81–89.

Civil, M., & Quintos, B. (2009). Latina mothers' perceptions about the teaching and learning of mathematics: Implications for parental participation. In B. Greer, S. Mukhopadhyay, S. Nelson-Barber, & A. Powell (Eds.), *Culturally responsive mathematics education* (pp. 321–343). New York, NY: Routledge.

Cummins, J. (2000). *Language, power and pedagogy: Bilingual children in the crossfire.* Tonawanda, NY: Multilingual Matters.

Díez-Palomar, J., Menéndez, J. M., & Civil, M. (2011). Learning mathematics with adult learners: Drawing from parents' perspectives. *Revista Latinoamericana de Investigación en Matemática Educativa (RELIME), 14*(1), 71–94.

Gándara, P., & Orfield, G. (2012). Why Arizona matters: The historical, legal, and political contexts of Arizona's instructional policies and U.S. linguistic hegemony. *Language Policy, 11*, 7–19. https://doi.org/10.1007/s10993-011-9227-2.

González, N. (2005). Theoretical underpinnings. In N. González, L. Moll, & C. Amanti (Eds.), *Funds of knowledge: Theorizing practice in households, communities, and classrooms* (pp. 25–46). New York: Routledge.

González, N., Moll, L., & Amanti, C. (Eds.) (2005). *Funds of knowledge: Theorizing practice in households, communities, and classrooms.* New York: Routledge.

Hodge, L. L., & Cobb, P. (2016). Two views of culture and their implications for mathematics teaching and learning. *Urban Education*. First published date: 11 April, 2016. https://doi.org/10.1177/0042085916641173.

Knijnik, G. (1993). An ethnomathematical approach in mathematical education: A matter of political power. *For the Learning of Mathematics, 13*(2), 23–25.

Menéndez, J. M., Civil, M., & Mariño, V. (2009, April). *Latino parents as teachers of mathematics: Examples of interactions outside the classroom.* Paper presented at the annual meeting of the American Educational Research Association (AERA), San Diego, CA.

Moll, L. C., Amanti, C., Neff, D., & González, N. (2005). Funds of knowledge for teaching: Using a qualitative approach to connect homes and classrooms. In N. González, L. Moll, & C. Amanti (Eds.), *Funds of knowledge: Theorizing practice in households, communities, and classrooms* (pp. 71–87). New York: Routledge.

Paris, D. (2012). Culturally sustaining pedagogy: A needed change in stance, terminology, and practice. *Educational Researcher, 41*(3), 93–97. https://doi.org/10.3102/0013189X12441244.

Sheridan, T. E. (1995). *Arizona: A history.* Tucson, AZ: The University of Arizona Press.

Suárez-Orozco, C., Pimentel, A., & Martin, M. (2009). The significance of relationships: Academic engagement and achievement among newcomer immigrant youth. *Teachers College Record, 111*(3), 712–749.

Suárez-Orozco, C., & Suárez-Orozco, M. (2001). *Children of immigration*. Cambridge, MA: Harvard University Press.

Valdés, G. (1998). The world outside and inside schools: Language and immigrant children. *Educational Researcher, 27*(6), 4–18.

Wright, W. E. (2005). The political spectacle of Arizona's proposition 203. *Educational Policy, 19*, 662–700.

Chapter 4
The Double Continuity of Algebra

Al Cuoco and William McCallum

Abstract We consider Klein's double discontinuity between high school and university mathematics in relation to algebra as it is studied in both settings. We give examples of two kinds of continuities that might mend the break: (1) examples of how undergraduate courses in algebra and number theory can provide useful tools for prospective teachers in their professional work, as they design and sequence mathematical tasks, and (2) examples of how questions that arise in secondary pre-college mathematics can be extended and analyzed with methods from algebra and algebraic geometry, using both a careful analysis of algebraic calculations and the application of algebraic methods to geometric problems. We discuss useful sensibilities, for high school teachers and university faculty, that are suggested by these examples. We conclude with some recommendations about the content and structure of abstract algebra courses in university.

Keywords Klein · Double discontinuity · Algebra · Secondary school teaching

4.1 Introduction

The young university student found himself, at the outset, confronted with problems, which did not suggest, in any particular, the things with which he had been concerned at school. Naturally he forgot these things quickly and thoroughly. When, after finishing his course of study, he became a teacher, he suddenly found himself expected to teach the traditional elementary mathematics in the old pedantic way; and, since he was scarcely able, unaided,

A. Cuoco (✉)
Center for Mathematics Education, EDC, 43 Foundry Ave, Waltham
MA, USA
e-mail: acuoco@edc.org

W. McCallum
Department of Mathematics, The University of Arizona, Tucson, AZ, USA
e-mail: wmc@math.arizona.edu

© The Author(s) 2018
G. Kaiser et al. (eds.), *Invited Lectures from the 13th International Congress on Mathematical Education*, ICME-13 Monographs,
https://doi.org/10.1007/978-3-319-72170-5_4

49

to discern any connection between this task and his university mathematics, he soon fell in with the time honoured way of teaching, and his university studies remained only a more or less pleasant memory which had no influence upon his teaching.

—Felix Klein, *Elementary Mathematics from an Advanced Standpoint* (1932, p. 1)

Does Klein's oft-quoted description of what he called the double discontinuity, given over 100 years ago, still hold today? The recommendations of the Conference Board on the Mathematical Sciences (CBMS 2012) for the mathematical education of teachers in the US were formulated on the premise that there was still a problem to be solved in at least one of the directions of the discontinuity: the transition from the university coursework of a prospective teacher to their practice in teaching secondary school. Chapter 2 of that report discusses some of the evidence for that point of view. In the other direction, de Guzmán et al. (1998) describe the discontinuity from the point of view of a secondary school student entering university, and find it particularly strong for prospective teachers.

Given the apparent persistence of the problem, it is reasonable to wonder if the double discontinuity results in part from a double discontinuity in the subject matter itself of the courses that students take. In the US, Wu (2015) has described what he calls *textbook school mathematics* as a subject alienated from genuine mathematics. The US Common Core State Standards for Mathematics were motivated in part by a desire to lessen the distance between school mathematics and university mathematics. Such projects need stories of continuity. In this paper we consider continuities between secondary school and university mathematics in both directions, focusing on the subject of algebra.

4.2   From University to Secondary School

In this section we follow mathematical threads from a course in abstract algebra or number theory to secondary school mathematics.

A common topic of discussion among teachers is finding tasks that can be used to launch a new topic. They want the answers to be simple, so that computational overhead does not get in the way of the ideas.

For example, consider the topic of the law of cosines, to be introduced by the task of finding the measure of $\angle UQS$ in Fig. 4.1. This task has a particularly simple answer, 60°. Examples like this are prized by teachers; they are traded at department meetings and sought after online. The problem of finding such examples is a problem in task design. A teacher with university background in abstract algebra or algebraic number theory can apply that knowledge to task design and find general methods for constructing such examples. This can be viewed as a sort of applied mathematics for the profession of teaching, as described at the elementary level in Ball et al. (2008).

Fig. 4.1 What is the measure of ∠UQS?

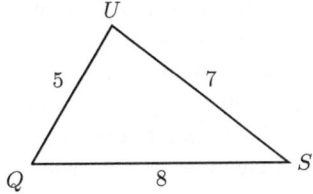

4.2.1 Pythagorean Triples

Before considering the task design problem of finding nice triangles with a 60° angle, we consider the simpler problem of finding nice triangles with a 90° angle, that is, nice right triangles. The triangle with side lengths 3, 4, and 5 is such a triangle because the lengths satisfy the Pythagorean identity:

$$3^2 + 4^2 = 5^2.$$

Another example is the triangle with side lengths 5, 12, and 13, because

$$5^2 + 12^2 = 13^2.$$

There are infinitely many such Pythagorean triples, that is, triples of positive integers (a, b, c) such that

$$a^2 + b^2 = c^2.$$

Diophantus of Alexandria developed, around 250 CE, a geometric method for generating such triples. In modern language, he realized that a rational point on the unit circle (the graph of $x^2 + y^2 = 1$), when written in the form $\left(\frac{a}{c}, \frac{b}{c}\right)$, produces a Pythagorean triple (Fig. 4.2):

$$\left(\frac{a}{c}\right)^2 + \left(\frac{b}{c}\right)^2 = 1 \quad \Rightarrow \quad a^2 + b^2 = c^2.$$

Fig. 4.2 The method of Diophantus

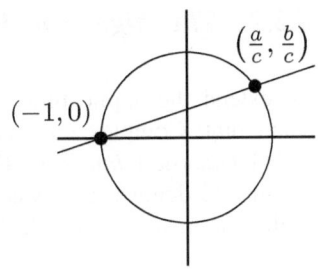

One can get such a rational point by forming a line with positive rational slope through the point $P = (-1, 0)$ and intersecting the line with the circle. The second intersection point will then be rational. Hence, it was known early on that there are infinitely many Pythagorean triples [for details, see Cuoco and Rotman (2013)].

In addition to this geometric method, there is an algebraic method for generating Pythagorean triples, using complex numbers and the observation that

$$x^2 + y^2 = (x + yi)(x - yi). \tag{4.1}$$

The sum of two squares can thus be written as the product of a complex number and its complex conjugate. So, if you want integers a and b so that $a^2 + b^2$ is a perfect square, you might write the sum of these two squares as

$$a^2 + b^2 = (a + bi)(a - bi)$$

and try to make each factor on the right-hand side the square of a complex number with integer real and imaginary parts. And it is within the scope of secondary school mathematics to show that if $a + bi = (r + si)^2$, then $a - bi = (r - si)^2$. So, for example,

$$(3 + 2i)^2 = 5 + 12i \quad \text{and}$$
$$(3 - 2i)^2 = 5 - 12i.$$

So,

$$\begin{aligned}
5^2 + 12^2 &= (5 + 12i)(5 - 12i) \\
&= (3 + 2i)^2(3 - 2i)^2 \\
&= ((3 + 2i)(3 - 2i))^2 \\
&= (3^2 + 2^2)^2 \\
&= 13^2,
\end{aligned}$$

and we have the Pythagorean triple $(5, 12, 13)$.

4.2.2 The Algebraic Method from a Higher Standpoint

To extend the applicability of this method, it helps to look at it from a higher standpoint. Complex numbers of the form $a + bi$ where a and b are integers are called *Gaussian integers*. The set of all Gaussian integers is denoted by $\mathbb{Z}[i]$, because $\mathbb{Z}$ denotes the system of ordinary integers that is the focus of much of arithmetic in school; so $\mathbb{Z}[i]$ is obtained from $\mathbb{Z}$ by adjoining i. Both $\mathbb{Z}$ and $\mathbb{Z}[i]$ are endowed with two operations—addition and multiplication. The properties of addition and multiplication that allow one to calculate with integers also hold in $\mathbb{Z}[i]$—order does not matter in addition or multiplication, multiplication distributes over addition, and so on. Formally, both systems are examples of *commutative rings*, and, in fact, $\mathbb{Z}$ is a subring of $\mathbb{Z}[i]$.

The complex conjugate of a complex number $z = a + bi$ is denoted by $\bar{z}$, so $\overline{a + bi} = a - bi$. This operation of multiplying a Gaussian integer by its complex conjugate is a map from $\mathbb{Z}[i]$ to $\mathbb{Z}$ called the *norm* and denoted by N:

$$N(z) = z\bar{z}.$$

The norm map has the following properties:

1. $N(zw) = N(z)N(w)$ for all Gaussian integers z and w.
2. Hence, if z is a Gaussian integer, then

$$N(z^2) = (N(z))^2.$$

3. If $z = a + bi$, $N(z) = a^2 + b^2$, a non-negative integer.

Put in the context of Pythagorean triples, item (3) shows that we are looking for Gaussian integers whose norms are perfect squares. Item (2) tells us how to do that:

To make $N(z)$ a square in $\mathbb{Z}$, make z a square in $\mathbb{Z}[i]$.

This gives an easily programmed algorithm for generating Pythagorean triples, giving secondary school teachers a useful tool for their lesson planning. Table 4.1, generated in a computer algebra system, shows $(r + si)^2 = a + ib$ and the corresponding norm c. The three integers a, b, and c form a Pythagorean triple.

4.2.3 Using Norms to Construct Triangles with a 60° Angle

We now return to the task design problem of finding examples like Fig. 4.1. The problem is to find a triple of positive integers (a, b, c) that are side-lengths of a triangle with a 60° angle.

Applying the law of cosines to the triangle in Fig. 4.3, we have

$$c^2 = a^2 + b^2 - 2ab \, \cos 60°$$
$$= a^2 + b^2 - 2ab \, \frac{1}{2}$$
$$= a^2 - ab + b^2$$

Table 4.1 $(r+si)^2$ and the resulting norm

	$s = 1$	$s = 2$	$s = 3$	$s = 4$
$r = 2$	3 + 4i, 5			
$r = 3$	8 + 6i, 10	5 + 12i, 13		
$r = 4$	15 + 8i, 17	12 + 16i, 20	7 + 24i, 25	
$r = 5$	24 + 10i, 26	21 + 20i, 29	16 + 30i, 34	9 + 40i, 41
$r = 6$	35 + 12i, 37	32 + 24i, 40	27 + 36i, 45	20 + 48i, 52
$r = 7$	48 + 14i, 50	45 + 28i, 53	40 + 42i, 58	33 + 56i, 65
$r = 8$	63 + 16i, 65	60 + 32i, 68	55 + 48i, 73	48 + 64i, 80

Fig. 4.3 A nice triangle

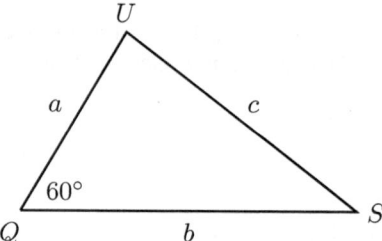

So, we want integers (a, b, c) so that

$$c^2 = a^2 - ab + b^2. \tag{4.2}$$

Examples of such triples are $(5, 8, 7)$ (corresponding to the example above) and $(15, 7, 13)$.

We will call such a triple an Eisenstein triple. We want to find a system analogous to $\mathbb{Z}[i]$ in which the right-hand side of (4.2) is a norm. Such an expression arises naturally in number theory courses that treat roots of unity.

Just as we can form the ring of Gaussian integers $\mathbb{Z}[i]$ by adjoining the fourth root of unity i to $\mathbb{Z}$, so we can form the ring of Eisenstein integers by adjoining the cube roots of unity, that is, the roots of $x^3 - 1 = 0$. Because

$$x^3 - 1 = (x - 1)(x^2 + x + 1),$$

the three roots are

$$\left\{ 1, \frac{-1 + i\sqrt{3}}{2}, \frac{-1 - i\sqrt{3}}{2} \right\}.$$

Let

$$\omega = \frac{-1 + i\sqrt{3}}{2},$$

and consider $\mathbb{Z}[\omega] = \{a + b\omega \mid a, b \in \mathbb{Z}\}$. This is a ring (the Eisenstein integers) with structural similarities to $\mathbb{Z}$ and $\mathbb{Z}[i]$ [details are in Cuoco and Rotman (2013)]. In particular, because

$$\omega + \overline{\omega} = -1 \quad \text{and}$$
$$\omega\overline{\omega} = 1,$$

a direct calculation shows that

$$\begin{aligned} N(z) &= (a + b\omega)\overline{(a + b\omega)} \\ &= (a + b\omega)(a + b\overline{\omega}) \\ &= a^2 - ab + b^2. \end{aligned}$$

Hence, the same mantra applies:

To make $N(z)$ a square in $\mathbb{Z}$, make z a square in $\mathbb{Z}[\omega]$.

Once again, teachers have a method for generating Eisenstein triples (see Table 4.2).

4.2.4   What Is to Be Learned from This?

At one level, the methods given above could be viewed as no more than charming tricks of no great consequence in mathematics education. It is certainly true that

Table 4.2 $(r + s\omega)^2$ and the resulting norm

	$s = 1$	$s = 2$	$s = 3$	$s = 4$
$r = 2$	$3 + 3\omega$, 3			
$r = 3$	$8 + 5\omega$, 7	$5 + 8\omega$, 7		
$r = 4$	$15 + 7\omega$, 13	$12 + 12\omega$, 12	$7 + 15\omega$, 13	
$r = 5$	$24 + 9\omega$, 21	$21 + 16\omega$, 19	$16 + 21\omega$, 19	$9 + 24\omega$, 21
$r = 6$	$35 + 11\omega$, 31	$32 + 20\omega$, 28	$27 + 27\omega$, 27	$20 + 32\omega$, 28
$r = 7$	$48 + 13\omega$, 43	$45 + 24\omega$, 39	$40 + 33\omega$, 37	$33 + 40\omega$, 37
$r = 8$	$63 + 15\omega$, 57	$60 + 28\omega$, 52	$55 + 39\omega$, 49	$48 + 48\omega$, 48
$r = 9$	$80 + 17\omega$, 73	$77 + 32\omega$, 67	$72 + 45\omega$, 63	$65 + 56\omega$, 61
$r = 10$	$99 + 19\omega$, 91	$96 + 36\omega$, 84	$91 + 51\omega$, 79	$84 + 64\omega$, 76

secondary school mathematics teachers can get by without them. However, when viewed not as methods but as examples of a certain sensibility, they acquire greater significance. That sensibility is a tendency to view the mathematics learned in university as a useful tool in teaching secondary mathematics. This is a useful sensibility for high school teachers to have.

For example, the way of thinking about arithmetic in complex numbers as "algebra with i" with an extra simplification rule—exemplified above—is often discouraged in secondary school mathematics. But it has quite a solid pedigree in modern algebra and can provide a glimpse of the reduction technique used to construct splitting fields for polynomials. Another example: In polynomial algebra, being explicit about the interplay between formal and functional thinking (something that is often blurred in secondary texts) helps students develop an appreciation for the "two faces" of algebra (Weyl 1995).

More generally, major themes in algebra—structure, extension, decomposition, reduction, localization, and representation—can help teachers bring coherence and parsimony to the entire secondary school curriculum.

4.3   From Secondary School to University

Now we look at a couple of mathematical threads that go in the opposite direction, from secondary school to university. Or, since the exact boundary between secondary school and university varies from country to country, it might be better to consider these simply as examples that go from some point in secondary school to a more advanced point, be it in secondary school, university, or beyond.

4.3.1   Ptolemy's Theorem

Consider the following secondary school mathematics problem.[1]

Given a cyclic quadrilateral whose sides are 2, 3, 5, 6, find the length of the square of the diagonal which makes a triangle with sides of length 2 and 3 (see Fig. 4.4).

Because the quadrilateral is inscribed in a circle, $\angle ABC$ is supplementary to $\angle CDA$, so $\phi = \theta - 180$. Applying the Law of Cosines to both triangles formed by the diagonal AC, we get

$$x^2 = 4 + 9 - 2 \cdot 6 \cos \theta$$
$$= 25 + 36 + 2 \cdot 30 \cos \theta$$

---

[1] We are indebted to Dick Askey for suggesting the sequence of ideas developed here.

Fig. 4.4 A cyclic
quadrilateral

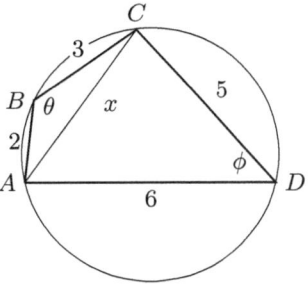

Eliminating $\cos\theta$ and solving for x yields $x = \sqrt{21}$. Now consider the same problem with a general quadrilateral, as in Fig. 4.5. The same method yields

$$x^2 = \frac{b^2cd + a^2cd + abc^2 + abd^2}{ab + cd}.$$

The expression on the right provides a wonderful opportunity for students to exercise what one might call algebraic insight. At first glance it is not obvious how to factor the numerator, but if one regroups the products in a way that shares the squared term with the other factors, it becomes easy to see:

$$x^2 = \frac{(bc)(bd) + (ac)(ad) + (ac)(bc) + (ad)(bd)}{ab + cd} = \frac{(ac + bd)(ad + bc)}{ab + cd}$$

Another exercise in algebraic insight is to imagine what the corresponding expression for y^2 would look like. One could repeat the calculation, or one could simply observe that y is in the same position with respect to (b, c, d, a) as x is with respect to (a, b, c, d). Therefore the formula for y^2 is obtained by performing the cyclic permutation $a \to b \to c \to d \to a$. Without actually writing the expression down, one can contemplate the effect of the permutation on the rightmost expression for x^2. The three parenthetical factors in that expression come in two types. Two of them, $ab + cd$ and $ad + bc$, are obtained by multiplying pairs of adjacent sides and adding the resulting products. One of them, $ac + bd$, is obtained by multiplying pairs of opposite sides and adding the resulting products. The permutation is going to swap the first two types and leave the second type unchanged. This has the effect of causing a lot of cancellation when you multiply x^2 and y^2. The swapped terms cancel each other out and we are left with

$$x^2y^2 = (ac + bd)^2,$$

or

Fig. 4.5 A general cyclic
quadrilateral

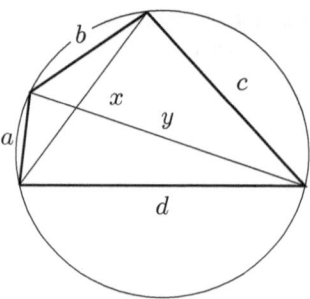

$$xy = ac + bd.$$

This is Ptolemy's theorem, a beautiful generalization of Pythagoras's theorem.

Theorem (Ptolemy) *In a quadrilateral inscribed in a circle, the product of the diagonals is the sum of the products of oppose sides.*

Pythagoras's theorem is the special case where the quadrilateral is a rectangle.

4.3.2 A Question from a Secondary School Class

This section is inspired by a story from the PROMYS for Teachers program at Boston University, recounted in Rosenberg et al. (2008). It starts with a question that could come up in a secondary school geometry class:

If two triangles have the same perimeter and same area, are they congruent?

It is natural to assume the answer is no, if only on the grounds that if the answer were yes it would be a well known theorem. However, it turns out to be surprisingly difficult to come up with counterexamples. From an advanced point of view, the reason for this is that the counterexamples live on a curve which, unlike the circle in Sect. 4.2.1, is not easy to parameterize. We briefly sketch the derivation of that curve here.

We want to parameterize the family of triangles with fixed perimeter p and fixed area A. The radius r of the inscribed circle of such a triangle is related to A and p by the equation

$$A = \frac{p}{2}r. \tag{4.3}$$

This can be seen by decomposing the triangle into three triangles with bases on the sides of the triangle and vertices at the incenter, and adding up their areas, taking note of the fact that the altitude of each of the smaller triangles is r (see Fig. 4.6). It follows from (4.3) that all the triangles in the family have the same in

Fig. 4.6 Triangle and
inscribed circle

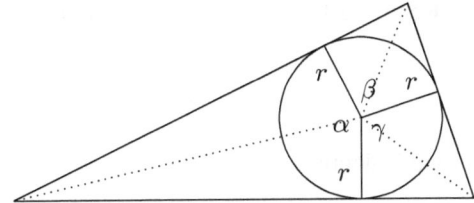

radius r, and they can all be circumscribed around a fixed circle, as in Fig. 4.6. We
consider the space of all triangles circumscribed around this circle, obtained by
varying the angles α, β, and γ and keeping the base horizontal. This is a two
parameter space, since the angles are constrained by the condition that they must
add up to 2π. Every triangle is similar to a triangle in this space, since every triangle
can be scaled to have inradius r. Our family of triangles with perimeter p and area A
is a curve within that space, defined by a constraint on the angles α, β, and γ, which
we now derive.

Another way of decomposing the triangle is to break it into 3 quadrilaterals
formed by the radii and the segments into which the sides are divided by the
perpendiculars from the incenter. Since the center of the inscribed circle is the
intersection of the angle bisectors, these quadrilaterals are divided into pairs of
congruent right triangles by the lines from the vertices to the center of the inscribed
circle (dotted in Fig. 4.6). Therefore these lines also bisect the angles α, β, and γ.
Adding up the lengths of the 6 line segments around the perimeter we get

$$p = 2r \left(\tan \frac{\alpha}{2} + \tan \frac{\beta}{2} + \tan \frac{\gamma}{2} \right). \tag{4.4}$$

From Eqs. (4.3) to (4.4) we see that, in our family of triangles with fixed area A
and fixed perimeter p, the sum of the tangents is also fixed;

$$\tan \frac{\alpha}{2} + \tan \frac{\beta}{2} + \tan \frac{\gamma}{2} = k, \quad \text{where } k \text{ is the constant } \frac{p^2}{4A}. \tag{4.5}$$

We can get an algebraic equation out of this by choosing parameters $x = \tan(\alpha/2)$ and $y = \tan(\beta/2)$. Since $\alpha + \beta + \gamma = 2\pi$, we have

$$\frac{\gamma}{2} = \pi - \frac{\alpha}{2} - \frac{\beta}{2},$$

so

$$\tan \left(\frac{\gamma}{2} \right) = \tan \left(\pi - \frac{\alpha}{2} - \frac{\beta}{2} \right) = -\tan \left(\frac{\alpha}{2} + \frac{\beta}{2} \right) = -\frac{x+y}{1-xy}.$$

Referring back to Eq. (4.5) we obtain, for fixed k, the equation

$$x + y - \frac{x+y}{1-xy} = k,$$

or equivalently

$$x^2y + xy^2 - kxy + k = 0. \tag{4.6}$$

This defines a curve in the xy-plane called an elliptic curve. Our original triangle gives a point on this curve; conversely, given a point on the curve with $x > 0$ and $y > 0$, we can reconstruct a triangle circumscribed around a circle of radius r with area A and perimeter p. Moreover, one can verify that that if x and y are rational numbers, then A and p are also rational numbers.

The method of finding rational points on elliptic curves using the secant method is well-developed in number theory and has a venerable history. We won't describe it further here, referring the interested reader to (Silverman and Tate 1994). We conclude with an example which is enjoyable to carry out by hand. The right triangle with sides 3, 4, and 5 corresponds to a point on the curve defined by (4.6) with $k = 6$. In fact there are six points, depending on which side you choose as base and how you orient the triangle: $(1, 2)$, $(2, 1)$, $(1, 3)$, $(3, 1)$, $(2, 3)$, and $(3, 2)$. Using the secant method one can find the rational point $\left(\frac{54}{35}, \frac{25}{21}\right)$ on this curve, which corresponds to the triangle with side lengths $\frac{41}{15}$, $\frac{101}{21}$, and $\frac{156}{35}$. Our method shows that this triangle has perimeter 12 and area 6, just like the $(3, 4, 5)$ triangle.

The journey does not stop here. The family of elliptic curves described here is closely related to the elliptic surfaced studied in (van Luijk 2007). Thus a journey that started in high school leads to the frontiers of research today.

4.3.3 *What Is to Be Learned from This?*

Again, we present these two examples as examples of a sensibility. Just as it is useful for high school teachers to view the mathematics learned in university as a useful tool in teaching secondary mathematics, it is useful for faculty at universities teaching prospective high school teachers to have the sensibility that the mathematics of high school can be mined for advanced examples in their courses. These two sensibilities are intertwined; together they could resolve the double discontinuity.

4.4 Implications for Teaching Abstract Algebra

The examples in Sect. 4.2 illustrate some of the many concrete applications of algebra, algebraic geometry, and number theory to the work of teaching mathematics at the secondary level, applications that are often missed in undergraduate courses and professional development programs. The examples in Sect. 4.3 illustrate ways in which high school mathematics can be applied to deep questions that show the utility of abstraction and of seeking structure in expressions. Courses in abstract algebra, in particular, would be much more useful to prospective teachers (and all undergraduate students, we claim) if they incorporated examples like these, examples that show how abstract methods provide useful tools for the day-to-day work of teaching and how questions and methods that live in high school mathematics can motivate some of those abstract methods.

We conclude with some suggestions for preservice courses in abstract algebra that we propose would contribute to closing the distance between school mathematics, university mathematics, and mathematics as it is practiced by mathematics professionals.

1. Abstractions should be capstones, not foundations—they should motivated with concrete examples whenever possible. This "experience before formality" is one of the hallmarks of mathematical work, and it is sometimes missing from dogmatic expositions of established mathematical results.
2. Groups should introduced in an historically faithful way, as part of an introduction to the Galois theory of polynomial equations.
3. The structural similarities between between $\mathbb{Z}$ and $k[x]$ (k a field) should be a major focus. These are the two major systems developed in school mathematics, and their underlying structure (that of a principle ideal domain) can form a bridge between arithmetic and algebra.
4. The development should follow the historical evolution of the ideas, showing how algebra evolved from techniques for solving equations to a study of systems in which the "rules of algebra" hold.
5. Applications should include those that are foundational for high school teaching. First, such applications can enrich and bring coherence to the mathematics that students study; second, they can help teachers in their professional work, such as designing lessons and tasks and sequencing topics; finally, to quote from (CBMS 2012, p. 54), they can help teachers understand that

"the mathematics of high school" does not mean simply the syllabus of high school mathematics, the list of topics in a typical high school text. Rather it is the structure of mathematical ideas from which that syllabus is derived.

Earlier texts, like the celebrated (Birkhoff and MacLane 2008), met many of these principles, except for item 5 above. The text (Cuoco and Rotman 2013) is one example of a course that attempts to meet all of them.[2] Some of the applications included in that course are:

- Pythagorean triples and the method of Diophantus.
- A historical development of $\mathbb{C}$.
- The mathematics of task design.
- Periods of repeating decimals.
- Cryptography.
- Lagrange interpolation and the Chinese Remainder Theorem.
- Ruler and compass constructions.
- Gauss' construction of the regular 17-gon.
- The arithmetic of $\mathbb{Z}[i]$ and $\mathbb{Z}[\omega]$.
- Solvability by radicals.
- Fermat's Last Theorem for exponents 3 and 4.

These are a mere sample of the ideas that have direct application to the work of teaching. Again, from (CBMS 2012, p. 66), "It is impossible to learn all the mathematics one will use in any mathematical profession, including teaching, in four years of college—teachers need opportunities to learn further mathematics throughout their careers." Professional development programs that developed other applications—in geometry, analysis, and statistics, for example—could carry this program forward for career-long learning for practicing teachers.

References

Ball, D. L., Thames, M. H., & Phelps, G. (2008). Content knowledge for teaching: What makes it special? *Journal of Teacher Education, 59*(5), 389–407.

Birkhoff, G., & McLane, S. (2008). *A survey of modern algebra. AKP classics.* Natick, MA: A K Peters.

Conference Board on the Mathematical Sciences. (2012). *The mathematical education of teachers II.* Providence, RI and Washington, DC: American Mathematical Society and Mathematical Association of America.

Cuoco, A., & Rotman, J. (2013). *Learning modern algebra. MAA textbooks.* Washington, DC: Mathematical Association of America.

de Guzmán, M., Hodgson, B. R., Robert, A., & Villani, V. (1998). Difficulties in the passage from secondary to tertiary education. *Documenta Mathematica, 3,* 747–762.

Klein, F. (1932). Elementary mathematics from an advanced standpoint. *Arithmetic, algebra, analysis.* London: Macmillan.

Rosenberg, S., Spillane, M., & Wulf, D. B. (2008). Delving deeper: Heron triangles and moduli spaces. *Mathematics Teacher, 101*(9), 656–663.

Silverman, J., & Tate, J. (1994). *Rational points on elliptic curves. Undergraduate texts in mathematics.* New York: Springer.

---

[2]Rotman died in 2016. Joe was a mathematician and teacher *extraordinaire*.

van Luijk, R. (2007). An elliptic K3 surface associated to Heron triangles. *Journal of Number Theory, 123*(1), 92–119.

Weyl, H. (1995). Part I. Topology and abstract algebra as two roads of mathematical comprehension. *The American Mathematical Monthly, 102*(5), 453–460.

Wu, H. (2015). *Mathematical education of teachers, part I: What is textbook school mathematics?* http://blogs.ams.org/matheducation/2015/02/20/mathematical-education-of-teachers-part-i-what-is-textbook-school-mathematics/#sthash.3Gzvh1Hp.dpbs. Accessed January 4, 2016.

Chapter 5
A Friendly Introduction to "Knowledge in Pieces": Modeling Types of Knowledge and Their Roles in Learning

Andrea A. diSessa

Abstract Knowledge in Pieces (KiP) is an epistemological perspective that has had significant success in explaining learning phenomena in science education, notably the phenomenon of students' prior conceptions and their roles in emerging competence. KiP is much less used in mathematics. However, I conjecture that the reasons for relative disuse mostly concern historical differences in traditions rather than in-principle distinctions in the ways mathematics and science are learned. This article aims to explain KiP in a relatively non-technical way to mathematics educators. I explain the general principles and distinguishing characteristics of KiP, I use a range of examples, including from mathematics, to show how KiP works in practice and what one might expect to gain from using it. My hope is to encourage and help guide a greater use of KiP in mathematics education.

Keywords Knowledge in pieces · Conceptual change · Complex systems

5.1 Introduction

5.1.1 Overview

Knowledge in Pieces (KiP) names a broad theoretical and empirical framework aimed at understanding knowledge and learning. It sits within the field of "conceptual change" (Vosniadou 2013), which studies learning that is especially difficult. While KiP began in physics education—in particular to provide a deeper

---

Some parts of this chapter are based on text in diSessa (2017), "Knowledge in Pieces: An evolving framework for understanding knowing and learning," in T. G. Amin and O. Levrini (Eds.), Converging perspectives on conceptual change: Mapping an emerging paradigm in the learning sciences, Routledge. Used by kind permission of Taylor and Francis/Routledge.

A. A. diSessa (✉)
University of California, Berkeley, USA
e-mail: disessa@berkeley.edu

G. Kaiser et al. (eds.), *Invited Lectures from the 13th International Congress on Mathematical Education*, ICME-13 Monographs,
https://doi.org/10.1007/978-3-319-72170-5_5

understanding of the phenomenon of "prior conceptions" (misleadingly labeled as "misconceptions"; Smith et al. 1993)—it has since engaged other areas, such as mathematics, chemistry, ecology, computer science, and even views of race and racism (Philip 2011).

I aim to produce a relatively non-technical introduction to KiP that can be understood by those who are not experts in the field of conceptual change, KiP's "home discipline." I emphasize breadth and "big ideas" over depth, while still pointing in the direction of KiP's distinctive fine structure and technical precision. A longer but still general introduction to KiP for those who want to pursue these ideas more deeply is diSessa et al. (2016).

Before beginning discussion in earnest, I would like to make two points about my strategy of exposition. First, the initial examples I give will be from physics, KiP's "home turf." I beg the (mathematical) reader's indulgence in doing so, but it allows me to select some of the best and most accessible examples of KiP analyses, where its core features are transparent, and where some competitive advantages over contrasting points of view are easiest to see. These examples are at the high-school level, so I do not expect them to be too much of a conceptual challenge. Mathematical examples will follow in Sects. 5.3 and 5.4. Second, with respect to mathematical examples, there are, of course, perspectives in the mathematics literature that bear on the same topics. While I will mention some of these (see comments and references in Sects. 5.3 and 5.4), careful comparative analysis is too complex for the scope of this paper. Readers who already know the relevant perspectives from mathematics education, of course, should be prepared to elaborate their own comparisons and conclusions.[1]

KiP is essentially epistemological: It aims to develop a modern theory of knowledge and learning capable of comprehending both short-term phenomena—learning in bits and pieces (hence the name, Knowledge in Pieces)—and long-term phenomena, such as conceptual change, "theory change," and so on. It aims to build a solid two-way bridge between, on the one hand, theory, and, on the other hand, data concerning learning and intellectual performance. "Two-way" implicates that (a) the theory is strongly constrained by and built out of observation, but also that (b) the theory can "project" directly onto what learners actually do as they think and learn, giving general meaning to their actions. KiP is, thus, a reaction against theories that are a priori, very high level, and consequentially are difficult to apply to the messiness of real-world learning.

KiP shares important features with two major progenitors. The first is Piagetian and neo-Piagetian developmental psychology, epitomized in mathematics education by Les Steffe, Ernst von Glasersfeld, Robbie Case, and many others.

---

[1]While I provide specific hints later for more detailed comparisons on a per-topic basis, probably the most effective single hint I can provide for reader-developed comparisons is to consider (a) whether work on the same topic identifies intuitive pre-cursor ideas in detail (few do), including their productive as well as problematic nature, and (b) whether data analysis includes extensive examination and explanation of students' in-process thinking, in addition to long-term comparisons. The presentation of distinctive KiP themes, just below, elaborates these points.

The core unifying feature of KiP with this work is constructivism, the focus on how long-term change emerges from existing mental structure. The second progenitor is cognitive modeling, such as in the work of John Anderson (e.g., his work on intelligent tutoring of geometry), or Kurt vanLehn (e.g., his work on students' "buggy" arithmetic strategies). The relevant common feature with KiP in the case of cognitive modeling is accountability to real-time data. A key distinctive feature of KiP, however, is its attempt to combine *both* long-term and short-term perspectives on learning. Piagetian psychology, in my view, was never very good at articulating what the details of students' real-time thinking have to do with long-term changes. In complementary manner, I judge that cognitive modeling has not done well comprehending difficult changes that may take years to accomplish.

I now introduce a set of interlocking themes that characterize KiP as a framework. These will be elaborated in the context of examples of learning phenomena to illustrate their meaning in concrete cases and their importance.

Complex systems approach—KiP views knowledge, in general, as a complex system of many *types* of knowledge elements, often having many particular *exemplars* of each type. Two contrasting types of knowledge are illustrated in the next main section.

Learning is viewed as a transformation of one complex system into another, perhaps with many common elements across the change, but with different organization. For example, students' *intuitive knowledge* (see the definition directly below) is fluid and often unstable, but *mature concepts* must achieve more stability through a broader and more "crystalline" organization, even if many of the same elements remain in the system. The pre-instructional "conceptual ecology" of students must usually be understood with great particularity—essentially "intuition by intuition"—in order to comprehend learning; general properties go only so far. A number of such particular intuitions will be identified in examples.

I use the terms "intuitive" and "intuition" here loosely and informally to describe students' commonsense, everyday "prior conceptions." However, consistent with the larger program, I will introduce a technical model of a very particular class of such ideas that has proven important in KiP studies.

A modeling approach—The learning sciences are still far from knowing exactly how learning works. It is more productive to recognize this fact explicitly and to keep track of how our ideas fail as well as how they succeed. Concomitantly, KiP builds models, typically models of different types of knowledge, not a singular and complete "theory of knowledge and learning," and the limits of those models are as important (e.g., in determining next steps) as demonstrated successes.

Continuous improvement—A concomitant of the modeling approach is a constant focus on improving existing models, and, sometimes, developing new models. In fact, the central models of KiP have had an extended history of extensions and improvements (diSessa et al. 2016). It is a positive sign that the core of existing models has remained in tact, while details have been filled in and extensions have been produced to account for new phenomena.

I call the themes above "macro" because they are characteristic of the larger program, and they are best seen in the sweep of the KiP program as a whole.

In contrast, the "micro" themes, below, can be relatively easily illustrated in many different contexts, which will be seen in the example work presented below.

A multi-scaled approach—I already briefly called out the commitment to both short-term and long-term scales of learning and performance phenomena, a *temporally multi-scaled approach*. Most conceptual change research, and, indeed, a lot of educational research, is limited to before-and-after studies, and there is almost no accountability to process data, to change as it occurs in moments of thinking.

A systems orientation also entails a second dimensional scale. Complex systems are built from "smaller" elements, and indeed, system change is likely best understood at the level of transformation and re-organization of system constituents. So, for example, the battery of "little" ideas, intuitions, which constitute "prior conceptions," can be selected from, refined, and integrated in order to produce normative complex systems, normative concepts. Since normative concepts are viewed as systems, their properties as such—both pieces and wholes—are empirically tracked. I describe a focus on both elements and system-level properties as *structurally multi-scaled*.

Richness and productivity—This theme is not so much a built-in assumption of KiP, but it is one of the most powerful and consistent empirical results. Naïve knowledge is, in general, rich and escapes simple characterizations (e.g., as isolated "misconceptions," simple false beliefs). Furthermore, learning very often—or always—involves recruiting many "old" elements into new configurations to produce normative understanding. This is the essence of KiP as a strongly constructivist framework, and it is one of its most distinctive properties in comparison to many competitor frameworks for understanding knowing, learning, and conceptual change. diSessa (2017) systematically describes differences compared to some contrasting theories of conceptual change. In my reading, assuming richness and productivity of naïve knowledge is comparatively rare, but certainly not unheard of, in mathematics, just as it is in science.

Diversity—An immediate consequence of the existence of rich, small-scaled knowledge is that there are many dimensions of potential difference among learners. Each learner may have a different subset of the whole pool of "little" intuitions, and might treat common elements rather differently. KiP may be unique among modern theories of conceptual change in its capacity to handle diversity across learners.

Contextuality—"Little" ideas often appear in some contexts, and not others. Furthermore, as they change to become incorporated into normative systems of knowledge, the contexts in which they operate may change. So, understanding *how knowledge depends on context* is core to KiP, while it is marginally important or invisible in competing theories. This focus binds KiP with situative approaches to learning ("situated cognition"). See Brown et al. (1989) for an early exposition, and continuing work by such authors as Jean Lave and Jim Greeno.

5.1.2 Empirical Methods

KiP is not doctrinaire about methods, and many different ones have been used.

Two modes of work are, however, more distinctive. First, KiP has *the development and continuous improvement of theory* (models) at its core. We in the community articulate limits of current models, encourage the refinement of old models and the development of new ones, when necessary.

Theory development, in turn, usually requires the richest data sources possible in order to synthesize and achieve the fullest possible accountability to the details of process. This is opposed to data that is quickly filtered and reduced to a priori codes or categories. In practice, *microgenetic* or *micro-analytic* study of rich data sources of students thinking (e.g., in clinical interviews) or learning (full-on corpora of individual or classroom learning) have been systematically used in KiP not only to validate, but also to generate new theory. See Parnafes and diSessa (2013) and the methodology section of diSessa et al. (2016). This kind of data collection and analysis is strongly synergistic with design-based research (diSessa and Cobb 2004), and iterative design and implementation of curricula—along with rich real-world tracking of data in concert with more cloistered and careful "break-out" studies of individuals—have been common.

I now proceed to concretize and exemplify the generalizations above with respect both to theory development and empirical work. I will boldface themes from the above list, as they are relevant. As mentioned, I start with examples having to do with physics, but then proceed to mathematics.

5.2 Two Models: Illustrative Data and Analysis

In this section I sketch the two best-developed and best-known KiP models of knowledge types. As such, the section illustrates KiP as a **modeling approach**. While both models are both **temporally** and **structurally multi-scaled**, the first model, p-prims, emphasizes smaller scales in time and structure. The second, coordination classes gives more prominence to larger scales.

5.2.1 Intuitive Knowledge

P-prims are elements of intuitive knowledge that constitute people's "sense of mechanism," their sense of which happenings are obvious, which are plausible, which are implausible, and how one can explain or refute real or imagined possibilities. Example p-prims are (roughly described): increased effort begets greater results; the world is full of competing influences for which the greater "gets its way," even if accidental or natural "balance" sometimes exists; the shape of a situation determines the shape of action within it (e.g., orbits around square planets are recognizably square). Comparable ideas in mathematics are that "multiplication makes numbers bigger" (untrue for multipliers less than one); a default assumption that a change in a given quantity generally implies a similar change in a related quantity (more implies more; less implies less, whereas, in fact, "denting" a shape may decrease area but increase circumference); and "negative numbers cannot apply to the real world" (what could a negative cow mean?). In the rest of this section, I will discuss physics examples only.

We must develop a new model for this kind of knowledge because, empirically, it violates presumptions of standard knowledge types, such as beliefs or principles. First, classifying p-prims as true or false (as one may do for beliefs or principles) is a category error. P-prims *work*—prescribe verifiable outcomes—in typical situations of use, but always fail in other circumstances. Indeed, when they will even be brought to mind is a delicate consequence of context (**contextuality**, both internal: "frame of mind"; or external: the particular sensory presentation of the phenomenon). So, for example, it is inappropriate to say that a person "believes" a p-prim, as if it would always be brought to mind when relevant, and as if it would always be used in preference to other ways of thinking (e.g., other p-prims, or even learned concepts). Furthermore, students simply cannot consider and reject p-prims (a commonly prescribed learning strategy for dealing with "misconceptions"). Impediments to explicit consideration are severe: There is no common lexicon for p-prims, and people may not even be aware that they have such ideas. Furthermore, "rejection" does not make sense for ideas that usually work, nor for ideas that may have very productive futures in learning (see upcoming examples).

Example data and analysis: J, a subject in an extended interview study (diSessa 1996), was asked to explain what happens when you toss a ball into the air. J responded fluently with a completely normative response: After leaving your hand, there is only one force in the situation, gravity, which slows the ball down, eventually to reverse its motion and bring it back down.

Then the interviewer asked a seemingly innocuous question, "What happens at the peak of the toss?" Rather than responding directly, J began to reformulate her model of the toss. She added another force, air resistance, which is changing, "gets stronger and stronger [as if to anticipate an impending balance and overcoming; see continuing commentary] to the point where when [sic] it stops." But then, she introduced yet another force, an upward one, which is equal to gravity, "in equilibrium for a second" at the top, before yielding to gravity. Starting anew, she provided a source for the upward force: It comes from your hand, and it "can only last so long against air and against gravity." In steps, she further decided that it's just gravity that is opposing the upward force, not air resistance, and gradually she reformulated the whole toss as a competition where the upward force initially overbalances gravity, reaching an equilibrium at the top, and then gravity takes over.

The key to understanding these events is that the interviewer "tempted" J to apply intuitive ideas of balancing and overcoming; he asked about the peak because the change of direction there looks like overcoming, one influence is getting weaker, or another is getting stronger. J "took the bait" and reformulated her ideas to include conflicting influences: The downward influence is gravity, but she struggled a bit to find another one, first trying air resistance, getting "stronger and stronger," but then introducing an upward force that is changing, getting weaker and weaker. This is a striking example of **contextuality**: J changed her model entirely after focusing attention on a particular part of the toss that suggested balancing. However, more surprises were to come.

Over the next four sessions, the interviewer continually returned to the tossed ball, providing increasingly direct criticism. "But you said the upward force is gone at the peak of the toss, and also that it balances gravity there. How can it both be zero and also balance gravity?" Over the last two sessions, the interviewer broke clinical neutrality and provided a computer-based instructional sequence on how force affects motion, including the physicist's one-force model of the toss. At the end of the instructional sequence, J was asked again to describe what happens in the toss. Mirroring her initial interview but with greater precision and care, she gave a pitch-perfect physics explanation. But, when asked to avoid an incidental part of her explanation (energy conservation), J reverted to her two-force model. So, we know that J exhibits not only surprising **contextuality** in terms of what explanation of a toss she would give, but that contextuality, itself, seems strongly persistent, a core part of her conceptual system.

After the completion of interviewing sessions, J reflected that she knew that it would appear to others that she described the toss in two different ways, and the "balancing" one might be judged wrong. But she felt both were really the same explanation.

Salient points: The dominant description of intuitive physics in the 1990s was that it constituted a coherent theory (see diSessa 2014, for a review and references), and the two-force explanation of the toss was a perfect example. External agents (the hand) supply a force that overcomes gravity, but is eventually balanced by it, and finally overcome. The KiP view, however, is that the "theory" only appears in particular situations (e.g., when overcoming is salient). Indeed, J did not seem to have the theory to start, but constructed it gradually, over a few minutes. **Contextuality** is missing from the then "conventional" view; "theories" comparable to Newton's laws don't come and go depending on what you emphasize in a visual scene. J's case is particularly dramatic since she never relinquished her intuitive ideas, even while she improved her normative ones. Instead, situation-specific saliences continued to cue one or the other "theory" of the toss. The long-term stability of an instability (the shift between two models of a toss) shows an attention to **multiple temporal scales** that is unusual in conceptual change studies but critical to understanding J's frame of mind. What happened in a moment each time it happened (shifting attention and corresponding shift in model of the toss), nonetheless continued to happen regularly over months of interviewing. Such critical phenomena test the limits of observational and analytical methods. For example, before and after tests are very unlikely even to observe the phenomenon. Attributing "misconceptions" categorically to a subject—"J has the non-normative dual force model of a toss"— fails to enfold this essentially **multi-scaled** and highly contextual analysis of J.

Another subject in the same study, K, started by asserting the two-force model of the toss. However, this subject reacted to similar re-directions of her attention concerning her explanation by completely reformulating her description to the normative model. She then observed that she had changed her mind and explained the reasons for doing so. The two-force model was then gone from the remainder of her interviews.

Ironically, a standard assessment employing first responses would classify J as normative, and K as "maintaining the naïve theory." Rather, K was a very different individual who could autonomously correct and stabilize her own understanding. J, in contrast, alternated one- and two-force explanations, and didn't really feel they were different. KiP methodologies did not assume simple characterization of either student's state of mind (**richness**), and they could also therefore better document and understand their differences (**diversity**). Neither J nor K would be well characterized by their initial responses. J, and not K, was deeply committed to a balancing view of many aspects of physics, even if both found balancing salient and significant in some cases.

Some lessons learned: The knowledge state of individuals is complex, and assessments cannot presume first responses will coherently differentiate them. The assumption of coherence in students' understanding is plainly suspect; J consistently maintained both the correct view and the "misconception," even in the face of direct instruction. The interviewer, knowing that fragile knowledge elements like p-prims are important, primed one (balancing, at the peak), and saw its dramatic influence. P-prims explain a lot about the differences and similarities between J and K (both used balancing, but J had a much greater commitment to it), but not everything. In continuing study (diSessa et al. 2002), we discovered that J showed an unusual and often counterproductive view of the nature of physics knowledge, which K did not. Modesty is the best policy: The complex conceptual ecology of students needs continuing work (**continuous improvement**).

One lesson learned here is that p-prims behave very differently than normative concepts. In terms that might be familiar to mathematics education researchers, p-prims provide a highly articulated version (specific elements whose use and contextuality can be examined across many circumstances) of a student's "concept image" (Tall and Vinner 1981). We need a different model to understand substantial, articulate and context-stable ideas, something roughly akin to "concept definition," but something that, in my view, uses KiP to better approach the cognitive and learning roots of expertise.

5.2.2   Scientific Concepts

Coordination classes constitute a model aimed at capturing central properties of expert concepts.

According to the coordination class model, the core function of concepts is to read out particular concept-relevant information reliably across a wide range of circumstances, unlike the slip-sliding activation of p-prims. Figure 5.1 explains.

Figure 5.2 shows the primary difficulty in creating a coherent concept. All possible paths from world (or imagined world) to concept attributes must result in the same determination. This is called *alignment*, and it is a property of the whole system, not of any part of it.

A physics example of lack of alignment is that students will sometimes determine forces by using intuitive inferences ("An object is moving; there must be a force on it."), and sometimes by "formal methods" ("An object is moving at a constant speed; according to Newton's third law, there is no net force on it."). A mathematical example is that students may deny that an equator on a sphere with three points marked on it is a triangle, even if they have agreed that any part of a great circle is a "straight line," and that a triangle is any three connected straight line segments.

Coordination classes are large and complex systems. This is structurally unlike p-prims, which are "small," simple, and relatively independent from one another. Alignment poses a strict constraint on all possible noticings (e.g., noticing F_1 or F_2 in Fig. 5.2) and all possible inferences (e.g., I_1 and I_2): All paths should lead to the same determination. That is, there is a *global* constraint on all the pieces of a

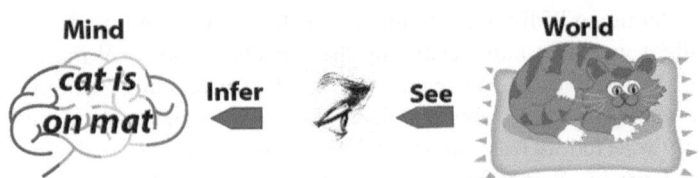

Fig. 5.1 Coordination classes allow reading out information relevant to concepts, here illustrated by "location," from the world. The readout happens in two stages. (1) "See" or "notice" involves extracting *any* concept-relevant information: "The cat is *above* the mat," and "The cat is *touching* the mat." (2) "Infer" draws conclusions specifically about the relevant information (location) using what has been seen: "The cat is *on* the mat."

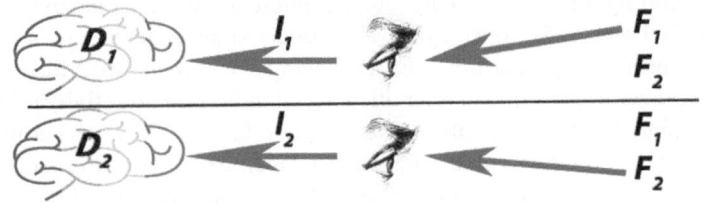

Fig. 5.2 In situations where multiple features (F_1, F_2) are available, different choices of what to observe may lead to different inferences (I_1, I_2) and potentially contradictory determinations (D_1, D_2) of the "same" information

coordination class, which makes the model essentially **multi-scaled**. In this case, multi-scaled refers to the **structure** of the knowledge system—pieces and the whole system—rather than to its **temporal** properties, which were emphasized with J.

I will not belabor a full taxonomy of parts of coordination classes, but, because it is relevant to an example from mathematics (Sect. 3.1), I note that a coordination class needs to include *relevance*, in addition to *noticings* and *inferences*. Relevance means that a coordination class needs to "know" when a concept applies and when information about it *must* be available. If you are asked about slope, there *must* be some available information about "rise" and "run," and it behooves one to attend to that information.

Dufresne et al. (2005) provided an accessible example of core coordination class phenomena. They showed two groups of university students, engineering and social science majors, various simulated motions of a ball rolling along a track that dipped down, but ended at its original height. They asked which motion looked most realistic. Subjects saw the motions in two contexts: one that showed only the focal ball, and another that also showed a simultaneous and constant ball motion in a parallel, non-dipping path. The social scientists' judgments of the realism of the focal motion remained nearly the same from the one- to two-ball situation. But, the engineers showed a dramatic shift, from preferring the correct motion to preferring another motion that literally no one initially believed to be realistic. In the two-ball case, engineers performed much worse than social scientists!

Using clinical interviews, the researchers confirmed that the engineers were looking at ("noticing") different things in the different situations. Relative motion became salient with two balls, changing the aspects of the focal motion that were attended to. In the two-ball presentation, a kind of balancing, "coming out even" dominated their inferences about realism. The very same motion that they had resoundingly rejected as least natural became viewed as most realistic.

Lessons learned: Scientific concepts are liable to shifts of attention during learning, and thus different (incoherent) determinations of their attributes. This is an easily documentable feature of learning concepts such as "force," and there is every reason (and some documentation) to believe this is also true for mathematical concepts. So, people must learn a variety of ways to construe particular concepts in various contexts, ways that are differentially salient in various conditions, yet all determinations must "align." Again, this local/global coherence principle shows KiP's attention to **multiple scales of conceptual structure**.

It is only mildly surprising that the "culprit" inference here is a kind of balancing, as implicated in J's case. So, once again, a relatively small-scaled element, similar to balancing p-prims, plays a critical role. Balancing is a core intuitive idea, but it also becomes a powerful principle in scientific understanding (**productivity**). Changes in kinetic and potential energy do always *balance out*. In this case, engineering students have elevated the importance and salience of balancing compared to social scientists, but have not yet learned very well what exactly balances out, and when balancing is appropriate (relevance). Certain p-prims are thus learned to be powerful, but they have not yet taken their proper place in understanding physics. Incidentally, this analysis also accounts for a very surprising difference (**diversity**) between different classes of students—engineers and social scientists.

P-prims and coordination classes are nicely complementary models. Within coordination class theory, p-prims turn out to account for certain problems (mainly in terms of inappropriate inferences), but they also can lie on good trajectories of learning, in constructing the overall system. Balancing is a superb physical idea, but naïve versions of balancing need to be developed precisely and not overgeneralized. Linearity is a comparable idea in mathematics. It is a wonderful and powerful idea, but it does not work, for example, for functions in general. $Sin(a + b)$ is not $sin(a) + sin(b)$. As balancing and linearity develop, they both need to be properly coordinated with checks and other ways of thinking.

5.3   Examples in Mathematics

This section displays some mathematical examples. The field of KiP analyses in mathematics is less rich than for physics, and overall trends are less well scouted out. But, to give a sense of what KiP looks like in mathematics and to encourage further such work is a primary goal of this article.

5.3.1   The Law of Large Numbers

Joseph Wagner (2006) used the main ideas of coordination class theory to study the learning of the statistical "law of large numbers": The distribution of average values in larger samples of random events hews more closely to the expected value (long-term average) than for smaller samples. In complementary manner, smaller samples show a greater dispersion; a greater proportion of their averages will be far from the expected value. So, if one uses a sample of 1000 coin tosses, one is nearly assured that the sample will have an average close to 50% heads and 50% tails. A sample of 10 tosses can easily lead to averages of, say, 70% heads and 30% tails. In the extreme case, a single toss, one is guaranteed of "averages" that are as far as possible from the long-term average: one always gets 100% heads, or 100% tails.

Wagner discovered that students often showed canonical coordination class difficulties during learning. Many had exceedingly long trajectories of learning, corresponding to learning in different contexts of use of the law of large numbers. In more technical detail, thinking in different contexts typically involves different knowledge (different noticings and different inferences), which may need to be acquired separately for different contexts. Furthermore, reasoning about the law in each context must *align* in terms of "conceptual output" (e.g., what is the relevant expected value) across all contexts. In short, **contextuality** is a dramatic problem for the law of large numbers, and systematic integrity (a **large-scale structural property**—in fact, the central-most large-scale property of coordination classes) is hard won in view of the **richness** of intuitive perspectives that may be adopted local to particular contexts (**small-scale structure**; think p-prims).

I present an abbreviated description one of Wagner's case studies to illustrate. Similar to the case of J, this is a fairly extreme case, but one in which characteristic phenomena of coordination class theory are easy to see. In particular, we shall see that learning across a wide range of situations appears necessary. The law of large numbers might not even appear to the learner as relevant to some situations, or it might be applied in a non-aligned way, owing to intuitive particulars of the situations. I sketch the subject's learning according to diSessa (2004), although a fuller analysis on most points and a more extensive empirical analysis appear in Wagner (2006).

The subject, called M ("Maria" in Wagner 2006), was a college freshman taking her first course in statistics. Wagner interviewed her on multiple occasions throughout the term (methodologically similar to J's study), and used a variety of near isomorphic questions involving the law of large numbers. The questions asked whether a small or large sample would be more likely to produce an average within particular bands of values, bands that include the expected value, or bands that are near or far from it. Would you choose a small or large sample if you wanted to get an average percentage of heads in coin tosses between 60 and 80% of the tosses? The law of large numbers says you would want a smaller number of tosses; in contrast, a very large number of tosses is almost certain to come out near 50% heads.

We pick up M's saga after she learned, with some difficulty, to apply the law of large numbers to coin tosses. Just after an extensive discussion of the coin situation, the interviewer (Jo) showed M a game "spinner," where a spun arrow points to one of 10 equal angular segments. Seven of the segments are blue, and three are green. Jo proceeded to ask M whether one would want a greater or lesser number of spins if one wanted to get an average of blues between 40 and 60 percent of the time.

M: OK. ... Land on blue? ... Well, 70% of the // of that circle is blue. Yeah. Seventy percent of it is blue, so, for it to land between 40 and 60 percent on blue, then, I would say there really is no difference. [She means it doesn't make a difference whether one does few or a lot of spins.]

Jo: Why?

M: Because if 70% of the // the circle, or, yeah, the spinner is blue, so ... it's most likely going to land in a blue area, regardless of how many times I spin it. It kinda really doesn't matter. It's not like the coins...

M is saying that she does not see the spinner situation as one in which the law of large number applies. The coordination class issue of *relevance* defines one of her problems. The larger data corpus suggests that a significant part of the problem is that M does not see that the concept of expected value applies to the spinner. She knows that in one spin, 70% of the time you will get blue, and 30% of the time you will get green. She reasons pretty well about "chances" for individual spins. But she simply does not believe that the long-term average, the expected percentage of blues or greens, exists. She "sees" chances, but does not infer from them a long-term average, nor even appear to know that a long-term average exists in this case.

Jo showed M a computer simulation of the spinner situation and proposed to do an experiment of plotting the result (histogram) of many samples of a certain number of spins. Would the percentages of blue pile up around any value, the way coin tosses always pile up around 50%? M was reluctant to make any prediction at all. But she very hesitantly suggested that the results might pile up around 70%. When the simulation was run, M was evidently surprised. "It *does* peak [pile up] around 70!!"

Here, we are at a disadvantage because we know much less about the relevant p-prims (or similar knowledge elements) that are controlling M's judgments, unlike the fact that, for J, the interviewer suspected balancing might provoke a different way of thinking about the toss, or that Dufresne et al. found that "balancing out" also sometimes controlled engineers' judgments about the realism of depicted motions of rolling balls. A good coordination class analysis demands a better analysis than the data here allow. However, a hint was offered earlier in the conversation when Jo pressed M to explain how the spinner differed from coins. M reported, "The difference, uh, between the coins and this [spinner] is that, in every toss, in the coin, I know that there's a … 50% chance of getting a head, 50% chance of getting a tail." But with a spinner, "It's just not the same." Although M cannot put her finger on the difference, it seems plausible that she sees the 50–50 split of a coin flip to be *inherent in the coin*, "in every toss…," while the spinner arrow, per se, does not visibly (to her) have 70–30 in its very nature. An alternative or contributing factor involves the well-known fact concerning fractions that students seem conceptually competent first with simple ones, like ½. But, again, there is not enough data to distinguish possibilities.

Independent of the reason, the big picture relevant to coordination classes is that M simply does not see the spinner as essentially similar to coins. The *relevance* part of her developing coordination class is the most obvious problem. In particular, she doesn't naturally see an expected value as relevant to (nor determinable for) spinners. This case has a happy ending because the empirical (computer simulation) result was enough to convince M that expected value existed in the spinner case, and she began to reason more normatively about Jo's questions. To summarize, there was a conceptual **contextuality** that prevented using the same pattern of reasoning, the law of large numbers, in different situations. M needed to learn that expected value existed for spinners, and that it related to the "chances" concerning a single case in the same way as for coins: The long-term expected average is the same as the "chances" for a single case.

The final case of contextuality I report (there are many others!) concerns the average height of samples of men, corresponding to men in the U.S. registering for the military draft at small or large post offices. If the average height in the U.S. is 5 ft 9 in., would a small or large post office (small or large sample) be more likely to find an average height for one day of more than 6 ft? At first, M had no idea how to answer the question. Pressed, she offered an uncertain reference to larger sets of numbers having smaller averages. The law of large numbers was, again, invisible to her in this context.

Jo improvised yet another context. Would you rather take a big or small sample of men at a university in order to find the average height? M was quick and confident in her answer. A larger sample would be "more representative,"[2] "more accurate." Arguably, the sampling context evoked a memory or intuition that larger samples are "better." Having made the connection to this intuition, M applied it relatively fluently to the post office problem.

The reason "representativeness" and "accuracy" were cued in the university sampling situation and not previously might not be clear. But M did not mention these intuitive ideas in any previous problems, and, once cued, she took those ideas productively into new contexts. The combination of **contextuality** and **productivity**, shown here, is highly distinctive of KiP analyses. Some intuitions, even if they are not usually evoked, can be useful if, somehow, they are brought to the learner's attention.

The next example is among the first applications of KiP to mathematics (a decade earlier than Wagner's work), and the final one is among the latest (a decade later than Wagner).

5.3.2   Understanding Fractions

Smith (1995) did an investigation of student understanding of rational numbers and their representation as fractions according to broad KiP principles. He began by critiquing earlier work as (a) using a priori analysis of dimensions of mathematical competence, and also (b) systematically assessing competence according to success on tests. Instead, he proposed to look at competence directly in the specific strategies students use to solve a variety of problems. In particular, he did an exhaustive analysis of strategies used by students during clinical interviews on a set of fractions problems that was carefully chosen to display core ideas in both routine and novel circumstances. Smith looked most carefully at the strategies used by students who could be classified as "masters" of the subject matter. So, his intent was to describe the nature of achievable, high-level competence by looking directly at the details of students' performance.

The results were surprising in ways that typify KiP work. Masters used a remarkable range of strategies adapted rather precisely to particulars of the problems posed. While they did occasionally use the general methods that they had been taught (methods like converting to common denominators or converting to decimals), general methods appeared almost exclusively when none of their other

---

[2]Kahneman and Tversky (1972) provide a now-canonical treatment of statistical "misconceptions," including representativeness. However, their theoretical frame is very different from KiP. **Productivity**, in particular, is missing, unlike the cited role of representativeness in M's learning. These authors maintain that, to learn, intuitions must be excluded, and formal rules must be followed without question. Pratt and Noss (2002) provide a KiP-friendly treatment of statistical intuitions.

methods worked. A careful look at textbooks suggested that it was unlikely that many, if any, of the particular strategies had been instructed. Student mastery seems to transcend success in learning what is instructed.

In net, observable expertise is: (a) "fragmented" (**contextual**) in that it is highly adapted to problem particulars; (b) **rich**, composed of a wide variety of strategies; and (c) significantly based on invention, rather than instruction. The latter two points suggest **productivity**, the use of rich intuitive, self-developed ideas, and that that richness is maintained into expertise, in contrast to what conventional instruction seems to assume.

One can summarize Smith's orientation so as to highlight typical KiP strategies, which contrast with those of other approaches:

- avoiding a priori or "rational" views of competence in favor of directly empirical approaches: Look at what students do and say about what they do.
- couching analysis in terms of knowledge systems (**a complex systems approach**) of elements and relations among them (e.g., particular strategies were often, but not always, defended by students by reference to more general, instructed ways of thinking).
- discovering that the best student understanding, not just intuitive precursors, is rich (many elements), diverse, and involves a lot of highly particular and con-textually adapted ideas (**contextuality**). Thus it is in some ways more similar to pre-instructional ideas than might be expected.

Smith did not use the models (p-prims, etc.) that later became the recognizable core of KiP. But, still, the distinctiveness of a KiP orientation proved productive. I believe this is an important lesson, that, independent of technical models and details, KiP's general principles and orientations can provide key insights into learning that are not available in other perspectives. Newcomers to KiP might do well to start their work at this level, and move to more technical levels when those details come to seem sensible, and when and if the value of technicalities becomes palpable.

5.3.3   Conceptual and Procedural Knowledge in Strategy Innovation

The relationship of procedural to conceptual knowledge is a long-standing, important topic in mathematics education. There is a general agreement that one should strike a balance between these modes. However, at a more intimate level, the detailed relations should be important. What conceptual knowledge is important, when, and how? It is known that students can (e.g., Kamii and Housman 2000) and do (e.g., Smith's work, above) spontaneously innovate procedures. How might conceptual knowledge be important to innovation, specifically what knowledge is important, and what is the nature and origin of those resources?

Levin (2012) studied strategy innovation in early algebra. Her study involved a student who started with an instructed guess-and-check method of solving problems like: "The length of a rectangle is six more than three times the width. If the perimeter is 148 ft., find the length and width." Over repeated problem solving, this student moved iteratively, without direct instruction, from guess-and-check to a categorically different method: a fluent algorithmic method that mathematicians would identify as linear interpolation/extrapolation. One of the interesting features of the development was that intuitive "co-variation schemes," more similar to calculus (related rates) than anything instructed in school, rooted his development (**productivity**). Indeed, his development could be traced through six distinct levels of co-variation schemes, progressively moving from qualitative (the "more implies more" intuition, but in a circumstance where it is productive), toward more quantitatively precise, general, and "mathematical-looking" principles.

In order to optimally track and generalize this student's progress, Levin extended the coordination class model to what she calls a "strategy system" model, demonstrating the generative and evolving nature of KiP (**continuous improvement**). Her model maintained a focus on perceptual categories ("seeing" in Fig. 5.1), and inferential relations (e.g., co-variation schemes). But there were also theoretical innovations: Typically more than one coordination class is involved in strategy systems. General conceptions (inferences) specifically supported procedural actions in particular ways.

In addition to the core co-variational idea, a cluster of intuitive categories, such as "controller," "result," "target," and "error" played strongly into the student's development. All in all, Levin's study showed the surprising power of intuitive roots—ones that may never be invoked in school—and provided a systematic framework for understanding their use in the development of procedural/conceptual systems.

5.3.4　Other Examples

In addition to what was presented above, I recommend a few other examples of KiP work that will be helpful for mathematics education researchers with different specialties in order to understand the KiP perspective. Andrew Izsák's has developed an extensive body of work using KiP to think about learning concerning, for example, area (Izsák 2005), and early algebra (Izsák 2000). Similarly Adiredja (2014) treated the concept of limit from a KiP perspective. Adiredja's analysis is important in the narrative of this article in that it takes steps to comprehend learning of the topic, limits, at a fine grain-size, including the productivity and not just learning difficulties that emerge from prior intuitive ideas. The work may be profitably contrasted with that of Sierpinska (1990) and Tall and Vinner (1981) on similar topics.

5.4   Cross-Cutting Themes

In this final section, I identify KiP's position and potential contributions to two large-scale themes in the study of learning in mathematics and science.

5.4.1   Continuity or Discontinuity in Learning

I believe that one of the central-most and still unsettled issues in learning concerns whether one views learning as a continuous process or a discontinuous one. In particular, how do we interpret persistent learning problems that appear to afflict students for extended periods of time? In science education, so-called "misconceptions" or "intuitive theories" views treat intuitive ideas as both entrenched and unproductive. They are assumed to be unhelpful—blocking, in fact—because they are simply wrong (Smith et al. 1993). In mathematics education, one also finds a lot of discussion about misconceptions (e.g., concerning graphing, Leinhardt et al. 1990) and also about the essentially problematic nature of "intuitive rules" such as "more implies more" (Stavy and Tirosh 2000). But, more often than in science, researchers implicate discontinuities of form, rather than just content. For example, Sierpinska (1990) talks about basic "epistemological obstacles," large-scale changes in "ways of knowing." Vinner (1997) talks about "pseudo-concepts" as bedeviling learners, and some interpretations of the distinction between process and object conceptualizations in mathematics (Sfard 1991) put process forms as inferior to conceptions that are at the level of objects (not necessarily Sfard's contention). Or, the transition from process to object modes of thinking is always intrinsically difficult. Tall (2002) emphasizes the existence of discontinuities possibly due to deep-seated brain processes ("the limbic brain;" sensory-motor thinking). Along similar lines (as anticipated in footnote 2), Kahneman and Tversky's view of difficulties in learning about chance and statistics relies on so-called "dual process" theories of mind. (See Glöckner and Witteman 2010, for a review and critical assessment.) Instinctive (intuitive) thinking must be *replaced* with a categorically different kind of thinking based on a conscious and explicit rule following.

On the reverse side, mathematics education researchers sometimes have supported the productivity of intuitive ideas (e.g., Fischbein 1987), and, most particularly, constructivist researchers have pursued important lines of continuity between naïve and expert ideas (Moss and Case 1999, is, in my view, an exceptional example from a large literature). However, very few studies approach the detail and security of documentation of elements, systems of knowledge, and processes of transformation of the best KiP analyses.

The issues are too complex and unresolved for a discussion here, but KiP offers a view and accomplishments to support a more continuous view of learning and to critique discontinuous views. For example, both experts and learners use intuitive ideas, even if their knowledge is different at larger scales of organization.

Gradual organization and building of a new system need not have any essential discontinuities: There may not be any chasm separating the beginning from the end of a long journey. It is just that, before and after, things may look quite different. A core difficulty in learning might simply involve (a) a mismatch between our instructional expectation concerning how long learning should take and the realities of the transformation, and (b) a lack of understanding of the details of relevant processes. KiP offers unusual but tractable and detailed models of small-scale, intuitive knowledge that can support its incorporation into expertise, and methodologies capable of discovering and carefully describing particular elements. These issues are treated in more detail in Gueudet et al. (2016).

5.4.2 Understanding Representations

To conclude, I wish to mention two KiP-styled studies concerning the general nature of representational competence—central to mathematical competence—and the roles of intuitive resources in learning about representations.

Sherin (2001) undertook a detailed study of how students use and learn with different representational systems (algebra vs. computer programs) in physics. One of Sherin's key findings was that p-prim-like knowledge mediates between real-world structure ("causality") and representational templates. For example, the idea of "the more X, the more Y" (e.g., more acceleration means greater force) translates into the representational form "$Y = kX$" (e.g., $F = ma$). Sherin's work will be most interesting to mathematics education researchers interested in how representations become meaningful in thinking about real-world situations (modeling), how such situations bootstrap understanding of mathematical structure, and the detailed role that intuitive knowledge plays in these processes. This work builds on similar earlier work by Vergnaud (1983), but in distinctly KiP directions.

Finally, diSessa et al. (1991) studied young students' naïve resources for thinking about representations. In contrast to misconceptions-styled work, we uncovered very substantial expertise concerning representations. However, the expertise was different than what is normally expected in school. It had more to do with the generative aspects of representation (e.g., design and judgments of adequacy) and less to do with the details of instructed representations. This repository of intuitive competence is essentially ignored in school instruction, an insight shared with a few (e.g., Kamii and Housman 2000), but not many, mathematics education researchers.

References

Adiredja, A. (2014). *Leveraging students' intuitive knowledge about the formal definition of a limit* (Unpublished doctoral dissertation). Berkeley, CA, University of California.

Brown, J. S., Collins, A., & Duguid, P. (1989). Situated cognition and the culture of learning. *Educational Researcher, 18*(1), 32–42.

diSessa, A. A. (1996). What do "just plain folk" know about physics? In D. R. Olson & N. Torrance (Eds.), *The handbook of education and human development: New models of learning, teaching, and schooling* (pp. 709–730). Oxford, UK: Blackwell Publishers, Ltd.

diSessa, A. A. (2004). Contextuality and coordination in conceptual change. In E. Redish & M. Vicentini (Eds.), *Proceedings of the International School of Physics "Enrico Fermi": Research on Physics Education* (pp. 137–156). Amsterdam: ISO Press/Italian Physics Society.

diSessa, A. A. (2014). A history of conceptual change research: Threads and fault lines. In K. Sawyer (Ed.), *Cambridge handbook of the learning sciences* (2nd ed., pp. 88–108). Cambridge, UK: Cambridge University Press.

diSessa, A. A. (2017). Conceptual change in a microcosm: Comparative analysis of a learning event. *Human Development, 60*(1), 1–37.

diSessa, A. A., & Cobb, P. (2004). Ontological innovation and the role of theory in design experiments. *Journal of the Learning Sciences, 13*(1), 77–103.

diSessa, A. A., Elby, A., & Hammer, D. (2002). J's epistemological stance and strategies. In G. Sinatra & P. Pintrich (Eds.), *Intentional conceptual change* (pp. 237–290). Mahwah, NJ: Lawrence Erlbaum Associates.

diSessa, A. A., Hammer, D., Sherin, B., & Kolpakowski, T. (1991). Inventing graphing: Meta-representational expertise in children. *Journal of Mathematical Behavior, 10*(2), 117–160.

diSessa, A. A., Sherin, B., & Levin, M. (2016). Knowledge analysis: An introduction. In A. diSessa, M. Levin, & N. Brown (Eds.), *Knowledge and interaction: A synthetic agenda for the learning sciences* (pp. 30–71). New York, NY: Routledge.

Dufresne, R., Mestre, J., Thaden-Koch, T., Gerace, W., & Leonard, W. (2005). When transfer fails: Effect of knowledge, expectations and observations on transfer in physics. In J. Mestre (Ed.), *Transfer of learning: Research and perspectives* (pp. 155–215). Greenwich, CT: Information Age Publishing.

Fischbein, E. (1987). *Intuition in science and mathematics: An educational approach*. Dortrecht, The Netherlands: Kluwer Academic.

Glöckner, A., & Witteman, C. (2010). Beyond dual-process models: A categorisation of processes underlying intuitive judgement and decision making. *Thinking & Reasoning, 16*(1), 1–25.

Gueudet, G., Bosch, M., diSessa, A., Kwon, O. N., & Verschaffel, L. (2016). *Transitions in mathematics education*. In G. Kaiser (Series Ed.), ICME-13 topical surveys. Switzerland: Springer International.

Izsák, A. (2000). Inscribing the winch: Mechanisms by which students develop knowledge structures for representing the physical world with algebra. *Journal of the Learning Sciences, 9*(1), 31–74.

Izsák, A. (2005). "You have to count the squares": Applying knowledge in pieces to learning rectangular area. *Journal of the Learning Sciences, 14*(3), 361–403.

Kahneman, D., & Tversky, A. (1972). Subjective probability: A judgment of representativeness. *Cognitive Psychology, 3*, 430–454.

Kamii, C., & Housman, L. (2000). *Young children reinvent arithmetic* (2nd ed.). New York, NY: Teachers College Press.

Leinhardt, G., Zaslavsky, O., & Stein, M. K. (1990). Functions, graphs, and graphing: Tasks, learning, and teaching. *Review of Educational Research, 60*(1), 1–64.

Levin (Campbell), M. E. (2012). *Modeling the co-development of strategic and conceptual knowledge during mathematical problem solving* (Unpublished doctoral dissertation). Berkeley, CA, University of California.

Moss, J., & Case, R. (1999). Developing children's understanding of the rational numbers: A new model and an experimental curriculum. *Journal for Research in Mathematics Education, 30*(2), 122–147.

Parnafes, O., & diSessa, A. A. (2013). Microgenetic learning analysis: A methodology for studying knowledge in transition. *Human Development, 56*(5), 5–37.

Philip, T. (2011). A "knowledge in pieces" approach to studying ideological change in teachers' reasoning about race, racism and racial justice. *Cognition and Instruction, 11*(3), 297–329.

Pratt, D., & Noss, R. (2002). The microevolution of mathematical knowledge: The case of randomness. *The Journal of the Learning Sciences, 11*(4), 453–488.

Sfard, A. (1991). On the dual nature of mathematical conceptions: Reflections on processes and objects as different sides of the same coin. *Educational Studies in Mathematics, 22*(1), 1–36.

Sierpinska, A. (1990). Some remarks on understanding in mathematics. *For the Learning of Mathematics, 10*(3), 24–36.

Sherin, B. (2001). A comparison of programming languages and algebraic notation as expressive languages for physics. *International Journal of Computers for Mathematical Learning, 6*(1), 1–61.

Smith, J. P. (1995). Competent reasoning with rational numbers. *Cognition and Instruction, 13*(1), 3–50.

Smith, J. P., diSessa, A. A., & Roschelle, J. (1993). Misconceptions reconceived: A constructivist analysis of knowledge in transition. *Journal of the Learning Sciences, 3*(2), 115–163.

Stavy, R., & Tirosh, D. (2000). *How students (mis-)understand science and mathematics.* New York, NY: Teachers College Press.

Tall, D. (2002). Continuities and discontinuities in long-term learning schemas. In D. Tall & M. Thomas (Eds.), *Intelligence, learning and understanding: A tribute to Richard Skemp* (pp. 151–177). Flaxton QLD, Australia: Post Pressed.

Tall, D., & Vinner, S. (1981). Concept image and concept definition in mathematics with particular reference to limits and continuity. *Educational Studies in Mathematics, 12*(2), 151–169.

Vergnaud, G. (1983). Multiplicative structures. In R. Lesh & M. Landau (Eds.), *Acquisition of mathematics concepts and processes* (pp. 127–174). New York, NY: Academic Press.

Vinner, S. (1997). The pseudo-conceptual and the pseudo-analytical thought processes in mathematics learning. *Educational Studies in Mathematics, 34*(2), 97–129.

Vosniadou, S. (Ed.). (2013). *International handbook of research on conceptual change* (2nd ed.). New York, NY: Routledge.

Wagner, J. (2006). Transfer in pieces. *Cognition and Instruction, 24*(1), 1–71.

Chapter 6
History of Mathematics, Mathematics Education, and the Liberal Arts

Michael N. Fried

Abstract This paper considers how the history of mathematics, if it is taken
seriously, can become a mode of thinking about mathematics and about one's own
humanness. What I mean by the latter is that by studying the history of mathematics
rather than simply using it as a tool—and that means attempting to understand it as
an historian does—one becomes aware of how mathematics is something human
beings do that therefore informs our human identity. In this way, the history of
mathematics in mathematics education has the potential to make us fuller human
beings, which is at the heart of the educational tradition known as the "liberal arts."
By considering the nature of the liberal arts, we may understand better the meaning
of the history of mathematics in mathematics education and, indeed, the meaning of
mathematics education tout court.

Keywords History of mathematics · Humanistic mathematics · Liberal arts
Whiggism

This paper concerns the history of mathematics and mathematics education.
I should say from the start that I will not display results from empirical research
showing how the history of mathematics is good for this or that. This is not because
I belittle such research. Not at all. However, much of that research treats the history
of mathematics as a tool, to use the phrase Jankvist (2009) has popularized. Again, I
have no objection to questions about tools and utility. Indeed, the last part of my
lecture concerning the liberal arts is in some way a matter of profound utility.
Nevertheless, emphasizing the use of the history of mathematics, as I pointed out in
my 2001 paper on the subject (Fried 2001), draws us away from the meaning of the
history of mathematics in mathematics education. It is that—the meaning of the
history of mathematics in mathematics education as something to study rather than
to use—that I wish to elaborate here.

M. N. Fried (✉)
Ben Gurion University of the Negev, Beersheba, Israel
e-mail: mfried@bgu.ac.il

© The Author(s) 2018   85
G. Kaiser et al. (eds.), *Invited Lectures from the 13th International Congress
on Mathematical Education*, ICME-13 Monographs,
https://doi.org/10.1007/978-3-319-72170-5_6

The paper will comprise four parts. By way of introduction, I will say a few words about D. E. Smith, whose importance both in the field of mathematics education—not the least because of his involvement in ICMI, the organization behind this conference—and in the history of mathematics is undeniable. Next, I will discuss the nature of history and its character as a discipline. Following that, I will make the point that a non-historical tendency enters mathematics teaching when history is viewed as something to be used only, and that that leads to a kind of dilemma for the teacher who has a serious interest in history. This third section will end, however, by suggesting that that very dilemma can provide us with an opportunity to review what we really mean by mathematics education or, rather, by the mathematically educated person. Finally, I will turn briefly to the old idea of the liberal arts. Taking the term *artes liberales* literally, these are the arts of a "free human being," or, better, of a fully human being. Thinking about the liberal arts in connection to the history of mathematics, I will claim, has the potential of bringing us back to a mathematics education aiming to make our students more fully human. And with that we may obtain insight into the meaning both of the history of mathematics in mathematics education and mathematics education itself.

6.1   By Way of Introduction: David Eugene Smith

Despite its apparent distance from mainstream empirical research in mathematics education, the subject of my paper is, I believe, appropriate for the ICME community. For one, there are several other sessions in the conference centered on historical ideas. But, more than that, the history of mathematics was a central preoccupation of David Eugene Smith (1860–1944), whose remarks in *L'Enseignement Mathématique* in 1905 set into motion the creation of ICMI in the first place. For this reason alone it is worth saying a few words about Smith. But thinking about Smith and his views also brings us directly into the set of ideas I wish to develop in this paper.

As most of you probably know, the 1905 article in *L'Enseignement Mathématique* was a response to an inquiry (published in the same volume of the journal) concerning the reforms necessary for the teaching of mathematics. Smith was only one respondent among others including such luminaries as Gino Loria and Emile Borel. It was Smith's view, though, that what was urgently needed was an *international* organization dedicated to questions on mathematics teaching. And, partly in response to Smith's proposal, the ICMI was created three years later. But more pertinent to my subject in this paper, in that same short piece in 1905, Smith also took the opportunity to express his views about the importance of the history of mathematics. He said that, regarding the training of mathematics teachers, besides knowing integral and differential calculus, the teacher, "…also ought to know, in a precise way, the historical development of subjects being taught, why were they were taught, how were they presented in different places" (Smith 1905, p. 470, my translation).

Smith had already stated a similar position more than once in his earlier work, *The Teaching of Elementary Mathematics* (Smith 1902). I would add too, that, like me, Smith is critical in this work about the motive of utility in mathematics education. He tells us that there are two main motives for teaching arithmetic:

> ...arithmetic, like other subjects is taught either (1) for its utility, or (2) for its culture. Under the former is included the general "bread-and-butter value" of the subject and its applications; under the latter, its training in logic, its bearing upon ethical, religious and philosophical thought. (p. 20)

He says that the utility motive favored by the "mechanical teacher" (as he puts it, together with another expression, the "machine teacher") is overrated and that it is the cultural motive that should be developed. On the other hand, by a "cultural motive" he seems mostly to mean a motive towards thoughtful and reflective learning. Thus, just a few lines after the sentence just quoted, he emphasizes that "[arithmetic] has cultural value because, if rightly taught, it trains one to think closely and logically and accurately" (p. 20)—which one might say is a more profound utility and more important for human life than the mechanical operations necessary for the day-to-day work of a storekeeper.

It is remarkable that D. E. Smith does not set the history of mathematics as an integral part of the "cultural motive." The history of mathematics would seem to be at the very heart of culture. For, whatever else it may mean, "culture" surely embraces at its core the doings and productions of human beings in a certain place and time. What makes this even more astonishing is that Smith was extraordinarily learned in the history of mathematics and wrote voluminously on the subject; he also amassed a collection of thousands of manuscripts and books related to the history of mathematics that was legendary (see Swetz and Katz 2011; Donoghue 1998). And in the book *On Teaching Elementary Mathematics*, which I have been referring to, Smith's arguments are, in a very pointed and explicit way, based on historical evidence. Almost from start to finish, Smith the historian of mathematics is at work. So how do we explain this seeming paradox?

The answer, I believe, can be discerned in the way that Smith argues that the history of mathematics should have a place in mathematics education. For he earnestly believes that, as I have already stated, and provides two justifications (Smith 1902). The first is based on the "parallelism argument," that is, that the development of an individual parallels the development of mathematics itself. As Smith himself suggests, this is an old argument. To be sure, it has had a long history before Smith, as one discovers in Schubring's thorough and deep works on the parallelism idea (e.g., Schubring 1978). It was also the driving argument for Toeplitz, whose "genetic approach" was laid out in 1926 (see Fried and Jahnke 2015), and it is still a potent argument for incorporating the history of mathematics in mathematics teaching (see Furinghetti and Radford 2008; Thomaidis and Tzanakis 2007). Here is how Smith states the position:

> ...the child learns somewhat as the world learns. This does not mean that the child must go through all of the stages of mathematical history—an extreme of the "culture-epoch" theory; but what has bothered the world usually bothers the child, and the way in which the

world has overcome its difficulties is suggestive of the way in which the child may overcome similar ones in his own development. (pp. 42–43)

Smith's second argument is that history of mathematics serves as a kind of filter
allowing one to see clearly what has proven important and fruitful and what turned
out to be effete and not worth pursuing. In Smith's words:

> ...the history of the subject [he is speaking specifically about the history of arithmetic, but
> the argument is general] gives us a point of view from which we can see with clear vision
> the relative importance of the various subjects, what is obsolete in the science, and what the
> future is likely to demand. (p. 43)

Despite Smith's immense factual knowledge of the history of mathematics and
his wide reading and scholarship, these arguments, I claim, presuppose a certain
view of mathematics in which mathematics, at bottom, is an unhistorical subject;
that is, it is one unaffected in any essential way by time and place or what we might
call culture. This can be seen in Smith (1921) presidential address to Mathematics
Association of America (Smith 1921). He called it *"Religio Mathematici,"* the
"religion of a mathematician," after Sir Thomas Browne's *Religio Medici*, which
was Browne's spiritual testament of his own identity as a doctor. So Smith's
"Religio Mathematici" is Smith's credo concerning the nature of mathematics and,
from it, one can infer with little trouble his credo concerning the history of
mathematics. Among other things, he writes:

> One thing that mathematics early imparts, unless hindered from so doing, is the idea that
> here, at last, is an immortality that is seemingly tangible,–the immortality of a mathematical
> law...The laws of the Medes and Persians, unchangeable though they were thought to be,
> have all perished; the canons that bound Egyptian activities for thousands of years exist
> only in the ancient records, preserved in our museums of antiquity...But in the midst of all
> these changes it has ever been true, it is true today, it shall be true in all the future of this
> earth, and it is equally true throughout the universe whether in the algebra of Flatland or in
> that of the space in which we live, that $(a + b)^2 = a^2 + 2ab + b^2$. (p. 341)

Mathematics does not change in this view, though it may be not be revealed all
at once. Its history, therefore, cannot be a history of change and development;
despite missteps here and there, it is rather the progressive unveiling of the
immortal truth—true everywhere "throughout the universe."

Thus, it is not surprising that, for Smith, history should reveal a kind of natural
direction of ideas and that its course should be consistent with that of an individual's intellectual development. It is not marked by the arbitrariness or idiosyncrasy
one finds in artistic creation and thus cannot be "cultural" in the way one expects
the history of art or literature to be. Thus Smith contrasts mathematics with the laws
of Persia and the canons of Egypt: If mathematics has a historical aspect, it is a
different kind of history from that of Egypt or Persia.

Moreover, the usefulness of the history of mathematics in mathematics education, its capacity to be a tool, comes from its unchanging character. Unlike the
norms of Egypt or Rome, it will always be relevant, and its history will open our
eyes and provide a measure of the importance of things according to an immortal,
unchanging scale. It is for this reason that, as Smith says in the passage quoted

above, "the history of the subject gives us a point of view from which we can see with clear vision the relative importance of the various subjects, what is obsolete in the science, and what the future is likely to demand."

These comments about Smith serve to bring out several points, all of which are central to any considerations regarding the history of mathematics and mathematics education. First, how one conceives the history of mathematics is not a *direct* result of one's learnedness; Smith's positions were not the result of his not knowing enough about dates, thinkers, and texts. Second, the role one assigns to the history of mathematics in mathematics education is inseparable from one's conception of the nature of mathematics. Third, in a similar way, how one conceives a cultural motive in mathematics teaching is connected to how one conceives mathematics; in particular, it is connected to its historical or non-historical character. Fourth, the non-historical character of the history of mathematics is in fact what allows it to be a tool, whether for guiding the teaching of individual students or for guiding the design of a curriculum.

To these I would like to add the converse of the last point, namely, that when one asks the history of mathematics to be a tool in mathematics teaching, one forces the history of mathematics to be non-historical. I will have to justify that claim, although it can be said immediately that to the extent that history is either what has been or the disciplined account of what has been, the student of history is something other than the user of a tool; therefore, treating history as a tool is ab initio contrary to history. However, before one even begins to talk about what it is to be non-historical, one should have a sense of the historical. So having encountered Smith's *Religio Mathematici* let us look at what might be called *Religio Historici*.

6.1.1 Religio Historici

Of course the word *religion* ought to be uttered with a smile. There is no single dogma to which all historians ascribe when it comes to their craft. Still, one can say that, if not quite a religion, there is at least a historical orientation: a set of pre-occupations recognizable by almost all historians despite considerable disagreement as to how one should pursue those preoccupations.

The question of sources is one such preoccupation, and in the case of intellectual history, of which the history of mathematics is an example, these are chiefly original texts. The centrality of original texts as a way of incorporating the history of mathematics in mathematics education was, accordingly, emphasized by Laubenbacher et al. (1994), for example. This has remained central in historical work, though its objective character has been challenged, for example, in Carr's *What Is History?* whose first chapter concerns what he called "the cult of facts" (Carr 1967, p. 9). Carr's position was that history cannot be removed from the historian's perspective on the past. This does not mean that the study of original texts is outdated. Far from it. Objections by Carr and many others keep alive the question of the meaning of sources.

Having a perspective on the past is connected to what is arguably the most important preoccupation of history, namely, the past itself or, rather, the relationship between the past and present. No doubt it is the past that jumps to one's mind when one hears the word *history*. However, a view of the past and history are not synonymous: It matters very much exactly *how* one considers the past.

Michael Oakeshott (1901–1990), in his various writings on history, has pressed the point that the "'historical past' denotes a distinguishable mode of the past" (Oakeshott 1999, p. 9). In his first and most famous book, *Experience and Its Modes* (1933), he says that in fact there are "certain pasts [that] may be dismissed at once as alien to history" (Oakeshott 1933, p. 102). There is, he says, among others, a remembered past or autobiographical past, a fancied past, and a practical past. It is the last of these that he considers most opposed to the historical past, and, accordingly, the one that brings out the nature of the historical past.

The practical past is a past whose entire mode of being is that of something involved with the present; it is derived and inspired by the present, important if important to the present, pursued if it is significant for our present concerns or even if it allows a way to escape them. As Oakeshott puts it:

> Wherever the past is merely that which preceded the present, that from which the present has grown, wherever the significance of the past lies in the fact that it has been influential in deciding the present and future fortunes of man, wherever the present is sought in the past, and wherever the past is regarded as merely a refuge from the present—the past involved is a practical, and not an historical past. (Oakeshott 1933, p. 103)

Our ordinary human day-to-day lives are so much directed to the present it is difficult to think of the past in any other way and perhaps impossible to engage in thinking of the past in a way that we utterly forget the present. Yet the historical past, the object of historical inquiry, involves, as Oakeshott says elsewhere, "a redirection" of this kind of activity "inherent in a human life" (Oakeshott 1999, p. 127). The past may be like the present, in fact, in that it must be in some way like the present if it is to be understood at all; however, the activity of history involves the attempt, even if it is ultimately doomed to failure, to see the past as other than the present, to see, as he says, "… the past as past, and with each moment of the past in so far as it is unlike any other moment" (Oakeshott 1933, p. 106).

The insidious side of Oakeshott's practical past is that it presents itself as historical. This is less so, for example, in the case of a fancied past or legends of yore: Serious people rarely take a Disney world of knights and unicorns as the real thing. But a past viewed from the perspective of the present is not always questioned. It was for this reason that Herbert Butterfield (1900–1979) wrote his famous book, *The Whig Interpretation of History* (Butterfield 1931/1951), just two years earlier than the work of Oakeshott that we have been referring to. It was a book about historians—Butterfield refers to it in the preface, with a smile no doubt, as a book about "the psychology of historians" (p. vi)—and it is presumably addressed to historians. As for the term, "Whig history," he says:

What is discussed is the tendency in many historians to write on the side of protestants and Whigs, to praise revolutions provided they have been successful, to emphasise certain principles of progress in the past and to produce a story which is the ratification if not the glorification of the present. (p. v)

And Butterfield, like Oakeshott, points to the essential character of a historical account in trying to grasp the otherness of the past in the attempt to see the past for itself:

...the chief aim of the historian is the elucidation of the unlikenesses between past and present and his chief function is to act in this way as the mediator between other generations and our own. It is not for him to stress and magnify the similarities between one age and another, and he is riding after a whole flock of misapprehensions if he goes to hunt for the present in the past. (p. 10)

6.2   History of Mathematics and Mathematics Education

This kind unhistorical history is particularly tempting in the history of mathematics, precisely because of the "religio mathematici" that Smith espoused. For if one takes mathematics to be essentially unchanging and immortal, then at bottom there is no difference between past and present. Therefore, with that in mind, one may freely translate the mathematics of the past into a modern idiom and use the present unabashedly as a guide to the past. The Whig perspective would, in that light, be completely unobjectionable and, undoubtedly, enlightening. Thus, in his well-known polemical article in 1975, Unguru declared that "Whig history, a dead horse nowadays—one would like to believe—in most branches of history, is alive and thriving in the history of mathematics, where its dangers are no less real than in the more traditional types of intellectual history" (Unguru 1975, p. 86).

Well aware of these difficulties, Grattan-Guinness (2004a, b) suggested two approaches to treating the mathematics of the past, history and heritage. These are distinguished by their guiding questions: History asks, "What happened?" or "Why did N happen?"; heritage asks "How did we get here?" The answer, Grattan-Guinness playfully points out, is more often than not via "the royal road to me." Grattan-Guinness is perfectly willing to say that "heritage resembles Whig history, the seemingly inevitable success of the actual victors, with predecessors assessed primarily in terms of similarities with the dominant position" (Grattan-Guinness 2004b, p. 171).

There is no doubt Grattan-Guinness's history/heritage dichotomy can be a useful tool for analyzing how mathematics of the past is treated; however, it does not adequately explain what it is we truly learn from heritage as opposed to history, and it does not bring out explicitly enough how these different approaches to the past are in fact different views of the past itself and place one in a different relation to the past. For they are truly different relations, even incompatible ones, as I think even Grattan-Guinness would have to admit. For this reason, in my book on Apollonius with Sabetai Unguru, we spoke about going through a historical door or a mathematical door (Fried and Unguru 2001, p. 404ff). Each door leads into a very different world: "The mathematical and the historical approaches are antagonistic. Whoever breaks and enters typically returns from his escapades with other spoils than the peaceful and courteous caller" (p. 406).

It should not be thought that the accusation of Whiggism or of an adherence to some other form of non-historical history is an accusation of being unlearned. It is not about not knowing enough history. As I have already mentioned, Smith was immensely learned. Clifford Truesdell (1919–2000) who was, among other things, the editor of Euler's collected works, was also a tremendously learned man, yet he was decidedly Whiggish in his approach to history, or at least happy to view the past as "practical past." Thus, he could write, for example, that

> one of the main functions [the history of mathematical science] should fulfill is to help scientists understand some aspects of specific areas of mathematics about which they still don't fully know. What's more important, it helps them too. By satisfying their natural curiosity, typically present in everybody towards his or her own forefathers, it helps them indeed to get acquainted with their ancestors in spirit. As a consequence, they become able to put their efforts into perspective and, in the end, also able to give those efforts a more complete meaning. (in Giusti 2003, p. 21)

It is clear from his writings that Truesdell felt truly that the figures of the past were, as Littlewood famously said, merely "fellows of another college" (Hardy 1992, p. 81). I emphasize the learnedness of Smith and Truesdell only to bring out that the meaning of history is to be found in how one approaches the past: It is not a direct function of how many names and dates one can recite. This is crucial not only for one who desires insight into the history of mathematics, but more importantly for those of us who are teachers who desire to use the history of mathematics to inform our teaching.

The two cases are not symmetrical. Historians of mathematics and those who wish simply to learn from history will gain by engaging with mathematical texts and thoughts from the past while giving cognizance to the meaning of history, the meaning of the past, and the meaning of thinking about the past. And that, as I have argued, requires actively avoiding the present and treating it as an unproblematic guide to the past. History, in fact, is not quite history without that.

But mathematics teachers as mathematics teachers have other unavoidable concerns. They have a curriculum to follow; they may have national examinations or some other kinds of large-scale examination for which they are obliged to prepare their students. A brave and bold teacher might decide, despite everything, to put aside such external constraints in order to treat the history of mathematics in a spirit of *religio historici*; however, that spirit, with its demand that the present be suspended, cannot be an imperative for mathematics teachers. It is not just that it is unnecessary: It conflicts with other imperatives. For the mathematics that mathematics teachers teach—the kind of mathematics laid out in the Common Core, for example—is crucial for the present and has more than historical import. No one can deny the kinds of approaches, techniques, and ideas that belong to mathematics of the present are genuinely useful in the sciences, engineering, and industry and that they are genuinely interesting, enlightening, and often beautiful. A teacher placing an emphasis on such mathematics cannot be condemned.

The external demands made on teachers and, perhaps, their own legitimate commitment to teaching the mathematics of the present—the mathematics needed in applications and in modern science—makes it easier to put history aside than put modern mathematics aside. But if there are mathematics teachers who nevertheless aim to bring history of mathematics into their teaching, they must, *because* of those external demands, economize by making history fit their other concerns. They must show its relevance to subjects already being taught or, alternatively, show its relevance to general mathematical thinking: what Smith called "culture." They must make history of mathematics useful. But in doing so, they are led almost ineluctably into precisely the "practical history" of Oakeshott or "Whiggist history" of Butterfield, which I described above: the past in the service of the present.

It is easy to find examples written by educators equally ignorant of mathematics education research and the history of mathematics, but such examples prove very little. On the other hand, it is not difficult to find papers where this is not the case. I have a paper in my files, for instance, called, "The history of mathematics as a pedagogical tool: Teaching the integral of the secant via Mercator's projection" (Haverhals and Roscoe 2010). I choose to highlight it because (1) it presents a good way of introducing the Mercator projection as it is used in geography and presents a pleasant way of teaching the integral of the secant function and showing its relevance and (2) its authors show an awareness of some of the literature being written in mathematics education research on history of mathematics (for example, Janvist 2009; Siu 2007; Ernest 1988) and a willingness to confront it. Yet, when it comes down to it, true to their title, history is not so much a subject to study as it is a tool to use and, as such, can be freely adapted for educational use. Thus, referring to the difficulties of original texts whose importance I have already mentioned, they say, "…student difficulty in confronting historical text can be alleviated by careful and thoughtful presentation that is at once historically accurate while educationally streamlined toward an intended goal, in this case, an understanding of the integral of the secant" (p. 354).

The authors have clear priorities. The priorities are not unreasonable. Haverhals and Roscoe are interested in teaching a subject, in this case the integral of the secant function, and have found a good example with historical color: It is an example that has the potential of holding the interest of the students and making their authors' task something more than an empty exercise. I emphasize this to make it clear that the authors' priorities are indeed priorities: They are to a great extent given in advance by commitments to a standard calculus curriculum. So if historical material is used, its use must be subordinated under the demands of such a modern curriculum. If the authors were to ignore the modern notion of the integral and the functional way of thinking, they could be more historical but undoubtedly less successful in achieving the teaching goals of their mathematics lessons. In effect, they are compelled to adopt a Whiggist perspective. For this reason, when I began thinking about the question of incorporating the history of mathematics in mathematics teaching, I stated the deliberations of a mathematics teacher interested in the history of mathematics as a kind of dilemma:

> ...if one is a mathematics educator, one must choose: either (1) remain true to one's commitment to modern mathematics and modern techniques and risk being Whiggish, i.e., unhistorical in one's approach, or, at best, trivializing history, or (2) take a genuinely historical approach to the history of mathematics and risk spending time on things irrelevant to the mathematics one has to teach. (Fried 2001, pp. 397–398)

At the time I wrote that, I thought the dilemma was inescapable and, much to my dismay, damning to almost any serious attempt to bring the history of mathematics into mathematics teaching. This was not for lack of good historical material. Writers such as Victor Katz, Frank Swetz, and the late John Fauvel, among others, work from the IREM team in France, and the many papers presented at the HPM-ESU conferences have all supplied plentiful historical sources and discussions for thinking about the history of mathematics. But as long as mathematics education was committed to the modern mathematical ideas necessary for the needs of modern life, as so many of the general documents pertaining to mathematics education declare—as, for example, when the NCTM's *Principles and Standards for School Mathematics* describes itself to be "the first set of rigorous, college, and career readiness standards for the 21st century" (NCTM 2016)—any use of history, I thought then, would have to be marked by the selection, abridgment, and organization of material for the modern ends that Butterfield underlined as the hallmark of Whig history.

But mathematics education both as an enterprise and as the focus of mathematics education research is not so rigidly defined: Its commitments are not written in stone. This means that, in principle at least, rather than asking how to adapt the history of mathematics to a fixed mathematics education with a predetermined set of commitments, one can ask how mathematics education might be conceived so that the history of mathematics plays an essential part in it, that is, where it is something to be studied and thought about. In this regard, the dilemma I set out in 2001 should be viewed more as a challenge to redefine mathematics education informed by the history of mathematics than as a criticism of past efforts.

It would be unfair to say there has been until now no thought in this direction. A survey of many of the ideas can be found in Fried (2014), and many of these predate 2001. It is beyond the scope of this lecture to summarize the work of all the scholars described there, but, as a sample, I would certainly mention the work of Radford, whose ideas on the semiotic and cultural core of mathematics education (for example, Radford 2015) have inspired many of us here; there is Jankvist and Kjeldsen's (2011) work within the context of the Danish competencies framework, which includes the history of mathematics; and then we have Jahnke's hermeneutic approach (e.g., Jahnke 2000), in which mathematics learners are conceived as interpreters, so that the reading of historical texts becomes a way of exploring one's own mathematical identity. My only contribution has been to sharpen the question to which these are possible answers.

6.3   The Liberal Arts

To characterize concretely mathematics education in which history forms an integral part, I could mention, for example, the centrality of original texts. I referred to this above in connection to the work of Laubenbacher et al. (1994). Or I could speak about some of the proposals connected to Jahnke's work cited above. But, since time is short, I would rather end with a much more general picture: one inspired by a concomitant history of the history of mathematics, namely, the history of education itself. What I have in mind specifically are the liberal arts.

It is true that the liberal arts come specifically from the history of education in the West, but, as I said, my intention is to use them as an image, a general picture, not a promotion of the West or a denigration of any other tradition. This tradition, however, is still very much a part of students' lives. Students who receive a BA or MA degree have, by name at least, received degrees in the liberal arts, for those are precisely the arts intended by Bachelor of *Arts* and Master of *Arts*. Even more, the MA, *Magister Artium*, is the qualification to *teach* the liberal arts. And here I say as an aside that whether or not it is right to think of history in terms of memory, it is certainly true that it treats the kind of forgetfulness that occurs when original meanings are lost in the light of modern transformations. Students who write the letters BA and MA after their names are oblivious, more often than not, to the traditions buried under their degrees.

The tradition of the liberal arts is a long one, although its systematization into an educational scheme became most clearly articulated in late antiquity. The usual scheme involved seven liberal arts. Of these, three were, in a sense, connected to language. These formed the *trivium*, the "three ways," of grammar: the nature of letters, words, sentences—rhetoric, or the artful use of words—and logic—the formation of arguments. The other four arts, contrary to the modern tendency to set mathematics and science apart from the liberal arts, were explicitly mathematical. These mathematical arts formed the *quadrivium*, the "four ways": the arts of arithmetic, geometry, astronomy, and music. I should mention that there were other schemes for the liberal arts; for example, in Varro's lost treatise, *The Nine Books of Disciplines*, written during the first century BCE, there were, besides these seven arts, two more: architecture and medicine (see Wagner 1986, p. 15ff). Nevertheless, in time the *trivium* and *quadrivium* became more or less the canonized scheme (see Fig. 6.1).

It is important to understand that the arts were thought of as a system—particularly the mathematical arts: a system no less unified than a system of the world. Thus, Proclus (412–485 CE), recalling the Pythagorean tradition, tells us that the *quadrivium* was structured so that arithmetic and geometry corresponded to the basic division between multitude and magnitude, with arithmetic being about numbers in themselves, pure multitudes, and geometry being about magnitudes at rest, figures that do not move. Music was derivative from arithmetic, being the study of numbers taken relatively to other numbers, such as the harmonic divisions of a string, while astronomy was geometry in motion (see Fig. 6.2).

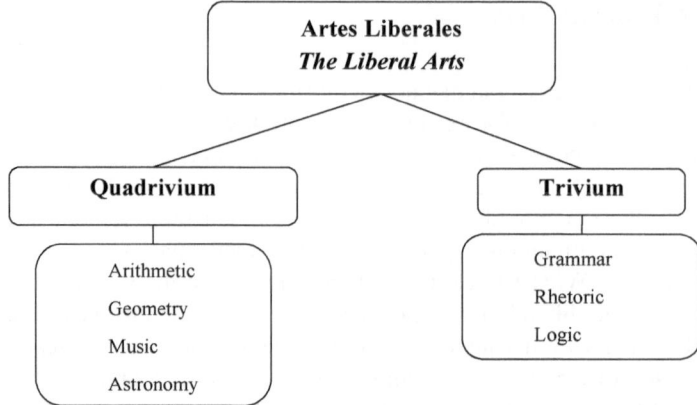

Fig. 6.1 The liberal arts

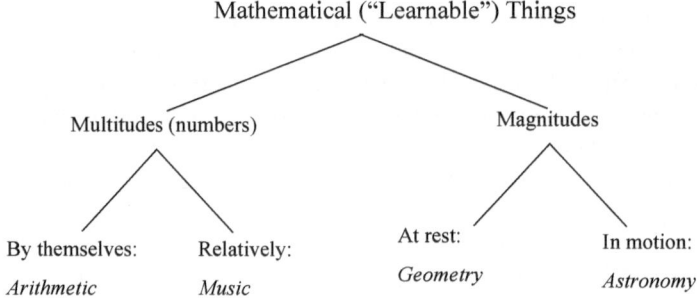

Fig. 6.2 The Pythagorean division of mathematics

One can disagree with the organizational principle; however, what is important is that there was an organizational principle at all, an attempt to present education as a reflection of a whole.

That the learnable things, the *mathēmata*, reflected the whole universe was almost certainly the result of Pythagorean doctrines. Whether the Pythagoreans had much use for the arts in the *trivium* is not nearly as clear, even though, surely, later Pythagoreans, such as Nichomachus of Gerasa, from whom we know much about Pythagorean teachings, were well versed in grammar, rhetoric, and logic. They could hardly express themselves so well without rigorous training in the trivium.[1]

---

[1]There was among especially the Church Fathers, Augustine and Jerome, for example, a certain tension between the rigor of this training and its connection to classical oratory on the one hand and unadorned inner spirituality on the other. One can detect an echo here of a similar tension today in mathematics education, namely, that between rigorous training in mathematical techniques and procedures and intuitive and original thinking.

But there is another sense in which the liberal arts reflected a whole, where the full complement of the arts was essential. Indeed, in Martianus Capella's early fifth century CE allegory, *The Marriage of Philology and Mercury*, the seven liberal arts, seven sisters, bestow the gifts that sanctify the marriage of the maiden Philology and Mercury, gifts which make the pair one. In general, our sources are largely "encyclopedias" written around this time, works by Boethius, Cassiodorus, and Isadore of Seville. This was the time when the Roman Empire was breaking down, and the need for inner coherence rooted in the tradition under threat was felt acutely.

The word *encyclopedia* is significant. It is a combination of two Greek words, *engkuklios* and *paideia*; the first, coming from *kuklos*, a circle, means what happens over and over, regularly, or common to all. The second may be translated variously as culture, upbringing, or, education, so that *engkuklios paideia* is something like the education common to all. The word *paideia* is truly the difficult—and therefore most important—word here. Jaeger (1945) required three thick volumes to explain it. But one can say that (1) *paideia* was rooted in the literature and thought of one's tradition—here the translation "culture" is apt and (2) it was meant to be carried throughout life, so that (3) it was very much an expression of being a human in the fullest sense of the word, thus the Latin translation of *paideia* came to be, revealingly enough, *humanitas* (see Marrou 1982, p. 218).

It is not by chance of course that encyclopedias, with their connection to the classical idea of *paideia*, should be the place where the liberal arts were discussed. For the study of the liberal arts was inseparable from the idea of *paideia*, even identified with it. They were called liberal arts because, like *paideia*, they were directed towards human beings who are not slaves but who are free to pursue a life allowing them to be fully what they are as human beings. In the history of *paideia*, this was particularly true in the post-Socratic period. Thus Jaeger (1945) writes:

> ...it is Socrates' idea of *the aim of life* which marks the decisive point in the history of *paideia*. It threw a new light on the purpose and duty of all education. Education is not the cultivation of certain abilities; it is not the communication of certain branches of knowledge.... The real essence of education is that it enables men [in gender-insensitive language of 1945] to reach the true aim of their lives.... This effort cannot be restricted to the few years of what is called higher education. Either it takes a whole lifetime to reach its aim, or its aim can never be reached. Therefore the concept of *paideia* is essentially altered; and education, in the Socratic sense, becomes the effort to form one's life along lines which are philosophically understood, and to direct it so as to fulfill the intellectual and moral definition of man. In this sense, man was born for *paideia*. It is his only real possession. (Jaeger 1945, Vol. II, pp. 69–70, emphasis original)

At this point, one might complain that although I have been looking back at a chapter in the history of education, I seem to have departed from history as such— and it is the history of mathematics in mathematics education that is my topic. The objection is actually more acute than one might think, for while the seven liberal arts were not fixed in the past, as a pointed out earlier, history was never considered one of them. On the other hand, today the liberal arts considered central to the "humanities" and have become the locus of historical study, while mathematics, traditionally always part of the liberal arts, has nearly become excluded from them.

The new place of history as the prime liberal art is partly the result of a change in the understanding of history, at least since the time of 17th–18th century figures such as Giambattista Vico (1668–1744) and certainly since Hegel. The idea that we ourselves may be historically constituted, that history might represent for human life the clearest kind of truth, has driven the modern idea of history. It has also made the aim of the liberal arts—the exploration and fulfillment of our human life as free, thinking beings—in some quarters the aim of history as well. Thus Collingwood could say:

> [The historian's knowledge] is not either knowledge of the past and therefore not knowledge of the present, or else knowledge of the present and therefore not knowledge of the past; it is knowledge of the past in the present, the self-knowledge of the historian's own mind as the present revival and reliving of past experiences. (Collingwood 1993, p. 175; see also Fried 2007)

The place of history as the central liberal art was certainly challenged by, for example, Thomas Huxley in the second half of the 19th century. Huxley doggedly made the case that science should be at the heart of education; indeed, one of his essays, which makes this point, is called "A Liberal Education; and Where to Find It" (1868, in Huxley 1899, pp. 76–110). Calls to increase science education today, with their tacit belittlement of the humanities, I believe, echo Huxley's well-meant sentiments. Moreover, to return to Smith, with whom I began, it may be said that his use of the word *culture* without history stems from the same sentiment that impelled Huxley to speak about liberal education with science at the center, more paleontology than history, to use one of Huxley's own points.

But I think it is fair to say that culture, for us, not only in general and in history but also in mathematics education itself, is impossible to untie from history and a view of ourselves as historical beings. Mathematics education in this light can look back to the liberal arts in which mathematics was central and in which their place in defining a full human life can be informed by the historical sense of a human life. The problem therefore of the history of mathematics and mathematics education in this way becomes a challenge to rethink mathematics education in terms of the liberal arts and the attempt to see ourselves more clearly and more deeply as the beings that created mathematics. If the history of mathematics is taken as a tool, then it must be taken in the way the liberal arts were tools: arts to be used but also a source on which we reflect about mathematics and ourselves.

6.4 Concluding Words

History, as we said, has become central to the liberal arts, while mathematics has become excluded. The history of mathematics is one way of restoring mathematics as a liberal art. Conversely, thinking of mathematics as a liberal art opens the way to the history of mathematics becoming an essential part of mathematics education. It allows mathematics to assume a place in a human life that is taken as an integral whole in a world that is taken as an integral whole. As in all history (as Collingwood has said) and in all the liberal arts, mathematics education, in this light, becomes a way of reflecting on ourselves.

Reflecting about ourselves and our human capabilities is humanism. It must be understood, however, that the liberal arts, which can be taken as another term for humanism, are not a dogma. In the same way, history and tradition, while being formative, are not binding. As one of my own teachers, Eva Brann, liked to point out, in thinking about tradition, one must remember that the Latin word *tradere* means both "to pass on" but also "to betray" (Brann 1979, p. 64). Searching for a whole, either of the world or of tradition or of traditions, a search that ultimately cannot be consummated, brings us thus to an openness to our own incompleteness —and therefore, to our own potentialities. We look back at our own foundations in history and by recognizing that truly, using Gadamer's (2006) language, we look beyond our own horizon.

References

Brann, E. T. H. (1979). *Paradoxes of education in a republic*. Chicago: Chicago University Press.

Butterfield, H. (1931/1951). *The whig interpretation of history*. New York: Charles Scribner's Sons.

Carr, E. H. (1967). *What is history?* London: Penguin Books.

Collingwood, R. G. (1993). *The idea of history*. Oxford: Oxford University Press.

Donoghue, E. F. (1998). In search of mathematical treasures: David Eugene Smith and George Arthur Plimpton. *Historia Mathematica, 25*, 359–365.

Ernest, P. (1988). *The impact of beliefs on the teaching of mathematics*. Paper presented at ICME VI, Budapest, Hungary.

Fried, M. N. (2001). Can mathematics education and history of mathematics coexist? *Science & Education, 10*, 391–408.

Fried, M. N. (2007). Didactics and history of mathematics: Knowledge and self-knowledge. *Educational Studies in Mathematics, 66*, 203–223.

Fried, M. N. (2014). History of mathematics and mathematics education. In M. Matthews (Ed.), *History, philosophy and science teaching handbook* (Vol. 1, pp. 669–705). New York: Springer.

Fried, M. N., & Jahnke, H. N. (2015). Otto Toeplitz's 1927 paper on the genetic method in the teaching of mathematics. *Science in Context, 28*(2), 285–295.

Fried, M. N. & Unguru, S. (2001). *Apollonius of Perga's Conica: Text, context, subtext*. Leiden, The Netherlands: Brill Academic Publishers

Furinghetti, F., & Radford, L. (2008). Contrasts and oblique connections between historical conceptual developments and classroom learning in mathematics. In L. English (Ed.), *Handbook of international research in mathematics education* (2nd ed., pp. 626–655). Mawah, NJ: Lawrence Erlbaum.

Gadamer, H. G. (2006). *Truth and method* (2nd ed., J. Weinsheimer & D. G. Marshall, Trans.). New York: Continuum.

Grattan-Guinness, I. (2004a). History or heritage? An important distinction in mathematics and for mathematics education. *The American Mathematical Monthly, 111*(1), 1–12.

Grattan-Guinness, I. (2004b). The mathematics of the past: Distinguishing its history from our heritage. *Historia Mathematica, 31*, 163–185.

Giusti, E. (2003). Clifford Truesdell (1919–2000), historian of mathematics. *Journal of Elasticity, 70*, 15–22.

Hardy, G. H. (1992). *A mathematician's apology*. Cambridge: Cambridge University Press.

Haverhals, N., & Roscoe, M. (2010). The history of mathematics as a pedagogical tool: Teaching the integral of the secant via Mercator's projection. *The Montana Mathematics Enthusiast, 7* (2&3), 339–368.

Huxley, T. H. (1899). *Science and education*. London: Macmillan and Co., Limited.

Jaeger, W. (1945) *Paidea: The ideals of Greek culture (Vol 3*, G. Highet, Trans.). New York: Oxford University Press.

Jahnke, H. N. (2000). The use of original sources in the mathematics classroom. In J. Fauvel & J. van Maanen (Eds.), *History in mathematics education: The ICMI study* (pp. 291–328). Dordrecht: Kluwer Academic Publishers.

Jankvist, U. T. (2009) A characterization of the "whys" and "hows" of using history in mathematics education. *Educational Studies in Mathematics, 71*(3), 235–261.

Jankvist, U. T., & Kjeldsen, T. H. (2011). New avenues for history in mathematics education: Mathematical competencies and anchoring. *Science & Education, 20*, 831–862.

Laubenbacher, R., Pengelley, D., & Siddoway, M. (1994). Recovering motivation in mathematics. In *Teaching with original sources, UME trends 6*. Available at the website: http://www.math. nmsu.edu/~history/ume.html. Accessed December 31, 2015.

Marrou, H. I. (1982). *A history of education in antiquity* (G. Lamb, Trans.). Madison, WI: University of Wisconsin Press.

NCTM. (2016). *Principles and standards for school mathematics*. At the website: http://www. nctm.org/Standards-and-Positions/Principles-and-Standards/. Accessed January 15, 2016.

Oakeshott, M. (1933). *Experience and its modes*. Cambridge: At the University Press.

Oakeshott, M. (1999). *On history and other essays*. Indianapolis, IN: Liberty Fund, Inc.

Radford, L. (2015). Early algebraic thinking: Epistemological, semiotic, and developmental issues. In S. J. Cho (Ed.), *The Proceedings of the 12th International Congress on Mathematical Education* (pp. 209–227). Cham: Springer.

Schubring, G. (1978). *Das genetische Prinzip in der Mathematik-Didaktik* [The genetic principle in mathematics education]. Stuttgart: Klett.

Smith, D. E. (1902). *The teaching of elementary mathematics*. New York: The Macmillan Company.

Smith, D. E. (1905). Réformes à accomplir dans l'enseignement des mathématiques: Opinion de M. Dav.-Eug. Smith. *L'Enseignement Mathématique, 7*, 469–471.

Smith, D. E. (1921). Religio Mathematici: Presidential address delivered before the Mathematical Association of America, September 7, 1921. *The American Mathematical Monthly, 38*(10), 339–349.

Siu, M. K. (2007). No, I don't use history of mathematics in my class. Why? In F. Furinghetti, S. Kaijser, & C. Tzanakis (Eds.), *Proceedings HPM2004 & ESU4* (pp. 268–277). Uppsala: Uppsala Universitet.

Swetz, F. J., & Katz, V. J. (2011). Mathematical treasures—The David Eugene Smith collection. *MAA: Convergence*. Available at the website: http://www.maa.org/press/periodicals/ convergence/mathematical-treasures-the-david-eugene-smith-collection. Accessed December 11, 2015.

Thomaidis, Y., & Tzanakis, C. (2007). The notion of historical 'parallelism' revisited: Historical evolution and students' conceptions of the order relation on the number line. *Educational Studies in Mathematics, 66*(2), 165–183.

Unguru, S. (1975). On the need to rewrite the history of Geek mathematics. *Archive for History of Exact Sciences, 15*, 67–114

Wagner, D. L. (1986). *The seven liberal arts in the middle ages*. Bloomington: Indiana University Press.

Chapter 7
Knowledge and Action for Change Through Culture, Community and Curriculum

Linda Furuto

Abstract At the 1984 International Congress on Mathematical Education (ICME-5), Ubiratan D'Ambrosio envisioned the creation of a global society where "mathematics for all" reached an unprecedented dimension as a social endeavor by questioning the equilibrium of mathematics education (1986, p. 6). To respond to the challenge three decades later, I will present a contemporary perspective by re-examining the sociocultural role of mathematics education in the schooling process. I will specifically discuss how knowledge and action for change are achieved through intersections of culture, community and curriculum in an ongoing process of navigating and wayfinding in Hawai'i and the Pacific. This will be accomplished by developing new theoretical insights into honoring and sustaining non-Western cultural systems and practices through examples in mathematics teacher education. In doing so, I will highlight diverse funds of teaching and learning that are grounded in a shared commitment to equity, empowerment and dignity.

Keywords Curriculum and instruction · Community · Culture-based education Equity

7.1 "Mathematics for All"

Three decades after the 1984 International Congress on Mathematical Education (ICME-5) where D'Ambrosio spoke about a vision of creating a society where "mathematics for all" reached an unprecedented dimension as a social endeavor (p. 6), we have come to understand that mathematics education is going through one of the most critical periods in its long recorded history since Western classical antiquity (Bishop 1988; Boaler 2002). The U.S. National Science and Technology Council's Federal Science, Technology, Engineering and Mathematics (STEM)

L. Furuto (✉)
University of Hawai'i at Mānoa, Honolulu, USA
e-mail: lfuruto@hawaii.edu

© The Author(s) 2018   103
G. Kaiser et al. (eds.), *Invited Lectures from the 13th International Congress on Mathematical Education*, ICME-13 Monographs,
https://doi.org/10.1007/978-3-319-72170-5_7

Education 5-Year Strategic Plan (2013) states, "We don't want our kids just to be consumers of the amazing things that science generates; we want them to be producers as well. And we want to make sure that those who historically have not participated in the sciences as robustly—girls, members of minority groups—are encouraged…this means teaching proper research methods and encouraging young people to challenge accepted knowledge" (p. 1). The current era emphasizes the role of an interconnected global society undergoing changing social, educational, political and economic conditions. Mathematics education has a direct role in influencing the equilibrium of achievement, particularly in traditionally underrepresented and underserved populations (Palhares and Shirley 2013; Weiss and Miller 2006).

To respond to the challenge of re-examining the sociocultural role of mathematics education in the schooling process, this article will discuss how knowledge and action for change are achieved through intersections of culture, community and curriculum in an ongoing process of navigating and wayfinding in Hawai'i and the Pacific (Furuto 2016; Tuhiwai Smith 1999). The research goal is to develop new theoretical insights into honoring and sustaining non-Western cultural systems and practices through examples in mathematics teacher education that are grounded in a shared commitment to equity, empowerment and dignity (Rosa et al. 2016). The underlying premise is that "Mathematics is powerful enough to help us build a civilization with dignity for all, in which iniquity, arrogance, violence and bigotry have no place, and in which threatening life, in any form, is rejected. School ethnomathematics practices encourage the respect, solidarity and cooperation with others. It is thus associated with the pursuit of peace" (D'Ambrosio 2004, p. ix).

7.1.1   Ethnomathematics and Ecological Systems Theory

Defined as the intersection of historical traditions, sociocultural roots, political dimensions and linguistics, among others, ethnomathematics encourages the investigation and adaptation of these concepts within and outside of classrooms around the world (Greer et al. 2009; PREL 1995). The term ethnomathematics was introduced by D'Ambrosio in 1977 to foster an "awareness of the many ways of knowing and doing mathematics that relates to the values, ideas, notions, procedures and practices in contextualized environments" (Rosa et al. 2016, p. 1). By drawing on the assets and backgrounds of our students and communities, we acknowledge the importance of strengths-based approaches in accessing diverse funds of teaching and learning experiences (Hall 1993; Maton et al. 2003).

At the 2016 International Congress on Mathematical Education (ICME-13), Barton expanded on the ethnomathematics program and invoked the ecological systems theory. Ecological systems theory seeks to bridge the divide between science and the humanities in order to help students achieve responsible creativity and ethical citizenship (Bronfenbrenner 1979). In this system, "justice and education are part of a larger environment in which there is more than one 'way of knowing' resulting in a diversity of knowledge" (Barton 2017, p. 3). Barton continued, "The extent to which we free mathematics and mathematics education from society and culture is the extent to which we are absolving ourselves from responsibility to others and to our world. It frees us from social and cultural responsibility. Ultimately, this makes us amoral" (2017, p. 3). In other words, to do mathematics is to engage in processes of understanding and fulfilling our civic and moral responsibilities.

7.2 Culture, Community and Curriculum

7.2.1 Theoretical Frameworks

Over the past three decades, research has emerged to support equity and empowerment, especially in communities traditionally underrepresented and underserved in mathematics education. Some of the important theories that frame the discussion on curriculum and instruction include the following: culturally relevant pedagogy (Ladson-Billings 1995), culturally congruent pedagogy (Au and Kawakami 1994), culturally compatible pedagogy (Jacob and Jordan 1987), engaged pedagogy (Hooks 1994), everyday pedagogies (Nassir 2008), critical care praxis (Rolón-Dow 2005), and most recently, culturally sustaining pedagogy (Alim and Reyes 2011; Cammarota 2007; Irizarry 2007; Paris 2012; Winn 2011). According to Paris (2012), "The term culturally sustaining requires that our pedagogies be more than responsive of or relevant to the cultural experiences and practices of young people—it requires that they support young people in sustaining the cultural and linguistic competence of their communities while simultaneously offering access to dominant cultural competence" (p. 95).

Research is critical in the democratic struggle toward principles of social justice in our schools and society (Alim et al. 2011; Hill 2009; Kirkland 2011). The literature encourages pedagogical, curricular and teaching innovations in sustaining and extending the richness of the past in the current struggle to overcome deficit theories and strive toward strengths-based approaches in education (Chang and Lee 2012; Morrell 2004; Paris 2011). According to Rosa et al. (2016), the main foundation of an ethnomathematics program is embracing these types of diverse instructional practices and pedagogy that are integral as we move toward equity.

As an example of ethnomathematics, Math in a Cultural Context emerged from ethnographic work with Yup'ik elders and teachers (Lipka et al. 2005). The macro themes that evolved in this work concern positive changes in relationships, both in the classroom and between the classroom and community, pride in identity and culture, and ownership of knowledge. This curriculum is locally and culturally based while meeting both the State of Alaska's cultural standards and the standards of the National Council of Teachers of Mathematics (2000). Importantly, "this curriculum holds great promise to improve Alaska Natives students' mathematical understanding while bridging the culture of the schools to that of the community. It also can be viewed as a way to tap rich Indigenous cultural heritages, thus liberating from the legacy of colonial education and the restrictive pedagogical forms it prescribes" (Lipka et al. 2005, p. 368).

7.2.2   Connections to Hawai'i and the Pacific

Ethnomathematics has fundamental connections to Hawai'i and the Pacific. Research in Native Hawaiian communities demonstrates the importance of culturally sustaining pedagogy as a means of engaging and empowering students and their families in the learning process (Furuto 2014; Kana'iaupuni et al. 2010). According to Kana'iaupuni et al. (2017), "Embracing the emancipatory potential of culture-based education is a 'win' for everyone in our increasingly plurilingual, pluricultural society, who will benefit from the assets found in Indigenous knowledge, values, and stories as models of vitality and empowerment through which we can all progress" (p. 334). By drawing on Indigenous wisdom and 21st century knowledge, we have opportunities to re-examine the schooling process of Native Hawaiian and all students.

A tradition that has run deep in the Indigenous peoples of Hawai'i and the Pacific for over 2000 years is open ocean, deep sea voyaging by celestial navigation without GPS systems, compasses, clocks or sextants (Baybayan et al. 1987; Finney et al. 1986; Goetzfridt 2008). Traditional wayfinding is done by the rising and setting of the sun, moon, stars, ocean swells, winds, currents, birds and principles of mathematics. Over time, knowledge of these traditional wayfinding techniques dwindled and nearly disappeared. However, in the past 40 years, traditional wayfinding has experienced a revival across the Pacific, especially in Hawai'i.

When the navigation renaissance began in the early 1970s by the Polynesian Voyaging Society (PVS), Native Hawaiians and others voyaged to prove that purposeful migration occurred across the Pacific (Goetzfridt 2008; Kyselka 1987; PVS 2016). Now, with the tradition of wayfinding revived and thriving, the voyages allow new generations of students to honor and sustain knowledge, culture and values through education. The Polynesian Voyaging Society's prototype canoe Hōkūle'a ("star of gladness") has sailed over 150,000 nautical miles, and inspired a revival of voyaging and Indigenous practices around the world (Finney et al. 1986; Furuto 2014). With the guidance of master navigator Mau Piailug and PVS founders Herb Kāne, Ben Finney and Tommy Holmes, among others, Hōkūle'a has spawned a legacy of more than 25 deep sea voyaging canoes birthed across 11 Pacific Island nations.

Hōkūle'a's most recent voyage was to circumnavigate the globe from 2013 to 2017 with a mission to mālama honua, which is to "care for island earth" and all people and places as 'ohana ("family"). It is a culture of caring for our students, schools, communities and homes. The author was on the first international leg from Hawai'i to Tahiti as apprentice navigator and education specialist, and subsequent voyages to Samoa, Olohega (Swain's Island), Aotearoa (New Zealand),

South Africa, Virginia, Washington, D.C. and New York City, sailing with leaders such as the Archbishop Desmond Mpilo Tutu, His Holiness the 14th Dalai Lama and United Nations (UN) Secretary General Ban Ki-moon. At the 2014 UN Small Island Developing States Conference in Samoa, Ban Ki-Moon presented PVS with a message in a bottle evocative of D'Ambrosio's challenge at ICME-5, "I am honored to be part of Hōkūle'a's Mālama Honua Worldwide Voyage. I am inspired by its global mission. As you tour the globe, I will rally more leaders to our common cause of ushering in a more sustainable future and a life of dignity for all."

In this next section, I will highlight examples in mathematics teacher education through the lenses of ethnomathematics and voyaging that honor Indigenous wisdom and 21st century connections. Powell and Frankenstein (1997) urge, "As we more clearly understand the limits of our educational practice, we will increase the radical possibilities of our educational action for liberatory change. Thus, we feel the most important area for ethnomathematical research to pursue is the dialectics between knowledge and action for change" (p. 327). Through culture, community and curriculum, we have witnessed firsthand how this is possible.

7.3 Knowledge and Action for Change

Knowledge and action for enduring, transformational change comes from essential understandings gained by working with and learning from the populations we are endeavoring to serve. According to Jaworski et al. (1999), "Inservice providers cannot just 'deliver' a course, or a workshop, or a session. They must become part of the learning community, to live with the teachers and the learners and the realities of their situation. In doing so, inservice providers will necessarily influence and be influenced by that situation, and be an intimate part of any research the inservice providers might be engaged in as part of the development work" (p. 12). This is what we have strived to do as we have brought voyages back to land, especially in Hawai'i and the Pacific.

7.3.1 Educational Context in Hawai'i and the Pacific

Hawai'i is the only statewide school district in the U.S., and operates a single public higher education University of Hawai'i System. As such, we have a unique opportunity to reach our schools through partnerships and strategic alignment. The long standing achievement gap of Native Hawaiian students in the Hawai'i State Department of Education and University of Hawai'i System represents a significant concern, and one that key stakeholders are committed to resolving (HIDOE 2017; UH IRO 2017). The connections from early childhood education through graduate studies (P–20) inspire meaningful, relevant and sustainable promising practices (Waitzer and Paul 2011; Weiss and Miller 2006).

In Fall 2013, the Polynesian Voyaging Society's Promise to Children was authored by educational leadership in Hawai'i and the Pacific, including the Hawai'i State Department of Education (HIDOE) Superintendent and the University of Hawai'i System (UHS) President who participated as crew members on the Mālama Honua Worldwide Voyage. According to the Promise to Children, "We believe that the betterment of humanity is inherently possible, and we believe our schools, from early childhood education through advanced graduate studies, are a powerful force for good. This is the voyage of our lifetimes...the University of Hawai'i's 10 campuses have active programs and projects to achieve this goal such as...ethnomathematics learning" (p. 3). This alliance spans early childhood education through graduate studies (P–20), public and private sectors, and invites new partners to achieve collective impact (Bryk et al. 2011; Kania and Kramer 2011).

As a result of P–20 collaborations, the HIDOE Office of Hawaiian Education created learning outcomes that all K–12 students will achieve by graduation. Nā Hopena A'o (2015) is a product of the HIDOE's Mālama Honua Worldwide Voyage efforts to inform policy implementation at the statewide level. It is a framework to develop the skills, behaviors and dispositions of Hawai'i's unique context, and to honor the qualities and values of the Indigenous language and culture of Hawai'i. Nā Hopena A'o (HĀ) reflects the HIDOE's core values and beliefs in action throughout the public educational system of Hawai'i to develop the competencies that strengthen a sense of belonging, responsibility, excellence, aloha, total well-being and Hawai'i in ourselves, students and others (HIDOE 2015). With a foundation in Hawaiian values, language, culture and history, HĀ supports a holistic learning process to guide the entire school community. The purpose of this policy is to provide a comprehensive outcomes framework to be used by those who are developing the academic achievement, character, physical and social-emotional well-being of students to the fullest potential (HIDOE 2015, 2017).

Similarly, the University of Hawai'i System (UHS) is the sole provider of public higher education in Hawai'i, and is comprised of 10 campuses. It is committed to improving the social, economic and environmental well-being of current and future generations, and services the needs of students not just in Hawai'i but throughout the Pacific, particularly in U.S. affiliated entities. Approximately 25% of the student population is Native Hawaiian (UH IRO 2017). The UHS Strategic Directions 2015–2021 guides the university's priorities to achieve systemwide outcomes, along with measurable goals and the ability to effectively monitor progress over time (UHS 2015). Interwoven in the strategic directions are two key imperatives embraced within the UHS mission—a commitment to being a foremost Indigenous-serving institution and advancing sustainability. With the Mālama Honua Worldwide Voyage as a catalyst, the UHS is firmly committed to advancing these directions in concert with its core values of academic rigor, excellence, integrity, service, aloha and respect (UHS 2015). According to the UHS, "There are powerful motivations for the University of Hawai'i to be supportive of its Indigenous population: some of its campuses sit on ceded lands; negative Native Hawaiian social and economic statistics exist; and inequity of success amongst its native and non-native students are factors that demand attention. There are many reasons…However, the best reason is because it is the right thing to do" (2012, p. 26).

7.3.2 Preparing Teachers as Leaders

The College of Education at the University of Hawai'i's flagship Mānoa campus is the ideal vehicle to help achieve P–20 knowledge and action for change. The UH Mānoa College of Education directs online and in-person teacher preparation programs, professional development, curriculum development and research projects across Hawai'i and U.S. affiliated Pacific Islands. It produces more than 65% of Hawai'i's teaching force and leads U.S. affiliated Pacific Islands in providing educational programs and professional development (UHM COE 2016). The underlying mission is to "envision a community of educators who provide innovative research, teaching and leadership in an effort to further the field of education and prepare professionals to contribute to a just, diverse and democratic society. The College aims to enhance the well-being of the Native Hawaiian people and others across the Pacific Basin through education" (UHM COE 2016, p. 2). The UH Mānoa College of Education is well-equipped to achieve UHS priorities, which include, "Continue improving P–20 education by establishing collaborative initiatives" (UHM COE 2016, p. 2).

The Ethnomathematics Institute is housed at the UH Mānoa College of Education, and the author is the principal investigator. Now in its 9th year, the project is an effort to address issues of equitable and quality education through culturally sustaining pedagogy grounded in the ethnic, cultural, historical, epistemological and linguistic diversities of Hawai'i and the Pacific. We bring together research institutions and community-based organizations to support yearlong professional development for P–20 inservice educators (note: for the first five years when the author was an Associate Professor of Mathematics at the University of Hawai'i—West O'ahu, the focus was on undergraduate students).

The three main objectives of the Ethnomathematics Institute are to: (1) explore promising practices in diverse, high needs populations in alignment with national and state standards, such as the Mathematics Common Core State Standards, Next Generation Science Standards and Nā Hopena A'o; (2) prepare teachers as leaders to provide instruction and professional development in ethnomathematics in their schools and communities through high-quality learning that is relevant, contextualized and sustainable; and (3) strengthen campus-community partnerships to build a research consortium within Hawai'i and the Pacific. In addition to classroom learning, place- and culture-based learning occur throughout the Hawaiian islands.

For example, students and teachers sail on Polynesian Voyaging Society canoes to perpetuate the art and science of traditional voyaging and the spirit of exploration. Through experiential, hands-on curriculum development, they inspire their communities to respect and care for themselves, each other, and their natural and cultural environments. Some of the experiments performed on land and sea to link Indigenous wisdom and 21st century knowledge include: plankton tows and identification, water quality research, marine debris collection and identification, hydroponic food growing, marine mammal acoustics and fish DNA identification. These mathematics and science experiments were designed and implemented with support from the HIDOE and UH Mānoa School of Ocean and Earth Science and Technology, spanning the disciplines of oceanography, geology and geophysics, marine biology, agriculture and zoology.

The project is guided by the shared HIDOE and UHS values of belonging, responsibility, excellence, aloha, total well-being and Hawai'i. We honor our students by connecting the classroom to the local ecological, cultural, historical and political contexts in which schooling itself takes place. Following an ethnomathematics lesson implemented by one of the teacher leaders on Polynesian Voyaging Society canoes, a 12th grade student reflected, "Papahānaumoku/Haumea has given birth to our world. She helps to provide our food and materials needed to survive. So in return, we as the children must take care of the land by making sure our lifestyle is balanced and keeps Papahānaumoku hau'oli loa (very happy) by understanding related rates of change in the tides, caring for the canoes that teach us analytic geometry, and using the Cartesian coordinate system to cultivate native plants that crew members take on voyages" (H. Barbieto, personal communication, December 1, 2015).

Over the past nine years, the Ethnomathematics Institute has grown through successes, challenges and lessons learned. For the first five years at the University of Hawai'i—West O'ahu, performance measures included a 1400% increase in the number of students enrolled in mathematics courses as the general student body population grew from 940 students in 2007 to 2361 students in 2013 (UH IRO 2017). This led to the development of 11 new mathematics courses tied to institutional learning outcomes, accreditation and graduation requirements, all of which are grounded in ethnomathematics. Over the past four years, the Ethnomathematics Institute transitioned into a yearlong professional development program for P–20 inservice educators, and has had participation of educators from all 15 complex areas and seven districts of the HIDOE. This has formed an integrated statewide network that demonstrates commitment to improving learner outcomes, particularly in traditionally underrepresented and underserved populations.

Next steps include institutionalizing the grant-funded Ethnomathematics Institute as a new academic program at the University of Hawai'i at Mānoa beginning in Fall 2018. As we work to become a stronger Indigenous-serving UHS, a new academic graduate-level program focused on preparing P–20 inservice educators to develop curriculum using both Western and non-Western approaches appeals to our diverse populations. The vision of preparing P–20 teachers as leaders to provide ethnomathematics instruction in their schools and communities strengthens the educational pipeline in alignment with the UHS Strategic Directions 2015–2021 and HIDOE Nā Hopena A'o. As stated by UHS President Lassner, "The new ethnomathematics graduate-level program will become a model for other programs interested in creating alternate pathways towards traditional academic goals" (D. Lassner, personal communication, April 27, 2017).

Research conducted by Bishop (1988) provides structure in the process of institutionalization. He asserts, "Of particular significance are the ideas that all cultural groups generate mathematical ideas, and that 'Western' mathematics may be only one mathematics among many…we must recognize the complex layers" (p. 179). We look forward to new developments guided by promising practices, and we know from quantitative and qualitative data that transformational change is occurring through the empowerment of individuals, schools, societies and nations. D'Ambrosio's declaration of a global society where "mathematics for all" reaches an unprecedented dimension as a social endeavor is not just a vision but a growing reality.

7.4 Further Discussion

Through developing new theoretical insights into honoring and sustaining non-Western cultural systems and practices, we have learned that we cannot change the winds but we can change our sails. When we change our sails, we often arrive not necessarily where we think we need to be, but exactly where we are supposed to be.

As a PVS apprentice navigator, I initially thought my responsibility was to arrive at a destination according to the sail plan. On the Mālama Honua Worldwide Voyage leg around the Samoan Islands, we planned to visit a number of islands but we were not able to due to the directional winds. Since it was necessary to return early to Pago Pago, American Samoa, we were able to interact with about 20 schools on Tutuila Island that were not in the original sail plan. Following our education presentation at Matatula Elementary School, the class expressed their appreciation and an 8-year old child stood and stated in the matai ("chief") language, "Thank you for teaching us what is not written in our textbooks" (I. Lagi, personal communication, September 30, 2015). The children remind us why we do the things we do, and they are the reason why we are voyaging the frontiers of education. Together we are writing the textbooks of island earth, and mathematics education is a powerful lens.

In conclusion, three decades after the 1984 International Congress on Mathematical Education (ICME-5), we have increasingly hopeful responses to the challenge of re-examining the equilibrium of mathematics education. Knowledge and action for change are continuing to be achieved through strengthening intersections of culture, community and curriculum in Indigenous wisdom and 21st century learning. In Hawai'i and the Pacific, we explored these through navigating and wayfinding traditions, and the Ethnomathematics Institute being institutionalized as a new academic program at the University of Hawai'i at Mānoa.

As we reflect on our educational visions and calls to action, I am inspired by the 'ōlelo no'eau ("Hawaiian proverb") shared by International Commission on

Mathematical Instruction (ICMI) President Ferdinando Arzarello to open the 2016 International Congress on Mathematical Education (ICME-13). "'A'ohe hana nui ke alu'ia—No task is too big when done together by all" (Pukui 1993, p. 18). Through storms and calm seas, we will change our sails as necessary and continue to remain steadfast in our firm commitment to equity, empowerment and dignity for all.

References

Alim, H., Ibrahim, A., & Pennycook, A. (2011). *Global linguistic flows: Hip hop cultures, youth identities, and the politics of language*. London: Routledge.

Alim, H., & Reyes, A. (2011). Complicating race: Articulating race across multiple social dimensions. *Discourse & Society, 22*, 379–384.

Au, K., & Kawakami, A. (1994). Cultural congruence in instruction. In E. Hollins, J. King, & W. Hayman (Eds.), *Teaching diverse populations: Formulating knowledge base* (pp. 5–23). Albany: SUNY Press.

Barton, B. (2017). Mathematics, education, and culture: A contemporary moral imperative. In G. Kaiser (Ed.), *The Proceedings of the 13th International Congress on Mathematical Education* (pp. 35–43). New York: Springer.

Baybayan, C., Finney, B., Kilonsky, B., & Thompson, N. (1987). Voyage to Aotearoa. *The Journal of the Polynesian Society, 96*(2), 161–200.

Bishop, A. (1988). *Mathematical enculturation: A cultural perspective on mathematics education*. Ann Arbor: Kluwer Academic Publishers.

Boaler, J. (2002). Learning from teaching: Exploring the relationship between reform curriculum and equity. *Journal for Research in Mathematics Education, 33*(4), 239–258.

Bronfenbrenner, U. (1979). *The ecology of human development*. Cambridge: Harvard University Press.

Bryk, A., Gomez, L., & Grunow, A. (2011). *Getting ideas into action: Building networked improvement communities in education*. http://archive.carnegiefoundation.org/pdfs/elibrary/bryk-gomez_building-nics-education.pdf. Accessed April 24, 2017.

Cammarota, J. (2007). A social justice approach to achievement: Guiding Latina/o students toward educational attainment with a challenging, socially relevant curriculum. *Equity and Excellence in Education, 40*, 87–96.

Chang, B., & Lee, J. (2012). "Community-based?" Asian American students, parents, and teachers in the shifting Chinatowns of New York and Los Angeles. *AAPI Nexus, 12*(2), 18–36.

D'Ambrosio, U. (1986). Socio-cultural bases for mathematical education. In M. Carss (Ed.), *Proceedings of the 5th International Congress on Mathematical Education* (pp. 1–6). New York: Springer.

D'Ambrosio, U. (2001). *Ethnomathematics link between traditions and modernity*. Rotterdam: Sense Publishers.

D'Ambrosio, U. (2004). Preface. In F. Favilli (Ed.), *Ethnomathematics and mathematics education* (pp. v–x). Pisa: Tipografia Editrice Pisana.

Finney, B., Kilonsky, B., Somseon, S., & Stroup, E. (1986). Re-learning a vanishing art. *Journal of the Polynesian Society, 95*(1), 41–90.

Furuto, L. (2014). Pacific ethnomathematics: Pedagogy and practices in mathematics education. *Teaching Mathematics and its Applications: International Journal of the IMA*. https://doi.org/10.1093/teamat/hru009.

Furuto, L. (2016). Lessons learned: Strengths-based approaches to mathematics education in the Pacific. *Journal of Mathematics and Culture, 10*(2), 55–72.

Goetzfridt, N. (2008). *Pacific ethnomathematics a bibliographic study*. Honolulu: University of Hawai'i Press.

Greer, B., Mukhodpadhyay, S., Powell, A., & Nelson-Barber, S. (2009). *Culturally responsive mathematics education*. New York: Routledge Press.

Hall, S. (1993). Cultural identity and diaspora. In P. Williams & L. Chrisman (Eds.), *Colonial discourse and post-colonial theory: A reader*. London: Harvester Wheatsheaf.

Hawai'i State Department of Education. (2015). *Nā Hopena A 'o*. http://www.hawaiipublicschools.org/DOE%20Forms/NaHopenaAoE3.pdf. Accessed April 24, 2017.

Hawai'i State Department of Education. (2017). *School data and reports*. http://www.hawaiipublicschools.org/VisionForSuccess/SchoolDataAndReports/Pages/home.aspx. Accessed April 24, 2017.

Hill, M. (2009). *Beats, rhymes and classroom life: Hip-hop pedagogy and the politics of identity*. New York: Teachers College Press.

Hooks, B. (1994). *Teaching to transgress*. New York: Routledge.

Irizarry, J. (2007). Ethnic and urban intersections in the classroom: Latino students, hybrid identities, and culturally responsive pedagogy. *Multicultural Perspectives, 9*(3), 21–28.

Jacob, E., & Jordan, C. (1987). Moving to dialogue. *Anthropology and Education Quarterly, 18*, 259–261.

Jaworski, B., Wood, T., & Dawson, A. (1999). *Mathematics teacher education: Critical international perspectives*. New York: Routledge Press.

Kana'iaupuni, S., Ledward, B., & Jensen, U. (2010). *Culture-based education and its relationship to student outcomes*. Honolulu: Kamehameha Schools, Research & Evaluation.

Kana'iaupuni, S., Ledward, B., & Malone, N. (2017). Mohala i ka wai: Cultural advantage as a framework for Indigenous culture-based education and student outcomes. *American Educational Research Journal, 54*(1), 311–339.

Kania, J., & Kramer, M. (2011). *Collective impact*. http://www.ssir.org/articles/entry/collective_impact. Accessed April 24, 2017.

Kirkland, D. (2011). Books like clothes: Engaging young Black men with reading. *Journal of Adolescent & Adult Literacy, 55*, 199–208.

Kyselka, W. (1987). *An ocean in mind*. Honolulu: University of Hawai'i Press.

Ladson-Billings, G. (1995). Toward a theory of culturally relevant pedagogy. *American Education Research Journal, 32*, 465–491.

Lipka, J., Hogan, M., Webster, J., Yanez, E., Adams, B., Clark, S., & Lacy, D. (2005). Math in a cultural context: Two case studies of a successful culturally based math project. *Anthropology & Education Quarterly, 36*(4), 367–385.

Maton, K., Schellenbach, C., Leadbeater, B., & Solarz, A. (2003). *Investing in children, youth, families and communities: Strengths-based research and policy*. Washington, DC: American Psychological Association.

Morrell, E. (2004). *Becoming critical researchers: Literacy and empowerment for urban youth*. New York: Peter Lang.

Nassir, N. (2008). Everyday pedagogy, lessons from basketball, track, and dominoes. *Phi Delta Kappan*, (March), 529–532.

National Council of Teachers of Mathematics. (2000). *Principles and standards for school mathematics*. Reston: National Council of Teachers of Mathematics.

National Science and Technology Council. (2013). *Federal science, technology, engineering, and mathematics (STEM) education 5-year strategic plan*. Washington, DC: Office of Science and Technology Policy Executive Office of the President.

Pacific Region Educational Laboratory. (1995). *Pacific standards for excellence in teaching, assessment and professional development*. Honolulu: Pacific Mathematics and Science Regional Consortium.

Palhares, P., & Shirley, L. (2013). The role of ethnomathematics in mathematics education. *Revista Latinoamericana de Etnomatemática, 6*(3), 4–6.

Paris, D. (2011). *Language across difference: Ethnicity, communication, and youth identities in changing urban schools*. Cambridge: Cambridge University Press.

Paris, D. (2012). Culturally-sustaining pedagogy: A needed change in stance, terminology, and practice. *Educational Researcher, 41*(3), 93–97.

Polynesian Voyaging Society. (2013). *Promise to children.* Honolulu: Polynesian Voyaging Society.

Polynesian Voyaging Society. (2016). *The story of Hōkūleʻa.* http://www.hokulea.com/voyages/our-story/. Accessed April 24, 2017.

Powell, A., & Frankenstein, M. (1997). *Ethnomathematics: Challenging Eurocentrism in mathematics education.* Albany: SUNY Press.

Pukui, M. K. (1993). *ʻŌlelo noʻeau: Hawaiian proverbs and poetical sayings.* Honolulu: Bishop Museum Press.

Rolón-Dow, C. (2005). Critical care: A color(full) analysis of care narratives in the schooling experiences of Puerto Rican girls. *American Educational Research Journal, 42*, 77–111.

Rosa, M., D'Ambrosio, U., Orey, D., Shirley, L., Alangui, W., Palhares, P., & Gavarrete, M. (2016). *ICME-13 topical surveys: Current and future perspectives in ethnomathematics.* New York: Springer.

Tuhiwai Smith, L. (1999). *Decolonizing methodologies: Research and Indigenous peoples.* New York: Zed Books.

University of Hawaiʻi Institutional Research Office. (2017). *MAPS enrollment projections.* http://www.hawaii.edu/iro/maps_release.php. Accessed April 24, 2017.

University of Hawaiʻi at Mānoa College of Education. (2016). *Currents.* Honolulu: University of Hawaiʻi at Mānoa College of Education.

University of Hawaiʻi System. (2012). *Hawaiʻi papa o ke ao.* http://www.hawaii.edu/hawaiipapaokeao/overview/. Accessed April 24, 2017.

University of Hawaiʻi System. (2015). *University of Hawaiʻi strategic directions 2015–2021.* http://blog.hawaii.edu/strategicdirections/files/2015/01/StrategicDirectionsFINAL-013015.pdf. Accessed April 24, 2017.

Waitzer, J., & Paul, R. (2011). Scaling social impact, when everybody contributes, everybody wins. *Innovations, 6*(2), 143–155.

Weiss, I., & Miller, B. (2006). *Developing strategic leadership for district-wide improvement of mathematics education.* Lakewood: National Council of Supervisors of Mathematics.

Winn, M. (2011). *Girl time: Literacy, justice, and the school-to-prism pipeline.* New York: Teachers College Press.

Chapter 8
The Impact and Challenges of Early Mathematics Intervention in an Australian Context

Ann Gervasoni

Abstract This paper explores the design and longitudinal effect of an intervention approach for supporting children who are mathematically vulnerable: the *Extending Mathematical Understanding* (EMU)—Intervention approach. The progress over three years of Grade 1 children who participated in the intervention was analysed and compared with the progress of peers across four whole number domains. The findings show that participation in the EMU program was associated with increased confidence and accelerated learning that was maintained and extended in subsequent years for most children. Forty per cent of children were no longer vulnerable in the year following the intervention, and others were vulnerable in fewer domains. Comparative data for non-EMU participants highlights the wide distribution of mathematics knowledge across all children in each grade level. This explains why classroom teaching is so complex and highlights the challenges teachers face in providing inclusive learning environments that enable all students to thrive.

Keywords Mathematics difficulties · Mathematics intervention
Inclusion · Whole number concepts · Mathematics assessment

8.1 Introduction

Leaders of school systems throughout the world voice concern about the phenomenon of children who experience difficulty with learning school mathematics and seek insight about how to overcome this situation. Ensuring that all children thrive mathematically is recognised as important for children's future citizenship and opportunities for work and further education, and ultimately for contributing to the economic and cultural prosperity of a society. The International Committee for Mathematics Education (ICME) actively supports research and development in this area. It commissioned a survey team to examine the state of the art with respect to

A. Gervasoni (✉)
Monash University, Melbourne, Australia
e-mail: Ann.Gervasoni@monash.edu

© The Author(s) 2018   115
G. Kaiser et al. (eds.), *Invited Lectures from the 13th International Congress on Mathematical Education*, ICME-13 Monographs,
https://doi.org/10.1007/978-3-319-72170-5_8

the *Assistance of students with mathematical learning difficulties—How can research support practice?* (Scherer et al. 2016). The survey results highlight shifting paradigms in understanding mathematics learning difficulties, and found that approaches for identifying and assisting children are moving away from medical models towards more inclusive approaches.

In Australia, there is increased attention placed on the early identification of children who experience difficulty with learning school mathematics, and on providing these children with access to research-informed interventions to enable their learning. One approach is the *Extending Mathematical Understanding* (EMU) intervention program (Gervasoni 2004, 2015), developed by the author, and implemented in hundreds of Australian schools as part of a whole school approach to enabling mathematics learning for all (Fullen et al. 2006; Gervasoni et al. 2010). This approach and the outcomes are explored in this paper. It is anticipated that the insights gained from examining this intervention approach may contribute to the international discussion about the type of resources and strategies that can enable children to thrive mathematically. A particular focus is considering the longitudinal effect of the intervention approach for children who initially failed to thrive when learning school mathematics, and any implications for providing inclusive classroom environments that enable all children to learn mathematics successfully.

8.2   Failure to Thrive When Learning Mathematics

Currently there are contested views for explaining why children initially fail to thrive when learning school mathematics and for describing the phenomena of children who experience difficulty with learning mathematics. Gervasoni and Lindenskov (2011) argue from a social justice perspective, that these children have 'special rights' in mathematics education because historically they have not had access to high-quality mathematics programs and instruction. These students fall into two groups. The first group comprises those children who are visually or hearing impaired, or who have physical or intellectual impairments such as Down syndrome. The second group are those who underperform in mathematics due to their exclusion from quality mathematics learning and teaching environments that are necessary for them to thrive mathematically. Underperformance in mathematics is too often due to issues associated with equity and quality. Gervasoni and Lindenskov (2011) argue that many students in this first group have been directly excluded from opportunities and educational pathways in learning mathematics because mathematics was deemed an inappropriate field of study for them (e.g., Faragher et al. 2008; Feigenbaum 2000). The second group of students participate in mathematics classes, but they do not receive the quality of instruction or experience that enables them to thrive mathematically (Gervasoni and Sullivan 2007; Lindenskov and Weng 2008). These students are indirectly excluded from mathematics education.

The perspective and principles underpinning the development and implementation of the EMU intervention approach are that all children have the right to access high quality mathematics education, at their local school, that enables them to thrive. This implies that all children can learn mathematics successfully given the necessary resources, environment, and teaching. However, it is important to acknowledge that providing high quality mathematics learning environments for all students is a struggle that may take some time to achieve. For example, in Tanzania, universal primary education has just become a reality for six-year-old children. First grade classes may include more than 100 students with access to few mathematics learning opportunities and teaching resources (Gelander et al. 2017). Even the most expert mathematics teacher may struggle to teach 100 children effectively in this environment. In contrast, Australia has the economic resources to provide six-year-old children with class sizes of 20–25 children and primary school teachers with at least 4-year degrees in teacher education. Australian children begin primary school at the start of the year in which they turn five, and have access to 15 h per week of pre-school education in the year prior to beginning school. However, even in this environment, not all Australian children thrive when learning mathematics at school and may be considered mathematically vulnerable. The term *vulnerable* is widely used in population studies (e.g., Hart et al. 2003), and refers to students whose environments include risk factors that can lead to poor developmental outcomes.

8.3 The Extending Mathematical Understanding (EMU) Intervention Approach

The EMU intervention approach is based on a social constructivist view of learning (Cobb et al. 1992) and the principle that all children can learn mathematics given access to the necessary resources, environment, and teaching. This view contrasts with many intervention approaches that consider mathematics learning difficulties from a medical or psychological paradigm rather than as a social construct. Magne (2003) extensive review of the literature on special educational needs in mathematics found that a medical model was adopted in the majority of studies surveyed; this positions mathematics difficulties as innate deficiencies as opposed to a socially, culturally, and politically constructed facet of identity and experience (Scherer et al. 2016). In the EMU approach, teachers focus on designing rich learning environments for all students that are responsive to differences in how children learn. The approach recognises that teachers need to be expert at understanding how individual children learn mathematics, and how they can advance this learning. This calls for a high level of professional knowledge. In Australia, primary classroom teachers are generalists and typically teach every curriculum area. This means that their initial teacher education rarely includes the depth of knowledge required to deal with the complexity of providing a truly inclusive mathematics classroom.

Schools using the EMU approach first concentrate on enabling all classroom teachers to increase their mathematical pedagogical content knowledge (PCK), and their ability to design and implement inclusive learning environments. This involves providing professional learning opportunities to develop each teacher's knowledge and confidence. These schools also employ an EMU specialist teacher whose role is to provide three levels of support for vulnerable students and classroom teachers (Gervasoni 2015), including an EMU intervention program for the most vulnerable students in Grade 1. Level 1 EMU support provides classroom teachers with advice about how to best advance a child's mathematics learning, and this is supplemented by an individual learning plan that outlines the learning goals and experiences that may boost a particular child's learning. Level 2 support provides this same advice plus in-classroom support (e.g., peer teaching, coaching, small group teaching) for children during mathematics lessons. Level 3 includes a small group EMU intervention program for prioritised students. This daily withdrawal program is coordinated with the classroom mathematics program. EMU specialist teachers complete a 36-hour course (at Masters level) that focuses on assessment of children's knowledge and dispositions, mathematical pedagogical content knowledge, and instructional design that maximises mathematics learning for all. This professional learning program recognises that specialist mathematics teaching knowledge is one important factor in enabling all students to thrive mathematically (Hill et al. 2005). Specialists also complete at least 25 h of field-based learning associated with teaching the EMU intervention program, and a program of professional reading prior to being accredited. Ongoing accreditation requires that EMU specialists engage in two days of ongoing professional learning each year.

Level 3 EMU support provides Grade 1 children who are mathematically vulnerable with an intervention program that aims to accelerate their learning and increase their confidence, so that ultimately they can thrive in the regular classroom environment. The intervention is also possible for older students, depending on the available resources in a school. During this intervention, groups of three children (6-year-olds) participate in 30-min lessons 5 days per week for a total of 10–20 weeks (i.e., 50–100 lessons), depending on their progress. The lessons are designed and customised for each student because of the diverse range of knowledge and difficulties noted amongst those who are mathematically vulnerable (see Table 8.1). Gervasoni and Sullivan (2007) found that it was rare to find two students with the same difficulties. The theoretical underpinnings of the EMU intervention program, the teaching approach, and lesson structure are described in detail in Gervasoni (2004), and also in the accompanying book for specialist teachers (Gervasoni 2015). In brief, each 30-min lesson focuses on: whole number learning with an introduction to build connections with recent learning (2-min); activities to develop children's understanding of quantities and numerosity, including place value and counting knowledge(8-min); mathematical investigations and open tasks involving the four operations, with an emphasis on the development of heuristic arithmetic and reasoning strategies (15-min); reflection on learning (5-min); and assignment of a daily home task (usually a mathematics game) to involve parents in the children's learning. Specialist teachers are encouraged to be responsive to what

Table 8.1 Combinations of whole number domains in which children participating in the 2010 EMU program were vulnerable in 2010 (pre-EMU) and 2011 (post-EMU)

Number of domains	Vulnerable in counting	Vulnerable in place value	Vulnerable in add and subtraction	Vulnerable in multi and division	Grade 1 pre-EMU vulnerable	Grade 1 pre-EMU total	Grade 2 post-EMU vulnerable	Grade 2 post-EMU total
0					2	2 (5%)	17	17 (41%)
1	✓				2	9 (21%)	3	10 (24%)
		✓			0		5	
			✓		4		1	
				✓	3		1	
2	✓	✓			1	13 (31%)	4	7 (17%)
	✓		✓		2		1	
	✓			✓	7		0	
		✓	✓		0		1	
		✓		✓	2		1	
			✓	✓	1		0	
3	✓	✓	✓		1	14 (33%)	2	5 (12%)
	✓	✓		✓	4		0	
	✓		✓	✓	8		2	
		✓	✓	✓	1		1	
4	✓	✓	✓	✓	4	4 (10%)	2	2 (5%)
Total					42	42 (100%)	41	41 (100%)
Pre EMU	29 (69%)	17 (41%)	17 (41%)	30 (71%)				
Post EMU	12 (34%)	18 (44%)	8 (20%)	7 (17%)				

they learn about each student during each lesson. The pedagogical approach encourages children to use concrete models to assist with their construction of new knowledge, and teachers prompt children to simulate, imagine, and describe solutions derived from using these concrete models. Children are also expected, and supported, to explain their thinking and strategies, and to develop confidence and positive mind-sets. The course for EMU specialist teachers emphasises these pedagogical approaches, and detailed explanations and illustrative examples are provided in the 220 page *Extending Mathematical Understanding: Intervention* book for teachers (Gervasoni 2015).

8.4 Using Growth Point Profiles to Identify Children Who May Benefit from an Intervention Program

While the aim of inclusive mathematics education is for each child to thrive with their learning in their classroom, it is also necessary to identify those children who are not thriving and to change their experience of learning. Sometimes a more intensive learning experience than what is available in a regular classroom is helpful in the short term. This is the intent of the EMU intervention program.

The EMU approach begins with the proposition that the classroom and specialist teachers need to deeply understand each child's current mathematical understanding and strategies. This process is facilitated through reference to a framework of mathematics growth points that help teachers recognise children's current understanding in four whole number domains, and guide their teaching. The growth point framework in the EMU approach was developed during the *Early Numeracy Research Project* (Clarke et al. 2002) and further refined in 2013 (Gervasoni et al. 2011). The processes for validating the growth points, the associated assessment interview items, and the comparative achievement of students are described in full in Clarke et al. (2002) and have been reported widely (e.g., Clarke 2001, 2013). From a research and evaluation perspective, the framework of growth points also enables children's learning progress to be measured. The growth points do not represent an assessment score, but rather describe a child's current knowledge in reference to the set of research-informed progressions in children's developing knowledge. This is a common approach in Australia (Bobis et al. 2005). The idea is that the growth points guide teachers about how they might respond to a child's current knowledge, and then provide the resources and teaching to extend their learning.

School communities using the EMU approach organise for classroom teachers to assess all children in their class using the *Mathematics Assessment Interview* (Gervasoni 2011). This one-on-one assessment requires the teacher to sit with each child, and to observe and probe their thinking and strategies for solving mathematics tasks, until they have a deep understanding of the extent of the child's current whole number knowledge. Based on children's strategies and responses, the detailed assessment script leads children through different tasks in nine whole number, measurement, and geometry domains, just like a *choose your own adventure story*. Following the assessment, the teacher analyses each child's responses to determine their growth point profile, and identifies whether any children are mathematically vulnerable. In the EMU intervention context, the four whole number domains (Counting, Place Value, Addition and Subtraction Strategies, and Multiplication and Division Strategies) have been shown to be the most reliable for identifying children who were mathematically vulnerable. In contrast, children's performance in the measurement and geometry domains was much less predictable (Gervasoni 2004). This emphasis on the whole number domains for identifying and prioritising children who may benefit from an EMU intervention program does not diminish the importance of children's measurement and geometry learning. Rather, it has been found that children have typically learnt measurement and geometry knowledge successfully in the classroom environment.

To illustrate the nature of the growth points (see Clarke et al. 2002; Gervasoni 2015 for detailed descriptions) and the wide distribution of knowledge in any grade level, Fig. 8.1 shows the 2016 Multiplication and Division Strategies distribution of growth points for all 21,884 primary students in a region of the New South Wales.

The wide distribution of growth points in every grade level was evident also in each of the other three whole number domains. For each grade level, it is clear that there is a group of students who are failing to thrive in comparison to their peers. It is also clear that responding to this wide distribution of knowledge is highly complex for teachers, and requires the classroom teacher to be highly skilled in customising mathematics lessons and teaching. An inference from these data is that children on the lowest growth points in a class may be marginalised, with the teacher struggling to provide the resources necessary to enable all to learn.

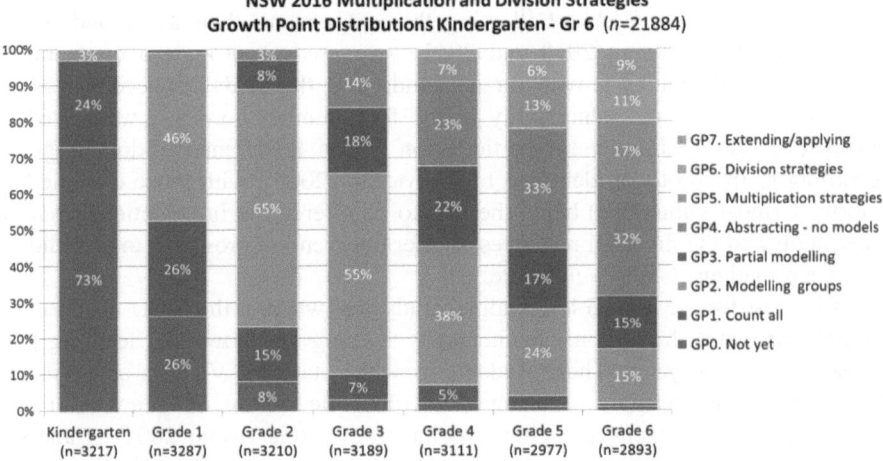

Fig. 8.1 Multiplication and Division Strategies growth point distributions for children in Kindergarten to Grade 6 in 2016

The EMU approach uses children's growth point profiles in the four whole number domains to prioritise children for Level 3 EMU intervention. For example, in the domain of Multiplication and Division Strategies, Grade 1 children who do not yet use count-all strategies (growth point one) to solve multiplicative problems are considered mathematically vulnerable in this domain, and they qualify for specialist EMU intervention support (Gervasoni 2004). This is because their current inability to use count-all strategies for simple multiplicative tasks excludes them from engaging with typical Grade 1 learning activities.

8.5 Progress of Students Who Participated in an EMU Intervention Program

Several studies have investigated the impact of the EMU intervention program (e.g., Clarke et al. 2002; Gervasoni 2004). During the *Bridging the Numeracy Gap in Low SES and Aboriginal Communities Pilot Project* (BTNG) (Gervasoni et al. 2011, 2012), the longitudinal progress of six-year-old students who participated in an EMU intervention program in 2010 was measured over three years using the ENRP growth point framework. The participants in the BTNG research all belonged to socially disadvantaged communities, as classified by the Australian Government, and formed two groups. The first group was the 42 Grade 1 children who, in 2010, took part in an EMU intervention program for 10–20 weeks. The second comparison group comprised all 2545 Grade 1 (6-year-old) to Grade 4 (9 year-old) children who attended the schools involved in the study during 2010–2011. All these schools employed an EMU Specialist Teacher who was able to provide support for children who were mathematically vulnerable, and all children in these schools were assessed by their classroom teachers using the *Mathematics Assessment Interview* (MAI). Following this assessment, their associated whole number growth points in 2010 and 2011 were used to provide a comparative measure of mathematics growth for all children in the study. Children's growth point profiles, and any vulnerability in the four number domains, were used to prioritise Grade 1 children for participation in the EMU intervention program, according to the protocol identified by Gervasoni (2004), with those classified as Priority 1 (most vulnerable) being the first to be offered the intervention. None of the schools had the financial resources to offer intervention programs to all children who were mathematically vulnerable.

Of interest for the BTNG longitudinal study was whether the EMU intervention program accelerated children's mathematics learning and how their learning progressed over three years. These students from the states of Victoria and Western Australia were all the most mathematically vulnerable students in their class, based on their MAI assessment and growth point profiles (Gervasoni 2004). Table 8.1 shows the number of whole number domains and the combination of domains for which these 42 children were vulnerable at two time points: (1) at the beginning of

Fig. 8.2 The percentage of EMU participants who were vulnerable in 0, 1, 2, 3 or 4 whole number domains in 2010 and 2011

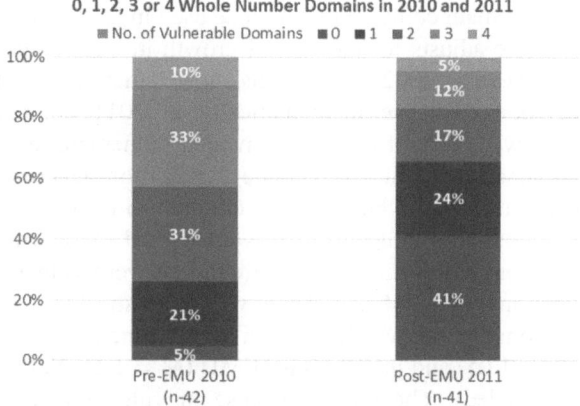

Grade 1 and before they began the EMU intervention program; and (2) 12 months later, at the beginning of Grade 2 and after these children had completed both the EMU Program in the previous year and the long summer holiday. Figure 8.2 shows the decrease in the number of domains for which children were vulnerable after the 2010 EMU Program. It is important to note that the growth points used for identifying children as vulnerable in Grade 2 (2011) were increased by one growth point in each domain to account for median growth across 12 months. For example, children beginning Grade 1 were identified as vulnerable in Counting if they did not reach Growth Point 2. At the beginning of Grade 2, children were identified as vulnerable in Counting if they did not reach Growth Point 3.

The data suggest that these children were a diverse group. Pre-EMU, some were vulnerable in only one domain (21%), some in two (31%) or three domains (33%), but only four students (10%) were vulnerable in all four domains. This is consistent with the findings of Clarke et al. (2002) during the *Early Numeracy Research Project*. Further, the combinations of domains for which the students were vulnerable prior to participation in the EMU intervention program varied. Clearly, there was no one pattern to describe students who were mathematically vulnerable. This highlights the complexity of teaching, and the need for teachers to be expert at understanding children's current knowledge and in designing learning environments that are personalised to enable all children to learn.

From an evaluation perspective, the EMU program aims for no children to be mathematically vulnerable at the end of the program or, as a minimum, for children to be vulnerable in fewer domains. To measure the effect of the EMU program for decreasing the domains for which children are mathematically vulnerable, children's EMU growth point data were collected by the new classroom teacher at the beginning of Grade 2, that is, after the long summer holidays. The data shown in Table 8.1 and Fig. 8.2 show that in 2011, after experiencing the EMU Program in 2010, the vast majority of students were vulnerable in fewer domains, and almost half were no longer vulnerable at all. It is important to note that the growth points used to identify a student as vulnerable in 2011 were one growth point higher in each domain compared with those used in 2010 and before the EMU Program. This increase adjusts for the typical growth in knowledge across one year of schooling (Clarke et al. 2002). The results suggest that most of the children were in a stronger position to thrive mathematically in 2011 compared with the previous year. However, 17% of children remained vulnerable in 3 or 4 domains. The challenge remained for their 2011 Grade 2 classroom teacher to create the learning experiences that would enable these children to thrive.

A paired-samples t-test was conducted to determine whether the change in the number of domains for which children were vulnerable before and after the intervention was significant. There was a statistically significant decrease in the number of domains for which children were vulnerable in 2010 before EMU (M = 2.20, SD = 1.05) and in 2011 after EMU (M = 1.15, SD = 1.24), t (40) = 6.13, $p < .001$ (two-tailed). The mean decrease in vulnerable domains was 1.05 with a 95% confidence interval ranging from 0.703 to 1.394. The eta squared statistic (.48) indicated a large effect size. While the study design does not enable the claim to be made that the EMU intervention program was the sole cause of this decrease in vulnerability, the intervention program was likely to be a contributing factor.

8.6 Longitudinal Impact on Mathematics Knowledge and Growth Points Over Three Years

The data presented in the previous section highlight that participation in the EMU program was associated, typically, with an acceleration in some children's whole number learning and a decrease in the number of domains in which they were vulnerable. However, it is also important to determine whether this learning was maintained and extended in the following years. To evaluate the longitudinal progress of students who participated in a Grade 1 EMU intervention program only in 2010 (but not in subsequent years), the EMU group's growth point distributions in 2010–2013 for each domain were calculated and compared with the progress of all students in the entire cohort (including EMU students). Figure 8.3 shows the growth point distributions for the two groups for the Multiplication and Division Strategies domain from 2010 to 2013. Note that due to the BTNG project ending in 2011, longitudinal data for students in the comparison 2012 Grade 3 cohort and the 2013 Grade 4 cohort were unavailable, so available 2011 data for all Grade 3 and Grade 4 students in the same schools were used to illustrate the distributions that might be expected of the 2012 Grade 3 and 2013 Grade 4 cohorts. An asterisk indicates this use of 2011 Grade 3 and Grade 4 data in Figs. 8.3, 8.4, 8.5 and 8.6. Only growth point data for children in the EMU intervention group were available for collection in 2012 and 2013 due to a limited extension of the BTNG project to enable these children's longitudinal progress to be measured.

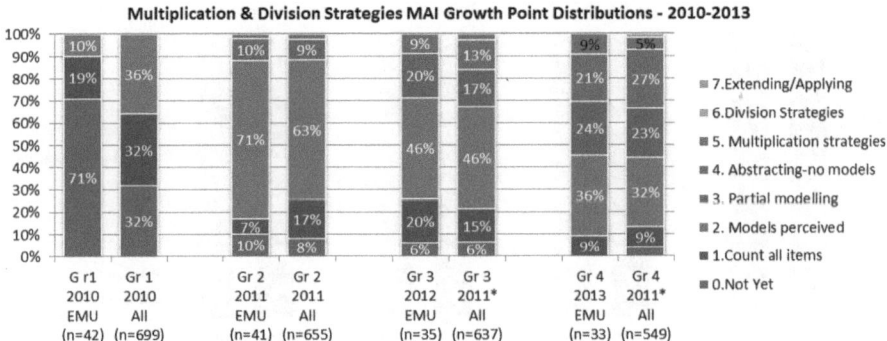

Fig. 8.3 2010–2013 multiplication and division growth point distributions (beginning of the year) for the 2010 EMU group, and comparison data for all Grade 1–Grade 4 students

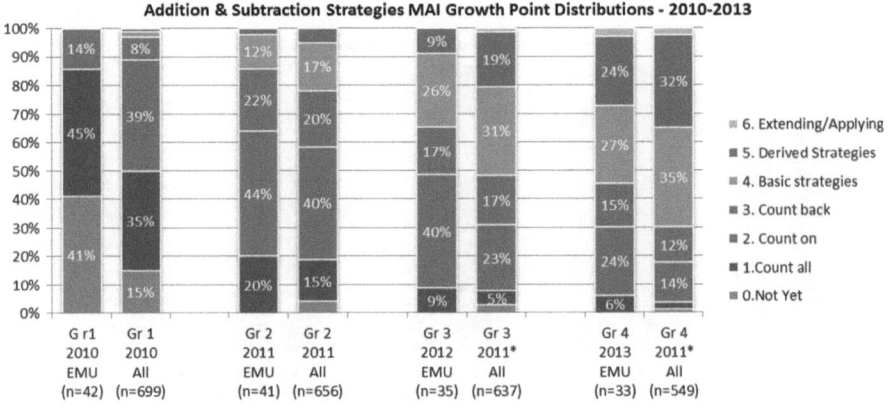

Fig. 8.4 2010–2013 addition and subtraction growth point distributions (beginning of the year) for the 2010 EMU group and comparison data (2010–2011) for all Grade 1–Grade 4 students

Figure 8.3 shows that the spread of Multiplication and Division Strategies growth points for the 2010 EMU group was substantially different to their peers, with 71% of the EMU group on Growth Point 0 (GP0) compared with only 32% of their peers. However, the EMU group made substantial growth by 2011. It is noticeable how similar the spread of growth points are in 2011 for both the EMU group and their peers (*All*) in Grade 2. This finding was also apparent for the other whole number domains (see Figs. 8.4, 8.5 and 8.6). These data suggest that one effect of the EMU Program in 2010 was an acceleration of whole number learning to the extent that the EMU group's growth point distribution at the beginning of 2011 (Grade 2) mirrored that of their peers. Nevertheless, while some EMU students progressed two or three growth points in each domain across 2010–2011, some remained vulnerable when they reached Grade 2 and remained on the lowest

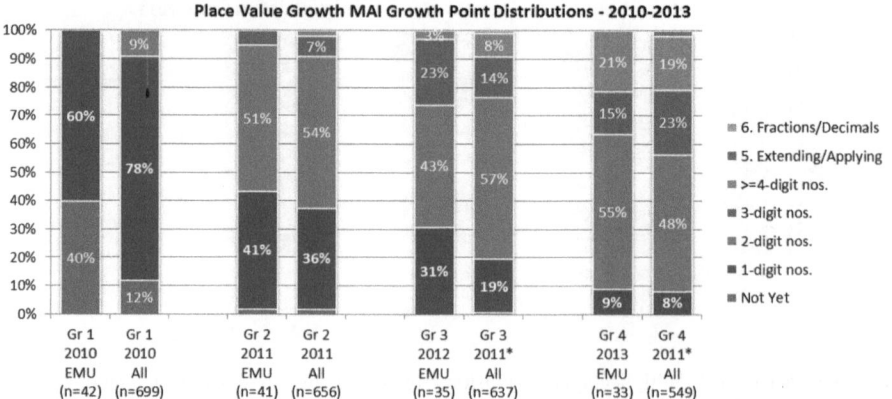

Fig. 8.5 2010–2013 place value growth point distributions (beginning of the year) for the 2010 EMU group and comparison data (2010–2011) for all Grade 1–Grade 4 students

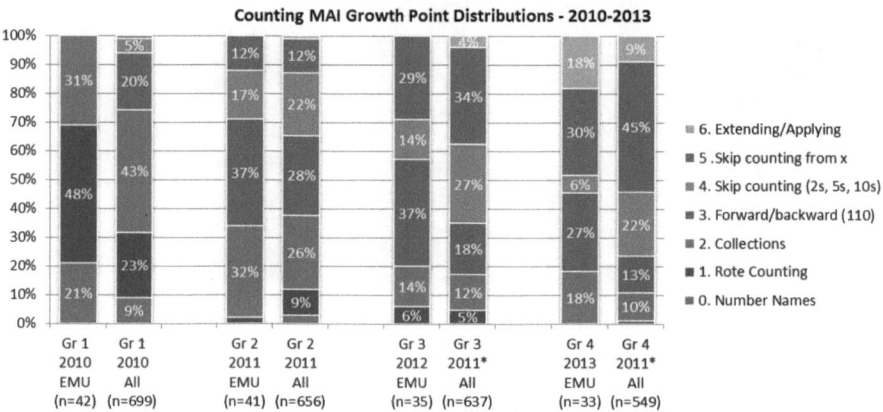

Fig. 8.6 2010–2013 counting growth point distributions (beginning of the year) for the 2010 EMU group and comparison data (2010–2011) for all Grade 1–Grade 4 students

growth points (GP0 and GP1). Such vulnerable students, including those who do not participate in the Grade 1 intervention program, are of concern and may benefit from additional assistance throughout Grade 2.

Overall, it is clear from the data presented in Figs. 8.3, 8.4, 8.5 and 8.6 that the majority of students participating in the EMU intervention program made accelerated progress in each domain by the beginning of Grade 2 (2011). It is also important to consider whether their progress continued or faded in 2011–2013 when they no longer had the opportunity afforded by an intervention program. A comparison of the Grade 2–Grade 4 Multiplication and Division growth point distributions (Fig. 8.3) for the 2010 EMU group suggests that their learning was maintained during this period, but that the rate of progress for many students was less in the following years when they no longer received additional support from a specialist teacher. Transition from one growth point to the next growth point in the framework represents a significant step in a student's development that may take 12 months to achieve, as opposed to smaller steps in learning that are noticeable day by day (Clarke et al. 2002). Figure 8.3 shows that, typically, the children in the highest quartile of the EMU cohort distribution (on GP3 and GP4) at the beginning of Grade 2 (2011) progressed one additional growth point from 2012 to 2013 in the Multiplication and Division Strategies domain, but EMU students in the lowest quartile distribution (on GP0 and GP1) of the cohort distribution made less progress, on average. Encouragingly, the rate of progress of students in the 2010 EMU group from 2011 to 2013 was consistent with the progress of their peers (All). Children on the lowest growth points struggled to make progress in 2011 and may not have benefitted greatly from their classroom experiences in Grade 2.

Figure 8.4 shows children's progress in the Addition and Subtraction Strategies domain. The EMU group made strong progress from 2010 to 2011 after the period of the intervention program but, although their learning was maintained in subsequent years, the rate of progress reduced.

It is of interest to examine the progress of both the EMU Group and their peers (All) from Grade 2 to Grade 3 (2011–2012). On average, comparison students in the upper 50% of the Grade 2 distribution were likely to progress at least one growth point by Grade 3 (e.g., GP3–GP4 or GP4–GP5), but the learning for most EMU students in the bottom half of the Grade 2 distribution has stagnated by Grade 3 on GP1 and GP2. That is, 64% of the EMU students were on GP1 or GP2 at the beginning of Grade 2 and 49% were still on GP1 or GP2 at the beginning of Grade 3. It appears that being able to move from using a count-on strategy for addition (GP2) to a count-back strategy for subtraction (GP3) is a difficult progression for many students.

In the Place Value domain, mean growth for children is just less than one growth point per year (Clarke et al. 2002). Examining the data in Fig. 8.5 for Grade 2, Grade 3 and Grade 4 children in both the EMU and comparison groups shows that the median was GP2 (understanding 2-digit numbers) in all these distributions. From year to year, this was less growth than might be expected and suggests that many children's place value knowledge was stagnating from Grade 2 through to Grade 4. Further, comparisons between the growth from Grade 2 to Grade 3 for both the EMU group and their peers (All) suggest that, on average, students beginning Grade 2 on GP1 or GP2 (success with tasks involving 1-digit and 2-digit numbers respectively), were highly likely to remain on these growth points one year later. These findings suggest that learning opportunities in Grade 2 and Grade 3 classrooms were insufficient for all students in Place Value. This is an issue for school systems to investigate.

Inspection of the 2010 EMU group's progress in the Counting domain (Fig. 8.6) suggests that, on average, learning was accelerated across Grade 1 for EMU students during the intervention period. There was substantial change in the proportion of EMU students in the lowest two growth points (GP0 and GP1) from Grade 1 to

Grade 2. Further, there were proportionally fewer children in the EMU group on these growth points in Grade 2 than among the *All* students group.

The data in Fig. 8.6 suggest that progress for some EMU students was likely to stagnate across Grade 2 and Grade 3, particularly if children began the year on Growth Point 2 (can count at least 20 objects) or Growth Point 3 (can count by ones past 109 and back from 24). In stark contrast, the EMU children who began on Growth Point 2 at the beginning of Grade 1, on average, were likely to progress to at least Growth Point 3 one year later. This suggests that instruction in Grade 2 and Grade 3 may not have been sufficiently focused on supporting students to learn to skip count (GP4 and GP5). Skip counting, or the teaching of skip counting, appears to present a barrier that prevents some students from progressing beyond GP3.

8.7   Impact of EMU Intervention on Children's Confidence for Learning Mathematics

Another goal of the EMU intervention program is to develop children's positive dispositions for learning mathematics. However, this aspect of children's learning was not investigated during the BTNG project described earlier. To gain some insight on any impact of the EMU intervention program in this regard, 127 EMU specialist teachers in New South Wales who completed their course in 2016 were surveyed. They were asked, "What key changes have you observed in your students as a result of their participation in the EMU program?" Seventy-nine of the teachers (62%) noted changes in children's confidence. Several responses that illustrate the impact of the EMU intervention program on children's confidence include the following:

> The increase in the students' confidence has been the biggest change. The students are more willing to participate in classroom mathematics.

> Confidence and engagement/enjoyment in maths has been a huge change - these kids are now approaching maths with a positive mindset and (as reported back from the classroom teachers following the course) these students are now much more likely to 'have a go' in their classrooms, persevere with their learning if they find it difficult, and look for different ways/strategies to solve a problem.

> I could not believe the speed with which they showed improvement. After 2 weeks they were more confident in sharing their strategies and ideas and it carried over into the classroom too. Teachers have reported that the students in EMU were more confident in Maths and used a wider variety of strategies than other students. The students taught their classmates some of their activities and became 'the experts'!

> The confidence that my students now demonstrate is fantastic. They are so much more engaged in numeracy in their classrooms and love to contribute ideas and explain their thinking.

Another strong theme in the responses was the positive change in children's engagement in classroom mathematics learning. This is also evident in the responses above. These data suggest that the EMU intervention program, as perceived by the specialist intervention teachers, had a positive effect on children's dispositions for learning mathematics, as well as on their mathematics knowledge and problem solving. Further, the data suggest that this increased confidence may have transferred to children's experience of learning mathematics in their classrooms.

8.8 Issues Related to Effective Intervention Approaches

The data presented earlier demonstrate that the mathematics learning of most children who participated in the EMU intervention program increased across the year, and that this learning was mostly maintained and extended in the subsequent three years. Further, most children gained in confidence and had more positive attitudes to learning mathematics, as noted by their classroom teachers. Thus, the majority of children were more strongly positioned to thrive mathematically in their classroom environment following participation in an EMU intervention program. This impact was further demonstrated by many EMU participants no longer being mathematically vulnerable or being vulnerable in fewer domains, with associated increased mathematics knowledge and confidence to bring to the classroom learning environment. Thus, the EMU program is likely to have assisted children to benefit more fully from their classroom mathematics learning in subsequent years. However, some children remained vulnerable. It is possible that these children may benefit from ongoing support from a specialist teacher who can advise the classroom teacher about the type of experiences and teaching adjustments that can enable their learning. Indeed, a specialist teacher can play an important role across the school in providing professional learning opportunities and advice for classroom teachers as they work towards providing a more effective and inclusive environment for mathematics learning.

The longitudinal data that described the progress of EMU children, alongside the whole cohort of students for comparison, clearly demonstrated that there were points when many children's mathematics learning stalled for 12 months or longer. This finding suggests that the classroom teachers may have needed further support to improve their pedagogical content knowledge to respond productively to all children's current mathematical understanding, and to provide experiences that enabled growth for all. Inclusive mathematics education calls upon teachers to provide learning experiences and teaching based on what children currently know, within the framework of a curriculum document. The wide distribution of knowledge in any one classroom across multiple mathematical domains, as demonstrated in Figs. 8.3, 8.4, 8.5 and 8.6, highlights the complex situation that teachers face when designing inclusive mathematics learning environments. It may be that the teaching approaches used by the EMU specialist teachers in the intervention program may be beneficial for classroom environments also. This would include: providing tasks that can be differentiated to enable children to engage in different ways and levels; adopting teaching strategies that enable children to use concrete models to assist with problem solving in new topic areas; encouraging children to simulate and describe their actions with concrete models; and providing opportunities for children to discuss their mathematical conjectures. Identifying effective teaching approaches that respond to the wide variation in children's mathematical knowledge is an important topic for ongoing research and development.

The increased learning for children in the EMU intervention group was noticeable when comparing the growth point spread for EMU students and their peers when the children reached Grade 2. Indeed, by Grade 2 (2011), and again in Grade 3 (2012) and Grade 4 (2013), the growth point distributions of both groups were very similar, in contrast to the marked differences observed between the two groups in 2010. However, an important issue apparent in the EMU group's Grade 2, Grade 3, and Grade 4 growth point distributions was that learning for some students seemed to stall across these grades; generally students in the top quartile of the growth point distribution were most likely to progress. This finding suggests that classroom mathematics teaching for Grade 2–Grade 4 students may not be sufficiently differentiated to enable all students to thrive. The analyses also suggest that not all Grade 1 children who participated in an EMU intervention program experienced accelerated learning in all whole number domains. Longitudinally, one-quarter of the EMU group reached the highest growth points found in the 2011, 2012 and 2013 distributions for *All* students, while a proportion of the EMU group remained mathematically vulnerable in subsequent years. These EMU children progressed, but remained mathematically vulnerable as the curriculum demands increased. Further insights are needed about effective strategies to assist these students.

Participation in the EMU intervention program was associated with most Grade 1 children progressing their whole number learning beyond the one growth point anticipated in each domain across one school year. This was true even for the children who began on the lowest growth points. It was also apparent that their learning was maintained over subsequent years, although some students' learning progression stagnated. This stagnation in learning was noted also for students in the *All* students comparison cohort. Profitable areas for further research and development are: (a) seeking insight into why some students make less progress during an intervention program than others, and (b) designing classroom instruction for Grade 2–Grade 4 students that is more inclusive and better enables mathematics learning for all. It may be beneficial for an EMU specialist teacher to be more available to advise Grade 2–Grade 4 teachers about how to refine curriculum and customise teaching to enable all to learn. It is also likely that some students may benefit from more specialised mathematics teaching beyond Grade 1, and also that classroom teaching in Grade 2–Grade 4 may need to be more responsive to students' individual learning needs. The importance of teachers' mathematical knowledge for teaching has been increasingly recognised as a key to achieving desired learning outcomes for all students (Hill et al. 2005).

8.9 Conclusion

The research and experiences presented in this chapter suggest that participation in a Grade 1 EMU intervention program was associated with accelerated mathematics learning for most students and that this learning was generally maintained and extended in the following three years. It appears that the EMU program also results in children gaining confidence as learners of mathematics. A review of the *Maths Recovery Program* (Smith et al. 2013) concluded that mathematics intervention programs must be coordinated with, rather than isolated from, the classroom mathematics program. This conclusion is supported by Clements et al. (2013) who further claim that interventions in the early years need to be scaled-up in subsequent years to be most effective for students. The longitudinal data presented in this chapter highlight that although most EMU intervention participants' learning was maintained in subsequent years, some students stalled in their learning at various points in the three years after the EMU Program concluded. This supports the findings of Smith et al. (2013) and Clements et al. (2013) that interventions need to be co-ordinated with classroom programs and scaled-up in subsequent years for their impact to be extended beyond the intervention period. Following an intervention program, if children continue to experience the same conditions under which they were marginalised and excluded from learning mathematics in the first place, then their learning may again be disrupted. It is possible that specialist intervention teachers have a role to play in scaling-up intervention in subsequent years through supporting classroom teachers to provide an inclusive mathematics learning environment in which all students can thrive.

Acknowledgements The research reported in this paper was funded by the Australian Government as part of the Bridging the Numeracy Gap in Low SES and Indigenous Communities Project. The author acknowledges gratefully the contribution of the research team and all participating teachers, parents, students, and school communities.

References

Australian Curriculum, Assessment and Reporting Authority [ACARA]. (2012). *Australian curriculum: Mathematics*. Sydney: ACARA. Retrieved from http://www.australiancurriculum. edu.au/mathematics/curriculum/f-10?layout=1.

Bobis, J., Clarke, B., Clarke, D., Thomas, G., Wright, R., Young-Loveridge, J., et al. (2005). Supporting teachers in the development of young children's mathematical thinking: Three large scale cases. *Mathematics Education Research Journal, 16*(3), 27–57.

Clarke, D. (2001). Understanding, assessing and developing young children's mathematical thinking: Research as powerful tool for professional growth. In J. Bobis, B. Perry, & M. Mitchelmore (Eds.), *Numeracy and beyond: Proceedings of the 24th Annual Conference of the Mathematics Education Research Group of Australasia* (Vol. 1, pp. 9–26). Sydney: MERGA.

Clarke, D. (2013). Understanding, assessing and developing children's mathematical thinking: Task-based interviews as powerful tools for teacher professional learning. In A. M. Lindmeier & A. Heinze (Eds.), *Proceedings of the 37th Conference of the International Group for the*

Psychology of Mathematics Education, Mathematics Learning Across the Life Span (Vol. 1, pp. 17–30). Kiel, Germany: PME.

Clarke, D., Cheeseman, J., Gervasoni, A., Gronn, D., Horne, M., McDonough, A., et al. (2002). *ENRP final report*. Melbourne: ACU.

Clements, D., Sarama, D., Wolfe, C., & Spitler, M. (2013). Longitudinal evaluation of a scale-up model for teaching mathematics with trajectories and technologies: Persistence of effects in the third year. *American Educational Research Journal, 50*(4), 812–850.

Cobb, P., Yackel, E., & Wood, T. (1992). A constructivist alternative to the representational view of mind in mathematics education. *Journal for Research in Mathematics Education, 23*(10), 2–33.

Faragher, R., Brady, J., Clarke, B., & Gervasoni, A. (2008). Children with down syndrome learning mathematics: Can they do it? Yes they can! *Australian Primary Mathematics Classroom, 13*(4), 10–15.

Feigenbaum, R. (2000). Algebra for students with learning disabilities. *The Mathematics Teacher, 93*(4), 270–276.

Fullan, M., Hill, P., & Crévola, C. (2006). *Breakthrough*. Thousand Oaks: Corwin Press.

Gelander, G. P., Rawle, G., Karki, S., & Ruddle, N. (2017). EQUIP-Tanzania impact evaluation midline issue note 1: The changing context for teacher in-service training—Reflections on equip-Tanzania's experience. Tanzania: Oxford Policy Management. Retrieved from http://microdata.worldbank.org/index.php/catalog/2838.

Gervasoni, A. (2004). *Exploring an intervention strategy for six and seven year old children who are vulnerable in learning school mathematics* Unpublished doctoral dissertation. Bundoora: La Trobe University.

Gervasoni, A. (2011). Exploring the whole number knowledge of children in grade 1 to grade 4: Insights and implications. In T. Dooley, D. Corcoran, & M. Ryan (Eds), *Mathematics Teaching Matters: Proceedings of the 4th Conference on Research in Mathematics Education* (pp. 168–178). Dublin: St Patrick's College Drumcondra.

Gervasoni, A. (2015). *Extending mathematical understanding: Intervention*. Ballarat, Australia: BHS Publishing.

Gervasoni, A., & Lindenskov, L. (2011). Students with 'special rights' for mathematics education. In B. Atweh, M. Graven, W. Secada, & P. Valero (Eds.), *Mapping equity and quality in mathematics education* (pp. 307–323). The Netherlands: Springer.

Gervasoni, A., & Sullivan, P. (2007). Assessing and teaching children who have difficulty learning arithmetic. *Educational & Child Psychology, 24*(2), 40–53.

Gervasoni, A., Parish, L., Hadden, T., Livesey, C., Bevan, K., Croswell, M., et al. (2012). The progress of grade 1 students who participated in an extending mathematical understanding intervention program. In J. Dindyal, L. P. Cheng, & S. F. Ng (Eds.), *Mathematics Education: Expanding Horizons: Proceedings of the 35th Annual Conference of the Mathematics Education Research Group of Australasia* (pp. 306–313). Singapore: MERGA Inc.

Gervasoni, A., Parish, L., Hadden, T., Turkenburg, K., Bevan, K., Livesey, C., et al. (2011). Insights about children's understanding of 2-digit and 3-digit numbers. In J. Clark, B. Kissane, J. Mousley, T. Spencer, & S. Thornton (Eds.), *Mathematics: Traditions and [new] Practices. Proceedings of the 23rd Biennial Conference of The Australian Association of Mathematics Teachers and the 34th Annual Conference of the Mathematics Education Research Group of Australasia* (Vol. 1, pp. 315–323). Alice Springs: MERGA/AAMT.

Gervasoni, A., Parish, L., Upton, C., Hadden, T., Turkenburg, K., Bevan, K., et al. (2010). Bridging the numeracy gap for students in low SES communities: The power of a whole school approach. In L. Sparrow, B. Kissane, & C. Hurst (Eds.), *Shaping the Future of Mathematics Education: Proceedings of the 33rd Annual Conference of the Mathematics Education Research Group of Australasia* (pp. 202–209). Fremantle: MERGA.

Hart, B., Brinkman, S., & Blackmore, S. (2003). *How well are we raising our children in the North Metropolitan Region?* Perth: Population health program, North Metropolitan Heath Service.

Hill, H. C., Rowan, B., & Ball, D. L. (2005). Effects of teachers' mathematical knowledge for teaching on student achievement. *American Educational Research Journal, 42*(2), 371–406.

Lindenskov, L., & Weng, P. (2008). Specialundervisning I matematik—kan det ikke være lige meget? In S. Tetler & S. Langager (Eds.), *Specialunderisning i skolen* (pp. 211–230). Copenhagen: Gyldendal.

Magne, O. (2003). *Literature on special educational needs in mathematics: A bibliography with some comments*. Educational and psychological interactions (4th Ed.), 124. Malmö, Sweden: School of Education.

Scherer, P., Beswick, K., DeBlois, L. Healy, L., & Optitz, E. (2016). Assistance of students with mathematical learning difficulties: How can research support practice? *ZDM Mathematics Education, 48*, 633–649. https://doi.org/10.1007/s11858-016-0800-1.

Smith, T. M., Cobb, P., Farran, D. C., Cordray, D. S., & Munter, C. (2013). Evaluating math recovery: Assessing the causal impact of a diagnostic tutoring program on student achievement. *American Educational Research Journal, 50*(2), 397–428.

Chapter 9
Helping Teacher Educators in Institutions of Higher Learning to Prepare Prospective and Practicing Teachers to Teach Mathematics to Young Children

Herbert P. Ginsburg

Abstract Research shows that young children possess surprising mathematical abilities and can benefit from Early Mathematics Education, which can lay a sound foundation for mathematics learning. Yet institutions of higher education generally provide their students with inadequate preparation in teaching mathematics to young children. To ameliorate this unfortunate situation, I have been working with colleagues on development of a comprehensive set of materials that teacher educators—usually professors and instructors in institutions of higher learning—can use in their teaching, either live or online. This paper describes a framework for training teacher educators and their students and presents an account of materials that can be used to promote understanding of the relevant mathematics, mathematical thinking of young children, and the kind of formative assessment that can be useful for teachers.

Keywords Professional development · Early mathematics education
Mathematical thinking · Formative assessment · Higher education

9.1 Introduction

In this paper, I describe the need for rich programs of teacher preparation in Early Mathematics Education (EME). Quality EME can lay a sound foundation for mathematics learning and also satisfy children's curiosity about numbers, shapes, and other mathematical topics. Yet, I argue, institutions of higher education generally do not prepare their students adequately for EME. They often fail to provide prospective and practicing teachers with an understanding of young children's mathematical thinking and the pressing need to foster it. To ameliorate this unfortunate situation, I have been working with colleagues on development of a

H. P. Ginsburg (✉)
Teachers College Columbia University, New York, NY, USA
e-mail: ginsburg@tc.edu

© The Author(s) 2018   135
G. Kaiser et al. (eds.), *Invited Lectures from the 13th International Congress on Mathematical Education*, ICME-13 Monographs,
https://doi.org/10.1007/978-3-319-72170-5_9

comprehensive set of materials that teacher educators—usually professors and instructors in institutions of higher learning—can use in their teaching, either live or online. Before introducing what we have developed, and the rationale for it, I describe the urgent need for an improved EME as well as some challenges we must overcome to implement it. I argue that it is clear that young children possess surprising mathematical abilities and can benefit from EME, if only the teacher educators and their students are properly prepared to understand and implement it. Then I describe a framework for training teacher educators and their students and present an account of the new materials we are developing with the collaboration of colleagues.[1]

9.2   The Need for EME

Many countries around the world stress the need for strong and extensive programs of EME. In the U.S., "Early childhood education has risen to the top of the national policy agenda with recognition that ensuring educational success and attainment must begin in the earliest years of schooling" (Cross et al. 2009). Latin American countries express similar concerns: "We have to do something [about EME], especially in the countries that were the land of brilliant civilizations like the Mayans and the Incas, who made important scientific and mathematical contributions" (Bosch et al. 2010, p. 5). Improving early mathematical competence has become a major priority around the world (Platas et al. 2016).

Why this focus on EME? One reason is that it has become increasingly clear that early proficiency in mathematics is a good predictor of academic success in later years (Duncan et al. 2007) and even college attendance (Duncan and Magnuson 2011). Further, "A causal relationship has been identified between early mathematical proficiency and later individual economic well-being and broader economic growth in countries including Kenya, Tanzania, Ghana, and Pakistan..." (Platas et al. 2016, p. 164).

We would add another important reason for EME, namely that failing to provide it is a disservice to young children. They are curious about mathematical ideas and want to learn. Contemporary developmental and educational research (Sarama and Clements 2009) shows that young children develop a relatively powerful informal mathematics as well as the capacity to acquire rather sophisticated foundational math skills. From the earliest days of infancy, they develop an everyday mathematics of some power and complexity. Infants can identify a collection containing more objects than another contains. Parents can easily confirm that babies prefer

[1]This work was done in collaboration with Megan Franke, Linda Platas, and Deborah Stipek, all of whom are members of the Development and Research in Early Mathematics Education (DREME) project, generously funded by the Heising-Simons Foundation. We are grateful for the Foundation's support. I also want to acknowledge the contributions of my students, Ma. Victoria Almeda, Bona Lee, Myra Luna-Lucero, Colleen Oppenzato, Colleen Uscianowski, and Eileen Wu.

more food to *less*. Babies approve of the parent *adding* food, and make clear their displeasure when some is *taken away*.

From about 2 years to 6, children engage in even more complex mathematical activities. Knowing a fair number of counting words, they develop ideas about "how many?" They learn, sometimes without adult assistance, to assign the counting words, one at a time, in order, to one and only one object. As they grow older, they gradually learn that the last number in the count sequence indicates the set's cardinal value and thus answers the "how many" question (Baroody and Dowker 2003).

Young children also enjoy their everyday mathematics. It is part of their intellectual life. It satisfies their curiosity. They do not need EME to make them ready for learning mathematics. Their everyday mathematics is real mathematics, involving thinking and exploration as well as the necessary memorization (for example of the numbers from one to ten).

Given young children's ability and potential to learn, promoting their mathematical learning should be a critical component of high quality early childhood education. Although preparation for the future is vital, we should also help children thrive in the present by providing them with the appropriate mathematical food for thought. If the focus is mainly on the future, one result may be high stakes testing, which may have the effect of deadening teaching and learning during the early years. But a focus on the present will not only respond to children's current interests but also help prepare them for future school success.

Given its importance (in the present as well as in the future), early mathematics proficiency is alarmingly inadequate around the world. "[A]t least 250 million primary-school age children around the world are not able to read, write or count well enough to meet minimum learning standards (Center for Universal Education at Brookings and UNESCO Institute for Statistics 2013, p. 1), including those who have spent at least four years in school. Further, "Of the more than 800 millions 0 to 6 year old children in the world, less than a third benefit from early childhood education programmes" (Lillemyr et al. 2001, p. 1).

Socio-Economic Status (SES) and ethnic differences in early mathematics proficiency exist in many countries. In the U.S., low-income and African-American children perform more poorly than middle-income and mainstream children (Denton and West 2002; Love and Xue 2010).

Given the low levels of mathematics performance, especially in lower-SES and minority children, and given the importance of early mathematics learning, many education authorities around the world (Bosch et al. 2010) have called for the widespread, even universal, implementation of high quality mathematics education for young children by at least the age of 4 years, especially for low-income children at risk of school failure (Cross et al. 2009) and of attending failing schools (Ginsburg et al. 2008). Meeting the educational needs of young children requires many different kinds of contributions, including:

- Political: The public needs to decide that early education is a social priority and should be universally available.

- Economic: The public needs to devote adequate public funding for classrooms, teachers and assistants, meals, and related needs.
- Resource Development: Educators need to create rich materials for children's mathematics learning.
- Teacher Education: Institutions of higher learning need to prepare teachers for EME and school systems need to support teachers who are engaged in implementing it.
- Research: Educators and psychologists need to understand the basic processes of EME and evaluate their efficacy.

Fortunately, many efforts along these lines are already underway. As we saw, many education authorities have called for extensive EME and are devoting funds to pay for it. Educators and researchers and others have developed and evaluated rich mathematics curricula (Casey et al. 2004; Ginsburg et al. 2003; Griffin 2004; Sarama and Clements 2004; Sophian 2004; Starkey et al. 2004).

One area that has received relatively little attention is pre-service education. Prospective teachers are seldom given adequate preparation in EME at the level of higher education (Ginsburg et al. 2014). The Development and Research in Early Mathematics Education (DREME) project aims to provide teacher educators—professors, instructors, and others—with effective EME pedagogy and materials that can be used in a flexible manner, in different courses, to prepare their students, prospective teachers, for EME.

9.3   A Guide for Teacher Educators

William James had it right over a hundred years ago (1899) when he wrote: "Psychology is a science, and teaching is an art; and sciences never generate arts directly out of themselves. An intermediary inventive mind must make the application, by using its originality" (James 1958, pp. 23–24). That is, although science may provide insights into children's mathematical thinking and learning, teachers need to use their "intermediary inventive mind[s]" to construct understandings of individual children and ways of teaching them. Our overall goal is to help teacher educators to prepare and support thoughtful, critical minded students.

9.3.1   What Do We Teacher Educators Want Our Students to Know?

The teaching of virtually any subject is complex, difficult and intellectually challenging. To teach well, our students need to learn a great deal about several interesting topics.

9.3.1.1 The Mathematics

It is self-evident that teachers need to understand what they attempt to teach. The problem, however, is that many students, and adults generally, do not realize that the mathematics young children need to learn is not simple. Indeed this basic mathematics—including whole number, shape and space—is far more complex than many adults recognize, precisely because it deals with basic ideas. Addition of the whole numbers, for example, is much more than remembering number facts, like the sum of 2 and 3. Addition also involves a set of fundamental mathematical ideas. For example, the child should know that the sum of any two numbers (other than zero) must be greater than each; that the order of adding makes no difference; that the sum indicates the total number of objects; that addition can be used to model certain real situations; and that methods for calculating a sum can and should make sense. These ideas are not hard for adults to learn, but many prospective teachers do not think about early addition in these terms and hence cannot teach the subject effectively.

9.3.1.2 The Development of Mathematical Thinking

How can anyone teach effectively without understanding the students to be taught? Would any teacher want *not* to understand her students? Teachers of young children in particular need to understand their thinking because it often differs from our own. The idea of distinctive child thought was one of the central points of Piaget's theory (Piaget and Inhelder 1969). We are surprised, and sometimes amazed and amused, to encounter a child who thinks that moving around a group of objects, without adding or taking away any of them, results in a change in the group's number, because we do not see the world in the same manner. Effective EME requires teachers to take a cognitive leap from their own ways of thinking, in order to understand the child's. In a sense, teachers need to overcome their own egocentrism to see the child's.

The psychology of children's mathematical thinking is very rich. Inspired by Piaget's theory (Piaget 1952), contemporary research has illuminated key aspects of everyday and formal mathematics learning (Sarama and Clements 2009). We now know in great detail how children think about the topics central to EME: basic number, shape, space, pattern, and measurement. The research has also plotted the developmental trajectories of thinking related to these topics. It is important to note that current research goes far beyond general Piagetian ideas about broad stages of development, like pre-operational and concrete-operational thought. Current research offers insight into the development of specific aspects of mathematical thinking, both everyday and schooled. This was not something that Piaget tried to do. Surely our students need to understand the details of student thinking and how it develops.

9.3.1.3 Formative Assessment and Understanding the Individual

Formative assessment is the process of collecting information that enables the teacher to understand individual children and to use what has been learned to improve instruction (Heritage 2010). Formative assessment is different in several ways from summative assessment, like achievement tests, year-end testing, or other forms of high-stakes assessment that use standard tests to focus mainly on children's achievement and mastery. By contrast, formative assessment employs flexible, and deliberately non-standardized methods, primarily clinical interviews and observations, to focus not only on performance, which of course is important, but also on what underlies it, namely children's ideas and strategies, knowledge of which can be used to shape teaching. Formative assessment is relevant for teachers, whereas summative assessment provides them with little actionable information.

It is important to note that the target of formative assessment is the *individual* child, not the average or *prototypical* child pictured by a developmental trajectory. Yes, 4-year-olds can be characterized in general as doing so and so. This is valuable information, but there is variation within the group, so that the prototype may not fully apply to the individual. Given the fact of widespread individual differences, the teacher needs to understand and assess the individual child.

9.3.1.4 Pedagogical Goals and Methods

Students need to learn that the overarching pedagogical goal is to produce meaningful learning in which children synthesize what they already know, their everyday mathematics, with the more powerful formal mathematics developed over the years in different cultures, from the Indian to the Arabic to the Western.

Vygotsky (1986) put the matter thus: "In working its slow way upward, an everyday concept [everyday mathematics] clears a path for the scientific concept [formal mathematics] and its downward development. It [the scientific concept] creates a series of structures necessary for the evolution of a concept's more primitive, elementary aspects, which give it body and vitality. Scientific concepts, in turn, supply structures for the upward development of the child's spontaneous concepts toward consciousness and deliberate use… The strength of scientific concepts lies in their conscious and deliberate character. Spontaneous concepts, on the contrary, are strong in what concerns the situational, empirical, and practical" (p. 194).

In other words, the goal of our pedagogy should be help the child develop a meaningful synthesis of the personal, which offers "body and vitality," and the formal, which is conscious and deliberate. The synthesis allows the child to "own" the resulting mathematical knowledge.

Accomplishing this goal requires several pedagogical approaches, and our students need to understand them all. Students need to appreciate the appropriate roles of free play, exploration, projects, guided instruction, group discussion, verbalization of ideas, memorization, and curriculum. These are all useful to the extent that they promote effective EME, each in their own ways. Thus free play can excite interest in mathematical ideas, but intentional teaching may be necessary for the child to understand them in depth.

9.3.2 Overcoming Negative Feelings

Unfortunately, many students, at least in the U.S., have negative feelings about mathematics. They experience anxiety about learning, doing, and teaching mathematics. Indeed, some students say that they chose the profession of early childhood education so that they would not have to teach mathematics. They may transmit negative feelings to the children they teach (Beilock et al. 2010). Clearly one of the teacher educator's goals must be to help students overcome their math anxiety. It's also true that some teacher educators may feel this kind of anxiety. To you, I can only say that the course materials may help you to overcome yours as well.

9.4 The DREME Modules

Our DREME project[2] has, to date, produced five modules, namely basic number (including counting words, enumeration, and cardinality); geometry (including shape and space); operations (including addition/subtraction and division into fair shares); pattern (including growing patterns); and measurement (informal and exact). In each module, the materials include:

- Readings on aspects of EME, some specially created for these modules,
- Explanations of the basic mathematics,
- Accounts of children's mathematical thinking and learning, with accompanying videos,
- Analyses of teaching, with accompanying videos,
- Guides to assessment, with accompanying videos,
- Guides to picture book reading from a mathematical point of view, with videos,
- Activities for higher education classroom use, with accompanying videos, and
- Vignettes of adult experiences related to EME.

Our general pedagogical approach is this: As much as possible, TEs should help adult students learn in the same way that we want teachers to help children learn. TEs need to engage students in active learning that bridges the gap between theory and practice, and that gives personal meaning to the concepts learned in the course.

We designed the modules for flexible use. TEs can use one component of a module in a course on science education; several components in a math methods course; or another component in a general introduction to Early Childhood

---

[2]http://prek-math-te.stanford.edu/.

Education. I now illustrate the modules, along with some methods of my own, through a personal use case: an account of my own teaching.

9.5   My Course

My course, *The Development of Mathematical Thinking* meets once a week, for 1 h and 40 min, over a period of 14 or 15 weeks, and also includes a web based component. The course (despite its name) uses many of the DREME materials to focus not only on children's mathematical thinking but also on early childhood pedagogy. Of course, few TEs will want to teach the course exactly the way I do, but they may choose DREME activities that help them meet their own goals and are consonant with their backgrounds. Indeed, this flexibility is exactly the goal of our DREME project.

Almost all of my students are prospective teachers at the Masters level in a Department of Curriculum and Teaching. Clearly these students are different from undergraduate education majors, but they do share an interest in teaching young children. The course (or the equivalent) is highly recommended for Early Childhood Education students.

9.5.1   Who Are You?

At the outset, like many TEs, I am interested in learning about the students taking the course. I have used several methods to learn about their interests, their background, and their feelings towards mathematics and EME. In my experience it is particularly useful to explore their anxieties about the course, particularly because some students may have been reluctant to take it in the first place.

The DREME materials include an activity dealing with student feelings. It begins with students reading a vignette called *But, I'm not good at math!* (Platas electronic document-a), which describes student fears and anxieties about teaching mathematics to a young child. The reading is followed by an activity, *"Engaging Mathematics" Activity*, (Platas electronic document-c), that helps students address and understand their feelings, beliefs, and attitudes, and thus begin the process of developing a positive approach to teaching mathematics.

I conduct a similar classroom discussion activity. During the first class I initiate a discussion of their ideas about and feelings toward EME. My intent is to help students understand that others may also have had traumatic (not too strong a word) experiences in their mathematics classes and still fear learning and teaching it. I also want them to know that I am concerned about their feelings and will try to help. I say that they should be patient because over the course of the semester, as they discuss the mathematics and children's responses to learning it, they will begin to shed their fears and in fact will find EME intriguing. I ask the students to send me, before the next class, a reflection on this activity. To insure confidentiality, I do not reveal the authors of the following typical quotes:

> I was concerned that I was going to be the only person that wasn't good at math coming into the course.

> Overall, this week's [class] made me realize that I am not alone in my feelings of discomfort surrounding teaching math in the classroom.

> I, myself, have had a negative outlook towards math and have always tried to avoid dealing with anything to do with math.

9.5.2 What Concerns You?

As a TE, I want to monitor my students' learning throughout the semester. Of course, I talk with them during class, ask questions, and the like, but sometimes students do not feel comfortable revealing their ideas or concerns to others in a group setting, particularly if the class is large.

For this reason, I require students to submit on our course website, after each weekly session, a short, ungraded reflection on what they learned in the previous class. Usually students are urged to discuss anything they found important or arguable. Occasionally I may ask them to discuss a specific issue. Before the next class, I read all of the reflections and send brief comments to each student. Then, at the outset of the next class, I show (via a projected PowerPoint) and discuss with the students parts of 4 or 5 reflections carefully selected to raise important issues. Here are some examples:

1. "For the past two classes now, what has really stood out to me is how much one can analyze from just 1–1.30 min of a video. Even the smallest details, from a glance to a slight hesitation, are indicative of something much larger. I wonder, though, how this would be possible in a classroom with 10 or 15 children, each one making several such gestures that indicate a thought, feeling or strategy. If only life had a pause and rewind button, then I could effectively analyze the mathematical thinking occurring in the classroom without missing a beat."
2. "I think symbols can give children different feelings depending on the scenarios in which they are presented. For example, exposing a child to symbols in a classroom setting and telling them that they must memorize these symbols can come off as very stressful and hard. But showing children symbols in everyday experiences can be less stress inducing and even fun, while also giving the children a visual representation of what certain symbols can mean; thus making the symbols less foreign for them to understand. I think that even taking children on walks as a field trip, allowing them to experience these symbols (and ask questions about such if they are so inclined) could be a great learning experience for young children living in any area."

3. "I think the thing that struck me the most was the highly detailed critique that we did of even just the first few pages of the storybook [name deleted]. It's just so surprising how many flaws that we were able to come up with as a class in about 20 min that this book had, and that such flaws were not taken into account prior to publishing this children's book."

As is evident, the student reflections provide the TE with opportunities to discuss and expand upon interesting issues. Thus the TE can use the first reflection to discuss a very real dilemma, namely how difficult it is for a teacher to attend to the activities of some 10 or 15 children, and to suggest possible solutions (for example, a plan to spend 5 min carefully observing each of 4 students a day). The TE can use the second reflection to dig more deeply into the nature of mathematical symbols and methods to help children learn them in meaningful and enjoyable ways. This reflection also allows students to draw conclusions about the differing value of activities depending on their context. The TE can use the third reflection to raise the issue of strategies for storybook reading, including how to deal with pages that are unclear or incorrect.

I have discovered that the post-class reflection offers the students a distinctive form of digital intimacy. After seeing that I take the reflections seriously and that the class discussions are interesting and useful, students begin to write lengthier reflections and to reveal in them insecurities and areas of ignorance, as well as questions about the readings, and comments on my lectures or other class events. Also, students get excited (and I think feel pride) when their reflections have been chosen for class discussion. The reflections provide me with insights into student needs, confusing remarks I may have made, and issues I had not considered. Sometimes a reflection teaches me something new about the subject matter—the children and also the mathematics.

9.5.3   Learning About the Math

Recognizing that many students are fearful of mathematics, the DREME modules contain several approaches to teaching number and geometry.

One approach is traditional. We have created some readings on the relevant mathematics, including a short piece on *The Mathematics of Counting*, (Platas electronic document-b) and *What Young Children Know and Need to Learn about Number* (Ginsburg electronic document-f), that discuss the mathematics to be learned in relation to children's existing knowledge.

DREME also offers classroom activities in which TEs can have adult students solve some mathematics problems, reflect on what they learned, and relate their learning to children's (Franke electronic document). This method helps teachers to see the world of mathematics from the learner's point of view.

Finally, I urge TEs to make the acquaintance of *Professor Ginsboo* (Ginsburg electronic document-e), who presents students with a hopefully amusing and engaging introduction to basic mathematical ideas from the perspective of a child trying to learn them. We find that some students and TEs find this approach useful, and others don't know what to make of it. We intend to investigate the appeal and effectiveness of the good Professor.

9.5.4 Learning About Children's Thinking

Students need to learn about the general trajectory of children's thinking during the early childhood years and beyond. The 3-year-old child's concept of "how many?" is very different from that of the 5-year-old's. Fortunately, researchers have contributed enormous insights into the nature and development of mathematical thinking during this age range.

One way for students to learn about the trajectory is to read relevant papers. Although this is valuable, our DREME project offers what we think are more engaging written accounts with built in videos. We call them "Thinking Stories" because they use a narrative form to present key aspects of mathematical thinking. One such thinking story describes a young boy's understanding of "how many?" as revealed by clinical interviews conducted when he was 3-, 4-, and 5-years of age (Ginsburg electronic document-a). This story, a kind of longitudinal case study, is particularly dramatic because it involves videos of the same child throughout.

Videos of young children are memorable. Former students have told me years later that they still remember several course videos. The DREME project offers on its website many videos that can be used to illustrate children's thinking and learning. You and your students can make such videos as well.

9.5.5 Assessment

TEs need to help students to understand individual children's behavior and thinking. Students need to learn to observe carefully, to think critically about what they see, develop reasoned interpretations, and use those interpretations to guide teaching. I use several methods for accomplishing these goals.

9.5.6 Analyzing Videos

One approach, which I use in almost all class sessions, is to engage the class in the analysis of videos, usually involving clinical interviews and observations of behavior. I use the "pedagogy of the video clip," which works as follows.

Imagine that you are in front of your higher education classroom. The topic is *Counting*. You say that the class will discuss a video illustrating key aspects of both the child's counting skill and knowledge, and the interviewer's method. The students' task is to examine the video carefully, develop an interpretation of what the child does and does not understand about counting, and consider how this interpretation can guide teaching. The students also examine the interviewer's clinical interview technique.

The video you use meets several essential requirements. First and foremost, the content illustrates children's thinking and learning of the mathematical topic of interest. Further, the video is intriguing, attractive, dramatic, and sometimes even funny. It is not too long (usually under two minutes). It grabs students' attention and animates what they read for their assignments. The video is a kind of intellectual manipulative with which students can explore children's thinking and construct a meaningful understanding of it.

Yet by itself the video, no matter how wonderful, is not sufficient. You, as the instructor, should not simply show the entire video, tell the students what you think it means, and then go on to the next topic. Instead you need to use and exploit the video as effectively as possible to promote your students' careful observation, analytic thinking, judicious interpretation, and consideration of productive instructional activities.

The essence of the pedagogy of the video clip is to:

- Show your students carefully selected segments of the video;
- Help students observe carefully;
- Repeat a segment or part of it to clarify observations and hypotheses;
- Ask students for interpretations supported by evidence;
- Challenge their interpretations;
- Encourage students to justify their interpretations;
- Ask students to evaluate classmates' interpretations;
- Do not allow students to get away with vague generalities, like "She is in Piaget's preoperational stage" or "His behavior is developmentally appropriate";
- Use flexible questioning to reveal the thinking behind your students' interpretations;
- Encourage the students to discuss and challenge interpretations, offer possible alternative hypotheses, and propose instructional approaches based on what has been learned about the child's understanding;
- And finally, help your students to relate the lessons learned from the video to academic papers and ideas concerning children's mathematical thinking and learning.

Examining videos in this fashion may be more effective than observing children directly or reading about children's thinking. A video is worth many more than a thousand words. The video allows you to view and review, to go forward and backwards in time, and to engage in deliberate and unhurried contemplation of the evidence. By comparison, direct observation is ephemeral and does not afford the kind of careful study provided by videos of thinking and learning. Hence, the DREME project offers a video-based introduction to the pedagogy of the video-clip (Ginsburg electronic document-d).

9.5.7 Clinical Interview

One of the most effective methods of formative assessment is the clinical interview, in which the adult uses flexible questioning to uncover what the child knows. Piaget's experience with the development of Binet's IQ test taught him that standardized tests are not particularly effective in uncovering cognitive processes; he felt that answers to IQ test items were ambiguous and required further examination by means of the clinical interview (Piaget 1976), which typically begins with a task chosen by the adult. Then the interviewer observes the student's response, behavior, affect and anything else that might be relevant, and develops an interpretation of the processes underlying the student's behavior. To check the interpretation, the interviewer develops new tasks as appropriate, follows up with questions designed to elicit thinking, and in general follows the student's thought process to where it leads. The interviewer sometimes challenges the student's response ("But Johnny said that 2 and 2 is really 5") to assess its stability and the student's confidence in it. The interviewer may sometimes employ gentle hints to help the student overcome a difficulty ("Can you use your fingers to help you figure it out?"). The interviewer continues these non- standardized investigative maneuvers, which must be constructed on the spot, in real time, until enough evidence has been obtained to support a reasonable interpretation of the student's behavior. If done well and guided by appropriate theory, clinical interviews can provide deep insights into thinking and can reveal strengths and weaknesses that otherwise may go undetected (Ginsburg 1997).

The students become familiar with the clinical interview almost each week as we discuss videos, most of which involve an adult interviewing a child. Although most of the analysis centers on children's thinking, we often examine, in some detail, the interviewer's technique.

Students are also required to complete a mid-term assignment in which they conduct and video record a clinical interview with a child. At this point, the students are impressed by the power of the interview to reveal hidden features of a child's thinking. But they are also apprehensive at the prospect of conducting an interview themselves, especially when they are required to video-record it and use clips from the video to justify their analyses. I tell the students to be brave and to go out and do the interview, and that everyone makes mistakes in interviewing, which is a complex skill.

In addition to cheerleading, I provide the students with various DREME materials: a general guide to interviewing, which we call math-thinking conversations, along with a sample interview protocols describing a series of questions and prompts that the student might use or modify (Ginsburg electronic document-b);

and a voice over analysis of a video recording showing a teacher's excellent interview with her preschool student (Ginsburg electronic document-c).

The details of the interview assignment are worth describing in detail. The student is required to write on the class website a paper that shows what the interview revealed about the child's thinking. The paper (available only to the instructor) begins in a standard way with an overall introduction to the problem, a very brief review of the literature, a statement of the goal, a description of the initial protocol (which the student is encouraged to modify as necessary in response to the child's responses), and a description of the child and the testing conditions (for example, the interview was conducted in a quiet part of the classroom). All of this is conventional.

But the next part, the results, takes advantage of a video technology that allows students to make short clips from the interview and insert them where appropriate to buttress their interpretations. The students provide an interpretation and then justify it with a carefully selected video clip of the child's response. The students essentially have to provide an argument and justify it with evidence from the video. I see this as promoting students' critical thinking skills—their use of evidence to justify an argument, or to show that the interpretation is not clear and requires further evidence to resolve. The assignment is also designed to help students avoid vague interpretations, empty concepts, and fuzzy thinking. When I grade the papers I can not only read what the students wrote but also view their videos, which may or may not support their arguments. Below is an example of a student response (Fig. 9.1).

Clinical interviewing can be a rewarding activity for the child as well as the interviewer. My student Catherine Rau wrote this about a 4-year-old whom she had interviewed a few weeks earlier.

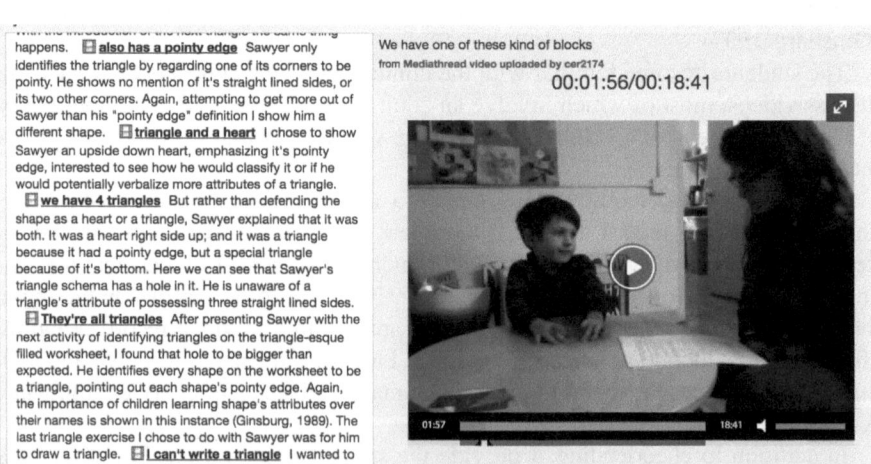

Fig. 9.1 Student assignment

But it wasn't until Sawyer approached me at school randomly, the other week, to ask when we would be doing math again together, that I knew I absolutely had to interview him. He told me that he really liked learning math with me and that he wanted to do more fun math games and for even longer this time. ... I couldn't help but smile and agree to do more math with him....

So the student can use the results of the interview to understand individual children and the children can enjoy the process of being interviewed. They seem to appreciate the attention of an adult who is interested in their thinking—how rare is that!—and the opportunity to solve interesting problems. For both adult and child, this is a win-win situation, clearly different from most high stakes assessments, which too often are lose-lose.

9.5.8 Pedagogy

The course naturally considers pedagogy. We have sessions and readings on the nature and value of pedagogy and curriculum for EME. My general approach is "constructivist," but I argue that it is important to consider the specifics of teaching, namely the particular maneuvers and methods that teachers use for different purposes (Ginsburg and Amit 2008). The argument is that teachers need to use a variety of methods to achieve the goal of children's meaningful construction of knowledge. Under some circumstances and for certain topics, free exploration is effective. At other times, direct instruction is desirable. For example, teaching the first 10 number words requires giving students basic practice that can facilitate memorization. At other times, the teacher may tell the child what to do—"I want you to make a color pattern with these blocks." There is a time for "telling" (Schwartz and Bransford 1998). The most important point for our students is to avoid being doctrinaire, and instead to attempt to match a teaching method with the needs of a child. If free exploration is what that child needs, fine. But if he needs drill to further the construction of knowledge, that is fine too.

Our DREME project makes available some teaching videos on its web site. In my class, we use analytic techniques, like the pedagogy of the video clip, to examine interesting cases of teaching.

Students are often skeptical about using curriculum for EME, even though several are available. I try to convince the class that using a planned curriculum is not necessarily "developmentally inappropriate." Because it is not the job of a higher education course to teach a particular curriculum, we examine examples of curriculum activities—from manipulatives to textbooks.

The course final project is designed to integrate what students have learned about children's thinking and about effective teaching and interviewing. Students engage in and video-record a three-part exercise in which they interview a child on a particular mathematical topic, teach what the child does not understand about the topic, and finally interview the child again to see what was learned. All of this provides the material for web-based papers in which students analyze the child's knowledge at each point, as well as their own teaching and interviewing.

9.5.9   Picture Books

Everyone loves children's picture books like *Goodnight Moon* (Brown and Hurd 2007). Parents snuggle up with their children at the end of the day to read books like these, sometimes over and over again. Teachers enjoy reading picture books to the whole class, at circle time, and also to the individual child or small groups of children. Children may want to read books over and over again (taxing parents' and teachers' literary patience). How many times has Moon gone to sleep?

Most picture books involve art and narrative. Some present simple exposition, not stories, as in the case of a book displaying different kinds of vehicles, from bicycles to fire trucks. But all these books for young children use pictures.

Reading books of this type can promote children's mathematics learning. Consider three types of picture books: those in which the mathematics is explicit, those in which significant mathematics is implicit, and everything else.

Explicit mathematics books are written for the express purpose of teaching children mathematics and may even contain a reference to mathematics in their titles, as in the case of counting books like, *Anno's Counting Book* (Anno 1977). Many books clearly written to teach mathematics do not have such titles, like *Rooster's Off to See the World* (Carle 1987), in which a different number of animals joins Rooster on each page, and children are invited to add up the total number of animals.

Other picture books were not written to teach children mathematics. However, they still address significant mathematical concepts in the narrative and illustrations. A well-known example is the classic story, *Goldilocks and the Three Bears,* which involves size comparisons (for example, the big bear, the medium-sized bear and the little bear) and correspondences (little bear with little bed and so on) that are crucial to the plot.

Finally, we have the rather large category of "everything else." The basic idea is that because a picture book page typically has objects or abstract shapes arranged in space, an adult reader can always ask the child to count them, or talk about their location (for example, "The hat is on top of his head"). In other words, adults can interject math conversation in virtually every single book ever written because "Mathematics is all around us" (or at least we can find mathematics all around us if we have a mind to), including in picture books. At the same time, we should not ruin an interesting story by interjecting math for its own sake. Occasionally, the adult might point to some math in an "everything else" story, but in general it's not a good idea to do so: the primary goal should be to enjoy and explore the story.

Given the popularity of picture books, and given many adults' fear and dislike of mathematics, reading storybooks can be a kind of stealth weapon in teaching mathematics. For this reason the DREME project offers materials designed to help students learn how to read storybooks to further mathematics learning.

I have found that an effective way to begin is to jump right in and read a picture book together as a whole-class activity. While ultimately the students will be presented with a guide to help them analyze books on their own, this introduction can be exploratory. The book is basically one well worth reading: it has many positive qualities, like an interesting story, but may also include problematic text or illustrations, thus providing opportunities for an interesting whole-class discussion. After reading the book once without commentary, we examine it page by page. I ask students to look at the text, the illustrations, and point out interesting and problematic features.

After students have analyzed a picture book together, I introduce a DREME guide, *How to Use Picture Books for Young Children to Teach Math* (Oppenzato et al. electronic document), which is designed to help students analyze books and determine their suitability and usefulness for teaching mathematics to young children. For example, the Guide encourages students to examine carefully the relation between the picture and the text on each page, the accuracy of the mathematics presented, and other important topics. Once armed with the *Guide*, the students can be broken into small groups to compare and contrast a pair of books. Students are asked to consider possibilities for using different pages to promote mathematics learning.

For example, *Ten Red Apples* (Hutchins 2000) begins by showing ten apples on a tree. Then apples are removed from the tree, one by one, by different animals, much to the annoyance of the farmer. At one point, there are three left and then one is stolen, although the result is not shown on the page. The students can discuss how a teacher might use this page to talk about subtraction. The students learn that it wise for the teacher to begin with a series of questions that can be used both to probe and promote children's understanding of the mathematical concepts.

We should also want our students to understand that reading books with mathematics content is in many respects no different from reading other books. In both cases, the primary goal is to enjoy and learn from the books. Also, in both cases, the adult reader should employ "dialogic reading," that is, reading that engages adult and child in a dialog around reading. The adult asks questions about the book, encourages the child's attention and participation, and in general takes the child on an intellectual adventure. Most likely, your students will have studied dialogic reading in classes on literacy. In any event, the guide (Oppenzato et al. electronic document) presents the major principles of dialogic reading in the context of picture books with significant focus on math. At the same time, the adult reader should not ask so many questions that they get in the way of enjoyable book reading.

Finally, after analyzing books as a whole class and in small groups, the students select a picture book and plan a lesson around it. Later they can conduct a final class project, involving interviewing and teaching, as described earlier, on picture book reading.

9.6 Conclusion

As I noted at the outset, the success of Early Mathematics Education depends in good measure on the professional development of prospective and practicing teachers in our institutions of higher learning. Teacher Educators have the opportunity and responsibility to train students to be thoughtful, sensitive, and effective guides of their children's mathematics learning. The DREME project aims to help Teacher Educators to seize the opportunity and fulfill their responsibility. We invite colleagues to share our materials and collaborate in our efforts to prepare prospective teachers to engage in exciting and meaningful mathematics education for all children.

References

Anno, M. (1977). *Anno's counting book*. New York: Crowell.

Baroody, A. J., & Dowker, A. (Eds.). (2003). *The development of arithmetic concepts and skills: Constructing adaptive expertise*. Mahwah, NJ: Lawrence Erlbaum Associates, Publishers.

Beilock, S., Gunderson, E., Ramirez, G., & Levine, S. (2010). Female teachers' math anxiety affects girls' math achievement. *Proceedings of the National Academy of Sciences, 107*(5), 1860.

Bosch, C., Álvarez Díaz, L., Correa, R., & Druck, S. (2010). *Mathematics education in Latin America and the Caribbean: A reality to be transformed* (Vol. 4). Rio de Janeiro and Mexico City: ICSU-LAC/CONACYT.

Brown, M. W., & Hurd, C. (2007). *Goodnight moon*. New York, NY: HarperCollins.

Carle, E. (1987). *Rooster's off to see the world*. Natick, MA: Picture Book Studio.

Casey, B., Kersh, J. E., & Young, J. M. (2004). Storytelling sagas: An effective medium for teaching early childhood mathematics. *Early Childhood Research Quarterly, 19*(1), 167–172.

Center for Universal Education at Brookings and UNESCO Institute for Statistics. (2013). *Toward universal learning: What every child should learn*. Retrieved from Washington, DC.

Cross, C. T., Woods, T. A., & Schweingruber, H. (Eds.). (2009). *Mathematics learning in early childhood: Paths toward excellence and equity*. Washington, DC: National Academy Press.

Denton, K., & West, J. (2002). *Children's reading and mathematics achievement in kindergarten and first grade*. Washington, DC: National Center for Education Statistics.

Duncan, G. J., Dowsett, C. J., Claessens, C., Magnuson, K., Huston, A. C., Klebanov, P., ... Japel, C. (2007). School readiness and later achievement. *Developmental Psychology, 43*(6), 1428–1446.

Duncan, G. J., & Magnuson, K. (2011). The nature and impact of early achievement skills, attention skills, and behavior problems. In G. J. Duncan & R. J. Murnane (Eds.), *Whither opportunity: Rising inequality, schools, and children's life chances* (pp. 47–69). New York, NY: Russell Sage.

Franke, M. L. (electronic document). *Counting collections*. Retrieved from http://prek-math-te.stanford.edu/counting/counting-collections-overview.

Ginsburg, H. P. (1997). *Entering the child's mind: The clinical interview in psychological research and practice*. New York: Cambridge University Press.

Ginsburg, H. P. (electronic document-a). *Ben learns how many*. Retrieved from http://prek-math-te.stanford.edu/counting/ben-learns-how-many.

Ginsburg, H. P. (electronic document-b). *Math thinking conversations*. Retrieved from http://prek-math-te.stanford.edu/overview/math-thinking-conversations.

Ginsburg, H. P. (electronic document-c). *How to analyze a child's counting*. Retrieved from http://prek-math-te.stanford.edu/counting/how-analyze-childs-counting.

Ginsburg, H. P. (electronic document-d). *The pedagogy of the video clip*. Retrieved from http://prek-math-te.stanford.edu/overview/pedagogy-video-clip.

Ginsburg, H. P. (electronic document-e). *Ben learns how to count*. Retrieved from http://prek-math-te.stanford.edu/counting/ben-learns-how-count.

Ginsburg, H. P. (electronic document-f). *What young children know and need to learn about counting*. Retrieved from http://prek-math-te.stanford.edu/counting/what-children-know-and-need-learn-about-counting.

Ginsburg, H. P., & Amit, M. (2008). What is teaching mathematics to young children? A theoretical perspective and case study. *Journal of Applied Developmental Psychology, 29*(4), 274–285.

Ginsburg, H. P., Greenes, C., & Balfanz, R. (2003). *Big math for little kids*. Parsippany, NJ: Dale Seymour Publications.

Ginsburg, H. P., Hyson, M., & Woods, T. A. (Eds.). (2014). *Preparing early childhood educators to teach math: Professional development that works*. Baltimore, MD: Paul H. Brookes Publishing Co.

Ginsburg, H. P., Lee, J. S., & Boyd, J. S. (2008). Mathematics education for young children: What it is and how to promote it. *Society for Research in Child Development Social Policy Report-Giving Child and Youth Development Knowledge Away, 22*(1), 1–24.

Griffin, S. (2004). Building number sense with number worlds: A mathematics program for young children. *Early Childhood Research Quarterly, 19*(1), 173–180.

Heritage, M. (2010). *Formative assessment: Making it happen in the classroom*. Thousand Oaks, CA: Corwin.

Hutchins, P. (2000). *Ten red apples*. New York, NY: Greenwillow Books.

James, W. (1958). *Talks to teachers on psychology: And to students on some of life's ideals*. New York: W. W. Norton & Company.

Lillemyr, O. F., Fagerli, O., & Søbstad, F. (2001). *A global perspective on early childhood care and education: A proposed model*. Retrieved from Paris.

Love, J. M., & Xue, Y. (2010). *How early care and education programs 0–5 prepare children for Kindergarten: Is it enough?* Paper presented at the Head Start's 10th National Research Conference, Washingon, DC.

Oppenzato, C., Uscianowski, C., Almeda, V., & Ginsburg, H. P. (electronic document). *Using picture books: Counting*. Retrieved from http://prek-math-te.stanford.edu/counting/using-picture-books-counting.

Piaget, J. (1952). *The child's conception of number* (C. Gattegno & F. M. Hodgson, Trans.). London: Routledge & Kegan Paul Ltd.

Piaget, J. (1976). *The child's conception of the world* (J. Tomlinson & A. Tomlinson, Trans.). Totowa, NJ: Littlefield, Adams & Co.

Piaget, J., & Inhelder, B. (1969). *The psychology of the child* (H. Weaver, Trans.). New York: Basic Books, Inc.

Platas, L. M. (electronic document-a). *But, I'm not good at math!* Retrieved from http://prek-math-te.stanford.edu/counting/im-not-good-math-activity.

Platas, L. M. (electronic document-b). *The mathematics of counting*. Retrieved from http://prek-math-te.stanford.edu/counting/mathematics-counting-0.

Platas, L. M. (electronic document-c). *"Engaging Mathematics" Activity*. Retrieved from http://prek-math-te.stanford.edu/counting/engaging-mathematics-activity.

Platas, L. M., Ketterlin-Geller, L. R., & Sitabkhan, Y. (2016). Using an assessment of early mathematical knowledge and skills to inform policy and practice: Examples from the early grade mathematics assessment. *International Journal of Education in Mathematics, Science and Technology, 4*(3), 163–173.

Sarama, J., & Clements, D. H. (2004). Building blocks for early childhood mathematics. *Early Childhood Research Quarterly, 19*(1), 181–189.

Sarama, J., & Clements, D. H. (2009). *Early childhood mathematics education research: Learning trajectories for young children*. New York: Routledge.

Schwartz, D. L., & Bransford, J. D. (1998). A time for telling. *Cognition and Instruction, 16*(4), 475–422.

Sophian, C. (2004). Mathematics for the future: Developing a head start curriculum to support mathematics learning. *Early Childhood Research Quarterly, 19*(1), 59–81.

Starkey, P., Klein, A., & Wakeley, A. (2004). Enhancing young children's mathematical knowledge through a pre-kindergarten mathematics intervention. *Early Childhood Research Quarterly, 19*(1), 99–120.

Vygotsky, L. S. (1986). *Thought and language* (A. Kozulin, Trans.). Cambridge, MA: The MIT Press.

Chapter 10
Hidden Connections and Double Meanings: A Mathematical Viewpoint of Affective and Cognitive Interactions in Learning

Inés M. Gómez-Chacón

Abstract This paper poses methodological questions concerning the evaluation of emotion in the process of mathematical learning where the interaction between emotion and cognition occurs. These methodological aspects are considered not only from the perspective of educational psychology but from that of mathematics education. Some epistemological and ontological aspects, which are considered central to the cognition-affect interplay, are noted. Special attention is given to the notion of cognitive-affective structure as a dynamic system. The interplay between cognition and affect in mathematics is viewed through the concepts of local and global affect and using a mathematical working space model. A model of this interplay is illustrated with research examples, enabling us to move from descriptions of cognition-affect at an individual level to the explanation of the tendency of a group. The non-linear modelling of emotion is reflected in the affect-cognition local structure.

Keywords Structures of affect · Affective and cognitive interactions
Epistemology and emotions · Mathematical working space · Learning

10.1 Introduction

Current advances in philosophy of science (Brun et al. 2008), social neuroscience for education (Immordino-Yang and Damasio 2007), and cultural approaches in social psychology (Harré 2009) have been highlighting interconnections between cognition and emotion, which frequently allow for emotions to contribute to the growth of knowledge. The demand for research that identifies these interconnections derives from the benefit that knowledge about them can impart to the design of learning environments.

I. M. Gómez-Chacón (✉)
Universidad Complutense de Madrid, Madrid, Spain
e-mail: igomezchacon@mat.ucm.es

© The Author(s) 2018
G. Kaiser et al. (eds.), *Invited Lectures from the 13th International Congress on Mathematical Education*, ICME-13 Monographs,
https://doi.org/10.1007/978-3-319-72170-5_10

Responding to this need to characterize these interconnections, we focus on specific mathematical knowledge. This chapter highlights some precise aspects of the epistemological and ontological dimension that this analysis of the cognition-affect interplay entails. We consider that the epistemic meaning of the emotions must be studied according to the specific characteristics of the epistemology of mathematical knowledge and under a dynamic interrelation approach (a dynamic system of affect). Recent reviews of affect (e.g., Pepin and Rösken-Winter 2015) have highlighted the need to make explicit the combined nature of cognition, motivation, and emotion on mathematical work in dynamic affect systems. Here, we propose viewing these systems in a holistic manner. This means an understanding of a system by examining the linkages, interactions, and relationships between the elements that compose the entirety of the system. Typologies of cognition and affect structures are identified. In describing these structures in our empirical studies, we found characterizations of affect that depend on the nature of the context (local–global) and on the kinematics of the mathematical processes involved (static–dynamic) as well as on the dynamic movement of the nature of the phenomenon at an individual or group level. Some dynamic affect systems may have properties that can only be studied at the higher emergent level in a group, so patterns of interaction that characterize the affective systems of individuals and collectives need be observed.

Specifically, we are looking to make progress in the understanding of the following aspects of the interaction between cognition-affect in mathematics:

(a) The conceptualization of structures of this interaction;
(b) The affective-cognitive reference system given by specific mathematical knowledge, which includes the distinction between mathematical knowledge and appraisal processes; and
(c) The complexity of explicitness of dynamic systems involved in this interaction, from the uniqueness of the individual patterns of reasoning to the characterization of the tendency of a group.

Section 10.2, Theoretical Fundamentals, deals with some aspects of what epistemologically characterizes the affective-cognitive system in mathematical reasoning and considers a model of analysis with its own categories. Sections 10.3 and 10.4 present the study by which the local affect-cognitive structure has been determined, using this model of analysis. A final section addresses the conclusions of the study and makes suggestions for developmental aspects in future research.

10.2   Theoretical Fundamentals

In order to tackle the intricacy of the subject under investigation, a number of theoretical considerations are employed to establish a consistent interpretative framework: epistemological dimension and the affective-cognitive reference system model.

10.2.1 *Affective-Cognitive Reference System: The Zig-Zag Path in Mathematical Reasoning*

We begin with the assumption that no affective behavior is devoid of cognition. Some authors use terms such as *affective schemes* or *cognitive-affective schemes* (Schlöglmann 2005) in an attempt to study this interaction in greater depth. Far from contradicting the aforementioned basic assumption, the acknowledgement of "affective structures" confirms that they are isomorphs of cognitive structures and the result of intellectualization (Piaget 1981), which exists whenever feelings are structured. In fact, structure and the workings of cognition and affect are indivisible in all behavior. Nevertheless, I note that there are authors such as Goldin (2000) who see this interaction as a representational system; in his works he leans more towards a separation of structures of affect and cognition, following the work of Zajonc (1980).

Maintaining a dialogue concerning cognition-affect interaction entails bearing in mind matters relating to the singularity of individual reasoning patterns as well as social interaction. An initial insight is that mathematical reasoning does not follow a straight line but, as Lakatos (1976) contends, a zig-zag course.

> Discovery does not go up or down, but follows a zig-zag path: prodded by counterexamples, it moves from the naïve conjecture to the premises and then turns back again to delete the naïve conjecture and replace it with the theorem. Naïve conjecture and counterexamples do not appear in the fully fledged deductive structure: the zig-zag of discovery cannot be discerned in the end-product. (Lakatos 1976, p. 42)

Affect is essential to the self-regulation and self-reflection that takes place in the course of reasoning. Self-assessment of personal competence, affective response, and self-regulation are keys to problem solving. In this regard, Lakatos notes that perseverance is needed to surmount the cognitive and affective difficulties arising in "conscious guessing, because it comes from the best human qualities: courage and modesty" (Lakatos 1976, p. 30).

Motion and e(motion)

The epistemological view of Lakatos challenges a simplistic view of objectivity in mathematical knowledge. In each theory (scientific or mathematical), the subjective dimension, either as a psychological process or a sociological process, is inexorably involved. For Lakatos, mathematical thinking does not develop monotonically: "Informal, quasi-empirical, mathematics does not grow through a monotonous increase of the number of indubitably established theorems but through the incessant improvement of guesses by speculation and criticism, by the logic of proofs and refutations" (1976, p. 5) in a dynamic zig-zag motion (trajectories).

In this characterization of mathematical knowledge, motion (hence, e(motion)) is essential. The word *(e)motions* or simply *motions* are used to denote mental states not fully described by a formalized language: They are individual instances of subjective, conscious experience. The "motions of the cognitive-affective interplay" are inner motions, interior movements consisting of thoughts, imaginings, emotions, inclinations, desires, feelings, repulsions, and attractions. To identify this interaction involves becoming sensitive to these movements, reflecting on them, and understanding where they come from and where they lead us. The word *motion* conveys the movement and purpose inherent in the meaning of the word *emotion*; in contrast, the word *emotion* conveys only a sense of spontaneous change and movement without a definite purpose.

In this light it may be said that the continuous process underlying motion holds and connects the stages that occur in the movement of problem solving. We invite the reader to consider the phenomenon of interaction as a key element of this movement. Moreover, affect is considered not only as energy (in the sense in which Piaget distinguished between affectivity as energy and cognition as structure) but as a structure, although a structure that is not static but dynamic (see Sect. 10.2.2).

10.2.2   Affective-Cognitive Reference System Model

Since our first study in 1997 on affect and cognition (Gómez-Chacón 2000a, b; Gómez-Chacón and Figueral 2007), we have been interested in the methodological aspects that would allow us to establish an analytical model. In this respect our major claim is that in order to understand cognitive-affect interplay in the acting individual at a particular moment, it is necessary to attain knowledge of that individual at different levels: individual, group, society, and integrally as holistic. This knowledge would capture aspects that model the dynamics of cognitive-affect using some constructs: structures of (local and global) affect, cognitive dimension (valuation processes and mathematical processes), and meta-affect.

Conceptualization of affect and structure

Prior to describing these constructs (categories) we need to clarify the meaning of the terms *affect* and *structure*.

We wish to note that the term affect has a different meaning if it is used by educational psychology or by mathematics education. Pekrun and Linnenbrink-Garcia (2012) note that the term affect in psychological emotion research refers more specifically to emotions and moods; here, we use the term in the broad sense prevalent in mathematics education (e.g., Goldin 2014; Pepin and Rösken-Winter 2015). We share the view of those who within mathematics education regard the importance of emotions as being partly or mainly through their connection with attitudes, beliefs, and values.

Affect is understood here as a notion of a higher order that includes all of the above as a phenomenon. Affect is defined as a "quality power status variable", with duration and intensity at the level of consciousness.

Regarding the meaning of structure, in every structure definition, the concept structure refers to following elements: (a) a whole, (b) the parts of this whole, and (c) the relationship between these parts. In the 50s, mathematical structure was defined as "a specific set of relations or laws describing the functions of a phenomenon that can be represented by a model" (Bastide 1962, p. 14). In this light, we understand structure as a system of interconnected elements in which a change in one element necessarily causes changes in other elements. This type of approach has been expressed in positions such as that of Piaget in the analysis of mathematical structures, defining structure as a system of transformations, which cites fullness (a whole), transformation, and self-regulation as important characteristics. The phenomenon of invariance is associated to the transformation. For many structuralist thinkers, this principle of invariance appears as a key element.

In our case we seek to confirm specific patterns, specific cases, and simple rules that give a typical structure or rhythmic sequences in this interaction between cognition and affect in the individual and explore whether these same patterns are extended to a group of individuals. The structure definition that we use here is dynamic (a dynamic system), and it corresponds to the predominant use of the term today.

We consider that the importance of capturing structure models, as already mentioned, or what other colleagues have expressed as "cognitive-affective schemes," resides not only in identifying schemes but also in the potential created for the recursive construction for the understanding of that scheme. Many of the individual actions are performed unconsciously. However, many of the operations through which we assemble our experiential world can be explored and the knowledge attained can help make learning different, and perhaps better.

Structures of affect: local and global

Local affect-cognition structure occurs at the micro and individual level. It is defined as the understanding of the affective reactions of students towards mathematics by observing and knowing the stages in the process of change of emotional reactions during problem solving and detecting cognitive processes associated with positive or negative emotion. It consists in representing the information on emotional reactions that have an impact on conscious processing. This allows us to establish productive affective pathways. Affective pathways are sequences of (local) emotional reactions that interact with cognitive configurations in problem solving.

Global affect-cognition structure occurs at the medium and macro (individual, group, and society) level and is understood as a result of these factors:

1. The summary of the pathways followed by the individual in the local affective dimension. These pathways are established with the cognitive system and they contribute to the construction of the general structures of one's self-concept as well as the beliefs about mathematics and the learning of mathematics.
2. The interactions and social-cultural influences on individuals and how that information is internalized and shapes their belief systems. Two aspects to take into account: the social representations of mathematical knowledge and the socio-cultural identity of subjects. The features that the students' identities have in their context are equivalent to a network of meanings that will be manifested in the learning of mathematics. These meanings will throw light on our search for a greater understanding of the global configuration of the affective aspect, their way of knowing and reacting affectively to the learning of mathematics, and their way of constructing belief systems and the knowledge of these.

To sum up, the term *local affect* includes emotional states and mood states but also their moment-by-moment interactions with cognition, the social environment, the emotions of others, and the individual's traits. Global affect includes trait emotions as well as stable structures that incorporate emotions—not only attitudes, beliefs, and values but constructs such as mathematical self-identity (Gómez-Chacón 2000a, b; Gómez-Chacón and Figueral 2007).

Cognitive dimension

In this study we use the term *cognitive* broadly. On the one hand, this refers to the extensive use of processes of evaluation (cognitive appraisal; Lazarus 1991) and, on the other hand, to the characterization of the subjects' personal meanings of the cognitive dimension of the heuristic that acts in the solution of problems (mathematical cognitive processes) (Goldin 2004; Gómez-Chacón 2000a, b, 2015; Schoenfeld 1994).

Certain distinctions between "knowledge" and "evaluation/appraisal" are essential in our study. The need for this differentiation arises not only from our studies but from studies that have expressed the need to make a distinction between knowledge and appraisal (Lazarus 1991). We consider that there are certain features with emotional implications that may be particular to mathematics as compared with other school subjects, and they are related to the cognitive dimension. For instance, the circulation between different types of mathematical reasoning (discursive, instrumental, and visual) and the difficulty of transitions between them or the necessity of formal language, which involves specific semiotic systems and representation. As we will present in Sect. 10.3, the analysis of the cognitive dimension of mathematical work requires a specific model. In this section we present the mathematical working space framework (MWS) model in order to go deeper into the analysis of local affect and cognition interplay.

We can also speak to the differentiation of cognitive demands according to concepts of the difference between mathematical knowledge in general and contextual mathematical knowledge. The former includes establishing attitudes, beliefs and intuitions about oneself, while the latter is active in a particular situation or with a specific content whose impact on emotion can be very different. Regarding cognitive appraisal, this will be referred to as the generation of personal meanings and how the valuation of that knowledge makes it potentially emotional, i.e., how the situation globally affects relevance for the person, in relation to the goals, and to resource management.

Meta-affect or meta-emotion

Notions such as 'meta-affect' or 'meta-emotion' are required to refer to affect about affect or to affect in cognition that is about affect. In this way they serve to monitor affect both through cognition and affect. It's referred to meta-emotional understanding and meta-emotional skills. It shows how meta-affect arises in the formation of an individual's cognitive and affective schemes (Gómez-Chacón 2000a, b, 2015; DeBellis and Goldin 2006; Malmivuori 2006; Schlöglmann 2005).

Cognitive understanding of affect enables individuals to control their actions in affective situations. Successful handling of affective situations stabilizes affect schemata and consequently beliefs through simulation as a cognitive window to emotions. Prior research has shown that the stability of an individual's beliefs is closely related to the interaction among belief structures. These include not only affect (feelings and emotions) but also and especially meta-affect (emotions about emotional states, emotions about cognitive states, thinking about emotions and cognitions, and regulation of emotions; Gómez-Chacón 2000b). These findings reveal the personal and social dimensions of the affective constructs and self-control of emotions.

The section that follows deals with determining the local affect-cognition structure. The main reasons for doing so are that the local affect-cognition structures offer a profile of global affect structure of the subjects and also because, as we indicated at the beginning, the results of this basic research could be easily integrated in classroom practice.

10.3 Determining the Local Affect-Cognitive Structure

To empirically illustrate this section we will take a recently developed study about affective pathways and interactive visualization in the context of technological and professional mathematical knowledge (Gómez-Chacón 2012, 2015). For a period of three years, a study with university students with degrees in mathematics who were possible future secondary school teachers (98 students, 65 female and 33 male) was carried out. In this study, a teaching experiment was developed through problem solving. A questionnaire was composed of six non-routine problems about geometric locus to be solved using GeoGebra. The problems required the solver to use a proposed chain of various steps of visual processing (technical, deductive, and analytical) in order to find the solution. Each of the 98 students solved the six problems. In order to identify both types of cognitive processes and emotions, data were collected from the subjects' problem-solving protocols as well as with two questionnaires: one on beliefs and emotions about visual reasoning completed at the outset and another on the interaction between cognition and affect in a technological context filled in after each problem was solved.

As introduced in Sect. 10.2.2, when we focus on the local affective-cognitive interaction as a research goal, we are trying to "capture" (explain and model) the precise mathematical elements of both cognitive processes and appraisal processes; patterns, routines, and dynamic changes in the affective pathways of each individual; and the transition of local affective-cognitive local structure to global affective-cognitive structure in the individual. Explaining and modelling these specific aspects involves both methodological and theoretical options. To describe these aspects and illustrate them with the results of empirical research the following points were established:

(a) Considerations for the analysis of the cognitive mathematical dimension (Sect. 10.3.1)
(b) Modeling the local structure of affect in the individual: routines and bifurcations (Sect. 10.3.2)
(c) Modeling local affect structure in a group (Sect. 10.4).

10.3.1   Considerations for the Analysis of the Cognitive Mathematical Dimension

Following the term cognitive described in Sect. 10.2, two aspects need to be characterized: the cognitive dimension of the mathematical visualization processes and the processes of cognitive appraisal.

Relative to the cognitive dimension, the mathematical working space (MWS) model (Gómez-Chacón and Kuzniak 2013; Gómez-Chacón et al. 2016), together with the instrumental dimension, are used to describe the complexity involved in applying technology to the mathematical and cognitive aspects of geometric tasks. Within the mathematical working space framework, cognitive and epistemological levels need to be articulated to ensure a coherent and complete geometric work.

Epistemological and cognitive levels are defined in terms of three geneses— semiotic-figural, instrumental, and discursive (see Fig. 10.1)—to guarantee complete and consistent geometric work. The cognitive plane is introduced to describe the cognitive activity of a single user. In this model, the idea of three cognitive processes involved in geometrical activity is adapted from Duval (2005): (a) a visualization process connected to the representation of space and material support, (b) a construction process determined by instruments (ruler, compass, etc.) and geometrical configurations, and (c) a discursive process that conveys argumentation and proofs. Both planes, cognitive and epistemological, need to be articulated in order to ensure a coherent and complete geometric work. This articulation assumes the presence of some transformations that are possible to define through three fundamental geneses represented in the diagram below (Fig. 10.1).

In the analysis of the problem solution we examined some key aspects of how both figural and instrumental developments are involved in the learning process in a computer environment. An understanding of the visualization processes identified which ones are associated with patterns of use, with structuring information by sign operations, or with a heuristic function that allows the user to anticipate and plan actions and modes of validation.

The study of the local affect-cognitive structure addresses the so-called intra- and inter-plane zig-zag paths in mathematical reasoning. These epistemological aspects of this point were introduced in detail in Sect. 10.2. There the conception of interaction cognition and the complexity of treating such interaction were expressed. As additional information, the interested reader can consult Gómez-Chacón et al. (2016), where the nonlinearity or zig-zag motion was given in detail, with an example of the interaction between demonstrative reasoning and mathematical attitudes by two secondary school students.

Regarding zig-zag paths in mathematical reasoning, in the presented study we focus on the plane that is associated with figural and instrumental genesis. Movements comprising inter-genesis and intra plane transitions were analyzed at the local level (see Table 10.1). Image typology and the use of visualization were analyzed as well. They are categorized conceptually: the use of visualization as a

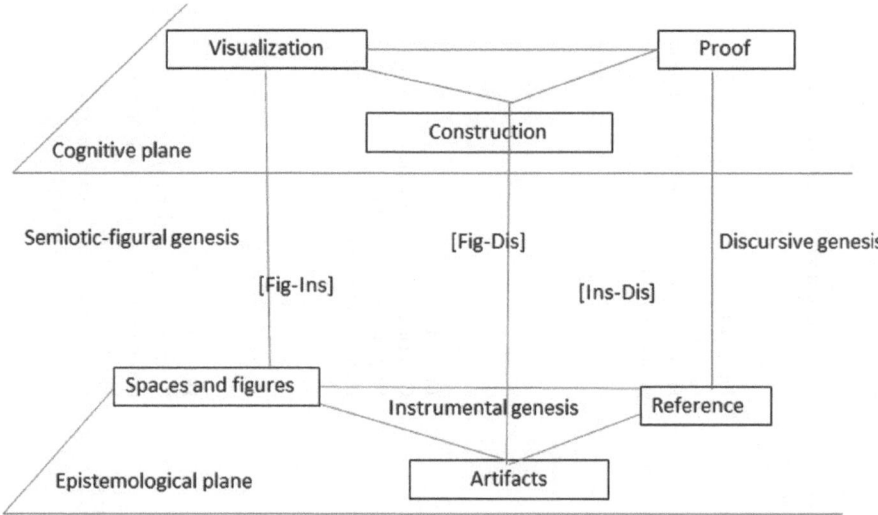

Fig. 10.1 Geometric working space, geneses, and vertical planes in an ideal MWS

reference and its role in mathematization and the heuristic function of images in problem solving.

Finally, referring to the processes of cognitive appraisal, different patterns of appraisal are analyzed (pleasantness and goal-path obstacle) and it is determined whether a given belief (beliefs about visualization and beliefs about technology) can elicit different emotions from different individuals.

10.3.2   Modeling the Local Structure of Affect in the Individual: Routines and Bifurcations

Establishing patterns of interaction of cognition and affect requires analysis at the microscopic level of individuals and all data sources. Below we present an analysis in the case study for the following problem:

> The ladder: The top of a 5-m ladder rests against a vertical wall, and the bottom rests on the ground. Define the locus generated by midpoint *M* of the ladder when it slips and falls to the ground. Define the locus for any other point on the ladder.

This analysis will endeavor to answer these research questions: What kind of cognitive affective pathways can be described? What are the influences that helped these students to stay on—or get back on—an enabling pathway of affect instead of sliding down to anxiety, fear, and despair?

Table 10.1 Part of the analysis of Alberto's solution to the ladder problem reported by the subject in his protocol, zig-zag reasoning in inter-genesis, inter-plane transitions and emotions

Method description, including visualization	Typology of the use of representation/image/	Local affect/ emotions	Reasoning pathway in inter-plane transitions
(1) In the first place, I made a representation of the problem on the paper. I tried to look for a way to solve it analytically but I do not find any. I reflect on the possible relationships between the triangles that the ladder gradually generates as it slides downward against the wall and the floor without being able to get anywhere (2) I think about the answer will it be a straight line, an ellipse, or a circle? (3) I left the problem for another day. I was thinking about it while I was doing other activities. I trusted my subconscious to continue to job	Drawing (of patterns and lines/figure) Analytical (Search for mental image (specific figure/illustration and dynamic image) Search for mental image and strategy. Mental block specific illustration and dynamic image	Curiosity Confusion Puzzlement Mental block	**Fig → Disc** Reference Pursuit of discursive genesis visual-figural justification only
(4) I cook up the problem with excitement and hope. I experiment with a pen and an elastic rubber rolled on its middle part. It seems to form an arc of a circle. A t least, I already have an idea (5) I start to work with GeoGebra. After trying some construction with straight lines, I notice that the ladder was a segment of length 5 allows me to make a construction based on a circle of radius 5 that runs the y-axis	Physical manipulation, kinetics (Kinesthetic manipulation) Mental image: identification mathematical object Technological manipulation with the computer Representing radius of the circle (Specific illustrations)	Confidence Perseverance-motivation	**Ins → Fig** Configuration (instrumentation) Shape deconstruction by manipulation-kinetics
(6) I generate a slider t and I define the centre of the circle C = (0, 1) The slider will shrink from 5 to 0. It is zero when the ladder lies on the ground. Point B represents the intersection of the circle and the x-axis	Interactive image generation, slider (Analogical)	Confusion Excitement and hope	**Dis → Lis → Fig** Structural analysis using tools, shape, and dimension to come up with a proposal Instrumental genesis Significant units in the interactive image generation to find the explanation graphically

A possible establishment of patterns was analyzed for each subject:

1. Exposed beliefs and beliefs in action about visual thought and emotional reacting that can be generated.
2. Coincidences in the typologies of use of visualization and associated emotion.
3. Valuation made about the events that stimulate feelings: local affect. In this case, we have concentrated on processes related to visualization and technological use and the zig-zag reasoning in inter-genesis and inter-plane transitions: Figurative → Instrumental → Discursive (Fig → Ins → Disc).

We will use Alberto's Case as an example. Alberto is university student with a mathematics degree and is a visualizing individual. His enjoyment of mathematical visualization is closely intertwined with the evolutionary conception of mathematics. He considers visual reasoning to be essential in problem solving. He defines his own pathway of affect-cognition for the ladder problem (Fig. 10.2).

In Table 10.1, we have an extract of the cognition-affect relations in the problem according with the theoretical frame, where relationships between affect, cognition, and epistemological consideration are evidenced. An in-depth analysis of the problem-solving protocol for this exercise and the affective-cognitive pathway reported showed that this subject was able to describe and control emotions and identify causes. Three types of affective perspective were identified. First, Alberto always tried to find an answer even when in doubt or blocked. Alberto was continuously active, which is one way that many students cope with stress. Second, he was able to walk away from the problem, aware of the role of the subconscious in mathematics (See number (3) in first column of Table 10.1). Third, he struck a balance between the combination of graphic geometric thought and analytical task solving (transitions between Dis-Ins planes). These three behaviors were indicative of interaction between the cognitive-affective system and self-control. The description of emotions revealed that, from both a mathematical and a technical-instrumental perspective, self-confidence, stimulus, and joy were associated with the reproduction of physical forms and the visual/perceptive control implicit in a command of ancillary mathematical objects.

This type of microscopic analysis was performed for the whole teaching experiment (six problems). This allows each student to describe pathways in different problems. Comparison of the pathways of the six problems (Table 10.2) in the teaching experiment allows us to model the local structure of affect in the individual that shapes a more stable structure we call the global structure.

Curiosity → Confusion → Puzzlement /Mental block → Confidence → Perseverance-

motivation → Confusion/*Exhilaration → Confidence → Confidence -Joy →*

Joy→Perceived beauty → Satisfaction→ GLOBAL AFFECT

Fig. 10.2 Affective pathways reported for the problem by Alberto

Table 10.2 Summary of Alberto's affective-cognitive pathways in questionnaire problems

Problem	Pathway type	Cognitive-emotional processes							Global affect
Pr-1	Path 3 (R3)	Curiosity	Stimulus-motivation	Joy-happiness	Satisfaction	Pleasure-joy			Positive self-concept
	Cognitive process	Search for mental image	Search for mental image imagination	Selection of one strategy	Full construction, obtaining a final solution	Checking of the solution			
Pr-2	Path 1 (R1)	Curiosity	Confusion	Perplexity	Stimulus-motivation	Placer	Joy-happiness	Satisfaction	Self-concept+
	Cognitive process	Search for mental image	Representation	Representation-semiotic-understanding	Intuition of the solution	Identification math object. Get representation	Specific illustration and dynamic image	Full construction, obtaining a final solution	
Pr-3	Path 3 (R3)	Curiosity	Confidence	Joy	Perplexity	Happiness	Pleasure-joy	Specific representation with interactivity (analogical)	Self-concept+
	Cognitive process	Search for mental image understanding problem	Visual reasoning identifying mathematical knowledge drawing on paper and pencil	Representation, drawing on paper and pencil (patterns and figure analytical)	Illustration and dynamic image with GeoGebra	Specific representation with interactivity (analogical)		Realization and checking of the solution	
Pr-4	Path 3 (R3)	Curiosity	Confusion	Perplexity puzzlement Mental block	Confidence	Perseverance-motivation	Confusion Excitement Hope	Confidence	Self-concept+
	Cognitive process	Reading and understanding problem	Search for mental image	Search for mental image	Search for mental image	Search for mental image and strategy Mental block specific illustration and dynamic image	Kinesthetic learning Mental image Identification mathematical object	Technological manipulation with computer Representing circle	Joy-happiness / Satisfaction
								Specific representation with interactivity (analogical)	Specific representation with interactivity (analogical) — Analytical-visual Memorized formulaic typology
Pr-5	Path 3 (R3)	Curiosity	Confidence	Certainty-confidence	Satisfaction	Specific representation with interactivity (analogical) Full construction from scratch			Self-concept+
	Cognitive process	Reading and understanding problem	Understanding of the problem Search for mental image	Technological manipulation with computer. Representing	Satisfaction				
Pr-6	Path 3 (R3)	Curiosity	Confidence	Certainty-confidence	Satisfaction	Checking of the solution graphically			Self-concept+
	Cognitive process	Understanding of the problem Mental block with analytical reasoning	Understanding of the problem	Using strategic and mathematical process					

In Alberto's case, it can be seen that the cognitive processes of visual reasoning and negative emotion interaction occur in identifying strategies of interactive representation and processing of certain representations, where the student has to put into play the identification of parametric variations. He is a student who has a fluid use of images: concrete, kinesthetic, and analogic. This student recognizes an overall positive self-concept structure when working with computer mathematics.

In short, this kind of analysis allows us to identify patterns in the individual and between individuals in relation to their local cognitive affective structure:

1. In the individual, the summary of the pathways followed by the individual in the local affective dimension in different problems allow us to identify invariance and variances that occur in their local structure and that shape a more stable structure we call global structure.
2. Among individuals, variations in local structure according to individuals. This type of data analysis allows us to identify in-depth profiles of students with varied characteristics: gender, achievement in mathematics, beliefs, display style, and emotions. The affective pathways they reported were compared in order to glean information on meta-emotion and visualization. The comparison revealed: (a) the use of visualization and associated emotion and (b) the dependence of their emotional self-control on their individual perception, which was influenced by style, disposition, type of activity or skill, instrumental command, and belief system around technology-aided mathematical learning. For more detailed information see Gómez-Chacón (2012, 2015).

10.4 Modeling Local Affect Structure in a Group

In Sect. 10.3 we have focused on the understanding of the cognition-affect interaction in the individual. In this section we are going to take a step further: we will try to see how to make the leap from the characterization of individuals to the characterization of the group. To carry out this characterization, we ask the following questions: What are the differences in a subject's choice of pathway? What information on meta-emotion and visualization can be gleaned from the productive affective pathways reported by students in locus problems? Of these, which allow characterization of the tendency of a group?

Methodologically, in the affect dimension the characterization of a group has been solved by quantitative studies, mainly based on surveys. Here, we would like to raise other methodological forms that are based on qualitative measures and on quantitative behavior modeling collected in a qualitative way. In our most recent work we have worked with implicative statistical analysis models or models based on fuzzy logic (Gómez-Chacón 2015, 2017). In this paper I describe the first models.

10.4.1 Implicative Data Analysis

In this research, together with the qualitative analysis (Sect. 10.3.2), an implicative analysis (Gras et al. 1997) was performed in order to explore the structure in cognition and affect interactions for the group. Gras's implicative statistical method has been conducted by using software called CHIC (Classification Hiérarchique Implicative et Cohésitive). At the descriptive level, it can be used to detect a certain degree of stability in the structuring, while for predictive purposes, it provides the grounds for assumptions. This statistical analysis was then used to establish rules of association for data series in which variables and individuals were matched in order to define trends in sets of properties on the grounds of inferential, non-linear measurement.

Defining categories

Two types of analyses were conducted in this study. The first was exploratory, descriptive, and interpretational, involving mainly inductive data analysis, with categories and interpretation building on the information collected (Sect. 10.3.2). This analysis used a qualitative approach based a cross-check of the solutions performed by three researchers. The following categories were defined:

1. Emotion associated with visual reasoning in the ladder exercise: P4EviP (like), P4EviN (dislike), P4EviM (mixed emotions), and P4viInd (indifferent).
2. Instrumental difficulties: The focus in this category was on two types of difficulties arising around the six problems (Fig. 10.3). Typology 1: Static constructions (discrete) (DT1P4). Typology 2: Incorrect definition of the construction (DT2P4).
3. Initial problem visualization: VisiP4.
4. Beliefs about visual reasoning: BeviP (positive), BeviN (negative).
5. Preferences and emotions around the use of visualization: EviP (like), EviN (dislike), EviInd (indifference).
6. Beliefs about computer-aided learning: BeGeoP (positive), BeGeoN (negative).

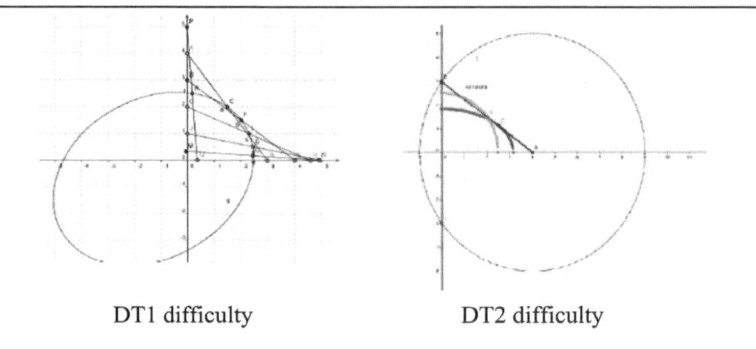

DT1 difficulty   DT2 difficulty

Fig. 10.3 Examples of subjects' difficulties with the ladder exercise

7. Emotions concerning computer use: EGeoP (like), EGeoN (dislike), EGeInd (indifference).
8. Affective-cognitive pathways: R1 (mainly positive emotions), R2 (mainly negative emotions), R3 (subject-formulated, as described in Fig. 10.2 and Table 10.2).

All categories were compiled and coded in a matrix for implicative analysis performed using CHIC software. The identification of possible links among affective-cognitive pathways, emotions, and meta-emotion was the subject of the analysis.

10.4.2 Results of the Modeling of Local Affect Structure in a Group

In this study, a similar response was received when the beliefs explored related to the use of dynamic geometry software as an aid to understanding and visualizing the geometric locus idea. All the subjects claimed to find it useful, and 80% expressed positive emotions based on its reliability, speedy execution, and potential to develop their intuition and spatial vision. They added that the tool helped them surmount mental blocks and enhanced their confidence and motivation. As future teachers they stressed that GeoGebra could favor not only visual thinking, but help maintain a productive affective pathway. They indicated that working with the tool induced positive beliefs towards mathematics itself and their own capacity and willingness to engage in mathematics learning (self-concept as a mathematics learner).

Table 10.3 summarizes the frequencies of pathways and emotions associated with visualization in the ladder problem. Mixed affective pathways were identified, with alternating negative and positive emotions and optimized self-control of emotions.

The question that was posed to study the mix of emotions and meta-emotion in greater detail was "What are the differences in a subject's choice of these three pathways?" The preliminary analysis showed that pathway R3 was largely self-formulated and contained a much greater mix of emotions (Sect. 10.3.2). In most cases, moreover, the trend was not as explicit as in R1 (positive) or R2 (negative). Rather, negative feelings (which were controlled) were attributed to certain stages of the visualization process and positive feelings to success in representing the desired images. A hierarchy study of R3 yielded some significant affective-cognitive implications respecting visual processes: R3P4 $\rightarrow$ 0.99 VisiP4 and R2P4 $\rightarrow$ 0.90 DT2P4.

Table 10.3 Percentage of affective-cognitive and emotional pathways associated with visualization in the ladder exercise

	R1	R2	R3	EviP	EviN	EviM	EviInd
Problem (%)	46	12	40	18	25	53	3

Fig. 10.4 Hierarchy tree

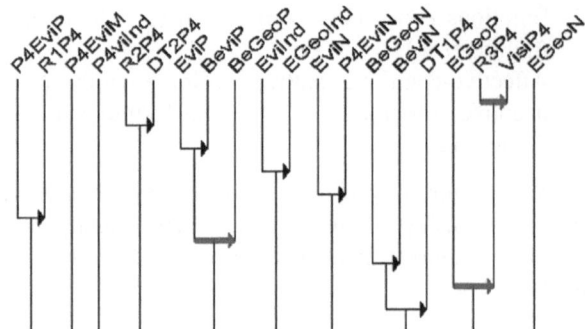

Three of the nine nodes obtained in the hierarchy tree were significant and identified the following groups (Fig. 10.4).

Group 1 (N (level 1, cohesion: 0.998) = (R3P4 VisiP4)), comprising over 40% of the initially visualizing subjects (in Problem 4) who indicated pathway R3 as the expression of their cognition-affect interaction. The most significant characteristic of these individuals was their positive feelings towards computers (use of GeoGebra (EGeoP) software).

Group 2 (N (level 7, cohesion: 0.276) = ((EviP BeviP) BeGeoP))), where the most prominent finding was that a belief in the use of GeoGebra was attendant upon a belief in and a preference for visual reasoning.

10.5   Conclusion

It has been argued, given the empirical evidence, that the link between cognition and affect is at the basis of the whole of mathematical activity. It has been noted, however, that similar investigations on the agenda of mathematics education are still very scarce. One of the main reasons for this shortfall is that it is difficult to carry out studies on affect: It is a question of the adequacy of theoretical and methodological frameworks.

We have tried to envisage both the epistemological and ontological keys inherent in the methodology designed to capture the interaction of cognition and affect in mathematics. Also, a study and research design has been shown that has resulted in an instrument that is significant and productive in addressing this goal.

The results of the research presented in Sects. 10.3.2 and 10.4 have shown the following findings: (1) Both cognitive mathematical processes and appraisal processes are key dimensions that explain the interplay between cognition and affect, and (2) the non-linear modeling of emotion is reflected in the affect-cognition local structure. It has been shown that the emotions in problem solving are not static.

They are dynamic events that can be on-going throughout a problem-solving session. The inspiration of the nonlinear model of emotion described here is based not only on socio-constructivist approaches (such as Mandler's theory; Mandler 1984) but on other emotion theories with a more holistic view of the individual or a perspective systemic of affectivity that we presented in Sect. 10.2.

Regarding the connections between cognition and emotion, two categories seem to appear within the study: emotions that have causes known by the person who experiences them and words that are used to denote the causes of the emotions. The analysis of these connections sought not only to determine the relations between variables but also to develop rule models as well as use a description of qualitative data that can enrich this relationship. Notice that the analysis of the group (Sect. 10.4) as a multiple variable analysis with the capability of decision trees enables us to go beyond simple one-cause, one-effect relationships and to discover and describe things in the context of multiple influences.

The results of studies where non-linear modeling of emotion is reflected in the affect-cognition local structure drive us to formulate an open question about the type of generated structure. We might ask whether or not the aforementioned affective-cognitive interactions have a so-called fractal structure; that is, are they basically similar (self-similar) at the mental and subjective, micro-social (interpersonal or small-group), and macro-social (international or intercultural) levels? As is well known, fractality is a property of a great number of natural and cultural phenomena which are intensely studied by dynamic systems theory (formerly called chaos theory) and of many biological, demographic, and economic processes. Fractal structure is generated by basically similar dynamics (algorithms) on different levels of functioning. The result is a characteristic so-called self-similarity (or scale-independence) of structural patterns on different levels of complexity. The notion of fractality permits a methodologically correct transfer of small-scale (e.g., individual or interpersonal) observations to large-scale (e.g., inter-group or international) processes and vice versa.

Notice that studies on the exploration of fractal structures have been developed in the understanding of mathematical concepts, but we do not meet them in the field of interplay of cognition and affect (e.g., Singer and Voica 2010). So, although the present study casts some light on this issue of the fractal structure, it is still a question open to debate. The discussion on fractality can be established in the global-local cognitive-affective perspective taken in this paper and in the methodological transfer of small-scale observations to large-scale processes.

In the results of the described study, a type of fractal construction could be seen in the patterns of use of visualization and associated emotion and in the meta-emotion and beliefs of the individuals. It has been shown that the cognitive processes of visual reasoning and negative emotion interaction occur in identifying strategies of interactive representation and processing of certain representations, where the student has to put into play the identification of parametric variations.

A distinction has been made between computer imagery and mathematical object imagery, and, through instrumental and dimensional deconstruction concepts, the existence of separate cognitive mechanisms for processing objects and relations among objects, where figurative and instrumental genesis processes are involved, has been shown.

In addition, the affective-cognitive dynamics can generally suffer sudden changes under special conditions. When the emotional tensions in mental or social systems reach a critical point, the dominant forms of feeling, thinking, and acting can suffer sudden global changes (called non-linear bifurcation). In this study, the control parameter that determines the moment of bifurcating was the meta-emotion and beliefs of subjects. Through implicative data analysis that took into account these small-scale observations as categories, it was possible to transfer to a large-scale process. As seen in Sect. 10.4, in individuals whose pathway is R3 (with alternating negative and positive emotions and optimized self-control of emotions) or individuals whose positive belief in the use of GeoGebra, both aspects may become the new relevant order-parameters (or nuclei of crystallization) around which the new global feeling-thinking-behaving patterns are organized.

The conceptualization of mathematical work can lead to an essential contribution to the methodology for the diagnosis of cognition and affect interaction. Regarding the cognitive dimension in appraisal processes, the categories of levels established have been useful for a global analysis, while the mathematical working space model has enabled a local look at how representations and images are produced. The transitions between figural, instrumental, and discursive processes have allowed us to characterize the dialectical process in the cognition and affect interaction between types of mathematical thinking. As we noted at the beginning, mathematical knowledge has specific characteristics that distinguish it from other areas of knowledge. Today, characterizing this is among the key issues around which we need to advance our knowledge.

Acknowledgements The paper's elaboration has been supported by a special action grant, EDU2013-44047-P, from the Government of Spain and a research grant Visiting Scholar Fellowship, University of California, Berkeley.

References

Bastide, R. (Ed.). (1962). *Senses et usages du terme "structure"*. Hague: Mouton.

Brun, G., Doğuoğlu, U., & Kuenzle, D. (2008). *Epistemology and emotions*. Hampshire, UK: Ashgate Publishing.

DeBellis, V. A., & Goldin, G. A. (2006). Affect and meta-affect in mathematical problem solving: A representational perspective. *Educational Studies in Mathematics, 63*(2), 131–147.

Duval, R. (2005). Les conditions cognitives de l'apprentissage de la géométrie: développement de la visualisation, différenciation des raisonnements et coordination de leur fonctionnements. *Annales de Didactique et de Sciences Cognitives, 10*, 5–53.

Goldin, G. A. (2000). Affective pathways and representation in mathematical problem solving. *Mathematical Thinking and Learning, 2*(3), 209–219.

Goldin, G. A. (2004). Problem solving heuristics, affect and discrete mathematics. *ZDM-Mathematics Education, 36*(2), 56–60.

Goldin, G. A. (2014). Perspectives on emotion in mathematical engagement, learning, and problem solving. In R. Pekrun & L. Linnenbrink-Garcia (Eds.), *Handbook of emotions and education* (pp. 391–414). New York: Routledge.

Gómez-Chacón, I. Mª. (2000a). *Matemática emocional. Los afectos en el aprendizaje matemático.* (Emotional mathematics. Affects in mathematics learning). Madrid: Narcea.

Gómez-Chacón, I. Mª. (2000b). Affective influences in the knowledge of mathematics. *Educational Studies in Mathematics, 43*, 149–168.

Gómez-Chacón, I. Mª. (2012). Affective pathways and interactive visualization in the context of technological and professional mathematical knowledge. *Nordic Studies in Mathematics Education, 17*(3–4), 57–74.

Gómez-Chacón, I. M. (2015). Meta-emotion and mathematical modeling processes in computerized environments. In B. Pepin & B. Rösken-Winter (Eds.), *From beliefs and affect to dynamic systems in mathematics education. Exploring a mosaic of relationships and interactions* (pp. 201–226). Switzerland: Springer.

Gómez-Chacón, I. Mª. (2017). Emotions and heuristics: The state of perplexity in mathematics. *ZDM-Mathematics Education, 49*, 323–338. https://doi.org/10.1007/s11858-017-0854-8.

Gómez-Chacón, I. Mª., & Figueral, L. (2007). Identité et facteur affectifs dans l'apprentissage des mathématiques. In *Annales de Didactique et de Sciences Cognitives* (Vol. 12. pp. 117–146). Strasbourg: IREM.

Gómez-Chacón, I. M., & Kuzniak, A. (2013). Geometric work spaces: Figural, instrumental and discursive geneses of reasoning in a technological environment. *International Journal of Science and Mathematics Education, 13*(1), 201–226.

Gómez-Chacón, I. Mª., Romero, I. Mª., & Garcia, Mª. M. (2016). Zig-zagging in geometrical reasoning in technological collaborative environments: A mathematical working space-framed study concerning cognition and affect. *ZDM-Mathematics Education, 48*(6), 909–924.

Gras, R., Peter, P., Briand, H., & Philippé, J. (1997). Implicative statistical analysis. In H. N. Ohsumi, N. Yajima, Y. Tanaka, H. Bock, & Y. Baba (Eds.), *Proceedings of the 5th Conference of the International Federation of Classification Societies* (Vol. 2, pp. 412–419). New York: Springer.

Harre, R. (2009). Emotions as cognitive-affective-somatic hybrids. *Emotion Review, 1*(4), 294–301.

Immordino-Yang, M. H., & Damasio, A. (2007). We feel, therefore we learn: The relevance of affective and social neuroscience to education. *Mind, Brain and Education, 1*(1), 3–10.

Lakatos, I. (1976). *Proofs and refutations: The logic of mathematical discovery.* New York: Cambridge University Press.

Lazarus, R. (1991). *Emotion and adaptation.* New York: Oxford University Press.

Malmivuori, M. L. (2006). Affect and self-regulation. *Educational Studies in Mathematics, 63*, 149–164.

Mandler, G. (1984). *Mind and body.* New York: Norton.

Pepin, B., & Rösken-Winter, B. (Eds.). (2015). *From beliefs and affect to dynamic systems in mathematics education. Exploring a mosaic of relationships and interactions.* Switzerland: Springer.

Pekrun, R., & Linnenbrink-Garcia, L. (2012). Academic emotions and student engagement. In S. L. Christenson, A. L. Reschly, & C. Wylie (Eds.), *Handbook of research on student engagement* (pp. 259–282). New York, NY: Springer.

Piaget, J. (1981). *Intelligence and affectivity: Their relationship during child development.* Palo Alto: Annual Reviews.

Schlöglmann, W. (2005). Affect and cognition—Two poles of a learning process. In C. Bergsten & B. Grevholm (Eds.), *Conceptions of mathematics. Proceedings of Norma 01* (pp. 215–222). Linköping: Svensk Förening för Matematikdidaktisk Forskning.

Schoenfeld, A. H. (1994). *Mathematical thinking and problem solving*. Lawrence Erlbaum. Hillsdale, NJ: Associates Inc.

Singer, F. M., & Voica, C. (2010). In search of structures: How does the mind explore infinity? *Mind, Brain and Education, 4*(2), 81–93.

Zajonc, R. B. (1980). Feeling and thinking: Preferences need no inferences. *American Psychologist, 35*, 151–175.

Chapter 11
The Role of Algebra in School Mathematics

Liv Sissel Grønmo

Abstract Algebra can be viewed as a language of mathematics; playing a major role for students' opportunities to pursue many different types of education in a modern society. It may therefore seem obvious that algebra should play a major role in school mathematics. However, analyses based on data from several international large-scale studies have shown that there are great differences between countries when it comes to algebra; in some countries algebra plays a major role, while this is not the case in other countries. These differences have been shown consistent over time and at different levels in school. This paper points out and discusses how these differences may interfere with individual students' rights and opportunities to pursue the education they want, and how this may interfere with the societies' need to recruit people to a number of professions.

Keywords Equal rights to education · Low emphasise on algebra
Daily life mathematics · Different profiles in mathematics education

11.1 Introduction

Algebra can be viewed as *a language of mathematics*. It is commonly accepted that competence in a countries language is essential for your opportunities in that country. The same may be said about algebra. Competence in algebra is essential for people across all types of education and professions where they use this language. To learn a country's language takes time, and it matures over time by intensive training through listening and by training to use it yourself, and it is

L. S. Grønmo (✉)
LEA: Research Group for Large-Scale Educational Assessment,
Department of Teacher Education and School Research,
Faculty of Educational Sciences, University of Oslo, Oslo, Norway
e-mail: l.s.gronmo@ils.uio.no

© The Author(s) 2018   175
G. Kaiser et al. (eds.), *Invited Lectures from the 13th International Congress on Mathematical Education*, ICME-13 Monographs,
https://doi.org/10.1007/978-3-319-72170-5_11

usually easier for young children to start learning it than for adults. To some extent the same can be said about algebra; except that in mathematics you learn arithmetic first, as arithmetic is the basis for algebra, so it seems reasonable to start with algebra after some fluency in arithmetic.

In a modern society, everyone goes to school for a long time; a school preparing them for being responsible citizens taking care of their own daily life as well as having a job to support themselves and to contribute to society. We have to ask what type of competence is it reasonable that school emphasises in our societies. Is it enough to teach them some arithmetic and statistics in mathematics to prepare them for their daily life, or do we have to put more effort into learning them the mathematical language algebra? A modern society needs a lot of people well educated in different types of technology such as computer science and engineering. A modern society faces problems related to the environment and economy. In all these domains, competence in the mathematical language algebra is essential. Algebra is an important tool for pursuing a profession in so many domains in our society. It is also important for all types of education in natural sciences as physics, biology, chemistry, or in mathematics itself. If you want to study geometry at a university, you need to be fluent in the language of algebra.

The school is responsible for giving students competence in algebra, and it is, for good reasons, part of school curriculum all over the world. Nevertheless, a number of analyses have shown that the emphasis on students learning algebra varies quite a lot around the world. This paper presents the results of a number of such analyses completed over the two last decades, based on data from different studies and at different levels in school. Drawing on these results, some consequences for individual students and societies not emphasizing the learning of algebra in their schools will be pointed out and discussed.

This include discussions of students equal rights to pursue all types of education and by that have the opportunity for a number of different positions in the society, possible reasons for the low emphasis on algebra in some countries, the relation between pure and applied mathematics, and also some reflection about teaching and learning algebra from the perspective that algebra is a language.

11.2   Different Profiles in Mathematics Education

Since it is commonly accepted that competence in algebra is an important tool for pursuing a number of types of education and profession in a modern high technology society, it may seem obvious that algebra should play a major role in school mathematics. However, a number of analyses based on data from several international large-scale studies have shown that there are great differences between countries when it comes to emphasis on algebra in school mathematics; in some countries algebra plays a major role, while this is not the case in other countries.

International assessment surveys such as TIMSS, TIMSS Advanced, and PISA (IEA 2017a; OECD 2017) aim at establishing reliable and valid scores for students' achievement which can be compared across countries or across groups of pupils within countries, and to relate achievement to various background and context variables that may give ideas about possible indicators for characterization of high performance in mathematics. There has also been an international comparative study of teacher education in mathematics (IEA 2017b; Tatto et al. 2012) collecting the same types of data for students in this type of tertiary education. All these studies also offer opportunities for secondary analyses to answer a number of other research questions. An important research question that has been asked is if it is reasonable to distinguish between different profiles of mathematics education in various countries or groups of countries, and to what extent such profiles seems to be consistent over time and at different levels in school. This paper will especially pay attention to the role of algebra in different groups of countries.

A number of analyses have been conducted based on data from all the studies mentioned above, looking for patterns in what type of content different countries seem to emphasis in mathematics in school. A method commonly used in these analyses is a type of cluster analysis looking for "item-by-country interactions" to investigate similarities and differences between countries or groups of countries across cognitive items. For more about these types of cluster analyses, see Olsen (2006). It has to be recognized that in these analyses, one is talking about *relative performance*. Countries at different levels of performance can therefore show equal patterns for the *type of mathematical content that is emphasized*, since the cluster analysis displays groupings of countries according to similarities in *relative response patterns* across items. Countries in the same group tend to have *relative* strengths and weaknesses in the same items. These types of analyses have been conducted on data from the first TIMSS-study in 1995, and later on data from a number of international studies at different levels in school and with different framework according to the type of mathematical competence that is measured in the study. The analyses of data from TIMSS 1995 concluded that the following clusters of countries formed meaningful profiles from a geographical, cultural or political point of view: English-speaking, German-speaking, East European, Nordic, and East Asian countries (Grønmo et al. 2004). In the following, the paper will concentrate on the four profiles that have revealed consistent profiles in a number of later analyses. The German-speaking profile will not be included because this profile has not been that consistent in later studies. Grønmo et al. also used the residuals in the matrix which was the basis for the cluster analysis in the previous section in order to identify items for which a certain group of countries achieved particularly well or badly. They concluded that typical for the items where East European and East Asian countries seemed to perform relatively best, was that they all focused on classical, pure and abstract mathematics such as algebra and geometry. Contrary to this finding, the Nordic group as well as the English-speaking group performed relatively better on items *closer to daily life mathematics* like estimation and rounding of numbers. The Nordic and the English-speaking groups also scored relatively low on items dealing with more classical abstract mathematics like fractions and algebra.

This type of analyses has later been conducted on data from TIMSS 2003, PISA 2003, TIMSS Advanced 2008, and TEDS-M 2008 (IEA 2017a, b; OECD 2017). The analyses have been conducted over an extended period of time, and there have been different countries participating in the studies which also influence the result. Nevertheless, *all these analyses* have concluded that it seems to be consistent patterns of countries clustering together in a Nordic group, an English-speaking group, an East European group, and an East Asian group (Grønmo et al. 2004; Olsen and Grønmo 2006; Grønmo and Olsen 2006a, b; Grønmo and Pedersen 2017; Blömeke et al. 2013) for mathematics in school. The analyses have been conducted at different levels in school, from lower secondary through upper secondary and even at the teacher education level of mathematics. The studies contributing to data for such analyses have a quite different framework for their testing, and different types of items for testing students' competence. PISA tests students' ability to solve problems presented mostly in some type of daily life context—or some in a more professional context (Wu 2009; Olsen and Grønmo 2006). The context is described with text, tables, and requires quite some reading, it also requires ability to relate and understand different types of information, before students use some mathematics to answer one or more questions. PISA does not have any items testing students' competence in pure algebra (ibid.). TIMSS in lower secondary school has items testing students in pure algebra, and items where algebra is tested in context, but less demanding when it comes to reading than items in PISA. TIMSS Advanced test students in a number of items in pure algebra, and some in context, but complexity in this study is in the mathematical domain, not in reading, as it is in PISA. TEDS-M test students' to become teachers in their understanding of pure algebra, in addition to also testing them from the perspectives of how to teach algebra in school (IEA 2017b). However, all these analyses give consistent results pointing out that it is meaningful to conclude that we have four different profiles in mathematics education that seem to be stable over time, at different levels in education and in different studies independent of the study framework or way of formulating the items of the tests.

Although the analyses reveal four different types of profiles, it is also meaningful to talk about two very different *types of profile*, a conclusion especially interesting from the *perspective of algebra in school*. One type of profile consists of East Asia and East Europe, the other type consists of the English-speaking and the Nordic countries. To summarise, even if there are *distinct differences between each of the four profiles*, we also find clear similarities between the two groups of countries we have linked as belonging to the *same type of profile*. The East Asian and the East European profiles are quite similar in the sense that both groups perform relatively better in traditional mathematical content areas like algebra than in mathematics more closely related to daily life such as data representation and probability. In the same way do the English-speaking and the Nordic profiles reveal similarities, both these groups of countries perform relatively better on data representation and probability and relatively worse in algebra.

The consistent difference between the two types of profiles according to algebra, based on a number of cluster analysis of data from TIMSS, TIMSS Advanced,

PISA and TEDS (IEA 2017a, b; OECD 2017) forms an important basis for discussions and conclusions in this paper.

Conclusions about differences according to *emphasis on algebra in different countries* are also supported by other types of analyses. Items in TIMSS, TIMSS Advanced and PISA have been re-categorized according to requirement of algebraic manipulation or not to be solved, then compared with students' success in different countries in solving the items (Hole et al. 2015, 2017).

11.3   Equal Rights to Education

It is interesting from several perspectives that some groups of countries like the Nordic and the English-speaking countries, emphasis teaching and learning of algebra less than other group of countries like East-Asian and East-European countries. It is reasonable to discuss possible consequences of such priorities in school, both for individual students as well as for society at large. If some countries do not give their students the opportunities to achieve the type of competence they need to be successful in todays modern societies, students from these countries will have disadvantages compared to students from other countries where this type of competence is achieved. It may also influence students' possibilities within each country, because if this type of competence is not achieved in school, students with highly educated parents, or economically well suited students may get some help outside school, and for that reason have a much better position to pursue a number of educations and important professions (Grønmo 2015).

Questions about what type of competence and knowledge students need in their life, daily life and in their professional working life; and questions about how the school can give them this type of competence seems therefore highly relevant. Algebra is not likely to be needed that much in daily life, but in a modern highly developed technological society it might be essential for students' possibilities for further education and for getting the job they want. It can also be argued that algebraic competency underpins higher level abstract reasoning, especially when it involves unknowns and generalised relationships. On this basis, algebraic competency is necessary or highly desirable for professional occupations including medicine, management and administrative occupations. In the PISA-study in 2015, 29% of the Norwegian students answered that they saw themselves at the age of 30 having a job categorized as based on some competence in technology, natural science or mathematics (Kjærnsli and Jensen 2016), while only 20% answered the same in 2006. According to the changes seen in our societies, it seems realistic that a high number of people will have these types of professions. OECD have pointed out an increasing need for more education related to science, technology and mathematics to give people a fair opportunity in our modern societies (OECD 2017). They also make strategy reports for a number of countries all over the world, about how to improve their educational system. In the Norwegian report, more competence in mathematics is pointed out as being of essential importance (OECD 2014).

The important issue discussed in this paper is that many professions require some type of competence in mathematics, and especially in the mathematical language algebra, since all these types of professions in one way or another use this mathematical language as a tool. This is true for educations and professions as engineering, economy, computer science, and natural sciences as physics, chemistry or biology. The responsibility for teaching students the algebra they need for further education and professions lies with the school. If this is not provided in school, it will influence student's possibilities to pursue a number of educations based on their home background (Grønmo 2015). This is not in accordance with the goal of equity for access to educations and later professional work that are a main goal in education in so many countries (Ibid.).

The school's responsibility for providing this type of knowledge to their students' is therefore closely related to students' equal rights to education in a changing world, and we have to take into account the direction of development in the society (OECD 2017). There is an ongoing discussion about the need for people with creativity and competence in how to handle changes in many countries including Norway. On the other hand, there seems to be less discussion about the need to emphasise students' learning of basic knowledge in the mathematical language algebra, needed in so many professions. I will argue that basic knowledge in algebra is more to be seen as complementary and necessary for being creative, rather than something opposing creativity. This may not be true for all types of societies, but at least for the highly developed technological society we have in many countries today. Without the language to develop technology and science, creativity is probably not very helpful. Algebra was probably not that important for so many fifty years ago as it is today. But taken into account the changes and challenges we are facing in a modern society (Ibid.), competence in algebra is essential and for that reason also an issue of importance from the perspectives of giving all students the possibility to pursue the education and job positions they want.

Failing to educate students to gain some fluency in the language of algebra may, for obvious reasons, have important consequences for the society, such as the shortage of people in a lot of professions and jobs. But the consequences for each individual student lacking this type of competence are no less serious. And it is especially thought provoking, that the low emphasis on the language algebra is most pronounced in the Nordic countries, well known for their emphasis on equal right for all citizens.

> During the last 50 years, the Nordic welfare state has been established as a unique model, with strong emphasis on equity of access to education of high level of quality. (Yang Hansen et al. 2014, p. 26)

The goal of equity of access to education of high quality may be more pronounced in the Nordic countries than in other countries, but many other countries around the world probably also support this goal.

Important characteristics of the Nordic welfare state, especially after the Second World War, have been free access to education and social mobility. After the war, education has in general been free of charge, at all levels, including college and university level. This has given people from all social classes in the society the possibility of pursuing all types of education. This means that it is no longer that important what type of background you have, economically or intellectually, and it has resulted in a large number of children from working class- and farmer class families having opportunities no one in their family ever had before. Based on the type of education people took, it was possible for people to join more or less any type of profession, and through this gain an influence in the society that their parents and grandparent could only dream of. This social mobility is probably also an important reason for the political stability seen in the Nordic countries today (Grønmo 2015).

But it is difficult to give equal right for all to educations and professions in the society, unless the school takes into account the competence needed for further education and professions today and in the future. We know that technological, economical, and natural science competencies are important today, and they are likely to be even more important in the future (OECD 2017). We also know that the mathematical language algebra is what students need to pursue many types of educations and professions in the society. Today, the Nordic countries seem to put a lot of effort into learning students *to use technology*, but little effort it seems to giving them the tools they need to be actively involved in *developing new technology*. This needs to be reflected upon from the perspective of students' equal right to education, and also from the perspective of social mobility and political stability. Also English-speaking countries seem to face problems by not emphasizing algebra throughout their school system, even if not to the same extent as in the Nordic countries.

In addition, since algebra is the language of generalization and the language of the relationships between quantities (Usiskin 1995) it is the basis for higher level abstract reasoning needed in all professions involving managerial decision making. So failing to develop algebraic competency among learners is denying them access to many occupations beyond science and technology, and is thus an obstacle to the human right of social mobility.

11.4   Reasons for Low Emphasis on Algebra

Analyses of different profiles in mathematics education around the world show notable differences between countries in how much emphasis is put on algebra. A consistent result of these analyses is that the Nordic and the English-speaking countries do not emphasize students' learning of pure mathematics as algebra, in opposition to countries in East Europe and East Asia (Olsen and Grønmo 2006; Grønmo and Olsen 2006a, b; Grønmo 2010). Why this is the case is an interesting question, probably related to development of curricula within individual countries, which again is influenced by the country's culture for school and education. Equal rights to education have been an important force for the development of school in many countries, and especially in the Nordic countries. Slogans as 'mathematics for all' have been part of this drive, as has the need for teaching all students the type of mathematics they need in daily life so as to be responsible, active citizens. The need for daily life mathematics has influenced discussion about content in mathematics in school. In particular, international comparative study PISA has highlighted the need for students to be able to use mathematics from such a perspective.

Such a daily life perspective is important, but it is also interesting to see to what extent this perspective has been especially influential in the Nordic and the English-speaking countries in opposition to countries in East Europe and East Asia. Based on data from PISA 2003 Olsen and Grønmo (2006) developed a classification system for analyzing this. All items in PISA were re-classified according to how close they were to "real world" or "daily life" mathematics as a way to further understand of the differences between the different profiles found in mathematics education. Their findings revealed that the profiles of the Nordic and the English-speaking countries were mainly accounted for by this variable, and that this variable had a higher degree of explanatory power than the aspects described in the framework of PISA when it came to understanding the clustering of countries in different profiles. The profile for the Nordic countries was strongly characterized by relatively high performance on items involving some sort of real world mathematics, and the same was true for the English-speaking countries. The East Asian and East European countries, however, achieved relatively lower on items categorized as some sort of real world mathematics. This result, consistent with Grønmo et al.'s (2004) findings, made it reasonable to conclude that real-world mathematics has been *a driving force* for school mathematics in the Nordic and in the English-speaking countries, in contrast to countries in East Europe and East Asia. Other researchers have also pointed out that an emphasis on everyday applications of mathematics has been an important driving force underlying changes in curriculum over the last decades (Mosvold 2009). The needs of mathematics for pupils in their daily lives have received more curricular attention than before, while more formal aspects of mathematics, such as algebra, have been reduced. From the mid-1980s, there has been a lot of discussion about the tendency to give more attention to daily life mathematics; see for example De Lange (1996) and Kilpatrick et al. (2005).

The findings presented here are consistent with those of other researchers in mathematics education, who have suggested that the mathematics curricula in the Nordic countries, as well as in the English-speaking countries, have been heavily influenced by an emphasis on real world mathematics and a daily life perspective on mathematics in compulsory school (Niss 1996; De Lange 1996; Gardiner 2004).

Olsen and Grønmo (2006) also found that there was a tendency for the pupils in the Nordic countries to perform relatively better on easier items, and with a non-significant tendency to achieve lower on items requiring accuracy in calculations. For the English-speaking countries the need for calculation was a significant factor indicating low relative achievement. This suggests that accuracy in calculations is not seen as that important in the Nordic and the English-speaking countries. To what extent an increased focus on daily life mathematics in Norway over the last decades has resulted in little attention being given to accuracy in calculation has also been discussed in several articles based on data from PISA and TIMSS (Bergem et al. 2005; Grønmo 2005; Grønmo and Olsen 2006a, b; Olsen and Grønmo 2006). It seems that it is not only algebra that is not emphasizes in some countries, but also emphasis on accuracy in calculation in arithmetic's seems to have been low.

Olsen and Grønmo (2006) concluded that on average the pupils in the Nordic countries performed relatively better on items with a realistic context, on items which included some sort of graphical material, on low difficulty items, and on items which did not include explicit algebraic expressions. Pupils in the Nordic countries also performed relatively better on items relating to probabilities and statistics in a daily life context (classified as Uncertainty in the PISA framework), and on items that tended to be of a more qualitative type which did not require any accuracy in performance of calculations. These results may also influence students' possibilities to learn algebra; since algebra may be seen as some type of generalization of arithmetic. Students' lack of competence in arithmetic is therefore likely to have a negative effect on students' learning of algebra (Brekke et al. 2000).

It is not only the content of curriculum that is important; it might be also the organization of the curriculum that might be problematic. In Norway are for example goals in the curriculum for grade 1 to grade 10 organized in three year blocs, while it is the teacher, the local school or community who decide what is to be emphasized each year (Utdanningsdirektoratet 2006). This seems to be one reason for why algebra is not emphasized in Norway. Algebra seems to be taken late, with teachers referring to algebra as abstract and therefore difficult to teach and difficult for students to learn. It might be abstract, which is part of why it is so powerful and useful in solving problems in so many different situations. It might also be argued in opposition to this, that since it is abstract, it is important to start learning it early so it can mature over time. Ordinary language, at least in written form, is also abstract; nevertheless, we start early in school so all shall be able to learn it.

Another important issue when it comes to what is emphasised in school is teacher education. The international comparative study TEDS-M 2008 of teacher education showed that the Nordic and the English-speaking countries do not emphasis algebra for their student teachers (Tatto et al. 2012; Grønmo and Onstad 2012; Blömeke et al. 2013). If teachers do not feel they are very competent in algebra themselves, it is understandable why they do not emphasis this mathematical content very much, or at least postpone it as long as possible according to what has to be done according to the curriculum.

You do not need very much algebra in daily life, but you do need it in many educations and professions. Researchers in Finland have warned about the problematic issue that even though Finland is a high achieving country in PISA, this does not reveal the whole truth about mathematics in their schools. Since PISA emphasis daily life mathematics, and not the type of mathematics needed for further education and professions, this study does not give the total picture of mathematics in their school.

> This conflict can be explained by pointing out that the PISA survey measured only everyday mathematical knowledge, something which could be - and in the English version of the survey report explicitly is - called "mathematical literacy"; the kind of mathematics which is needed in high-school or vocational studies was not part of the survey. No doubt, everyday mathematical skills are valuable, but by no means enough. (Astala et al. 2005)

A central question is therefore how to find a balance between what students need in their daily life and their needs to pursue further education and professions in the society. Equally important is a question about how mathematics for daily life and professions best can be implemented in school, taking all levels in school into account. These questions are by no means easy to answer, but they are far too important not to be asked and reflected upon.

11.5   Pure and Applied Mathematics

The discussions about what should constitute mathematics in compulsory school may be understood in the light of the considerable efforts and use of resources to develop education for all citizens in Western societies (Ernest 1991; Skovsmose 1994). The relationship between pure and applied mathematics has been part of these discussions, even if it has been argued that the distinction between pure and applied mathematics may not be very well founded from a historical point of view. Some of the main contributors in mathematics, as Newton, Fermat, Descartes, and Gauss among others, would probably not have recognised the distinction being made today between pure and applied mathematics—indicating that mathematics should be taught as a whole (Kline 1972). However, in a discussion about what should be the content of mathematics in school, this distinction does seem to be relevant and fruitful, as illustrated in the former analyses of different profiles in mathematics education. It also seems relevant in discussions of curriculum and curriculum changes for mathematics. As some have argued, an increasing focus on applied mathematics seems to have resulted in too little attention given to what we may call pure mathematics. Gardiner (2004) has argued that to apply mathematics you need some competence in traditional pure mathematics, and that it is a misunderstanding that teaching applied mathematics is *an alternative* to teaching pure mathematics, even if some seem to believe that. Grønmo (2005) and Grønmo and Olsen (2006a, b) also pointed to problems created by underestimating the importance of pure mathematics and that only emphasizing applied mathematics may be one possible reason for the low performance of Norwegian pupils in studies as TIMSS and PISA, especially on items involving algebra.

Figure 11.1 presents a commonly accepted model of the relationship between pure and applied mathematics taken from an influential United States policy document on standards in mathematics (National Council of Teachers of Mathematics 1989). PISA uses a slightly different form of this model (OECD 2003, p. 38). The right-hand side of the figure represents the mathematical world (what we may refer to as pure mathematics)—an abstract world with well-defined symbols and rules.

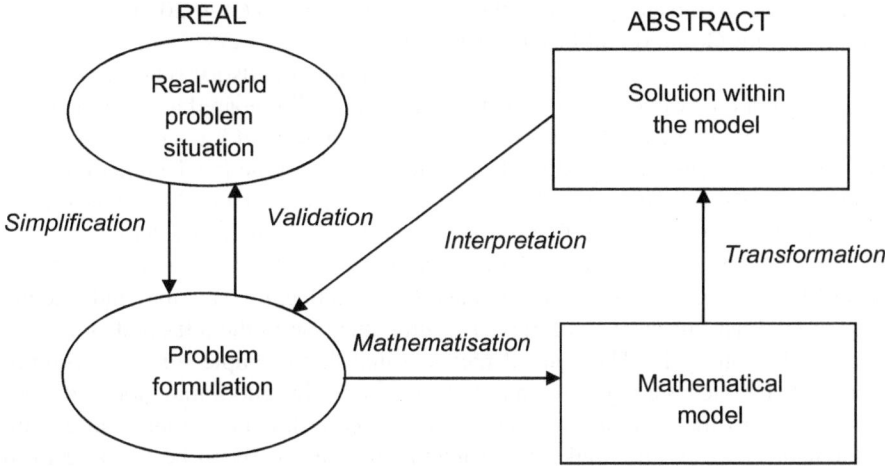

Fig. 11.1 The mathematisation cycle. *Source* National Council of Teachers of Mathematics (1989)

The left-hand side represents the real, concrete world, containing an infinite number of different contexts and situations. The context or situation presented may either be scientific or what might be called daily life. Working with pure mathematics, such as numbers or algebra out of any context, means working only on the right-hand side of the model. In applied mathematics, the starting point is intended to be a problem from the real world, which first has to be simplified, and then mathematized into a model representing the problem. School mathematics rarely starts with a real problem. What is presented as a problem for pupils has in almost every case already been simplified to make it accessible to them.

For any type of applied mathematics, the pupils need to have some knowledge of pure mathematics to find a correct mathematical solution. Applied mathematics can therefore be seen as more complex than pure mathematics, if the same mathematics is involved in the two cases. Gardiner (2004) argues extensively that even if the ability to use mathematics to solve daily life problems is a main goal for school mathematics, this cannot be seen as an alternative to basic knowledge and skills in pure mathematics. It may rather underline the pupils' need for being able to orient them in the world of pure mathematics as a necessary prerequisite to solving real world problems.

PISA aims at embedding all items in a context as close to a real-life situation as possible, while most items in TIMSS are pure mathematical items with no context, or items with a simplified context, as has long been the tradition in school mathematics. TIMSS therefore gives extensive information about pupils' knowledge in pure mathematics—or what may be called traditional school mathematics—while PISA mainly displays pupils' competence in solving items in a daily life context with the use of some mathematical knowledge—what may be referred to as applied mathematics, and usually in a rather complex context (Wu 2009; Olsen and Grønmo 2006; Grønmo and Olsen 2006a, b).

Many countries have as a goal that, on leaving compulsory school, all pupils should have a type of competence that makes them well prepared to solve daily life problems using mathematics. This has been seen as important for active citizens in a modern society, and has by some been referred to as functional numeracy (Niss 1994, 2003; De Lange 1996). The aim of PISA is to test pupils in this type of mathematical competence, defined in the study as Mathematical Literacy.

Countries representing the East European profile performed relatively better in TIMSS than in PISA in 2003 (Grønmo and Olsen 2006a, b). This may indicate that most of the East European countries give little attention to the left-hand side of the mathematisation cycle. The general message that this example serves to communicate is that concentrating only on pure mathematics in school may not be the best if the aim is to foster pupils who are mathematically literate, pupils who can use mathematics to solve the daily life problems they are likely to be exposed to. In contrast to the East European countries, countries representing the East Asian profile as for example Japan, are high achieving in both TIMSS and PISA. This may indicate that pure mathematics is emphasized in the mathematics curriculum in East Asian countries as for example Japan, at the same time as attention is given to the full cycle of applied mathematics.

A European country such as the Netherlands, also high-achieving in both studies in 2003, revealed some clear differences from Japan on performance levels in different topics in TIMSS. Comparing achievement in Japan and the Netherlands in Grade 8, the countries achieved equally well in the topics number, measurement, and data, while there were clear differences between these countries in their achievement levels in algebra and geometry. This indicates that even high-achieving countries may have pronounced differences in what they emphasise in their curriculum. Algebra and geometry seem to be *much more in focus in Japan* than in the Netherlands. But when it comes to achieving well in mathematical literacy, as tested in PISA, the Netherlands is doing just as well as Japan. Grønmo and Olsen (2006a, b) took this as an indication that the "basics" of most importance for daily life mathematics, are the fundamental concepts of number and operations with numbers.

The achievement in algebra has been low for a long time in Norway (IEA 2017a), which have been pointed out in several national report based on data from TIMSS (Grønmo and Onstad 2009; Grønmo et al. 2010, 2012; as well as in articles based on TIMSS studies (Grønmo 2010; Grønmo and Onstad 2013a, b). It is also worth noticing that despite the fact that Norway measured a general improvement in mathematics in grade 8 in TIMSS from 2011 to TIMSS 2015, there was a significant decrease in achievement in algebra (Bergem et al. 2016).

Even if algebra is not the most important content for applying mathematics to solve daily life problems, this content knowledge is highly relevant for those going into studies and professions in need of more mathematical competence. The conclusions pointed out in the discussion of Fig. 11.1 about the need for basic knowledge and skills to be able to apply mathematics is just as relevant for applying algebra as it is for applying number in mathematics problem solving. This aspect received more attention in the TIMSS 2007 report (Grønmo and Onstad 2009), referring to the problems in Norway in recruiting pupils to educational programs and professions requiring knowledge in algebra. In connection to this, the report also posed a critical question about what the Norwegian compulsory school offers to their most talented pupils in mathematics, the pupils who are most likely to be recruited to studies and professions in need of this type of mathematical knowledge (Grønmo et al. 2014).

One consequence of a growing focus on applied mathematics may be that problems arise if too little attention is given to pure mathematics. If pupils lack elementary knowledge and skills with numbers, this is important also for their *possibility to learn algebra*. It has been pointed out that problems pupils have learning algebra in many cases are caused by a *too weak basis in arithmetic* (Brekke et al. 2000). And as already underlined, if talented pupils are not given the opportunity to learn basic concepts and skills in algebra, it will probably lead to later problems in recruiting them to studies and professions in need of such knowledge (Grønmo et al. 2016).

11.6   How to Learn the Mathematical Language Algebra

There are different ways of learning a language, but to be fluent in a language, to be able to use it in a lot of different contexts, a good way is to experience using it in a lot of different situations, and to give students the opportunities to mature their competence over time. To learn algebra, as other languages, we have to take this into account. This indicates that since learning of algebra is essential for so many students in our societies, it is necessary to reflect this throughout the school system. Algebra is not only relevant for what we teach students in lower or upper secondary school, the basis for this language, as with spoken and written languages is laid much earlier. This paper has already pointed to how learning of arithmetic is an important basis for learning algebra (Brekke et al. 2000). But it is also necessary to discuss what part of algebra can be implemented even at lower levels in school.

The curriculum in a Nordic country as Norway put much more emphasize on algebra in the sixties and seventies than today. Textbooks showed that students were exposed to letters as X for a variable number already in grade 1 in the 50s and 60s, for students at the age of seven. The Norwegian curriculum from 1974 indicated that students should start learning about variables in a simple setting from the start of school, and that there should be a special focus on elementary algebra from grades 4–6, with consolidation and expansion of it in grade 7–9. This was a period that is referred to as modern mathematics, with a lot of emphasis on abstract mathematics at all levels in school. It is not a good solution to go back and copy this, because it probably went too far in the abstract and formal direction. But it seems troublesome that after that time there has been a long period with the opposite problem, too little emphasis on pure abstract mathematics as arithmetic and especially on algebra. Education, as many other things in life, is probably more about a balance between different aspects and goals. Some have referred to changes in school as jumping from one ditch to the other, or following the swing of a pendulum (Ernest 1989), but never finding the best balance between all the different and sometimes contradictory goals schools are supposed to handle.

If you want to improve something, a common way is to look at and learn from those who are good at it. However, the Nordic countries seems to have a strong tradition for mostly looking to each other since they share common values and ideology based on geography, culture and history (Bergem et al. 2016; Grønmo et al. 2016). For good reasons, the Nordic countries have pointed out that their societies are stable with a good social system for all citizens. But this is not the same as the Nordic countries being best in everything, nor a good reason for not looking to other countries in the world for improvement.

To do well in algebra, it seems more reasonable to look at some of the East-Asian countries, like Japan, Singapore or Hong Kong. These countries have a very different culture, but nevertheless, we can look at what they are doing, and pick up ideas about what is good and what is not, even if our cultural background is very different. Cultural similarities can be an advantage, but also a disadvantage if taken too far.

A country like Japan has, for example, produced some very interesting videos about their way of teaching algebra to middle school students (TIMSSVIDEO 1999). In this video, they use the differences between students' competences as a resource in their teaching, not a problem, activating their students in interesting discussions. It is well documented that East-Asian countries perform much better in algebra than Nordic and English-speaking countries. We have master students comparing textbooks in Norway and Singapore, especially looking at how they start to teach algebra (Karimzdeh 2014).

We also need to discuss at what level in school we should start teaching algebra to our students. Some countries start teaching formal algebra at early levels in school, while other countries prefer to formalize algebra much later. As already mentioned, it is also interesting to look back to the curriculum in our countries some decades ago, where algebra was also emphasized more than in our present curriculum. This was before the emphasis on daily life mathematics was supposed to, more or less; solve the problems in school mathematics (Gardiner 2004).

The school curriculum has to meet the needs of all students. Not only for those who will need algebra for an academic career, but also for those only needing daily life mathematics. This is the challenge we have to meet in our societies, based on students' needs and how to meet them in compulsory school, as well as in lower and upper secondary school. One source of answers is to look at world wide practices. Looking at countries with different types of educational systems and different cultures, to learn, to discuss and reflect upon their methods to find ways of improving our own system. We can probably learn from everybody, from different countries all over the world, at the same time as we keep in mind that just copying anybody is not the way to do it.

11.7   Summary and Further Research

As pointed out earlier in this paper (ref Sect. 11.5) numbers and algebra constitute important parts of pure mathematics. Working with numbers and algebra out of any context are pure mathematics, while using numbers or algebra to solve a problem in daily life or in professional life are applied mathematics. It is also underlined that to apply any type of mathematics, the students need to be competent in the type of pure mathematics they are supposed to apply (Gardiner 2004). It is especially for their professional life, higher education and work, students need algebra. To little emphasize on pure mathematics as algebra in school is therefore problematic from the point of view to give all students' equal opportunities to pursue the education or professions they want.

A number of analyses based on data from several international large-scale studies have shown that there are great differences between countries when it comes to the role of algebra in school mathematics. While countries in East Asia and East Europe emphasize students' learning of pure mathematics as algebra, this is not the case in the Nordic or English speaking countries. This is likely to interfere with students' rights and opportunities to pursue the education they want, which is especially thought provoking given the common consensus, especially in the Nordic countries, when it comes to students' equal right to education at all levels in school. The fact that education in general is free of charge is important characteristic of the Nordic welfare state. This is based on a consensus about the importance of an equal rights to education for all citizens and at all educational levels.

To what extent a country offers students the opportunity to learn algebra may also interfere with the society's possibilities to recruit people into a number of professions needed in a modern society. This is the case for professions in technology as engineering and computer science, in natural sciences as physics, chemistry, biology, and in studies of economy. Educations or professions using mathematics as a tool need students with basic competence in the mathematical language algebra. We also have to take into account that this type of competence is likely to be more important as our society become more dependent of technology, to solve problems related to environmental problems or problems in economics. It has also been suggested that all higher level professionals need the high level thinking skills that rest on a basic understanding of algebra. Countries can, as many already do, hire people from other countries if their school system does not educate enough people with the competence needed. But not educating the working force needed in a country is not a good solution, neither for individuals in the country, or for the country as a whole. Especially for rich countries spending a lot of resources on education, this seems to be an unsatisfactory or bad solution.

This paper has referred to a number of analyses drawing the same conclusions about the problems arising from the low emphasis algebra in several countries, and also pointed out some problems and perspectives to be discussed in solving this. Also some solutions such as looking to other countries that are successful in teaching and learning algebra have been mentioned. However, we need more research on how this can be done in a good way. More research projects comparing countries with different profiles in mathematics education seems to be one possibility, as a way to learn from each other. Copying other countries ways of dealing with algebra may be tempting, but probably not the best way forward. In a complex field like education, differences in culture are likely to play a major role we must take into account if we want to improve our own educational system. Cooperation and discussions between countries with different educational systems and cultures is more likely to be successful, especially if we include it as part of research about education. Earlier in this paper, reference was made to a tendency in the Nordic countries to mostly compare and discuss our systems in relation to other countries close to us, geographically or culturally. This is probably true for many countries in the world. But just as our societies become more dependent on all types of technology, countries in the world will also be more dependent on cooperation on a broader range, in education as in other fields, than in former times. There have been more references lately to the fact that we share one world together, and that we have to solve and take care of it together, especially when it comes to environmental or economical problems. But this may be just as true when discussing education all over the world, an important factor for solving most problems we are facing.

This paper is heavily based on a number of secondary analyses of data from different type of international comparative studies, as TIMSS, PISA, TIMSS Advanced and TEDS-M. To be able to conduct such type of analyses it is important that countries participate in several of these studies. No study, no matter how good the quality of the study is, can give the best answer on how to improve a country's educational system. All the studies referred to in this paper have quite different frameworks for what they want to test, as well as quite different ways of developing items to cover their frameworks. For this paper, the differences between TIMSS and PISA have played an important role, and giving researchers the possibility to ask and answer more questions about the role of algebra and daily life mathematics than any of these studies could answer alone. Participation in both TIMSS for compulsory school and TIMSS Advanced last year of upper secondary school has also given countries the possibility to see how what is emphasized at one level in school seems to influence other levels in school. In that way, a country will get a much broader view of their own educational system, and better information about how to improve education in their country.

Researchers in Finland have warned about the problem Finland is facing because they mostly only participated in PISA:

> A proper mathematical basis is needed especially in technical and scientific, biology included. The PISA survey tells very little about this basis, which should already be created in comprehensive school. Therefore, it would be absolutely necessary that, in the future, Finland would participate also in international surveys which evaluate mathematical skills essential for further studies. (Astala et al. 2005, https://matematiikkalehtisolmu.fi/2005/erik/PisaEng.html)

International comparative studies have produced a large databank that can be used to answer a high number of interesting research questions. Especially by combining analyses of data from different international comparative studies will give researchers unique possibilities for better answers to their research questions. This type of further research needs to be emphasized to improve education all over the world, which is also likely to influence our chances of solving the problems we are facing today and will face in the future.

References

Astala, K., Kivelä, S. K., Koskela, P., Martio, O., Näätänen, M., & Tarvainen, K. (2005). polytechnics, m. t. i. u. a. *The PISA survey tells only a partial truth of Finnish children's mathematical skills*. Retrieved from https://matematiikkalehtisolmu.fi/2005/erik/PisaEng.html.

Bergem, O. C., Grønmo, L. S., & Olsen, R. V. (2005). PISA 2003 og TIMSS 2003: Hva forteller disse undersøkelsene om norske elevers kunnskaper og ferdigheter i matematikk. *Norsk Pedagogisk Tidsskrift, 89*(1), 31–44.

Bergem, O. K., Kaarstein, H., & Nilsen, T. (2016). *Vi kan lykkes i realfag. Resultater og analyser fra TIMSS 2015*. Oslo: Universitetsforlaget.

Blömeke, S., Suhl, U., & Döhrmann, M. (2013). Assessing strengths and weaknesses of teacher knowledge in Asia, Eastern Europe and Western countries: Differential item functioning in TEDS-M. *International Journal of Science and Mathematics Education, 11*, 795–817.

Brekke, G., Grønmo, L. S., & Rosén, B. (2000). *KIM (Kvalitet i matematikkundervisningen): Veiledning til algebra* [Quality in mathematics teaching: A guide in algebra]. Oslo: Nasjonalt Læremiddelsenter.

De Lange, J. (1996). Using and applying mathematics in education. In A. J. Bishop, K. Clements, C. Keitel, J. Kilpatrick, & C. Laborde (Eds.), *International handbook of mathematics education* (Vols. 1–2). Dordrecht: Kluwer Academic Publishers.

Ernest, P. (1989). *Mathematics teaching: The state of the art*. London: Falmer Press.

Ernest, P. (1991). *The philosophy of mathematics education*. London: Falmer Press.

Gardiner, A. (2004). *What is mathematical literacy?* Lecture given at the ICME-10 conference, Copenhagen, Denmark, July 2004.

Grønmo, L. S. (2005). Matematikkprestasjoner i TIMSS og PISA [Mathematics achievement in TIMSS and PISA]. *Nämnaren, 32*(3), 5–11.

Grønmo, L. S. (2010). Low achievement in mathematics in compulsory school as evidenced by TIMSS and PISA. In B. Sriraman, C. Bergsten, S. Goodchild, G. Pálsdóttir, B. Dahl, & L. Haapasalo (Eds.), *The first sourcebook on Nordic research in mathematics education* (pp. 49–69). Charlotte, NC: Information Age Publishing.

Grønmo, L. S. (2015). *Cómo alcanzar la equidad de acceso a la educación un perfil nórdico en matemáticas* [How to achieve equity of access to education. A Nordic profile in mathematics]. Paper presented at PEDAGOGIA 2015, Havana, Cuba.

Grønmo, L. S., Hole, A., & Onstad, T. (2016). *Ett skritt fram og ett tilbake. TIMSS advanced 2015. Matematikk og fysikk i videregående skole*. Oslo: Cappelen Damm Akademisk.

Grønmo, L. S., Jahr, E., Skogen, K., & Wistedt, I. (2014). *Matematikktalenter i skolen-hva med dem?* [How do we threat student with a talent for mathematics?] Oslo: Cappelen Damm.

Grønmo, L. S., Kjærnsli, M., & Lie, S. (2004). *Looking for cultural and geographical factors in patterns of responses to TIMSS items*. Paper presented at the 1st IEA International Research Conference, May 11–13, 2004, Lefkosia, Cyprus.

Grønmo, L. S., & Olsen, R. V. (2006a). Matematikkprestasjoner i TIMSS og PISA: ren og anvendt matematikk [Mathematic achievement in TIMSS and Pisa: Pure and applied mathematics]. In B. Brock-Utne & L. Bøyesen (Eds.), *Å greie seg i utdanningssystemet i nord og sør*. Bergen: Fagbokforlaget.

Grønmo, L. S. & Olsen, R. V. (2006b). TIMSS versus PISA: The case of pure and applied mathematics. In *Proceedings at the 2nd IEA International Research Conference*, Washington, DC.

Grønmo, L. S., & Onstad, T. (2009). *Tegn til bedring. Norske elevers prestasjoner i matematikk og naturfag i TIMSS 2007* [Signs of improvement]. National report from TIMSS 2007. Oslo: Unipub.

Grønmo, L. S., & Onstad, T. (Eds.). (2012). *Mange og store utfordringer* [Many great challenges]. National report from TEDS-M 2008. Oslo: Unipub.

Grønmo, L. S., & Onstad, T. (Eds.). (2013a). *Opptur og nedtur. Analyser av TIMSS-data for Norge og Sverige*. Oslo: Akademika forlag.

Grønmo, L. S., & Onstad, T. (Eds.). (2013b). *The significance of TIMSS and TIMSS advanced. Mathematics education in Norway, Slovenia and Sweden*. Oslo: Akademica Publishing.

Grønmo, L. S., Onstad, T., Nilsen, T., Hole, A., Aslaksen, H., & Borge, I. C. (2012). *Framgang, men langt fram. Norske elevers prestasjoner i matematikk og naturfag i TIMSS 2011*. Oslo: Akademika forlag.

Grønmo, L. S., Onstad, T., & Pedersen, I. F. (2010). *Matematikk i motvind* [Mathematics against headwinds]. National report from TIMSS advanced 2008. Oslo: Unipub.

Grønmo, L. S., & Pedersen, I. F. (2017). *Do analyses of TIMSS advanced data confirm that countries have a similar cultural profile in mathematics at all levels in school?*

Hole, A., Grønmo, L. S., & Onstad, T. (2017). *Measuring the amount of mathematical theory needed to solve test items in TIMSS advanced mathematics and physics*. Paper presented at the 7th IEA International Research Conference, June 28–30, 2017, Prague, Czech Republic.

Hole, A., Onstad, T., Grønmo, L. S., Nilsen, T., Nortvedt, G. A., & Braeken, J. (2015). *Investigating mathematical theory needed to solve TIMSS and PISA mathematics test items*. Paper presented at the 6th IEA International Research Conference, June 24–26, 2015, Cape Town, South Africa.

IEA. (2017a). *TIMSS and PIRLS International Study Center*. US: Lynch School of Education, Boston College. https://timssandpirls.bc.edu. Retrieved May 15, 2017.

IEA. (2017b). *TEDS-M. Teacher Education and Development Study in Mathematics*. https://arc.uchicago.edu/reese/projects/teacher-education-and-development-study-mathematics-teds-m. Retrieved May 15, 2017.

Karimzdeh, A. (2014). *Algebra i norske og singaporske matematikklaereboker (Comparing algebra in Norwegian and Singaporian texstbooks)* (Master thesis). ILS, University of Oslo.

Kilpatrick, J., Hoyles, C., Skovsmose, O., & Valero, P. (2005). *Meaning in mathematics education*. New York: Springer.

Kjærnsli, M., & Jensen, F. (Eds.). (2016). *Stø kurs. Norske elevers kompetanse i naturfag, matematikk og lesing i PISA 2015*. Oslo: Universitetsforlaget.

Kline, M. (1972). *Mathematical thought from ancient to modern times*. Oxford: Oxford University Press.

Mosvold, R. (2009). Teachers' use of projects and textbook tasks to connect mathematics with everyday life. In B. Sriraman, C. Bergsten, S. Goodchild, G. Pálsdóttir, B. Dahl, & L. Haapasalo (Eds.), *The first sourcebook on Nordic research in mathematics education* (pp. 169–180). Charlotte, NC: Information Age Publishing.

NCTM. (1989). *Curriculum and evaluation standards for school mathematics*. Reston, VA: National Council of Teachers of Mathematics.

Niss, M. (1994). Mathematics in society. In R. Biehler, R. W. Scholz, R. Straesser, & B. Winkelmann (Eds.), *The didactics of mathematics as a scientific discipline*. Dordrecht: Kluwer Academic Publishers.

Niss, M. (1996). Goals of mathematics teaching. In A. J. Bishop, K. Clements, C. Keitel, J. Kilpatrick, & C. Laborde (Eds.), *International handbook of mathematics education* (Vols. 1–2). Dordrecht: Kluwer Academic Publishers.

Niss, M. (2003). Mål for matematikkundervisningen [Goals of mathematics teaching]. In Grevholm, B. (Ed.), *Matematikk for skolen*. Bergen: Fagbokforlaget.

OECD. (2003). *PISA 2003 assessment framework. mathematics, reading, science and problem solving. Knowledge and skills*. Paris: OECD Publications.

OECD. (2014). *OECD Skills strategy action report* Norway. Retrieved from http://www.oecd.org/skills/nationalskillsstrategies/OECD_Skills_Strategy_Action_Report_Norway.pdf

OECD. (2017). *PISA, Programme for International Students Assessment*. Accessible at http://www.oecd.org/pisa/publications/. Retrieved May 15, 2017.

Olsen, R. V. (2006). A Nordic profile of mathematics achievement: Myth or reality? In J. Mejding & A. Roe (Eds.), *Northern lights on PISA 2003—A reflection from the Nordic countries*. Oslo: Nordisk Ministerråd.

Olsen, R. V., & Grønmo, L. S. (2006). What are the characteristics of the Nordic profile in mathematical literacy? In J. Mejding & A. Roe (Eds.), *Northern lights on PISA 2003—A reflection from the Nordic countries*. Oslo: Nordisk Ministerråd.

Skovsmose, O. (1994). *Towards a philosophy of critical mathematics education*. Dordrecht: Kluwer Academic Publishers.

Tatto, M. T., Schwille, J., Senk, S. L., Ingvarson, L., Rowley, G., Peck, R., et al. (2012). *Policy, practice, and readiness to teach primary and secondary mathematics in 17 countries. Findings from the IEA teacher education and development study in mathematics (TEDS-M)*. Amsterdam: IEA.

TIMSSVIDEO (Producer). (1999). *Lesson JP3. TIMSS video study*. From http://www.timssvideo.com/49.

Usiskin, Z. (1995, Spring). Why is algebra important to learn? *American Educator*, 30–37. Retrieved on January 7, 2017 from https://www.researchgate.net/publication/240415845_Why_Is_Algebra_Important_to_Learn.

Utdanningsdirektoratet. (2006). The Norwegian Directorate for Education and Training. *The curriculum for the common core subject of mathematics*. https://www.udir.no/laring-og-trivsel/lareplanverket/finn-lareplan/#matematikk&englishundefined. Retrieved May 15, 2017.

Wu, M. (2009). *A critical comparison of the contents of PISA and TIMSS mathematics assessments*. Retrieved from https://www.researchgate.net/publication/242149776.

Yang Hansen, K., Gustafsson, J. E., & Rosén, M. (2014). School performance differences and policy variations in Finland, Norway and Sweden. In *Nothern lights on TIMSS and PIRLS 2011. Differences and similarities in the Nordic countries* (Vol. TemaNord 2014: 528, pp. 24–47). Norway: Nordic Council of Ministers.

Chapter 12
Storytelling for Tertiary Mathematics Students

Ansie Harding

Abstract This paper offers a narrative of ideas, events and opinions addressing the underexposed area of storytelling in tertiary mathematics. A short discussion on storytelling is followed by a brief account of the history of storytelling. Features of stories are discussed as well as options for when a story should be told and the requirements of a good story. The main thrust of the paper is a personal account of experiences of storytelling in a tertiary mathematics classroom. The study involves a large group of engineering students doing a calculus module. The storytelling discussed in this paper takes the form of a structured activity in a specific timeslot. Student feedback presents an unexpected angle, deviating from the intended purpose of entertain, inspire and educate, namely, giving a perception of caring from the teacher's side.

Keywords Storytelling · Mathematics · Tertiary students · Features of storytelling Mathematics stories

12.1 About Stories and Storytelling

Storytelling is part of every culture: It is an ancient art that has been practiced through millennia of human interaction. It ranges from a mother telling her child a story to theatrical storytelling on a stage. Storytelling often relies on the imagination and speaking ability of the storyteller and the listening ability of the audience. Stories have travelled and still travel all over the world, and commonalities in different cultures abound.

Stories come in many forms: they can be in written form, orally conveyed or visually depicted. Stories appear everywhere: in newspapers, on the internet, in magazines, on television and in discourse between people. Stories can be factual or fictional, stem from actual events or be the product of someone's imagination.

A. Harding (✉)
University of Pretoria, Pretoria, South Africa
e-mail: aharding@up.ac.za

© The Author(s) 2018
G. Kaiser et al. (eds.), *Invited Lectures from the 13th International Congress on Mathematical Education*, ICME-13 Monographs,
https://doi.org/10.1007/978-3-319-72170-5_12

They can range from centuries-old traditional folktales to accounts of current events. Young and old alike find appeal in stories, and it seems as if stories are built into our thinking, as it is easy to remember and to repeat when the time comes. Stories fuel conversation, are compelling and can be a source of entertainment.

Storytelling is basic to education. From a young age children are exposed to stories that open up new worlds for them and expose them to characters and situations from which they can learn moral lessons. Many of these stories come through generations and are brought to life for a new generation by the storyteller at each telling.

12.2   History of Storytelling

As the practice of storytelling stretches over many millennia, an accurate account would be ambitious and falls outside the scope of this narrative. For the sake of simplifying this history, we note several landmarks in the long history of story-telling. This account is a personal perspective formed over a long period and is sourced widely.

The origins of storytelling are lost in the mists of time, but most probably first took form in oral storytelling, which could date back as far as the time of the Neanderthal people. One can imagine people sitting around a fire and relating events of the day, perhaps telling of a narrow escape or a heroic encounter. Stories possibly travelled from clan to clan where the stories also conveyed news.

The first recorded stories date from around 35,000 BC, from when cave paintings show a recording of events, telling stories of people and animals. Amongst the cuneiform clay tablets dating from Babylonian times (around 2000 BC) is the story of Gilgamesh, a forerunner of the modern day superheroes. The Epic of Gilgamesh is considered the oldest piece of epic Western literature (*Ancient History Encyclopedia*, n.d.). Gilgamesh is widely accepted as a real person of superhuman capabilities.

Stories about real people evolved towards stories of imaginary characters. From Greece (500 BC) we inherited the fables of Aesop, presenting moral lessons for life. Homer's *Odyssey* (800 BC), with its mythological characters such as the one-eyed Cyclopes, is one of many Greek and Roman works on fictional characters. The story of Merlin and King Arthur (500 AD) and the *One Thousand and One Nights* stories are more of the early treasures of fictional stories.

Perhaps the biggest thrust in the history of storytelling is the advent of printed books. The first book printed with movable metal type is *Jikji* (an abbreviated title), a Korean Buddhist document, which dates from 1377. The Gutenberg Bible, printed by Johannes Gutenberg in 1450, was the first major book printed in Europe with movable metal type. The printing of books was the process that would make recorded stories accessible to a wide population.

The invention of storytelling machines dates from the early 1900s. A 1907 Lee de Forest company advertisement promised: "It will soon be possible to distribute grand opera music from transmitters placed on the stage of the Metropolitan Opera House … to almost any dwelling in Greater New York and vicinity." (TVTechnology, n.d., p. 1). This promise became a reality in 1910. The radio became a storytelling device that has a place in almost every household and has survived the times.

In 1890, Edison invented the kinetograph, technology that led to an enormous film and television industry. The first of the silent movies, *The Great Train Robbery*, appeared in 1903 and is all of about 10 min long. The year 1927 saw the first of the talkies, *The Jazz Singer*, in which characters could first be heard talking. Almost simultaneously, in 1926, Logie Baird gave life to television, perhaps the greatest storytelling device of all time.

The final milestone in the history of storytelling is internet storytelling. The new millennium has seen storytelling blossom by means of social media, blogs, Facebook, Twitter, YouTube and other platforms. People write fictitious stories but also relate events and experiences in their own lives, reminiscent of the first storytelling of the Neanderthal people, thus seemingly completing a full circle.

Through all the new modes of storytelling that have emerged through times, it is noticeable that despite new devices and modes appearing, the older ones remain. For example, there was talk that printed books would be replaced by e-books, but it has not happened. Both of these co-exist. Oral storytelling is another example of how the most ancient form of storytelling has survived. Every new mode has supplemented rather than replaced previous ones.

12.3 Literature on Storytelling in Education

Zazkis and Liljedahl (2009) have been instrumental in promoting storytelling in the mathematics classroom. The purpose of telling stories in the mathematics classroom, according to these authors, is to create an environment of imagination, emotion and thinking; to make mathematics more enjoyable and more memorable; to engage students in a mathematical activity; to make them think and explore; and, perhaps most importantly, to help them understand ideas and concepts. The ability that stories have for shaping and orientating the listener's feelings is mentioned by Zazkis and Liljedahl (2009) as a great power. These authors divide stories according to their function in the classroom and their potential for engagement into six categories: stories used to ask a question, stories accompanying a topic, stories for introducing an idea, stories intertwined with a topic, stories to explain a concept and stories used to introduce an activity. Friday (2014) claims that teachers have been storytellers for millennia but do not necessarily see themselves as storytellers. He admits that becoming a storyteller takes effort and inclination but that the effort makes it worth it. Hamilton and Weiss (2005) maintain that stories are the best gifts teachers can give their students because they can never be taken away: they belong to students forever.

Although Tobin's (2007) comprehensive study on using storytelling in tertiary education focuses on information technology (IT), it covers a broad scope of storytelling, listing formats, structure, uses and benefits and implementation with a comprehensive source of references to each of these. He states that that the use of storytelling is not "as well established or commonly accepted as the more traditional logical or scientific content-based lecture method" (p. 55). The article serves to stimulate interest in storytelling and concludes with a checklist of issues to be considered when using a storytelling approach. Other valuable resources in the higher education context that offers ways in which storytelling can be used effectively as a tool are presented by Alterio and McDrury (2002) and Kruyvenhoven (2009).

The term digital storytelling describes the practice of people who use digital tools to tell their story, involving some means of technology in storytelling, as opposed to face-to-face storytelling. The implementation and effects of digital storytelling in education have been discussed widely (for example, Heo 2009; Hull and Katz 2006; Ohler 2005; Robin 2005; Sadik 2008).

The purpose of using storytelling to bring the culture of the community—in our case, the culture of mathematics—into the classroom and making it part of students' awareness is captured by Harold Rosen, well-known engineer and educationalist, as quoted in Zipes (1995): "If the culture of the community is to enter the culture of the school, its stories must come too and, more profoundly perhaps, its oral storytelling traditions must become an acknowledged form of meaning making" (p. 1).

Scepticism about the value of storytelling in the corporate environment is voiced by Denning (2004) as he describes his journey in the business world. Executives thrive on analysis and although analysis "might excite the mind, it hardly offers a route to the heart" (p. 3). Denning claims that storytelling is the place to go "to motivate people not only to take action but to do so with energy and enthusiasm" (p. 3). The latter statement applies generally and also in education.

Huggins (2017) discusses the purpose of storytelling from the game design environment that has the mission to inform, inspire and entertain in order to channel teens' interest. The author of this chapter differently interprets the notion of informing as education, thus subscribing to the threefold purpose to entertain, inspire and educate.

12.4 Storytelling for Tertiary Mathematics

The task of teaching tertiary mathematics seems to be remote from the act of telling stories, and lecturers often shy away from this "juvenile" activity. This does not mean that storytelling does not happen in the tertiary mathematics classroom, but rather that it is often an informal rather than structured activity. Historical anecdotes are often woven into teaching, providing context to topics while bringing a moment of relaxation to the class. Mathematics in particular is embedded in a rich history and also relates to almost any other field in some way, thus offering ample material for storytelling. In this paper I would like to propose including storytelling into tertiary mathematics as a more structured activity.

The question of where to fit storytelling into a classroom experience has no definite answer, but there are various options. Probably the most common usage of storytelling is as an introduction to the lecture or a topic. The story is then related to the day's work or the topic at hand. The aim is to give context to the topic that will follow. My personal experience and opinion is that although there are positives to this method, there are also negatives. There is a measure of sugar coating involved in this practice, in that the pleasantness of the story is followed by the toughness of the mathematics, and unintentionally the lecturer is trying to soften the blow. For large classes of more than 100 students there is another concern, namely that the nature of a story is such that it is a trigger for discussion, conversation and sharing. There is no real opportunity for this, as it not only takes time but also disturbs the calm in the classroom and necessitates regaining its harmony.

Another option is using a story as a "by the way", weaving it into the teaching as a short anecdote or an amusing snippet. The intention is to vary the teaching and pace through the appeal that a story has. Although a commendable practice, the story could in this case come in an abbreviated form, so the richness of the characters and context of the story are not fully exposed. In other words, the potential of the story is curbed.

A third option of storytelling is to use it as a "commercial break" somewhere in the middle of the lesson, between topics. Student attention span is limited and pausing the teaching for the light entertainment provided by the story is an option. The story can be fleshed to suit in terms of characters and storyline. Calling it a commercial break distinguishes the story from the work. After the story, the second half of the lecture ensues. This practice can be successful but holds the same danger as an introductory story; namely, to regain the attention and focus of the lesson could be problematic as one has to deprive students of the repartee that follows a story.

The fourth option emerged for me as the most successful mode after many years of experimenting with different modes of storytelling in tertiary teaching. This mode is to use storytelling as a reward. I teach calculus to a large first-year group of engineering students of around 300 that consists of four contact sessions of 50 min each per week. The last 10 min of the last lecture of the week is dedicated to a full-fledged story. I take care in preparing a story that is mostly in low-tech oral form or sometimes centred around one or two slides. The idea is that we have worked hard on the week's study material and this is the reward. The story could be related to the work at hand but often is not. The mathematics link is always there, stronger sometimes than at other times. Students look forward to the story of the week and alert me as the time gets closer. Student reaction is the motivation; why this is a successful way of incorporating storytelling will be discussed subsequently.

For tertiary students a story needs care from the storyteller: care in preparation and care in presentation. The practice of telling a fleshed-out story demands searching and compiling of facts and anecdotes from various sources. Most important is that the storytellers make the stories their own. You need to be comfortable with both the facts and the storyline. You have to put soul into the storytelling, fleshing out the characters and bringing the story to life. I have compiled a long list of suitable stories as a resource but am continuously searching for new stories. Sources include the internet, books, listening to people, talking to people, using your imagination to embellish the story and always adding your own touch.

I have practiced storytelling for tertiary students for a number of years. It seems unlikely that a large group of engineering students would take to storytelling, yet it does happen and supports the point that storytelling is for all ages. The time spent on storytelling is little enough not to impact negatively on teaching time.

The question that most probably arises in the reader's mind is what is typical of such stories. A few of the stalwart stories are: "How long is a year?" "How a memory stick burned down the houses of parliament in England." "Pythagoras was possibly a plagiarist." "l'Hôpital's rule? No, Bernoulli's rule." "Why x represents the unknown." "Newton vs Leibniz." "Memorising the digits of pi." "De Moivre's story." "The Millennium problems." "The Fields medal." "Why there is no Nobel prize in mathematics". Two of these stories are included as examples at the end of the chapter.

12.5   Features of Storytelling

Although storytelling is a diverse activity, mostly influenced by the personality traits of the storyteller or author, there are recognizable elements. Firstly there are characters placed in a setting, which the storyteller has to flesh out to bring to life. Characters have ambitions or quests that they pursue or would like to pursue. The character then encounters some problem or some conflict ensues. Through a series of events, the story leads to an outcome or resolution and, most of the time, "they lived happily ever after". In all stories, the human element plays a major role. It provides the audience member with a human connection to the events or character through which the story is brought to life. When considering a story, I pay attention to the presence of these elements.

Compiling a good story is a skill to be cultivated. A collection of facts does not make a good story. My personal list of requirements for a story are: (1) The gist of the story has to be mathematics related in some way. (2) There has to be a human element, preferable a hero and anti-hero. (3) There has to be a flow of events in the story from the start, running through events towards a conclusion. (4) This story has to contain an element that will trigger reaction, be it humour, outrage or the unexpected.

12.6   Data Gathering

The students involved in this study were first-year engineering students doing a calculus module presented by the mathematics department. The teaching model was one of large-group teaching, with around 1500 students enrolled for the module taught by five lecturers. The study involves one group of around 300 students. The students were of mixed ethnic, socioeconomic and gender distribution, typical of a South African university. The study was conducted in 2016 over a semester of 14 weeks. After concluding the semester, while students were doing examination preparation, I posted an invitation on the Blackboard learning management system for students to share their opinions and views on the storytelling feature at the end of the last lecture of the week. A high response rate was not expected because of the timing. A total of 26 students responded, and it was noticeable that care was taken in responding.

The analysis of the data is based on the systematic methodology of grounded theory. Student responses were studied and anchors were identified in each response and coded. Codes of similar content allowed the data to be grouped. Broad groups of similar concepts were used to generate the following six categories of responses.

12.7 Feedback

The responses were overwhelming in volume and in detail. Not only did students express their appreciation, but they also suggested additional topics for stories and shared their own reading experiences. It was clear that the storytelling struck a chord and that students wanted to share their opinions. Many students did not stop at a few lines but continued to write a page or more about the value it had had for them. The prolific writing certainly came as a surprise.

As mentioned before, the mantra I adhere to is entertain, inspire and educate. The intention is to tell a story, consciously leaving the actual mathematics aside, that will lift their spirits through the entertainment element while at the same time weaving a picture that brings inspiration. Stories open up circumstances and events previously unknown to the students and in so doing educate them.

1. *Emotional impact*

What emerged is that students have a personal need to be included and recognised. The storytelling proved to have an unexpected positive emotional impact.

"Made me feel you are a parent of my own in the university environment because no one seems to care about students in university, unlike in high school."

"The fact that you tell us inspirational or even just fun stories … makes me calmer and more focused."

"Made me feel welcome in the class(room)."

"It is a way of saying: 'It's okay, you are in competent hands.'"

2. *Reward*

The intention to use the storytelling as a reward, although never stated, paid off, as students seem to have picked up on this. It also underlines the positive outcome of storytelling at the end of the week instead of at the start of a lecture.

"Gave us something to look forward to."

"They are a sort of reward for the week's work. A refreshing beer at the end of a long day."

"It was almost like an energy bar for the weekend."

3. *Motivation*

Motivation resulting from the storytelling was mentioned by students as a gain. Motivation links to inspiration, which was one of the three intentions of presenting storytelling.

"Motivates me to wake up that early for a lecture."

"Motivates me to go to even the last lecture of the week."

"I will have strength to study because I have a smile at the end of the lecture."

"It motivated me personally to pursue my studies more enthusiastically."

4. *Subject impact*

Although storytelling was presented as a separate activity to the formal lecturing, feedback showed that it had a definite impact on perceptions of the subject itself.

"The stories showed me that mathematics need not begin and end with difficult integrals and limits."

"I feel that a story is an immersive, simple way to get to know the skeleton of the work, before having to add the flesh to the bones."

"… it can truly inspire some students into delving deeper into mathematics."

"It made me feel like I was part of some 'maths family'."

5. *Appreciation*

A pleasant personal reward came from appreciation showed by students. This appreciation did not so much relate to the stories themselves but to the human side shown through storytelling. This is a significant finding, as the result is far removed from the intention of entertaining, inspiring and educating. Students perceive the lecturer as someone who cares and is approachable.

"Showed me that you as a lecturer put in effort to make the classes interesting for the students."

"It made you seem less of an almighty professor in front of the class and more of a teacher that actually cares about your class."

"The stories demonstrated that the lecturer was someone who actually (still) cares about her craft."

"… it is your peculiar signature move."

"It helped all of us know you were much more approachable than some others."

"It made you more approachable and added to your standing in the students' eyes."

"I viewed you more like a 'guardian' and less like a lecturer."

6. *Bigger picture*

The storytelling seems to have opened up horizons for students to give a wider view beyond the subject content.

"The stories remind us that there is life ahead of university as they portray general knowledge."

"It proved that there is more to maths than the calculus we studied in this course."

"… shows us how maths relates to the world around us."

"It inspires me not only to work for distinctions but to have a broader view such as inventing new things and being innovative in my career."

12.8 Critical Reflection

The chapter presents a personal storytelling journey that follows the thoughts of Gallagher (2011), who argues that storytelling is central to education research and maintains that storytelling as a narrative methodology is here to stay.

The storytelling discussed in this paper takes the form of a structured activity in a specific time slot. Feedback shows that the gain experienced by students is perhaps more on an emotional level than on a cognitive level. Students see the storytelling as an act of caring and as a contribution to their well-being from the lecturer's side. In this sense a fourth reason for telling a story is added to those given by Zazkis and Liljedahl (2009). We tell stories because we enjoy it, because the students like it and because we believe it is an effective instructional tool for teaching mathematics but also because it gives the student the sense of caring from the lecturer's side, a sense of giving beyond the subject content. This quote captures student perception best: "It shows you care about maths and you care about your students."

Being active in storytelling for a period leads to agreement with Friday (2014), who claims that teachers do not necessarily see themselves as storytellers and that becoming a storyteller takes effort but that the efforts make it worth it. Becoming a storyteller is a learning process, both in animating the story to the appeal of the audience and in searching for suitable stories. The benefit of delving into the history and characters of mathematics proves to be an ultimately enriching experience.

Scepticism, such as voiced by Denning (2004), about storytelling in a tertiary environment is not uncommon, and yet the enthusiasm and motivation encountered counteract any negativity.

My experience in storytelling has also led to the sobering realisation that it requires full buy-in from the storyteller regarding collecting stories, personalising them and presenting them in an enthusiastic manner.

12.9   Examples of Stories

To conclude the paper, two stories are presented as examples, each followed by an interpretation. The first story on how a memory stick burned down the houses of parliament is loosely related to mathematics and brought into the modern context through seeing tally sticks as the forerunner of the memory stick.

How a memory stick burned down the Houses of Parliament:

Tally sticks have been around for a long, long time. A tally stick is a piece of wood or bone on which notches are carved, mainly to remember things. So a tally stick is just a primitive memory stick. The oldest tally stick found dates back 35,000 years, found in a cave in the Lebombo Mountains on the border between South Africa and Swaziland. It shows 29 notches on a baboon bone that could point to the number of days in a lunar cycle. It is the first evidence of recorded counting.

In medieval Europe, tally sticks came to another use. With coins in short supply and the population largely unable to read and write, tally sticks were used to keep record of transactions. If you borrowed money or bought goods from me, we carved the amount in terms of notches on a tally stick. The tally stick was then split in half (hence "split tally") through the centre of the carving so that both halves showed the amount. Neither you nor I could add marks as the other had proof of the original transaction. One part was then slightly shortened and given to you the borrower, the longer part to me the lender. Hence the expression that the borrower had "the short end of the stick". The longer part was called the "stock" and the shorter part the "foil".

The tally stick system formed the basis of commerce in the British Empire until the 1600s, when the Bank of England was formed in which a paper system was followed. People found it difficult to let go of the tally system, and legislation was slow, as we all know that governments take their time. Charles Dickens was one of the people that canvassed against the use of tally sticks as an outdated practice. It was only in 1826 when the sticks were finally removed from circulation and stored in the Houses of Parliament. The basement was overflowing. In 1834, it was decided to get rid of the mass of tally sticks. Rather than give them away as firewood, it was decided to burn them in the two underfloor coal furnaces in the House of Lords.

Two guys, Joshua and Patrick, were assigned to do the job. They unfortunately chose to ignore the warning that the old building was a fire risk. They stuffed the furnaces with tally sticks all day long. The job was inspected early on, but left to them later in the day. The copper-lined brick flues overheated and during the late afternoon, as people were getting ready to go home, they noticed that the House of Lords chamber was smoky and unusually hot. Again this was ignored. Joshua and Patrick wanted to finish the job and they pushed on as lock-up time drew nearer. An hour later the place was ablaze, helped on by a gusting wind. It is believed that the overheated copper linings set the wooden wall panelling alight.

As expected, a multitude of spectators gather to witness a sight too spectacular to miss, lining the banks of the Thames, testing the crowd controlling skills of the police and the army. It was a huge disaster and the subject of a painting by JMW Turner. Only the foundations remained.

Interpretation: The story contains historical information, two characters of dubious intention and a disturbing outcome. The story has a beginning and end, it has a human element and an element of surprise. It aims towards creating an environment of imagination, emotion and thinking as well as towards to making mathematics more enjoyable and more memorable (Zazkis and Liljedahl 2009). The story also complies with the threefold purpose to entertain, educate and, in this case to a lesser extent, inspire. The story describes events caused by people's actions and can be seen a educating about historical events and about how the forerunner of our banking system worked. It also illustrates how mathematics is engrained in society.

The second story on the rule that is wrongly attributed to l'Hôpital is strongly embedded in mathematics and positioned in the years following Newton and Leibniz's formulation of calculus.

L'Hôpital's rule or Bernoulli's rule?

This story has two players: Guillaume l'Hôpital and Johann Bernoulli. Guillaume l'Hôpital was born in 1661, which makes him about 20 years younger than Newton. His family was considered to be nobility in France. Since childhood, l'Hôpital was passionate about mathematics. He briefly followed a military career because of his family background. He spent his days in the tent doing mathematics and soon found an excuse to quit. He then worked to become one of the best mathematicians in France.

Johann Bernoulli, six years younger than l'Hôpital, was part of the Bernoulli family, who produced six outstanding mathematicians over three generations. Johann's family were traders and he, along with his brother Jacob, did not want to take over the family spice business. They began studying mathematics together and, although successful, the two developed a rather jealous and competitive relationship, trying to outdo each other. After Jacob's early death from tuberculosis, Johann took over his brother's position as professor and merely shifted his jealousy toward his own talented son, Daniel. At one point, Johann published a book based on Daniel's work, even changing the date to make it look as though his book had been published before his son's.

When l'Hôpital was 30 years old and Bernoulli 24, they met by chance at a science meeting in Paris. Bernoulli had just arrived in Paris after giving lectures on the latest development in mathematics, namely Leibniz's differential calculus. Bernoulli liked l'Hôpital for his pleasant personality and l'Hôpital, on the other hand, quickly became intrigued by Bernoulli's knowledge on this new mathematics. Bernoulli agreed to give four lectures a week over a six-month period that l'Hôpital attended. After that, l'Hôpital managed to persuade Bernoulli to give him private lessons on his estate. Then l'Hôpital came with a proposition: He would start by paying Bernoulli 300 pounds. Bernoulli would sell his work and ideas to l'Hôpital and would keep quiet about the transaction. L'Hôpital could publish it as his own. Why Bernoulli agreed to this is not clear. Did he need the money or did the fact that he came from a tradesman's background and l'Hôpital from a nobleman's background make him obliged to be subservient?

Five years after first meeting Bernoulli, l'Hôpital published the first ever textbook on differential calculus. In the introduction, l'Hôpital acknowledges Leibniz and Johann Bernoulli as knowledgeable, but the impression was that the work was his own. He also acknowledges Newton as discoverer of calculus but says that Leibniz's notation was better. In Chap. 9 appears the rule now known as L'Hôpital's rule for a limit where both numerator and denominator tend to zero. This book was an enormous success. It was used for a long time, with new editions produced for more than 100 years. Bernoulli said nothing at first, but after L'Hôpital's early death eight years later (he died at age 43) he became more forceful in saying that the book was essentially his. His claims were not taken too seriously as he had been involved in many disputes. Towards the end of his life Bernoulli boasted of the money he had received from L'Hôspital, exaggerating the amount he had received.

Only in 1921 did a manuscript copy of the course given by Johann Bernoulli to l'Hôpital come to light, and it was seen how closely the book followed the course notes. It was only when the agreement between the two men came to light that more understanding of the events became possible. In fact Bernoulli had not been in a position to complain when l'Hôpital's book was published because of the agreement between them.

We should not judge l'Hôpital's procedure too harshly. L'Hôpital, being a nobleman, was accustomed to paying for the services of others. In fact, Bernoulli did a similar thing to his own son later. The bottom line is that the rule is still known as L'Hôpital's rule and not Bernoulli's rule.

Interpretation: The human element is again present in this story. The sequence of events leads to a surprising outcome and the listener is left wondering whether justice prevailed or not. The purpose of the story is to place L'Hôpital's rule in historical context but also to expose a situation of perceived injustice and thus to create an environment of imagination, emotion and thinking (Zazkis and Liljedahl 2009). The aspiration is also that the oral format does justice to shape the listener's feelings about the information that is communicated. In the categorisation of Zazkis and Liljedahl (2009), this is a story accompanying a topic.

References

Alterio, M., & McDrury, J. (2002). *Learning through storytelling in higher education*. New Zealand: Dunmore Press Limited.

Ancient History Encyclopedia. (n.d.). *Gilgamesh*. http://www.ancient.eu/gilgamesh/. Accessed August 25, 2017.

Denning S. (2004.) Telling tales. *Harvard Business Review, 82*(5), 122–129.

Friday, M. J. (2014). *Why storytelling in the classroom matters*. Edutopia. https://www.edutopia.org/blog/storytelling-in-the-classroom-matters-matthew-friday. Accessed July 28, 2017.

Gallagher, K. M. (2011). In search of a theoretical basis for storytelling in education research: Story as a method. *International Journal of Research & Method in Education, 34*(1), 49–61.

Hamilton, M., & Weiss, M. (2005). *Children tell stories: Teaching and using storytelling in the classroom.* Katovah: Richard C. Owen Publishers.

Heo, M. (2009). Digital storytelling: An empirical study on the impact of digital storytelling on pre-service teachers' self-efficacy and dispositions towards educational technology. *Journal of Educational Multimedia and Hypermedia, 18*(4), 405–428.

Huggins, S. (2017). Storytelling and young adults: An overview of contemporary practices. In J. M. Del Negro & M. A. Kimball (Ed.), *Engaging teens with story: How to inspire and educate youth with storytelling.* Santa Barbara: ABC-CLIO.

Hull, G. A., & Katz, M.-L. (2006). Crafting an agentive self: Case studies of digital storytelling. *Research in the Teaching of English, 41*(1), 43–81.

Kruyvenhoven, J. (2009). *In the presence of each other: A pedagogy of storytelling.* Toronto: University of Toronto Press.

Ohler, J. (2005). The world of digital storytelling. *Educational Leadership, 63*(4), 44–47.

Robin, B. R. (2005). *The educational uses of digital storytelling.* https://digitalliteracyintheclassroom.pbworks.com/f/Educ-Uses-DS.pdf. Accessed July 28, 2017.

Sadik, A. (2008). Digital storytelling: A meaningful technology-integrated approach for engaged student learning. *Educational Technology Research and Development, 56*(4), 487–506.

Tobin, P. K. J. (2007). Teaching IT through storytelling. *SACJ, 38,* 51–61.

TVTechnology. (n.d.). *Metropolitan opera to celebrate 100 years of live broadcasts.* http://www.tvtechnology.com/news/0086/metropolitan-opera-to-celebrate-years-of-live-broadcasts/227629. Accessed August 25, 2017.

Zazkis, R., & Liljedahl, P. (2009). *Teaching mathematics as storytelling.* Rotterdam: Sense Publishers.

Zipes, J. (1995). *Creative storytelling—Building communities/changing lives.* New York and London: Routledge.

Chapter 13
PME and the International Community of Mathematics Education

Rina Hershkowitz and Stefan Ufer

Abstract The International Group for the Psychology of Mathematics Education (PME) was founded in 1976 in Karlsruhe (Germany), during the ICME-3 Congress. Since 1977, the PME group has met every year somewhere in the world, since then, and has developed into one of the most interesting international groups in the field of educational research. In this paper, after a short introduction, we draw some main features of the unique essence of the PME as a research group. We focus on and analyse the change and development of the group's research over the past 40 years, and exemplify these changes and developments by tracing on a few main research lines. Based on specifics of PME research, we describe the more comprehensive lines of PME research, its change and progress in the past four decades.

Keywords The International Group for the Psychology of Mathematics Education
History of mathematics education · Research trends in IGPME · Theory in mathematics education research · Methods of mathematics education research

13.1 Introduction

13.1.1 Some General Features of PME

In the introduction to the "Handbook of Research on the Psychology of Mathematics Education" (PME 1976–2006) which was published in 2006 for the

R. Hershkowitz (✉)
Science Teaching Department, Weizmann Institute of Science,
76100 Rehovot, Israel
e-mail: rina.hershkowitz@weizmann.ac.il

S. Ufer
Department of Mathematics, Ludwig-Maximilians-Universität München,
Theresienstraße 39, 80333 Munich, Germany
e-mail: ufer@math.lmu.de

© The Author(s) 2018
G. Kaiser et al. (eds.), *Invited Lectures from the 13th International Congress on Mathematical Education*, ICME-13 Monographs,
https://doi.org/10.1007/978-3-319-72170-5_13

30 years of PME, the editors Gutierrez and Boero (2006), mentioned two reasons for the success of the PME group.

Their first reason is the human and scientific quality of the founding members —"The fathers": E. Fischbein, H. Freudenthal & R. Skemp and many others who shared both the important decision itself and bringing the PME into the existence.

We feel that this is the place to say some words in the memory of Efraim Fischbein (Fig. 13.1) who was in a sense the "inspiration spirit" of the PME organization. We borrowed these words from Tall (1998), who wrote after Fischbein's death:

> The 23rd meeting of the International Group for the Psychology of Learning Mathematics in Israel is touchingly the first in which we cannot be joined by our Founder President, Professor Efraim Fischbein, who left us on July 22nd 1998. It is a time of sadness, yes, but it is also a time for celebrating the achievements of this gentle man who is responsible for the existence of our organization. In particular, it is to him that we owe our focus on the psychology of learning mathematics. (Tall 1998)

Gutierrez and Boero see the second reason for the success of the PME group in 'the fact that the growth of "PME"… happened during the full development of mathematics education as a research domain, contributing to that development, but also profiting from it', (2006, p. 1).

We may conclude from the above that the editors of the book see a kind of symmetrical relationships between mathematics education research community in general and the PME group activity. We may also conclude that the main contribution of PME to mathematics education research and practice is being a—not necessarily always coherent—core group in the domain in which mathematics education researchers and practitioners, psychologists, and mathematicians from different countries and cultures may meet on an annual basis within a well-organized group and work together. The group created a constitution which has been adopted in the PME 1980 conference, and went through a few democratic changes from then.

Fig. 13.1 Efraim Fischbein, first PME president

13.1.2 PME Spirit Through the Lens of Its Goals, Conferences, Proceedings and Books

The major formal goals of PME, as they appeared in the PME constitution (1980–2016), frames its activity. They are:

> (i) to promote international contacts and exchange of scientific information in the field of mathematics education; (ii) to promote and stimulate interdisciplinary research in the aforesaid area; and (iii) to further a deeper and more correct understanding of the **psychological** and other aspects of teaching and learning mathematics and the implications thereof.

In spite of the diversity of research included in the PME, PME as a scientific organization and especially the annual conferences and the proceedings volumes, in which the annual conference contributions are published, have the following unifying characteristics:

- Democratization and freedom spirit is one of the characteristics of the International Group for the Psychology of Mathematics Education from its very beginning.
- This spirit encourages members of PME to present (orally and in writing) their empirical studies, as well as their hypothetical new thoughts and theories (partial or more complete), in a very early stage of their development. These presentations enable the researchers to get the critical and often wise feedback from their peers in PME, and to interweave what they had learned from it in the longitudinal thread of their work and also publications in the PME proceedings without the highly demanding procedures of scientific journals. This way, young researchers are able to "gain time and help" on their way to become mature researchers in their domain, working as members in a community and/or cooperate with others on individual or small group basis, rather than in isolation. Members of PME are quite aware to this advantage. E.g., the dedication which has been written on the new handbook for the 40 years of PME says (Gutierrez et al. 2016):

> *To the young researchers, throughout the world, who are the future of mathematics education research and of the PME community*

- An additional characteristic of the PME activity is the fact that there are and always were mathematics education researchers who are active in both the PME community, as well as in other international frameworks in mathematics education and educational research beyond. These people are "interactive pipes" among the various mathematical activity frameworks.

13.2 First Views on the Research Presented at PME

One focus of research in PME, among many others, from the very beginning has been conceptual understanding of the number system. When having a superficial glance at the papers from PMEs in the late 70s and early 80s, one might sometimes wonder if anything has changed at all since then. One illustrative example is the following: In the PME38 proceedings, we find a neatly designed, experimental study with pre-, post- and follow-up test by Heemsoth and Heinze (2014), titled "How should students reflect upon their own errors with respect to fraction problems?" It investigates, flatly speaking, if it is better for students to reflect on their errors in exercises or if it is better if they study the correct solutions of the exercises. In the proceedings of PME7, one finds a neatly designed, experimental study with pre-, post- and follow-up test by Swan (1983) titled "Teaching decimal place value. A comparative study of 'conflict' and 'positive only'" approaches. The study investigates, flatly speaking, if it is better in instruction to focus on the errors students' make in exercises, when learning decimals, or if it is better to focus on the correct solution of these exercises. Both studies come to the conclusion that, particularly in the long run, focusing on errors is more effective than an exclusive focus on the correct solution. So, nothing has changed? Well, of course important things have changed—when looking at the research in both periods in more detail; we see continuity, but also substantial development.

13.2.1 The Theoretical Basis That Is Used to Frame Findings

When we consider these two studies, we see important differences. The Swan study (1983) starts out from the point that "traditional courses" do not remedy students' misconceptions on fractions, and then proposes the two teaching styles as potential solutions to this problem. Afterwards, it presents a mathematical analysis of potential student errors in decimal arithmetic—which is actually mostly in line with what we can read in the literature today. Next, the sample, intervention methodology, and results are presented. At the end, the author hypothesizes that the "conflict" approach should be particularly effective to foster *conceptual knowledge* in contrast to *procedural knowledge*, connecting his study to a model of cognition that is not further specified. The Heemsoth and Heinze (2014) study can refer to the then more extensive literature on students' misconceptions about fractions and on learning from errors, but it also refers explicitly to a developmental learning model that includes conceptual and procedural knowledge. The authors interpret their results in view of this model and see some peculiarities, e.g. that the "error centred" (conflict) intervention showed immediate effects on procedural knowledge, while an effect on conceptual knowledge occurred only by the follow-up test. Our main point here is not if the results are conclusive. What is visible in the newer contribution explicitly—even though it might also have been done in the older one —is how theoretical models can be challenged by empirical data, and how questions regarding their power to describe mathematical learning processes can be derived from well-planned studies. In the sequel, we will contrast the accounts of numerical cognition in earlier and newer times of PME. However, the discussion of the theoretical basis for and the role of theories in mathematics education in general cannot be tackled here in detail (see English and Sriraman 2009).

In his PME2 paper, Noelting (1978) discusses a developmental model for proportional reasoning, based on Piaget and Inhelders' developmental stage models. He presents 23 tasks that required students to compare which of two mixtures of orange juice and water with respect to the "relative orange taste". He categorizes the answers of 321 students aged 6–16 into four developmental stages statistically, and provides a qualitative description of each stage. This study is one example of a number of studies from "early PME" that closely connected to general theories of cognitive development and learning going back to the Piagetian tradition. We see the criticism of these theoretical accounts reflected in later PME proceedings, for example in an analysis of students' conceptions of multiplication by Herscovics et al. (1983) in PME7, who indicate that some of Piaget and Szeminska's original findings might be due to the specific tasks used in their experiment. Even though no empirical data is presented in their contribution, the more critical stance towards the established psychological theories is visible. One direction this discussion took was to take into account the socio-cultural context in which learning and mathematical thinking take place (see Sect. 13.3.4).

Beyond this, descriptions of tasks and students' individual understanding, based on analyses of the underlying mathematical structures, have a tradition in and beyond PME (e.g., Carpenter et al. 2012). In the recent years, different perspectives, based on specific theories of numeric processing like Dehaene's triple code model (Dehaene 1992), and assumptions about how human process numbers are studied in detail. Under the umbrella term "natural number bias", for example, several studies follow the question if and under which conditions humans process fractions as one holistic magnitude, instead of processing their denominator and numerator separately (e.g. van Dooren 2016). These accounts use psychological theories of number processing (e.g., Dehaene 1992), as well as dual process theories of cognition which differentiate between quick default heuristics and more demanding analytic strategies. Of course, strategies which students use when dealing with fractions have been described in studies long before, primarily using self-reports. One possible focus of PME in the future might be to study the link between the existing descriptions of mathematical thinking and these specific models of number processing. This cannot only enrich our understanding of mathematical cognition, but also help interpreting students' strategy choice and provide means which advance students' use of mathematics.

13.2.2 Methods Used to Approach Questions

As the theoretical perspectives of research shift, this often has an impact on the research methods required or deemed relevant to study mathematical cognition and learning. With the strong basis on the ideas from Piaget's school, research in the early phases of PME focused primarily on individual thinking and learning processes, often in well-controlled settings such as clinical interviews or paper-and-pencil tests. Two developments can be distinguished from that point. Firstly, together with a perspective on learning which focused on the social and cultural embedding of mathematical cognition and learning, methods have been developed to study these phenomena in authentic, realistic settings, taking into account not only psychological, but also social phenomena that influenced mathematical learning and thinking (e.g., Cobb and Yackel 1996). Secondly, new theoretical models, such as those about number processing mentioned above, drove the application of methods recently, that had not been used before a lot. These "new" research methods, like eye-tracking and reaction time analysis or, less frequently applied, brain imaging methods, have been discussed in several recent group activities in PME, including a Research Forum on the role of neurocognitive research for mathematics education in PME39 (Tzur and Leikin 2015), and a working session on the use of eye-tracking technologies organized by Barmby et al. (2014) in PME38. A good example for a new method that found its way into PME this way is the choice/no-choice method to study students' strategy choice in different kinds of tasks (Luwel et al. 2009). Originally, it builds on the assumption that students chose their calculation strategies from a pre-defined set of strategies, and it allows to study strategy choice within this theoretical frame, well beyond the restrictions of usual self-report methods. It is also a good example that a new method need not be based on innovative technology—sometimes creative but systematic thinking is a good first step.

13.3 Development and Changes in PME Research on Mathematics Learning

13.3.1 General Features of Trends in This Research

We use the thread of the research on processes of mathematics learning as a second example for a more detailed demonstration of the development, the theoretical and methodological changes and milestones, in PME research.

While starting to search in PME's proceedings and books we realized that the reality is quite complex and it is not easy to describe how the focus changed, raised and fell along the 40 years of PME; paradigms, trends and fashions were changing in an evolutional way, were in use, and would almost vanish after a short time, or would become at least partially nesting in the following one or vice versa. In our view, the main reason for such an interweaving emerges from the following research situation:

> ...it seems that more than in the past, researchers today do not feel obliged to and/or satisfied with sticking to one methodological paradigm. Research trends in our area are nowadays characterized by flexibility and creativity in combining research methods and methodological tools, which fit the researchers' theoretical framework and meet their goals and needs to explain and answer some 'big questions' emerging from their explorations. (Hershkowitz 2009, p. 273)

In the following we will try to discuss a few research trends that became milestones in certain points during the PME 40 years. Those trends seemed to be the most dominant in the area of investigating processes of learning mathematics. This discussion is not a statistical survey. Our words express our view which is supported by our knowledge, memory and rereading in PME proceedings and in other publications. On our way we were helped by others who wrote on similar topics—"we were standing on the shoulders of giants" (Sreen 1990).

13.3.2 Learning as It Is Expressed in the Accumulation of Learners' Responses (as Individuals) to Purposeful Tasks in Tests and Questionnaires (Quantitative Research)

In this trend of research, the responses of each individual are analysed separately. However the accumulation of all responses draws a picture of the collective knowledge as a product at one point of time. But, it does not allow direct inferences on the processes of the knowledge construction; neither on the individual, nor on the collective knowledge construction. Yet, valuable information on achieved knowledge of the collective at one point of time is collected.

While searching in one of the first proceedings of PME, (e.g., the first part in the proceedings of the 1980 PME conference, edited by Robert Karplus), it can be seen that many of the contributions use a very popular quantitative methodological tool: The questionnaire. If we will examine the first part of the 1980 proceedings, there are mainly two patterns of research methodology of making use in questionnaires:

The first one can be described by the following pattern of research elements:

- Research question;
- Hypothesis;
- The rationale for the hypothesis;
- The methodology and methodological tool—the questionnaire;
- Big heterogeneous sample which fits the research question and the hypothesis;
- Results presentations and analysis;
- Comparison between the findings and the hypothesis;
- Answer to the research question;

- Interpretation and explanation;
- Some ideas for a follow-up research and/or for activities of practice.

The above is a kind of *Top-Down Research*, and many quite valuable studies were done according to such a pattern. A classic example is the seminal study done by Efraim Fischbein and Irith Kedem, which was published in the PME (1982) proceedings. In this study the researchers investigated the question:

> Does the high-school student, normally involved in courses of mathematics, physics etc. ... clearly understand that a formal proof of a mathematical statement confers on it the attribute of a priori, universal validity – and thus excludes the need for any further check?

The second pattern of research in which its main research tool is the questionnaire, is a pattern of *bottom-up research*, whose main line might be described as follows:

- Focusing on the mathematics learning topic to be investigated;
- A priori epistemic analysis of the knowledge students are expected to develop by learning the above topic;
- Constructing questionnaire;
- Defining the research sample and circulating the questionnaire;
- Analyses of the finding; Interpretations of the above, which on one side is often supported by previous well known studies and learning theories, but on the other side provides opportunities for the emergence of new theories (either partial or not);

This pattern is a *Bottom–up research,* as the findings and also the conclusions are concluded directly from the data, without the need to confirm or to refute a pre-given theory or hypothesis. As an example of this pattern we may read the contribution by Haseman (1980), which investigates difficulties of 7th graders in addition of fractions.

13.3.3   Theory in the Center

It is quite amazing that only three years after the 1980 Conference in Berkeley, we may find so many theoretical contributions in the Proceedings of the 1983 PME Conference in Israel. As for the 1980 conference, we should mention the theory concerning the differentiation between the *Concept Definition* and the *Concept Image*, suggested at PME 1980 by Vinner and Hershkowitz (1980) (and later by Tall and Vinner 1981).

Section B in the above proceedings with the title: "Learning Theories" (pp. 52–122) includes 12 contributions, with almost each of them discussing a theme in mathematics learning from a theoretical point of view, rather than having a discussion based on empirical data. Often some mathematical-pedagogical examples do appear, mostly as illustrations for the main ideas. If empirical data are mentioned (for example interviews with learners), they are summarized in a "meta way"

Fig. 13.2 Richard Skemp,
second PME president

without real examples. We assume that the lack of a written space (six pages per a contribution) is also a reason for this style of representing findings. At any case it leaves the impression that for these researchers theoretical ideas are much more important than the findings or their interpretations. Also if empirical findings appear, they mostly emphasize the end-products and less the observation and analysis of the learning processes.

A very typical example is the dominant theory in the late seventies and eighties by Richard Skemp (Fig. 13.2), the second PME President. Richard Skemp was a mathematician who later studied psychology (Skemp 1986) and drew on both these disciplines to explain understanding in mathematics. The main 'thrust' of his argument is that learners construct *schemata* to link what they already know with new learning. According to Skemp, mathematics involves an extensive hierarchy of concepts—we cannot form any particular concept until we have formed all the subsidiary ones upon which it is depends. Skemp stressed that instrumental and relational understandings are both ways of understanding. This is a distinction in a theory of understanding: ***Instrumental understanding***: a rule/method/algorithm' kind of understanding, which gives quicker results in the short term. ***Relational understanding***: a more meaningful understanding in which the pupil is able to understand the links and relationships which give mathematics its structure (which is considered more beneficial in the long term and aids motivation). Both are deemed important for mathematics learning (Skemp 1977).

13.3.4 Constructivism and Socio-cultural Approaches, as Catalysts for Classroom Research or Vice-Versa

One of the main theories which raised an intensive theoretical, empirical and practical interest as well as intensive debates at PME's community is *constructivism*, which might be considered as one of the PME's milestones in the late eighties. Looking at PME proceedings from 1987 in Montreal, we may learn about the place of the constructivism theory within the thinking, discussions and debates at the PME's eleventh conference. Confrey and Kazak (2006) described clearly this milestone, with its peak at the PME conference, in their chapter in the PME 30 book. The authors start from mentioning PME organization's main goal which expresses the need to integrate together *Mathematics Education and Psychology* (see the third goal of PME above). There is no doubt that the presentations at the 1987 Conference (especially the four plenaries) demonstrate the advantages as well as the difficulties in interweaving together the "P", "M" and "E" at the PME community. Confrey and Kazak describe main features of constructivism and explain how it became so popular. They wrote:

> As a grand theory constructivism served as a means of prying mathematics education from its sole identification with the formal structure of mathematics as the sole guide to curricular scope and sequence. (p. 306)

An extreme example of the above trend of mathematics education is the sequence of the SMSG text books (e.g., SGMG 1961) in which the "curriculum developers" were mostly mathematicians, who seemed to believe that if only the text book will be mathematically correct and the order of the various chapters will be constructed according to a mathematical logic, students will be able to learn the mathematical subject, no matter how abstract it is. Confrey and Kazak continued and said:

> The constructivism created means to examine mathematics from a new perspective, the eyes, mind and hand of the child... Constructivism evolved as researchers interests' in the child's reasoning went beyond a simple diagnostic view of errors, to understanding the richness of student strategy and approach. (p. 306)

Gerard Vergnaud, the PME third president, explained in his presentation at the 1987 PME Conference what he considered as the main goal of constructivism:

> As a matter of fact, our job, as researchers, is to understand better the processes by which students learn, construct or discover mathematics and help teachers, curriculum and test devisors and other actors in mathematics education to make better decisions. (p. 43)

This approach, which is related to the investigation of the learning process of the student as an individual, and was based on the belief that the learner has to construct his/her mathematics by him/herself, had its own research methodologies. For example, Steffe and his colleagues describe the methodologies they used for teaching experiments and for clinical interviews with an individual child, through which they built models of children's mathematics, meaning the mathematics which was created by the student. In these models they suggest ways to consider the role of interactions between the interviewer and the student and/or between the teacher and the student (Confrey and Kazak, p. 313).

Confrey and Kazak also explained why the constructivist approach held for quite a long time:

It took hold in practice because it addressed the two primary concerns of teachers: (1) students' weak conceptual understanding with over-developed procedures (relational vs. instrumental in Skemp's language) and (2) students' demonstrated difficulties with recall and transfer to new tasks.... (p. 306)

When mathematics education researchers' and practitioners' interest moved on to focus on processes of the collective's mathematics learning (mostly classroom research), constructivist psychological approach was not enough. The socio-cultural approaches, which were established and developed at the beginning mainly in the Soviet countries (Vygotsky 1978 and many others), were adopted by theoreticians and researchers in many areas all over the world and also by the mathematics education people as well as the PME community. E.g., "*The materialist psychology by Vygotsky*", was mentioned by Lerman (2006) in his chapter in the 30 PME handbook. Lerman (2006) wrote that the main elements of the theory are: "That development led by learning (Vygotsky 1986); that concepts appear first on the social plane and only subsequently on the individual plane; that the individual plane is formed through the process of internalization; Psychological phenomena are social events; Learning takes place in the *zone of proximal development* and pulls the child into their tomorrow; and motives are integral to all actions" (p. 350).

Coordination between the constructivist and the socio-cultural approaches (theories and methodologies), led to a deep investigation of what is going on in the mathematical classroom. As an example we cite Cobb and Yackel (2011):

...we differentiated between what we termed the social aspects of the classroom which included classroom social norms and the cognitive aspects which included students' mathematical reasoning.... We instead came to the view that any aspect of the classroom can be analysed from either social or a cognitive perspective. (p. 38)

They called this new theoretical framework *The Emergent Perspective*. The emergent perspective framework is a powerful framework for describing socio-cognitive development within a classroom and was established upon the need to better understand and interpret what was observed in the mathematics classroom.

13.3.5   Research in the Mathematics Classroom and the Mathematics that is Taught and Learned in the Classroom

For more than 20 years now mathematics educators and researchers have been discussing intensively teaching and learning practices like cooperative learning, interactions and argumentation in the various classroom settings, social norms, socio-mathematical norms and more. The discourse about these practices is often general and does not always relate much to what contents and structures of mathematics are the most appropriate for teaching and learning so that the above practices will be activated by needed mathematical contents and mathematical means. Yet, there were and still are projects and innovative curricula in different countries which are enriched by the vision of such practices and vice versa.

Realistic Mathematics Education (RME), which is a teaching and learning theory in mathematics education, was the vision and curriculum development of the Freudenthal Institute in the Netherlands. This theory influenced and has been adopted by a large number of countries all over the world (de Lange 1996). The vision of RME was led mostly by Freudenthal's view on mathematics learning (Freudenthal 1991). Two of his important points of views are: "mathematics must be connected to reality" and "mathematics as human activity". First, mathematics must be close to children and be relevant to everyday life situations. However, the word 'realistic', refers not just to the connection with the real-world, but also refers to problem situations which are real in students' minds. Second, the idea of mathematics as a human activity is stressed. Mathematics education organized as *a process of Guided Reinvention*, where students can experience a similar process compared to the process by which mathematics is being invented. The meaning of *invention* is steps in learning processes while the meaning of *guided* is the instructional environment of the learning process. The *reinvention process* can use concepts of mathematization (Freudenthal 1991, p. 41) as a guide.

13.3.6 Networking—Connecting Theoretical Approaches for Better Interpretation of Empirical Findings

In Sect. 13.3.4 we discussed the coordination between the constructivist and the socio-cultural approaches (theories and methodologies), for a deep investigation of what is going on in the mathematical classroom. In the current years researchers discuss *networking* which is engaged in connecting different theoretical/methodological approaches, where each of them is a framework underlying some trend/s of research. Coordinating them together in planning the study and in analysing the findings enables higher levels of interpreting the results, innovative understanding and insights (Prediger et al. 2008). We see the interest in networking an additional evidence for the trend of flexibility in choosing both, theoretical and methodological basis for research with the aim of better understanding and interpreting the meaning of research.

The work of the researchers (In alphabetic order—Dreyfus, Hershkowitz, Rasmussen and Tabach) is a good example representing the above trend. First, the researchers coordinated together two theoretical/methodological frameworks: The AiC (Abstraction in Context) framework which analysed construction processes of mathematical knowledge of individuals as well as small groups within an inquiry classroom. The second framework is the DCA (Documenting Collective Activity), whose aim is investigating processes of constructing mathematical knowledge within the whole class community. By the above coordination, the researchers were able to trace processes of knowledge constructing and knowledge shifts between and within the different settings in the working mathematics classroom along a whole lesson and more. While doing so they were able to reveal the active role of some students and the teacher in the shifts of knowledge. Currently (Hershkowitz et al. 2017) the group started to characterize the shifts of knowledge by additional step of networking. The group raised the question: In what way is the shifted knowledge creative? For searching the authors used the work of Lithner (2008) on creativity. This new study represents progress in terms of what Prediger et al. (2008) refer to as the local integration of different theoretical/methodological approaches.

13.4 Factors Influencing PME's Development—Examples from Research on Mathematics Teachers

Besides the perspectives on developments within PME, we will outline factors influencing PME from the outside, exemplified by research on mathematics teachers. This research area went through an extraordinary development in the last 30–40 years. Some researchers speak of an "explosion". Thus it is not possible to aggregate all the many individual single contributions within and outside PME into a consistent account of the development of the field and derive influencing factors. However, we will highlight some aspects we regard as most important.

13.4.1 The Development of Research on Teachers and Teaching in PME

Several papers have summarized the development of research on teachers and teaching in PME in the past (e.g., Hoyles 1992; da Ponte and Chapman 2006; Llinares and Krainer 2006; Jaworski 2011; Lin and Rowland 2016). In most contributions, the development of the field is described by three phases.

1. *Teachers getting recognized*: All these accounts agree that the teacher was not in the focus of PME research until the end of the 80s. Even though research from this phase has been criticized as simplistic and deficit-oriented, it has certainly played an important role in the formation of the research area.
2. *Towards a research area*: In her 1992 plenary lecture, Hoyles (1992) diagnoses two trends in PMEs work: A quantitative increase in contributions that focus on the "teacher as an integral – and crucial – facet of mathematics learning and a series of qualitative shifts as to how the teachers' role is conceptualized". If the teachers occurred in research contributions before, they seem to have played a side-role as the facilitator for students' development, while the student was at the centre of researchers' attention. At the time, teachers were increasingly recognized as a possible focus of research, initially with a restriction to teachers' beliefs, later on the relation of these beliefs to teachers' classroom practice. Studies addressed the question, if and how teachers' attitudes could be changed so that curricular innovations would be taken up. For the late 80s, Hoyles observed that research increasingly addressed how teachers' beliefs and actions

were connected to a specific classroom context as well as its broader cultural (e.g., national) embedding. Finally, she described the first developments of a research area focusing on teachers, including a reflection on theoretical perspectives and methodologies used.

3. *Differentiation of the field*: As Jaworski (2011) outlines in her historical account, the 1990s started a very active phase of teacher and teaching research in PME. In particular, she mentions three working groups, which met regularly for five years during PME conferences during this phase (Psychology of In-Service Education of Mathematics-Teachers, Research on the Psychology of Mathematics Teacher Development, Teachers as Researchers in Mathematics Education). Each of these groups produced a book volume by 1999, there were two PME plenaries on the topic (Hoyles 1992), and the Journal of Mathematics Teacher Education was founded under the lead of Tom Cooney in 1998. The topic was also taken with a special survey for ICME-10 in 2004 and the ICMI Study 15 *The professional Education and Development of Teachers of Mathematics*.

In their summary of the field for the 30 years PME volume in 2006, da Ponte and Chapman (2006) identify four main objects of study in teacher-related research on PME: Teachers' mathematical knowledge, Teachers' knowledge of mathematics teaching, Teachers' beliefs and conceptions, Teachers' practices. While all of these topics are still in place, Lin and Rowland (2016) put teacher knowledge in the centre of their contribution for the 40 year handbook ten years later, highlighting its role in PME research.

13.4.2 Trends Impacting the Development of Research on Teachers and Teaching

The development described above was influenced by different other trends in mathematics education and related fields.

Advent of so-called "socio-cultural approaches": Several authors offer explanations for the increased focus on the teacher in the late 80s and in the 90s. Lin and Rowland (2016) note that this development coincided with the so-called "social turn", meaning an increased focus on social context, in which students' mathematical thinking and development takes place (see Sect. 13.3.4). Mathematical thinking and learning cannot be considered as something that happens in the students' isolated mind, independently of external influences. This idea that the environment—most prominently the classroom and the teacher—have an influence is already visible in many early PME papers. As soon as these ideas spread, it was only natural to pay more attention to the role of the teacher.

Discussion on the situatedness of cognition: This trend relates to the discussion, to what extent cognition in general, or knowledge specifically is connected to specific situational contexts, or to what extent it may be considered as a more general disposition that can be activated in a variety of situations. A famous part of this discussion it the so-called Anderson-Greeno-debate, which went on over several papers in the *"Educational Researcher"*. In a joint paper (Anderson et al. 2000), the opposing groups propose a research agenda that pursues both approaches "vigorously", and argue for attempts to integrate the different understandings of learning and knowing into a comprehensive account in the future.

Within PME, researchers favouring the "situated approach" have often focused on teachers' practices in realistic situations, research initiatives from the "cognitive side" have tried to build up models that describe the knowledge that is necessary for the professional work of a teacher, and studied them often using paper-and-pencil tests. Apart from the different conceptualizations of knowledge in both approaches, each perspective has developed and often stuck to a specific set of research methods. The discussion seems to have split the research tradition into two parts. Even (2009) asks "Are the two perspectives compatible? Do they complement each other?" Based on a model of assessment proposed by Blömeke et al. (2015), Gabriele Kaiser illustrated one approach towards an integration of both views in her plenary lecture on PME38 (Kaiser et al. 2014). They propose to study teachers' knowledge not only with methods traditionally applied in the cognitive tradition, as paper and pencil tests, but also using complex, authentic assessment situations to observe teachers' practices systematically. However, the path towards the integration of both perspectives seems to be rocky. Lin and Rowland (2016) state after discussing these and related ideas: "The paradigmatic differences in conceptualizations of mathematics teacher knowledge [...] remain intact".

Parallel developments in teacher research in other areas: Krauss (2011) reviews the history of teacher related research in education in general, mostly focusing on developments that were not explicitly connected to PME. He describes four phases of teacher research in the past on the quest for identifying characteristics of "good teachers", in the sense of teachers who support their students' development successfully. While the first phase did not yield strong results on the effectiveness of teachers, the second and the third phase brought up substantial knowledge that can today be found in typical texts on instructional psychology, for example on the role of instructional clarity, prevention of disturbances, and adequate speed of instruction. Since 1985, Krauss (2011) describes an increasing interest in the teacher again, now with a focus on teacher characteristics that can be connected to teacher practices and student development theoretically and empirically. The historical narrative in the mathematics education tradition is that developments within mathematics education itself led to this "discovery of the teacher" at this time, and that the developments in general educational research on the role of the teacher were viewed with scepticism at that time, due to different theoretical perspectives on classroom learning. However, some PME members surely had contacts to the general education community. Given that the trend to focus on the teacher developed about the same time in both communities, it cannot be excluded that they were connected to a certain extent—be it with the goal of integrating or of delineating the different approaches in both research communities.

13.4.3   What Can We Learn from Research on Teachers and Teaching?

Discussing frameworks: Mathematics education is a quite young science, and thus it is not clear yet in many fields, which notions are best suited to describe the phenomena and problems we observe in mathematics teaching and learning. Diverse models of teacher knowledge have been discussed in the past (see Lin and Rowland 2016 for an overview). Lin and Rowland (2016) describe attempts to find relations between these frameworks and study their unique characteristics. Whether a "Mainstream Theory", as Lin and Rowland (2016) call it, is a realistic goal of research or merely a guiding ideal is still under debate. Ideas to compare and combine different theories have been proposed to deal with these multiple perspectives in design research (Artigue and Mariotti 2014). However, it will be very interesting to follow the development of our understandings of teacher knowledge and its effects towards increasing coherence, since this field has proven productive in generating a variety of frameworks in the past (Lin and Rowland 2016).

Struggling to integrate opposing views: As indicated above, the discussion about the nature of knowledge and cognition, as well as different methodological approaches, pose a major challenge for a field of research. Along them comes the danger of the field splitting into separate subfields, but also the chance to reach a deeper understanding of the field of study, be it teachers and teaching, or students' mathematical cognition. Of course we can admit that each perspective has something to contribute, but does this really increase our understanding? Sometimes it increases our joint confusion, since we arrive at different conclusions, even from the same data. If our goal is to further our joint understanding of mathematics learning and teaching, we will not get around trying to find a common basis to talk to each other.

Talk to your neighbour: Research on mathematics teachers and teaching has developed parallel, and more or less independently, with trends in neighbouring disciplines like psychology, education, and sociology. Research on teachers is perfect for such an intense discourse with these disciplines, since effective teaching is not a purely subject-related matter. Many PME contributions are discussing not only problems that are closely connected to the specifics of mathematics teaching and learning. They address also general aspects of learning, knowing, and cognition. The increased and ongoing contact with disciplines that deal with the same issues will stimulate the discussion and scientific progress on both sides and prevent divergent developments.

13.5   Epilog

We would like to end this paper, with a few sentences from Hans Freudenthal (Mathematician and educator, who left his deep traces on the mathematics education community, one of the PME's "fathers"). The following sentences are borrowed from his plenary at PME 1983.

Fig. 13.3 Hans Freudenthal

Freudenthal (Fig. 13.3) claimed that for him an education is a human activity, which is about learning and teaching as processes, taking place in a more or less organized way". He complains that in the late 70th and early 80th, *education* meant for many people *education research*, and many publications were concerned with **states** (his words) rather than **processes**. Many of the publications were in the style of: "before the treatment and after and what happened in between was indeed a treatment rather than a teaching-learning process." Freudenthal was also happy to tell the PME 1983 conference that: about a third of the contributions in the few PME proceedings which existed at 1983, "were concerned in what I (he) like to call education, that is learning and teaching as a process" (1983, p. 46). Over the years, PME members have taken Freudenthal's insightful comments to heart. It is our wish that the community continue to follow his inspirational path.

References

Anderson, J. R., Greeno, J. G., Reder, L. M., & Simon, H. A. (2000). Perspectives on learning, thinking, and activity. *Educational Researcher, 29*(4), 11–13.

Artigue M., & Mariotti, A. M. (2014). Networking theoretical frames: The ReMath enterprise. *Educational Studies in Mathematics, 85*, 329–355.

Barmby, P., Andrà, C., Gomez, D., Obersteiner, A., & Shvarts, A. (2014). The use of eye-tracking technology in mathematics education. In P. Liljedahl, C. Nicol, S. Oesterle, & D. Allan (Eds.), *Proceedings of the Joint Meeting of PME 38 and PME-NA 36* (Vol. 1, p. 253). Vancouver, Canada: PME.

Blömeke, S., Gustafsson, J.-E., & Shavelson, R. (2015). Beyond dichotomies. *Zeitschrift für Psychologie, 223*, 3–13.

Carpenter, T. P., Fennema, E., & Romberg, T. A. (Eds.). (2012). *Rational numbers: An integration of research*. London, UK: Routledge.

Cobb, P., & Yackel, E. (1996). Constructivist, emergent, and sociocultural perspectives in the context of developmental research. *Educational Psychologist, 31*(3/4), 175–190.

Cobb, P., & Yackel, E. (2011). Chapter 4. Introduction. In E. Yackel, K. Gravemeijer, & A. Sfard (Eds.), *A journey in mathematics education research—Insights from the work of Paul Cobb* (pp. 33–40). Berlin: Springer. https://doi.org/10.1007/978-90-481-9729-3.

Confrey, J., & Kazak, S. (2006). A thirty-year reflection on constructivism in mathematics education in PME. In A. Gutierrez & P. Boero (Eds.), *Handbook of research on the psychology of mathematics education* (pp. 305–342). Rotterdam: Sense Publishers.

da Ponte, J. P., & Chapman, O (2006). Mathematics teachers' knowledge and practices. In A. Gutierrez & P. Boero (Eds.), *Handbook of research on the psychology of mathematics education: Past, present and future* (pp. 461–494). Rotterdam: Sense.

de Lange, J. (1996). Using and applying mathematics in education. In A. J. Bishop, K. Clements, C. Keitel, J. Kilpatrick, & C. Laborde (Eds.), *International handbook of mathematics education: Part one* (pp. 49–97). Dordrecht: Kluwer Academic Publisher.

Dehaene, S. (1992). Varieties of numerical abilities. *Cognition, 44*(1), 1–42.

English, L., & Sriraman, B. (2009). *Theories of mathematics education: Seeking new frontiers.* Springer: Heidelberg.

Even, R. (2009). Teacher knowledge and teaching: Considering the connections between perspectives and findings. In M. Tzekaki, M. Kaldrimidou, & H. Sakonidis (Eds.), *Proceedings of the 33rd Conference of Psychology of Mathematics Education* (Vol. 1, pp. 147–148). Thessaloniki, Greece: PME.

Fischbein, E., & Kedem, I. (1982). Proof and certitude in the development of mathematical thinking. In A. Vermandel (Ed.), *Proceedings of the Sixth International Conference for the Psychology of Mathematics Education* (pp. 128–131). Antwerp, Belgium.

Freudenthal, H. (1983). Is heuristics a singular or plural? In R. Hershkowitz (Ed.), *Proceedings of the Seventh International Conference for the Psychology of Mathematics Education* (pp. 193–198). Rehovot, Israel: The Weizmann Institute of Science.

Freudenthal, H. (1991). *Revisiting mathematics education: China lectures.* Dordrecht: Kluwer Academic Publishers.

Gutierrez, A., & Boero, P. (2006). *Handbook of research on the psychology of mathematics education.* Rotterdam: Sense Publishers.

Gutierrez, A., Leder, G., & Boero, P. (2016). *The second handbook of research on the psychology of mathematics education.* Rotterdam: Sense Publishers.

Haseman, K. (1980). On the understanding of concepts and rules in secondary mathematics: Some examples illustrating the difficulties. In R. Karplus (Ed.), *Proceedings of the Fourth International Conference of the International Group for the Psychology of Mathematics Education* (pp. 68–74). Berkeley, California.

Heemsoth, T., & Heinze, A. (2014). How should students reflect upon their own errors with respect to fraction problems? In S. Oesterle, P. Liljedahl, C. Nicol, & D. Allan (Eds.), *Proceedings of the 38th Conference of the International Group for the Psychology of Mathematics Education and the 36th Conference of the North American Chapter of the Psychology of Mathematics Education* (Vol. 1, pp. 265–272). Vancouver, Canada: PME.

Herscovics, N., Bergeron, J. C., & Kieran, C. (1983). A critique of Piaget's analysis of multiplications. In R. Hershkowitz (Ed.), *Proceedings of the Seventh International Conference for the Psychology of Mathematics Education* (pp. 193–198). Rehovot, Israel: The Weizmann Institute of Science.

Hershkowitz, R. (2009). Contour lines between a model as a theoretical framework, and the same model as a methodological tool. In B. B. Schwarz, T. Dreyfus, & R. Hershkowitz (Eds.), *Transformation of knowledge through classroom interaction* (pp. 273–280). London, UK: Taylor & Francis, Routledge.

Hershkowitz, R., Tabach, M., & Dreyfus, T. (2017). Creative reasoning and shifts of knowledge in the mathematics classroom. *ZDM—The International Journal for Mathematics Education, 49,* 25–36.

Hoyles, C. (1992). Illuminations and reflections: Teachers, methodologies and mathematics. In W. Geeslin (Ed.), *Proceedings of the Conference of the International Group for the Psychology of Mathematics Education* (Vol. 3, pp. 263–286). Durham, NH: PME.

Jaworski, B. (2011). Situating mathematics teacher education in a global context. In N. Bednarz, D. Fiorentini, & R. Huang (Eds.), *International approaches to the professional development for mathematics teachers* (pp. 2–50). Ottawa: University of Ottawa.

Kaiser, G., Blömeke, S., Busse, A., Döhrmann, M., & König, J. (2014). Professional knowledge of (prospective) mathematics teachers: Its structure and development. In P. Liljedahl, C. Nicol, S. Oesterle, & D. Allan (Eds.), *Proceedings of the Joint Meeting of PME 38 and PME-NA 36* (Vol. 1, pp. 35–54). Vancouver, Canada: PME.

Krauss, S. (2011). Das Experten-Paradigma in der Forschung zum Lehrerberuf. In E. Terhart, H. Bennewitz, & M. Rothland (Eds.), *Handbuch der Forschung zum Lehrerberuf* (pp. 171–191). Münster: Waxmann.

Lerman, S. (2006). Socio-cultural research in PME. In A. Gutierrez & P. Boero (Eds.), *Handbook of research on the psychology of mathematics education* (pp. 305–342). Rotterdam: Sense Publishers.

Lin, F. L., & Rowland, T. (2016). Pre-service and in-service mathematics teachers' knowledge and professional development. In A. Gutiérrez, G. Leder, & P. Boero (Eds.), *The second handbook of research on the psychology of mathematics education* (pp. 483–520). Rotterdam: Sense Publishers.

Lithner, J. (2008). A research framework for creative and imitative reasoning. *Educational Studies in Mathematics, 67*(3), 255–276.

Llinares, S., & Krainer, K. (2006). Mathematics (student) teachers and teacher educators as learners. In A. Gutiérrez & P. Boero (Eds.), *Handbook of research on the psychology of mathematics education: Past, present and future* (pp. 429–459). Rotterdam: Sense Publishers.

Luwel, K., Onghena, P., Torbeyns, J., Schillemans, V., & Verschaffel, L. (2009). Strengths and weaknesses of the choice/no-choice method in research on strategy use. *European Psychologist, 14*(4), 351–362.

Noelting, G. (1978). The development of proportional reasoning in the child and adolescent through combination of logic and arithmetic. In E. Cohors-Fresenborg & I. Wachsmuth (Eds.), *Proceedings of the Second International Conference for the Psychology of Mathematics Education* (pp. 242–277). Osnabrück, Germany: Universität Osnabrück.

PME. (1980–2016). *Constitution of IGPME.* http://www.igpme.org/index.php/home. Accessed November 22, 2016.

Prediger, S., Bikner-Ahbahs, A., & Arzarello, F. (2008). Networking strategies and methods for connecting theoretical approaches: First steps towards a conceptual framework. *ZDM Mathematics Education, 40*, 165–178. https://doi.org/10.1007/s11858-008-0086-z.

SGMG (School Mathematics Study Group). (1961). *Mathematics for the elementary school.* https://catalog.hathitrust.org/Record/001883911. Accessed May 18, 2017.

Skemp, R. R. (1977). Relational understanding and instrumental understanding. *Mathematics Teaching, 77*, 20–26.

Skemp, R. R. (1986). *The psychology of learning mathematics* (2nd ed.). London: Penguin Books.

Sreen, L. A. (1990). *On the shoulder of giants—New approaches to numeracy.* Washington, DC: National Academic Press.

Swan, M. (1983). Teaching decimal place value: A comparative study of "conflict" and "positive only" approaches. In R. Hershkowitz (Ed.), *Proceedings of the Seventh International Conference for the Psychology of Mathematics Education* (pp. 211–216). Rehovot, Israel: The Weizmann Institute of Science.

Tall, D. (1998). *Efraim Fischbein, 1920–1998, Founder President of PME. A tribute.* http://homepages.warwick.ac.uk/staff/David.Tall/pdfs/dot1999b-fischbein-tribute.pdf.   Accessed November 30, 2016.

Tall, D. O., & Vinner, S. (1981). Concept image and concept definition in mathematics, with special reference to limits and continuity. *Educational Studies in Mathematics, 12*(2), 151–169.

Tzur, R., & Leikin, R. (2015). Interweaving mathematics education and cognitive neuroscience. In K. Beswick, T. Muir, & J. Wells (Eds.), *Proceedings of 39th Psychology of Mathematics Education Conference* (Vol. 1, pp. 91–124). Hobart, Australia: PME.

Van Dooren, W. (2016). Understanding obstacles in the development of the rational number concept—Searching for common ground. In C. Csikos, A. Rausch, & J. Szitanyi (Eds.), *Proceedings of the 40th Conference of the International Group for the Psychology of Mathematics Education* (Vol. 1, pp. 383–412). Szeged, Hungary: PME.

Vergnaud, G. (1987). On constructivism. In J. C. Bergeron, N. Herscovics, & C. Kieran (Eds.), *Proceedings of the Seventh Conference of the International Group for Psychology of Mathematics Education* (pp. 42–54). Montreal.

Vinner, S., & Hershkowitz, R. (1980). Concept image and common cognitive paths in the development of some simple geometrical concepts. In R. Karplus (Ed.), *Proceedings of the 4th Conference of the International Group for Psychology of Mathematics Education* (pp. 177–184). Berkeley: University of California.

Vygotsky, L. (1978). *Mind in society: The development of higher psychological processes.* Cambridge, MA: Harvard University Press.

Chapter 14
ICMI 1966–2016: A Double Insiders' View of the Latest Half Century of the International Commission on Mathematical Instruction

Bernard R. Hodgson and Mogens Niss

Abstract This paper concentrates on the latest five decades of the International Commission on Mathematical Instruction. We had the privilege of occupying leading positions within ICMI for roughly half the period under consideration, which has provided us with a unique standpoint for identifying and reflecting on main trends and developments of the relationship between ICMI and mathematics education. The years 1966–2016 have seen marked trends and developments in mathematics teaching and learning around the world, at the same time as mathematics education as a scientific discipline came of age and matured. ICMI as an organisation has not only observed these developments but has also been a key player in charting and analysing them, as well as in fostering and facilitating (some of) them. We offer, here, observations, analyses and reflections on key issues in mathematics education as perceived by us as ICMI officers, and as influenced by ICMI.

Keywords ICMI (International Commission on Mathematical Instruction)
History of ICMI · Mathematics education as a scientific discipline
Internationalisation of mathematics education

14.1 Introduction

The year 2016 marks more than a century of existence of the International Commission on Mathematical Instruction (ICMI) since its establishment in Rome in 1908. This paper concentrates on the last five decades of that period.

B. R. Hodgson (✉)
Université Laval, Quebec, Québec, Canada
e-mail: Bernard.Hodgson@mat.ulaval.ca

M. Niss
Roskilde University, Roskilde, Denmark

G. Kaiser et al. (eds.), *Invited Lectures from the 13th International Congress on Mathematical Education*, ICME-13 Monographs,
https://doi.org/10.1007/978-3-319-72170-5_14

During the years 1966–2016, ICMI witnessed and took note of marked trends and developments in mathematics teaching and learning around the world, in terms both of the socio-economic and institutional boundary conditions and of the diverse and multi-faceted practices of mathematics education. This half century is also the one in which mathematics education as a scholarly and scientific discipline came of age and matured. ICMI as an organisation has not only observed these developments but has also been a key player in charting and analysing them, as well as in fostering and facilitating (some of) them, for instance by way of conferences, studies or other activities.

It has been our privilege to having occupied leading positions in the Executive Committee (EC) of ICMI for roughly half the period of time under consideration, including those of consecutive Secretaries-General from 1991 to 2009. This has provided us with a unique platform from which we could identify and reflect on the main trends and developments of the relationship between ICMI and mathematics education from the perspective of two "insiders".

Our paper thus offers observations, analyses and reflections on key issues in mathematics education as perceived by us as ICMI officers, and as influenced by ICMI.

The history of ICMI and the roles played by some of its protagonists have also been subject of attention at recent International Congresses on Mathematical Education (ICMEs). This is reflected for instance by the regular lecture by Howson (2008) presented at ICME-10 (Copenhagen, 2004), as well as by the talk by Arzarello et al. (2008) at ICME-11 (Monterrey, 2008).

Although the focus of our paper will be on the years 1966–2016, we have found it necessary to provide a brief outline of ICMI's first 58 years, so as to set the stage for understanding and appreciating the target years. The paper is thus divided into four sections:

- 1908–1982: Foundation, (re)formation and "the first crisis" around ICMI;
- 1983–1998: Consolidation and expansion;
- 1999–2016: Calm waters, but with "a second crisis" around ICMI; and finally
- ICMI and the field of mathematics education.

14.2 1908–1982: Foundation, (Re)Formation and "The First Crisis" Around ICMI

Following a suggestion of the US mathematician and teacher educator David E. Smith made in the then recently created journal *L'Enseignement Mathématique* (Smith 1905, p. 469), ICMI was first established at the General Assembly of the 4th International Congress of Mathematicians (ICM) held in Rome in 1908, based on the following resolution:

> The Congress, recognizing the importance of a comparative study on the methods and plans of teaching mathematics at secondary schools, charges Professors F. Klein, G. Greenhill, and Henri Fehr to constitute an International Commission to study these questions and to present a report to the next Congress. (Lehto 1998, p. 13)[1]

This instigated what might be called the "Klein Era" of ICMI—from the name of ICMI's first President, Felix Klein (1849–1925), see Bass (2008)—characterised by activities focusing on curricular reflections and comparisons. The first host of results of the Commission's work, undertaken by mathematicians with educational interests, teachers of high reputation and institutional representatives, were presented at the ICM in Cambridge (UK), in 1912. The mandate of the Commission was extended and the work continued during WWI. By 1920, 310 reports (totalling more than 13,500 pages) had been produced from eighteen countries plus the so-called Central Committee, the ancestor of the EC—see Lehto (1998, p. 14) and Fehr (1920–21, p. 339). Even though the Commission was international and open to all countries, it was, in fact, highly Euro- and US-centric.

Because of difficulties in international relationships caused by WW1, the so-called "Central Powers" were excluded from the then newly established International Mathematical Union (IMU)—historically named the "Old IMU" in the parlance of Lehto (1998). Nevertheless, the mandate of the Commission was re-confirmed during the 1920s and 1930s, but activity was progressively reduced.

After WWII there was a strong desire to avoid international division, so all countries were invited to take part in the international mathematical collaboration. Thus IMU was re-established in 1951, and in 1952 ICMI was re-constituted as a sub-commission of IMU with the following brief, forming part of the Terms of Reference (and still in force today):

> The Commission shall be charged with the conduct of the activities of IMU, bearing on mathematical and scientific education, and shall take the initiative in inaugurating appropriate programmes designed to further the sound development of mathematical education at all levels and to secure public appreciation of its importance. (ICMI Terms 1954)[2]

The members of ICMI were then national representatives of IMU member states, plus an Executive Committee elected by the General Assembly at the ICMs.[3] During the years 1952–1966, ICMI gradually moved from "Old ICMI" style

---

[1]Original text: "Il Congresso, avendo riconosciuto la importanza di un esame accurato dei programmi e dei metodi d'insegnamento delle matematiche nelle scuole secondarie delle varie nazioni, confida ai Professori Klein, Greenhill e Fehr l'incarico di costituire un Comitato internazionale che studii la questione e ne riferisca al prossimo Congresso." (Castelnuovo 1909, p. 33)

[2]It is interesting that this brief asks for the furthering of "sound development of mathematical education" and the securing of "public appreciation", both of which are of a normative nature.

[3]This is in distinction to the current situation, where the *members* of ICMI are now countries, as was always the case with IMU. Hence the members of the ICMI EC are no longer considered as "members of ICMI". This change in the definition of ICMI membership was formalised in the 2002 revision of the ICMI Terms of Reference—see http://www.mathunion.org/icmi/icmi/icmi-as-an-organisation/terms-of-reference/. The members of the Commission, as in the original 1954 wording, now form the ICMI General Assembly.

actions, where mathematics education was predominantly seen as a "national business", to more international activities, marked by the concerns of individual actors on the stage and involving, along the road, mathematics educators ("didacticians of mathematics"). This evolution eventually lead to the emergence of an international mathematics education community, collaborating with organisations such as OEEC/OECD and UNESCO, which in turn gave rise to initiatives towards developing countries. One instance of this development was the launching in 1961 of the *Comité Interamericano de Educación Matemática* (CIAEM)—see Hodgson et al. (2013, pp. 911–913)—on the initiative of Marshall Stone (1903–1989), ICMI President for the term 1959–1962.

The 1950s saw an emerging interest in curriculum design and reform combined with approaches to teaching aligned with these reforms, whilst paying attention to contributions from psychology and general education (e.g., Jean Piaget and Jerome Bruner). One also began to gradually realise that (good) teaching is not the same as (good) lecturing. The establishment of the *Commission Internationale pour l'Étude et l'Amélioration de l'Enseignement des Mathématiques* (CIEAEM)—(Hodgson et al. 2013, pp. 910–911)—initiated by Caleb Gattegno and with early members including Gustave Choquet, Jean Dieudonné, Georges Papy and Piaget, also exerted an influence on ICMI's development. Among the founding members of CIEAEM were also André Lichnerowicz (1915–1998), ICMI President for the term 1963–1966, and Hans Freudenthal (1905–1990), Lichnerowicz' successor as ICMI President—André Delessert (1923–2010) served as Secretary-General under both Lichnerowicz and Freudenthal. During the presidencies of Stone and Lichnerowicz, ICMI became an agent for fostering and promoting the set-theory based New Math (or *mathématiques modernes*) in school curricula around the world. This can be seen in the first volumes of UNESCO's series *New Trends in Mathematics Teaching* (from 1966), published in collaboration with ICMI.

With this historical background in view, we now enter the first segment of the time span covered by this paper, 1967–1982.

A significant turning point in ICMI's life was the presidency of Hans Freudenthal (1967–1970). Even though this presidency lasted only one term, as was usual in those days, Freudenthal introduced so many new features into ICMI and to mathematics education that his influence lasted more than a decade after his presidency. So, it is fair to use the term the "Freudenthal Era"—in the spirit of Bass (2008)—for the years 1967–1980. His presidency marked a break away from New Math and—albeit slowly at first—from the dominance of research mathematicians in mathematics education that had been prevalent up till then. One of Freudenthal's most significant moves regarding mathematics education was the inauguration of the International Congresses on Mathematical Education (the ICMEs), the first of which was held in Lyon in 1969. At the same time, but not formally under the auspices of ICMI, he launched the world's first international journal of mathematics education *Educational Studies in Mathematics* (ESM) in 1968.[4] The developments leading to these decisions are captured in a resolution adopted at ICME-1:

> The theory of mathematical education is becoming a science in its own right, with its own problems both of mathematical and pedagogical content. The new science should be given a place in the mathematical departments of Universities and Research Institutes, with appropriate academic qualifications available. (Editorial Board of Educational Studies in Mathematics 1969, p. 284)

It would be wrong to say that these initiatives were received with applause by IMU. Secretary Otto Frostman wrote as follows to Freudenthal in December 1967 (Frostman 1967):

> On the ESM: "I must admit that I am not too happy about the new pedagogical journal. Do you really think that there is a market for two international journals of that kind (I do not)? If you are not satisfied with *L'Enseignement*, ICMI's official journal, perhaps it would be better to try to reform it."

> On ICME: "I can agree with very much of your criticism of the meetings of ICMI at the International Congresses [of Mathematicians], but I am not sure that ICMI should isolate itself from those who have, primarily, a scientific interest but who have, nevertheless, very often taken part in the discussions of ICMI."

One reason for such reactions from IMU might well have been that Freudenthal launched these initiatives without much interaction with IMU officials, so that IMU was often facing *faits accomplis* from ICMI. This constituted the first ICMI/IMU crisis. This is well captured by a comment of IMU President Henri Cartan, in reaction to an initiative taken by Freudenthal concerning ICME-2. In October 1970, right at the beginning of a letter to IMU Secretary Frostman, Cartan wrote: "Freudenthal me donne encore du souci" ("Freudenthal again causes me worries") (Cartan 1970). This time, Cartan was worried because Freudenthal wanted the outgoing ICMI Executive Committee to appoint the International Programme Committee for ICME-2 with only 2½ months left of his presidency. This inaugurated some tension between the ICMI President and the IMU leadership, arising again from time to time in the years to come.

However, these were also years with an abundance of initiatives on the part of ICMI. Quite a few of these initiatives were taken during the presidencies of James Lighthill (1924–1998)—ICMI President for 1971–1974 with Edwin Maxwell (1907–1987) as Secretary-General—and Shokichi Iyanaga (1906–2006)—1975–78 President with Yukiyoshi Kawada (1916–1993) as Secretary-General. In addition to sponsoring the ICMEs (ICME-2, 1972; ICME-3, 1976; and ICME-4, 1980—quadrennial except for the first interval), ICMI affiliated two Study Groups at ICME-3,

---

[4]ICMI's official organ since its inception in 1908, *L'Enseignement Mathématique* (launched in 1899), was never really a mathematics education journal, even though it did—and still does—publish education reports and papers from time to time. In the opinion of Freudenthal, in relation to the launching of ESM, the "contributions [of *L'Enseignement Mathématique*] on education were not pedagogical but organisatory and administrative." (Freudenthal 1967)

the *International Group for the Psychology of Mathematics Education* (PME) and the *International Study Group on the Relations between the History and Pedagogy of Mathematics* (HPM), instigated the so-called ICMI Regional Conferences, and held other ICMI-related symposia in Europe, Africa, India, Latin America, and Southeast Asia as an expression of the first outreach efforts of the Commission. Finally, ICMI established the *ICMI Bulletin* as a rather informal means of communication within the "ICMI family".

These developments of ICMI were concurrent with the emergence of mathematics education as a scientific and scholarly discipline, a field of systematic reflection and investigation. At the institutional level this was marked by the establishment—in addition to ESM, founded by an ICMI President—of the *Journal for Research in Mathematics Education* (JRME) and the *International Journal of Mathematical Education in Science and Technology* (IJMEST) in 1970, and of *For the Learning of Mathematics* (FLM) and *Recherches en Didactique des Mathématiques* (RDM) in 1980. Whilst secondary education received most of the attention in the first fifty years of ICMI, primary and tertiary education now entered the field of interest as well. The fourth volume of UNESCO's *New Trends in Mathematics Teaching* series, which appeared in 1979 (volume I had been published in 1966, II in 1970, and III in 1972), contained chapters on the goals of mathematics teaching (by Ubiratan D'Ambrosio), on applications (by Henry Pollak) and on algorithms (by Arthur Engel), which went beyond the teaching of established mathematical areas and topics.

The 1979–1982 term of Hassler Whitney (1907–1989) as President and Peter Hilton (1923–2010) as Secretary-General turned out to be a difficult one as far as relationships both with the IMU and within ICMI itself were concerned. In the minutes of an IMU Executive Committee meeting held in 1980, one can read: "The [IMU] EC expresses concern about the lack of communications between IMU and ICMI." And again in 1981: "Much concern concerning the difficulties that arose in the [ICMI] EC." (IMU EC Minutes 1980, p. 14, and 1981, p. 25)

The difficulties were to do with Whitney's wilfulness in his way of undertaking his office—for example the EC only rarely met—and with the fact that the EC seemed to think of Hilton's role as Secretary-General to be that of an office clerk rather than that of an organiser and decision making executive officer. At least this was the perception he expressed in a confidential letter to one the Ex-Officio members of the ICMI EC, IMU Secretary Jacques-Louis Lions: "It is clear to me that I was expected by some of my colleagues on the EC to act purely in a 'secretarial' capacity, (…) and that I could not exercise the influence I hoped to have from that position" (Hilton 1980). That perception had led Hilton to present his resignation from the Secretary-General's office. However, for reasons (yet) unknown this resignation did not materialise and Hilton finally remained as the ICMI Secretary-General till the end of his term.

The controversies were also due to the fact that members of the ICMI EC put forward as its candidate for the next President the Danish mathematics educator Bent Christiansen (1921–1996), ICMI Vice-President for two terms, since 1975. This was not well received by the IMU leadership. Thus the IMU President, the Swedish mathematician Lennart Carleson, in a letter to the ICMI EC at the end of 1981 wrote: "The [next] President should be a well-known mathematician with established interests in education" (Carleson 1981). Evidently, IMU officers thought that mathematics education was far too important to be left to the mathematics educators.

Eventually, IMU elected the French mathematician Jean-Pierre Kahane (1926-2017) as the next President and the British mathematician/mathematics educator Geoffrey Howson as the Secretary-General. Besides, Bent Christiansen was elected to a third term as Vice-President. The—perhaps implicit—mandate of Kahane and Howson was to put ICMI back on track, or at least—as can be seen in the video interview with Kahane made for the ICMI Centennial in 2008 (under "Interviews and film clips" on the *History of ICMI* site at http://www.icmihistory. unito/it)—to revitalise ICMI. This takes us to our next section.

14.3   1983–1998: Consolidation and Expansion

Kahane and Howson both served for two terms: 1983–1989. This was the first time, since the presidency of Klein, that an ICMI President was elected twice. Their terms represented a much wanted consolidation and stabilisation of the ICMI leadership after a number of years of turbulence and tension.

During the Kahane-Howson era, ICMI instigated significant new activities (some of which had been proposed in previous terms), above all the first series of the ICMI Studies, according to the following format:

- for each Study, the ICMI EC selected a theme, described in general terms, and appointed an International Programme Committee;
- the Programme Committee produced a Discussion Document to be circulated internationally, inviting written reactions;
- based on the written reactions, a rather small invited symposium/Study Conference was organised;
- based on the conference activities, a comprehensive Study Volume was written, typically with Kahane and Howson as the main authors. Sometimes also conference proceedings were put out.

The Study Volumes for the first five Studies (1–5), which were—as a deliberate choice—rather slim, were all published by Cambridge University Press in the *ICMI Study Series*. They were devoted to the following themes, four of which were already identified at the very outset of the Kahane-Howson term[5]:

---

[5]Howson (1982) presents the idea of ICMI Studies under the heading "Possibilities for future action" and describes the first four of these (but calling them "symposia"). In the report on ICMI for the year 1983, Howson (1983) uses the word "Studies".

- *The Influence of Computers and Informatics on Mathematics and Its Teaching* (Strasbourg, 1985)
- *School Mathematics in the 1990s* (closed seminar, Kuwait, 1986)
- *Mathematics as a Service Subject* (Udine, 1987)
- *Mathematics and Cognition* (no conference, written under the auspices of PME, published in 1990)
- *The Popularization of Mathematics* (Leeds, 1989).

[A report on these five Studies presented in 1990 at the Kyoto ICM can be found in Hodgson (1991).] The Studies can be seen as a reflection of needs pertinent to new issues and developments in mathematics education concerning technology, school mathematics, service subject, cognition and popularisation. In the first series, the Studies hadn't yet found a uniform format.

During the Kahane-Howson era, the *International Organisation of Women and Mathematics Education* (IOWME) in 1987 became the third ICMI Affiliated Study Group, and regional meetings continued to be supported, as was collaboration with UNESCO. The activities and roles of the representatives of member countries caused concern, as the links between many of them and the EC were frail or non-existent. So, the IMU General Assembly held on the occasion of the 1990 Kyoto ICM passed a resolution (#5) limiting the number of consecutive terms served by representatives—this came to be known as the "Kobe rule", from the name of the host city of the General Assembly:

> All Adhering Organizations are reminded that they should review their national representation on ICMI and that normally national representatives should not be asked to serve for more than two consecutive four-year terms. (IMU General Assembly 1990, p. 8)

Two ICMEs, both of which added new facets to the format and perspectives of the congresses, were held during their terms, ICME-5 (Adelaide, 1984) and ICME-6 (Budapest, 1988).

The range and scope of mathematics education as a field of research and development expanded considerably during the Kahane-Howson era. The educational levels dealt with expanded "downwards" to kindergarten and pre-school children, and "upwards" to tertiary programmes, especially those involving mathematics as a service subject. Also the public image and perception of mathematics and their influence on mathematics education received increasing attention, hence the Study on "popularisation".

With particular regard to research, foci moved from curriculum design and teaching to mathematics learning on the one hand, and to classroom communication in mathematics on the other. But new foci concerning mathematical substance per se gained momentum as well—such as problem solving, applications and modelling, and technology in mathematics education.

These different sorts of expansion led people to begin to systematically reflect on the nature of mathematics education research, not least the German mathematics educator Hans-Georg Steiner, who established a forum, *Theory of Mathematics Education* (TME), for discussing these issues—first at ICME-5 in 1984, leading to the paper (Steiner 1985), and in subsequent colloquia elsewhere (see also Hodgson and Rogers (2012) for comments about TME).

The following two ICMI EC terms, 1991–1998, for which the Spanish mathematician with a strong interest in mathematics education Miguel de Guzmán (1936–2004) and the Danish mathematician/mathematics educator Mogens Niss were President and Secretary-General, respectively, can be characterised as one of continued consolidation and expansion of ICMI and its activities along the lines established in the Kahane-Howson years. This era was one of continuity and calm reform, not one of abrupt changes and revolution, even though some dark clouds emerged at the end of the second de Guzmán-Niss term (see the next section).

One of the most significant changes during those years was the re-shaping of the ICMI Studies. First, their goals were clarified as being to provide a state-of-the art account and review of the *problématiques* and topics chosen for the Studies, for which developments in research were to receive increased emphasis. Moreover, there was an increased uniformisation of the Study formats as regards the nature and role of their main components (see above): International Programme Committee—Discussion Document—Study Conference—Study Volume. There was a growing and widening interest and participation in the Studies, which considerably expanded the "ICMI family". This was also meant to be stimulated by the fact that the relatively expensive Study Volumes were made available to individuals at reduced rates by agreements between ICMI and the publishers (first Kluwer, then Springer when they bought Kluwer). Unfortunately, however, these agreements were never as widely known or used as anticipated. Six Study Conferences were held during the Guzmán-Niss era, the resulting volumes being published in the *New ICMI Study Series* (Studies 6–11):

- *Assessment in Mathematics Education* (Calonge, 1991, resulted in two books)
- *Gender and Mathematics Education* (Höör, 1993)
- *What is Research in Mathematics Education and What are its Results?* (College Park, 1994)
- *Perspectives on the Teaching of Geometry for the 21st Century* (Catania, 1995)
- *The Role of the History of Mathematics in the Teaching and Learning of Mathematics* (Luminy, 1998)
- *The Teaching and Learning of Mathematics at University Level* (Singapore, 1998).

It was systematically attempted by the ICMI EC, during those years, to always have three Studies underway in different stages of completion at the same time: one for which the International Programme Committee has been appointed and the Discussion Document is in the process of being written; one for which the Study Conference is under planning; and one for which the Study Volume is being written and edited. (This resulted in having roughly three Study Conferences per four-year term of a given ICMI EC.) Along with the above-mentioned key purpose of an ICMI Study as being to capture and gauge the state-of-affairs and trends concerning pertinent issues and topics, the Studies also had a dual purpose, namely, for ICMI to identify, shape and facilitate work with new foci.

During the de Guzmán-Niss era the *World Federation of National Mathematics Competitions* (WFNMC) was accepted as a new Affiliated Study Group (1994). On the personal initiative of de Guzmán, announced in his Presidential address at ICME-7, a so-called Solidarity Programme and Fund were established. At the same time it was decided to include a 10% Solidarity Tax on ICME conference fees as part of a concerted effort to reach out to new places and groups in mathematics education, in both geographical, socio-economic and cultural terms. Moreover, efforts were made by the ICMI EC to stimulate the creation of Sub-Commissions of ICMI so as to provide a bridge between ICMI and its member states and to compensate for the sometimes insufficient functioning of some country representatives. Finally, the *ICMI Bulletin* was consolidated both in format and publishing regularity during those years.

The ICMEs held during the de Guzmán-Niss terms were ICME-7 (Québec, 1992) and ICME-8 (Sevilla, 1996). As a reflection of the general growth of ICMI activities and undertakings, the time line for deciding upon and planning the ICMEs became extended considerably, roughly 5–6 years in advance. And in that respect, controversies and conflicts sometimes began to arise.

A look at the concurrent development of mathematics education as a field reveals an extension of its radius of action to encompass

- assessment;
- history and philosophy of mathematics and their impact on mathematics education;
- teacher education and professional development;
- students' and teachers' beliefs and affect in mathematics;
- socio-cultural factors influencing mathematics teaching and learning;
- equity.

New international journals were established in those years, including the *Mathematics Education Research Journal* (MERJ—1989), *Nordisk Matematikdidaktik* (NOMAD—1993), *ZDM–Mathematics Education* (1997), the *Journal of Mathematics Teacher Education* (JMTE—1998), *Mathematical Thinking and Learning* (MTL—1999), whilst some "national" journals became increasingly international, as was the case with the *Journal für Mathematik-Didaktik* (JDM—1980). Moreover many new ideas for Studies were in the pipeline by the end of 1998.

14.4   1999–2016: Calm Waters, but with "A Second Crisis" Around ICMI

With the election by the 1998 IMU General Assembly of Hyman Bass (President) and Bernard Hodgson (Secretary-General), an ICMI leadership duo once again took office for what turned out to be two consecutive terms (1999–2006). According to the 2002 revision of the Terms of Reference for ICMI, "Secretary-General" was then instituted as the official title of what was previously named "Secretary"[6]—see Bass (2002). Right away the Bass-Hodgson era opened with some problematic issues that the new leadership had to deal with "on the first day in the office".

As part of the transition from the previous era, it was assumed that ICME-10 was going to be held in Brazil (the planning of ICME-9, Makuhari/Tokyo, 2000, was already well under way). However, already in December 1998, the incoming ICMI President received a letter, signed jointly by the Brazilian representative to ICMI and the President of the Brazilian Mathematical Society, speaking against the possibility of ICME-10 being hosted by Brazil. This point of view was presented as being "shared by the Council" of the Society—see Soares and Cordaro (1998). Since holding an ICME requires the concerted effort of all relevant parties in a country, including of course the research mathematicians, this was in effect a veto statement. So, the new ICMI leadership had to work hard for several months to find an alternative host country, eventually persuading the so-called Nordic Countries (Denmark, Iceland, Finland, Norway and Sweden) to expedite previously expressed ideas to host ICME-11 in that region. Eventually, Copenhagen was chosen as the venue for ICME-10. This course of events urged ICMI to develop a more closely monitored bidding process for future ICMEs, including a 7-year in advance "preliminary declaration of intention of presenting a bid to act as host." (Hodgson 2000, p. 14)

The other problematic issue, too, was a leftover from the previous term. At the 1998 ICM, held in Berlin, serious problems about the education section of the scientific programme occurred. Instead of accepting ICMI—IMU's commission for mathematics education—as the responsible body for the education activities of IMU's own ICM, the general Programme Committee for the congress designed these activities by itself. De Guzmán and Niss reacted vigorously to the IMU leadership, who agreed to sort things out for future ICMs together with the new ICMI EC. This was achieved during the Bass-Hodgson era.

The "ICM crisis" provided momentum to thoughts prevailing in some ICMI quarters about the justification of having ICMI as an organization living "inside" the IMU, leading to the question: "Should ICMI seek independence from IMU?". Michèle Artigue, at that time Vice-President of ICMI, later returned to this issue in her Presidential address at the ICMI Centennial Symposium held in Rome in 2008:

> Retrospectively this crisis was beneficial. It obliged the ICMI EC to deeply reflect about the nature of ICMI and what we wanted ICMI to be. This led us to reaffirm the strength of the epistemological links between mathematics and mathematics education. (Artigue 2008, p. 190)

---

[6]This change of nomenclature, to some extent of a trivial nature, is nonetheless related to the perception and understanding of the role attached to this position within the ICMI EC—see the Hilton episode discussed above.

So, there was, at the very outset of the Bass-Hodgson terms, an urgent need to re-establish a relationship of mutual understanding and respect between IMU and ICMI, and to reinvigorate links through concrete actions. This became a central objective of the ICMI EC. As a first step it was agreed that the ICMI President and Secretary-General would regularly be invited to the IMU EC meetings, whilst the IMU President or Secretary would attend ICMI EC meetings. However, the most marked outcome of the growing harmony and intensive collaboration between IMU and ICMI was a new constitutional foundation of ICMI as a commission of IMU. In fact a truly historic and unexpected change of the governance of ICMI took place during the years 2002–2006.

The 2002 IMU General Assembly requested a change in the election procedure of the IMU EC, introducing a Nominating Committee to produce a slate of proposals for the EC members. The first proposal of IMU was that this same Nominating Committee would also produce the slate for the ICMI election, but it was promptly stressed by the ICMI EC that this scheme would not pay sufficient attention to the specificity of ICMI and its community. The ensuing discussions eventually gave rise to an agreement between the IMU President John Ball and the ICMI EC, reached at the ICMI EC meeting during ICME-10 in 2004, leading to the introduction of a specific ICMI Nominating Committee whose task is to propose a slate to be voted on, not by the IMU General Assembly, but by the ICMI General Assembly. This major change of constitution was put before the following IMU General Assembly, held in 2006 in Santiago de Compostela Spain, which—after a rather fierce debate in which ICMI President Hyman Bass, in his capacity as a distinguished mathematician, played a crucial part—decided to adopt the proposed change. The first election according to the new scheme took place at ICME-11 in Monterrey in 2008, where the 2010–2012 EC was elected by the ICMI General Assembly.[7]

Another reflection of the improved relationship between ICMI and IMU was the so-called "Pipeline Project", launched in 2004 on the request of IMU in order to chart the supply and demand for mathematics students and personnel in educational institutions and in workplaces. One task was to provide data for decision making and for a better understanding of the situation internationally. Reports were presented at ICME-11 (2008) and ICM-2010.

Several other new initiatives were taken during the Bass-Hodgson era. For instance collaboration with UNESCO, which because of funding problems had been rather dormant during the previous era, was renewed. The two organisations thus collaborated on establishing the travelling exhibition "Experiencing Mathematics", which was launched at ICME-10 in 2004 and was thereafter visited by around 1 million pupils, students, teachers and parents in 50 cities in 20 countries. There were also actions towards reinforcing the links with *L'Enseignement Mathématique*,

---

[7]Consecutive 3-year terms of office for the 2007–2009 and 2010–2012 ICMI ECs allowed transferring the election year from the IMU to the ICMI General Assembly.

the official organ of ICMI since its inception in 1908, notably through a joint *L'Enseignement Mathématique*-ICMI symposium held in 2000 to celebrate the first one hundred years of the journal, established in 1899—see Coray et al. (2003).

At a late stage in the last term of de Guzmán and Niss, the ICMI EC had received a proposition to establish ICMI awards so as to recognise outstanding contributions to mathematics education research and development.[8] The proposition was carried over to the first Bass-Hodgson EC, which decided to establish two ICMI Awards, named after legendary ICMI Presidents: the Felix Klein Award, honouring lifetime achievement, and the Hans Freudenthal Award, honouring a major cumulative programme of research. These awards are awarded in odd-numbered years, from 2003 on.

The ICMI Study Series was continued during the Bass-Hodgson terms with Studies 12–17, thus pursuing the rhythm of having three Studies in progress at a given time:

- *The Future of the Teaching and Learning of Algebra* (Melbourne, 2001)
- *Mathematics Education in Different Cultural Traditions: A Comparative Study of East Asia and the West* (Hong Kong, 2002)
- *Applications and Modelling in Mathematics Education* (Dortmund, 2004)
- *The Professional Education and Development of Teachers of Mathematics* (Águas de Lindóia, 2005)
- *Challenging Mathematics in and Beyond the Classroom* (Trondheim, 2006)
- *Digital Technologies and Mathematics Teaching and Learning: Rethinking the Terrain* (Hanoi, 2006).

On the organisational side, the *International Study Group for Mathematical Modelling and Applications* (ICTMA) was adopted as a new Affiliated Study Group in 2003, and several ICMI Regional Conferences were held, including the *Conferencia Interamericana de Educación Matemática* (CIAEM—launched in the 1960s), the *ICMI-East Asia Regional Conference in Mathematics Education* (EARCOME—1998, but originating from a series started in 1978), *Espace Mathématique Francophone* (EMF—2000), based on the notion of a "region" being conceived in linguistic terms, and the *Africa Regional Congress of ICMI on Mathematical Education* (AFRICME—2005).

Having in mind to pave the way for a smooth transition to the new governance structure, a new EC was established by the 2006 IMU General Assembly for a 3-year term, 2007–2009. Former Vice-President Michèle Artigue was elected not only as the first female ICMI President ever,[9] but also as the first President in the "New ICMI" era, inaugurated in 1952, whose credentials are primarily based on the reputation as a mathematics educator rather than as a classical research

---

[8]The idea of having "a medal (or possibly two) to be awarded to someone who has made an outstanding contribution to mathematics education" had been raised earlier by ICMI Secretary-General Howson (1982, p. 8).

[9]But certainly not the last, as Jill Adler was elected ICMI President for 2017–2020 at ICME-13.

mathematician. To ensure continuity from the past to the future, Bernard Hodgson was exceptionally asked to serve as the Secretary-General for a third term, also a complete novelty since the time when Henri Fehr (1870–1954) served for decades as Secretary-General of the "Old ICMI".

This term saw further consolidation of established ICMI activities. Thus ICMI Studies 18 and 19 were conducted:

- *Statistics Education in School Mathematics: Challenges for Teaching and Teacher Education* (Monterrey, 2008), organised jointly with IASE, the International Association for Statistical Education
- *Proof and Proving in Mathematics Education* (Taipei, 2009),

and ICME-11 was held in Monterrey (2008), for the first time in a country outside what used to be called the "First World".

It was in this very eventful term that the ICMI Centennial 2008 was celebrated in Rome in a symposium organised in Palazzo Corsini, home of the Accademia Nazionale dei Lincei and the very birthplace of ICMI at the 1908 ICM. The centennial is commemorated in the symposium proceedings, edited by Menghini et al. (2008).

The year 2008 also saw the inception of a new ICMI project whose subsequent underpinnings are related to the thematic session on the *Legacy of Felix Klein* held at this congress. The Klein project stems from a proposal made by Vice-President Bill Barton, at the first meeting of the 2007–2009 ICMI EC, to foster the promotion of mathematics through the revisiting of Felix Klein's famous *Elementary Mathematics from an Advanced Standpoint*, originally published in 1908—see ICMI EC Minutes (2007, p. 1). The aims of the project are to produce resources for secondary teachers about contemporary mathematics, so as to help them make connections between their teaching and the field of mathematics as a living subject.

At that same ICMI EC meeting, a decision was made to launch an electronic newsletter for prompt, efficient and brief communication with the community, a project that had already been considered for a while, partially inspired by the *IMU-Net* initiated in 2003. The first issue of *ICMI News* appeared in December 2007. The aim of this new and "light" channel of communication was to complement the *ICMI Bulletin*, which retained interest in a long-term archival perspective, but whose size and scope, since the turn of the century, had become more ambitious while its appearance was more erratic, as only ten issues were published between 1999 and 2009. Less than a year after its launching, *ICMI News* had more than a thousand subscribers. Also in the year 2009, the project of updating the ICMI website, which had been in progress for a few years, was finally completed. (The initial version of the ICMI website, originally a mere page on the IMU server, went back to 1995.)

New Terms of Reference for ICMI were adopted in 2009. The Terms themselves, under the jurisdiction of the IMU EC, are accompanied by Guidelines, concerning some ICMI internal rules of operation. For instance the definition of the ICMI General Assembly and the voting rights therein are part of the Terms, whereas the details of the ICMI EC election procedure are under the jurisdiction of the ICMI General Assembly. With the 2009 revision of the Terms of Reference, the traditional notion of an Affiliated Study Group was extended and generalised to that of an *Affiliate Organisation*. The *International Group for Mathematical Creativity and Giftedness* (MSG) was accepted in 2011 as the sixth ICMI Affiliated Study Group, whilst existing multi-national mathematical education societies became affiliate organisations shortly after the adoption of the new scheme: CIAEM (2009), CIEAEM (2010), the *European Society for Research in Mathematics Education* (ERME—2010) and the *Mathematics Education Research Group of Australasia* (MERGA—2011).

A new EC was elected for a 3-year term, 2010–2012, at ICME-11 in Monterrey (2008) with Bill Barton as President and Jaime Carvalho e Silva as Secretary-General. Under their leadership, ICMI Studies 20 and 21 were conducted:

- *Educational Interfaces between Mathematics and Industry* (Lisbon, 2010), organised jointly with the International Council for Industrial and Applied Mathematics (ICIAM)
- *Mathematics Education and Language Diversity* (Águas de Lindóia, 2011).

ICME-12 was held in Seoul in 2012, and the Klein Project underwent considerable development.[10] Moreover it was decided at the 2011 EC meeting to launch the *Database Project*, with the ultimate goal of building a free access database of mathematics curricula from all over the world.

This EC also saw, in 2011, the inauguration of the IMU Secretariat in Berlin, where a position of ICMI Administrator had been established. This event turned out to be a major change in the daily maintenance of ICMI business thanks to the most welcome support thus provided to the work of the EC.

But probably the most significant new development in the Barton-Carvalho e Silva era was the launching in 2010 of the so-called CANP (*Capacity and Networking Project*), meant to stimulate outreach to developing countries by fostering networking amongst teachers, mathematics educators and mathematicians within a given region. The project emerged as a joint initiative of ICMI and UNESCO, spurred by the renewed collaboration that had started in the early 2000s with the exhibition "Experiencing Mathematics" (see above). As a result of the regular links that then arose between the two bodies, UNESCO invited in 2009 ICMI President Michèle Artigue to pilot the preparation of a White Paper on *Challenges in basic mathematics education* (UNESCO 2011). Inspired by the recommendations of that document, UNESCO proposed to ICMI during the year 2010 to organise an event in Africa aiming at "reinforcing teacher education

---

[10]At the time of writing, more than 20 "Klein Vignettes" have been produced, each being a short piece on a selected mathematical topic likely to be new to most secondary teachers and typically requiring some undergraduate mathematical knowledge. Some vignettes are available in different languages.

capacities, building synergies between communities, and reinforcing South-South collaboration" (Artigue 2017). Led by the incoming President Barton, the ICMI EC developed, jointly with UNESCO, the CANP model, of which the first event actually took place in Africa in 2011.

Five CANP workshops—each with a follow up meeting held about a year later —took place in the period 2011–2016 (which goes into the term of the next ICMI EC):

- two in Africa: Mali (2011) and Tanzania (2014);
- two in Latin America: Costa Rica (2012) and Peru (2016); and
- one in South East Asia: Cambodia (2013).

These workshops and follow up events gathered more than 400 participants from more than 25 developing countries in five regions—see Koch (2016), a preliminary report on CANP by the ICMI Administrator. CANP is considered within ICMI as a most successful endeavour, notably with regard to the improvement in the individual scientific capacity of the participants, as well as to the fostering of regional network building—data supporting this view are provided in (Koch 2016, 2017).

The organisation of CANP workshops could be considered somewhat expensive, as the average cost per workshop for ICMI is of the order of 50,000 € (in addition to funding and support from local sources). However the cost per participant for these two-week workshops is very low. Moreover more than 85% of these expenses up till now have been covered by special grants, mainly from IMU but also from the International Council for Science (ICSU) and UNESCO (Koch 2016). In spite of the substantial amounts involved, the ICMI EC clearly finds the cost worth the while because of the most significant outreach impact obtained.

After the two 3-year terms, it was now time to go back to the usual 4-year terms. At the ICMI General Assembly in Seoul, 2012, a new ICMI EC was elected for the term 2013–2016, with Ferdinando Arzarello as President and Abraham Arcavi as Secretary-General. A major decision made by the EC was the establishment, announced early in 2015, of a third ICMI Award, the Emma Castelnuovo Award honouring excellence in the practice of mathematics education, to be awarded every four years, starting in 2016 at ICME-13 (Hamburg). The Arzarello-Arcavi era also saw ICMI Studies 22 and 23:

- *Task Design in Mathematics Education* (Oxford, 2013)
- *Primary Mathematics Study on Whole Numbers* (Macau, 2015).

as well as the adoption by the EC of *Guidelines for conducting an ICMI Study* crystallising the goals and process of an ICMI Study (Arzarello et al. 2014, p. 83).

Regarding communication with the ICMI community, this new EC decided, at its very first meeting in 2013, to officially discontinue the production of the *ICMI Bulletin* (Arzarello et al. 2014, p. 92). The previous EC had supported in principle the importance of the role played by the *Bulletin*, but in practice no issue had been published during its term of office. The Arzarello-Arcavi EC aimed at improving the use of *ICMI News* as the main communication channel, notably by producing a more sophisticated version of the journal (still in a brief style), and also aimed at reinforcing the collaboration with *L'Enseignement Mathématique*.

14.5   ICMI and the Field of Mathematics Education

In this paper we have attempted to link a brief account of ICMI's organisational history to a portrayal of ICMI as a facilitator of international cooperation and collaboration in mathematics education in a broad sense. ICMI can be perceived as a body that reflects, and reflects on, important developments in mathematics education as a field of research and development. ICMI can also be perceived as a body that takes initiatives to identify new issues and needs in mathematics education and provides a platform for the exploration and unfolding of these issues and demands.

Thus, ICMI has engaged in a symbiotic relationship with mathematics education. However, it is important to understand that for the past half century, ICMI was never a body taking political stances on pertinent issues, e.g., by passing resolutions and making particular educational recommendations.[11] Nor was ICMI ever a managerial body that tried to "rule over" mathematics education. Since mathematics education is—of course—about mathematics, a key theme throughout those five decades has been to create and maintain strong and mutually respectful links between mathematics and mathematics education, and between research mathematicians and mathematics educators.

The half century covered by this paper has been an epoch of expansion and enlargement in a multitude of different respects. There is every reason to believe that thanks to new generations of concerned, committed and competent ICMI officers, new land will be reclaimed and charted during the next fifty years.

References

Artigue, M. (2008). ICMI: A century at the interface between mathematics and mathematics education. In M. Menghini, F. Furinghetti, L. Giacardi, & F. Arzarello (Eds.), *The first century of the International Commission on Mathematical Instruction (1908–2008). Reflecting and shaping the world of mathematics education* (pp. 185–198). Rome, Italy: Istituto della Enciclopedia Italiana.

Artigue, M. (2017). The origin of the CANP project. Email to Bernard Hodgson, 9 May. *IMU Archive* (uncatalogued documents).

Arzarello, F., Arcavi, A., & Koch, L. (2014). Quadrennial activity and financial report on ICMI activities in the period 1 January 2010–31 December 2013. *IMU Bulletin, 64*, 81–98.

---

[11]In the words of ICMI President Kahane, then speaking of the ICMI Studies, one should not expect, as a result of ICMI endeavour, "any ICMI-labelled solution to any educational problem" (Kahane 1990, p. 3).

Arzarello, F., Giacardi, L., Furinghetti, F., & Menghini. M. (2008). Celebrating the first century of ICMI (1908–2008): Some aspects of the history of ICMI. In *Proceedings of the 11th International Congress on Mathematical Education* (July 6–13, 2008), *Materials from ICME-11 Mexico* (regular lecture, pp. 65–93). http://www.mathunion.org/icmi/publications/icme-proceedings/materials-from-icme-11-mexico/. Accessed December 15, 2016.

Bass, H. (2002). A general change in nomenclature. *ICMI Bulletin, 51,* 13.

Bass, H. (2008). Moments in the life of ICMI. In M. Menghini, F. Furinghetti, L. Giacardi, & F. Arzarello (Eds.), *The first century of the International Commission on Mathematical Instruction (1908–2008). Reflecting and shaping the world of mathematics education* (pp. 9–24). Rome, Italy: Istituto della Enciclopedia Italiana.

Carleson, L. (1981). Letter to the members of the EC of ICMI, 22 December. *IMU Archive, Box 14C—International Commission on Mathematical Instruction, 1981–1982.*

Cartan, H. (1970). Letter to Otto Frostman, IMU Secretary, 15 October. *IMU Archive, Box 14B—International Commission on Mathematical Instruction, 1967–1980.*

Castelnuovo, G. (Ed.). (1909). *Atti del IV Congresso Internazionale dei Matematici* (Vol. I, Roma, Aprile 6–11, 1908). Rome, Italy: Tipografia della R. Accademia dei Lincei.

Coray, D., Furinghetti, F., Gispert, H., Hodgson, B. R., & Schubring, G. (Eds.). (2003). One hundred years of L'Enseignement Mathématique: Moments of mathematics education in the twentieth century. In *Proceedings of the EM-ICMI Symposium*, Geneva, 2000. Geneva, Switzerland: L'Enseignement Mathématique.

Editorial Board of Educational Studies in Mathematics. (Eds.). (1969). *Proceedings of the First International Congress on Mathematical Education* (International Commission on Mathematical Education [sic], ICMI). Dordrecht: D. Reidel [Also in *Educational Studies in Mathematics,* 2(1969), 135–418].

Fehr, H. (1920–21). La Commission internationale de l'enseignement mathématique de 1908 à 1920: Compte rendu sommaire suivi de la liste complète des travaux publiés par la Commission et les Sous-commissions nationales. *L'Enseignement Mathématique, 21,* 305–339.

Freudenthal, H. (1967). Letter to Otto Frostman, IMU Secretary, 20 December. *IMU Archive, Box 14B—International Commission on Mathematical Instruction, 1967–1980.*

Frostman, O. (1967). Letter to Hans Freudenthal, ICMI President, 2 December. *IMU Archive, Box 14B—International Commission on Mathematical Instruction, 1967–1980.*

Hilton, P. (1980). Letter to Jacques-Louis Lions, IMU Secretary, 24 September 1980. *IMU Archive, Box 14B—International Commission on Mathematical Instruction, 1967–1980.*

Hodgson, B. R. (1991). Regards sur les Études de la CIEM. *L'Enseignement Mathématique, s. 2, 37,* 89–107.

Hodgson, B. R. (2000). A call for bids for ICME-11. *ICMI Bulletin, 49,* 14–15.

Hodgson, B. R., & Rogers, L. F. (2012). On international organizations in mathematics education. *International Journal for the History of Mathematics Education, 7*(1), 17–27.

Hodgson, B. R., Rogers, L. F., Lerman, S., & Lim-Teo, S. K. (2013). International organizations in mathematics education. In M. A. Clements, A. Bishop, C. Keitel, J. Kilpatrick, & F. Leung (Eds.), *Third international handbook of mathematics education*, Chapter 28, Springer International Handbooks of Education 27 (pp. 901–947). New York, NY: Springer.

Howson, A. G. (1982). Minutes of a meeting held in Paris on 3 December 1982 between J.-P. Kahane, B. Christiansen and A. G. Howson. *IMU Archive, Box 14C—International Commission on Mathematical Instruction, 1981–1982.*

Howson, A. G. (1983). The International Commission on Mathematical Instruction: Past, present and future. *L'Enseignement Mathématique, s. 2, 29,* 348–350.

Howson, A. G. (2008). Klein and Freudenthal. In M. Niss (Ed.), *Proceedings of the 10th International Congress on Mathematical Education* (July 4–11, 2004), CD (regular lecture, 19 p.). Roskilde, Denmark: IMFUFA.

ICMI EC Minutes. (2007). Minutes of the meeting of the Executive Committee of the International Commission on Mathematical Instruction (London, June 13–16, 2007). *IMU Archive* (uncatalogued documents).

ICMI Terms. (1954). *Terms of reference of the International Commission on Mathematical Instruction.* http://www.mathunion.org/icmi/icmi/icmi-as-an-organisation/terms-of-reference/. Accessed December 15, 2016.

IMU EC Minutes. (1980). Minutes of the 41st meeting of the Executive Committee of the International Mathematical Union (Paris, April 8–9, 1980). *IMU Archive, Box 4D—Executive Committees, 1979–1982.*

IMU EC Minutes. (1981). Minutes of the 42nd meeting of the Executive Committee of the International Mathematical Union (Paris, April 27–28, 1981). *IMU Archive, Box 4D—Executive Committees, 1979–1982.*

IMU General Assembly. (1990). Report of the 11th general assembly of IMU. *IMU Bulletin, 32,* 3–9.

Kahane, J.-P. (1990). A farewell message from Jean-Pierre Kahane, the retiring President of ICMI. *ICMI Bulletin, 29,* 3–8.

Koch, L. (2016). *CANP review—First results.* Preliminary report to the ICMI EC (July 2016). http://www.mathunion.org/icmi/activities/outreach-to-developing-countries/canp-review/. Accessed January 31, 2017 (full report available by July 2017).

Koch, L. (2017). *Evaluation of the capacity and network project CANP 1–5 of the International Commission on Mathematical Instruction (ICMI)* (Thesis, Master of Business Administration in Higher Education and Research Management). University of Applied Sciences Osnabrück, Germany (Thesis under preparation).

Lehto, O. (1998). *Mathematics without borders: A history of the International Mathematical Union.* New York, NY: Springer.

Menghini, M., Furinghetti, F., Giacardi, L., & Arzarello, F. (Eds.). (2008). *The first century of the International Commission on Mathematical Instruction (1908–2008). Reflecting and shaping the world of mathematics education.* Rome, Italy: Istituto della Enciclopedia Italiana.

Smith, D. E. (1905). Opinion de David-Eugene Smith sur les réformes à accomplir dans l'enseignement des mathématiques. *L'Enseignement Mathématique, 7,* 469–471.

Soares, M. G., & Cordaro, P. D. (1998). Letter to Hyman Bass, ICMI President, 4 December. *IMU Archive* (uncatalogued documents).

Steiner, H.-G. (1985). Theory of Mathematics Education (TME): An introduction. *For the Learning of Mathematics, 5*(2), 11–17.

UNESCO. (2011). *Les défis de l'enseignement des mathématiques dans l'éducation de base.* Paris, France: UNESCO (English translation, 2012).

Chapter 15
Formative Assessment in Inquiry-Based Elementary Mathematics

Alena Hošpesová

Abstract The chapter presents findings related to Czech teachers' and pupils' difficulties with, opinions on, and needs associated with formative assessment, namely, peer assessment, in inquiry-based lessons. The research was conducted within the EU-funded Assess Inquiry in Science, Technology, and Mathematics Education project (ASSIST-ME). Six teachers of primary mathematics worked with researchers on inquiry tasks and methods of peer assessment and implemented them in their classrooms. The paper focuses mainly on (a) the interplay of teachers' intentions, subject matter, and learners in inquiry; (b) the teachers' role in supporting learning via (formative) assessment; and (c) the pupils' role in their own learning and the learning of peers. Significant phenomena in implementation of assessment were identified, namely, the importance of formulation of learning objectives; pupils' ability to decide about the correctness, identify the mistakes, and give supporting feedback to their peers; possible (and needed) support; and institutionalization of knowledge.

Keywords Formative assessment · Peer assessment · Self-assessment
Inquiry based mathematics education · Primary school level

15.1 Introduction

School assessment as a feedback tool and an important part of interaction among key actors in school education has for a long time been the subject of discussion in the Czech Republic because it influences the character of the entire system of teaching and learning. School assessment is closely related to school tradition and the culture of education. For this reason, the implementation of inquiry based approaches in mathematics education at school includes, among other problems, the challenges of having to comply with curriculum requirements, classroom

A. Hošpesová (✉)
University of South Bohemia, České Budějovice, Czech Republic
e-mail: hospes@pf.jcu.cz

© The Author(s) 2018
G. Kaiser et al. (eds.), *Invited Lectures from the 13th International Congress on Mathematical Education*, ICME-13 Monographs,
https://doi.org/10.1007/978-3-319-72170-5_15

management issues in responding to unexpected and uncertain situations, and the problem of conducting assessment. The objective of the chapter on the general level is to show what changes occurred in classrooms of teachers developing inquiry-based approaches and assessing their pupils' results and achievement, in other words, what the relationship between assessment and inquiry-based education in mathematics is.

The chapter reports the results of a research study on peer assessment carried out in the Czech Republic under the framework of the EU-funded Assess Inquiry in Science, Technology, and Mathematics Education project (ASSIST-ME). The project develops and studies formative and summative methods of assessment that support inquiry-based approaches in teaching science, technical disciplines, and mathematics. Based on analysis of what is known about summative and formative assessments of knowledge, skills and attitudes, the project team proposed a variety of combined methods of assessment. These methods were tested on primary and secondary schools in different countries across Europe (the Czech Republic, Denmark, Finland, France, Cyprus, Germany, Switzerland, and the United Kingdom). The research carried out was focused on creation of formative assessment methods that (1) fit into everyday classroom practice, (2) provide qualitatively oriented feedback of competence-oriented, inquiry-based learning processes, and (3) can be combined with existing summative assessment requirements and methods used in different educational systems.

The chapter is based on data from primary mathematics classrooms collected within the framework of the ASSIST-ME project in the Czech Republic and aims to answer the questions: What difficulties are faced in introducing formative assessment in primary mathematics classroom?

15.2 Background of the Study

15.2.1 Assessment

The issue of assessment has a long tradition at ICME, e.g., Topic Study Group "Assessment and Testing in Mathematics Education" (Suurtamm and Neubrand 2015) at ICME 12 focused on:

- Issues connected to the development of teachers' professional knowledge of assessment and their use of assessment in the mathematics classroom.
- Issues and examples related to the enactment of classroom practices that reflect current thinking in assessment and mathematics education (e.g., the use of assessment for learning, as learning, and of learning in mathematics classrooms).

This chapter contributes to debate the latter of the two topics.

Our study focused on *classroom-based assessment of pupils* (in the sense of definition published in OECD 2013). This assessment requires specification of the purpose for which the data are collected and interpreted, i.e., the purpose of assessing. This affects a number of decisions the teacher makes about (a) the data to be collected (e.g., whether systematically or occasionally), (b) interpretation of them, (c) communication about them, and (d) building further decisions on them (Black et al. 2004). With respect to how assessment is used and with what purpose it is carried out, two approaches are distinguished:

- Summative assessment in which the evaluator checks and summarizes what the pupil has learned. It may concern individual pupils, groups of pupils, or the whole population (for example, large-scale external tests and examinations).
- Formative assessment that supports the pupils' learning process. This involves the processes of data collection and interpretation that learners and their teachers use to make decision about the following: What have the pupils learned so far? Where is their learning aim? How can they be supported and assisted on their way to learning?

Black and Wiliam (2009) explained formative assessment in these words:

> Practice in a classroom is formative to the extent that evidence about student achievement is elicited, interpreted, and used by teachers, learners, or their peers to make decisions about the next steps in instruction that are likely to be better, or better founded, than the decisions they would have taken in the absence of the evidence that was elicited. (p. 9)

From this definition, the main characteristics of formative assessment can be summarized (see ARG 2002):

- Pupils play active roles in making decisions about their own learning. They can be expected to be able to channel their effort more efficiently if they know the objectives of their learning.
- The teachers' feedback includes advice on how to advance; it does not compare pupils with each other.
- Teachers use the information to make adaptations to their lessons in a way that gives their pupils more opportunity to learn.
- Dialogues between teachers and the pupils support the pupils' reflection on the learning process.
- Pupils develop self-assessment by taking part in determination of what will make them advance.

The idea of formative assessment has been discussed in Czech educational context since the 1990s, but it has only been slowly introduced in school practice (for more details, see Žlábkova and Rokos 2013). Slavík (2003) summarized that although several examples of good practice were offered, empirical research findings focused on problems in using the formative assessment methods were almost absent. Some forms of formative assessment are seen as embedded in common Czech teaching culture, such as immediate teacher feedback as a response to a pupil's problem solving (on-the-fly assessment) or less frequent written comments from teachers. Others are seen as difficult in mathematics on primary school level (e.g., structured classroom dialog or formative peer assessment).

Teaching/learning processes including formative assessment were depicted by Harlen (2014) as a repetitive cyclic process the learner in the center (see Fig. 15.1).

In inquiry-based education, formative assessment naturally penetrates the process of inquiry:

- The teacher formulates a short-term goal of a lesson or several lessons, i.e., the norm the pupil is expected to achieve. In mathematics education, the goal is often operationalized in the form of a problem or a set of problems (see Samková et al. 2015).
- In discussions with peers and the teachers, pupils communicate about how they understand the problems in question, how they solve them, and how they understand their classmates' solutions. If teachers do not find these discussions meaningful, they stimulates them by asking open questions; i.e., the teacher collects data during the activities being carried out, interprets them, and intervenes if necessary. In school mathematics, the teacher makes often conjectures about how the learner thinks.
- The teacher then formulates some recommendation, which Harlen (2014) refers to as "judgement," about the next steps. These steps are then carried out by the pupil in the subsequent process of learning.

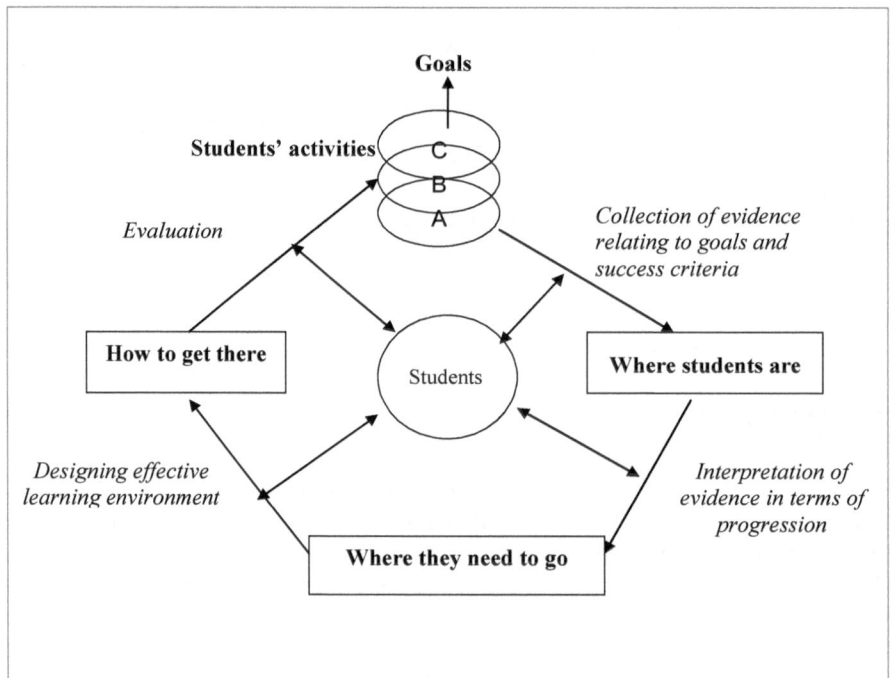

Fig. 15.1 Assessment for formative purposes (Harlen 2014, p. 6)

– Information the teacher gets about a pupil's activity is related to the short-term objectives of a lesson, or several lessons. This information constitutes the basis for making decisions about the next steps or about how to help the pupil carry out these steps. The goal is to support pupils' learning and to provide feedback on the progress they have made in their understanding or skills. It is at the same time feedback for the teacher, who can adapt tasks to the pupil and maximize the opportunity to learn. Empirical evidence has verified that formative assessment increases efficiency of learning, e.g., Black and Wiliam (1998) mentioned studies that stated that "improved formative assessment helps the (so-called) low attainers more than the rest, and so reduces the spread of attainment whilst raising it overall" (p. 3).

Peer assessment, which is the focus of our attention in this study, is understood here in accordance with Boud and Falchikov (2007, p. 132):

Peer assessment requires students to provide either feedback or grades (or both) to their peers on a product or a performance, based on the criteria of excellence for that product or event which students may have been involved in determining.

Slavík (2003) and others consider peer assessment to be a way to autonomous assessment, i.e., deeper reflection on one's own learning and its results "that learners use on their own, master it, that they understand to the needed extent, that they can explain or defend" (p. 14). Slavík (2003) stated that autonomous assessment partially develops and deepens in relation to self-assessment and partially in relation to assessment of others' performance (most likely of classmates', i.e., peer assessment) through which pupils learn to reflect on their work. Pachler et al. (2010) stressed that learner self-regulation is a core factor in formative assessment and that it is linked to motivation and emotional factors which affect learners' engagement.

Learning benefits are supposed for both pupils acting as assessor and as assessee, as well, because they both can bridge the gaps in their understanding of particular contents and get a more sophisticated grasp on their learning (Topping 2013). In addition, there has been great interest in upscaling formative assessment to change learning/teaching culture (OECD 2015).

15.2.2 Inquiry-Based Approach in Mathematics Education and Assessment

Formative assessment is especially important in the situation of inquiry-based education:

As is the case in the natural sciences, inquiry-based mathematics education refers to an education which does not present mathematics to pupils and students as a ready-built structure to appropriate. Rather it offers them the opportunity to experience:

- how mathematical knowledge is developed through personal and collective attempts at answering questions emerging in a diversity of fields, from observation of nature as well as the mathematics field itself, and,
- how mathematical concepts and structures can emerge from the organisation of the resulting constructions, and then be exploited for answering new and challenging problem (Artigue et al. 2012, p. 8).

Inquiry-based education in teaching mathematics helps not only to build inquiry-based attitudes in pupils but also to reinforce pupils' understanding of mathematical concepts and procedures. According to Donovan and Bransford (2005), it uses (a) a knowledge-centered lens, focusing attention on "what is taught (learning goals), why it is taught, and what mastery looks like"; (b) an assessment-centered lens, emphasizing the need to provide frequent opportunities "to make students' thinking and learning visible as a guide for both the teacher and the student in learning and instruction"; and (c) a community-centered lens, based on a culture of "questioning, respect, and risk taking," as well as the interaction of learners and teacher as central to the learning process (p. 13). It follows that in the process of inquiry, the roles of the pupil and the teacher and their responsibility for the teaching/learning process change. Primarily, it is the pupil/group of pupils who must be active when looking for information, estimating and guessing, making conjectures, and discovering solutions. When peer assessment is present, pupils must try to understand the solutions of others, comment on them, and give feedback. The teachers' role is to create the right conditions for this. They must create an environment that encourages cooperation, guide their pupils, support them in their search for unknown solving methods, and ask questions, such as "Why?" "How would you explain?" "Is it really so?" and "Do you know any similar problem/task?" The teacher must be proactive, support pupils' efforts, praise pupils' contributions (including giving feedback on mistakes the pupils have made) and must help their pupils advance in learning based on their own independent discoveries and interpretations.

Implementation of inquiry-based approaches brings a radical change to the whole process of education; starting with response to the demands of curricula, to problems that stimulate independent inquiries, to a change in the pupils' and teachers' roles in the teaching/learning process. Pupils and their teacher constitute a complex system with its own dynamics, conditions, and rules. The system can be illustrated by the schema in Fig. 15.2.

Inquiry at school can be depicted as a cycle:

Like scientific inquiry, mathematical inquiry starts from a question or a problem, and answers are sought through observation and exploration; mental, material or virtual experiments are conducted; connections are made to questions offering interesting similarities with the one in hand and already answered; known mathematical techniques are brought into play and adapted when necessary. This inquiry process is led by, or leads to, hypothetical answers—often called conjectures—that are subject to validation. (Artigue and Baptist 2012, p. 4)

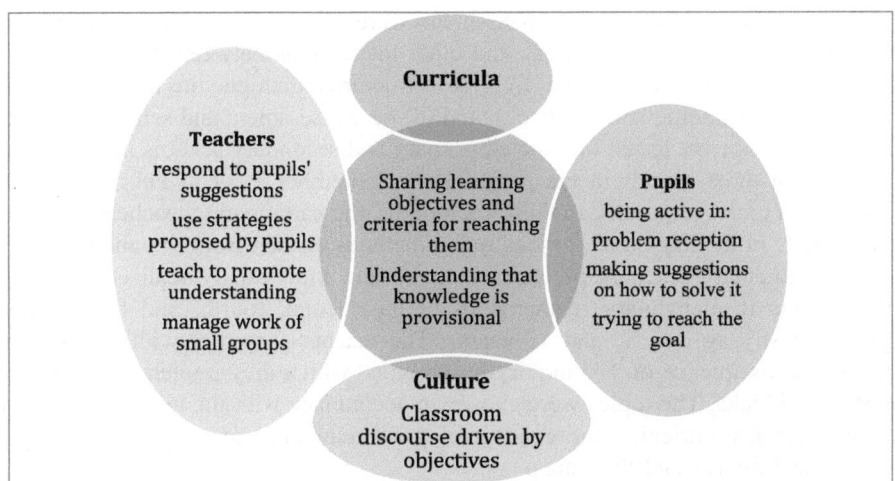

Fig. 15.2 Roles of different agents in the classroom in the course of pupils' independent inquiry (modified according Samková et al. 2015, p. 97)

In respect to this study, we have to better characterize the problems that initiated the inquiries. Openness of problems as the stimuli of inquiry has been requested already by Dewey (1938):

> The original indeterminate situation is not only "open" to inquiry, but it is open in the sense that its constituents do not hang together. A variety of names serves to characterize indeterminate situations. They are disturbed, troubled, ambiguous, confused, full of conflicting tendencies, obscure, etc. (p. 105)

In other words, problems can be interpreted in more ways and there are more correct ways of solving them and sometimes more correct answers. To solve the problem, the pupils discover (or better, rediscover) the ways of its solution. In accordance with their actual knowledge, they make experiments (mental, material, and virtual), observe similarities and differences, and compare them with their current knowledge. Adding more experience and knowledge to their network of knowledge, they restructure the existing. They make mistakes and learn from them (especially their own but also other people's mistakes).

15.3 Empirical Study

15.3.1 Goals and Organization of the Study

The goal of the research study presented here is to identify those phenomena that could be observed in a planned implementation of peer assessment in inquiry-based education in mathematics at the primary school level and, in particular, the difficulties faced in such an implementation.

Four methods of formative assessment were proposed in the frame of the ASSIST-ME project: (a) questions and other interactions between the teacher and their pupils conducted "on the fly", (b) structured dialogue in the classroom, (c) evaluation (grading and feedback), and (d) peer assessment and self-assessment. These methods were tested in selected primary and secondary schools in the Czech Republic in three rounds in the period from September 2014 to February 2016. Each round took six months. In these six months, the participating teachers assessed their pupils in inquiry-based lessons using methods chosen from the above list.

Six elementary teachers prepared in pairs and individually realized teaching experiments in elementary mathematics in the second, fourth, and fifth grades (pupils mostly aged 7, 9, and 10 years). The teachers together with researchers developed a sequence of 4–6 inquiry-based units, which they implemented mostly in 90-min blocks. The topics were chosen in accordance with the teaching plans of relevant grades: enriching the concept of great numbers and properties of plane geometrical figures and their area.

15.3.2   Preparation of the Educational Experiments

The educational experiments were carried out with the intention of creating a space for independent pupils' inquiry supported by peer assessment. Lesson planning was always carried out via joint discussion by the local working group (teachers and researchers), during which the goal was formulated and various options for its fulfilment were discussed. The discussion focusing on a priori didactical analysis of content (in the sense formulated by Brousseau 2002) identified the key concepts that became the goal of the teaching experiment.

The goals were operationalized in the form of "the problem of a lesson." We always were looking for problems that would stimulate inquiry. The main characteristics of these problems were their openness (see Dewey 1938). The following problems are two examples:

Problem 1 Find out how many lentils there are in a half-kilogram package.

Problem 2 Create instruction for your friends on how to determine the number of tiles needed to tile the triangle in Fig. 15.3.

The first problem is indeterminate in terms of the solution procedure. The goal in solving this problem is to cover the topic of great numbers in order to:

– understand that the basis of representing a number in the position system is the grouping of elements according to the base of the numeral system and the notation of the number of these groupings and
– perceive that the value of a figure in the notation of a number depends on its position.

Fig. 15.3 Figure in
assignment of tiling problem

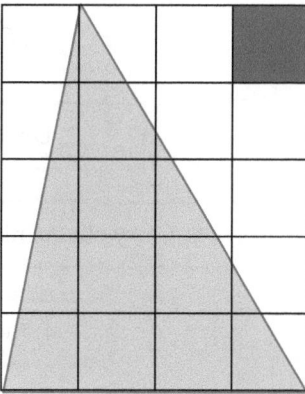

Pupils are familiar with grouping into tens and hundreds from counting to 100. It is very likely they have not come across a similar activity with larger sets at school or in their real life. The problem may also be solved by weighing.

The second problem prepares students for the introduction of the concept of the area of a triangle. Pupils may count whole tiles and their parts and may also discover that a triangle is one half of a 4 cm by 5 cm rectangle.

Each of the teachers then individually elaborated a plan for the lesson and realized them with their pupils. The plan was based in general on Polya's model of stages of problem solving (Polya 1945), supplemented by the peer-assessment stage (5) and reflection (6) (see schema in Fig. 15.4).

In most cases, a worksheet was created in which the pupils recorded a solution to the problem and provided peer assessment (an example of the worksheet for the topic of great numbers is included in the Appendix). Pupils described their solution procedure in the worksheets. After that they swapped the worksheets and provided each other written feedback on the solution. Having received comments from peers, each pupil (or a group of pupils) had the chance to revise the original solution with respect to the feedback they had received. In addition, each student also briefly responded in writing to the feedback they had received.

15.3.3 Data and Their Analysis

We have acquired a rich source of data from the 16 experiments that were implemented:

- Video recordings (2 cameras: the teacher's and one in the classroom) of lessons and their transcriptions
- Structured classroom observation protocols from the researchers
- Pupils' written productions (worksheets capturing problem-solving and assessment comments by classmates)

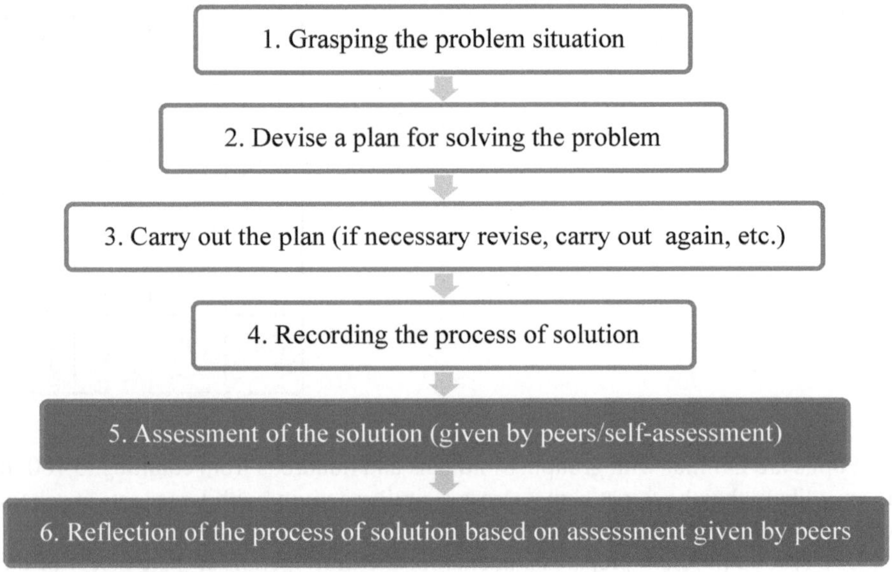

```
┌─────────────────────────────────────────┐
│      1. Grasping the problem situation     │
└─────────────────────────────────────────┘
                    ↓
┌─────────────────────────────────────────┐
│    2. Devise a plan for solving the problem   │
└─────────────────────────────────────────┘
                    ↓
┌─────────────────────────────────────────────────┐
│ 3. Carry out the plan (if necessary revise, carry out  again, etc.) │
└─────────────────────────────────────────────────┘
                    ↓
┌─────────────────────────────────────────┐
│     4. Recording the process of solution      │
└─────────────────────────────────────────┘
                    ↓
┌─────────────────────────────────────────────────┐
│ 5. Assessment of the solution (given by peers/self-assessment) │
└─────────────────────────────────────────────────┘
                    ↓
┌───────────────────────────────────────────────────────┐
│ 6. Reflection of the process of solution based on assessment given by peers │
└───────────────────────────────────────────────────────┘
```

Fig. 15.4 Schema of planning experimental education units

- Audio recordings of group work if it took place
- Semi-structured interviews with the teachers who realized the teaching experiments before the lesson to determine how they understood inquiry-based mathematics education and the role of formative assessment in it
- Audio recordings of discussions in local working groups during the preparation of lessons and their transcripts
- Short interviews with teachers after each lesson.

Data were analyzed qualitatively. Transcripts were coded via open coding. The codes were derived from the characteristics of formative assessment, the inquiry-based education, and perceived actions. Gradually, a list of codes were created, including codes for the activities of both teachers (e.g., assignment of the problem; explaining on the initiative of the teacher or pupil(s); monitoring the activities of pairs or groups of pupils; discussion with pupils; individuals, pairs, or groups giving feedback; questions to individuals, pairs, or groups; individuals requesting clarification; feedback directing solutions for the whole class, feedback directing solutions for individuals; reactions to pupils' explanations; general evaluation of pupils' work; assessment of pupils' work; putting knowledge into context with what has been previously learned; and indication of the importance of knowledge for the future) and pupils (clarification requests, comments on the problem, solving tasks in pairs or groups, a request for clarification during the solution of the problem, a request for equipment, loud comment on the solution of the problem solution summary, and reaction to summarize for the teacher).

Specific issues related to difficulties that teachers and pupils had with introducing and conducting assessment gradually emerged in the process of pupils' independent inquires, such as how to support peer assessment, what the role of the teacher in peer assessment is, and what difficulties pupils have with peer assessment.

15.4 Selected Findings and Discussion

We identified significant phenomena in implementation of assessment in the teaching experiments.

15.4.1 Formulation of Learning Objectives

It is essential for the teachers' planning and decisions in the lesson that the didactical objective to be achieved in the lesson should be clearly defined. This premise is often discussed in the materials for teachers. When teaching mathematics at the primary school level in the Czech Republic, teachers often use materials to support their teaching that define only the topic of the lesson and cover classroom management (what problems will be included in the lesson, how they will be arranged, and what form of classroom organization will be used). Teachers do not plan lessons with respect to what the pupils will learn but with the objective of correctly solving problems provided in textbooks, workbooks, and worksheets.

In inquiry mathematics education, the quality of feedback that pupils may gain from their solutions depends on the accuracy of the definition and the operationalization of the learning objective of the "inquiry." At the beginning of our sequence of lessons, the objectives defined in cooperation with the teachers were quite general (e.g., "get experience," "apply a known procedure in a new environment"). The experiments showed that it is essential for assessors to state the objective in terms of the expected pupil's performance.

This can be illustrated on the solution of Problem 2, in which the pupils were asked to formulate comprehensible instructions on how to determine the number of tiles needed to tile the triangle in Fig. 15.3. Assessors decided whether the instructions were clear and could be used for determining the number of tiles needed to tile other triangles. This assessment could have been initiated by a concrete question: Did the solvers determine correctly that 15 tiles are needed? But asking this question would not correspond to the defined learning objectives: Pupils gain pre-concepts that form the basis of measuring the area of a triangle, namely, an experience with filling in a triangle with, for instance, squares (i.e., by a selected unit). For that reason, we asked the assessor to determine the number of tiles in another triangle according to the instructions, and after that we asked questions: Is the instruction correct? Is the instruction clear?

15.4.2 Supporting Self-Assessment and Formative Peer Assessment

The pupils who participated in the experiments had before had the opportunity to occasionally do inquiries in mathematics under the guidance of their teachers, but this experimental education was their first experience with peer assessment. As mentioned above, to support the peer assessment we (the members of the local working group) designed the worksheets. The worksheets were used both for recording the solution of the problem and peer assessment. In the process of discussion of the content of the worksheet, we immediately realized the difficulties that peer assessment in inquiry-based education in mathematics create. The problems that the pupils deal with can usually be solved in various ways. Not all of these ways are directed to meet intentional educational goals, and for that reason it is difficult to formulate the rubrics in the worksheet to enable the pupils to assess the work of their classmates.

For example, in the Lentil problem (determine how many lentils there are in a half-kilogram package) we assumed that pupils would determine the weight (or volume) of a certain number of lentils or vice versa and then use a reasoning based on knowledge of proportionality. It would also be possible to use ways based on estimation. The specific solutions can vary in details. The uncertainty of the situation created difficulty in formulating the questions on the worksheet. We decide to distinguish the correct solution and ask the question: What did you like in the solution? In the solutions that the assessors considered erroneous, the instruction was: Recommend to your classmates how to get the correct number.

The pupils solved this problem in groups. The groups then assessed the work of another group using the questions from the worksheet. Table 15.1 presents three

Table 15.1 Several solutions of the lentil problem and its peer assessment

	Solvers (assessees)	Assessors
S1	First we determined that 5 g contains 80 grains. $500 \div 5 = 100$, $100 \times 80 = 8000$ There should be 8000 lentil grains in the package	☺ We like that they have the same principle as we have ☹ But weighing needs to be accurate
S2	First we determined the mass of the whole package. Then we calculated how many grains there are in one gram. We then calculated the problem and got 93,258 as the result	☹ We cannot assess this. They do not write what problem they were solving. Therefore we do not know how they calculated it
S3	We poured the grains into a large vessel and weighed it. We subtracted the mass of the vessel from the mass of the grains. We found out that the grains weighed 501 g. Then we found out that 20 grains weigh 1 g Calculation: $501 \times 20 = 10,020$	☺ The procedure was correct ☹ But our result was different

descriptions of the solution procedure (S1 correct, S2 incomplete, S3 incorrect) and their assessment by classmates.

Although the assessors of S2 expressed the error in the solution quite accurately, the assessees in the final discussion criticized the assessors not to give them supporting feedback.

In the following experiments, we split the solution into stages and asked for separate assessment of each stage. However, our questions still had to be general in order not to lead the pupils to a "correct" solution.

15.4.3   Correctness of Solution of the Problem and Peer Assessment

The greatest advantage of peer assessment lies in the fact that a classmate may often give a problem solver more comprehensible feedback because assessor and assessee have almost the same learning experience and speak the same language. In our study, we experienced two difficulties. The solvers found it very difficult to record their solution procedure (see examples in Table 15.1), and for this reason the assessors sometimes did not fully understand the solution. The second problem was related to the assessors, who were not always able to assess the correctness of the solution. Sometimes it was difficult for assessors to decide whether the problem had been solved correctly, in other words, to assess the individual steps of the solution procedure described by their classmates, and it was equally difficult to communicate this assessment in a comprehensible way. In S2 (Table 15.1), some steps that were not needed for the solution are described (the mass of lentils in the package was given in the assignment), while other steps were not described clearly enough to make a decision on their correctness. The assessors commented on the second part of the solution quite clearly. However, they did not comment on the fact that it was not necessary to determine "what the mass of the package was." Another problem was that number of lentils was approximately 10 times higher than the correct solution. The peers did not comment this fact because they were not sure of the correct answer.

Essential in inquiries are those erroneous contributions that move the solution forward. However, these are not often assessed by an evaluator who is familiar with similar methods of work as the solver. The situation becomes even more complicated if the solution is not described clearly and comprehensibly by the solver.

In our study, the solvers could react to the peer assessment at the conclusion of the whole solution process. Several answers to the question: "Did your friends' advice help you?" were negative, using a sad emoticon: "☹ It did not because if they write that the measuring should be accurate, we don't know how exactly." "No. Those who were checking our work must have lost their lentils." "☹ Saying the result was different is of no help." As the children grew more experienced with peer assessment, they grew more self-critical in these final comments: "We think it should have been briefer and we did not finish the manual because we had for-gotten." "We think it's all quite muddled." "We could do it better to make it comprehensible for everybody. But otherwise we think it's OK."

In the final interview, students answered a question: Do you think you did well in assessing your peer(s)? The majority of pupils' felt relatively competent in assessing their peers (58%), while about a quarter of the pupils responded that they did not work well. Others (17%) were not able to evaluate their work and responded that they did not know. Among reported difficulties, the most frequent was the lack of knowledge or skills necessary for correct assessment, which was associated either with uncertainty about the solution of the inquiry task or the criteria for the assessment. The pupils mentioned, for example, that "it was difficult to decide whether it is correct or not," "I did it differently and I am not sure that this could be realized," etc. Dealing with this uncertainty in the classroom is crucial for imple-mentation of peer feedback in a broader context.

15.4.4   Peer Assessment and Institutionalization of Knowledge

Formulation of the objectives of an inquiry is connected to the issue of institu-tionalization of the gained knowledge. We found out that at the end of inquiry-based lessons, the pupils expected an unequivocal decision on what had been done correctly. They expected the result of their solution—the discovered knowledge—to be shared by the group, critically discussed, and then accepted. The final summary was in the hands of the teacher. However, if the teacher had not stated the learning objective clearly enough, their summary was very vague ("You worked very nicely," "I am pleased with your work."). Our findings are in agree-ment with the theory of didactical situations. The need for the inclusion of an institutionalization phase was theoretically grasped by Brousseau and Balacheff (1997) and introduced in the model of the so-called a-didactical situation (a situ-ation in which the teacher let the pupils to discover part of mathematical knowl-edge). The presence (and necessity) of the institutionalization phase in independent problem-solving situations has been confirmed in the Czech educational context (Novotná and Hošpesová 2013).

15.4.5   Other Methods of Formative Assessment in Our Experiments

Although the Czech part of the project focused on peer assessment, we also monitored the implementation of other forms of formative assessment or action supporting it. On-the-fly   assessment   (immediate   corrective   feedback   and reinforcement) was present, especially in the inquiry phase. It is a part of the Czech culture of primary mathematics that the teacher corrects pupils' work while they are working. On-the-fly assessment can also be considered peer feedback. During the group discussions, the pupils considered the suggestions for the solutions presented by their members.

15.5 Concluding Remarks

The research question we were focusing on was: What difficulties have to be faced in introducing formative assessment in the primary mathematics classroom? Let us now summarize these difficulties from the point of view of the actors in education: teachers and pupils. The interviews with teachers after the realization of each lesson showed that the peer assessment during the inquiry-based tasks in mathematics is rather difficult and challenging for both teachers and pupils.

15.5.1 Formative Assessment and Teachers

The teachers reflected on their role, which they found even more important and difficult than they foresaw. The main problems they identified were time and resources demands (worksheets, assessment tools, and teacher assistant time). This is related to the issue of appropriate support that teachers need. At the beginning of experimental teaching, the teachers appreciated the worksheets for pupils. During and after the experiments, they reported that they preferred to see a more experienced colleague's teaching, real teaching-situation stories, a databank of tasks, training courses for teachers, and researchers' on-site support. Some training, therefore, should precede the implementation of formative assessment, and assistants in classes would also be helpful, as the implementation would be time consuming and not all teachers have proper readiness.

During the experimental teaching, the teachers realized that pupils frequently were not able to provide effective feedback to their peers, and they felt uncertain how to help them. After the experimental teaching, the teachers mentioned that the pupils were often not able to give meaningful feedback to the recipient that would provide enough hints about how to proceed further. They saw the importance of development of the pupils' assessment skills, which should help pupils not only in learning to assess (both self and peers), but also in mastering the curriculum.

The teachers also reflected that they had difficulties in summing up the lessons in a final whole-class discussion that could institutionalize the new piece of knowledge (Brousseau and Balacheff 1997).

Some teachers saw parents' views on learning as an obstacle to the implementation of formative assessment: "… only working with the pupils' book is seen as sound learning, anything else is seen as entertainment or relaxation," "… I do not know how I could defend the time we have spent on it and explain to parents that we did not practice enough of the tasks in the pupils' book."

Some teachers' comments were directly related to the realization of experimental lessons. In particular, written peer feedback seemed too difficult; the pupils were able to be more precise and detailed when speaking. In introducing formative (peer) assessment, it would be good to start with simpler tasks and with some task for training, e.g., working on a series of similar tasks and only at the end asking the pupils for peer feedback. The teachers also mentioned that the pupils would prefer to have an opportunity to see and discuss more solutions before assessing. Some of the teachers considered this to be a big issue, as they realized that pupils may need more time to think the assessment over.

We realized that it is quite essential to elaborate the learning progress structure of the inquiry task, but it is more or less impossible to do it for inquiry that has been initiated using an open problem. Summary of peer feedback and learning accomplished is important for "institutionalization of new knowledge," i.e., deepened understanding of the concept being studied.

15.5.2   Formative Assessment and Pupils

The trend is to make pupils responsible for their learning. Whether they are ready to accept this responsibility largely depends on whether they understand what the meaning of various school activities is and on their ability to recognize when to use which activity. This means pupils need the space to define and accept goals of their learning and the space to see whether they have actually achieved those goals. If pupils are given the chance to speak about their learning and reflect on the process of learning, higher order thinking skills and metacognition are developed.

Our experiments showed that pupils adapted to inquiry-based education and peer assessment very quickly. Pupils' willingness to inquire and to assess each other and their success in these activities developed as they gained experience. For both individuals and groups to become independent while solving problems and develop towards autonomous assessment is a gradual process that must be given enough space at schools. Our experience showed that formative assessment of peers' work is more productive if preceded by a discussion with the whole class on the possible solutions of the problem. This discussion reduces pupils' uncertainty associated with the correctness of the problem's solution, which is the main problem of peer assessment from the pupil's point of view.

15.5.3   Formative Assessment and Culture

The relationship between content (curriculum requirements), how the content is used (pedagogy), and assessment is well known: What is taught is influenced by how it is taught, and what is assessed influences both what is taught and how it is

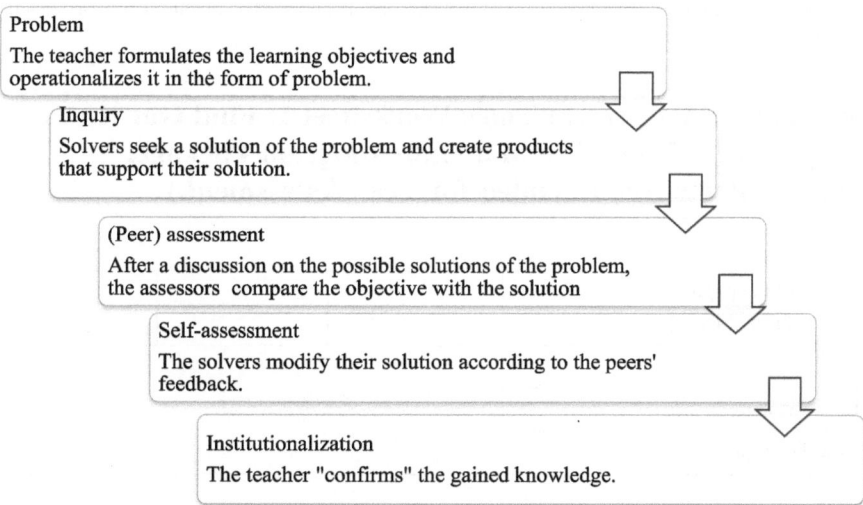

Fig. 15.5 The process of formation of autonomous assessment (*Source* Hošpesová and Žlábková 2016)

taught. It follows that if the curriculum is packed with content and assessment requires memorization of facts instead of conceptual understanding, learner-centered approaches including inquiry-based education will be out of place and there is no need to demand that pupils should be responsible for their learning.

In our project, the course of an inquiry-based lesson including formative assessment is expressed by the schema in Fig. 15.5. Although inquiry and formative assessment are both described as cyclical processes, in the lessons they have the nature of a linear process that starts with the assignment of the problem and ends with a teacher's summary.

Mathematics, with a greater extent of generic skills, is different from the science subjects, as teachers see open inquiry in math as more difficult to prepare and manage. Peer assessors feel somewhat less certain when providing feedback unless the institutionalization of learned knowledge precedes.

For primary-level group and pair activities, including inquiry and peer feedback, seem to be convenient. We recommended starting with formative assessment activities very soon: Even second-grade students can try them and learn a lot (though mostly about the feedback process itself).

The strong tradition of summative assessment in the Czech Republic calls for open discussion in broader contexts, including the general public (media, etc.), in order to slowly change the assessment culture.

Acknowledgements This research was supported by the Capacity, Collaborative Project no. 321428, entitled "Assess Inquiry in Science, Technology, and Mathematics Education" (under the Seventh EU Framework Program) and the Czech Science Foundation under project

No. 14-01417S, entitled "Enhancing mathematics content knowledge of future primary teachers via inquiry-based education."

Appendix 1: Assessment Tools Worksheet 1: Find Out How Many Lentils There Are in a Half-Kilogram Package. (Colored Parts Are Intended for Peer Assessment.)

Names of the pupils solving the problem:				

Estimation:

How did we proceed?

Result:

Names of pupils who assess the solution:				

Is the result correct?   ☺   ☹

☺ What did you like in the solution?

☹ Recommend your classmates how to get the correct number

Revision:

Did the advice from your classmates help you? How?   ☺   ☹

References

Artigue, M., Baptist, P., Dillon, J., Harlen, W., & Lena, P. (2012). *Learning through inquiry. The fibonacci project resources*. Retrieved from: http://fibonacci-project.eu.

Assessment Reform Group (ARG). (2002). *Assessment for learning: 10 principles*. Retrieved from: www.assessment-reform-group.org.

Black, P., Harrison, C., Lee, C., Marshall, B., & Wiliam, D. (2004). Working inside the black box: Assessment for learning in the classroom. *Phi Delta Kappan*. https://doi.org/10.1177/003172170408600105.

Black, P., & Wiliam, D. (1998). Assessment and classroom learning. *Assessment in Education, 5*(1), 1–74.

Black, P., & Wiliam, D. (2009). Developing the theory of formative assessment. *Educational Assessment, Evaluation and Accountability, 21*(1), 5–31.

Boud, D., & Falchikov, N. (Eds.). (2007). *Rethinking assessment in higher education: Learning for the longer term*. London: Routledge.

Brousseau, G., & Balacheff, N. (1997). *Theory of didactical situations in mathematics: Didactique des mathématiques, 1970–1990*. Boston: Kluwer Academic Publishers.

Dewey, J. (1938). *Logic: The theory of inquiry*. New York: Holt.

Donovan, M. S., & Bransford, J. D. (Eds.). (2005). *How students learn: Science in the classroom*. Committee on How People Learn: A Targeted Report for Teachers. National Research Council.

Harlen, W. (2014). *Assessment, standards and quality of learning in primary education*. York: Cambridge Primary Review Trust.

Hošpesová, A., & Žlábková, I. (2016). Assessment in inquiry based education in primary mathematics. In M. Flégl, M. Houška, & I. Krejčí (Eds.), *Proceedings of the 13th International Conference Efficiency and Responsibility in Education 2016, ERIE* (pp. 194–201). Prague: Czech University of Life Sciences.

Novotná, J., & Hošpesová, A. (2013). Students and their teacher in a didactical situation. A case study. In *Student voice in mathematics classrooms around the world* (pp. 133–142). Rotterdam: Sense Publishers,

OECD. (2013). *Synergies for better learning: An international perspective on evaluation and assessment*. OECD Reviews of Evaluation and Assessment in Education. Paris: OECD Publishing.

OECD. (2015). *Education at a glance 2015: OECD indicators*. Paris: OECD Publishing. http://dx.doi.org/10.1787/eag-2015-en.

Pachler, N., Daly, C., Mor, Y., & Mellar, H. (2010). Formative e-assessment: Practitioners cases. *Computers & Education, 54*(3), 715–721.

Polya, G. (1945). *How to solve it*. New Jersey: Princeton University Press.

Samková, L., Hošpesová, A., Roubíček, F., & Tichá, M. (2015). Badatelsky orientované vyučování matematice [Inquiry based mathematics education]. *Scientia in Educatione, 6*(1), 91–122 (in Czech).

Slavík, J. (2003). Autonomní a heteronomní pojetí školního hodnocení – aktuální problem pedagogické teorie a praxe [Autonomous and heteronomous understanding of assessment at school—Topical problem of educational theory and practice]. *Pedagogika, 40*(1), 5–25 (in Czech).

Suurtamm, C., & Neubrand, M. (2015). Assessment and testing in mathematics education. In S. J. Cho (Ed.), *The Proceedings of the 12th International Congress on Mathematical Education*. https://doi.org/10.1007/978-3-319-12688-3_58.

The Framework Educational Programme for Basic Education. (2007). Retrieved from http://www. msmt.cz/areas-of-work/basic-education-1?lang=2.

Topping, K. (2013). Peers as a source of formative and summative assessment. In J. H. Mac Millan (Ed.), *SAGE handbook of research on classroom assessment* (pp. 395–412). London: Sage.

Žlábkova, I., & Rokos, L. (2013). Pohledy na formativní a sumativní hodnocení žáka v českých publikacích [Formative and summative assessment in Czech publications]. *Pedagogika, 58*(3), 328–354 (in Czech).

Chapter 16
Professional Development of Mathematics Teachers: Through the Lens of the Camera

Ronnie Karsenty

Abstract The VIDEO-LM project (Viewing, Investigating and Discussing Environments of Learning Mathematics), developed at the Weizmann Institute of Science in Israel, is aimed at enhancing secondary mathematics teachers' reflection and mathematical knowledge for teaching. In the project, videotaped lessons serve as learning objects and sources for discussions with teachers. These discussions are guided by an analytic framework, comprised of six viewing lenses: mathematical and meta-mathematical ideas; goals; tasks; dilemmas and decision making; interactions; and beliefs. To assess and characterize the impact of the project, data was collected from 17 different implementations of in-service VIDEO-LM courses around the country conducted by facilitators specifically qualified for this pursuit. This paper reports on some of the findings, with particular reference to possible mechanisms that can explain the processes of change that teachers undergo.

Keywords Video-based professional development · Secondary mathematics teachers · Reflection · Mathematical knowledge for teaching

16.1 Introduction

Video has been used as a tool for teacher education and professional development (PD) for the past 50 years, however the focus and methods of its uses has changed considerably over time (Sherin 2004; see Fig. 16.1). Presently, the low cost of portable easy-to-use digitized video recording devices, combined with accessible means of editing and exchanging clips, has increased the dissemination of this technology within PD programs for mathematics teachers around the world [e.g., Mathe sicher können in Germany; the Problem-Solving Cycle and the Learning and Teaching Geometry programs in the USA (Borko et al. 2011); MILE in the Netherlands (Goffree and Oonk 2001); and Effective Mathematics Teaching and

R. Karsenty (✉)
Weizmann Institute of Science, Rehovot, Israel
e-mail: ronnie.karsenty@weizmann.ac.il

© The Author(s) 2018
G. Kaiser et al. (eds.), *Invited Lectures from the 13th International Congress on Mathematical Education*, ICME-13 Monographs,
https://doi.org/10.1007/978-3-319-72170-5_16

CBL in Australia (Clarke et al. 2013)]. Online video resources are now largely available to educators (MET in the USA and Teachers Media in UK are prominent examples) and international symposia are dedicated to the use of video in professional development for mathematics teachers (e.g., http://www.weizmann.ac.il/conferences/video-lm2014).

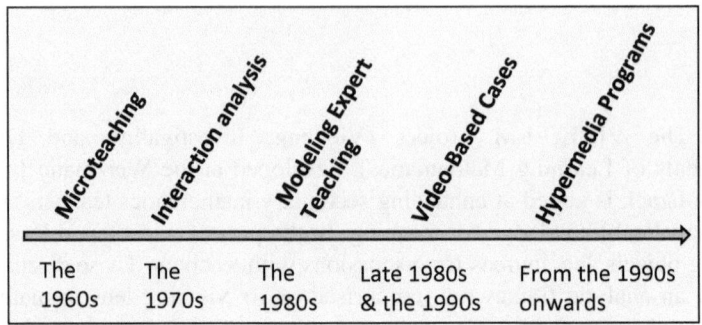

Fig. 16.1 Changes in video uses in teacher education (based on Sherin 2004)

The affordances of videotaped episodes as a source for teacher learning have been investigated in a growing number of studies (e.g., Brophy 2004; Borko et al. 2011; Coles 2014; Gaudin and Chaliès 2015; Nemirovsky and Galvis 2004; Santagata and Yeh 2013; Sherin and van Es 2009). Sherin and van Es (2009) claim that "teachers benefit from opportunities to reflect on teaching with authentic representations of practice" (p. 21); Brophy (2004) argues that video can introduce "the complexity and subtlety of classroom teaching as it occurs in real time" (p. 287); and Nemirovsky and Galvis (2004) suggest that "because of the unique power of video to convey the complexity and atmosphere of human interactions, video case studies provide powerful opportunities for deep reflection" (p. 68). All of these scholars emphasize the role of video as a window to the authentic practice of teaching, which allows teachers to focus on complex issues that may be unpacked through observing, re-observing, and reflecting on specific occurrences.

Three main directions can be identified within programs that use videotaped episodes from mathematics lessons as resources for teacher development. First, video is utilized for introducing new curricula, activities, pedagogical strategies, etc. This target is mainly implemented through supplying teachers with video cases that model and demonstrate how teaching the new curricula or using the pedagogical strategies may be enacted (e.g., Seago et al. 2010). A second direction is using videotaped lessons as a source for feedback and evaluation. Teachers watch videotapes from their own classrooms and discuss them with colleagues or instructors, often with the use of pre-constructed standard-based rubrics such as those developed by Danielson (2013) or Hill et al. (2008). The third direction is using videotaped episodes to enhance teachers' proficiency to notice, understand

and discuss students' mathematical thinking (Sherin et al. 2011), usually in the form of "video clubs" (van Es and Sherin 2008).

The VIDEO-LM Project (Viewing, Investigating and Discussing Environments of Learning Mathematics) is aimed at a fourth direction: the elaboration and use of tools for reflection on the mathematics teaching practice through the development of a productive language that supports deep peer conversations. The project also aims at promoting the development and enrichment of mathematics knowledge for teaching, in the sense defined by Ball et al. (2008).

In this paper, I describe the project and its theoretical roots. I then introduce the framework of analysis, called the six-lens framework, developed to achieve the project's aims. I present findings from an evaluative study conducted to assess the impact of the project. Finally, I suggest possible mechanisms that can explain the processes of change that teachers undergo.

16.2 The VIDEO-LM Project: Rationale, Theoretical Roots, and Framework

Teaching is known to be a rather lonely profession. Despite participation in professional communities, online forums, and other forms of communication and collaboration with other teachers, the reality is that the vast majority of teachers are the "solo adult actors" in their classrooms, where they spend the lion's share of their professional life. In many countries teachers seldom get the chance to watch their peers in action once the pre-service period is over. This is not merely a social deficit but also a barrier to certain processes of professional evolution embedded in peer learning in situ. The VIDEO-LM project, developed at the Weizmann Institute of Science in Israel, is a research-based PD program for secondary mathematics teachers that creates opportunities for teachers to watch whole lessons given by other teachers. The project uses a collection of videotaped lessons, which serve as learning objects and sources for discussions with teachers. Since teachers do not watch themselves, as is frequently done in video clubs, but rather observe videotaped lessons of unknown teachers, the videos are taken, in a sense, as "vicarious experiences" that allow for indirect exploration of one's own perceptions on the practice of mathematics teaching through the observation of "remote" teaching events. This is done in a supportive atmosphere that does not focus on evaluative feedback.

The project is rooted in two theories: Schoenfeld's (1998, 2010) Teaching in Context theory and the theoretical framework of Mathematical Knowledge for Teaching (MKT; Ball et al. 2008). According to the Teaching in Context framework (Schoenfeld 1998, 2010), teaching is goal-oriented; teachers strive to achieve various types of goals and are constantly modifying and changing their goals in correspondence with classroom realities. The theory asserts that teachers have a body of knowledge resources they can call upon for both expected and unexpected situations and that teachers, like everyone else, have a set of orientations, i.e., predispositions and beliefs about mathematics, about students, and about teaching.

This triad of goals, resources, and orientations monitors teachers' decision-making processes and shapes their choice of actions.

The MKT framework, proposed by Ball et al. (2008) and refined by Hill et al. (2008), is comprised of two categories, Subject Matter Knowledge and Pedagogical Content Knowledge, further divided into six sub-categories. This framework is valuable both as a conceptualization tool of the kind of knowledge we wish to enhance within a PD setting and as an analysis tool that allows a scrutinized look at what teachers are focusing on during PD sessions, as I shall demonstrate in Sect. 16.4.

16.2.1 The Six-Lens Framework

In light of these theoretical frameworks, we suggest that teachers can and should be actively involved in a deep reflection and analysis of their own (and others') goals, resources, and orientations and of their mathematical knowledge for teaching. Following previous initial experimentation with video-based discussions that centralize these ideas (Arcavi and Schoenfeld 2008), we designed a framework consisting of six analytical tools with which mathematics teachers can reflect on a videotaped lesson. We call these tools *lenses*, to emphasize their use as means of *observation*, in the dual sense of watching an occurrence but also commenting on it. Viewing a lesson through a certain lens implies shedding light on a specific feature of the mathematics teaching practice. Table 16.1 presents this six-lens framework (henceforth: SLF), consisting of the following components: mathematical and meta-mathematical ideas; goals; tasks and activities; interactions; dilemmas and decision-making; and beliefs about mathematics teaching. Table 16.1 outlines the focus of observation activities around each of these lenses, and exemplifies the sort of questions that direct discussions with teachers.

16.2.2 Features of Using SLF in Video-Based PD Sessions

The SLF framework was designed with a particular desired learning environment in mind. We envisioned a supportive and nonthreatening setting in which a group of teachers feels comfortable enough to elicit ideas and thoughts, while opportunities are created for deep reflection on practice. Our aim was that the activities of watching and analyzing videotaped lessons will lead to forming peer groups that are highly engaged in core issues of the mathematics teaching profession. Therefore, we explicitly defined the use of SLF in PD sessions around the following features and norms:

- SLF is *not evaluative* in nature and is not used for the purpose of providing feedback. In line with the works of Jaworski (1990) and Coles (2013), the use of SLF attempts to establish nonjudgmental norms of discussion through the redirection of highly evaluative comments into "issues to think about". This is closely connected to the next feature.

Table 16.1 The Six-Lens Framework (SLF)

Lenses for observing a videotaped mathematics lesson	The focus of activities around each lens	Examples of questions that direct teachers' discussions
Mathematical and meta-mathematical ideas	Scanning the space of relevant ideas, concepts, and procedures, as well as meta-mathematical ideas (e.g., one counter example is sufficient to refute a conjecture) that may be associated with the lesson's topic	• Which ideas did the filmed teacher bring forward in the lesson? Which ideas were left out? How can this decision be explained? • Which meta-mathematical notions were evident in the lesson?
Explicit and implicit goals	Attributing goals that may underlie the teacher's actions or decisions, on the basis of what was observed in the video. Rather than "scientifically verifying true goals", the aim is to sharpen awareness of different possible goals and negotiate the pros and cons of preferring certain goals over others.	• Try to identify the goals that you think the filmed teacher was attempting to achieve. Show evidence from the video to support your assertion. • Did you notice a moment when the teacher's goals have changed or a new goal was added? Why do you think this happened?
Tasks and activities	Conducting an "a posteriori task analysis": discussing features of the task and how it was enacted by the filmed teacher and students. Noticing if and when it develops differently than expected.	• Observe and document how the task is introduced and carried out and how the teacher addresses students' reactions. • What may be the benefits and pitfalls in bringing this task to class?
Interactions with students	Observing and analyzing if and how the filmed teacher poses further questions to those of the task; listens to (or ignores) comments or difficulties raised by students; manages discussions; delegates responsibilities in the process of knowledge generation.	• How does the filmed teacher navigate students' responses during the mathematical activity? What kind of questions does the teacher ask? Who gets permission to speak? • Characterize the teacher's feedback to students.
Dilemmas and decision-making	Uncovering situations of dilemma (i.e., when there is no evident optimal course of action) that the filmed teacher seemed to have faced during the lesson. Discussing the decisions taken in order to resolve these dilemmas, and their consequent tradeoffs.	• Did you notice a dilemma during the lesson? What did the teacher decide to do? Are there alternatives you can think of for this decision? • What may be the constraints and affordances of the teacher's choice and of the suggested alternative paths?
Beliefs about mathematics teaching	Eliciting orientations, beliefs and values that may be attributed to the filmed teacher on the basis of the video. Unpacking implicit messages that may be conveyed to students through the teacher's communications and actions.	• What may be the filmed teacher's views about the nature of mathematics as a discipline? • How does the teacher perceive his or her role? What may be the teacher's ideas about what "good mathematics teaching" is? What does the teacher think about the students' role as learners?

- An SLF-based discussion pre-accepts a basic working assumption that *the filmed teacher is acting in the best interest of his/her students*. Thus, observers are required to "step into the shoes" of the filmed teacher in an attempt to understand his/her goals, decisions, and beliefs, maintaining a respectful conversation. This viewpoint allows for deeper layers of reflection than those entailed in comments such as "she's doing it all wrong".
- *SLF does not pursue the demonstration of "best practice"*. This is intentional; we believe that for different teachers there may be different best practices and that these differences may be linked to personal, contextual, and cultural settings. Our aim is to choose lessons that can serve as springboards for meaningful discussions on different aspects of practice, rather than on alignment with criteria of what teaching should look like. In this sense, we adopted the term "better than best practice" coined by Lefstein and Snell (2014).
- SLF is *deeply rooted in the subject matter of mathematics* and shuns generic discussions on teaching. Issues of classroom management, the teachers' body language and other generic aspects are marginal, if not completely absent, during discussions around the screened lessons. Instead, SLF refers to what lies at the heart of mathematics teaching, such as mathematical concepts and ideas, meta-mathematical concerns, possible targets of mathematics lessons, and beliefs about mathematics teaching.

In addition, the following two choices regarding the use of SLF are important to mention:

- SLF is a *teacher-centered framework*, i.e., the focus is on the filmed teacher's actions and choices. Students' voices and actions are taken into account within the interaction lens; however, the lion's share of an SLF-based discussion is dedicated to what the teacher is doing. In this regard, SLF is significantly different from the *noticing* framework (Sherin et al. 2011) mentioned earlier.
- SLF does not refer to clips or short episodes edited from a lesson; rather, the units of analysis for teachers' discussions are *whole lessons*, in which a more comprehensive "story" can unfold, with a beginning, a development of a process, and a closure. This characteristic marks SLF as unique amongst other frameworks used in most PD programs.

16.3   Exploring Possible Gains of Video-Based Discussions Directed by the SLF Framework

16.3.1   *VIDEO-LM Courses for Secondary Mathematics Teachers*

During the 2012–13 academic year, we conducted two pilot courses for mathematics teachers. Based on this pilot, we refined the design of the course to obtain a model which has since then been implemented in 29 new PD courses (7, 8, and 14 in the academic years of 2013–14, 2014–15, and 2015–16, respectively). The course consists of 30 academic hours, usually configured as 7–8 monthly sessions of 4–4.5 h each, and is led by a VIDEO-LM facilitator who has been specifically qualified for this pursuit, in consultation with the development team. In each session, the teachers watch a videotaped mathematics lesson. Several modes of "watching and discussing" may apply, according to a predesigned session plan (e.g., watching together or in small groups, focusing on different lenses, watching the whole lesson uninterruptedly vs. breaking it to sequenced episodes). The collection of videotaped lessons (mostly filmed by the VIDEO-LM team in Israel and a few videos from Japan and USA with Hebrew subtitles) as well as supplementary materials such as the tasks used in the lesson and lesson graphs describing the flow of the lesson are available on the VIDEO-LM website.[1]

16.3.2 Research Aim and Questions

Following the growing demand for VIDEO-LM courses, we designed several studies aiming to explore possible impacts of these courses on the participating teachers. Two of the research questions (RQ) investigated were the following:

RQ I. What may be the gains of video-based teacher discussions around the SLF framework, in terms of the teachers' MKT?
RQ II. To what degree do VIDEO-LM sessions stimulate reflections and deep conversations about the teaching practice?

16.3.3 Data Collection and Analysis

Data was collected from the 17 VIDEO-LM professional development courses that were conducted during 2012–2015 at nine different sites in Israel. The analyses of data are still ongoing, and in this paper I report on selected findings from five courses. Details on these courses and the data collection means used appear in Table 16.2.

All participants were secondary school mathematics teachers with different levels of experience—from new teachers to experienced teachers. Participation was recognized by the Ministry of Education for accruing credential points for promotion. Although courses were somewhat different from one another, according to each facilitator's approach and the local dynamics of the group, all were aligned with the course model described above, and in all of them SLF was used as a base for peer discussions.

---

[1]http://adasha.weizmann.ac.il.

Table 16.2 Courses details and data collection means

Site	Year	Location	No. of participants	Written data (reflections or feedback questionnaires)	Documentation of sessions
(a)	2012–13	WIS	10	✓	Video of all sessions
(b)	2013–14	WIS	12	✓	Video of all sessions
(c)	2014–15	RTC, large city in the center of Israel	17	✓	Video or audio of several sessions
(d)	2014–15	RTC, town in the center of Israel	11	✓	Video or audio of several sessions
(e)	2014–15	RTC, town in the north of Israel	11	–	Video or audio of several sessions

WIS Weizmann Institute of Science; *RTC* Regional Teacher Centre

As shown in Table 16.2, the data collected included video and audio documentation of PD sessions and written reflections or feedback questionnaires submitted by teachers at the end of the course (these submissions were part of the course assignments; the decision whether to include written reflections or feedback questionnaires in the final assignments was left to the facilitator in each site). The analysis of the data was carried out using various qualitative content analysis methods. Each method was applied to selected parts of the data, according to both availability of data at the time of analysis and the target of the analysis. Two analysis methods that are relevant to findings reported in this paper are described below.

(1) In order to answer RQ I, we performed a sequence of steps as follows (Karsenty et al. 2015; Nurick 2015): Transcribing video or audio records of PD sessions; tracing all utterances of participants' associated with MKT (i.e., unpacking mathematical concepts or relating to teaching these concepts); grouping utterances into units of analysis that share similar ideas; using the units to form "discussion maps" that convey the evolution of knowledge throughout different parts of sessions (examples follow in the Findings section); and comparing utterances in the discussions before and after watching the video using the six MKT categories. This type of analysis was performed on data from Sites (a) and (b).

(2) In order to answer RQ II, we performed a sequence of steps based on grounded theory methods (Glaser and Strauss 1967) as follows (Karsenty and Schwarts 2016; Schwarts 2016): Reading all the documented material—both spoken and written—relating to a the same lesson watched in various sites and identifying common themes; categorizing participants' utterances by the themes identified;

defining major themes according to considerations of prevalence and interest, merging categories where necessary; building "theme narratives" in order to characterize teachers' reactions in each category; and reexamining the narratives in search for different types of reflections that may be identified. This type of analysis was performed on data from Sites (a), (c), (d) and (e).

16.4   Selected Findings

16.4.1   Growth of Mathematical Knowledge for Teaching

In the third session conducted at Site (a), teachers watched an episode from a lesson on the commutative and associative laws given in a seventh grade heterogeneous class. Prior to watching the video, teachers were asked to elicit any mathematical ideas that may be associated with the topic of the commutative and associative laws. They suggested a fairly wide range of ideas, from the simple fact that addition and multiplication satisfy both laws, while subtraction and division do not, through various models that demonstrate the laws, to efficient solutions of multi-term exercises using the laws. It appeared that most teachers perceived the topic as natural and intuitive for students, at least at the numerical level. The discussion was coded in terms of the MKT categories (Ball et al. 2008): Each unit of analysis was coded as reflecting Common Content Knowledge (CCK), Specialized Content Knowledge (SCK), Knowledge of Content and Teaching (KCT), etc. The following excerpt demonstrates a unit of analysis coded as KCS (Knowledge of Content and Students):

64   T1: In seventh grade it's difficult to construct a serious generalization, so you smooth it over to things that work or don't work. As ideas, the associative and commutative laws are too early for seventh grade and it's difficult to create learning

65   T2: There is use in it, applications. For example 99 + 3232 + 1. A student that looks at it intuitively will do it

66   T3: They will do it without us calling it the commutative law and generalizing it

The part of the discussion before the video was screened was formed into two "discussion maps", one of which is presented in Fig. 16.3. Each unit was colored according to its MKT categorization using the color key introduced in Fig. 16.2. The discussion map clearly shows that prior to watching the video, teachers mainly demonstrated pedagogical content knowledge. The other discussion map, not appearing herein due to space limitations, conveys the same conclusion.

In the videotaped episode, the teacher asked the class whether operations that satisfy the commutative law necessarily satisfy the associative law as well and vice versa. The students' spontaneous collective answer was "yes". The teacher then introduced operation tables that were shown to be counterexamples to this conjecture (see Fig. 16.4) and led a discussion resulting in the conclusion that the laws are not interdependent.

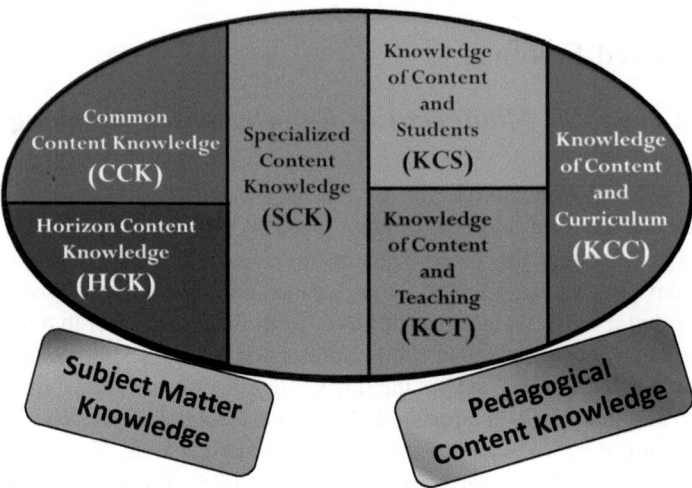

Fig. 16.2 Components of MKT (adapted from Ball et al. 2008, p. 403; For interpretation of the references to color in this figure, the reader is referred to the web version of this article)

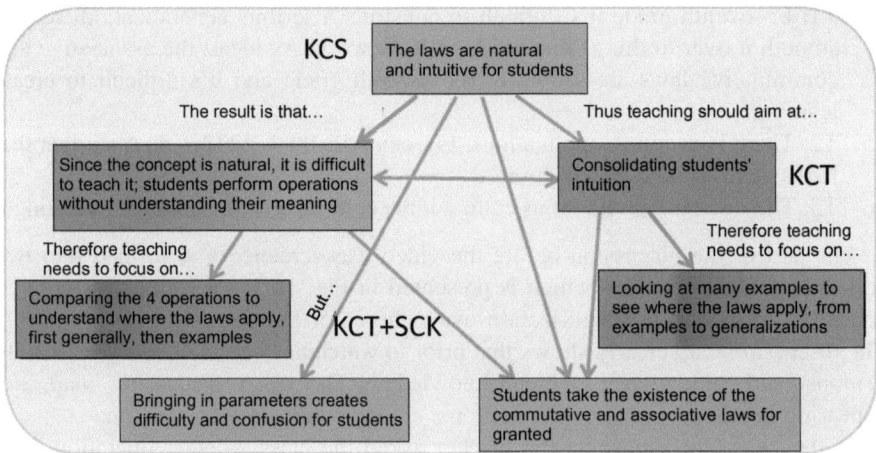

Fig. 16.3 One of the "discussion maps" describing the discussion before watching the video "The commutative and associative laws" (For interpretation of the references to color in this figure, the reader is referred to the web version of this article)

While watching the episode, each couple of teachers was requested to focus on one of the lenses comprising SLF. Then, in the plenary, observations were shared and discussed by all participants. On the whole, teachers were surprised by the episode, since the main mathematical idea raised by the filmed teacher was not considered by the group earlier. One of them described the teachers goal as "*undermining the perception that an operation can either satisfy both the associative and commutative laws or none of them*". The teachers used concepts from set theory to express this idea (see Fig. 16.5), noting that addition and multiplication are in the intersection of the commutative operations and the associative operations sets, while subtraction and division are in the complement of the union of these sets. While students might hold the misconception that the other possible two sets are empty, the lesson demonstrates that operations exist in all possible sets. Teachers also discussed the use of finite operation tables. Some teachers asserted that operations on small finite groups are not equivalent, mathematically and pedagogically, to operations defined on the real numbers. Thus, they challenged the group to find an operation, defined on the real numbers and relevant to students' school learning, for which only one of the laws holds. Eventually, two such

	Operation that returns the **first** number in the pair	Arbitrary operation on all possible pairs made of a, b, c
Operation table presented:	$\triangle$ 0 1 2 3 0 0 0 0 0 1 1 1 1 1 2 2 2 2 2 3 3 3 3 3	* a b c a a a b b a c b c b b a
Commutative law	✗	✓
Associative law	✓	✗

Fig. 16.4 Examples of operations discussed in the video "The commutative and associative laws"

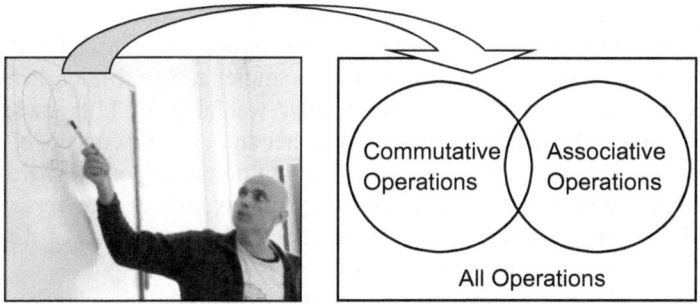

Fig. 16.5 A teacher presenting the mathematical idea of the episode using set theory

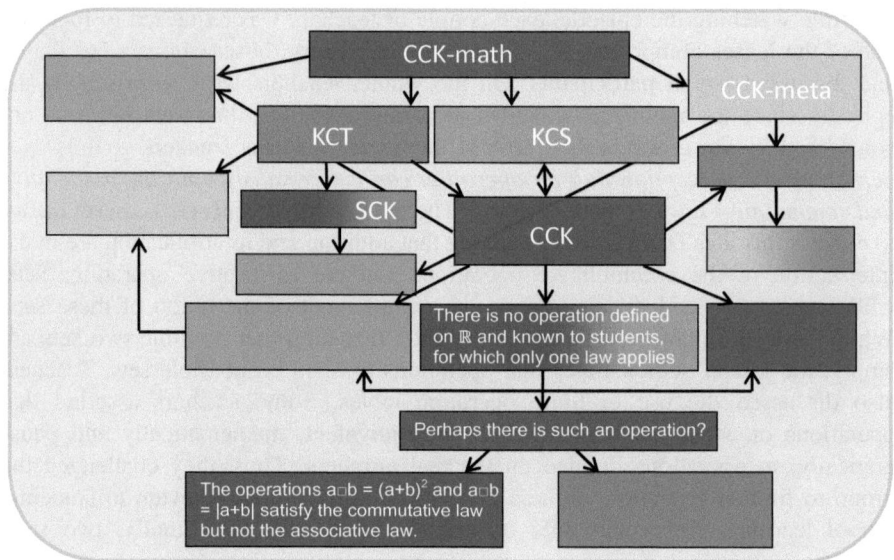

Fig. 16.6 A schema of the "discussion map" describing the discussion after watching the video "The commutative and associative laws" (For interpretation of the references to color in this figure, the reader is referred to the web version of this article)

examples were found: $a\square b = (a + b)^2$ and $a\square b = |a + b|$. In both cases the operation satisfies the commutative law but not the associative law.

This part of the discussion was also coded in terms of the MKT categories and formed into a discussion map. Figure 16.6 presents the schema of this map, illustrating the colored MKT categories (since space is limited, only several units are presented in words within this map). Comparing the discussion maps before and after the video was observed and analyzed by teachers, reveals that watching the video triggered a shift in the participants' utterances from pedagogical considerations towards the eliciting of more mathematical ideas, as was evident from the considerable increase in the units coded as Common Content Knowledge (CCK). In terms of quantification, the percentage of units coded as CCK before and after observing the video was 20 and 45%, respectively.

The findings from the case of "the commutative and associative laws" video are representative of other findings as well. For example, at Site (b), teachers explored various definitions of an inflection point, after watching an 11th grade Calculus class. In the video, the teacher discussed with her students the concept of concavity of functions, leading to the definition of inflection points as points where the graph changes from concavity upwards to concavity downwards or vice versa. This was then translated into a working tool, associating inflection points of $f(x)$ with the extreme points of $f'(x)$ or the zeros of $f''(x)$.

The video triggered a discussion about possible deficiencies of this tool, focusing on the following question: What about an inflection point where the first or

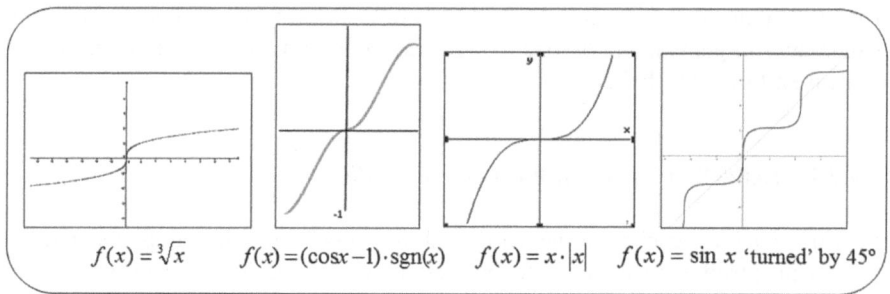

$$f(x) = \sqrt[3]{x} \qquad f(x) = (\cos x - 1) \cdot \mathrm{sgn}(x) \qquad f(x) = x \cdot |x| \qquad f(x) = \sin x \text{ 'turned' by } 45°$$

Fig. 16.7 Examples generated by teachers for functions $f(x)$ that have an inflection point at 0 but $f'(0)$ and/or $f''(0)$ do not exist

the second derivatives do not exist? The group became motivated to find counterexamples where $f(x)$ has an inflection point in x_0 but $f'(x_0)$ or $f''(x_0)$ do not exist, and found a graphic example but not an algebraic representation of such a function. Following the session, in an intense and rich email exchange, teachers found and shared different counterexamples, as described in Fig. 16.7.

As a result, the group reached a consensus about the accuracy of definitions of inflection points that are customarily presented in advanced calculus classrooms. The new collectively generated MKT also included valuable pedagogical suggestions offered by participants, such as the idea to have students find their own counterexamples to the "rule" that identifies inflection points with $f''(x) = 0$. Another opportunity to extend knowledge evolved during the session, when the goals of the videotaped teacher were discussed. Participants attempted to justify the teacher's choice of presenting an inaccurate working definition by ascribing to her two major considerations: Firstly, students may not be ready to grasp the correct definition, which requires advanced thinking, and, secondly, functions such as $x \cdot |x|$ are not included in the curriculum and in the final exams. This part of the discussion opened a debate on more general questions: How far should teachers go beyond what is delimited by the curriculum? To what extent are we allowed to "sacrifice" mathematical rigor in favor of our students' immediate practical interests?

To sum up this section, the cases analyzed above suggest that SLF-based peer discussions around videotaped lessons can be a powerful tool for prompting the growth and refinement of relevant mathematical knowledge for teaching.

16.4.2 Enhancement of Reflection on the Mathematics Teaching Practice

In this section I will demonstrate, through representative examples, how viewing videotaped mathematics lessons of unknown teachers, using lenses included in

SLF, contribute to the development of rich reflection on the practice of teaching mathematics in general and on one's own practice in particular. I will focus here on two lenses: tasks and beliefs.

16.4.2.1 Reflecting Through the Lens of Tasks

The video enables teachers to watch a "task in action", how it is implemented, the nuances in introducing it, how students attempt to solve it, and how the teacher addresses the students' reactions. We refer to this as an "a posteriori task analysis", which may be very different from the somewhat limited "a priori analysis", i.e., examining the same task as it appears in a written text. This turned out to be a very engaging activity in VIDEO-LM courses. For example, when we investigated what teachers talked about in sessions around a Japanese video, in which a challenging geometrical problem was given to eighth grade students, we found that 29.3% of the teachers' talk was devoted to the task, its characteristics, affordances, and limitations, how it was presented and how students handled it (Karsenty and Schwarts 2016; Schwarts 2016). This collective analysis led many teachers to relate to the kinds of tasks and problems they use in their own classrooms, as illustrated in the following teacher citations, taken from PD sessions or from written reflections submitted after the course:

- "There's an embarrassment here, do I surprise my students at all, occasionally? It's difficult to deal with this embarrassment [...] Seeing this unusual problem raises the question of how many times do I do that, and what it tells about the way I teach" (PD session, Site a)
- "Many times I try to select problems that are unique, special [...], it's not always simple, sometimes I have them from last year, sometimes I find them accidently [...], and then once you do the irregular stuff, the other problems they can handle" (PD session, Site c)
- "There are beautiful proofs using areas, but in fact we actually never do them" (PD session, Site e)
- "Watching this Japanese lesson left me with frustration, that I as a teacher mainly teach technique, solving algorithms and not much beyond that, I feel chained to the time constraints. Or is this just an excuse for not being creative?" (written reflection).
- "I'm in my 22nd year of teaching, and I look at this thing and I know that I'm taking this today [...] I'm not going to be this teacher in this classroom but I definitely leave here asking myself what I'm going to do with it tomorrow, in my classes" (PD session, Site a).

Talking about tasks and their implementation in class may also evoke reflections about risks that teachers take (or refrain from taking) when choosing tasks for students. Following a discussion on a videotaped lesson on sequences, given in a low-track class, one of the teachers wrote in his reflection:

In this session teachers occasionally raised doubts (that I also feel sometimes) about the ability of students to deal with the tasks we give them. The one who phrased it in the best way was Sam, who said "I don't have the courage to throw my students into it, just like that, on their own…". I think that this is the heart of the matter, it is *us* who don't have courage to let them strive. If *we* dare a little more, so will they. (written reflection, emphasis in original)

In this case, the teacher raises considerations of what he calls "courage," related to selecting tasks that students may struggle with. In another case, teachers talked about selecting tasks that are challenging for *teachers*. The conversation below took place in Site (d):

677   T1: What does a teacher do if he just now opened the textbook, saw some tasks, tried to solve them and did not succeed. Does he take it to class? […]
685   Facilitator: Do I take to my classroom something that I cannot solve?
686   T2: Of course not! Are you kidding me?
687   T3: Surely not
688   T1: I don't know, maybe yes
689   T2: What [do you mean] yes?
691   Facilitator: Why?
692   T1: Why? Because if I come to class with the approach of "let's learn together" …
693   T2: Let's think together?
694   T1: Let's think together, here, there are certain things that I too…

In both cases, clearly the discussion through the lens of *tasks* is interrelated with the teachers' *beliefs* regarding their role as teachers, although this interrelation remains implicit. This connection is not surprising; the issue of how teachers' beliefs shape their practice has been widely studied (e.g., Schoenfeld 1998; Li and Moschkovich 2013). Thus, we acknowledge that in fact the use of most of the lenses comprising SLF (i.e., goals, tasks, interactions, and dilemmas) is likely to be guided by the beliefs teachers hold. This is one of the main sources for our decision to explicitly include conversations about beliefs in VIDEO-LM courses, or in other words, to incorporate the lens of beliefs as one of the six lenses. In the next section I elaborate on possible gains of using the lens of beliefs.

16.4.2.2 Reflecting Through the Lens of Beliefs

Facilitating discussion about beliefs is a delicate matter; for many teachers, this theoretical construct is foreign, thus it needs to be carefully presented. As shown in Table 16.1 above, some the questions we focus on within this lens are: On the basis of the observed teacher's actions, what may be her views about the nature of mathematics as a discipline? How does she perceive her role? What may be her ideas about what "good mathematics teaching" is? The exercise of inferring and attributing beliefs to another teacher is not a trivial one. However, it often triggers catalytic comments, especially in later stages of the course, when teachers begin to internalize the SLF language and connect the analysis to their own practices. This was demonstrated vividly in one of the PD sessions at Site (b). The topic of the lesson watched was sketching, for a given function $f(x)$, the graphs of $e^{f(x)}$ and $\ln f(x)$. The teaching in this lesson was frontal, with the teacher's tight control over the development of the mathematical knowledge. Students appeared to be highly engaged in the questions posed by the teacher, who never left her position near the board. In the discussion, one teacher said:

> The lesson really challenges our beliefs. [...] If you'd ask me at the beginning, before watching the video, what... how should a lesson look like, I would have said many nice things [...] such as you need to have a discussion, you need to have shared thinking, students should experiment right and wrong things, you need to have interaction in the class, and dynamics, and then suddenly I see something that... doesn't have these things - there's no discussion, or just a very short one, and I'm looking at it and I say 'what a beautiful lesson!' [...] so now I have an internal conflict, really, I have an internal conflict, because on the one hand everything I know about teaching is missing here, but on the other hand I like what I see. So I'm trying to settle this dissonance, so I say okay, maybe it's class dependent, maybe it's students dependent.

This citation indicates that, when given the opportunity to directly speak about beliefs, teachers may re-inspect their most deep convictions and practices and confront the complexities of teaching. This may or may not lead to changes in one's own beliefs, or in one's practice, but it increases teachers' awareness to various decisions they make, which are often left implicit.

16.5　Discussion

VIDEO-LM professional development courses provide opportunities for secondary mathematics teachers to watch authentic lessons and discuss them in a supportive and non-evaluative environment. In the previous section, I presented indications of the development and refinement of mathematical knowledge for teaching among courses participants, as well as enhancement of focused reflection on various aspects of the mathematics teaching practice. One of the interesting questions to be raised in light of these findings concerns the mechanisms by which such developments may take place. In this specific context, I define "mechanisms" as "*actions, thinking processes, or behaviors* occurring during the activities of watching a videotaped lesson and engaging in an SLF-based discussion". Accordingly, the aim is to identify and characterize mechanisms that possibly enable, or account for, observed outcomes of reflection and knowledge growth that are associated with participation in VIDEO-LM courses. Pointing to such explanatory mechanisms is an elusive pursuit, as it is difficult to determine a causal connection between certain features of a PD activity and observed products of the PD. Nevertheless, several mechanisms can by mentioned as a starting point for further exploration:

I. *Using an explicit "language" and a multi-focused tool.* The SLF framework and norms can be seen as a new language that teachers get acquainted with. The explicitness of SLF and its presence in all sessions function as an organizer of experience, in the sense described in classic psycholinguistics: "Language enables us to extract from the fleeting mass of phenomena the common elements or qualities essential for our experience, and to give them permanence" (Hörmann 1979, p. 11). This possible mechanism is reflected, for instance, in the following citation from the written feedback of a participant in Site (a):

> These are really tools that now I use to look at lessons, and also when I plan lessons [...] everything suddenly has names, selecting tasks as well. There are many kinds of spectacles that now became natural to me.

II. *Comparing and contrasting.* Comparison to others is a powerful mechanism, encountered by people on a daily basis (Mussweiler et al. 2004). Although such comparisons can often be unproductive, situations in which a subtle comparison to other professionals is triggered carry an opportunity to reflect on one's goals and decisions. VIDEO-LM's agenda does not include direct comparisons, yet these are apparently unavoidable, and in most PD conversations teachers switch back and forth from analyzing actions of the videotaped teacher to self-inspections of their own teaching, as shown, for instance, in Sect. 16.4.2.1. In some of the written feedbacks we found even "meta-reflections" on this process, for example:

> During the video watching and discussions [...] I found myself engaged in questions: Where do I stand? What would I have done? How come I never thought of this? [...] In what ways am I different? What should I keep? What should I change?

III. *Intentional stepping into another person's shoes.* This mechanism is explicitly present in SLF-based discussions, as described in Sect. 16.2.2. We invite teachers to infer and attribute goals, dilemmas, and beliefs to the filmed teacher; Rather than evaluating the teacher, they are requested to seek possible reasons for certain decisions made. One of our facilitators developed a unique strategy for this request: A chair is put in the front of the room, and whoever wants to offer an analysis of a specific occurrence is asked to sit in that chair and speak in a first person voice, attempting to adopt the perspective of the teacher in the video. This unusual stance has a considerable influence on participants, as illustrated in the following citation by a teacher who was also a regional teacher mentor employed by the Ministry of Education:

> [It] completely changed the nature of my observations on teachers' lessons [...] all of the conversation, the conversation that I hold now with a teacher, after visiting his classroom for observation, is more like "what's your motivation, and what brought you [to do this], and what were your considerations", and it leads to a different kind of meaningful conversations. [...] Something changed, even in the way I observe.

IV. *Postponing judgment.* In the first PD sessions, the facilitators establish this norm almost "forcefully"; Instead of judgmental comments about the filmed teacher's decisions, participants are asked to consider alternative paths and their consequent tradeoffs. Later on in the course, this norm seems to be internalized as an almost automatic mechanism, and judgmental viewpoints are replaced with the need for mindful decisions, as reflected in the following feedback:

> We all teach fine, the point is to understand what you're doing, why do you do it, and do you really agree with what you decided to do. If you agree, fine, but if you don't – go and fix it! But be aware of what you did. I never thought about that.

V. *Discovering collective wisdom.* Hearing opinions expressed by peer teachers, rather than by "authorities" such as facilitators or researchers, seems to have the potential of convincing teachers to consider a change in their own opinion. We encountered an interesting example of this mechanism in the case of Daniela (pseudonym), a teacher who argued passionately that the Japanese lesson could never be successfully duplicated in an Israeli classroom. She nevertheless decided to try it in her classroom and reported back in the next session on its overwhelming success. When asked later why she decided to act against her intuition, she said:

> The fact that people were in favor. I'm trying to figure out if, let's say, everyone was against it, would I still want to try this lesson? Probably the fact that there were other people that said… that supported this lesson and said "it might be a good thing, it might be beneficial." […] Yeah, it definitely reinforced it […] Other opinions that upset me is actually a fascinating thing, to try them, because, again, who says I'm in this place that is guaranteed? The minute this opinion was strengthened by opinions of the participants, and people justified their stance, so I was even more interested to check this out.

VI. *Exposure to a variety of styles and methods.* The videotaped lessons observed during a typical VIDEO-LM course are varied in terms of teaching styles, approaches to teaching core subjects in the curriculum, use of technology, and more. Possibly, this diversity serves as an eye opener by itself and has an impact on teachers' readiness to elicit ideas of their own and reflect on their practice. The following citation from a written reflection illustrates this:

> It's a pleasure to look at different teachers and diverse teaching styles that often were a mirror to my own conduct and sometimes were a source for inspiration and pondering.

I end this paper with a last citation, taken from a teacher's written reflection, that conveys the spirit of VIDEO-LM and the kind of teacher learning we aspire to nurture within it:

> Theoretically, I know that there is an infinite variety of teachers that I can regard as "good teachers" and still they will be different from one another, and in various decision crossroads they may take totally opposite decisions. However,

each time I witness this it is a refreshing discovery, and I feel that slowly slowly it wears out my inherent belief that there are absolute "rights" and "wrongs" in teaching too.

Acknowledgements This study was supported by the Israel Science Foundation, Grant #1539/15, and by the Israel Trump Foundation for Science and Mathematics Education, Grant #7/143. I wish to thank my colleague and friend Abraham Arcavi, with whom I co-direct the VIDEO-LM project, for a most valuable and productive partnership.

References

Arcavi, A., & Schoenfeld, A. H. (2008). Using the unfamiliar to problematize the familiar: The case of mathematics teacher in-service education. *Canadian Journal of Science, Mathematics, and Technology Education, 8*(3), 280–295.

Ball, D. L., Thames, M. H., & Phelps, G. (2008). Content knowledge for teaching: What makes it special. *Journal of Teacher Education, 59*(5), 389–407.

Borko, H., Koellner, K., Jacobs, J., & Seago, N. (2011). Using video representations of teaching in practice-based professional development programs. *ZDM—The International Journal of Mathematics Education, 43*(1), 175–187.

Brophy, J. (Ed.). (2004). *Using video in teacher education.* The Netherlands: Elsevier.

Clarke, D., Hollingsworth, H., & Gorur, R. (2013). Facilitating reflection and action: The possible contribution of video to mathematics teacher education. *Sisyphus—Journal of Education, 1*(3), 94–121.

Coles, A. (2013). Using video for professional development: The role of the discussion facilitator. *Journal of Mathematics Teacher Education, 16*(3), 165–184.

Coles, A. (2014). Mathematics teachers learning with video: The role, for the didactician, of a heightened listening. *ZDM—The International Journal of Mathematics Education, 46*(2), 267–278.

Danielson, S. (2013). *The framework for teaching evaluation instrument* (2013 ed.). Princton, NJ: The Danielson Group.

Glaser, B. G., & Strauss, A. L. (1967). *The discovery of grounded theory: Strategies for qualitative research.* New York: Aldine.

Gaudin, C., & Chaliès, S. (2015). Video viewing in teacher education and professional development: A literature review. *Educational Research Review, 16*, 41–67.

Goffree, F., & Oonk, W. (2001). Digitizing real teaching practice for teacher education programmes: The MILE approach. In F. L. Lin & T. J. Cooney (Eds.), *Making sense of mathematics teacher education* (pp. 111–145). The Netherlands: Springer.

Hill, H. C., Blunk, M., Charalambous, C., Lewis, J., Phelps, G. C., Sleep, L., et al. (2008). Mathematical knowledge for teaching and the mathematical quality of instruction: An exploratory study. *Cognition and Instruction, 26*, 430–511.

Hörmann, H. (1979). *Psycholinguistics: An introduction to research and theory.* New York: Springer.

Jaworski, B. (1990). Video as a tool for teachers' professional development. *British Journal of In-Service Education, 16*(1), 60–65.

Karsenty, R., Arcavi, A., & Nurick, Y. (2015). Video-based peer discussions as sources for knowledge growth of secondary teachers. In K. Krainer & N. Vondrová (Eds.), *Proceedings of the 9th Congress of the European Society for Research in Mathematics Education* (pp. 2825–2832). Prague: ERME.

Karsenty R., & Schwarts, G. (2016). *Enhancing reflective skills of secondary mathematics teachers through video-based peer discussions: A cross-cultural story.* Paper presented in

TSG-50 of the 13th International Congress on Mathematical Education. Hamburg, Germany, July 2016.

Lefstein, A., & Snell, J. (2014). *Better than best practice: Developing teaching and learning through dialogue*. New York: Routledge.

Li, Y., & Moschkovich, J. N. (2013). *Proficiency and beliefs in learning and teaching mathematics—Learning from Alan Schoenfeld and Günter Törner*. Rotterdam: Sense Publishers.

Mussweiler, T., Rüter, K., & Epstude, K. (2004). The ups and downs of social comparison: Mechanisms of assimilation and contrast. *Journal of Personality and Social Psychology, 87*(6), 832.

Nurick, Y. (2015). *The crystallization of mathematical knowledge for teaching of high school teachers in video-based peer discussions* (Unpublished master's thesis). Weizmann Institute of Science (in Hebrew).

Nemirovsky, R., & Galvis, A. (2004). Facilitating grounded online interactions in video-case-based teacher professional development. *Journal of Science Education and Technology, 13*(1), 67–79.

Santagata, R., & Yeh, C. (2013). Learning to teach mathematics and to analyze teaching effectiveness: Evidence from a video-and practice-based approach. *Journal of Mathematics Teacher Education, 17*(6), 491–514.

Schoenfeld, A. H. (1998). Toward a theory of teaching-in-context. *Issues in Education, 4*(1), 1–94.

Schoenfeld, A. H. (2010). *How we think: A theory of goal-oriented decision making and its educational applications*. New York: Routledge.

Schwarts, G. (2016). *Characterizing video-based peer discussions of Israeli mathematics teachers watching a Japanese lesson* (Unpublished master's thesis). Weizmann Institute of Science (in Hebrew).

Seago, N., Jacobs, J., & Driscoll, M. (2010). Transforming middle school geometry: Designing professional development materials that support the teaching and learning of similarity. *Middle Grades Research Journal, 5*(4), 199–211.

Sherin, M. G. (2004). New perspectives on the role of video in teacher education. *Advances in Research on Teaching, 10*, 1–27.

Sherin, M. G., Jacobs, V. R., & Philipp, R. A. (Eds.). (2011). *Mathematics teacher noticing: Seeing through teachers' eyes*. New York: Routledge.

Sherin, M. G., & van Es, E. A. (2009). Effects of video participation on teachers' professional vision. *Journal of Teacher Education, 60*(1), 20–37.

van Es, E. A., & Sherin, M. G. (2008). Mathematics teachers learning to notice in the context of a video club. *Teaching and Teacher Education, 24*, 244–276.

Chapter 17
Powering Knowledge Versus Pouring Facts

Petar S. Kenderov

Abstract Many problems related to the real world admit a mathematical description (i.e., a mathematical model) based on what is studied at school. Solving the mathematical model, however, often requires a higher level of mathematics, and this is the reason for not including such problems in the curriculum. We present several problems of this kind and propose solutions to their mathematical models by means of widely available dynamic mathematics software (DMS) systems. For some of the problems, it is possible to directly use the in-built functionalities of the DMS and to construct a computer representation of the problem that allows exploring the situation and obtaining a solution without developing a mathematical model first. Using DMS in this way can broaden the applicability of school mathematics and increase its appeal. The ability of students to solve problems with the help of DMS has been tested by means of two types of competitions.

Keywords Mathematical modelling · Inquiry education · Computational thinking

17.1 Introduction

There are two partially contradicting trends in high school mathematics education. On one hand, we want mathematical knowledge to be based on a solid logical base (rigor). On the other hand, we want this knowledge to be rich both in content and applications. These two trends cannot always (and easily) be reconciled (De Lange 1996). One of the reasons for this contradiction is the fact that only a few problems related to practice allow mathematically pure and complete treatment with the traditional rigor. The demonstration of patterns of logical thinking is time consuming and often related to simplified mathematical content that does not properly reflect the unavoidable complexity of the real world. The formulation of a

P. S. Kenderov (✉)
Institute of Mathematics and Informatics, Bulgarian Academy
of Sciences, Sofia, Bulgaria
e-mail: pkend@math.bas.bg

© The Author(s) 2018   289
G. Kaiser et al. (eds.), *Invited Lectures from the 13th International Congress
on Mathematical Education*, ICME-13 Monographs,
https://doi.org/10.1007/978-3-319-72170-5_17

mathematical model for a real-life situation cannot be based on rigor only. Dropping out some features and keeping only the most essential ones in the mathematical model requires skills that have little to do with rigor, and this is an obstacle for the inclusion of complex real-life situations in the mathematics curriculum. Furthermore, there are many problems related to practice (some of which will be considered below) that can be equipped with a reasonable mathematical model based on what is studied at school. The corresponding model may be a system of equations, an optimization problem, or something else of a mathematical nature. Solving this mathematical model, however, with the traditional rigor within the frame of the school mathematics is not always possible. It may require a higher level of mathematical knowledge, for instance, advanced calculus and/or numerical methods for approximation of the exact solution. This is another reason for avoiding the consideration of genuine real-life applications within the school mathematics. However, with the appearance of powerful and widely accessible dynamic mathematics software (DMS) systems it became possible to reduce, at least partially, the mentioned contradiction between rigor and applications. Solving a model can be performed by means of DMS. As mentioned in Hegedus et al. (2017) "This leaves more time for essential mathematical skills, e.g., interpreting, reflecting, arguing and also modeling or model building for which there is mostly no time in traditional teaching" (p. 20). With the help of technology, it is possible to offer to students much more demanding mathematical content and interesting applications (Hoyles and Lagrange 2009). Such a change would drastically increase the realm of real-life problems that can be considered in school. We do not have in mind only the traditional application of computers where a mathematical model of the problem is solved by a computer; in addition, some examples will be described below where the standard in-built operations ("buttons") of the DMS system can be used directly to make a computer representation of the problem without first writing the formulas of a mathematical model. This DMS representation of the problem will be called a "computer model of the problem." By means of this model and the in-built functionalities of DMS (such as dragging, measuring distances, and areas), the solution of the problem can be found with a reasonable degree of precision. This direct DMS modelling of the problem as well as the mathematical modelling of the problem, followed by a DMS-assisted solution, are in the focus of this paper, which is mainly oriented toward problem solving. Both types of modelling support the most natural way of knowledge acquisition: by experimenting, by formulating and verifying conjectures, by discussing with peers, and by asking more experienced people. In a nutshell, the technology provides the opportunity to learn mathematics by inquiry. This refers not only to what happens (and how it happens) in class but also to extracurricular activities that provide a fruitful playground for building mathematical literacy and cultivating elements of computational thinking (Freiman et al. 2009). Another advantage of using technology in this way is that much larger and more operational mathematical content could be given to the students at an earlier age.

Later in the paper, several problems are considered for which it is easy to assign a proper mathematical model based on school mathematics but whose solution with the necessary rigor while remaining in the frame of school mathematics is relatively difficult (or at least not easy). On the other hand, these models are easily solvable by means of DMS systems. This way of problem solving opens further opportunities for inquiry and cultivates the elementary computational thinking skills of the students, thus *powering* (in the sense of "adding power to") their existing knowledge and skills. Problems such as the ones considered below and the inquiry-based approach to their solving can make the mathematics studied in school more applicable and more appealing in contrast to the now prevailing *pouring* of mathematical facts. The ability of school students of different ages to solve such problems has been tested by means of two online competitions called VIVA Mathematics with Computer and Theme of the Month. The participants' scores show that the use of DMS for problem solving is gradually gaining popularity in Bulgaria. The students are interested in this approach and many are capable of using it. The problems considered next have been used in these competitions.

17.2 The Sample of Problems

The problems in this section illustrate the differences in the uses of the models we consider in this paper: a *mathematical model* which can (or cannot) be solved in the traditional way, a *mathematical model* allowing a simple DMS supported solution, and a *computer model* (direct DMS representation of the problem). Each of the problems is easy to formulate as a mathematical model but not so easy to solve with the usual rigor within school mathematics. On the other hand, an approximate DMS-assisted solution is readily available, or a computer model of the problem is easy to construct by means of which the problem can be solved even at the earlier stages of secondary education.

The Parking Entrance Problem

This problem is a further elaboration of one of the *Problems of the Month* used in the European Project MASCIL (http://www.mascil-project.eu/). We present both the computer modeling, which is amenable for younger students using DMS, and the pencil-and-paper mathematical modeling, which requires rather advanced knowledge of mathematics.

Problem 1 A vehicle (car, baby carriage, or wheel-chair) with a wheelbase b (the distance between the centers of the wheels) and clearance (ride height) c is to be moved from the street to the basement of a house over a slope of γ degrees (Fig. 17.1; $\gamma = 20°$). Is this possible without damaging the bottom of the vehicle? (Fig. 17.2)

The answer to this problem depends on the concrete values of the parameters b, c, and γ. A steeper slope γ is more likely to cause damage to the vehicle. Damage will occur also if b is big enough. The clearance c is also decisive. The interplay between these parameters is not simple and the usual intuition does not help much. The heavy scratches on the surface of many "sleeping policemen" (speed bumps)

Fig. 17.1 Moving a vehicle from the street to the basement

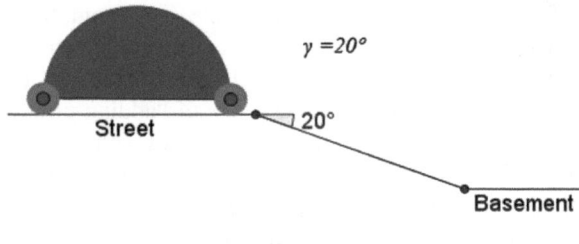

$\gamma = 20°$

Fig. 17.2 The lowest flat part of the vehicle can hit the "vertex" at the beginning of the slope

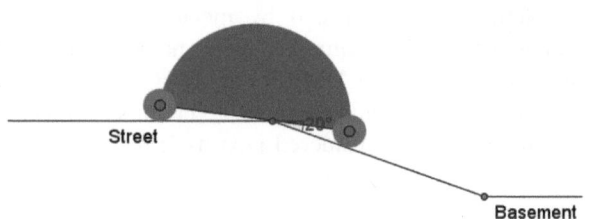

on the streets indicate that problems similar to this one are important. Further, for the sake of simplicity, we will depict the vehicle only by its two wheels (circles of radius c centered at A and B respectively) and the segment AB (the wheelbase) connecting the wheels. Both the computer model of this problem and its mathematical model rely on the very basic geometric fact that the opposite angles formed by two intersecting lines are equal (angles α in Fig. 17.3). Figure 17.3 shows the collision situation when the vertex at the beginning of the slope hits the bottom of the vehicle at some point C from the segment AB. The second arm of the angle β on Fig. 17.3 is the tangent from C to the front wheel.

A collision occurs only if $\alpha + \beta < \gamma$ (the front wheel is no longer rolling on the slope). This suggests the idea for the computer model that is visualized on Fig. 17.4. The numbers b, c, and γ are entered in the model as parameters (sliders in GeoGebra). Using the built-in operations of GeoGebra, one constructs a segment AB of length b and two circles (the wheels) of radius c centered at A and B and takes an arbitrary point C on AB that is outside the two wheels. Further, tangents from C to these circles are drawn as shown in Fig. 17.4 and, finally, the angles α and β are measured by the corresponding operation in GeoGebra.

Fig. 17.3 Collision situation

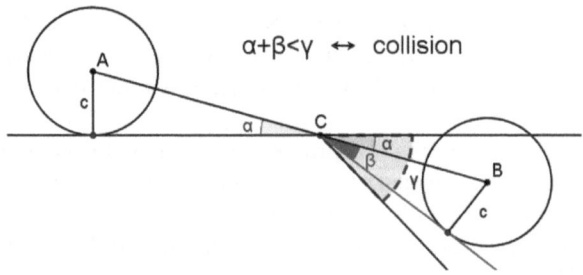

$\alpha + \beta < \gamma \leftrightarrow$ collision

Fig. 17.4 A computer model
for the parking problem

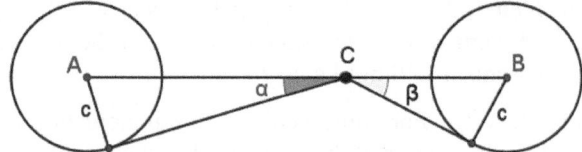

The sum $\delta = \alpha + \beta$ is a function of the position of the point C. By moving
(dragging) point C along AB and observing the change of δ, one can establish
experimentally that the function δ attains its minimum at the point M, which is the
middle of AB. If this minimum is bigger or equal to γ, the vehicle could be parked
safely in the basement. Otherwise a collision occurs. This observation confirms the
intuitive expectation that the middle M of the segment AB is the critical and most
vulnerable point. If it passes above the slope vertex, the vehicle can be parked
safely in the basement. This observation also shows that even a simpler computer
model can solve the problem. Note that if C and M coincide, then $\alpha = \beta$, and the
condition for non-collision takes the form $2\alpha \geq \gamma$. Given the numbers b, c, and γ,
one finds the middle M of the segment AB, draws the tangents from M to the two
wheels, and measures the angle δ between these tangents (Fig. 17.5). If $\delta \geq \gamma$, the
vehicle can be moved safely. If $\delta < \gamma$ there will be a collision and moving it without
damage becomes impossible.

The second computer model solution of this problem is completely amenable for
students at earlier stages of secondary education. In contrast, as we will now see, the
mathematical model of the problem requires knowledge of inverse trigonometric
functions, and the classical solution uses some elements of calculus. Denote by x the
length of the segment CA in Fig. 17.4. Then $\alpha = \arcsin\frac{c}{x}$ and $\beta = \arcsin\frac{c}{b-x}$. One has
to find the minimum of the function $\delta(x) = \arcsin\frac{c}{x} + \arcsin\frac{c}{b-x}$ in the interval
$[c, b - c]$ (this is the interval where the function $\delta(x)$ is well-defined; we implicitly
assume here that $b > 2c$). By finding the zeros of the derivative of $\delta(x)$, one can derive
that the minimum of this function is attained for $x = \frac{b}{2}$ and solve the problem.

Here are some tasks for further inquiry with the computer or the mathematical
model of this problem:

Problem 1.1 What is the steepest slope (in degrees) that a baby carriage with
$b = 130$ cm and $c = 12$ cm can overcome without troubles?

Fig. 17.5 A simpler
computer model for the
parking problem

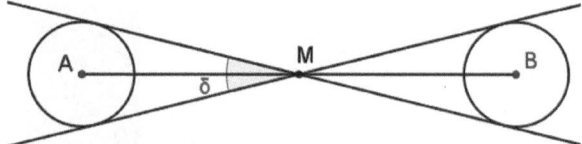

Problem 1.2 If the slope to the basement is 20° and the wheelbase of the car is $b = 290$ cm, what is the smallest radius of the wheels such that moving the car to the basement will not be a problem?

Problem 1.3 For some vehicles, the bottom line is different from the line connecting the centers of the wheels. Also, the front wheels and the rear wheels are not always of the same radius (Fig. 17.6). Develop a computer and a mathematical model for the exploration of the dangers for moving such vehicles down slopes.

One could further explore the parking problem by means of the more realistic computer model developed by Toni Chehlarova. The corresponding GeoGebra file is available at http://cabinet.bg/content/bg/html/d22178.html (last visited December 2016).

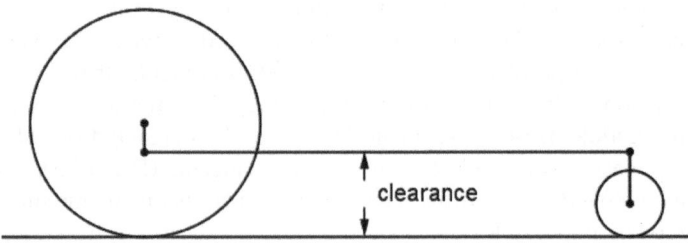

Fig. 17.6 A model with different wheels

The Cylindrical Container Problem

Problem 2 Two thirds of the volume of a closed cylindrical can of radius 5 cm (Fig. 17.7) is filled with some liquid. What is the height of the liquid if the can is laid horizontally?

The problem seems to be three dimensional but could be easily reduced to a two dimensional one. In the horizontal position, two thirds of the circle area of the can base are covered by the liquid. Hence, the problem is reduced to finding a horizontal chord AB (Fig. 17.8) in a circle of radius 5 cm with center at O that cuts off a circular segment (slice) of area one third of the total area of the circle.

Fig. 17.7 The cistern problem

Fig. 17.8 Cutting a circular
segment with an area one
third of the circle area

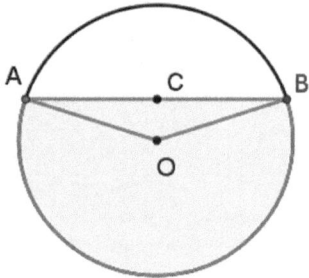

This can be done in different ways. The in-built operations of the DMS can be used to find the area of the circular sector outlined by the segments OA, OB, and the arc from B to A (in the counterclockwise direction) and the area of the triangle AOB. The difference between the two areas is the area of the circular segment we are looking for. If the horizontal chord AB is made movable (the DMS takes care of the dynamics and automatically re-calculates the areas), a position for the chord AB can be found such that the area of the circular segment is one third of the area of the entire circle. If C is the middle of the chord AB at this position, then the height of the liquid in the horizontal can is equal to the radius of the can base (5 cm) plus the length of the segment CO (which can be measured by the functionalities of the DMS). In our case, an approximate value for the height of the liquid is 6.32 cm. The computer model just developed allows exploration of similar situations with other cylindrical cans (the radius of the can could be made changeable, the part of the can volume which is filled with liquid in vertical position can change, etc.).

We will now proceed to a mathematical model of the problem. For the sake of generality (and since this will not introduce further complications), we will denote the radius of the can base by r. Let α be the measure (in radians) of the angle in the circular sector considered above. The area of this sector is $\frac{\alpha}{2}r^2$. The area of the triangle OBA is $\frac{1}{2}r^2 \sin \alpha$. Hence, the angle α that corresponds to a circular segment with area equal to one third of the area of the circle has to satisfy the equation $\frac{\alpha}{2}r^2 - \frac{1}{2}r^2 \sin \alpha = \frac{1}{3}\pi r^2$. Equivalently, $\alpha - \sin \alpha - \frac{2}{3}\pi = 0$. As we see, the mathematical model of this problem is an exotic equation. School mathematics does not deal with such equations, and this seems to be the reason for not including this important cistern problem in the curriculum. The numerical/graphical solution of this model by DMS, however, is available. The graph of the function $f(x) = x - \sin x - \frac{2}{3}\pi$ is depicted in Fig. 17.9. The point A has been constructed as the intersection of the graph of f and the x-axis. The first coordinate of A gives the angle we are looking for: $\alpha = 2.60533$ (the precision of 5 digits after the decimal point is taken here arbitrarily; it can be increased or decreased).

The length of the segment OC corresponding to this α and $r = 5$ can be calculated: $OC = r \cos \frac{\alpha}{2} = 1.32465$. For the height of the liquid in the horizontal position of the can, we obtain 6.32465.

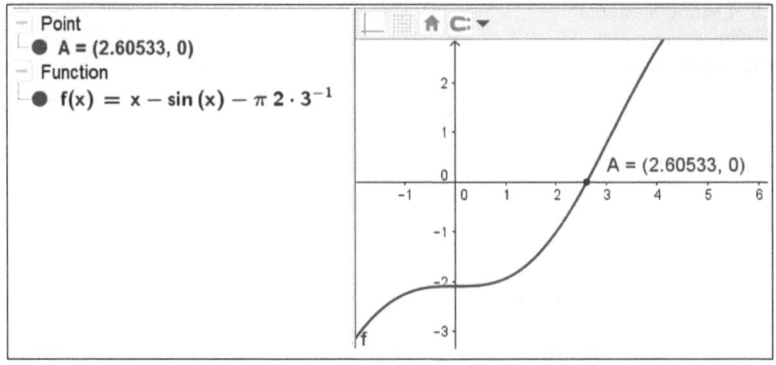

Fig. 17.9 Graphical presentation of the function $f(x)$

If the angle α is measured in degrees, the area of the circular sector is $\frac{\alpha}{360}\pi r^2$. Correspondingly, the equation from which the angle α will be determined has the following appearance:

$$\frac{\alpha}{180}\pi - \sin\alpha - \frac{2}{3}\pi = 0$$

For further inquiries with either the computer model or with the mathematical model, one could consider the following related problems:

Problem 2.1 A horizontally laid cylindrical tank with diameter 200 cm and length 500 cm is partially filled with petrol so that the level of the petrol is 80 cm. How many liters of petrol are there in the tank?

Problem 2.2 If the height of the can from Problem 2 is 24 cm, how much additional liquid should be poured into it in a horizontal position so that the level of the liquid is elevated by 1 cm? If after the addition of the liquid the can is turned into vertical position, what is the height of the liquid level?

Problem 2.3 If the height of the can from Problem 2 is 24 cm, how much liquid should be removed from it so that in a horizontal position the liquid level drops down by 1 cm?

Problem 2.4 A heavy metal ball of radius 4 cm is placed into an empty vertically placed can of radius 5 cm and height 25 cm. Then liquid is poured into the can until its level reaches 20 cm and then the can is sealed. What would the liquid level be, if the can is laid horizontally (see Fig. 17.10)?

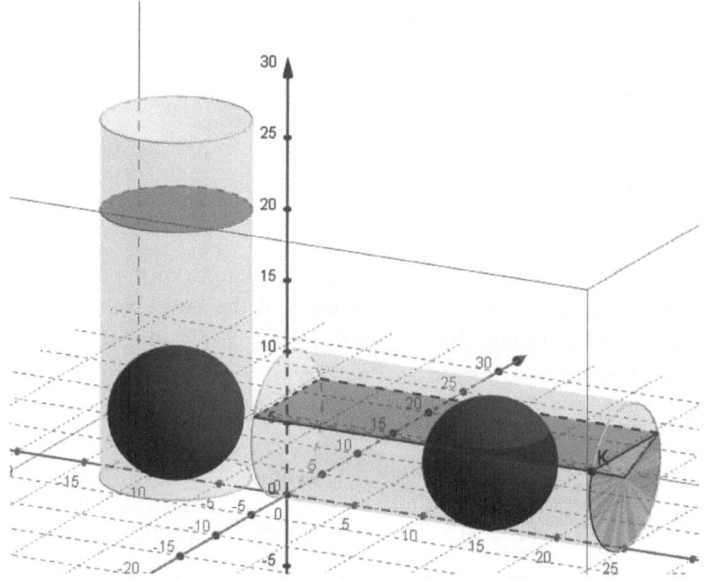

Fig. 17.10 In a horizontal position, a part of the ball might be above the liquid surface

The Conical Container Problem

This problem is a well-known mathematics exercise for university students. It can be settled by means of calculus or by a mathematical trick with inequalities. We present the mathematical model and demonstrate that by means of a DMS the problem can be considered and solved in school.

Problem 3 A circular sector of measure α (in degrees) has been cut out from a circular plastic sheet of radius l with center O (Fig. 17.11). From the remaining part, a right circular cone is made by sticking (gluing) the cuts (Fig. 17.12). What is the size of angle α (in degrees) for which the volume of the resultant cone is maximal?

The mathematical model of this problem is based on the well-known formula for the volume V of the cone: $V = \frac{\pi R^2}{3} h$. Here R is the radius of the cone base and h is

Fig. 17.11 Cutting a sector

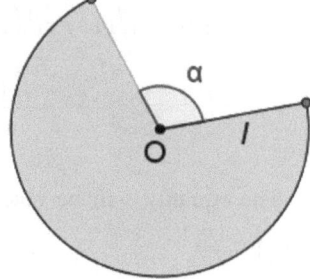

Fig. 17.12 Gluing a cone

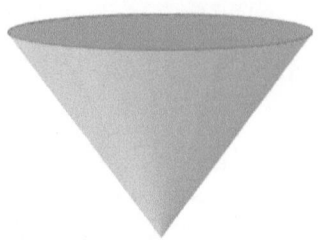

cone's height. Since α is measured in degrees, the length of the arc of the removed circular sector is $\frac{\alpha}{360} 2\pi l$. Therefore, the length of the cone base circumference is what remains after the cutting: $2\pi l - \frac{\alpha}{360} 2\pi l$. Hence, $2\pi l - \frac{\alpha}{360} 2\pi l = 2\pi R$. It follows that the radius R can be expressed as function of $x = \frac{\alpha}{360}$: $R = l(1 - x)$. Further, it follows from Pythagoras's theorem that $h^2 = l^2 - R^2 = l^2\left(1 - (1 - x)^2\right)$. i.e., $h = l\sqrt{1 - (1 - x)^2}$. Thus, the volume of the cone is $V = \frac{1}{3}\pi l^3 (1 - x)^2 \sqrt{1 - (1 - x)^2}$. The essence of the problem, its mathematical model, is to find a number x, $0 \leq x \leq 1$, for which the function $f(x) = (1 - x)^2 \sqrt{1 - (1 - x)^2}$ attains its maximal value. Once again we see that the derivation of the mathematical model is based on school mathematics. Solving this model however requires more advanced mathematics. Using calculus one can find the extremal values of this function f by finding the zeros of its derivative. These zeros are $x = 1 - \frac{\sqrt{2}}{\sqrt{3}}$, $x = 1$ and $x = 1 + \frac{\sqrt{2}}{\sqrt{3}}$. The last of these numbers is outside the interval $[0, 1]$ and is not relevant for our considerations. The value $x = 1$ corresponds to a minimum for f because $f(1) = 0$. Therefore the maximum of f is attained at $x = 1 - \frac{\sqrt{2}}{\sqrt{3}}$ and the value of f at this point is equal to $\frac{2}{3}\sqrt{\frac{1}{3}}$.

There is a nice trick which allows solution of this mathematical model by means of the well-known inequality between the arithmetic mean and the geometric mean of any non-negative numbers a, b, and c: $\sqrt[3]{abc} \leq \frac{a+b+c}{3}$. It is known also that equality is attained in this inequality if and only if $a = b = c$. Applying this inequality for $a = b = \frac{(1-x)^2}{2}$, $c = 1 - (1 - x)^2$, we get

$$f(x) = \sqrt{(1 - x)^4 \left(1 - (1 - x)^2\right)}$$
$$= 2\sqrt{\frac{(1 - x)^2}{2} \frac{(1 - x)^2}{2} \left(1 - (1 - x)^2\right)} \leq 2\sqrt{\left(\frac{1}{3}\right)^3} = \frac{2}{3}\sqrt{\frac{1}{3}}.$$

The equality will be reached when $\frac{(1-x)^2}{2} = \left(1 - (1 - x)^2\right)$. This again yields $x = 1 - \frac{\sqrt{2}}{\sqrt{3}}$.

If at all, calculus and the mentioned trick with the inequality are available only at the last stages of school mathematics. With the help of DMS, however, the mathematical model of this problem can be solved by younger students. It is possible to draw the graph of the function $f(x)$ and see where its maximum is. The graph of the function $f(x)$ can be seen in Fig. 17.13.

It is clear from this picture that the function f has two maxima. Only the one in the interval $[0, 1]$ on the x-axes is of interest for us. The DMS (GeoGebra) allows observation of the coordinates of a point A, which moves along the graph of the function. When A is dragged to the highest point in the graph, its first coordinate will be equal to the value of x we are looking for. In Fig. 17.13, this is the point $A = (0.18, 0.38)$. If the precision of calculations is increased, one gets $x = 0.1835$, which is a very good approximation of $x = 1 - \frac{\sqrt{2}}{\sqrt{3}}$. This value of x corresponds to $\alpha \approx 66.06°$, and the latter value could be accepted as a reasonable solution to Problem 3.

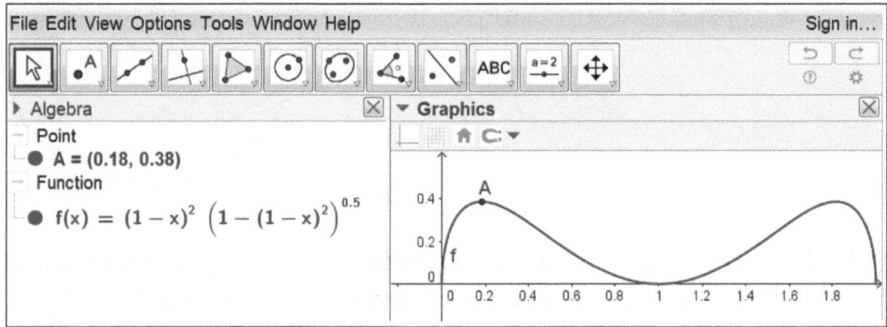

Fig. 17.13 The graph of the function $f(x)$

The Ice Cream Container Problem

The next problem is a challenge for pencil-and-paper technology, even for university students. With the help of DMS it is completely amenable for school students.

Problem 4 An ice cream container (as depicted in Fig. 17.14) is to be made of a circular plastic sheet of radius l with center O by cutting and gluing (sticking). The cutting and gluing operations allowed and the order in which they are performed are:

(a) Cut a circular sector of measure α (in degrees) from the plastic sheet (Fig. 17.15) and, by gluing, make from it a cone that will serve as the lower part of the ice cream container.
(b) Cut off from the remainder (Fig. 17.15) a full circular sector of radius t (this number t is to be specified later) and glue a cut cone (truncated cone) that will serve as the upper part of the ice cream container.

For what size of α will the ice cream container have largest volume?

Fig. 17.14 The ice-cream
container

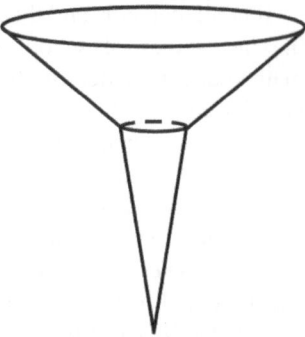

Fig. 17.15 The cutting and
gluing process

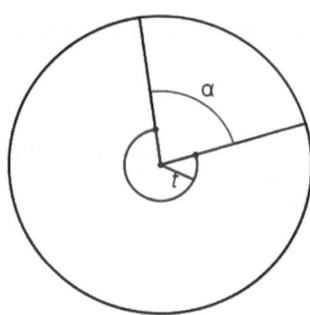

The length of the arc of the circular sector of measure α is $\frac{2\pi l\alpha}{360}$. The cone made of
this sector will have a radius r of the base determined from the equation $2\pi r = \frac{2\pi l\alpha}{360}$,
i.e., $r = lx$ where $x = \frac{\alpha}{360}$.

The radius t of the full circular sector mentioned in (b) is determined in such a
way that the upper circle of the lower cone fits the lower circle of the upper
truncated cone: $\frac{(360-\alpha)}{360} 2\pi t = 2\pi r$. Hence $t = \frac{r}{1-x}$. Note that the length of the gen-
eratrix of the truncated cone obtained in (b) is $l - t$. The resultant container is
depicted in Fig. 17.16. As in Problem 3, we see that the radius R of the upper
circle of the truncated cone is $R = (1 - x)l$. The altitude h_1 of lower cone is
determined by Pythagoras's theorem: $h_1^2 = l^2 - r^2 = l^2(1 - x^2)$. The volume of the
lower cone is

$$V_1 = \frac{\pi}{3} l^3 x^2 \sqrt{1 - x^2}.$$

The altitude h_2 of the truncated cone is determined similarly (using Pythagoras's
theorem):

Fig. 17.16 Geometry behind
the mathematical model

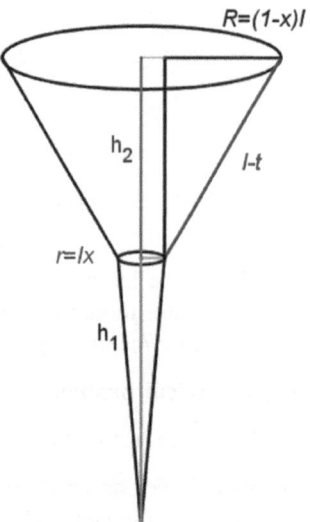

$$h_2 = \frac{l(1 - 2x)}{1 - x} \sqrt{2x - x^2}.$$

The volume V_2 of the truncated cone is $V_2 = \frac{\pi}{3} h_2 (R^2 + Rr + r^2)$ where $R = (1 - x)l$, $r = lx$, $R^2 + Rr + r^2 = l^2 \left((1 - x)^2 + (1 - x)x + x^2 \right) = l^2 \left((1 - x)^2 + x \right)$.

Hence $V_2 = \frac{\pi}{3} l^3 \frac{1-2x}{1-x} \sqrt{2x - x^2} (1 - x + x^2)$. The volume of the ice cream container is $V = V_1 + V_2$. We note here that x must belong to the interval $[0, \frac{1}{2}]$. This follows from the fact that the number $t = \frac{r}{1-x} = \frac{lx}{1-x}$ cannot be bigger than l.

Finding the maximum of V by means of calculus is a challenge. With the help of a DMS it can be found, as in the previous problem, that the maximal value of V is attained for $x \approx 0.23088$, which corresponds to $\alpha \approx 83.12°$.

Here are some problems for further inquiry:

Problem 4.1 What is the minimal radius l of the initial circle from which the ice cream container is produced in the above way so that its volume is at least 200 cm^3?

Problem 4.2 A bucket (the far right of Fig. 17.17) with a circular base of radius $r = 10$ cm has to be made from a circular plastic sheet of radius $l = 60$ cm with center O by cutting and gluing (sticking). The cuts that are allowed and the order in which they are performed are:

(a) Cut circles centered at O (i.e., concentric with the initial circle).
(b) Cut from the remainder a radial segment of measure α (in degrees).

For what size of α will the volume of the bucket be the largest?

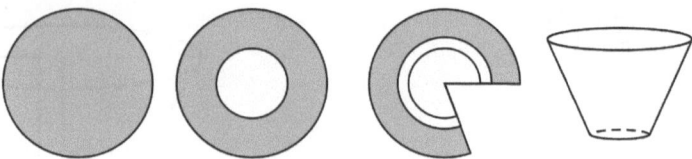

Fig. 17.17 The construction of a bucket

What is the largest possible volume of the bucket?

A computer model for Problem 4 was developed by Toni Chehlarova. It can be found at http://cabinet.bg/content/bg/html/d22582.html (visited December 2016).

A geometrical problem

This is the last of the sample problems:

Problem 5 For an arbitrary triangle *ABC*, denote by *D*, *E*, and *F* its orthocenter, incenter, and the centroid, correspondingly (Fig. 17.18). Are there triangles *ABC* for which the area of the triangle *DEF* is bigger than the area of the triangle *ABC* itself?

This problem deviates in style from the previously considered problems. It contains a research-like component that is suitable for work on a project by the students. The computer model for this problem is easy to construct. The in-built operations of GeoGebra can be used to construct the orthocenter, the incenter, and the centroid of an arbitrary triangle. Using the "finding area of a polygon" command, the areas of the triangles *ABC* and *DEF* are calculated and displayed on the monitor. Due to the dynamic functionalities of GeoGebra, this computer model of Problem 5 allows to explore many triangles (by dragging some of the vertices *A*, *B*, and *C*). Playing with the vertices can experimentally establish that for some obtuse triangles *ABC* the answer to the question in Problem 5 is positive.

Note that this computer model solution of the problem does not require knowledge of more advanced mathematics (trigonometry, analytical geometry, etc.). It relies on the knowledge of the basic notions involved (orthocenter, incenter,

Fig. 17.18 The orthocenter
D, the incenter *E*, and the
centroid *F* of the triangle *ABC*

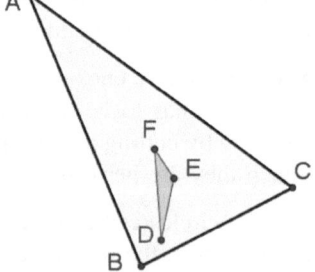

and centroid), on acquaintance with the functionalities of GeoGebra, and on some modeling skills.

Problem 5.1 For an arbitrary triangle *ABC*, find the area of the triangle with vertices at the orthocenter, the circumcenter, and the centroid of *ABC*.

Exploring this task with the corresponding computer model can show that the required area is always zero and, therefore, the three points are collinear (they lie on the famous Euler line of the triangle *ABC*).

The following simplified form of Problem 5 was given as one of the tasks in the competition VIVA Mathematics with Computer.

Problem 5.2 Given is a triangle *ABC* (by its sides or by the coordinates of its vertices; see Fig. 17.18). Find the area of the triangle with vertices at the orthocenter *D*, the incenter *E*, and the centroid *F* of the triangle *ABC*.

17.3 The Competitions Viva Mathematics with Computer and Theme of the Month

In order to examine the attitudes of Bulgarian students to problems like those in the previous section and to test the students' ability to solve such problems, two online competitions named VIVA Mathematics with Computer (VIVA MC) and Theme of the Month (TM) were launched in 2014 with the financial support of VIVACOM, a major telecommunication operator in the country (https://www.vivacom.bg/bg). The VIVA MC competition is for students from Grade 3 to Grade 12 and has two rounds. The first round is conducted twice during the academic year (in December and April) and is with open access. The second round takes place in September or early October and is only for the best performers in the December and April editions of the first round from the previous academic year. Pre-registration is needed at the VIVAcognita portal (http://vivacognita.org/) for participation in VIVA MC. Each registered student chooses how to participate in the competition: from any place with internet access by desktop, tablet, or laptop. On a fixed day and time every participant gets access for 60 min to a worksheet that contains 10 tasks corresponding to the participant's age group. The easier tasks are equipped with several possible answers. i.e., these are multiple-choice questions. The participant is expected to select the correct answer on the basis of performing some mathematical operations. The majority of the remaining tasks require a decimal number (usually up to two digits after the decimal point) as an answer that has to be entered in a special answer field. To find this answer, the student has to make a computer model of the task and explore it with the functionalities of DMS. Some of the most difficult tasks are accompanied by a file (a computer model) that solves a similar problem, and participants must modify the files accordingly in order to solve the tasks assigned to them. The number of points given for the answer to a task depends on both how close the student's answer is to the one calculated by the jury and/or by the author of the task and the difficulty of the problem. The maximum possible score is 50 points. There are no restrictions concerning the use of resources: books, internet search, advice from specialists, etc. More information about this competition can be found in Chehlarova and Kenderov (2015). In April 2016, there were 474 participants while in December 2016 the number of participants was 1321. In both cases there were five age groups (two grades per group). An impression of the degree to which the participants were capable of solving problems with the help of DMS can be gained by the overview of their scores presented in Tables 17.1 and 17.2.

Students' scores in solving the problems from Sect. 17.2 were similar. Problem 5.2 from Sect. 17.2 was proposed as a last (presumably most difficult) task in the very first edition of VIVA MC (December 2014) to 207 students from Grades 8 to 12. The lack of experience with such problems and the short time to work on the problems (60 min) is clearly seen from the obtained results: About half of the students (48%) did not enter any answer for this task, 13% provided precise answer, and 2% gave an answer with satisfactory precision. The cylindrical container problem (Problem 2 from Sect. 17.2) was given to 317 students from Grade 8 to Grade 12 at the December 2015 edition of VIVA MC. An auxiliary DMS file was provided in order to facilitate the exploration of the problem. Only 13% provided an answer with sufficiently high precision. The answers of a further 37% were given with satisfactory precision. The general feeling has been that with every new edition of VIVA MC the performance of the participants improves, though rather gradually.

The other competition, TM, is conducted monthly. A theme of five tasks related to a common mathematical idea is published at the beginning of the month on the abovementioned portal (vivacognita.org). The tasks are arranged in the direction of increasing difficulty. The participants are expected to solve the problems and send responses online by the end of the month. Some of the problems are accompanied by auxiliary DMS files which allow the students to explore the mathematical

Table 17.1 Scores of participants in April 2016 competition VIVA MC

Grades in a group	3 and 4	5 and 6	7 and 8	9 and 10	11 and 12
Number of participants	146	142	79	67	40
Participants with 35–50 points	80	49	2	9	1
Participants with 20–34 points	44	59	15	21	16
Participants with 10–19 points	19	24	22	23	5

Table 17.2 Scores of participants in December 2016 competition VIVA MC

Grades in a group	3 and 4	5 and 6	7 and 8	9 and 10	11 and 12
Number of participants	449	385	268	123	86
Participants with 35–50 points	180	27	7	12	11
Participants with 20–34 points	147	146	24	29	28
Participants with 10–19 points	75	114	84	37	20

problem, find suitable properties, try out different strategies, and find a (usually approximate) solution. To solve the more difficult tasks from the theme, the students have to adapt the auxiliary files from previous problems or to develop their own files for testing and solving the problem. Each problem brings at most 10 points (depending on the degree of preciseness of the answer). The maximum total score is 50 points. Usually there are hundreds of visits to the site where the theme is published. Only dozens, however, submit solutions. The theme for February 2015 was related to the parking problem (Problem 1 from Sect. 17.2). Seventeen participants submitted their solutions. Seven received between 35 and 50 points and two received between 20 and 34. Much better were the results from the theme from September 2015, which was related to conical containers (Problems 3, 4, and 4.2 from Sect. 17.2). Sixteen students submitted their solutions, with 14 scoring between 41 and 50 points and one scoring 34 points. The results of the first several runs of TM are published in Kenderov et al. (2015) and Chehlarova and Kenderov (2015).

After the April 2017 edition of VIVA MC, the participants (more than 500) were asked to fill in a questionnaire and submit it to organizers. Of the 143 participants who returned the questionnaire, 95.51% said they liked the event. Here are some of their responses:

The problems are interesting because they require logical thinking.
I like it because I could use GeoGebra for each problem.
The contest is nice since I don't feel pressed when solving the problems.
The questions are at the right level for me.
It is interesting and helps me develop.
I find the problems entertaining.
It was easy for me to understand the formulation of the problem by means of the dynamic file I could use.
Every problem is interesting in its own way.
I like the fact that I can explore while solving the problem.
I like the parking entrance problem because it is something you could face in the real world.
This relatively modest feedback confirms the expectation that providing the students with appropriate exploration tools can increase their awareness of both the beauty and the applicability of mathematics.

Acknowledgements The author is indebted to T. Chehlarova and E. Sendova for the long-standing collaboration in promoting the mathematics with computers idea and for the fruitful remarks and suggestions related to this paper. In particular, the final version of the title of the paper was suggested by E. Sendova. The author is indebted also to Dj. Kadijevich for pointing some relevant literature sources. Special thanks are due also to anonymous reviewers whose critical notes helped clarify and improve the paper.

References

Chehlarova, T., & Kenderov, P. (2015). Mathematics with a computer—A contest enhancing the digital and mathematical competences of the students. In E. Kovacheva & E. Sendova (Eds.), *UNESCO international workshop: Quality of education and challenges in a digitally networked world* (pp. 50–62). Sofia, Bulgaria: Za Bukvite, O'Pismeneh.

De Lange, J. (1996). Using and applying mathematics in education. In A. J. Bishop et al. (Eds.), *International handbook of mathematics education* (pp. 49–97). Dordrecht: Kluwer.

Freiman, V., Kadijevich, D., Kuntz, G., Pozdnyakov, S., & Stedřy, I. (2009). Technological environments beyond the classroom. In E. J. Barbeau & P. Taylor (Eds.), *Challenging mathematics in and beyond the classroom: The 16th ICMI study* (pp. 97–131). New York: Springer.

Hegedus, S., Laborde, C., Brady, C., Dalton, S., Siller, H.-S., Tabach, M., et al. (2017). *Uses of technology in upper secondary mathematics education. ICME-13 topical surveys.* Cham, Switzerland: Springer Open.

Hoyles, C., & Lagrange, J.-B. (Eds.). (2009). *Mathematics education and technology—Rethinking the terrain.* New York: Springer.

Kenderov, P., Chehlarova, T., & Sendova. E. (2015). A mathematical theme of the month—A web-based platform for developing multiple key competences in exploratory style. *Mathematics Today, 51*(6), 305–309.

Chapter 18
Mathematical Problem Solving in Choice-Affluent Environments

Boris Koichu

Abstract This chapter presents a proposal for an exploratory confluence model of mathematical problem solving in different instructional contexts. The proposed model aims at bridging the knowledge of how problem solving occurs and the knowledge of how to enhance problem solving. The model relies of the premise that a key solution idea to a problem is constructed as a result of shifts of attention stipulated by the solver's individual resources, interaction with peers, or with a source of knowledge about the solution. The exposition converges to the conclusion that successful problem solving is likely to occur in choice-affluent learning environments, in which the solvers are empowered to make informed choices of a challenge to cope with, problem-solving schemata, a mode of interaction, an extent of collaboration, and an agent to learn from. The theoretical argument is supported by an example from an empirical study.

Keywords Mathematical problem solving · Shifts of attention
Choice-affluent environments

18.1 Introduction

The centrality of problem solving in doing and studying mathematics is broadly recognized in the mathematics education research community (e.g., Halmos 1980; Mason 2016a; Schroeder and Lester 1989). Research on problem solving keeps growing, and many approaches to translating the developed problem-solving

B. Koichu (✉)
Weizmann Institute of Science, Rehovot, Israel
e-mail: bkoichu@technion.ac.il

© The Author(s) 2018
G. Kaiser et al. (eds.), *Invited Lectures from the 13th International Congress on Mathematical Education*, ICME-13 Monographs,
https://doi.org/10.1007/978-3-319-72170-5_18

frameworks and accumulated research results into recommendations for practice have been articulated (e.g., Schoenfeld 1983, 1985; Felmer et al. 2016). In particular, the professional literature suggests various specifications of "good" mathematical problems (e.g., Lappan and Phillips 1998), characterizations of problem-solving classrooms (e.g., Engle and Conant 2002; Lampert 1990; Schoenfeld et al. 2014), and sets of principles for teaching for and through problem solving (Cai 2010; Heller and Hungate 1985; Koichu et al. 2007a; Lester 2013; Lester and Cai 2016; Schoenfeld 1983). In many cases, the recommendations are presented as generalized reflections on successful classroom practices, experiments, or series of experiments (e.g., Koichu et al. 2007a, b; Lester 2013; Schoenfeld 1992). In some cases, the recommendations are based also on theories of problem-solving architecture (e.g., Ambrus and Barczi-Veres 2016; Clark et al. 2006) or decision making (e.g., Schoenfeld 2013).

It is indicative, however, that recent reflections of the state of the art tend to emphasize lacunas and open questions rather than the accomplishments of problem-solving research as a servant of mathematics instruction (Mason 2016a, b; Lester 2013; Schoenfeld 2013; Vinner 2014). Vinner (2014), for instance, questions the feasibility and even the relevance of problem solving for exam-oriented school education. Mason (2016a) reminds us that a variety of factors should be taken into account in order to make problem-solving instruction feasible in school and university settings. In his words, "all aspects of the human psyche, cognition, affect, behavior, attention, will and metacognition or witnessing must be involved" (p. 110). He then characterizes a research approach attempting to isolate particular features of problem solving as simplistic and unlikely to bring the desired change. In the same volume (Felmer et al. 2016), Mason (2016b) suggests that the crucial yet not sufficiently understood issue for adopting a problem-solving approach to mathematics teaching is the *when-issue*, that is, the issue of "*when* to introduce exploratory tasks, *when* to intervene, and in what way" (p. 263).

Lester (2013) acknowledges that research on mathematical problem solving has provided some valuable information about problem-solving instruction, but argues that the progress is slow and, generally speaking, insufficient. As one of the explanations of "this unfortunate state of affairs" (p. 251), Lester (2013) reiterates the claim that he and Charles made 25 years ago (Lester and Charles 1992): Research on mathematical problem solving remains largely atheoretical. Lester (2013) then argues that the comprehensive framework for research on problem-solving instruction proposed by Lester and Charles (1992) is still worth pursuing. Likewise, Schoenfeld (2013) reflects on the gains and limitations of a problem-solving framework that he authored 30 years ago (Schoenfeld 1985). He then suggests that the current challenge is to advance from a *framework* for examining problem solving to a *model* that would specify the theoretical architecture of this activity, i.e., would say "what matters" in problem solving (Schoenfeld 2013, p. 17), explain "how decision making occurs within that architecture" (p. 17) and theorize "how ideas grow and can be shared in interaction" (p. 20).

Stimulated by the aforementioned calls, the goal of this chapter is to present a particular proposal for an exploratory model of problem solving that would bridge our knowledge of how problem solving occurs with the knowledge of how to support and enhance problem solving in instructional settings.[1] The proposed model is confluence, namely, it consolidates ideas from several conceptual and theoretical frameworks. The consolidation is pursued by means of a strategy that is referred to as *networking theories by iterative unpacking*. In brief, this strategy consists of sequencing theoretical developments so that at each step of theorizing one theory serves as an overarching conceptual framework, in which another theory, either existing or emerging, is embedded in order to elaborate on the chosen elements(s) of the overarching theory. Mason's theory of shifts of attention (Mason 1989, 2008, 2010) serves as the overarching conceptual framework of the proposed model. Throughout the chapter, the model is illustrated by consideration of a single geometry problem, which is analyzed theoretically and then based on empirical evidence. Thoughts about possible implications of the model are shared in the concluding section.

18.2 The Proposed Model at a Glance

The proposed model is schematically presented in Fig. 18.1. The model is referred to as the shifts and choices model (SCM) in the rest of the chapter.

The inner part of Fig. 18.1 concerns the process that might be termed, with reference to Pólya (1945/1973) or Schoenfeld (1992), as a heuristic search embedded in the planning phase of problem solving. The main query associated with this part of the SCM is, simply stated: Where can a solution to a challenging problem come from? A more elaborated formulation of the query is as follows: Through which activities and resources does a problem solver construct a pathway of shifts of attention towards an invention of a key solution idea to a mathematical problem?

The outer part of Fig. 18.1 concerns a configuration of choices available to a problem solver. This part of the SCM deals with the following query: What choices is a problem solver empowered to make when constructing or co-constructing a chain of shifts of attention towards invention of a key solution idea to a mathematical problem that he or she perceives as a challenge? Among endless conscious and unconscious choices that individuals face when solving problems on their own or with others, the model takes into account the following: a choice of a challenge to be dealt with, a choice of schemata for dealing with a challenge, a choice of the mode of interaction, a choice of an extent of collaboration, and a choice of an agent to learn from.

As Fig. 18.1 shows, a *key solution idea* notion is in the core of the SCM. It is a solver-centered notion. Namely, a key solution idea is a *heuristic idea* that is

---

[1]In a way, this paper synthesizes and develops ideas that have been presented separately in Palatnik and Koichu (2014, 2015) and Koichu (2015a, b, 2017).

Fig. 18.1 Schematic presentation of the proposed model (SCM)

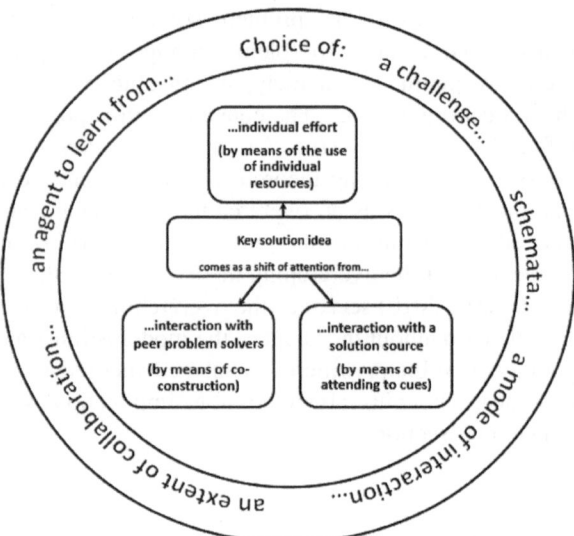

invented by the solver and evokes the conviction that it can lead to a *full solution* to the problem. A full solution is referred to as a solution that, to the solver's knowledge, would be acceptable in a situation in which it is communicated. Furthermore, Raman (2003) explains the heuristic idea notion as follows[2]: it is "an idea based on informal understandings, e.g., grounded in empirical data or represented by a picture, which may be suggestive but does not necessarily lead directly to a formal proof" (p. 322). Note that not any heuristic idea is a key solution idea. An in-depth discussion of an *idea* notion is beyond the scope of this chapter. It is sufficient to mention here that the Oxford Dictionary defines *idea* as "a thought or suggestion as to a possible course of action" (https://en.oxforddictionaries.com/definition/idea). Accordingly, a heuristic idea notion can operationally be treated as a piece of content-level mathematical discourse (see Sfard 2007) suggestive as to a possible way of solving the problem. An elaborated example is presented below.

The SCM relies on three premises:

1. Even when a problem is solved in collaboration, it has a *situational solver*: an individual who invents and eventually shares its key solution idea.
2. A key solution idea can be invented by a situational solver as a shift of attention in a sequence of his or her shifts of attention when coping with the problem.
3. Generally speaking, a solver's pathway of the shifts of attention is stipulated by choices the solver is empowered to make and by enacting the following types of resources:

---

[2]See also Koichu et al. (2007a) and Liljedahl et al. (2016) for detailed discussions of approaches to conceptualizing heuristics.

    (i)   individual resources,
    (ii) interaction with peer solvers who do not know the solution and struggle in their own ways with the problem or attempt to solve it together, and
    (iii) interaction with a source of knowledge about the solution or its parts, such as a textbook, an internet resource, a teacher, or a classmate who has already found the solution but is not yet disclosing it.

These three possibilities are intended to embrace all frequent situations of problem solving in instructional settings. Needless to say, the possibilities can be employed separately or can complement each other in problem solving.

18.3   Elaboration on the Elements of the Proposed Model

Discussion of the elements of the SCM in this section is supported by consideration of the two-circle problem (Fig. 18.2). The reader is invited to approach it before continuing reading.

18.3.1   *Invention of a Key Solution Idea as a Shift of Attention*

Mason's theory of shifts of attention had initially been formulated as a conceptual tool to dismantle constructing abstractions (Mason 1989) and was then extended to the phenomena of mathematical thinking and learning (Mason 2008, 2010). Palatnik and Koichu (2014, 2015) adapted the theory as a tool for analyzing insight problem solving. To characterize attention shifts, Mason (2008) considers *what* attended to by an individual and *how* it is attended to. To address the "how" question, he distinguishes five different *ways of attending* or *structures of attention*.

According to Mason (2008), *holding the wholes* is the structure of attention where a person is gazing at the whole without focusing on particulars. This is what probably happens when a reader flashes a glance at Fig. 18.2. *Discerning details* is a structure of attention in which one's attention is caught by a detail that becomes distinguished from the rest of the elements of the attended object. For example, one's attention can be caught by the segments *EF* and *GH* or by triangle *MEF* in Fig. 18.2. Mason (2008) suggests that "discerning details is neither algorithmic nor logically sequential" (p. 37). *Recognizing relationships* between the discerned elements is a development from discerned details that often occurs automatically: It refers to specific connections between specific elements. Say, for instance, that when gazing at the central part of Fig. 18.2, one notices that segments *EF* and *GH* look equal and can be considered sides of a quadrilateral *EFHG*. The *perceiving properties* structure of attention is different from the *recognizing relationships* structure in a subtle but essential way. In the words of Mason (2008): "When you

Two extrinsic circles are given. From the center of each circle, two tangent segments to another circle are constructed. The points of intersection of the tangent lines with the circles define two chords, *EF* and *GH* (see drawing). Prove that *EF* = *GH*.

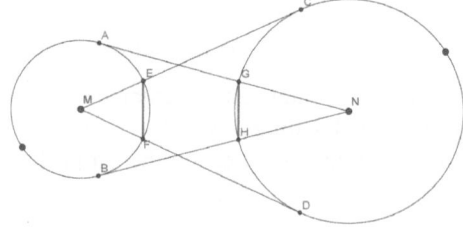

Fig. 18.2 The two-circle problem (translated from Sharygin and Gordin 2001, No. 3463)

are aware of a possible relationship and you are looking for elements to fit it, you are perceiving a property" (p. 38). In our example, one can draw the segments *EG* and *FH*. The perceived property would be "*EFHG* is a rectangle." Finally, *reasoning on the basis of perceived properties* is a structure of attention in which selected properties are attended to as the only basis for further reasoning. For example, one might consider what needs to be proved for sides *EG* and *FH* in order to prove that *EFHG* is a rectangle (see Fig. 18.3).

Palatnik and Koichu (2014, 2015) added a "why" question to Mason's "what" and "how" questions about attention: Why do individuals make shifts from one object of attention to another in the way that they do? One way of addressing this question is related to the obstacles embedded for the solver in attending to a particular object and to continuous evaluation of potential gains and losses of the choice to keep attending to the object or shift the attention to another one.

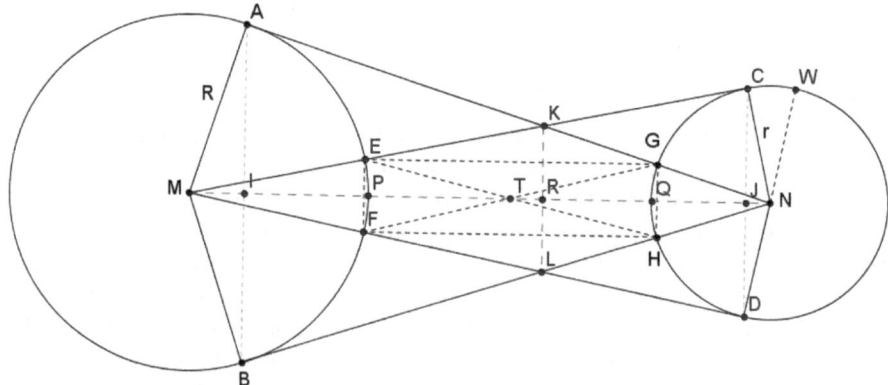

Fig. 18.3 Auxiliary construction for the two-circle problem

For example, one can try proving that *EFHG* is a rectangle by applying the available schemata associated with the rectangle notion, such as: it is enough to prove that *EFHG* is a parallelogram with a right angle, so let's try proving first that *EF* = *FH*. To prove this conjecture, it is enough to prove that $\Delta EKG \cong \Delta FLH$ and so on. (The reader is invited to check that this reasoning line is not particularly productive). At some point one can decide that enough attention has been given to *EFHG* and consider another object. A shift is likely to be mediated by mathematical resources within the reach for the solver. Our imaginary solver might think: "What else can be done? How can the congruency of two segments be proved? Maybe, it is worth including the segments into some pair of triangles and prove that they are congruent. Are *EFM* and *GHN* congruent? Apparently not. Should I stop considering *EF* and *GH* for a while and focus on their halves, *EP* and *GQ*?" Such a shift may seem trivial (especially if one knows the solution), but in fact it is not.

In a while, the solver might consider *EP* and *GQ* to be the sides of the right-angle triangles *MEP* and *NGQ*. The solver can then perceive the following property: $\Delta MEP \sim \Delta MCN$ and $\Delta NGQ \sim \Delta NAM$. There is a gap, however, between perceiving a property and choosing it as the only basis for further reasoning. Indeed, the solver should somehow realize that the triangle similarity can help in proving that two segments are equal. If our imaginary solver arrives this far, she has a good chance of inventing a key solution idea to the problem. One such idea consists of the observation that *EP* and *GQ* can be expressed through the same elements, *R*, *r* and *MN*, based on the aforementioned triangle similarities. This idea can be developed into a full solution to the problem: $EP = \frac{r \cdot R}{MN}$ from the first similarity and $GQ = \frac{r \cdot R}{MN}$ from the second similarity, consequently, *EF* = *GH*, QED.

The presented imaginary scenario suggests how the process of inventing a key solution idea can be seen in terms of the solver's shifts of attention. Generally speaking, the solver attends to the objects embedded in the problem formulation and mentally manipulates them by applying available schemata. The process is goal directed, but particular shifts can be sporadic. However, it should be noted that the presented scenario is neither complete nor compelling. More should be done in order to realistically characterize one's problem solving process as a chain of shifts of attention. In particular, the specificity of problem solving in socially different educational contexts should be considered. This is done in the next three subsections.

18.3.2   Shifts of Attention in Individual Problem Solving

The lion's share of the data corpus that underlies the development of the foremost problem-solving frameworks (e.g., Schoenfeld 1985; Carlson and Bloom 2005) consists of cases of individual problem solving. Carlson and Bloom (2005) consider four phases in individual problem solving by an expert mathematician: orientation, planning, executing, and checking. The model also includes a sub-cycle, "conjecture-test-evaluate," and operates with various problem-solving attributes (this notion is due to Schoenfeld 1985), such as conceptual knowledge, heuristics, metacognition, control, and affect. Generally speaking, Carlson and Bloom's framework offers a kit of conceptual tools that can be used for producing thick descriptions of individual problem solving. These tools enter the SCM as tools for addressing "how" and "why" questions about the shifts of attention.

For example, when our imaginary problem solver individually coped with the two-circle problem, she first directed her attention to proving that a particular quadrilateral is a rectangle and then shifted her attention to proving the similarity of two pairs of triangles. The pre- and post-stages of the shift can be described as two "conjecture-test-evaluate" sub-cycles within the planning phase. The shift itself can be characterized in terms of her mathematical, heuristic, and affective resources.

18.3.3 Shifts of Attention When Interaction with Peers Is Available

While studying problem-solving behaviors in small groups of students, Clark et al. (2014) extended Carlson and Bloom's (2005) framework by introducing two new categories/codes. They termed them *questioning* and *group synergy*. The former category was introduced in order to give room in the data analysis to various questions (for assistance, for clarification, for status, and for direction) that the participants had asked. The latter category appeared to be necessary in order

> to capture the combination and confluence of two or more group members' problem-solving moves that could only occur when solving problems as a member of a group. … A key characteristic of this group synergy code is that it leads to increased group interaction and activity, sometimes in unanticipated and very productive ways. (Clark et al. 2014, p. 10–11)

Indeed, when a possibility to collaborate with peers is available to problem solvers, their shifts of attention can be stipulated by inputs of the group members, especially when the inputs are shared in some common problem-solving space in a non-tiresome way. Peer interaction can increase one's chances to produce a key solution idea, but can also be overwhelming or distracting. In particular, when nobody in a group knows how to solve the problem, the other members' inputs of potential value are frequently undistinguishable for the solver from inputs of no value.

Schwartz et al. (2000) deeply explored the cognitive gains of two children who failed to solve a problem individually, but who improved when working in peer interaction. They distinguished between the *two-wrongs-make-a-right* and *two-wrongs-make-a-wrong* phenomena. The mechanisms of co-construction behind the two-wrongs-make-a-right phenomenon were the mechanisms of disagreement, hypothesis testing, and inferring new knowledge through challenging and conceding. These mechanisms might be involved in those cases of collaborative problem solving in which group synergy led the participants in Clark et al.'s (2014) study to "very productive ways" (p. 11) of solving the given problems. To further explore the phenomenon of group synergy, it seems necessary to me to acknowledge that the above mechanisms can become active on condition that at least sometimes solvers shift their attention from an object that they are exploring to an object attended to by a peer.

Consider as an example once more the two-circle problem, but this time based on real data. The data were collected from a two-year experiment conducted in a class of 17 regular (i.e., not identified as gifted) 10th grade students. During the experiment, many difficult problems were offered to the students to solve over the course of 5–7 days for each problem in an environment combining classroom work and work from home. The work from home was supported by an online discussion forum at Google+. The forum devoted to the two-circle problem was active for 4 days and contained 230 entries. Three different solutions were finally produced, including the one presented in Sect. 18.3.1. An excerpt from the beginning of the forum is presented in Fig. 18.4.

Evidently, the forum participants are still far from any productive heuristic idea. Some of them are at the orientation stage and, generally speaking, are occupied by creating initial drawings. Maya and Shira begin to develop the direction that "*EFHG* is a rectangle," which, as we know, is a dead end. The excerpt is suggestive about the following phenomena: The students independently choose different objects of attention (e.g., a kite-looking quadrilateral, a pair of triangles) and then share what they do. Sometimes the attempts to get somebody's attention are successful, and sometimes they are not. The students occasionally choose to explore the same object together. The excerpt is also suggestive about the aforementioned mechanisms of productive peer interaction.

How can such interactions influence an individual pathway of shifts of attention? Let me address this question with particular focus on one student, Maya. According to the teacher, Maya has neither been an active student in a classroom nor a successful student in terms of mathematics exams. However, Maya was one of the most active participants in the two-circle problem forum. She initiated six out of 34 discussion threads, replied to 17 threads, uploaded two drawings and a scanned hand-written solution. Based on her reflective questionnaire, we know that she devoted about 5 h to the problem during 4 days and that for about 3 h she worked outside the forum. A part of the Maya's devious pathway of shifts of attention is presented on Fig. 18.5.

The key solution idea of Maya was similar to the idea presented in Sect. 18.3.1, but she considered different pairs of similar triangles, $\Delta KNL \sim \Delta GNH$ and $\Delta KML \sim \Delta EMF$ (see Fig. 18.3). To prove their similarity, Maya first proved that $KL \| EF \| GH$. This was the major challenge for her. She then concluded the proof by consideration of proportions stemming from these similarities, in conjunction to a proportion based on the "bridging" pair of similar triangles, $\Delta MAK \sim \Delta NCK$.

Maya was a situational problem solver in the group. It is of note that Maya's interactions with the peers were mostly around the objects of attention chosen by her. She switched her attention to the objects suggested by the other students only

Teacher (Day 1, 04:37): Good week, everybody! Today we begin to work on the third task of the project [posts the two-circle problem]. Good luck! [one "like"]

Shira (Day 1, 22:14): Look guys if you understand anything [in the drawing on the right side]. But this is, in general, the direction that I took :-) [two "likes"]

Zila (Day 1, 22:16): The truth is that I also thought about this way as the beginning, but how can you develop it?

...

Zila (Day 1, 23:49): How the quadrilateral (looks like a kite at Shira' drawing), you said that it is a parallelogram?

Shira (Day 2, 07:22): I think it does not matter what the drawing looks like. I just did not draw exactly :-). But this is, in general, the direction, and you're invited to continue :-)

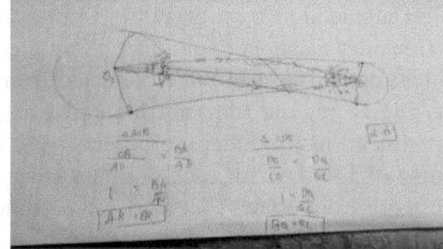

Maya (Day 2, 19:11): I began working on the exercise only now and your auxiliary construction is exactly what I drew, even without looking! ^_____^ [this sign roughly means "I am proud of myself"]. I thought to use the segment between the centers and the distance from the center of the circle, I've not really begun, just a sort of conjecture :) [one "like"]

Zila (Day 2, 19:12): [Maya], did you reach the conclusion that this is a parallelogram (there is a good chance that it is!)

Maya (Day 2, 19:27): So far I've drawn a sketch, and I'll begin thinking in a moment.

Shira (Day 2, 21:20): The teacher, Maya, and I talked by phone for about 15 min about the exercise and thought and thought and got something. We first understood that the key points are those points that define the given arcs. So, the auxiliary construction should be from them or they should be included in two triangles so that it will be possible to prove their congruence. [one "like"]

Teacher (Day 2, 21:26): Excellent. It is worth adding to your previous ideas. . .

Maya (Day 2, 21:40): I invented this idea with Shira! In general, the idea is to get to the rectangle *EFHG*. I thought about how to prove the congruence of triangles *EKG = FLN* [six "likes"].

Zila (Day 2, 21:42): Maya, they are congruent.

Maya (Day 2, 21:52): How???

Zila (Day 2, 21:53): According to the calculation of the angles.

Zila (Day 2, 21:53): Or, in fact, they are not.

Maya (Day 2, 22:01): Hhhhhhhh... I am sitting already for two hours on this exercise. . . . If it would be that simple, I'd already be successful.

Fig. 18.4 An excerpt from the beginning of the two-circle problem forum at Google+

MN and the radii (*with Shira*) →Points defining arcs *GH* and *EF* → *EFHG* → *EKG = FLN* → *KL* → Possible similarity of *GHN* and *CMD*, *ANB* and *EMF* → Similarity of *MAK* and *KCN*, *LDN* and *MLB* → *MAN* (Meirav's idea, which appears to be a dead-end for Maya) → return to *EFHG* → Return to Arcs *GH* and *EF* → *MN*, chords *GH* and *EF* → *KL*, similarity of *NHG* and *EMF*, *KNL* and *MKL* → ??? → solution

Fig. 18.5 Maya's pathway of shifts of attention in solving the two-circle problem

occasionally. Interestingly, Maya never explicitly acknowledged that the ideas published on the forum had helped her. However, some steps in her solution can be traced back to the ideas suggested and explored by the other participants.[3]

18.3.4   Shifts of Attention When Interaction with a Solution Source Is Available

The option to interact with a source of knowledge about a key solution idea to a problem can drastically change a pathway of one's shifts of attention, up to the point that the entire process can stop being a problem-solving process and become a solution-comprehending process. The proposed model seeks to encompass only the situations in which a solution source can be present as a provider of cues to the solution or as a convenient storage of potentially useful facts, but not as a source of *telling* the solution. Such situations are common, for instance, when a teacher orchestrates a classroom problem-solving discussion by favoring some of the students' ideas over the others. In this way, the problem is usually solved before the bell rings. A danger in this situation, however, is that solvers may be deprived of inventing the solution themselves or being misled by a deceptive feeling, such as "we solved the problem with the teacher, so next time I will be able to do so alone."

When a source of knowledge about the solution is present but does not give the solution, the solver may attempt to extract it from the source (e.g., see questions for assistance and questions for direction in Clark et al. 2014). In some cases, one's shifts of attention may occur as a straightforward result of such attempts. In other cases, a shift may occur as a result of a conflict that emerges when more knowledgeable and less knowledgeable interlocutors assign different meanings to the same assertions (cf. Sfard 2007 for commognitive conflict).

For example, the assertion "triangle similarity is a good idea" can either pass unnoticed in the group discourse or be a trigger for solvers to shift their object of attention. The effect of the assertion would depend on who it has come from, a regular member of the group or a teacher or a peer who acts as if she has already solved the problem. In one case, the assertion can be perceived as "it is possible that similarity helps"; in another, "I've tried it and it helped"; and in yet another, "this is the direction approved by the authority." In any case, the perceived meaning of the assertion does not necessarily match the intended meaning. In line with Sfard (2007), one can suggest that a conflict of meanings can either hinder the communication or help the solver to progress.

Let me illustrate this suggestion by an additional excerpt from the two-circle problem forum. This time I focus on an episode from the fourth day of the forum, when Maya announced that she solved the problem. As one can see (Fig. 18.6),

---

[3]Unfortunately, there is no room in this chapter for presenting the entire story and the method of its SCM-driven analysis. It will be done elsewhere (Koichu and Harris, in preparation).

Maya: I succeeeeeded!!!!!!!!! And the teacher was right—it is so simple that I'd like to die. When I saw that I got nothing from [consideration of] this rectangle [*EFHG* on Fig. X.3], I stopped and changed the direction. I went to proportions and solved the problem in several theorems, but anyway the long way towards the rectangle helped me to prove that *KL* is parallel to *GH* and to *EF*, and this is what helped me with the proportions. I'll now organize everything and upload the solution ☺.

Meirav: But I've tried a lot of times to find some parallelograms and it did not work out. A hint?

Maya: Look at Thaleses…

Meirav: Thanks!!

Alina: How did you prove that *KL* is parallel to *GH* and *EF*?

Zila: She used the theorems about tangent lines and triangle similarity. Is it right?

Maya: Oh, it took years….

Fig. 18.6 Interactions with Maya as a source of knowledge about the solution

when Maya announced her success, she was perceived by the classmates as a source of knowledge about the solution. This fact essentially shaped the interaction. The students did not ask Maya to share the full solution—they knew that this would be against the rules of the forum—but they sought direction. Meirav asked for a hint. Alina inquired about a particular solution step. Zila assumed the role of translator of Maya's ideas. Interestingly, the situation is not as festive as it probably looks. First, when Maya actually published her solution, it appeared to have a logical flaw. She succeeded in producing a mathematically valid proof only after polishing her reasoning in interaction with two forum participants. Second, the hint provided by Maya in response to Meirav's request sounds as if the theorem of Thales might help, but this was not the case. Neither Maya nor other students used this theorem. Apparently, "Thaleses" was an informal tag for the idea "use proportions." Third, Zila's suggestion that Maya used "the theorems about tangent line and triangle similarity" was not helpful at best and misleading at worst. Maya indeed used these theorems, but not in proving that $KL\|EF\|GH$.

The point is that interaction with a source of knowledge about the solution affected the students' pathways of shifts of attention. It seems that an interplay of different meanings assigned to the same statements (e.g., "Thaleses") was an indispensable characteristics of such an interaction.

18.3.5 Shifts of Attention and Choices That Problem Solvers Are Empowered to Make

The argument presented in the previous subsections can be condensed into the following sentence: One's pathway of shifts of attention when solving a mathematical problem is essentially stipulated by choices that he or she is empowered to make. At a glance, this sentence echoes a description of a problem-solving process offered by Poincaré (1908/1948): A problem-solving process consists of a multi-stage pathway of conscious and unconscious steps towards the minimalistic choice of a "proper" combination of ideas out of a huge number of possible combinations. Let us recall however that the Poincaré description concerns only the choices of solution steps and applies to mathematicians. Accordingly, it concerns problem solving having very specific characteristics: The choice of a problem is made by its solver, the solver has immensely rich mathematical resources and is confident in his or her mathematical ability, and has virtually unlimited time and motivation for pursuing the problem. Few of these characteristics hold for mathematical problem solving in instructional settings, but let me argue that Poincaré's idea of choice can usefully be stretched.

Let us come back to the case of Maya once more. This time I wish to summarize the choices available to Maya when solving the Two-Circle Problem. As the presented excerpts suggest, Maya acted in a situation in which she could choose her solution moves. Moreover, she was empowered to make many additional choices such as whether to attempt to solve the given problem, follow problem-solving ideas of her peers, or merely ignore the challenge, as some of her classmates did; whether to work independently or with her peers; who to communicate with and when; which ideas to respond to and when; which and whose ideas to include into her own reasoning line; which ideas to share and how; and how much time to devote to the problem. It seems that only two choices were not up to Maya: Which problem to solve and when. In the described episode, these choices have been made for her.

The presented situation was particularly rich with opportunities for the students to choose. It is possible to recall or imagine instructional situations in which a configuration of choices available to the problem solvers would be different and include more or less choices. In fact, any instructional situation involving problem solving can be thought of in terms of choices that mathematics teachers empower their students to make, either intentionally or not.

18.4 Pedagogical Uncertainty and Choice-Affluent Environments

The diversity of choices involved in problem solving in instructional settings is immense, as are the diversity of individual pathways of shifts of attention. Being aware of this, we must acknowledge a fundamental role of pedagogical uncertainty: We can never know in advance which pathways of the shifts of attention the students construct when solving problems; thus, we will probably never be able to formulate universal recommendations as to how to organize a problem-solving classroom so that it would fit the individual needs and traits of each student. In the other words, we can probably never produce a satisfactory answer to Mason's (2016b) "when" question, which he posits as *the* question of problem-solving instruction: "*when* to introduce exploratory tasks, *when* to intervene, and in what way" (p. 263). In yet other words, there are probably no configuration or configurations of teaching decisions that would be optimal for enhancing problem solving for all.

There is an alternative, however, to the perennial search for such configurations of decisions. Its roots can be traced to seminal work of Dewey (1938/1963), who substantiated the idea that students must be involved in choosing what they learn and how. I term this alternative as *constructing choice-affluent learning environments*. By a choice-affluent learning environment, I mean an environment in which students can at different times choose the most appropriate (1) challenge to pursue, from solving a difficult problem to comprehending a worked-out example; (2) mathematical tools and schemata for dealing with the challenge; (3) extent of collaboration, from being actively involved in exploratory discourse with peers of their choice to being independent solvers; (4) a mode of interactions, that is, whether to talk, listen, or be temporarily disengaged from the collective discourse, as well as whether to be a proposer of an idea, a responder to the ideas by the others, or a silent observer; and (5) agent to learn from, that is, the opportunity choose whose and which ideas are worthwhile of their attention.

How feasible are choice-affluent environments in school reality? One example of such an environment, an asynchronous problem-solving forum characterized by exploratory discourse, was presented above. An additional example involving engagement of students in long-term mathematics research projects has been presented elsewhere (Palatnik and Koichu 2015; Koichu 2017). Furthermore, it can be argued that even a lesson in a classroom that has a time constraint can be a choice-affluent environment.

For example, one characteristic of what Liljedahl (2016) termed the *thinking classroom* is the use of vertical surfaces (e.g., whiteboards or blackboards) as media that substitute for notebooks or working sheets that traditionally lie on the student desks. In such a classroom, students are given time to solve mathematical problems in a small group while standing and writing on the vertical surfaces instead of sitting and writing in their notebooks. Accordingly, students all have access not only to the content of the whiteboard of their own small groups, but can also see what is written on the other groups' whiteboards. I had a chance[4] to observe the following phenomenon in such a lesson: When a small group felt that they were in progress, they paid little attention to the work of the other groups. But when the students felt that they were stuck, some of them looked over the other groups' work without directly interacting with the members of these groups with the hope of getting a useful cue or evaluating their progress in comparison with the other groups' progress. In this way, they got what can be called *non-intrusive assistance*. Furthermore, the students engaged themselves in interactions exactly when *they* needed them and not when the teacher decided for them that they needed them. In terms of the definition of a choice-affluent environment, the students in Peter Liljedahl's class were empowered to make choices (2), (3), and (5) above.

---

[4] I thank Peter Liljedahl for the opportunity to attend such a class during my visit to Simon Fraser University in 2016.

18.5 Concluding Remarks

Developing a model of mathematical problem solving that would be applicable to different educational contexts is motivated by several causes. First, with few exceptions, the existing problem-solving frameworks utilize different conceptual tools for exploring problem solving in socially different educational contexts. Second, the foremost frameworks (Carlson and Bloom 2005; Schoenfeld 1985) are comprehensive within the problem-solving contexts within which they have emerged, but it is sometimes difficult to apply them to additional contexts. Third, the central issue of connecting our knowledge of how problem solving occurs and how to enhance this activity in instructional settings is still underdeveloped (e.g., Schoenfeld 2013).

In this chapter, a particular way of constructing a model of mathematical problem solving is presented. The proposed model capitalizes on the Mason's (1989, 2008, 2010) theory of shifts of attention (which was initially developed for other reasons) and consideration of choices that problem solvers are empowered to make. Hence, the model has been named the shifts and choices model, or the SCM. In fact, the model is a confluence and embeds theoretical tools from the existing problem-solving frameworks and theories. As mentioned, the model is only exploratory. Its use as a research tool is stipulated by the availability of research methodologies for identifying shifts of attention in socially different problem-solving contexts. In part, such methodologies are available from past research, but they should be further developed. The use of the model as a pedagogical tool depends on further unpacking the mechanisms underlying the choices that problem solvers make in different instructional situations and on research on how a teacher's decisions affect these choices. Little research has been conducted in this direction so far.[5]

Furthermore, the choice-affluent learning environment notion is introduced in this chapter. It is important to note that I do not argue for the claim that the more choices that are left to the students, the better. I rather argue for being aware of the fundamental role of pedagogical uncertainty related to Mason's (2016b) "when" question, for being aware of configurations of choices that we, mathematics teachers, inevitably create for our students and for being aware of the complexity and sensitivity of the student pathways of shifts of attention when they are engaged in problem solving.

I choose to end this chapter by mentioning some of the questions that I have had a tendency to ask myself as a mathematics educator ever since I have begun looking at my own lessons through the lenses provided by the SCM: What student choices do I tend to support? What student choices am I aware of? How do my students choose what they choose? How can I support the desirable choices without

---

[5]A study by Flowerday and Schraw (2000) on teachers' beliefs about instructional choices is an exception and a useful step towards understanding how teachers construct choices for their students.

choosing for the students? Which choices should I leave to the students? To what extent were my lessons during the last week choice affluent? My best hope in relation to this chapter is that some of the readers would find some of these questions worth thinking about.

Acknowledgements I am grateful to the Program Committee of ICME-13 for the invitation to give a talk at the conference. The empirical example presented in this chapter is taken from a study supported, in part, by the Israel Science Foundation (grant# 1596/13, PI B. Koichu). The opinions expressed here are those of the author and are not necessarily those of the Foundation.

References

Ambrus, A., & Barczi-Veres, K. (2016). Teaching mathematical problem solving in Hungary for students who have average ability in mathematics. In P. Felmer, E. Pehkonen, & J. Kilpatrick (Eds.), *Posing and solving mathematical problems. Advances and new perspectives* (pp. 137–156). Switzerland: Springer.

Cai, J. (2010). Helping students becoming successful problem solvers. In D. V. Lambdin & F. K. Lester (Eds.), *Teaching and learning mathematics: Translating research to the elementary classroom* (pp. 9–14). Reston, VA: NCTM.

Carlson, M., & Bloom, I. (2005). The cyclic nature of problem solving: An emergent multidimensional problem-solving framework. *Educational Studies in Mathematics, 58*(1), 45–75.

Clark, K., James, A., & Montelle, C. (2014). We definitely would't be able to solve it all by ourselves, but together...: Group synergy in tertiary students' problem-solving practices. *Research in Mathematics Education, 16*(2), 306–323.

Clark, R. E., Kirschner, P. A., & Sweller, J. (2006). Why minimal guidance during instruction does not work: An analysis of the failure of constructivist, discovery, problem-based, experimental, and inquiry-based teaching. *Educational Psychologist, 41*(2), 75–86.

Dewey, J. (1938/1963). *Experience and education* (reprint). New York: Collier (Original work published 1938).

Engle, R. A., & Conant, F. R. (2002). Guiding principles for fostering productive disciplinary engagement: Explaining an emergent argument in a community of learners classroom. *Cognition and Instruction, 20,* 399–483.

Felmer, P., Pehkonen, E., & Kilpatrick, J. (Eds.). (2016). *Posing and solving mathematical problems. Advances and new perspectives*. Switzerland: Springer.

Flowerday, T., & Schraw, G. (2000). Teacher beliefs about instructional choice: A phenomenological study. *Journal of Educational Psychology, 92*(4), 634.

Halmos, P. (1980). The heart of mathematics. *American Mathematical Monthly, 87*(7), 519–524.

Heller, J., & Hungate H. (1985). Implications for mathematics instruction of research on scientific problem solving. In E. A. Silver (Ed.), *Teaching and learning mathematical problem solving: Multiple research perspectives* (pp. 83–112). Hillsdale, NJ: Erlbaum.

Koichu, B. (2015a). Towards a confluence framework of problem solving in educational contexts. In K. Krainer & N. Vondrová (Eds.), *Proceedings of the 9th Conference of the European Society for Research in Mathematics Education* (pp. 2668–2674). Prague, Czech Republic: Charles University.

Koichu, B. (2015b). Problem solving and choice-based pedagogies. In F. M. Singer, F. Toader, & C. Voica (Eds.), *Electronic Proceedings of the 9th International Conference Mathematical Creativity and Giftedness* (pp. 68–73). Sinaia, Romania (ISBN: 978-606-727-100-3). Retrieved December 12, 2016 from http://mcg-9.net/pdfuri/MCG-9-Conference-proceedings.pdf.

Koichu, B. (2017). On mathematics with distinction, a learner-centered conceptualization of challenge and choice-based pedagogies. *The Mathematics Enthusiast, 14*(1–3), 517–540.

Koichu, B., Berman, A., & Moore, M. (2007a). Heuristic literacy development and its relation to mathematical achievements of middle school students. *Instructional Science, 35*, 99–139.

Koichu, B., Berman, A., & Moore, M. (2007b). The effect of promoting heuristic literacy on the mathematic aptitude of middle-school students. *International Journal of Mathematical Education in Science and Technology, 38*(1), 1–17.

Lampert, M. (1990). When the problem is not the question and the solution is not the answer: Mathematical knowing and teaching. *American Educational Research Journal, 27*(1), 29–63.

Lappan, G., & Phillips, E. (1998). Teaching and learning in the connected mathematics project. In L. Leutzinger (Ed.), *Mathematics in the middle* (pp. 83–92). Reston, VA: NCTM.

Lester, F. (2013). Thoughts about research on mathematical problem-solving instruction. *The Mathematics Enthusiast, 10*(1–2), 245–278.

Lester, F. K., & Cai, J. (2016). Can mathematical problem solving be taught? Preliminary answers from 30 years of research. In P. Felmer, E., Pehkonen, & J. Kilpatrick (Eds.), *Posing and solving mathematical problems. Advances and new perspectives* (pp. 117–136). Switzerland: Springer.

Lester, F. K., & Charles, R. I. (1992). A framework for research on mathematical problem solving. In J. P. Ponte, J. F. Matos, J. M. Matos, & D. Fernandes (Eds.), *Issues in mathematical problem solving and new information technologies* (pp. 1–15). Berlin: Springer.

Liljedahl, P. (2016). Building thinking classrooms: Conditions for problem-solving. In P. Felmer, E. Pehkonen, & J. Kilpatrick (Eds.), *Posing and solving mathematical problems. Advances and new perspectives* (pp. 361–386). Switzerland: Springer.

Liljedahl, P., Santos-Trigo, M., Malaspina, U., & Bruder, R. (2016). *Problem solving in mathematics education. ICME-13 topical surveys.* Berlin: Springer International Publishing.

Mason, J. (1989). Mathematical abstraction as the result of a delicate shift of attention. *For the Learning of Mathematics, 9*(2), 2–8.

Mason, J. (2008). Being mathematical with and in front of learners: Attention, awareness, and attitude as sources of differences between teacher educators, teachers and learners. In B. Jaworski & T. Wood (Eds.), *The mathematics teacher educator as a developing professional* (pp. 31–56). Rotterdam/Taipei: Sense Publishers.

Mason, J. (2010). Attention and intention in learning about teaching through teaching. In R. Leikin & R. Zazkis (Eds.), *Learning through teaching mathematics, mathematics teacher education* (Vol. 5, pp. 23–47). The Netherlands: Springer.

Mason, J. (2016a). Part 1 reaction: Problem posing and solving today. In P. Felmer, E., Pehkonen, & J. Kilpatrick (Eds.), *Posing and solving mathematical problems. Advances and new perspectives* (pp. 109–116). Switzerland: Springer.

Mason, J. (2016b). When is a problem...? "When" is actually the problem! In P. Felmer, E. Pehkonen, & J. Kilpatrick (Eds.), *Posing and solving mathematical problems. Advances and new perspectives* (pp. 263–287). Switzerland: Springer.

Palatnik, A., & Koichu, B. (2014). Reconstruction of one mathematical invention: Focus on structures of attention. In P. Liljedahl, C. Nicol, S. Oesterle, & D. Allan (Eds.), *Proceedings of the 38th Conference of the International Group for the Psychology of Mathematics Education* (Vol. 4, pp. 377–384). Vancouver, Canada: PME.

Palatnik, A., & Koichu, B. (2015). Exploring insight: Focus on shifts of attention. *For the Learning of Mathematics, 2*, 9–14.

Poincaré, H. (1908/1948). *Science and method.* New York: Dover (Originally published in 1908).

Pólya, G. (1945/1973). *How to solve it.* Princeton, NJ: Princeton University Press.

Raman, M. (2003). Key ideas: What are they and how can they help us understand how people view proof? *Educational Studies in Mathematics, 52*(3), 319–325.

Schoenfeld, A. H. (1983). *Problem solving in the mathematics curriculum. A report, recommendations, and an annotated bibliography.* USA: The Mathematical Association of America.

Schoenfeld, A. H. (1985). *Mathematical problem solving.* Orlando, FL: Academic Press.

Schoenfeld, A. H. (1992). Learning to think mathematically: Problem solving, metacognition, and sense-making in mathematics. In D. Grouws (Ed.), *Handbook for research on mathematics teaching and learning* (pp. 334–370). New York, NY: Macmillan.

Schoenfeld, A. H. (2013). Reflections on problem solving theory and practice. *The Mathematics Enthusiast, 10*(1–2), 9–34.

Schoenfeld, A. H., Floden, R. E., & The Algebra Teaching Study and Mathematics Assessment Project. (2014). *An introduction to the TRU Math document suite.* Berkeley, CA & E. Lansing, MI: Graduate School of Education, University of California, Berkeley & College of Education, Michigan State University. Retrieved from: http://ats.berkeley.edu/tools.html.

Schroeder, T., & Lester, F. (1989). Developing understanding in mathematics via problem solving. In P. Traffon & A. Shulte (Eds.), *New directions for elementary school mathematics: 1989 yearbook* (pp. 31–42). Reston, VA: NCTM.

Schwartz, B., Neuman, Y., & Biezuner, S. (2000). Two wrongs may make a right… if they argue together! *Cognition and Instruction, 18*(4), 461–494.

Sfard, A. (2007). When the rules of discourse change, but nobody tells you: Making sense of mathematics learning from a commognitive standpoint. *The Journal of the Learning Sciences, 16*(4), 565–613.

Sharygin, I. F., & Gordin, R. K. (2001). *Collection of problem in geometry. 5000 problems with solutions* (Sbornik zadach po geometrii. 5000 zadach s otvetami). Moscow: Astrel (in Russian).

Vinner, S. (2014). The irrelevance of research mathematicians' problem solving to school mathematics. In Koichu, B. Reflections on problem solving. In M. N. Fried & T. Dreyfus (Eds.), *Mathematics & mathematics education: Searching for common ground. Advances in mathematics education* (pp. 113–135). The Netherlands: Springer.

Chapter 19
Natural Differentiation—An Approach to Cope with Heterogeneity

Günter Krauthausen

Abstract Teachers in their classes always have to cope with heterogeneity, and that by no means is a new problem. In Germany e.g. plenty of (mostly pedagogical) publications from the midst 1970s until today offer brilliant advice for several kinds of differentiation. How then can it be that after forty years, heterogeneity and differentiation are still called a ›mega issue‹? Could it be that those traditional kinds of differentiation are admittedly to be considered or necessary, but not sufficient—and if: why? This paper will discuss questions like these aiming to bring together crucial issues for (primary) math education in heterogeneous classes, like standards for mathematical practice, standards for mathematical content, social learning with and from each other, and heterogeneity. Main theoretical concepts are substantial learning environments (Wittmann in Educational Studies in Mathematics 15(1):25–36, 1984; Wittmann in Educational Studies in Mathematics 48(1):1–20, 2001a; Wittmann in Proceedings of the Ninth International Congress on Mathematical Education. Kluwer Academic Publishers, Norwell, MA, 2004) and natural differentiation (Wittmann and Müller in Grundkonzeption des Zahlenbuchs. Klett, Stuttgart, 2012; Krauthausen and Scherer in Ideas for natural differentiation in primary mathematics classrooms. Vol. 1: The substantial environment number triangles. Wydawnictwo Uniwersytetu Rzeszowskiego, Rzeszòw, 2010a; Krauthausen and Scherer in Motivation via natural differentiation in mathematics. Wydawnictwo Universytetu Rzeszowskiego, Rzeszów, pp. 11–37, 2010b; Krauthausen and Scherer in Natürliche Differenzierung im Mathematikunterricht – Konzepte und Praxisbeispiele aus der Grundschule. Kallmeyer, Seelze, 2014).

Keywords Natural differentiation · Heterogeneity · Substantial learning environments · Number triangles

G. Krauthausen (✉)
University of Hamburg, Hamburg, Germany
e-mail: krauthausen@uni-hamburg.de

© The Author(s) 2018
G. Kaiser et al. (eds.), *Invited Lectures from the 13th International Congress on Mathematical Education*, ICME-13 Monographs,
https://doi.org/10.1007/978-3-319-72170-5_19

19.1 Heterogeneity and Differentiation—A Traditional Problem with Traditional Answers?

Math teachers in their classes have to cope with heterogeneity every day of their professional lives. It is by no means a new problem. And it is still as much a theoretical challenge as it is a practical one. This paper tries to illuminate the perspective, concepts and experiences from Germany. Since the early 1970s there have been plenty of publications in that country addressing differentiation (e.g. Bönsch 1976; Geppert and Preuß 1981; Klafki and Stöcker 1976; Winkeler 1976), when it was already called a ›mega issue‹, just as it is still called these days. For primary schools, normally just ›inner differentiation‹ has been taken into account. That means methods to be used within a classroom and not splitting up the class in (supposed) homogeneous groups for a longer time. Today's references in Germany often quote the same theoretical and methodical concepts of inner differentiation, though new terms may have been created (cf. Bönsch 2004; Paradies and Linser 2005). Some of those traditional methods are:

- social differentiation: single work, partner work, group work, …
- differentiation by teaching methods: course-like formats, projects, …
- differentiation by media: textbook, worksheets, manipulatives, digital media, …
- quantitative differentiation: same amount of time for different workload/amount of content, or different amount of time for identical workload/amount of content
- qualitative differentiation: objectives and tasks with different levels of difficulty.

This list looks like a current offer of in-service courses. Actually, it dates back to a booklet from Winkeler (1976), which may raise two questions: (1) Is there no namable progress since then? Why else would differentiation still be so prominent on the agenda? (2) And why is that so?

For sure, those traditional recommendations should not be devalued per se, nor can be claimed that they are non-effective. But obvious problems can possibly hinder or prevent what is actually intended. Four examples for that:

The idea of ›difficulty‹

Declaring a task as difficult/moderate/easy—a common practice found in workbooks from publishers or on self-made worksheets by teachers—necessarily comes up against limiting factors:

(a) A level of difficulty varies not just between different students, but also with the same student, at different times, and even with the same task (cf. Selter and Spiegel 1997). Difficulty is a question of subjective valuation and not an objective concern.

(b) A *felt* grade of difficulty depends on diverse considerations:

- complexity of the demands of calculation (kind and size of numbers etc.),
- involved arithmetic operations (addition and subtraction often seem to be easier to do than multiplication and division),

- demands of cogitation, strategic comprehension, process-related competency, linguistic understanding of the task, amount of required (oral or written) text production or documentation, etc.

(c) The level of a task's difficulty cannot be measured just by the formal-syntactic steps that the solution requires.

All these aspects clearly relativize the sometimes stated claim (by workbooks and even more by digital media) that a learning offer would *automatically* adjust its level of difficulty to the capability or demands of the individual learner.

Individualization and social learning

Put to practice, the postulate of individualization sometimes even results in the abolition of social learning, the learning with and from each other. But individualization does not mean that each student should deal with his very own, individual, and different tasks or even topics. That kind of misunderstanding leads to scenarios that evokes pictures of open-plan offices: Students working at their desks, dispatching different things, and no substantial communication about the things they individually deal with. In this case, social learning is often seen in a mainly pedagogical sense or as a question of classroom management, aiming at implementing effective rules and rituals within and for lessons in order to make classroom relations affable, friendly and non-threatening. No doubt that this all is important, too.

But if restricted to that, argumentation—that is the *communication of minds on shared contents*—turns to become nearly impossible, because there *are no* shared experiences with a common content. Some teachers even take pride in that by declaring: »We abolished those common plenary phases in favor of a thorough individualization«. In doing so, individualization is made absolute and actually leads to the isolation of the learners.

But social learning is not independent of *contents*, and it is reliant on *communication*. And that, according to Bakhtin (1981, 1986), means at least two voices engaging in persuasive discourses about shared contents. Teaching mathematics means to foster the internalization of multivoiced dialogical thinking. In contrast to transmission models (Shannon and Weaver 1949), Bakhtin postulates that *multivoicedness* of communication. And then, heterogeneity comes into play not as an obstacle, but as a source of *cognitive pluralism* to evoke multivoiced discourses (Wertsch 1991).

The importance of collaborative working in groups is a well known basic assumption. But more than students with essentially similar ways of thinking and contributing each a piece of the whole, here it is a matter of students with *truly different* ways of thinking. A heterogeneous classroom in a *natural* way can provide qualitatively different voices. In addition to that, Bakhtin's (1981) rent metaphor may be helpful: A voice, an utterance can just ›rent‹ meaning instead of owning a fixed meaning as it is assumed in an authoritative discourse (Wertsch 1991). Tenants are individuals (students), renter is the community they are part of (e.g. a classroom). And progressing the metaphor (cf. Hollenstein 1997): Renting implies options of influence. What is used cannot remain unchanged. In a multivoiced discourse meanings are steadily modified. In that sense providing and using meanings can create new meaning.

Arbitrariness and wasted thoughtfulness

›Open learning‹ and ›free work‹ are sometimes interpreted as leaving it to the students themselves which contents they would like to deal with. This may harbor the risk of arbitrariness, namely if learning needs are confused with students' desire to deal with whatever they fancy. Instead, the *teacher* him- or herself is responsible for …

- identifying and choosing mathematically substantial contents,
- the didactic design of so called *substantial learning environments* (sensu Wittmann 2001a) and
- keeping in mind far-reaching didactic and subject-matter goals as well as process-related competency (communication, argumentation, problem solving, representation, modelling; cf. KMK 2005).

This requires specific professional competency and cannot just be handed over to elementary students. Even an autonomously learning and high-performing child needs sound support when (s)he meets the zone of his or her proximal development (Vygotsky 1978). The teacher, on the one hand, is responsible for leading the child to its individual limits. On the other hand, (s)he must offer the child sound impulses in order to push those limits more and more forwards. Delving into mathematical structures of the learning contents in that sense does not happen automatically, rather a well-considered encouragement is necessary.

This does not deny the requirement to gradually qualify children for autonomy and self-reliance regarding their own learning process. But about it is necessary to contemplate *where, when* and with *which prerequisites* which degrees of freedom are meaningful, important, and rational for the child.

What about mathematics …?

In Germany, theoretical and conceptual discussions concerning heterogeneity and differentiation were mostly dominated by organizational and methodical questions. In addition to that, most publications originated from a pedagogical point of view. This neglects the essential importance of the subject matter, in this case mathematics, and its specifics.

Meanwhile, several proposals for learning environments in mathematics education were developed where desirable forms of differentiation can take effect— because, in a sense, it is implemented in the topic *itself* (e.g. Hengartner 2006; Hirt and Wälti 2009; Wittmann and Müller 1992, 2017).

19.2 Modified Requirements and Potential Risks

Traditional approaches as a matter of principle are limited. They come along with modified or increased requirements regarding education, school, or society:

- *Increased range of heterogeneity*: The gap between low and high achievers has perceptibly expanded over the years. In one and the same classroom there may sit students whose proficiency may spread over three school years.
- *Inclusion*: In 2008, the Convention of the United Nations on the rights of persons with disabilities came into effect. Implementing its demands for schools and mathematics education is far from being trivial. Until this very day it requires development efforts in order to provide »[e]ffective individualized support [...] in environments that maximize academic and social development, consistent with the goal of full inclusion« (UN-Convention 2014, p. 36).
- Traditional kinds of inner differentiation may be helpful and needed. But evidently they are not sufficient. It still lacks a crucial element in order to make perceptible progress in coping with heterogeneity.

The limited range and efficiency of traditional kinds of inner differentiation as well as the varied requirements mentioned above, and finally yet importantly, the demands of actual Common Core State Standards (NGA Center/CCSSO 2010; KMK 2005) involve potential risks: Teachers can feel left alone when trying to bridge the gap between fitting learning processes individually for *all* students *as a basic principle* and fulfilling the requirements of the standards. More than a few teachers in Germany complain that they do not have convincing and effective tools for that.

Consequences can be observed in classes, when the terms differentiation or individualization as popular catch cries are understood rather ambiguously. Their meaning remains mostly unexamined in terms of effective classroom practice, while being still expected to serve as a universal secret weapon for optimizing students' proficiency in math classes.

Additionally, the wide-spread availability and the familiarity with *traditional* kinds of differentiation may cause their application for anything and everything, even in situations where they come up against limiting factors. In cases of perceived helplessness teachers may settle for the mere semblance of what can be called ›modern teaching‹—potentially ending in »open communication with closed mathematics« (Steinbring 1999). In that case, differentiation and individualization become mere labels. But because that practice at least looks ›modern‹, teachers can live with it—more or less, despite an awkward aftertaste.

19.3   Natural Differentiation—A Redefined Answer

The concept of natural differentiation (cf. Wittmann 2001a; Krauthausen and Scherer 2014) intends to fill the gaps of traditional inner differentiation, in particular by ...

- orientating actions of differentiation explicitly towards the *specifics of mathematics*,
- doing justice to the different areas of *responsibilities* for teachers and for students,
- ensuring degrees of *freedom* for individual learning processes,
- laying great emphasis on guaranteeing common *social learning* in the sense mentioned above (multivoiced discourses).

There is no unambiguous and comprehensive ›definition‹ of what natural differentiation encompasses. The constitutive characteristics of natural differentiation are embedded in theoretical concepts of math education which are widespread (not only) in German schools and teacher education, namely discovery learning (Bruner 1961; Winter 2016), productive practicing (Winter 1984; Wittmann 1992), and substantial learning environments (Wittmann 1984, 2001a, 2004).

19.3.1   Constitutive Features

Due to text length constraints, they can just briefly be sketched here (cf. Fig. 19.1), followed by a short example (Figs. 19.2 and 19.3; cf. more concrete examples in Krauthausen and Scherer 2010a, b, 2014):

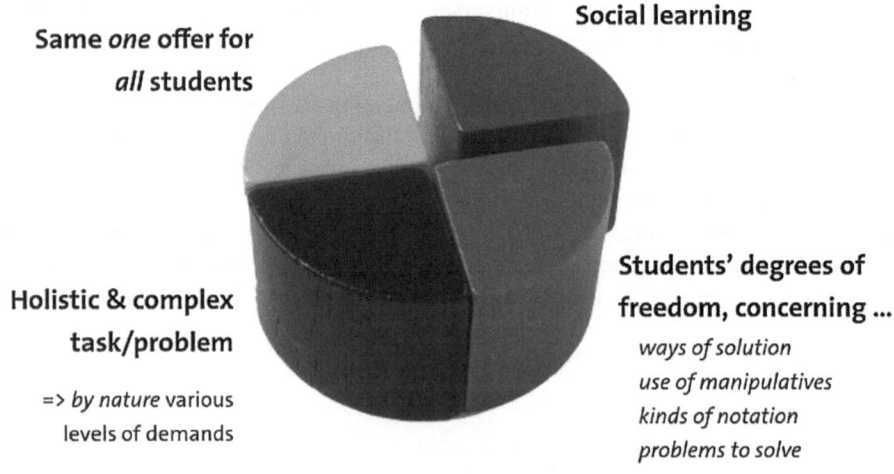

Fig. 19.1 The constitutive components of natural differentiation

Fig. 19.2 Number triangle

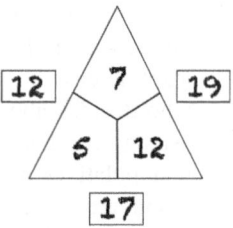

1. *Same one offer for all*: All students get the same offer like e.g. a common task or problem. In contrast to traditional differentiation, there is no need for a vast number of additional worksheets or ›special tasks‹ for different levels of capabilities.
2. *Holistic and sufficiently complex*: Because holistic content includes more meaning than isolated parts, this offer must not be split up into several isolated pieces. It must be *holistic* with regards to the content (so, not meant here is the pedagogical sense of ›head, heart and hand‹). This facilitates access for learners of *all* capabilities. The task or problem may not fall below a specific amount of complexity and mathematical substance. This may startle teachers at first or make them sceptical—especially with respect to low achievers who (in traditional differentiation) have been fostered by applying the principle of small and smallest steps and the principle of isolated difficulties. But holistic and complex problems *naturally* allow to develop a momentum of their own, giving room for the inherent dynamism of the topic.

   It is helpful to carefully distinguish between *complex* and *complicated*, which are not necessarily identical! Complexity does not by nature make things more complicated; but complexity can rather facilitate an overview, allow assorting (despite not yet mastering all the details of the content), and a personal access at a certain point for the individual student. On the contrary, if the learning environment or the task is too narrow, the accesses are too limited, possibly to just the one which successfully leads to the (one and only) solution. This kind of challenging and complex learning environments (in contrast to common isolated tasks) are not only an advantage for better learning students (cf. Scherer 1999).

It is important to emphasize in particular: The sound realization of these first two features of natural differentiation is specifically the *teacher's* responsibility. It cannot be delegated to elementary students, as some questionable teaching methods may suggest. Because what is needed here, is a professional background in several respects: Knowledge of mathematical content and mathematics pedagogical content in accordance with the design of well-considered substantial learning environments (cf. Wittmann 2004) as well as of goals concerning the content and the process for elementary math education (and beyond). A substantial learning environment, developed on the basis of a structural-genetic didactic analysis (Wittmann 2013), *then* offers a sufficiently complex frame, including meaningful, reasonable and beneficial degrees of freedom, that means for …

3. *Students' degrees of freedom*: Those first two features mentioned above, *by nature* (naturally) imply different levels of aspiration and difficulty within such a learning environment, without determining them in advance. It is not the teacher who decides about the grade of difficulty to actually work on, but the student, asking him-/herself: Which ways could I follow for a solution? Which aids or manipulatives may be helpful? How could I argue? Which kinds of documentation are at my disposal? Which levels of argumentation are plausible or adequate? (cf. Example in Sect. 19.3.2)

4. *Social learning*: The postulate of social learning from and with each other is fulfilled in a *natural* way as well, since it makes sense *by the content itself*: Because if the whole group has worked on the same problem (though on different levels), then it is obvious to share the various approaches, experiences and solutions. Everybody knows what is on the agenda and what is talked about. Everybody has the opportunity to link his/her own experiences with those of others. And the multivoiced discourse (Bakhtin 1986; Wertsch 1991) can serve as a thinktank to create meaning. Compared with that, traditional differentiation with separated worksheets requires that each student at once makes him-/herself familiar with the pretty different topics presented … if (s)he is at all motivated and capable of that in a final plenary.

»All students will be confronted with alternative ways of thinking, different techniques, variable conceptions, independent from their individual cognitive level. Rigid inner differentiation is more likely to just complicate this opportunity. […] So, the various, individually organized ways of solution also have an impact on affective, emotional areas. They leave a cognitive scope to students which can facilitate their identification with the learning demands. In this way, the direct experience of autonomy can lead to motivation and interest« (Neubrand and Neubrand 1999, p. 154 f., transl. GKr; also cf. Freudenthal 1974, p. 66 ff.).

19.3.2 An Example: Number Triangles

A well-known topic in nearly every German mathematics textbook for primary schools (grade 1–4) are number triangles. They consist of three interior fields, filled with one number each, and three exterior fields with the particular sum of the corresponding adjacent interior fields (cf. Fig. 19.2).

Number triangles could be used just as a container for any addition or subtraction tasks, simply chosen by chance. Traditional differentiation mostly offered different worksheets with easy/medium/difficult number triangles, allocated to the students by the teacher or chosen by the students themselves. After completing the fields the results just were compared, and the task was done. But this would not at all savor what is actually inherent in number triangles (cf. Wittmann 2001b).

Here are just some tasks or problems around number triangles going beyond simple addition and subtraction (cf. Krauthausen and Scherer 2010a, b, 2014):

- *Practicing the number triangles rule*: In order to make oneself acquainted with the rule, students have to fill in some number tringles. This is more interesting with an additional focus: »Make number triangles with your own numbers—three number triangles you would call easy, three you would call difficult, and three ›special‹ ones« (The latter turned out to be a very interesting question because the term special is so vague!). »And in each case write down why you think so« (to be discussed in the plenum …).
- *Discovering and describing pattern*s: Three filled out number triangles with an inherent pattern are given, two empty ones left to be filled. »Work out and continue! What do you discover? Write down your explanation« (to be discussed …).
- *Generating own patterns*: »Make number triangles with your own patterns and describe them.«
- *Generating patterns for others*: »Design number triangles with different patterns to be continued and described by your partner.« And: »Write down descriptions of different patterns for your partner to be filled into and continued in number triangles.«
- *Moving counters*: »Place a counter into one of the interior fields of a completed number triangle (= increasing this field by 1). Now let the counter move conjointly around the interior fields. What do you discover?« Or: »Two of those counters move clockwisely, one counter moves counter clockwisely. What do you discover?«
- *Number triangles with numbers from the multiplication tables*: Partly completed number triangles just contain numbers from the multiplication tables. »What do you discover? Explain …« (The inherent distributive law can be justified by patterns of counters.)
- *Even/odd exterior fields*: This example will be explained below (cf. Fig. 19.3).
- *Three exterior fields given*: »How can you find the numbers for the interior fields?« (there are different ways, with and beyond just trial and error)
- *Sums of the interior/exterior fields*: »Have a closer look at sums of interior fields and sums of exterior fields. What do you discover? Explain …« (This problem also offers some hints for the task mentioned before.)

How do number triangles with problems like these serve the constitutive features of natural differentiation? As an example, Fig. 19.3 shows the upper part of a corresponding worksheet which was tested in many classes (the lower part just offered empty number triangles for investigations).

Same one offer for all students

The whole class got this same worksheet with claims from Mandy and John. No different tasks and ›special‹ worksheets for ›special‹ needs.

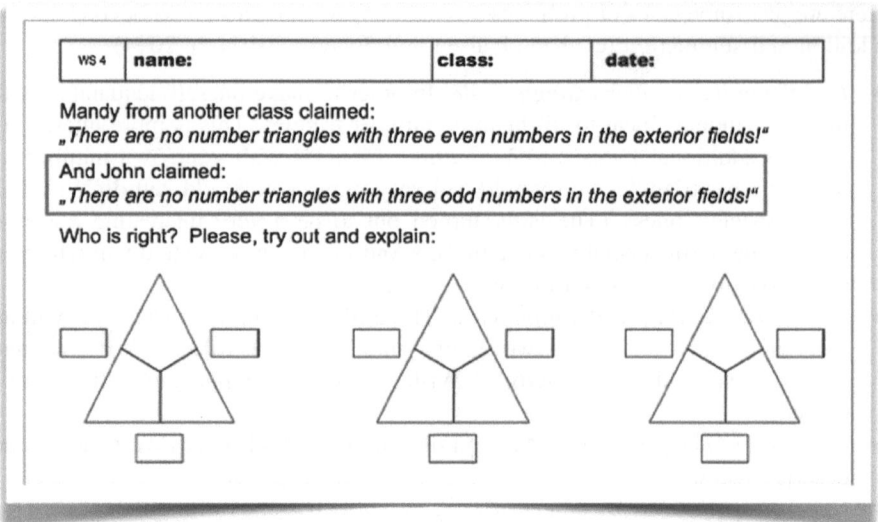

Fig. 19.3 The even/odd-problem

Holistic and sufficiently complex

Mandy's and John's claims by nature offered various levels of demands, ranging from more or less trial-and-error approaches (calculating several number triangles) to arguing with number properties in a more general way.

Students' degrees of freedom

It was possible for the students to choose own numbers for their investigations (arithmetical approach). Or they used abbreviations like *e*(ven) or *o*(dd), a pre-algebraical approach. Mandy's utterance could be disproved by just one counter-example. This may be found with or without a case discrimination (three even numbers or three odd numbers in the interior fields). John's claim could also be investigated with concrete numbers or pre-algebraically. Several bidirectional transitions between those approaches could be observed within classrooms.

Some students used counters or other manipulatives to explain their arguments, others worked just on the symbolic level. Some fourth-graders were not settled for the fact that there are no triangles with three odd exterior numbers (John's claim). So they felt free to change the triangle format to a number square with four interior fields—and for that John was proved right. Then they investigated more cases, ending up in the general utterance that John's claim is true for all cases where there is an even number of interior fields. Others argued that John is right because they used rational numbers in the interior fields.

Social learning

All those different approaches and argumentations *by nature* suggested common discussions in a plenary phase: Is it allowed to use rational numbers? Why or why not, who tells? What is the definition of even numbers—a number that can be split up evenly, into two equal halves? »But that is the case with 7 = 3.5 + 3.5! So seven is even?!« This utterance of a fourth-grader opened a substantial discussion among the students of that class ...

19.4 Demands for Teacher Education

Natural differentiation is not a magic wand, evolving its efficiency automatically or in the sense of self-evident, unstudied etc. The term ›natural‹ (in its colloquial meaning) might connote this, but in clear contrast to that the concept of natural differentiation is understood in the sense explained above. And then it can be a pretty powerful tool for math educators. But it is reliant on diverse prevailing circumstances. Just two of them will be shortly introduced here as they emerge for teacher education.

Mathematics content knowledge

Due to text length constraints it cannot explored here in detail what kind of content knowledge is needed for elementary teachers (cf. e.g. Loewenberg Ball 2003; Osana et al. 2006). But for sure, the content knowledge of elementary teachers has to be expanded, as TEDS-M has shown (cf. Blömeke and Delaney 2012). For a general characterization a postulate from Freudenthal can be helpful. For teachers' mathematics content knowledge he required the same as for the mathematics that they have to teach: It has to be diversely related (Freudenthal 1978, p. 71 f.). Mathematics content knowledge for prospective teachers should not just consist of isolated collections of facts (cf. above: ›Even numbers can be divided into two equal parts‹), but rather make the manifold interconnections and structural relations transparent and available. And it includes *attitudes* as well.

Math education has to enable students to realize the demands of the subject matter. Teachers might succeed in this even better, the more they themselves feel an aptitude to the contents of their teaching. »The art of teaching has to convey the claim of the content. It is generally agreed that we also have to know something about the learners [...]. It is a given that we have to reflect the order of presentation [...], the arrangement of our teaching. But all of this remains hollow without love for the content, without a steady effort to do justice to it [...]. The shift of pedagogical interest, away from contents and instead to psychology or methods, seems questionable to me. I ask myself, how can contents be imparted by people who know how to present them, but who themselves do not feel the demands involved. How can somebody who is not interested in a topic, make this topic interesting for somebody else?« (Schreier 1995, p. 14 f.; transl. GKr).

Methodical competency

A fundamental aspect is the ability to stimulate and to maintain a shared communicative exchange among the learners in the sense of mathematical discourses. This »is not an easy task—neither for the teacher as a moderator, nor for the students, who at first will have to learn a more self-directed communication; and for that they are entitled to professional support by their teacher« (Krauthausen and Scherer 2014, p. 82; transl. GKr). Some teachers may confuse a plenum with rigid ›chalk and talk method‹, a teacher-centered approach from the front of the classroom and generally associated with an antiquated understanding of classroom practice. But in fact, there are several good reasons for a plenary phase, e.g. content-related ones.

A common plenum by no means involves a revitalization of an outdated traditional« method. Instead it gets *a new function* with the main goal of deepening the content-related demands. It is just this *newly customized* plenary phase which first and foremost allows a deeper incursion into the mathematical core. Because students hardly will and can do that by themselves (e.g. in the range of traditional differentiation), a higher point of view is needed for that. In other words: A professional moderation by the teacher is as a matter of fact not just indispensable, but also much more possible than at other times when students work actively (alone or in small groups) on a problem. A sound moderation in that sense certainly belongs to the most demanding tasks of teaching. Because a deep understanding of content-knowledge is needed, as well as a sure instinct for the right moment and the appropriate impulse—and all that in real-time, spontaneously, and without time to contemplate. The teacher in the example mentioned above did that quite well (cf. transcripts in Krauthausen and Scherer 2010a).

Another reason would be a sociological perspective on the role of mathematical discourses about shared contents. Miller (2006, p. 200 ff.) considers social discourses a compulsory factor in modern learning (cf. Bakhtin). According to that, learning can only happen as desired (that means: effectively and sustainably), if learners enter a shared argumentation—about the *process of generating knowledge*, and not first and foremost about the completed products (cf. Krauthausen and Scherer 2014). »Only in collective discourses the learners involved will be able to develop argumentative contexts for generating new insights and new knowledge [...] and that by moments of reciprocal differences, misunderstandings and irritations« (Schülke and Söbbeke 2010, p. 21, transl. GKr; cf. also Schülke 2013).

Therefore both the following demands are essential for teaching:

(a) Content must be expressed in *language*—*as a matter of principle* there must be communication (= emphasis on *language*); and ...

(b) *Content* must be expressed in language—so, there must be content involved (= emphasis on *content*), not just talking about anything.

Connecting both meanings in a fruitful way is one of the special and most demanding tasks for teachers moderating such kind of argumentative discourses in a sound and child-oriented way—mathematically substantial, elaborately expressed and effective.

Moderation competence, too, is a complex concept and cannot be discussed here in its entirety. That's why just a few catch words will be mentioned in order to mark the direction and to hint at some facets of the bundle of skills (cf. examples in Krauthausen and Scherer 2014):

- *Imperative of influence*: It is a misunderstanding (and in a sense a failure to render assistance) that students should discover all and everything just by themselves, and teachers would have nothing to do but observe. Didactic responsibility includes exerting influence. The question indeed is what that means, and how it is done. In his famous paper ›Taboos of the Teaching Profession‹ Adorno says: »Success as a teacher is apparently due to the absence of any kind of predictive influence and relinquishing persuasion« (Adorno 1965, p. 491; transl. GKr). Adorno does not argue against influence, but against predictive influence, e.g. by too deterministic and prescribing lesson plans which then are strictly executed. Possible deviations from prescription are rebalanced by means of Bauersfeld's funnel pattern (Bauersfeld 1983; Voigt 1984). This, of course, is not what Adorno had in mind.

- *Reserve*: Teachers have to control their own ›missionary enthusiasm‹. They must not tell and explain their students everything immediately. »To reveal something to a child what it could find out by himself is not just bad teaching, it is a crime« (Freudenthal 1971, p. 424; transl. GKr).

- *Monitoring the learning process and analytical listening*: Once again Freudenthal explains the difference to just occasionally watching students' activities: »I called it intelligent observing. Not recording photographically. Before you start observing you have to know what to pay attention to. On the other hand, you must not know this too exactly, because then you will *just* see what you want« (Freudenthal 1978, p. 162; transl. GKr).

- *Authentic curiosity*: Genuine curiosity for what a child knows and how (s)he thinks as well as authentic, true and no staged enthusiasm are a fundamental tenor of teachers who want to foster and support mathematical discourses with and among their students. Their comments and answers are not just classified as wrong or right or (counter-)productive, but helpful for the teacher to understand even better what and how the students think.

- *Encouragement to express oneself and turn towards others*: All students must have secure confidence that they can express all their thoughts, assumptions, even ventured ideas, free of sanctions and without the prospect of hasty evaluations.

- *Manifold repertoire of questions*: Especially valuable are ›higher order‹ questions and impulses which initiate new/distinct/variable thinking as well as autonomous/reasoning/inferential thinking.

- *Probing into the subject matter again*: This is to encourage students to dig deeper into the subject. To repeat questions does not at all mean that an answer must have been wrong.

- *Having a break*: Productive discourses sometimes need a break. Not in order to interrupt the thinking process, but to pause for thought. Short moments of silence—caused e.g. by speechless astonishment, by surprise, by hesitation, by skepticism—should be experienced as productive, not as embarrassing blankness which ought to be filled as soon as possible with a strange comment or a displacement activity.

19.5 Conclusion

Natural differentiation as a specific kind of inner differentiation offers opportunities to design the learning of heterogeneous students in a way that is more productive and more sustainable for all. It is natural, because heterogeneous groups of learners *by nature* evoke and foster multivoiced discourses expressing truly different ways of thinking on truly different levels. And it is natural, because complex and holistic problems, like substantial learning environments, *by nature* allow a momentum of their own, giving room for the inherent dynamism of the *content*. And it is natural because it is the *learner* who can make use of his/her degrees of freedom in several respects in a designated frame of a mathematical substantial learning environment.

The special prospects in particular lie in the following attributes of the concept:

- Emphasizing the *specifics of mathematics* and consciously valuating the demands of the content as well as the *social learning* postulate, especially via moderated discourses about shared experiences with working on a common learning environment.
- Emphasizing an *integrative* access to content-related and process-related mathematical competency (KMK 2005).
- No claim as an *all-in-one* tool for the whole range of mathematical learning and teaching (though rather likely for its major part). Practicing basic facts or introducing a specific procedure may require other methods.
- Availability of numerous *appropriate learning environments* (sensu Wittmann 2004) for substantial hands-on activities that meet the demands of proper natural differentiation.

References

Adorno, T. W. (1965). Tabus über den Lehrberuf. *Neue Sammlung, 5*, 487–498.
Bakhtin, M. M. (1981). *The dialogic imagination. Four essays*. Austin: University of Texas Press.
Bakhtin, M. M. (1986). *Speech genres and other late essays*. Minneapolis: University of Minnesota Press.

Bauersfeld, H. (1983). Kommunikationsverläufe im Mathematikunterricht. Diskutiert am Beispiel des ›Trichtermusters‹. In K. Ehlich & J. Rehbein (Eds.), *Kommunikation in Schule und Hochschule* (pp. 21–28). Tübingen: Narr.

Blömeke, S., & Delaney, S. (2012). Assessment of teacher knowledge across countries: A review of the state of research. *ZDM—The International Journal on Mathematics Education, 44*, 223–247.

Bönsch, M. (1976). *Differenzierung des Unterrichts: Methodische Aspekte* (3rd ed.). München: Ehrenwirth.

Bönsch, M. (2004). *Intelligente Unterrichtsstrukturen. Eine Einführung in die Differenzierung* (3rd ed.). Baltmannsweiler: Schneider.

Bruner, J. S. (1961). The act of discovery. *Harvard Educational Review, 31*(1), 21–32.

Freudenthal, H. (1971). Geometry between the devil and the deep sea. *Educational Studies in Mathematics, 3*(3/4), 413–435.

Freudenthal, H. (1974). Die Stufen im Lernprozeß und die heterogene Lerngruppe im Hinblick auf die Middenschool. *Neue Sammlung, 14*, 161–172.

Freudenthal, H. (1978). *Vorrede zu einer Wissenschaft vom Mathematikunterricht.* München: Oldenbourg.

Geppert, K., & Preuß, E. (1981). *Differenzierender Unterricht – konkret. Analyse, Planung und Gestaltung. Ein Modell zur Reform des Primarbereichs* (2nd ed.). Bad Heilbrunn: Klinkhardt.

Hengartner, E. (2006). Lernumgebungen für das ganze Begabungsspektrum: Alle Kinder sind gefordert. In E. Hengartner, U. Hirt, & B. Wälti (Eds.), *Lernumgebungen für Rechenschwache bis Hochbegabte. Natürliche Differenzierung im Mathematikunterricht* (pp. 9–15). Zug: Klett & Balmer.

Hirt, U., & Wälti, B. (2009). *Lernumgebungen für den Mathematikunterricht in der Grundschule: Begriffsklärung und Positionierung.* Seelze-Velber: Klett & Kallmeyer.

Hollenstein, A. (1997). Kognitive Aspekte sozialen Lernens. In K. P. Müller (Ed.), *Beiträge zum Mathematikunterricht* (pp. 243–246). Hildesheim: Franzbecker.

Klafki, W., & Stöcker, H. (1976). Innere Differenzierung des Unterrichts. *Zeitschrift für Pädagogik, 22*(4), 497–523.

KMK, Kultusministerkonferenz. (Ed.). (2005). *Bildungsstandards im Fach Mathematik für den Primarbereich. Beschluss vom 15.10.2004.* Neuwied: Luchterhand Wolters Kluwer.

Krauthausen, G., & Scherer, P. (Eds.). (2010a). *Ideas for natural differentiation in primary mathematics classrooms. Vol. 1: The substantial environment number triangles.* Rzeszòw: Wydawnictwo Uniwersytetu Rzeszowskiego.

Krauthausen, G., & Scherer, P. (2010b). Natural Differentiation in Mathematics (NaDiMa). Theoretical backgrounds and selected arithmetical learning environments. In B. Maj, E. Swoboda, & K. Tatsis (Eds.), *Motivation via natural differentiation in mathematics* (pp. 11–37). Rzeszów: Wydawnictwo Universytetu Rzeszowskiego.

Krauthausen, G., & Scherer, P. (2014). *Natürliche Differenzierung im Mathematikunterricht – Konzepte und Praxisbeispiele aus der Grundschule.* Seelze: Kallmeyer.

Loewenberg Ball, D. (2003). *What mathematical knowledge is needed for teaching mathematics?* Remarks prepared for the Secretary's Summit on Mathematics, U.S. Department of Education, February 6, 2003, Washington, D.C.

Miller, M. (2006). *Dissens. Zur Theorie diskursiven und systemischen Lernens.* Bielefeld: transcript.

Neubrand, J., & Neubrand, M. (1999). Effekte multipler Lösungsmöglichkeiten: Beispiele aus einer japanischen Mathematikstunde. In C. Selter & G. Walther (Eds.), *Mathematikdidaktik als design science* (pp. 148–158). Leipzig: Klett Grundschulverlag.

NGA Center/CCSSO—National Governors Association Center for Best Practices, & Council of Chief State School Officers. (2010). *Mathematics standards.* http://www.corestandards.org. Accessed March 11, 2016.

Osana, H. P., Lacroix, G. L., Tucker, B. J., & Desrosiers, C. (2006). The role of content knowledge and problem features on preservice teachers' appraisal of elementary mathematics tasks. *Journal of Mathematics Teacher Education, 9*, 347–380.

Paradies, L., & Linser, H. J. (2005). *Differenzieren im Unterricht* (2nd ed.). Berlin: Cornelsen Scriptor.

Scherer, P. (1999). Mathematiklernen bei Kindern mit Lernschwächen. Perspektiven für die Lehrerbildung. In C. Selter & G. Walther (Eds.), *Mathematikdidaktik als design science* (pp. 170–179). Leipzig: Klett Grundschulverlag.

Schreier, H. (1995). Unterricht ohne Liebe zur Sache ist leer. Eine Erinnerung. *Grundschule, 6,* 14–15.

Schülke, C. (2013). *Mathematische Reflexion in der Interaktion von Grundschulkindern. Theoretische Grundlegung und empirisch-interpretative Evaluation.* Münster: Waxmann.

Schülke, C., & Söbbeke, E. (2010). Die Entwicklung mathematischer Begriffe im Unterricht. In C. Böttinger, K. Bräuning, M. Nührenbörger, R. Schwarzkopf, & E. Söbbeke (Eds.), *Mathematik im Denken der Kinder. Anregungen zur mathematikdidaktischen Reflexion* (pp. 18–28). Seelze-Velber: Kallmeyer.

Selter, C., & Spiegel, H. (1997). *Wie Kinder rechnen.* Leipzig: Klett Grundschulverlag.

Shannon, C. E., & Weaver, H. (1949). *The mathematical theory of communication.* Urbana: University of Illinois Press.

Steinbring, H. (1999). Offene Kommunikation mit geschlossener Mathematik? *Grundschule, 3,* 8–13.

UN-Konvention – Beauftragte der Bundesregierung für die Belange behinderter Menschen (Ed.). (2014). *Convention of the United Nations on the rights of persons with disabilities.* Berlin.

Voigt, J. (1984). Der kurztaktige, fragend-entwickelnde Mathematikunterricht. Szenen und Analysen. *Mathematica didactica, 7,* 161–186.

Vygotsky, L. S. (1978). *Mind in society: Development of higher psychological processes.* Cambridge, MA: Harvard University Press.

Wertsch, J. V. (1991). *Voices of the mind. A sociocultural approach to mediated action.* London: Harvester Wheatsheaf.

Winkeler, R. (1976). *Differenzierung: Funktionen, Formen und Probleme* (4th ed.). Ravensburg: Otto Maier.

Winter, H. (1984). Begriff und Bedeutung des Übens im Mathematikunterricht. *mathematik lehren, 1*(2), 4–16.

Winter, H. (2016). *Entdeckendes Lernen im Mathematikunterricht. Einblicke in die Ideengeschichte und ihre Bedeutung für die Pädagogik* (2nd ed.). Wiesbaden: Springer.

Wittmann, E. C. (1984). Teaching units as the integrating core of mathematics education. *Educational Studies in Mathematics, 15*(1), 25–36.

Wittmann, E. C. (1992). Üben im Lernprozeß. In E. C. Wittmann & G. N. Müller (Ed.), *Handbuch produktiver Rechenübungen: Vom halbschriftlichen zum schriftlichen Rechnen* (pp. 175–186). Stuttgart: Klett.

Wittmann, E. C. (2001a). Developing mathematics education in a systemic process. *Educational Studies in Mathematics, 48*(1), 1–20.

Wittmann, E. C. (2001b). *Drawing on the richness of elementary mathematics in designing substantial learning environments.* http://www.mathematik.uni-dortmund.de/didaktik/mathe2000/pdf/rf4-2wittmann.pdf. Accessed May 01, 2017.

Wittmann, E. C. (2004). Developing mathematics education in a systemic process. In H. Fujita, Y. Hashimoto, B. R. Hodgson, P. Yee Lee, S. Lerman, & T. Sawada (Eds.), *Proceedings of the Ninth International Congress on Mathematical Education* (pp. 73–90). Norwell, MA: Kluwer Academic Publishers.

Wittmann, E. C. (2013). Strukturgenetische didaktische Analysen – die empirische Forschung erster Art. In G. Greefrath, F. Käpnick, & M. Stein (Eds.), *Beiträge zum Mathematikunterricht* (pp. 1094–1097). Hildesheim: Franzbecker.

Wittmann, E. C., & Müller, G. N. (1992). *Handbuch produktiver Rechenübungen. Band 2: Vom halbschriftlichen zum schriftlichen Rechnen*. Stuttgart: Klett.

Wittmann, E. C., & Müller, G. N. (2012). *Grundkonzeption des Zahlenbuchs*. Beilage auf CD-ROM zum Materialband Zahlenbuch 2 (pp. 158–173). Stuttgart: Klett.

Wittmann, E. C., & Müller, G. N. (2017). *Handbuch produktiver Rechenübungen. Band 1. Vom Einspluseins zum Einmaleins* (2nd ed.). Stuttgart: Klett.

Chapter 20
Changes in Attitudes Towards Textbook Task Modification Using Confrontation of Complexity in a Collaborative Inquiry: Two Case Studies

Kyeong-Hwa Lee

Abstract This study examined how two middle school mathematics teachers changed from being reluctant to modify tasks in mathematics textbooks to having positive attitudes about textbook task modification. In order to successfully coordinate a curriculum revision with the textbooks they use, mathematics teachers need to be able to use their in-depth understanding of the intentions of both the revision and textbooks to modify and implement tasks appropriately. The two middle school teachers' cases in this study showed that it is possible to change teachers' negative attitudes about modifying tasks in mathematics textbooks if they explicitly understand the complexity in mathematics teaching and go through a sequence of activities that help them understand the revised curriculum in detail, interpret and modify textbook tasks, and implement the modified tasks and reflect on their implementation.

Keywords Mathematics textbook · Textbook task modification
Complexity map · Collaborative inquiry · Professional development

20.1 Introduction

Teacher researcher- or teacher-led inquiry communities have been increasingly viewed as promising for professional growth and development of theory and practice in mathematics education (Lin and Cooney 2001; Dowling 2013; Jaworski 2003; Slavit and Nelson 2010; Robutti et al. 2016; Goodchild 2008; Goodchild et al. 2013). Researchers have reported that teachers both deepen content knowledge and pedagogical knowledge by learning ways of teaching and develop their understanding of how to facilitate students' conceptual understanding through collaborative work with their colleagues and researchers (e.g., Sullivan et al. 2012;

K.-H. Lee (✉)
Seoul National University, Seoul, South Korea
e-mail: khmath@snu.ac.kr

© The Author(s) 2018   343
G. Kaiser et al. (eds.), *Invited Lectures from the 13th International Congress on Mathematical Education*, ICME-13 Monographs,
https://doi.org/10.1007/978-3-319-72170-5_20

Cooper et al. 2006). However, with the popularity of community approaches, pitfalls have also arisen such as limited local school resources, shortages of qualified teachers in distressed areas, and stress related to performance on high stakes testing (Ledoux and McHenry 2008). In addition, a number of tensions produced from the complex nature of relationships among members in communities have suggested that one cannot be purely optimistic about such collaborative work (Martin et al. 2011). Therefore, there is a need to unpack the tensions and complexities involved in community approaches along with the learning opportunities among teachers and researchers by recognizing the genuine perspectives and needs of teachers (Schwarz 2001).

Textbook modification is a common procedure used in collaborative approaches by teachers and researchers for professional development (Bao and Stephens 2013; Boston and Smith 2011; Zaslavsky 1995). Through textbook modification, teachers can learn how to redesign textbook tasks and how to teach mathematics differently. In order to successfully coordinate a curriculum revision with the textbooks they use, mathematics teachers need to be able to use their in-depth understanding of the intentions of both the revision and textbooks in order to modify and implement tasks appropriately. However, studies have shown that a number of mathematics teachers merely follow textbooks as they are written (Manouchehri and Goodman 1998; Choe and Hwang 2004, 2005). A deeper understanding is needed of the reason that mathematics teachers place textbooks in a rather fixed position of high authority. Professional development programs giving opportunities for teachers to reflect on this passiveness towards textbook modification may help teachers to consider textbooks to be a type of curriculum material that can be evaluated, interpreted, and redesigned prior to and during lessons (Drake and Sherin 2006; Lloyd 1999; Remillard 2005). In this study, teachers were invited to co-learning activities with the researcher that focused on textbook task modification (TTM) for professional development. Although the community was initiated by the researcher, the teachers participated voluntarily in the whole collaboration and all members had equal status (Hospesovà et al. 2006). To recognize the teachers' genuine perspectives on and needs for TTM, teacher narrative analysis was used. This article will describe and discuss why and how two middle school mathematics teachers who participated and had a voice in all phases of the research process (Sullivan et al. 2012; Goodchild 2008; Jaworski 2003) changed their initial attitudes about modifying tasks in mathematics textbooks from negative to positive.

20.2   Learning by Collaborative Work on TTM

A professional learning community (PLC) of teachers and educators facilitates teacher and researcher learning through collaboration, conversation, and inquiry (Jaworski 2003; Goodchild et al. 2013). The ways in which the participants in a PLC interact is closely linked to the roles they play within the community (Robutti et al. 2016). The three elements of practice—*engagement, imagination,* and *alignment of participant*—can be employed as essential norms in a PLC (Jaworski 2003) for lesson study. Firstly, members of a community can be engaged in the activities of analyzing and modifying textbook tasks and applying and reflecting on the modified tasks (Bao and Stephens 2013; Boston and Smith 2011; Coe et al. 2010). The purpose of analyzing textbook tasks is to gain insights into their intended mathematical and pedagogical meaning through revealing specific learning goals and making distinctions between task features, such as context-based and open-ended tasks (Sullivan et al. 2012). In addition, predicting student misconceptions and errors can be done at this stage. Finally, possible dimensions and possible ranges of variation (Watson and Mason 2006) of concepts, procedures, or representations embedded implicitly or explicitly in tasks can be described using task analysis. Based on these detailed analyses, we move to the modification stage, in which we make judgments on modifications in detail. We prepare supplementary tasks to help those students who cannot begin the given task or those who complete tasks in a very short time. Applying this stage includes not only implementation of modified tasks but also improvisational adaptation of tasks based on in-the-moment decisions in reaction to students' responses.

Secondly, imagination, which requires participants in the community to disengage by moving back and looking at the engagement through the eyes of an outsider (Wenger 1998, p. 185), can be considered in collaborative work between teachers and researchers. In order to take a step back and look at the big picture of teachers' engagement related to textbook task use, understanding teachers' perspectives by asking the following questions is useful. Why do you think teachers should analyze textbook tasks? What does it mean to analyze? What are the criteria and methods for analyzing? What was the most important thing you learned from the experience of analysis? Have you had any experience in modifying textbook tasks? When, why, and how did you do any modification? How did you understand the results of implementing modified tasks? What were your main concerns both when you were modifying textbook tasks and after you implemented the modified tasks? What was the most important thing you learned from the experience of modifying and implementing the tasks?

When answering the above questions, teachers can regard themselves as subjects who interpret and modify textbook tasks (Remillard 2005) and not as subjects who use textbook tasks with few or no modifications. The above questions are to recall and describe teachers' perceptions of the complexity of textbook use in their everyday classrooms. Teachers' answers to these questions may be represented as a map showing key issues and concerns; this will be called a *complexity map* in this study. A complexity map is defined as a diagram that is made using teachers' answers to the above questions and teachers' perceptions of complexity in textbook use. A complexity map is an ongoing process that makes it possible to examine teachers' concerns and understand the indirect effects that influence the direction and degree of task modification and implementation. In other words, a complexity map can be a communication tool that teachers initiate or lead the creation of that represents various components and the relationships among those components that teachers have perceived while they were analyzing, modifying, and implementing textbook tasks. A complexity map by an individual teacher can reveal a part of a *subjective scheme*, a kind of mental framework by which objectively given structures of information such as tasks and narratives in textbooks are understood and interpreted (Otte 1986). A complexity map may vary with respect to educational environments, teaching cultures, and value systems. Mathematical emphasis, referring to the mathematics knowledge and practices that are valued (Remillard et al. 2014, p. 739), may differ among teachers who have different perceptions of the complexities in textbook use (Ben-Peretz 1990; Heaton 2000; Sherin and Drake 2009). Furthermore, teachers' different complexity maps may have different influences on students' learning opportunities (Schmidt 2007; Valverde et al. 2002; Grouws and Smith 2000; Stein et al. 2007).

By constructing a complexity map as a communication tool in collaborative works, *critical alignment* (Goodchild et al. 2013) can be pursued. Teachers and researchers can create a *special synergy* by means of critical alignment mediated by complexity maps as suggested in Goodchild et al. (2013). One possible aim for both teachers and researchers in a PLC for textbook modification can be engagement of students in mathematics learning by modifying textbook tasks. The other possibility is doing research and drawing implications on textbook task use in classrooms. Teacher-researcher collaboration to achieve these aims can be facilitated by critical alignment between both participants. Complexity maps can be viewed as a set of problems to be solved or set of constraints to be overcome by collaborative treatments. We can see a complexity map as a window to look at practices from teachers' perspectives. Using a complexity map, we can invite teachers to reflect on their current practices and to highlight partial or general complexity when considering TTMs. If critical alignment between practice- and research-based awareness in teaching with textbook tasks can be developed and rooted in the learning community, then knowledge can grow in practice.

20.3 Mathematics Curriculum and Mathematics Textbooks in Korea

The place that curriculum and textbooks take in mathematics education can vary from country to country. In the Korean context, the curriculum and textbooks are regarded as having high authority by teachers, parents, and students. The mathematics curriculum in Korea is developed by a committee that is sanctioned at the national level. Mathematics curriculum reform has been conducted based on discussions about the objectives, content, and methods of mathematics education. When this study was conducted, we used the curriculum that was revised in 2009 (Ministry of Education, Science, and Technology [MEST] 2009). In the 2009 Curriculum, mathematical problem solving, communication, and reasoning were considered to be crucial process standards for nurturing mathematical creativity, and it was strongly recommended that these three aspects be realized in teaching and learning (MEST 2009). In addition, the 2009 Curriculum suggested finding and exploring real-life contexts that are familiar to students and accepting student intuition and an informal approach. The 2009 Curriculum's structure has five sections: characteristics, objectives, content, teaching and learning methods, and evaluation. In the characteristics section, school mathematics is described as a subject that deals with mathematical concepts, principles, and rules to be explored in various contexts; develops logical thinking; cultivates the ability to observe and interpret various phenomena; and develops an understanding of how to use various methods to solve problems. MEST (2009) puts particular emphasis on the development of mathematical literacy:

> The in-depth understanding and application of mathematical concepts, including problem-solving ability, are essential in learning diverse content successfully and are also necessary to increase one's skills and ability to solve problems as a democratic citizen. Moreover, mathematical knowledge and thinking methods act as an intellectual driving force in the development of human civilization and are necessary in the rapidly changing information-based society. (p. 5)

The objectives for the three school levels, primary, middle, and high school, are set by integrating perspectives from relevant research studies as well as the aims and the requests of the *noosphere* (Chevallard and Bosch 2014). For example, the objective for middle school mathematics is

> to obtain the basic knowledge and understand the functions of mathematics, to cultivate the ability to think mathematically and communicate in order to create practical solutions to social and natural phenomena and problems, and to cultivate a positive attitude toward mathematics. (MEST 2009, pp. 8–9)

Even terms and notations that should be included in textbooks and lessons are presented in the curriculum. In addition to teaching and learning methods, recommendations for didactic transposition of content are presented. For example, in the functions section for seventh grade, the following recommendations are included: (a) Use a daily-life context where one quantity changes as another quantity changes and (b) teach the concept of functions at an intuitive level. In the evaluation section, a great deal of emphasis is put on conducting assessments in order to provide useful cognitive and definitive suggestions that can help students' learning and well-rounded development and improve teaching practices. Considering the level of students' mathematical knowledge is also explicitly mentioned, and abiding by the content presented in the curriculum documents is suggested. A variety of types of evaluation, such as formative and summative evaluations, is suggested as well. This systematically organized intended curriculum influences Korean mathematics textbook development and mathematics teaching in classrooms.

Textbook writers make an effort to realize the reform ideas prescribed in the curriculum by following its terms and the notations, instructions for teaching specific content areas, and aspects to be emphasized in evaluation. In general, Korean textbooks are written based on thorough interpretation of the content, the teaching and learning methods, and the evaluation policies in the intended curriculum. Therefore, even though there are various kinds of secondary school textbooks in Korea, they have a lot in common in many aspects. The coverage and the depth of content areas, teaching and learning strategies, and assessment systems in textbooks are almost the same. Therefore, studies have shown that it is very natural for teachers follow the national curriculum and their textbooks in planning lessons (Choe and Hwang 2004, 2005). Teachers' tendency to follow the intended curriculum explains the difficulty of opening a discussion on the necessity of TTM in professional development programs about TTM. In this study, making each mathematics teacher's complexity perception explicit is used as a strategy to open a discussion on the necessity of TTM.

20.4 Research Context

As a large part of research project, 88 middle school mathematics teachers first participated in a six-hour professional development program on textbook-task use. Seventeen of these teachers then voluntarily participated in a second professional development program on TTM that lasted eight months. At the beginning of this advanced professional development program, a survey was administered to examine the teachers' experiences with and attitudes toward TTM where they were asked to choose between three items: (1) I have experience with modifying text-book tasks, (2) I am not willing to modify textbook tasks, and (3) I want to learn more about modifying textbook tasks if there is any follow-up program. Based on the survey results, two teachers who had experienced TTM but had negative atti-tudes toward it were selected for the purpose of this study in order to examine why these teachers had negative attitudes toward TTM, the reasons behind their negative attitudes, and why and how they changed their perspective on TTM over time. The two teachers, Euna and Miyeong, were both female. Euna was in her early 30s and had six years of teaching experience but did not have any textbook-writing expe-rience. She taught a seventh grade class consisting of approximately 34 students at a large, low-achieving public middle school for boys located in a large metropolitan city. The other teacher, Miyeong, was in her mid-40s, had 17 years of teaching experience, and had textbook-writing experience. Miyeong taught an eighth grade class of approximately 33 students at a large, low-achieving public school for boys and girls located in a large metropolitan city. Miyeong was widely recognized as an expert teacher and was actively involved in enhancing students' interest in mathematics.

The data were collected in three ways: teachers' narratives, which were used to get insights about their perceptions and beliefs (Connelly and Clandinin 1990; Schwarz 2001); discussions; and classroom observations of their mathematics classrooms. First, teachers were asked to write free narratives on the following topics: previous TTM experiences; a brief explanation about why they were not willing to do TTM; key roles, affordances, and constraints of textbook tasks; complexities in task use, teaching, and learning; tensions and dilemmas experienced in teaching and participating in PD programs; and what they learned from TTM and its implementation. Teachers' narratives were used in discussion meetings as prompts. Discussion meetings were conducted 12 times from March to October. After selecting key issues to be discussed in these meetings, the meetings were facilitated to discuss the issues systematically. Another source of data was classroom observation. Euna's four lessons, where the topics were the concept of a function, graphs of functions, and applications of functions, were observed. Miyeong's five lessons, where the topics were events, relative frequency, the concept of probability, and probability calculation, were also observed.

20.5   Findings

The relationships between teachers and the curriculum seem to be relatively simple in the Korean mathematics education context, as mentioned in the earlier section. In particular, the two teachers, Euna and Miyeong, clearly presented their negative attitude to TTM in the beginning. The reasons for their prior thoughts on TTM will be reported first. Then, how complexity maps showing the teachers' recognition of their practices were drawn and utilized over PLC meetings will be described. The significant tensions and challenges faced by the teachers and how the teachers changed their attitudes to TTM can provide interesting insights into the ways in which the teachers incorporated the reform principles prescribed in the revised curriculum into their teaching practices.

20.5.1   Rationales for Negative Attitudes Toward TTM

Both teachers in this study had negative attitudes toward TTM at the beginning of the PLC activity. Their participation in the prior workshop for six hours and their volunteering to participate in the PLC meetings for eight months showed that they did not just blindly follow textbooks and that they were at least somewhat interested in TTM. Euna viewed textbooks as effective tools for teaching mathematics, a view similar to those of the teachers in Choe and Hwang's studies (2004, 2005). She used textbooks without modifications because she felt that textbook tasks (a) are effective in deepening students' mathematical understanding, (b) are systematically sequenced and have appropriate scope, and (c) reflect new visions and recent research trends in the mathematics curriculum and teaching and learning of mathematics. Her positive evaluation of textbooks can be attributed to her trust in the process of textbook development and its authors in Korea. In describing her trust in textbooks, she said:

> Textbooks are usually written by a team of mathematicians, mathematics educators, and experienced mathematics teachers. They must incorporate new visions, appropriate content,

and new teaching and thinking methods into the textbook they are writing. Why not follow the textbook? It is an *optimal solution* [emphasis added] for teaching and learning mathematics in our environment.

The idea of optimal solution that Euna described is closely related to the way of organizing content in Korean mathematics textbooks. Korean mathematics textbooks are structured in a deductive way in the sense that their basic structure starts with the definition and ends with its application via some explanations, with worked out examples and drills in the meantime. Before the body of the content begins, an interesting opening is provided that gives a context in which the learning content can be related to what students are familiar with.

In summary, Euna's negative attitudes toward TTM can be attributed to her satisfaction with current textbooks and her belief that it is more effective to follow textbooks. The dilemma Euna faced is associated with her perception of unchanging practices with the revised curriculum. For her, an optimal solution has nothing to do with the newly emphasized competences such as communication, reasoning, and problem solving. She was invited to reconsider what would be optimal for her interpretations and implementations of textbook tasks in order to find another optimal solution.

Similar to Euna, Miyeong also trusted textbooks. However, she provided different reasons for her negative attitudes toward TTM. Having 17 years of teaching experience, she was widely recognized as an expert teacher. Being sensitive to curriculum revisions, she fully understood what had been revised in the curriculum. She had rich experience with modifying textbook tasks. Despite all of this, she was reluctant to modify textbook tasks because she thought that it was much easier to design new tasks rather than modify the existing tasks in textbooks:

> There is a saying that "revolution is easier than reform." Likewise, I prefer to design new tasks instead of TTM. I am free to use various contexts and knowledge from a variety of fields such as film, travel, finance, and history when designing new tasks. However, this only applies to extra classes, such as work done after finishing units or after-school programs, but not to regular classes. TTM is not easy to implement because we need to cover the limited range of concepts and procedures that are contained in the curriculum. If the advantages of TTM for teaching specific content in the curriculum are clear, it would be okay.

Even if textbook tasks were to be modified, it would be important for Miyeong to maintain the original learning goals and content as intended in the textbooks; however, it was difficult to create learning opportunities that were better than the textbook tasks. The reason for this limitation was that she was concerned with being able to attain the original learning goal and was sensitive to potential changes to what can be learned using modified tasks. The rationale behind her negative attitudes toward TTM included: (a) TTM does not guarantee good learning and teaching and (b) TTM is a challenging and risky job. While explaining that TTM is a challenging and risky job, Miyeong mentioned cases where the main focus of instruction is on non-mathematical issues, for example, cases that mainly focus on students' interests.

It is noteworthy that both teachers had negative attitudes toward TTM because they trusted textbooks more than their unsuccessful TTM. The rationales behind the teachers' negative attitudes toward TTM show that one should not interpret merely following textbooks as a signal of teachers' resistance to being independent of textbook use. Instead, teachers' resistance to TTM can be viewed as their faithful implementation of the curriculum. Having identified the fact that the two teachers greatly value faithful implementation of curriculum, we naturally moved to discuss the complexities faced by the teachers and were able to draw complexity maps.

20.5.2   Complexities Perceived by the Two Teachers

Faithful implementation of the curriculum has proved challenging, since the mathematical emphases and pedagogical approaches included in the curriculum materials are difficult to carry out (Lloyd 1999; Remillard 2005). Moreover, Korean mathematics teachers are under intense pressure to raise or maintain students' test scores while at the same time realizing the reform ideas described in the curriculum documents (Lee 2010; Park 2004). Although reasonable rationales for their negative attitude to TTM were discussed, the two teachers recognized the necessity of TTM, which led them to volunteer to participate in the PLC on TTM. The initial complexity maps were constructed based on the teachers' perceptions about complexity in teaching. Among various fragmentary components, those that were perceived first became the discussion topics and were reflected in the initial complexity maps. For example, Euna expressed her uncomfortable feeling about the fact that some students did their homework from cram schools during mathematics classes or independent study time. It was also difficult for her to see that her students' parents mainly focused on their children's test scores. These issues came out in Euna's narratives and during discussion meetings and were reflected in her initial complexity map. In her narratives and PLC meetings, Miyeong focused more on potential mathematical meanings and the structures of particular concepts that should be highlighted in her classes. The two teachers' perceptions about complexity in their teaching practices were different, which might explain their different understandings of the challenges and dilemmas of TTM (Leder et al. 2006; Goldin et al. 2011; Gates 2006).

Table 20.1 shows the categories of the types of complexity that Euna and Miyeong initially perceived: student-related complexity (SRC), mathematics-related complexity (MRC), and external complexity (EC). Several differences in their experiences, including teaching experience and experience with writing textbooks, could be connected to the difference in the initial complexity that each teacher perceived. Miyeong was able to recognize the complexities in her implementation well and competently explain each complexity. On the other hand, Euna's order of priorities was EC, SRC, and MRC, and she rarely recognized the relationship among the complexities. Unlike Miyeong, Euna expressed difficulties with teaching mathematics and spent a considerable amount of time expressing such difficulties as complexities and making the connection to task modification. Before Euna discussed issues with Miyeong, Euna was not sure how to reflect some of the complexities in TTM. For instance, regarding the issue of cram schools, Euna was very concerned about the interruptions made by students who attended cram schools but did not come up with specific solutions, whereas Miyeong identified the issue of cram schools as "bad habits in learning mathematics." This indicates that cooperation with colleagues, particularly with more experienced teachers, is important in developing mathematics teachers' expertise, as previous research has evidenced (Garet et al. 2001; Desimone et al. 2002).

Euna frequently mentioned cram schools, which she identified as one of the external complexities. Considering that approximately 30% of her students attended cram schools and learned content in advance, Euna had difficulty in designing tasks and planning lessons for those students.

> One thing I never feel easy about is the interruption made by the students who go to cram schools. They already know the content before I teach it. If these students dominate the dialogue among students by employing the mathematics concepts that are considered the learning goal, then other students lose learning opportunities. Inquiry-based learning is not easily pursued in this situation.

Teaching was also very complex for Euna because parents were very interested in and passionate about their children's education, and they tended to evaluate and supervise her teaching. Among SRCs, Euna mentioned the range in students' achievement levels most frequently. As curriculum revision had been conducted so frequently, Euna felt the pressure to change things in her practice. This indicates that she was trying to understand the revised curriculum and reflect it in her practice. Another MRC is evaluation. Assessment-focused mathematics education has been characterized as one of the main features of mathematics education in East

Table 20.1 Initial complexities that Euna and Miyeong individually perceived

Type of complexity	Complexities Euna initially perceived	Complexities Miyeong initially perceived
Student related (SRC)	• A wide range of achievement • Negative attitude • Passive attitude	• A wide range of achievement • Negative attitude • Changing students • Bad habits in learning mathematics • Learning anxiety pressure
Mathematics related (MRC)	• Frequently changed curriculum • Assessment	• Unfamiliar terminology • Hierarchy of mathematics • Abstract mathematics • Not differentiated curriculum • Assessment
External (EC)	• Cram schools • High level of parents' enthusiasm • Large class size	• Cram schools

Asia (e.g., Park 2004; Leung 2001), and Euna was also very concerned about preparing students for assessments while teaching. She commented that she taught in ways that "emphasized the content and form of knowledge reflected in assessments but did not focus as much on knowledge that would not be assessed." This may have influenced her task designing and lesson planning, which made her teaching complex. Miyeong had more factors related to MRC.

For Miyeong, teaching was complex because of issues involving the nature of mathematics, such as "the hierarchy of mathematics" and "abstract mathematics." Other factors such as "unfamiliar terminology," "not differentiated curriculum," and "assessment" were complexities related to the limitations of school mathematics. In terms of SRC, beyond "a wide range of achievement" and "negative attitudes (toward mathematics and mathematics learning)" that Euna perceived, Miyeong thought that the complexity was attributed to "different students," "bad habits in learning mathematics," and "learning anxiety or pressure." Among these, Miyeong provided the following explanation for "different students":

> We never have the same students; on the contrary, teachers need to be prepared to satisfy new students every year. The students I have this year are quite different from those I taught last year. Hence, I had to learn what and how they learned and can learn in my classes day by day. I had to develop how to teach mathematics in many ways.

The changes in the cognitive aspects of students mentioned above were not the only difficulties Miyeong experienced. She said in the later discussion that she had difficulties with task design and lesson planning in terms of students' different dispositions and attitudes toward mathematics. For instance, she said that "students in the past waited quietly, but students nowadays do not wait at all, so it is difficult to include problems that require in-depth investigation from the students." When Miyeong explained about "bad habits in learning," she associated it with cram schools and assessment. For example:

> Many students seek recipes for solving problems with the goal of getting high scores on the tests. Thus, creating tasks that do a good job of enhancing the students' learning is very important. Otherwise, students may develop bad habits, such as depending on strange recipes without understanding the necessary mathematics behind them. Cram schools are places where students learn such irrelevant strategies. I have tried to tackle this issue in different ways, but it is not easy to resolve.

Whenever Miyeong referred to "bad habits in learning," Euna thought that this was related to her EC and agreed with Miyeong. For example, Euna said, "Right, I had students who bragged about weird methods learned from cram schools, but I could understand them," and emphasized the interruptions made by these students.

Even though the initial complexity maps were not clear in meaning, they played roles in forming discourses between the teachers and the researcher about practices. It was especially helpful for me as a researcher to understand the teachers' circumstances and the practices they used in teaching mathematics. This was effective in decreasing the asymmetry between the teachers and the researcher that is generally caused by the researcher presenting certain theories or results of previous studies (Goodchild 2008) and achieving equal status between practitioners and theorists (Hospesovà et al. 2006). Discussions that were dominated by discourses initiated by the teachers rather than by the researcher led to teachers having greater curiosity about theories and previous studies. For example, there were many discussions led by Miyeong about abstraction in mathematics and the structure of mathematical knowledge, and at the end of these discussions, the teachers asked questions about mathematics teaching-learning theory and related research results.

Although teachers did not reach a level where they were finding and discussing theories and research results themselves, they tended to appreciate each component in a complexity map from a theoretical standpoint as they related them to TTM. They also tried to understand the relationships between various components in the complexity map. In addition, teachers attempted to connect what they understood from various research results to analyzing and modifying textbook tasks. These collaborative activities led teachers to deepen complexities of teaching and to utilize complexity maps as a tool to form a productive discourse that can be helpful in analyzing and modifying textbook tasks. It provided teachers with opportunities to take some theoretical perspectives into consideration for resolving their teaching dilemmas. Gradually, the two teachers began to play researcher roles in the sense that they contributed to the elaboration of some research questions to be examined using implementations of their own TTMs. The two teachers became deeply involved in the collaborative inquiry, which resulted in the development of profound content knowledge and pedagogical content knowledge (Jaworski 2007; Darling-Hammond and Richardson 2009).

It is worth mentioning that the two teachers found the necessity for and ways of implementing TTM after a few discussions about their complexity maps. They started talking about their unsuccessful TTM experiences and tried to see those from a distance in order to link them to a particular theoretical perspective that would give them ideas for improvements. By introducing relevant studies, the teachers could relate to their previous experiences with TTM and the researcher identified potential research questions to collaboratively tackle in the later stages. In the meantime, the teachers determined the affordances and constraints of textbook tasks they wanted to highlight or overcome in the classroom. For example, Euna said,

> Textbook tasks are ideal for average or low-achieving students because they introduce the standard method. They specify the detailed steps, so I do not need to provide any extra explanation. However, they are limited in that they do not consider that various students are at different achievement levels. I can try TTM when targeting students from different achievement levels.

She further tried to find the affordances and constraints of textbook tasks based on her perception of the instructional reality (Zhao et al. 2006):

> Textbook tasks are too simple and stereotyped to deepen conceptual understanding. As many students use the secret recipes, they think that they are good at mathematics as long as they are dealing with textbook tasks. I would like to design tasks that cannot be solved by their methods but provide rich conceptual understanding.

Fig. 20.1 The shared
complexity map

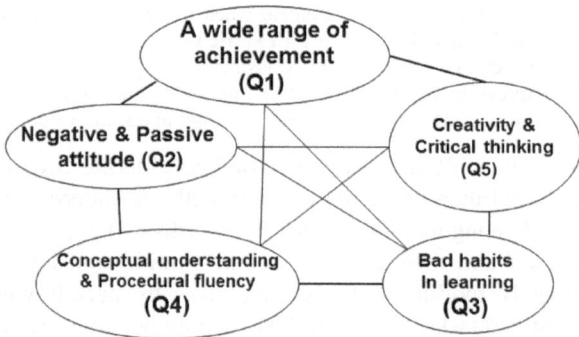

Both teachers emphasized that it was a very meaningful experience to develop expertise while considering a complexity map and analyzing textbook tasks. Opportunities to discuss the initial complexities that each teacher perceived, to share teaching practices, and to analyze the affordances and constraints of textbook tasks served to develop expertise for both teachers. This is evidence of closing the gap between theory and practice in mathematics teaching and learning based on teacher-researcher collaborations. In the end, all members in the PLC constructed a *shared complexity map* to tackle five core questions (see Fig. 20.1):

Q1. How can we consider a wide range of student achievement?
Q2. How can we consider students' negative and passive attitudes towards mathematics and mathematics learning?
Q3. How can we change students' bad learning habits using TTM?
Q4. How can we enhance conceptual understanding with procedural fluency using TTM?
Q5. How can we engage students in creative and critical thinking using TTM?

20.5.3 Two Teachers' Use of the Shared Complexity Map in the TTM Process

Euna tried TTM on 109 tasks. Of the 109 modified tasks, 35 (about 32%) had a cognitive level higher than the original tasks, 66 (about 61%) maintained the same cognitive level as the original tasks, and 8 (about 7%) had a lower cognitive level than the original tasks. Euna was developing a sensibility about changes in the cognitive demands of a task by implementing TTM over time. Her main focus was to increase the participation of various students, especially those with low achievement. This became a turning point for her and helped her become more active in implementing TTM.

> I will never forget the moment when I discovered Sucheol working very hard on a task I was able to offer him using TTM. He used to be like a ghost in the previous classes. At this moment, however, he was so visible to everyone, including me. I was like, like... I cannot express the emotion I felt at that moment. It was one of the best moments in my teaching career. Since then, I have had no doubt about the value of TTM.

Euna's TTM process included considering SRC explicitly and implementing it first and then implicitly and indirectly considering EC and MRC (see Fig. 20.2).

Miyeong modified 43 textbook tasks using the process of first modifying the task herself and then doing a second modification after a discussion. Miyeong additionally modified 74 textbook tasks by herself without discussion. Of the 117 modified tasks, 82 (about 70%) had higher cognitive demands than the originals, 26 (about 22%) maintained the original cognitive demands, and 9 (about 8%) had lower cognitive demands than the originals. Miyeong was very sensitive to the changes of cognitive demands of tasks at the beginning of TTM. Unlike Euna, who focused on SRC, Miyeong mainly considered MRC, consequently increasing the cognitive demands of the tasks beyond those in the textbook. Euna consistently identified SRC, but Miyeong first considered MRC and made opportunities for low-achieving students to participate if possible. Miyeong's turning point in cultivating a positive attitude toward TTM was quite different from Euna's:

> It was quite striking for me to see doubtful contexts or prompts included in textbook tasks. For example, the warm-up task for teaching probability using a 14-faced die was far from the fundamental idea of experimental probability. The first prompt was rolling the die 20 times, which is irrelevant to the meaning of experimental probability. Doing this may provide an opportunity to think about the "law of large number," but that is not the focus in this unit. Suddenly I felt embarrassed, as I did not consider that ever before and just asked students to complete the table. Why that kind of prompt was there without proper intent and why did I just follow it? Why wasn't I aware of it?

Miyeong's TTM process included considering MRC first and then SRC while explicitly considering EC. Figure 20.3 illustrates this process and its dynamics.

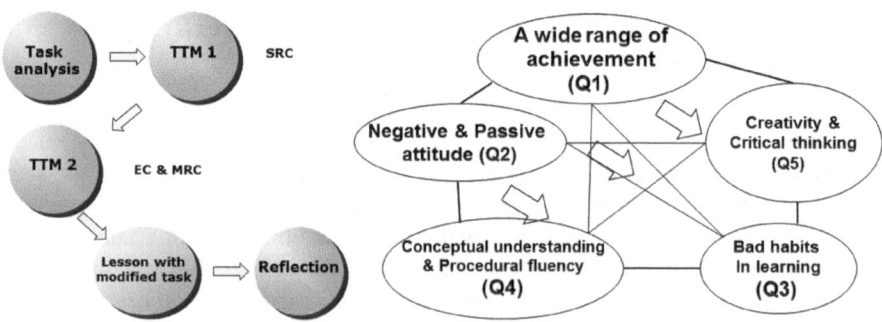

Fig. 20.2 Euna's TTM process and dynamics in the shared complexity map

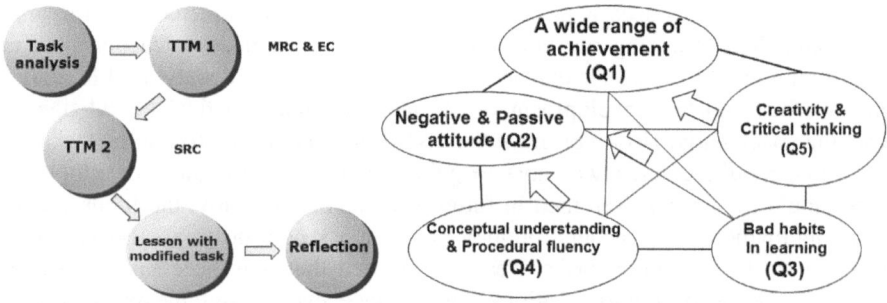

Fig. 20.3 Miyeong's TTM process and dynamics in the shared complexity map

20.6   Conclusion

Unlike what has been shown in previous studies (e.g., Drake and Sherin 2006), the two middle school teachers in this study trusted the contents and structure of MEST-authorized mathematics textbooks rather than having been influenced by the way they learned and experienced mathematics when they were students. The teachers were concerned that TTM itself did not promote effective lessons or reform lessons and that it would sometimes result in bad modifications that were worse than textbook tasks. In order to successfully coordinate the revised curriculum with the textbooks they use, mathematics teachers need to be able use their in-depth understanding of the intentions of both the revision and textbooks in order to modify and implement tasks appropriately. The two middle school teachers' cases in this study showed that it is difficult to change teachers' passive attitudes about modifying tasks in mathematics textbooks unless they go through a sequence of activities that help them understand the revised curriculum in detail, interpret and modify textbook tasks, and implement the modified tasks and reflect on their implementation.

Instead of making teachers learn theoretical concepts or the results of previous studies that are provided to them by a researcher, this study progressed by encouraging the teachers to verbalize the complexities they perceived, recognize them as problems, and find solutions through modifying textbook tasks. This process was very challenging because the teachers and the researcher sometimes either had different meanings for the same terms or used ambiguous terms. Over time, however, examining the complexity became helpful in clarifying the common goal of modifying tasks for effective instruction and finding their implementation strategies. This shows that mathematics teachers can play an active and key role in constructing and operating a learning community with other mathematics teachers. In other words, the use of a complexity map played a significant role in critical alignment of the intent and the implementation of class using the modified textbook tasks, as shown by Goodchild et al. (2013).

From their experiences with modifying textbook tasks, implementing modified tasks, and then reflecting on what they had done, mathematics teachers had an opportunity to understand the essence of school mathematics deeply. In this study, I did not focus on the development of the teachers' mathematical knowledge for teaching specifically. However, from the discussions about modifying the tasks and the actual modification, I was able to gain evidence of substantial development. This should be closely examined in future studies. In this study, the focus was on changing mathematics teachers' prejudices about task modification, especially their absolute trust in textbook tasks. Teachers learned that it is necessary to modify textbook tasks according to situations that they face, even though the textbook tasks were developed based on previous studies. Moreover, when they implemented modified tasks in their class, they learned how to observe whether they were helpful in revealing the essence of school mathematics and whether the lesson was facilitated efficiently. The two teachers in this study mentioned that for a long time they had recognized the importance of reflecting on lessons, but they did not know what aspects of lessons should be reflected upon or how to reflect. They stated that through participating in this professional development program, they had learned why they should think about the textbook tasks when they prepared lessons, how and in what ways they could modify the tasks, how to observe when they implement the modified tasks in class, and how they could draw improvement ideas for the next class. The professional development in this study changed the asymmetric discourse structure between teachers and researcher to a symmetric one that was different in quality from ones that add several theories or results of studies to the teachers' previous perspective. This kind of professional development is significant since it has transformative potential (Price 2001), in that teachers can continue to reflect and improve their reality for themselves.

References

Bao, L., & Stephens, M. (2013). Using a modified form of lesson study to develop students' relational thinking in years 4, 5 & 6. In V. Steinle, L. Ball, & C. Bardini (Eds.), *Mathematics education: Yesterday, today and tomorrow (Proceedings of the 36th Annual Conference of the Mathematics Education Research Group of Australasia)* (pp. 74–81). Melbourne: MERGA.

Ben-Peretz, M. (1990). *The teacher-curriculum encounter: Freeing teachers from the tyranny of texts*. USA: SUNY Press.

Boston, M. D., & Smith, M. S. (2011). A 'task-centric approach' to professional development: Enhancing and sustaining mathematics teachers' ability to implement cognitively challenging mathematical tasks. *ZDM, 43*(6–7), 965–977.

Chevallard, Y., & Bosch, M. (2014). Didactic transposition in mathematics education. In S., Lerman (Ed.), *Encyclopedia of mathematics education* (pp. 170–174). The Netherlands: Springer.

Choe, S.-H., & Hwang, H.-J. (2004). A study on implementation of the seventh mathematics curriculum at the elementary school level. *School Mathematics, 6*(2), 213–233.

Choe, S.-H., & Hwang, H.-J. (2005). A study on the seventh national curriculum at the secondary school level. *School Mathematics, 7*(2), 193–219.

Coe, K., Carl, A., & Frick, L. (2010). Lesson study in continuing professional teacher development: A South African case study. *Academica, 42*(4), 206–230.

Connelly, F. M., & Clandinin, D. J. (1990). Stories of experience and narrative inquiry. *Educational Researcher, 19*(5), 2–14.

Cooper, T. J., Baturo, A., & Grant, E. (2006). Collaboration with teachers to improve mathematics learning: pedagogy at three levels. In *Proceedings of the 30th Conference of the International Group for the Psychology of Mathematics Education* (Vol. 2, pp. 361–367) Prague, Czech Republic: Charles University.

Darling-Hammond, L., & Richardson, N. (2009). Research review/teacher learning: What matters. *Educational Leadership, 66*(5), 46–53.

Desimone, L. M., Porter, A. C., Garet, M. S., Yoon, K. S., & Birman, B. F. (2002). Effects of professional development on teachers' instruction: Results from a three-year longitudinal study. *Educational Evaluation and Policy Analysis, 24*(2), 81–112.

Dowling, D. (2013). *Hungary for calculation: Developing approaches to calculation in the new curriculum using Hungarian methodology as our inspiration (NCETM CTP4213)*. Sheffield: NCETM. Retrieved from https://www.ncetm.org.uk/files/20365123/CTP4213+Final+Report.pdf.

Drake, C., & Sherin, M. G. (2006). Practicing change: Curriculum adaptation and teacher narrative in the context of mathematics education reform. *Curriculum Inquiry, 36*(2), 153–187.

Garet, M. S., Porter, A. C., Desimone, L., Birman, B. F., & Yoon, K. S. (2001). What makes professional development effective? Results from a national sample of teachers. *American Educational Research Journal, 38*(4), 915–945.

Gates, P. (2006). Going beyond belief systems: Exploring a model for the social influence on mathematics teacher beliefs. *Educational Studies in Mathematics, 63*(3), 347–369.

Goldin, G. A., Epstein, Y. M., Schorr, R. Y., & Warner, L. B. (2011). Beliefs and engagement structures: Behind the affective dimension of mathematical learning. *ZDM, 43*(4), 547–560.

Goodchild, S. (2008). A quest for 'good' research. In B. Jaworski & T. L. Wood (Eds.), (2008). *The mathematics teacher educator as a developing professional* (pp. 201–220). The Netherlands: Sense Publishers.

Goodchild, S., Fuglestad, A. B., & Jaworski, B. (2013). Critical alignment in inquiry-based practice in developing mathematics teaching. *Educational Studies in Mathematics, 84*(3), 393–412.

Grouws, D. A., & Smith, M. S. (2000). NAEP findings on the preparation and practices of mathematics teachers. In E. A. Silver & P. A. Kenney (Eds.), *Results from the seventh mathematics assessment of the national assessment of educational progress* (pp. 107–139). Reston, VA: National Council of Teachers of Mathematics.

Heaton, R. M. (2000). *Teaching mathematics to the new standard: Relearning the dance* (Vol. 15). New York: Teachers College Press.

Hospesovà, A., Machàckovà, J., & Tichà, M. (2006). Joint reflection as a way to cooperation between researchers and teachers. In *Proceedings of the 30th Conference of the International Group for the Psychology of Mathematics Education* (Vol. 1, pp. 99–103). Prague, Czech Republic: PME.

Jaworski, B. (2003). Research practice into/influencing mathematics teaching and learning development: Towards a theoretical framework based on co-learning partnerships. *Educational Studies in Mathematics, 54*(2–3), 249–282.

Jaworski, B. (2007). Theory and practice in mathematics teaching development: Critical inquiry as a mode of learning in teaching. *Journal of Mathematics Teacher Education, 9*(2), 187–211.

Leder, G. C., Pehkonen, E., & Törner, G. (Eds.). (2006). *Beliefs: A hidden variable in mathematics education?* (Vol. 31). Berlin: Springer Science & Business Media.

Ledoux, M. W., & McHenry, N. (2008). Pitfalls of school-university partnerships. *The Clearing House: A Journal of Educational Strategies, Issues and Ideas, 81*(4), 155–160.

Lee, K. H. (2010). Searching for Korean perspective on mathematics education through discussion on mathematical modeling. *Research in Mathematics Education, 20*(3), 221–239.

Leung, F. K. (2001). In search of an East Asian identity in mathematics education. *Educational Studies in Mathematics, 47*(1), 35–51.

Lin, F. L., & Cooney, T. (Eds.). (2001). *Making sense of mathematics teacher education.* Berlin: Springer Science & Business Media.

Lloyd, G. M. (1999). Two teachers' conceptions of a reform-oriented curriculum: Implications for mathematics teacher development. *Journal of Mathematics Teacher Education, 2*(3), 227–252.

Manouchehri, A., & Goodman, T. (1998). Mathematics curriculum reform and teachers: Understanding the connections. *The Journal of Educational Research, 92*(1), 27–41.

Martin, S. D., Snow, J. L., & Franklin Torrez, C. A. (2011). Navigating the terrain of third space: Tensions with/in relationships in school-university partnerships. *Journal of Teacher Education, 62*(3), 299–311.

Ministry of Education, Science, and Technology. (2009). *Mathematics curriculum.* Seoul, Korea.

Otte, M. (1986). What is a text? In B. Christiansen, A. G. Howson, & M. Aile (Eds.), *Perspectives on mathematics education* (pp. 173–203). The Netherlands: Springer.

Park, K. M. (2004). Factors contributing to Korean students' high achievement in mathematics. In *Korea sub-commission of ICMI. The report on mathematics education in Korea* (pp. 85–92).

Price, J. N. (2001). Action research, pedagogy and change: The transformative potential of action research in pre-service teacher education. *Journal of Curriculum Studies, 33*(1), 43–74.

Remillard, J. T. (2005). Examining key concepts in research on teachers' use of mathematics curricula. *Review of Educational Research, 75*(2), 211–246.

Remillard, J. T., Harris, B., & Agodini, R. (2014). The influence of curriculum material design on opportunities for student learning. *ZDM, 46*(5), 735–749.

Robutti, O., Cusi, A., Clark-Wilson, A., Jaworski, B., Chapman, O., Esteley, C., et al. (2016). ICME international survey on teachers working and learning through collaboration. *ZDM, 48* (5), 651–690.

Schmidt, W. H. (Ed.). (2007). *Characterizing pedagogical flow: An investigation of mathematics and science teaching in six countries.* Berlin: Springer Science & Business Media.

Schwarz, G. (2001). Using teacher narrative research in teacher development. *The Teacher Educator, 37*(1), 37–48.

Sherin, M. G., & Drake, C. (2009). Curriculum strategy framework: Investigating patterns in teachers' use of a reform-based elementary mathematics curriculum. *Journal of Curriculum Studies, 41*(4), 467–500.

Slavit, D., & Nelson, T. H. (2010). Collaborative teacher inquiry as a tool for building theory on the development and use of rich mathematical tasks. *Journal of Mathematics Teacher Education, 13*(3), 201–221.

Stein, M. K., Remillard, J., & Smith, M. S. (2007). How curriculum influences student learning. In F. K. Lester (Ed.), *Second handbook of research on mathematics teaching and learning* (Vol. 1, pp. 319–369).

Sullivan, P., Clarke, D., & Clarke, B. (2012). *Teaching with tasks for effective mathematics learning* (Vol. 9). Berlin: Springer.

Valverde, G. A., Bianchi, L. J., & Wolfe, R. G. (2002). *According to the book: Using TIMSS to investigate the translation of policy into practice through the world of textbooks.* Berlin: Springer.

Watson, A., & Mason, J. (2006). *Mathematics as a constructive activity: Learners generating examples.* UK: Routledge.

Wenger, E. (1998). *Communities of practice: Learning, meaning, and identity.* Cambridge: Cambridge University Press.

Zaslavsky, O. (1995). Open-ended tasks as a trigger for mathematics teachers' professional development. *For the Learning of Mathematics, 15*(3), 15–20.

Zhao, Q., Visnovska, J., Cobb, P., & McClain, K. (2006). Supporting the mathematics learning of a professional teaching community: Focusing on teachers' instructional reality. In *Annual Meeting of the American Educational Research Association Conference,* San Francisco, CA.

Chapter 21
How Can Cognitive Neuroscience Contribute to Mathematics Education? Bridging the Two Research Areas

Roza Leikin

Abstract This paper, which describes neurocognitive studies that focus on mathematical processing, demonstrates the value that both mathematics education research and neuroscience research can derive from the integration of these two areas of research. It includes a brief overview of neuroimaging research related to mathematical processing. I base my claim that cognitive neuroscience and mathematics education are still two tangent areas of research on a brief comparison of these two fields, with a particular spotlight on research goals, conceptions, and tools. Through a close look at several studies, I outline possible directions in which mathematics education and educational neuroscience can capitalize on each other. Mathematics education can contribute to the stages of research design, while neuroscience can validate theories in mathematics education and advance the interpretation of the research results. To make such an integration successful, collaboration between mathematics educators and neuroscientists is crucial.

Keywords Mathematics education research · Cognitive neuroscience
Educational neuroscience · Mathematical processing

21.1 Introduction

In this paper, I analyze the potential contribution of neurocognitive research to the theory of mathematics education and exemplify some of its implications. This analysis is motivated by the following three observations:

First, there is no consensus among researchers that neuroscience has relevance for education. Educational neuroscience is seen as an emerging discipline with its roots in cognitive neuroscience and its focus on applying the findings of neuroscience to education and posing educational questions to be pursued in neurosci-

R. Leikin (✉)
Department of Mathematics Education, RANGE Center - Research and Advancement
of Giftedness and Excellence, University of Haifa, Haifa, Israel
e-mail: rozal@edu.haifa.ac.il

© The Author(s) 2018
G. Kaiser et al. (eds.), *Invited Lectures from the 13th International Congress
on Mathematical Education*, ICME-13 Monographs,
https://doi.org/10.1007/978-3-319-72170-5_21

entific investigation (Geake 2009). In 1998, Byrnes and Fox suggested that brain research findings might have useful applications in education. Since then many researchers have supported this view with several theoretical hypotheses and have attempted to link neurocognitive empirical findings with the development of educational theory and practice. However, Bowers (2016) argued that there are still no examples of neuroscience motivating new and effective teaching methods, and further asserted that neuroscience is unlikely to improve teaching in the future.

Second, whereas some researchers (e.g., De Smedt et al. 2010), underscored the importance of "balanced dialogue" between neuroscience and education, Turner (2011) argued that this relationship is imbalanced, with a clear dominance of neuroscience (Clement and Lovat 2012). Furthermore, while De Smedt et al. (2010) maintained that neuroscience does not replace the need for behavioral studies, because behavioral studies may be needed to test conclusions drawn from neuroscientific observations, even in this argument behavioral studies and neuroscientific studies are not presented symmetrically. On the contrary, I argue that studies in the field of neuroscience can and should help in testing conclusions drawn from behavioral research in mathematics education. Consequently, research goals and research questions in neurocognitive research can be determined by the results of behavioral research, while behavioral studies can inform neuroscientific research vis-à-vis task design and research interpretations.

Third, mathematics education and cognitive neuroscience are two tangent areas of research. Even though a relatively large number of neurocognitive studies have been performed in the field of numerical cognition, these studies are rooted in cognitive psychology and are not connected to the findings of mathematics education research. Consequently, they use somewhat different terminology and have little impact on the processes of learning primary mathematics in school. Furthermore, only a small number of studies in cognitive neuroscience are currently exploring brain processing associated with (relatively) advanced mathematical concepts while these are rarely connected to theories in mathematics education.

Three notes:

(1) This paper does not provide a broad and detailed meta-analysis of research in the field or detailed descriptions of the studies observed, and it intentionally omits technical details related to the data collection and data analysis procedures of the reviewed research. Instead, it attempts to simplify complex information, present examples to illustrate the main ideas, and propose some directions through which research in cognitive neuroscience can contribute to the development of mathematics education as a scientific field.

(2) This paper does not address eye-tracking, a promising and interesting neuroscientific area of research. Implementation of eye tracking in mathematics education—e.g., in analysis proof reading (e.g., Andrá et al. 2015), exploration of problem-solving strategies (e.g., Obersteiner et al. 2014), and even examination of creative problem solving (e.g., Muldner and Burleston 2015)—has been developing exponentially. For example, the PME-40 conference included

a relatively large number of presentations that applied eye tracking to the examination of mathematical processing at different levels (Csíkos et al. 2016).

(3) A glossary of technical terms in cognitive neuroscience can be found in the "Cognitive neuroscience and Mathematics Learning" special issue of *ZDM-Mathematics Education 42*(6) (Grabner et al. 2010a, b).

21.2   Mathematics Education and Educational Neuroscience are Two Tangent Areas

Even though neuroscientific research in the field of mathematical processing is making progress, some limitations to this research are still evident. De Smedt and Grabner pointed out that neuroscientific research is mostly performed with adult participants and requires better ecological validity. That is, many studies are performed in laboratory settings which are not similar to classroom settings in which students cope with tasks at different levels of mathematical challenge. In this section I argue that mathematics education and educational neuroscience are two tangent areas and illustrate this argument with the results of a brief search performed in several research outlets in the fields of cognitive neuroscience and mathematics education.

21.2.1   Publications on Mathematical Processing in Cognitive Neuroscience Journals

De Smedt and Grabner (2015) stress that "in the past decade, there has been a tremendous increase in neuroscience research on mathematics learning" (p. 2), while "the field of mathematics learning has been proposed as an ideal workspace for making applications of neuroscience to education" (p. 3).

I present herein a brief summary of publications in three journals in the field of cognitive neuroscience. I have chosen these particular journals according to their goals and scopes, which all include educational publications: *Frontiers in Human Neuroscience, Neuropsychologia: An International Journal in Behavioral and Cognitive Neuroscience*, and *Trends in Neuroscience and Education*. An analysis of a number of publications related to mathematical processing (at different levels of mathematics) during 2012–2016 demonstrated that despite researchers' growing interest in this area (reflected in *Trends in Neuroscience and Education*), the number of neuroscientific studies related to mathematical processing is very small. The search was made using the following key words: mathematics, arithmetic, numerical cognition, numerical operations, dyscalculia, algebra, calculus, and geometry. After this search, the papers were downloaded and compared in order to count only once those papers that came up repeatedly in the searches using different key words. The percentage of papers published in these journals varies significantly. During this period, in contrast to *Frontiers in Human Neuroscience*, where less than 1% of papers (19 of 2417) presented original research dealing with mathematical processing, 33% (12 of 36) of original research papers in *Trends in Neuroscience and Education* included reports associated with research in mathematical processing. Overall, across the three journals, about 2.4% of all publications (including original research papers, review papers, and commentaries) were focused on various aspects of mathematical processing. No less interestingly, among the 105 articles (of 4375) in the same three journals, 92 papers (87%) addressed numerical processing (including 18 articles on dyscalculia). Only a handful of studies explored brain activity related to mathematics studies in school.

21.2.2 Neurocognitive Studies Published in Journals in Mathematics Education

For analysis of publications in mathematics education journals, also during 2012–2016, I chose *Educational Studies in Mathematics* (*ESM*), *Journal for Research in Mathematics Education* (*JRME*), *Mathematical Thinking and Learning* (*MTL*), and *Journal of Mathematical Behavior* (*JMB*). The search, with a focus on original research papers, was conducted using the following key words pertaining to neuroscientific methodologies: EEG, ERP, fMRI, fNIRS, and eye tracking. I found only one publication, by Inglis and Alcock (2012), in *JRME*. This paper presented an investigation comparing expert and novice approaches to reading mathematical proofs using eye tracking methodology.

There are several possible explanations for the results of this search. First, I may have overlooked some publications (my apology if this is the case) and if so I would be glad to receive information from authors and readers who are familiar with such publications. Second, of the mathematics education researchers who consider these journals to be venues for publication of their findings, only a small number employ neuroscientific methodology in their studies. Third, those who do such research usually collaborate with neuroscientists, who prefer publishing their manuscripts in neuroscientific journals.

Fourth, there is another side of this coin. In my experience, reviews from neuroscientists and mathematics educators in response to the same publication had different foci of attention, and the requirements for revisions were contradictory to some extent. One of the central issues here is that, as I mentioned in the introduction, the implications of neuroscientific research for mathematics education are not straightforward, and it is difficult to explain the connections in a convincing way. The other central issue, which I describe in Sect. 21.4.1 of the paper, is the difference in theoretical frameworks and, correspondingly, in terminology and interpretation of findings, in the two fields. These issues are illustrated by the response that my colleagues and I received from the editor of one of the leading mathematics education journals (which I will call *X*) justifying why the paper was rejected without sending it to reviewers:

> Articles published in *X* journal pertain to the teaching or learning of mathematics and advance research in this area.... Although I read your paper with great interest, its findings do not move the field of research in mathematics education forward in clearly identifiable ways. (Editor)

The positions of the researchers in mathematics education and educational neuroscience are not contradictory but complementary, and thus bridges built between mathematics education and educational neuroscience can contribute meaningfully to the development of both fields. I argue that making connections between the two fields is a challenging task, and reviewers in both fields have to take greater care in presenting arguments that are compelling for researchers from a different discipline. In what follows, I demonstrate that research methods and tools are one of the reasons for tangency of the research in the two fields.

21.2.3 Special Issues in Mathematics Education

Fortunately, three special issues devoted to neuroscientific research related to mathematics education were published in two mathematics education journals. Two special issues were published in *ZDM—Mathematics Education*: one was "Cognitive Neuroscience and Mathematics Learning," edited by Grabner, Ansari, Schneider, De Smedt, Hannula, and Stern in 2010, and another was "Cognitive Neuroscience and Mathematics Learning: Revisited After Five Years," edited by Grabner and De Smedt in 2016. Another special issue edited by Anderson, Love, and Tsai, "Neuroscience Perspectives for Science and Mathematics Learning in Technology: Enhanced Learning Environments," was published in 2014 in the *International Journal of Science and Mathematics Education* (*iJSME*).

Most of the original papers in the *iJSME* special issue edited by Anderson et al. (2014) used eye-tracking techniques in research on processing related to mathematical and scientific concepts and processes. A paper by Norton and Deater-Deckard (2014) is one of two papers related to studies that investigated brain functioning associated with mathematical problem solving. The researchers take a neo-Piagetian approach to mathematical learning of fractions with computer games in order to frame two studies involving the use of EEG and FMRI techniques. Based on the neuroimaging data, the authors arrive at conclusions about the memory and attention mechanisms involved at different task levels.

The two *ZDM* special issues revealed a significant increase in the variety of topics under investigation. The 2016 *ZDM* special issue contains articles on an impressive diversity of topics—including fraction comparison, geometry, arithmetic, and artificial symbol learning, to name just a few. In comparison to the state of the art in 2010, a more diverse set of questions pertaining to mathematics education is being investigated from a cognitive neuroscience perspective. This represents significant and exciting progress (Ansari and Lyonsi 2016, p. 380).

There was also an obvious shift in the methodology used—from fMRI investigations only to studies that employed a variety of neuro-cognitive techniques: fMRI, EEG (ERD), and ERP. In 2010, the *ZDM* special issue included eight original research papers, three overview manuscripts, and a glossary of terms. In 2016, the special issue included nine papers presenting original research and three commentary papers. As a critique, Ansari and Lyonsi (2016) pointed out that most of the studies published in both special issues presented experiments that had adults as the research participants and included "well-controlled psychological experiments, but their connections to the educational context and the mathematics classroom are unclear" (p. 380). This argument supports my observation that these two research areas are still tangent. However, I am certain that mathematics educators can find a wealth of exciting and useful information in these studies that can help in understanding the underlying processes of mathematical cognition, problem solving under different stress conditions, and neuroimaging aspects related to intuitive rules (see Sect. 21.4 in this paper).

21.3 Neuroimaging Research Associated with Mathematical Processing: A Brief Overview of Issues Mathematics Education Research Does not Address

Neuroimaging research focuses on the underlying brain structures (the magnitude of brain activation as well as brain topographies) associated with different types of mental activities in different population groups. A variety of neuroimaging techniques (for definitions, see Grabner et al. 2010a, b) allow researchers to obtain high-quality information on both temporal and spatial brain activity associated with different kinds of information-processing, including mathematical processing at different levels in individuals with varying levels of abilities. For example, the event-related brain potentials (ERP) technique offers high temporal resolution over the course of problem solving due to a precise reflection of perceptive and cognitive mechanisms. ERPs are electrophysiological measures that reflect changes in the electrical activity of the brain in relation to external stimuli and/or cognitive processes. These measures provide information about the process in real time, before the appearance of any external response (Neville et al. 1993). Another major technique is functional magnetic resonance imaging (fMRI), which offers high spatial resolution and enables the detection of differences in processing that are not evident from behavioral and ERP measures alone, thereby potentially leading to a more comprehensive understanding of the underlying processes and brain structures involved.

21.3.1   Localization of Brain Activation Associated with Mathematical Processing

As mentioned, neuroimaging research focuses on localization of brain activation associated with mathematical processing and its relationship to general cognitive abilities (e.g., memory and attention). One example can be seen in the triple code theory of numerical knowledge, which emphasizes the role of the parietal cortex in number processing and arithmetic calculations (Dehaene et al. 2003) and identifies three regions of the parietal cortex that have been linked to the different functions connected to number processing. The horizontal intraparietal sulcus (HIPS) has been found to be involved in calculations; the posterior superior parietal lobule (PSPL) has been linked to the visuospatial and attention aspects of number processing (Dehaene et al. 2003); the angular gyrus (AG) has been found to be involved in the verbal processing of numbers and in fact retrieval (Grabner et al. 2009). Additionally, the parietal cortex has been found to be associated with word-problem solving (Newman et al. 2011), algebraic equations (Sohn et al. 2004), and geometry proof generation (e.g., Anderson et al. 2011).

Another example can be found in studies that show that the posterior superior parietal cortex is involved in visuospatial processing, including the mental representations of objects and mental rotations (Zacks 2008), while the frontal cortex has been linked to attention-control processes (Badre 2008) and working memory (Gruber and Von Cramon 2003). Solving of (relatively) advanced mathematical problems, such as calculus integrals, was found to activate a left-lateralized cortical network (Krueger et al. 2008).

Research on mathematical problem solving associated with different representations of mathematical objects is also a focus of neuroscientists. For example, different brain areas are known to be involved in recalling different representations of the functions (verbal vs. equation representations) and are thus connected to different cognitive processes involved in the corresponding mathematical processing (Sohn et al. 2004). Lee et al. (2007) compared brain activation in diagrammatic and equation representations for mathematical word problems and found that both modes of representation were associated with activation of areas linked to working memory and quantitative processing (the left frontal gyri and bilateral activation of HIPS). However, the symbolic representation activated the posterior superior parietal lobules (PSPL) and the precuneus. These findings suggest that the two representation modes impose different attention demands (symbolic representation being more demanding).

21.3.2   Individual Differences Reflected in Structural and Functional Characteristics of Brain Activation

Neurocognitive research also focuses on individual differences. Neuroimaging studies demonstrate the neural correlates of mathematical difficulties and disabilities (e.g., developmental dyscalculia; Butterworth et al. 2011). At the other end of the continuum, research has also demonstrated connections between intelligence and brain activity related to different cognitive tasks. Neuroimaging research shows that intelligence is associated with the reciprocity of several brain regions within a widespread brain network (Colom et al. 2010; Desco et al. 2011). Another branch of neurocognitive research focuses on the relationship between intelligence and the extent of induced brain activity during cognitive task performance (Jausovec and Jausovec 2000). These studies have led to the formulation of the neural efficiency hypothesis of intelligence, which states that "brighter individuals display lower (more efficient) brain activation while performing cognitive tasks" (Neubauer and Fink 2009, p. 1004). The neural efficiency phenomenon has also been shown to be related to individuals' expertise in a given field (in our case, excellence in mathematics) (e.g., Grabner et al. 2006). At the same time, task difficulty has an effect: The neural efficiency phenomenon is revealed in easy to moderately difficult tasks, whereas when it comes to performing difficult and challenging tasks, more intelligent individuals exhibit higher brain activity (e.g., Neubauer and Fink 2009).

21.4   Mathematics Education and Educational Neuroscience Can Capitalize on Each Other

21.4.1   Goals, Terms, and Tools in the Two Fields of Research

In the last decade, several publications have been devoted to the various theories in mathematics education (e.g., a volume edited by Sriraman and English 2007). Some debate on the existence and essence of the theories in the field is to be expected. Silver and Herbst (2007) argued that "the development of the grand theory of mathematics education is not simply attainable but desirable for organizing the field" (p. 4), whereas Sriraman and English (2007) questioned the feasibility of creating such a grand theory due to the mathematical, social, and cultural contextualization of mathematics teaching and learning. In our review of the volume (Leikin and Zazkis 2012) we suggested that research in mathematics education is integrated in general education research in two ways. On the one hand, mathematics education is informed by more general theories such as, for example, cognitive sciences, sociology, and anthropology. On the other hand, the recent mathematics education research findings can inform and extend general educational theories. In this paper, I argue that while (in the meantime) mathematics education is not well informed by neuroscience research, and findings of mathematics education research are rarely used in neuroscience research, the integration of the two research areas can empower each of them.

Cognitive research in mathematics education has a variety of foci of attention and research methods. These studies include, but are not limited to, learning and understanding of mathematics as related to problem solving, proofs, proving and argumentation, and defining and exemplification. Special attention is given to investigation and modeling activities, while substantial attention is devoted to difficulties and misconceptions, as well as to expertise, creativity, and giftedness. The *Handbook of Research Design in Mathematics and Science Education* (Kelley and Lesh 2000) emphasizes research designs that are intended to radically increase the relevance of research to actual practice. Examples of such research designs include: teaching experiments, clinical interviews, analyses of videotapes, action research studies, ethnographic observations, software development studies, and computer modeling studies (Kelley and Lesh 2000, p. 18). Schoenfeld (2000) highlighted two main purposes of research in mathematics education. One is a theoretical objective directed at better understanding the nature of mathematical processing as it pertains to thinking, teaching, and learning. The second is an applied objective; that is, to use such understanding to improve mathematics instruction, which ultimately helps realize mathematical giftedness and encourages mathematical creativity. Schoenfeld (2000) stressed that models and theories in mathematics education must have explanatory and predictive power, possess a broad scope, and allow replicability.

As noted in Sect. 21.3 of this paper, neuroimaging research focuses on the underlying brain structures (magnitude of brain activation as well as brain topographies) associated with different types of mental activities in varying population groups. Interestingly, De Smedt and Grabner (2015) identified three types of applications of neuroscience to education: neuro-understanding, neuro-prediction, and neuro-intervention. Neuro-understanding is based on the capacity of neuro-scientific research to deepen understanding of mathematical processing at the biological level and thus to inform mathematics education theories regarding typical and atypical development of mathematical competencies. Neuro-prediction opens opportunities to use neuroimaging results to predict learning trajectories. Neuro-intervention includes both (1) the use of brain imaging data to analyze the impact of education on the neural circuitry underlying development of mathematical knowledge and (2) the effect of neurophysiological interventions on mathematical performance or learning. An analysis of exemplary studies of each type can be seen in De Smedt and Grabner (2015).

An interesting connection between the two fields of research can be seen in the parallel between Schoenfeld's (2000) call for the explanatory and predictive powers of the theories in mathematics education and the neuro-understanding and neuro-prediction types of applications of neuroscience to education. In turn, neuroscience has a strong potential for increasing the explanatory and predictive powers of mathematics education theories as well as examining the power of different educational interventions using neuro-intervention Type 1 mentioned above.

I also would like to suggest the verification power of neuroscience studies, and later in this paper I will illustrate an example of a study (Anderson et al. 2014) that can be categorized as being of a neuro-verification type, even though the authors did not connect their results with mathematics education theories.

Table 21.1 outlines a comparison between the research goals and methodologies in the two fields. I do not include references in the table since each row could include at least a dozen references in each column.

Obviously, neuroscience research on mathematical processing and cognitive research in mathematics education are complementary. They have many features in common, and each field can provide information that cannot be attained by research methodologies in other fields. Clearly, mathematics education research does not address biological data of the kind that is provided by neuroscience. However, behavioral data, collected over long periods of time—related to analysis of processes of mathematical creation (Hadamard 1945), solving mathematical problems of varying complexity and classroom communication and classroom discourse— still are not a part of neuroscience research (excluding eye tracking methodology, which seems to come close to the field of mathematics education research, as mentioned in Note 2 in the Introduction section).

In what follows, I analyze examples of several neurocognitive studies on relatively advanced mathematical processing that suggest interesting and rather clear connections between mathematics education research and neurocognitive research, and I go on to explain these connections. I also provide two examples from a large-scale research project entitled "Multidimensional Investigation of Mathematical Giftedness" performed by the research group of Haifa University's Research and Advancement of Giftedness and Excellence Center (RANGE; Leikin et al. 2013).

Note that I do not provide examples of studies in the fields of number sense and arithmetic. One of the latest comprehensive reviews of neurocognitive studies in this field can be seen in Kaufman et al. (2015). Additionally, de Freitas and Sinclair (2015) provided a critical review of neurocognitive studies on number sense with special attention devoted to studies of dyscalculia. They argued that neurocognitive research, in contrast to mathematics education research, deployed images of numbers with an emphasis on cardinality rather than ordinality and concluded that there is a need for new kinds of neurocognitive research. I take a less critical view, suggesting that integration of the two fields can enable both to benefit from each other.

21.4.2   Between Pólya and Neuroscience: Discovering the Structure of Mathematical Problem Solving

Pólya's works (1945/1973) in mathematics education are among the most influential ones in the field of problem solving. His four-step approach to heuristically solving problems included understanding the problem, devising a plan,

Table 21.1 Brief comparison of cognitive studies in mathematics education and studies in cognitive neuroscience associated with mathematical processing

	Cognitive studies in mathematics education	Studies in cognitive neuroscience associated with mathematical processing

Goals: better understanding of	• Skills, expertise, difficulties in
	  – Numerical operations
	  – Problem solving processes, proving, defining, exemplifying, investigating
	  – Logical, critical, creative thinking
	  – Conceptual understanding
	  – Teaching, learning, communication
	  – Teacher knowledge and skills
	• Individual differences
		• Brain activation associated with
		  – Mainly numerical processing (Subsidizing, enumeration, approximation, comparison, arithmetic operations)
		  – Problem solving (mainly in arithmetic)
		  – Training
		  – Neuro-stimulation
		• Individual differences
		• Domain-general cognitive abilities involved in mathematical processing

Mathematical topics	Broad range of topics, concepts, and properties from elementary to university mathematics	Number sense and arithmetic (mainly) Relatively advanced mathematics (a small number of studies)
Different representations	• Numerical, graphical, algebraic, pictorial, verbal • Translations between different representations	• Numerical magnitude representation (mainly) – Concrete quality, verbal, number line • Symbolic versus pictorial (few)
Research participants	• K-12 • University students • Research mathematicians	• Adults (mostly university students) • Children (a small number of studies) • Research mathematicians (few studies)

Research conditions	• Laboratory/clinics
	• Field experiments,
	  – Design experiment
	  – Teaching experiment
	• Ethnographical research
		• Laboratory
		  – e.g.: MRI (fMRI), EEG (ERP, ERD), fNIRs,
		  – GSR
		  – Eye tracking

Research tools	• Tests
	  – Written, oral, computerized
	• Interviews
	  – Individual, collective
	• Observations
	• Written questionnaires
	• Self-reports
		• Tests
		  – Computerized (e.g., E-Prime)
		• Self-reports

Tasks	• Open/closed • Multiple solutions • Multiple choice • Differ in conceptual density	• Very short • Multiple choice • Yes/no

(continued)

Table 21.1 (continued)

	Cognitive studies in mathematics education	Studies in cognitive neuroscience associated with mathematical processing

Measures	• Behavioral
	  – Correctness
	  – P-S/proving strategies
	  – Critical reasoning, creative thinking
	  – Communicative collaborative and processes
	  – Teachers' knowledge and competences
		• Behavioral
		  – Correctness (%)
		  – Reaction time
		• Neurocognitive
		  – Magnitude of the brain activation
		  – Brain topographies
		• Cognitive
		  – Connections between mathematical processing of different types and basic cognitive functions associated with these

carrying out the plan, and looking back. Schoenfeld (1992) suggested somewhat more detailed stages of problem solving that included reading, analyzing, exploring, planning, implementing, and verifying. Pólya and Schoenfeld demonstrated that a close look into these stages can distinguish experts from non-experts in problem solving when the participants are required to cope with a non-standard problem—one for which they do not have a ready-to-use procedure.

Without any connection to Pólya (1945/1973) and Schoenfeld (1992), Anderson et al. (2014) conducted neuroimaging (fMRI) research that was aimed at discovering the stages of mathematical problem solving, the factors that influence the duration of these stages, and how these stages are related to the learning of a new mathematical competence. This study demonstrated that participants went through five major phases when solving a class of problems: (1) Define Phase, where they identified the problem to be solved, characterized by activity in visual attention and default network regions; (2) Encode Phase, where they encoded the needed information, characterized by activity in visual regions; (3) Compute Phase, where they performed the necessary arithmetic calculations, associated with activity in regions active in mathematical tasks; (4) Transform Phase, at which they performed any mathematical transformations, characterized by activation of mathematical and response regions; and (5) Respond Phase, at which they entered an answer, associated with activation in motor regions. Two features distinguished the mastery trials during which participants came to grasp a new problem type. First, the duration of late phases of the solution process increased. Second, there was increased activation in the rostro-lateral prefrontal cortex (RLPFC) and angular gyrus (AG) regions associated with metacognition. This indicates the important contribution of reflection to successful learning.

Obviously, the stages identified by Anderson et al. (2014), which go beyond the task design, are in harmony with the stages devised insightfully by Pólya and Schoenfeld in their works: the *define* and *encode* phases correspond to the *reading* and *analyzing* stages in Schoenfeld's terms, or to *understanding the problem* in Pólya's terms. The *compute* and *transform* phases correspond to *carrying out the plan* (Pólya). The *respond* phase corresponds to the *looking back* or *verifying* stages (of Pólya and Schoenfeld, respectively). Anderson and colleagues provided biological validation for the big ideas of mathematics education researchers and, in this sense, theirs can be considered a neuro-validation study. At the same time, it provides us with further information about the basic cognitive abilities (visual attention, visual encoding, and motor skills), which are very often overlooked in mathematics education literature. This connection to cognitive processes can be helpful in gaining a better understanding of the effectiveness of educational practices as they are connected to specific cognitive traits. Thus, this study is also of a neuro-understanding type. Moreover, from the point of view of a mathematics educator, connecting the work of Anderson et al. to other works of mathematics educators related to learning and teaching equations can have an added value for the interpretation of the behavioral research results achieved through individual or collective interviews and relevance to the educational practices.

21.4.3 Mathematical Expertise: Connections Between Advanced Mathematical Processing with Language and Number Sense

Experts are usually characterized by consistently superior performance on a specified set of representative tasks (Ericson 1996), while expertise reflects a varying balance between deliberate practice and innate differences in capacities and talents. Experts usually have more robust mental imagery, more numerous images, and the ability to make flexible use of different images and focus their attention on appropriate features of problems (Carlson and Bloom 2005). Experts differ from novices in the problem-solving strategies they employ (Schoenfeld 1992). There is consensus that professional mathematicians are experts in mathematics. Poincare (1908) linked the activity of a mathematician to mathematical creation that requires a feeling of mathematical order and mathematical intuition, which, in his opinion, cannot be possessed by everyone. Still, research on mathematical expertise in school students is rare. To the best of my knowledge, the connections between expertise at earlier ages and expertise in research mathematicians remain unexplored.

Neurocognitive research by Amalric and Dehaene (2016) demonstrated connections between numerical processing and relatively advanced mathematical thinking. The researchers performed an investigation into the neuronal origins and consequences of mathematical expertise. They employed fMRI with 15 expert mathematicians and 15 non-mathematicians who had comparable educational backgrounds. The participants were asked to evaluate the correctness of mathematical and non-mathematical statements. The non-mathematical statements referred to general knowledge and could be meaningful or meaningless, while the mathematical statements referred to domains of higher level mathematics: geometry, analysis, algebra, and topology. No differences were found in the cortical network activated (a) for all four domains of mathematics examined and (b) in the expert mathematicians when reacting to meaningful vs. meaningless mathematical statements. At the same time, the study revealed a contrast of brain activation measured during the reflection on mathematical statements versus activation associated with reflecting on non-mathematical statements. A direct comparison of the groups revealed that parietal and frontal activation during reflection on mathematical statements was only present in the group of expert mathematicians. The experiment demonstrated that the brain regions employed by expert mathematicians during their reflection on mathematical statements are located outside areas typically associated with language. The findings contradicted previous findings of studies on numerical cognition, which had demonstrated connections between activation evoked by numerical processing and by language. The research by Amalric and Dehaene (2016) shows that language may play a role in the initial acquisition of mathematical competencies and that brain activation during elementary numerical processing and higher level mathematics are connected; they thus demonstrated that advanced mathematical processing is connected to symbolic number processing.

The connection between advanced mathematical processing and number sense can develop awareness of the importance of nurturing mathematical minds from early stages of development. Additionally, these findings can lead to a hypothesis stating that early competencies associated with number processing and numerical operations can constitute predictors of later mathematical expertise and, probably, of mathematical talent. This hypothesis, supported by some self-reports by mathematicians (e.g., Tao 1992) about their first steps of success in mathematics, requires a longitudinal systematic investigation.

21.4.4 Starting from Theories in Mathematics Education to Enrich Them

Only a small number of neuroimaging experiments are rooted in theories of science and mathematics education. For example, Babai et al. (2016) explored the effect of mode (discrete/countable vs. continuous perimeters) and the order of presentation on elementary schoolchildren's performance on the "comparison of perimeters" task. They found that providing students with the opportunity to overcome difficulties by altering the mode or order of presentation may lead to improved student performance on the tasks. Their fMRI brain-imaging findings point to two factors that are involved in solving the task correctly: inhibitory control mechanisms and salience. The authors claim that the fMRI brain-imaging results corroborate, validate, and support behavioral findings and, as such, they contribute theoretically and practically to the understanding of reasoning processes and to improved teaching.

Another interesting research study was performed by Tzur and Depue (2014), who examined how task design, rooted in a constructivist theory of learning and thinking, may impact adults' brain processing of numerical comparisons. They examined four independent variables: number comparison—whole numbers or unit fractions $(1/n)$, task sequencing—cueing first by a number or by an operation; distance between the two compared numbers—large $(1 > 8?)$ or small $(7 > 5?)$, and testing occasion—pre/post a purely conceptual teaching intervention. The study showed that each independent variable had a significant impact on reaction time, whereas the error rate remained invariant. The authors suggest implications for mathematics education and cognitive-neuroscience with rethinking distance effect and the need to amend the limiting notion of fractions as equal-parts-of-whole. Tzur (2015) took it one step further by illustrating differentiated circuitry for comparing whole numbers and unit fractions in support of the hypothesis that a fraction is not merely a simple extension of a whole number.

21.4.5 Mathematical Giftedness: Designing a Neurocognitive Study Based on Mathematics Education Theories

Our research group in the RANGE Center at the University of Haifa conducted a study aimed at gaining a better understanding of mathematical giftedness (Leikin et al. 2013). It was motivated by the observation that the evaluation of individuals presented as over-performers or who excel in the field of mathematics is not an easy matter due to the lack of strong definitions of the phenomenon of mathematical giftedness. We also argued that the development of tools for evaluation of individual abilities (especially high abilities) in the field of mathematics is insufficient and that applying brain research to the study of mathematical giftedness seems to be of importance to the attainment of an operative definition of mathematical giftedness and, consequently, to the development of tools that enable researchers to identify mathematical giftedness.

Several distinctions were introduced in the study: First, based on theories of gifted education (e.g., Milgram and Hong 2009), a distinction was made between levels of intelligence ("general giftedness," G, determined by IQ scores higher than 130) and levels of expertise ("excellence in mathematics," EM, determined by high scores in secondary school mathematics). This was applied in the sampling procedure, whereby four research groups were designed by a varying combination of EM and G characteristics. Second, based on the theories of mathematics education, a distinction was made between the translations of different representations of mathematical objects required by the task (Kaput 1998) and different areas of mathematics (i.e., algebra and geometry), together with a third distinction between learning-based and insight-based tasks; these distinctions were implemented in the design of the research tools. The task design was determined by Pólya's (1973) theory of problem-solving strategies. Three strategies—understanding the task conditions, understanding the question, and verifying the results—constituted the stages of the task design and the corresponding cognitive processes.

The study design led to some exciting discoveries: The distinction between general giftedness and expertise in mathematics proved to be powerful in understanding that these two characteristics, even though interrelated, are different in nature. It was also obvious that using behavioral measures only is insufficient and sometimes misleading.

For example, coping with a function-related task that required translation between graphical and algebraic representations of the functions (Waisman et al. 2014) both at the behavioral (accuracy and reaction time for correct responses) and electrophysiological levels (amplitudes, latencies, and scalp topographies of brain activity identified using the ERP procedure) was affected by students' level of mathematical expertise, with significantly higher accuracy of responses and significantly shorter reaction times among non-gifted students only. Interestingly, students who excelled in school mathematics but were not identified as being generally gifted exhibited the highest electrical brain activity as compared to all the other groups of students. That is, for (relatively) simple learning-based mathematical problems, mathematical expertise appeared to be the main factor that influenced problem-solving performance.

When comparing ERP measures associated with performance of function-based tasks and tasks that involved geometrical inferences (Leikin et al. 2014), we found differences both in the magnitudes and topographies of brain activation at the stage of answer verification (Fig. 21.1). Based on these results, we argue that problem

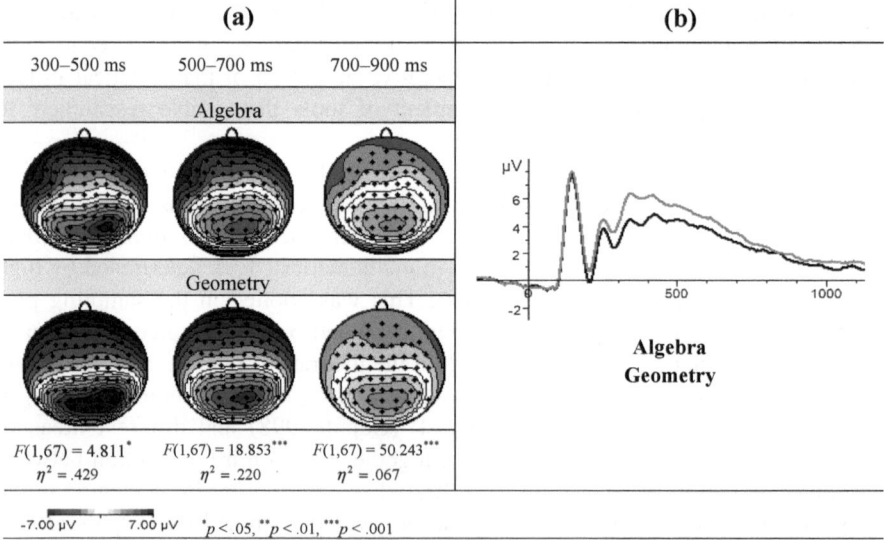

Fig. 21.1 **a** ERP topographies in the three time frames at the answer verification stage. **b** Amplitudes in the posterior regions at the answer verification stage

solving in algebra and geometry, even when requiring a similar translation between visual to symbolic representation, is associated with variant patterns of brain activity as related to different underlying cognitive processes.

The distinction between learning-based and insight-based problems led to additional surprising results (Leikin et al. 2016): behavioral and neurocognitive measures led to somewhat contradictory findings, and thus neurocognitive characteristics provided essential information that was hidden in the behavioral analysis. We found that, contrary to our research hypothesis, expertise in school mathematics affected behavioral measures associated with the insight-based test only, while general giftedness affected accuracy of the responses on both tests. At the same time, as hypothesized, expertise in mathematics and its interaction with general giftedness affected ERP measures associated with solving learning-based problems only, while ERP measures associated with solving insight-based problems appeared to be affected mainly by the G factor.

Our findings of stronger activation of the PO4–PO8 electrode site (Fig. 21.2) matched findings of Jung-Beeman et al. (2004), who demonstrated the increased activation of the PO8 electrode being associated with the "Aha!" moment. Thus, in Leikin et al. (2016) we raised a hypothesis that our findings on increased activation at the PO4–PO8 electrode site associated with solving insight-based problems indicate that mathematical insight is a specific characteristic unique to generally gifted students. Moreover, our findings expanded upon the previous findings by Jung-Beeman et al. (2004) by showing that students who excel in school mathematics exhibit increased activation of the PO4–PO8 electrode site when they are presented with learning-based mathematical tasks. This finding led to the hypothesis (that opens a window for future studies) that this increase in the absolute ERP

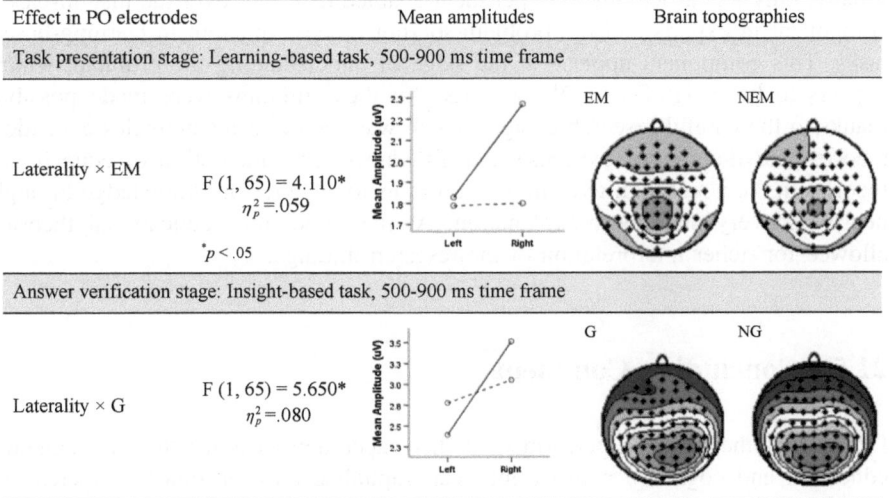

Fig. 21.2 ERP topographies and amplitudes at the stage of answer verification

amplitudes at the PO4–PO8 electrodes is linked to the ability of experts to predict the problem question based on the problem givens (Schoenfeld 1992). This prediction can also be considered an insight-based process related to learning-based tasks reflected in our findings and, thus, can be considered to be evidence of the insight-related component of problem solving by experts at the stage of understanding the problem.

We connected our findings regarding the strength of electrical potentials evoked in different groups of participants when solving learning-based tasks to the neural efficiency effect (see Sect. 21.3 in this paper and critique by Ansari 2016). We suggest that our findings showed no neuro-efficiency associated with solving insight-based tasks for either G or EM, due to the high difficulty of the task. Furthermore, the effects of the G characteristic on the cortical topographies associated with solving insight-based problems were explained by the presumably different problem-solving strategies applied by students with different levels of intelligence.

Clearly, our neurocognitive experiment not only validated our initial hypotheses, which were not always supported by behavioral data, but also led to new insights and new hypotheses. First, mathematical expertise and general giftedness are not equivalent constructs. We hypothesized that both of these characteristics are necessary conditions for mathematical giftedness. Second, we realized that externally similar algebra and geometry problems are based on different underlying cognitive processes, as reflected in different brain activation when solving the problems. Third, only ERP data allowed us to develop an understanding that success in solving insight-based problems is a function of general intelligence and is not attained by school mathematical expertise. We believe that further behavioral research is needed to ascertain to what extent classroom culture determines these findings or whether mathematical insight is an innate characteristic of the gifted. Finally, only a neurocognitive experiment enabled us to discover the insight-based component in experts and the problem-solving process inherent in learning-based tasks. This component appears at the stage of understanding the problem, which appears to be insightful for the experts. All these findings were made possible thanks to the careful research design, which was rooted in mathematics education theories and theories of expertise and giftedness. The integration of educational theories in the neurocognitive study allowed us to enhance our knowledge through neuro-discovery and neuro-explanation. At the same time, educational theories allowed for richer interpretation of the research findings.

21.5   Concluding Comments

I hope that the analysis performed in this paper demonstrates that mathematics education and cognitive neuroscience can capitalize on each other by increasing validity of findings and mutually providing more substantiated interpretations of findings. Mathematics education can clearly contribute to research design, and neuroscience can validate (or perhaps refute) theories in mathematics education and, later, advance the interpretation of research results. Mathematics education initially developed as a branch of cognitive psychology; neurocognitive investigation can enrich mathematics education by contributing to our understanding of the underlying cognitive processes involved in different types of mathematical performance and by explaining the roots of success and difficulties in mathematics learning, proving, problem solving and creative, intuitive, and critical reasoning. To successfully integrate the fields, collaboration between mathematics educators and neuroscientists is crucial. This collaboration should be symmetrical to allow reciprocal enhancement and further development of these two fields of research and, eventually, to allow implementation of the resulting findings in educational practice.

References

Amalric, M., & Dehaene, S. (2016). Origins of the brain networks for advanced mathematics in expert mathematicians. *Proceedings of the National Academy of Sciences (PNAS)*, *113*, 4909–4917).

Anderson, J. R., Lee, H. S., & Finchama, J. M. (2014). Discovering the structure of mathematical problem solving. *NeuroImage*, *97*, 163–177.

Anderson, O. R., Love, B. C., & Tsai M.-J. (Eds.). (2014). Neuroscience perspectives for science and mathematics learning in technology-enhanced learning environments. *International Journal of Science and Mathematics Education*, *12*(3), 669–696.

Andrá, C., Lindström, P., Arzarello, F., Holmqvist, K., Robutti, O., & Sabena, C. (2015). Reading mathematics representations: An eye tracking study. *International Journal of Science and Mathematics Education*, *13*(2), 237–259.

Ansari, D., & Lyonsi, I. M. (2016). Cognitive neuroscience and mathematics learning: How far have we come? Where do we need to go? *ZDM Mathematics Education*, *48*, 379–383.

Babai, R., Nattiv, L., & Stavy, R. (2016). Comparison of perimeters: Improving students' performance by increasing the salience of the relevant variable. *ZDM*, 1–12.

Bowers, J. S. (2016). The practical and principled problems with educational neuroscience. *Psychological Review*, *123*(5), 600–612.

Butterworth, B., Varma, S., & Laurillard, D. (2011). Dyscalculia: From brain to education. *Science*, *332*, 1049–1053.

Byrnes, J. P., & Fox, N. A. (1998). Minds, brains, and education: Part II. Responding to the commentaries. *Educational Psychology Review*, *10*(4), 431–439.

Csíkos, C., Rausch, A., & Szitányi, J. (Eds.). (2016). *Proceedings of the 40th Conference of the International Group for the Psychology of Mathematics Education*. China: PME.

De Freitas, E., & Sinclair, N. (2015). The cognitive labour of mathematicsdis ability: Neurocognitive approaches to number sense. *International Journal of Educational Research*, *1103*, 9.

De Smedt, B., Ansari, D., Grabner, R. H., Hannula, M. M., Schneider, M., & Verschaffel, L. (2010). Cognitive neuroscience meets mathematics education. *Educational Research Review*, *5*(1), 97–105.

De Smedt, B., & Grabner, R. H. (2015). Applications of neuroscience to mathematics education. In R. Cohen Kadosh & A. Dowker (Eds.), *The Oxford handbook of numerical cognition*. Oxford: Oxford University Press.

Dehaene, S., Piazza, M., Pinel, P., & Cohen, L. (2003). Three parietal circuits for number processing. *Cognitive Neuropsychology, 20*(3–6), 487–506.

Desco, M., Navas-Sanchez, F. J., Sanchez-González, J., Reig, S., Robles, O., Franco, C., & Arango, C. (2011). Mathematically gifted adolescents use more extensive and more bilateral areas of the fronto-parietal network than controls during executive functioning and fluid reasoning tasks. *Neuroimage, 57*(1), 281–292.

Grabner, R. H., Ansari, D., De Smedt, B., & Hannula, M. M. (2010a). Glossary of technical terms in cognitive neuroscience. *ZDM-Mathematics Education, 48*(3), 461–463.

Grabner, R. H., Ansari, D., Koschutnig, K., Reishofer, G., Ebner, F., & Neuper, C. (2009). To retrieve or to calculate? Left angular gyrus mediates the retrieval of arithmetic facts during problem solving. *Neuropsycholog*ia, *47*(2), 604–608.

Grabner, R. H., Ansari, D., Schneider, M., De Smedt, B., Hannula, M. M., & Stern, E. (2010b). Cognitive neuroscience and mathematics learning. Special Issue of *ZDM-Mathematics Education, 48*(3).

Grabner, R. H., & De Smedt, B. (2016). Cognitive neuroscience and mathematics learning— revisited after five years. Special Issue of *ZDM-Mathematics Education, 48*(3).

Grabner, R. H., Neubauer, A. C., & Stern, E. (2006). Superior performance and neural efficiency: The impact of intelligence and expertise. *Brain Research Bulletin, 69*(4), 422–439.

Hadamard, J. (1945). *The psychology of invention in the mathematical field.* New York: Dover Publications.

Inglis, M., & Alcock, L. (2012). Expert and novice approaches to reading mathematical proofs. *Journal for Research in Mathematics Education, 43*(4), 358–390.

Jausovec, N., & Jausovec, K. (2000). Correlations between ERP parameters and intelligence: A reconsideration. *Biological Psychology, 55*(2), 137–154.

Kaput, J. J. (1989). Linking representations in the symbol systems of algebra. *Research Issues in the Learning and Teaching of Algebra, 4,* 167–194.

Kaufman, L., Kucian, K., & von Aster, M. (2015). Development of the numerical brain. In R. Cohen Kadosh & A. Dowker (Eds.), *The Oxford handbook of numerical cognition.* Oxford: Oxford University Press.

Kelly, A. E., & Lesh, R. A. (2000). *Handbook of Research Design in Mathematics and Science Education.* Mahwah, NJ: Routladge.

Leikin, M., Waisman, I., & Leikin, R. (2013). How brain research can contribute to the evaluation of mathematical giftedness. *Psychological Test and Assessment Modeling, 55*(4), 415–437.

Leikin, M., Waisman, I., Shaul, S., & Leikin, R. (2014). Brain activity associated with translation from a visual to a symbolic representation in algebra and geometry. *Journal of Integrative Neoroscience, 13*(1), 35–59.

Leikin, R., Waisman, I., & Leikin, M. (2016). Does solving insight-based problems differ from solving learning-based problems? Some evidence from an ERP study. *Special issue on Neuroscience and Mathematics Education—ZDM—The International Journal on Mathematics Education, 48*(3), 305–319.

Muldner, K., & Burleston, W. (2015). Utilizing sensor data to model students' creativity in a digital environment. *Computers in Human Behavior, 42,* 127–137.

Neubauer, A. C., & Fink, A. (2009). Intelligence and neural efficiency. *Neuroscience and Biobehavioral Reviews, 33*(7), 1004–1023.

Neville, H. J., Coffey, S. A., Holcomb, P. J., & Tallal, P. (1993). The neurobiology of sensory and language processing in language-impaired children. *Journal of Cognitive Neuroscience, 5*(2), 235–253.

Norton, A., & Deater-Deckard, K. (2014). Mathematics in mind, brain and education: A neo-piagetian approach. *International Journal of Science and Mathematics Education, 12*(3), 647–667.

Obersteiner, A., Moll, G., Beitlich, J. T., Ciu, C., Schmidt, M., Khmelivska, T., & Reiss, K. (2014). Expert mathematicians strategies for comparing the numerical values of fractions— evidence from eye movements. In P. Liljedahl, S. Oesterle, C. Nicol & D. Allan (Eds.),

Proceedings of the 38th Conference of the International Group for the Psychology of Mathematics Education (Vol. 4, pp. 338–345). Vancouver, Canada: PME.

Poincare, H. (1908/1952). *Science and method.* New York: Dover Publications Inc.

Pólya, G. (1945/1973). *How to solve it.* Princeton, NJ: Princeton University.

Schoenfeld, A. H. (1992). Learning to think mathematically: Problem solving, metacognition, and sense-making in mathematics. In D. Grouws (Ed.), *Handbook for research on mathematics teaching and learning* (pp. 334–370). New York: MacMillan.

Schoenfeld, A. H. (2000). Purposes and methods of research in mathematics education. Notices of the *American Mathematical Society, 47*, 2–10.

Tao, T. C. S. (1992). *Solving mathematical problems: A personal perspective.* Geelong: Deakin University Press.

Tzur, R., & Depue, B. E. (2014). Conceptual and brain processing of unit fraction comparisons: A cogneuromathed study. *Proceedings of PME 38, 5*, 297–304.

Waisman, I., Leikin, M., Shaul, S., & Leikin, R. (2014). Brain activity associated with translation between graphical and symbolic representations of functions in generally gifted and excelling in mathematics adolescents. *International Journal of Science and Mathematics Education, 12*(3), 669–696.

Chapter 22
Themes in Mathematics Teacher Professional Learning Research in South Africa: A Review of the Period 2006–2015

Mdutshekelwa Ndlovu

Abstract In this chapter, I review and identify themes in in-service mathematics teacher professional development/learning research in South Africa over a 10-year period from 2006 to 2015. No less than 92 journal articles were reviewed. Nine themes were identified as characterising research during this period. Mathematical knowledge for teaching and pedagogical content knowledge were the two most dominant themes. Subject matter knowledge was the fourth and closely aligned to the first two. Curriculum knowledge was the third most frequently occurring research theme and was also closely aligned to the first two. Together the first four themes constituted 54% of the research output for this period, an indication of the centrality of practising teachers' professional knowledge of school mathematics. Under-researched themes included the integration of ICTs in mathematics education as well as impact studies that were apparently constrained by lack of funding for large-scale research.

Keywords In-service training · Mathematics teacher · Professional development Professional learning · Teacher knowledge

22.1 Introduction

The mathematics education situation in South Africa has been described as a crisis by many researchers based on the low performance and perception of the country in international benchmark studies. Principal examples of such studies include the Trends in Mathematics and Science Study (TIMSS), the Southern and Eastern Africa Consortium for Monitoring Educational Quality (SACMEQ) and the World Economic Forum's (WEF) annual Global Competitiveness Reports. Various

M. Ndlovu (✉)
Stellenbosch University, Stellenbosch, South Africa
e-mail: mcn@sun.ac.za

© The Author(s) 2018
G. Kaiser et al. (eds.), *Invited Lectures from the 13th International Congress on Mathematical Education*, ICME-13 Monographs,
https://doi.org/10.1007/978-3-319-72170-5_22

reasons have been proffered and chief among them is the apartheid legacy (e.g. Kaino et al. 2015) that has provided unequal educational resources and opportunities based on race, not only at school level but also at initial teacher education level. While the white minority received world-class education, the majority black underclass received substandard education that was dismissive of mathematics as it was being taught to the white master's hewers of wood and drawers of water. That psyche permeated the preparation of teachers in subtle ways. While white teachers were well prepared at universities to be graduates, black teachers were underprepared at under-resourced teacher education institutions that awarded mainly a three-year teaching diploma. There was no requirement for teacher education institutions to conduct research. In an effort to redress the imbalances of the past, the post-apartheid dispensation sought to upgrade historically underqualified teachers through formalized in-service teacher education programmes such as the Advanced Certificate of Education (ACE). It is no coincidence therefore that most of the in-service teacher education research in the period reported here is apparently dominated by involvement in this programme.

Initially, teacher education was incorporated into universities so that all newly qualified teachers could receive degrees (i.e., receive a relative educational qualification value of Level 14 [REQV-14], which consists of matriculation and 4 years of training) irrespective of where or who they were going to teach. Concomitantly, universities began to offer upgrading courses such as the ACE to those teachers who had received REQV-13 (Brown 2010). This provided an opportunity for universities to carry out systematic scrutiny (research) on teacher education programmes as part of their core business of research, teaching and service to the community. Apart from upgrading teachers in mathematics teaching skills, the ACE also became a vehicle for retraining in new subjects in the curriculum. Mathematical literacy was one such subject introduced in the Further Education and Training (FET) phase (Grades 10–12), as a compulsory alternative to pure mathematics in order to fulfil the mathematics for all policy adopted by the new democratic government. This subject was introduced with no experience around the world to draw from, thus attracting considerable research interest (e.g., Julie 2006; Bansilal et al. 2014).

Frequent changes in the school curriculum stimulated research interest in their own right in the broader education reform process. The rapid changes in the curriculum saw an evolution from the National Education Department (NATED) curriculum to outcome-based education (OBE) in the form of Curriculum 2005, the Revised National Curriculum Statements (RNCS) and the National Curriculum Statement (NCS) to the Curriculum and Assessment Policy Statements (CAPS). This is why I prefer to refer optimistically to a mathematics education system in transition or in search of an identity, rather than one in crisis, for a relatively young nation state.

The purpose of this paper is to analyse the research themes privileged in mathematics teacher professional development/learning (MTPD/MTPL) for the 10-year period from 2006 to 2015. To achieve this goal I will use some meta-analysis and the following overarching research question and sub-questions to guide the study:

22.1.1 Main Research Question

What were the main themes (issues) investigated by the researchers with respect to MTPL?

22.1.2 Sub-questions

(a) Where was the research surveyed published for the period under review?
(b) What volume of research was published in (accredited) journals over the period?
(c) Who were the research participants?
(d) What were the themes (issues) problematized in the research?

22.2 Methodology

Firstly, I conducted a literature search on Google Scholar using the following search phrases: 'mathematics teacher professional development in South Africa', 'in-service training for mathematics teachers in South Africa', 'mathematics teacher professional learning in South Africa' etc. I ticked the box for articles and unticked the boxes for 'case law', 'include patents' and 'include citations'. I selected the custom range 2006–2015 to limit the search to article publications during that period. I repeated the same procedure for the other two search phrases, yielding results of approximately the same number of pages each time: 100, 99 and 100 respectively. Figure 22.1 below shows the top and bottom screenshots for page 2 of 100 Google Scholar pages that were prompted when the first phrase was used for searching within the 2006–2015 range. Each of the first 99 pages contained a list of about 10 publications (articles, books, conference proceedings papers, etc.). The last page only had one publication listed on it. Table 22.1 shows the total number of journal articles that spoke to research on mathematics teacher professional development in South Africa for the 10-year period 2006–2015. I repeated the same procedure for the other search phrases and specifically looked for articles that had not been prompted by the preceding searches.

Secondly, I searched for publications in South African databases (e.g., Sabinet and Ebscohost) for accredited journals. I excluded conference publications to narrow down the search and ensure only articles involving more rigorous peer review were included. I excluded book chapters, as most appeared to be drawing from prior published journal articles or unpublished thesis work. I also excluded unpublished theses from this study, as they were not rigorously peer-reviewed work. However, many of the published articles were based on thesis work and hence publishable thesis work was indirectly included in that sense.

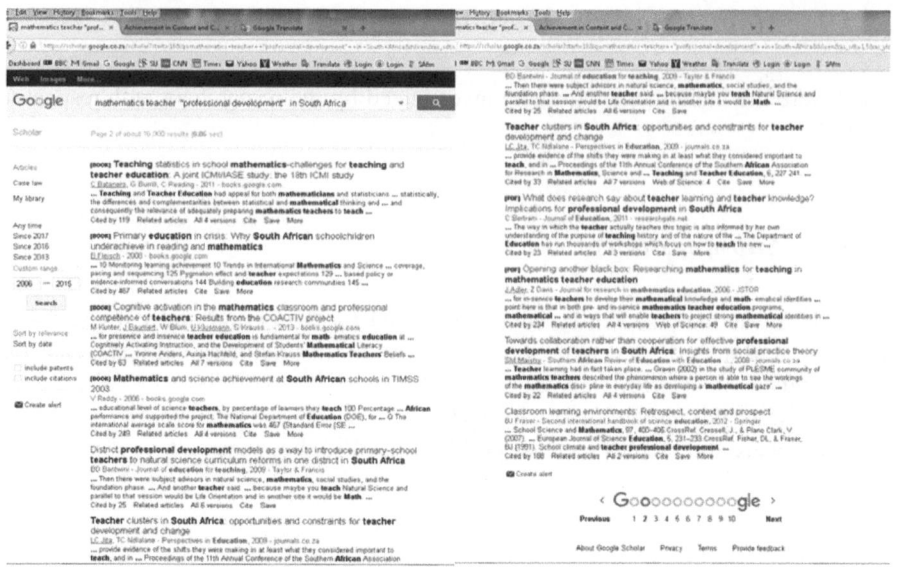

Fig. 22.1 Top and bottom screenshots of Google Scholar page 2 of 100 for the phrase 'mathematics teacher professional development in South Africa'

Table 22.1 Google Scholar results for the phrase 'mathematics teacher professional development in South Africa'

Pages	1–10	11–20	21–30	31–40	41–50	51–60	61–70	71–80	81–90	91–100	Total
Total	8	12	9	4	4	5	4	4	0	2	51

Thirdly, I scrutinized each article's title and abstract to confirm whether it was indeed about mathematics teacher professional development. Where details of participants were not clear from the title and abstract, I proceeded to the methodology section to confirm or refute whether it was a study on in-service teacher professional development/learning or not. I also discarded multidisciplinary studies that involved more than two subjects, as these tended to be more general and less specific to mathematics teaching. I included articles that involved mathematics and science teacher professional development/learning as the twin gateway subjects were frequently researched together (e.g., Mokhele and Jita 2012a, b; Jita and Mokhele 2012, 2014; Mokhele 2013). In the larger study, I categorised each article that related to teacher development or professional learning research according to the following sub-headings: author(s), year, journal, topic/theme, theoretical framework, purpose of research, research method, paradigm, participants and main findings. In this paper, I only deal with the main themes/issues explored. In all, 92 articles were included for analysis. Although this was not an exhaustive list, I considered it a representative sample of identified research studies.

22.3 Results

22.3.1 Main Journals Surveyed

Figure 22.2 shows the distribution by journal of the 92 articles reviewed. The journal that contained the largest share (32%) of mathematics teacher professional learning research articles was *Pythagoras*. This accredited journal is exclusively dedicated to mathematics education and the only one of its kind is South Africa. It is published by the Association of Mathematics Education of South Africa (AMESA), the largest mathematics education research association in the country and one of only two accredited mathematics education conferences. The second largest contributing journal is the *African Journal of Research in Mathematics, Science and Technology Education (AJRMSTE)* which contributed 21% of the total number of articles from the journals surveyed. This journal is published by the Southern African Association of Mathematics, Science and Technology Education (SAARMSTE). The SAARMSTE annual conference is the only accredited conference dedicated to mathematics, science and technology education. It draws its membership from the Southern African region.

The *South African Journal of Education* and the *South African Journal of Higher Education* respectively contributed 13 and 9% to the total number of articles selected for review. The category 'Other' included any other journal from which articles on MTPL in the sample were obtained, such as *Perspectives in Education (PiE)*, *Acta Academica*, *Education as Change*, *Journal of Educational Studies*, *Journal of Social Sciences*, and *Anthropologist*.

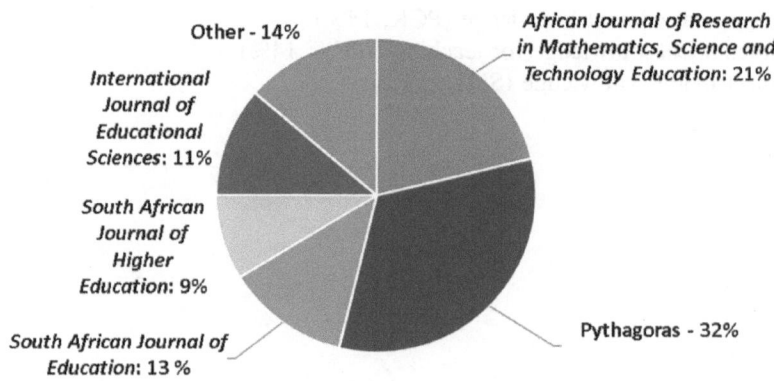

Fig. 22.2 Distribution of journals surveyed

22.3.2   Volume of Journal Publication Output

The graph in Fig. 22.3 shows the volume of publication output from year to year in this sample.

The graph shows that the research output experienced erratic growth from 2006 to 2010 before an acceleration or exponential growth pattern from 2010 to 2015. The overall picture is one of steady growth during the 10-year period.

22.3.3   Distribution of Research Participants

Figure 22.4 shows the distribution of research participants.

Almost three quarters of the research (72%) involved secondary school mathematics teachers as participants. Only 11% was on primary school teachers. In 11% of the articles, the research participants were both primary and secondary school teachers. In 6% of the articles, it was unclear at what level the participants were. There was not much evidence of Foundation Phase (Grades R-3) or Intermediate Phase (Grades 4–6) teachers participating in the research.

22.3.4   Main Research Themes

Figure 22.5 shows distribution of main research themes.
Nine themes were identified:

1. Pedagogical content knowledge (PCK; 14%)
2. Mathematical knowledge for teaching (MKT; 14%)
3. Subject matter knowledge (SMK; 13%)

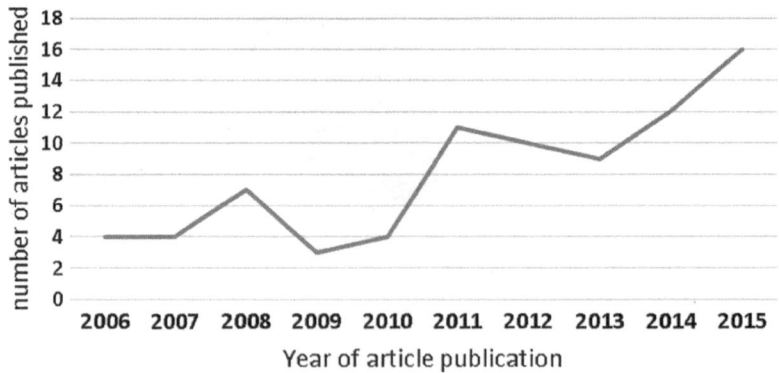

Fig. 22.3 Journal article output 2006–2015

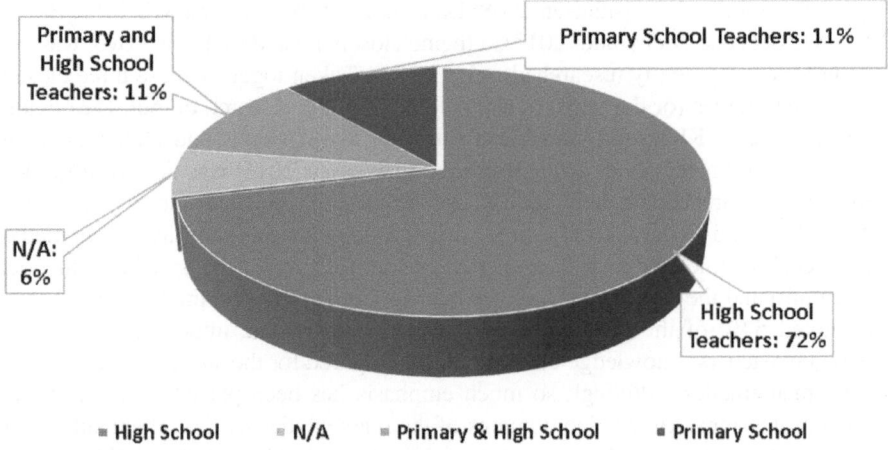

Fig. 22.4 Distribution of research participants

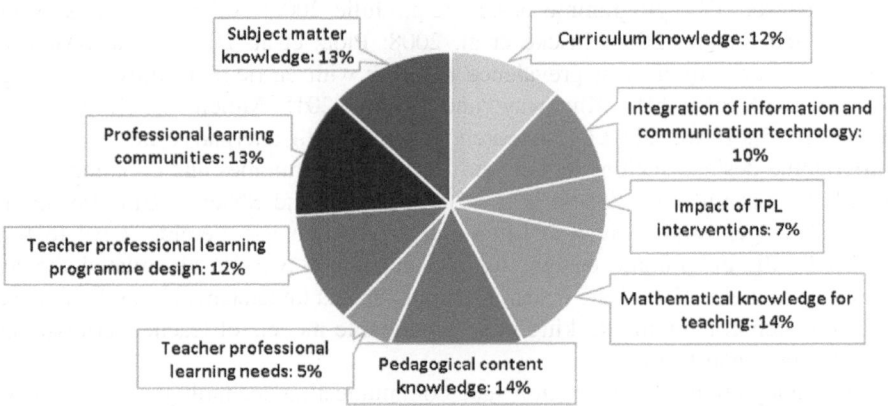

Fig. 22.5 Main research themes privileged

4. Curriculum knowledge (12%)
5. Teacher professional learning (TPL) programme design (12%)
6. Integration of information and communication technologies (ICTs; 10%)
7. Impact of TPL interventions (7%)
8. Professional learning communities (PLCs; 13%)
9. Professional learning needs (5%).

MKT (e.g., Adler and Davis 2006; Adler and Pillay 2007; Aldridge et al. 2009; Brijlal et al. 2012; Gierdien 2008; Mhlolo et al. 2012; Mudaly and Moore-Russo 2011; Tosavainen et al. 2013; Kazima et al. 2008; Kazima and Adler 2006; Bansilal 2014a, b; Lampern 2015) and PCK (e.g., Mhlolo and Schafer 2012; Brijlal 2014; Brodie and Sannie 2014) were the two most researched themes. SMK (e.g., Likwambe and Christiansen 2008; Bansilal 2011; Ndlovu and Mji 2012; Berger 2013; Wessels and Nieuwoudt 2013), a theme closely related to the first two, was tied for third most frequently researched (with PLCs). Taken together these three closely intertwined themes (or domains) constituted 41% of the research output. Curriculum knowledge (e.g., Khuzwayo and Mashiya 2015; Mwakapenda and Dhlamini 2009; Bansilal 2015; Shalem et al. 2013; Biccard and Wessels 2015) was tied for fifth most frequently researched theme (with TPL programme design) and closely aligned to MKT, SMK and PCK. Curriculum knowledge helps teachers to align their mathematics content knowledge and mathematics for teaching to the syllabus (intended curriculum) and the assessment (assessed curriculum). Together, the first four themes constituted 54% of the research output, which conveys the importance placed on practicing teachers' knowledge of mathematics required for the successful teaching of school mathematics. Although so much emphasis has been placed on these fundamental professional knowledge domains of the mathematics teacher, there still seems to be a long way to go in the attempts to solve the teacher knowledge problem, more so in an environment where the mathematics curriculum itself keeps on evolving (Paulsen 2015; Phoshoko 2015).

Formalised TPL programme design (e.g., Julie 2006; Adler and Davis 2006; Brown and Schafer 2006; Fricke et al. 2008; Plotz et al. 2012; Owusu-Mensah 2014) was tied for fifth in prevalence together with curriculum knowledge (e.g. Graven and Venkat 2014; Khuzwayo and Mashiya 2015; Molefe and Brodie 2010; Webb 2015). If we add PLC research, together with alternative non-formalised programme designs such as lesson study, cluster programmes and communities of practice, research (e.g., Brodie 2007, 2013; Brodie and Shalem 2011; Posthuma 2012; Pausigere and Graven 2014; Singh 2011; Ono and Fereira 2010) to formalised TPL programme design, we see that these two closely related themes combine to make 25% of the research output. This is understandable in the context of a research community seeking more sustainable models of teacher professional development and learning.

The integration of information and communication technologies into mathematics teaching (e.g., Stols et al. 2008, 2015a, b; Van der merwe and Van der Merwe 2008; Berger 2011; Van Staden and Van Westhuizen 2013; Gierdien 2014; Leendertz et al. 2015; Stols and Kriek 2011) came in seventh but with increasing intensity in the second half of the period under review. With the proliferation of ICTs in numerous forms and platforms, this is an encouraging sign. However, it still falls short of the rate at which ICT tools themselves are becoming increasingly ubiquitous. To this end, Blignaut et al. (2010) lament the lack of ICT competency among teachers in terms of basic ICT, as revealed by the Second International Technology Education Studies (SITES) of 2006.

Studies relating to the impact/effectiveness of TPL programme interventions, notably on student learning outcomes, were few (e.g., Frick et al. 2008; Ndlovu 2011a, b, c, d; Ndlovu 2014; Pourana et al. 2015). Similarly, studies that specifically focused on TPL needs were few and far between (e.g., Rakumako and Laugksch 2010; Wessels and Nieuwoudt 2011; Julie et al. 2011).

22.4 Discussion of Results

It is clear that the research emphasis focussed sharply on teachers' mathematical knowledge, be it as pure SMK or fused with pedagogy as in PCK or more specialised as mathematics for teaching (MFT) or mathematical knowledge for teaching (MKT/MKfT). Mathematical knowledge for teaching also re-appeared as mathematical literacy knowledge for teaching (MLK), Statistical knowledge for teaching (SKT) and more recently as mathematical discourse in instruction (MDI; e.g., Adler and Ronda 2015; Venkat and Adler 2012). Although this might be taken for granted in other countries, it appeared to be the core problem vexing the quality of mathematics teaching to the extent that teachers' mathematical discourses in action needed to be analysed in context—some form of situated cognition—and acted upon relevantly.

Lack of resources in disadvantaged schools included knowledge resources such as mathematics for teaching abilities of teachers and cultural resources such as the undermined/underutilised potential of indigenous languages in the teaching of mathematics, requiring re-sourcing of teachers (e.g., Adler 2012; Dicker 2015; Setati 2008). Research on the integration of ICT in mathematics education might have been impeded by the novelty of the technology itself, which caused in-service providers to also be co-learners in most instances. The paucity of research on ICT integration pointed to the possibility that very few in-service teacher educators were comfortable (i.e., digital natives) with the new tools. For example, Stols and Kriek (2011) asked the question 'why don't all maths teachers use dynamic geometry software in their classrooms?' They postulate that their beliefs could initially hinder their intention to use technology, but once exposed, the intention and the actual use of technology actually increases. However, more research is needed to establish the full range of impediments. For example, a strong possibility exists that as ICT proliferates, there may be many more practitioners who use technology than researchers think—it just has not been written about in journal articles.

Many mathematics teachers appeared to be out of their depth, especially in the secondary school sector. The majority of the studies sought to describe or unpack the nature of mathematical knowledge required for teaching and what mathematical difficulties underqualified teachers encountered even as they were being upgraded. Given the long history of neglect of the education of the majority, it was understandable that unless a completely new breed of teachers was trained overnight to replace the old cadres, the problems of under-equipped and inappropriately deployed teachers would persist.

The rapid changes in the curriculum made the design of teacher professional programmes a rather messy issue, as researchers had to attempt to fix a system that in itself was a moving target. Very often, changes in policy outstripped the supply of appropriately qualified or appropriately retrained teachers, as Mbekwa (2006) and other researchers highlighted, for example, in the case of mathematical literacy. A constant shortcoming of in-service teacher education was its general failure to influence the quality of education positively. The very few studies on impact may suggest a persistent lack of capacity or competencies as well as funding constraints among researchers to conduct large-scale randomised control trial (RCT) projects. Following the success of the East Asian countries in international benchmark tests, some research on professional learning communities and lesson study has surfaced as a prospectively more sustainable, teacher-driven (bottom-up) alternative. The notion of PLCs has been embraced at the policy level by both the Department of Basic Education (DBE) and the Department of Higher Education and Training (DHET) in the integrated strategic planning framework for teacher education and development (DBE and DHET 2011). However, the initiation of such communities of practice might still be a challenge as there may not be enough skilled personnel (e.g., professors) to stimulate them all over the country with equal zeal and skill as in the reported research. There might also not be enough resources nor enough time and incentives for participating teachers to embrace them wholeheartedly.

22.4.1 Implications for the Way Forward

The research has on balance led to greater local understanding and interpretation of pedagogical problems affecting the South African mathematics education conundrum. Whereas some researchers appear to bemoan the ineffectiveness of one-off or hit-and-run approaches to teacher professional learning (e.g., Jita and Mokhele 2012), formal in-service teacher education programmes such as the ACE reported here have also lacked consistency in standardising what is taught. The broad consensus appears to be that more long-term formats of teacher professional learning programmes may yield more long-lasting improvements. Whereas the pace of curriculum change has to be moderated, there is also the challenge for initial teacher education to produce more adaptable educators, perhaps partly by extending teaching experience and partly by investing in ICT infrastructure at higher education institutions so that the integration of ICT in mathematics can be more easily and naturally embraced in the field. This could be backed by improved research funding for its integration in schools, not only for mathematics but for all other subjects as well.

Figure 22.6 highlights the need for constant interaction between initial teacher education, continuous teacher professional learning and student learning outcomes.

Fig. 22.6 The interrelationship between TPL, student learning outcomes and initial teacher education

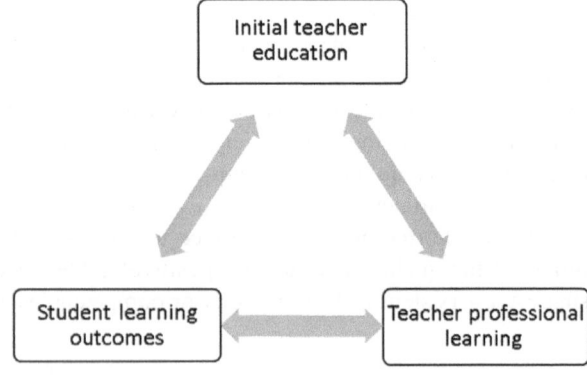

Furthermore, looking at the comparatively extensive volume of research on in-service mathematics teacher professional development and learning that South Africa has produced, it is worth recommending the consolidation of the funding model for higher education research that has been a major catalyst for research productivity. Since most research emanates from the higher education sector, it raises the prospects of the researchers to influence the re-curriculation of initial teacher education (ITE) so that the gap between ITE and the realities of the practicing South African mathematics teacher can be as narrow as possible.

References

Adler, J. (2012). Knowledge resources in and for school mathematics teaching. In G. Gueudet et al. (Eds.), *From text to 'Lived' resources, mathematics teacher education* (Vol. 7, pp. 3–22). Dordrecht: Springer.

Adler, J., & Davis, Z. (2006) Opening another black box: Researching mathematics for teaching in mathematics teacher education, *Journal for Research in Mathematics Education, 37*(4), 270–296.

Adler, J., & Pillay, V. (2007). An investigation into mathematics for teaching: Insights from a case. *African Journal of Research in Mathematics, Science and Technology Education, 11*(2), 87–102.

Adler, J., & Ronda, E. (2015). A framework for describing mathematics discourse in instruction and interpreting differences in teaching. *African Journal of Research in Mathematics, Science and Technology Education, 19*(3), 237–254.

Aldridge, J., Fraser, B., & Ntuli, S. (2009). Utilising learning environment assessment to improve teaching practices among in-service teachers undertaking a distance-education programme. *South African Journal of Education, 29,* 147–170.

Bansilal, S. (2011). Unpacking mathematical literacy teachers' understanding of the concept of inflation. *African Journal of Research in Mathematics, Science and Technology Education, 15*(2), 179–190.

Bansilal, S. (2014a). Exploring the notion of mathematical Literacy teacher knowledge. *South African Journal of Higher Education, 28*(4), 1156–1172.

Bansilal, S. (2014b). Mathematical literacy teachers' engagement with contextualised income tax calculations. *Pythagoras, 35*(2), 35–44.

Bansilal, S. (2015). An exploration of the assessment and feedback practices in a practical teaching intervention for in-service teachers. *International Journal of Educational Sciences, 8*(1-i), 23–35.

Bansilal, S., Mhkwananzi, T., & Brijlal, D. (2014). An exploration of the common content knowledge of high school mathematics teachers. *Perspectives in Education, 32*(1), 34–50.

Berger, M. (2011). A framework for examining characteristics of computer-based mathematical tasks. *African Journal of Research in Mathematics, Science and Technology Education, 15*(2), 111–123.

Berger, M. (2013). Examining mathematical discourse to understand in-service teachers' mathematical activities. *Pythagoras, 34*(1), 1–10.

Biccard, P., & Wessels, D. (2015). Student mathematical activity as a springboard to developing teacher didactisation practices. *Pythagoras, 36*(2), 1–9.

Blignaut, S., Els, C., & Howie, S. (2010). Contextualising South Africa's participation in the SITES 2006 module. *South African Journal of Education, 30*(4), 555–570.

Brijlall, D. (2014). Exploring the pedagogical content knowledge for teaching probability in middle school: A South African case study. *International Journal of Educational Sciences, 7* (3): 719–726.

Brijlall, D., Bansilal, S., & Moore-Russo, D. (2012). Exploring teachers' conceptions of representations in mathematics through the lens of positive deliberative interaction. *Pythagoras, 33*(2), 1–8.

Brodie, K., & Sanni, R. (2014). 'We won't know it since we don't teach it': Interactions between teachers' knowledge and practice. *African Journal of Research in Mathematics, Science and Technology Education, 18*(2), 188–197.

Brodie, K., & Shalem, Y. (2011). Accountability conversations: Mathematics teachers' learning through challenge and solidarity. *African Journal of Research in Mathematics, Science and Technology Education.*

Brodie, K. (2007). Dialogue in mathematics classrooms: Beyond question-and-answer methods. *Pythagoras, 66,* 3–13 (PCK).

Brodie, K. (2013). The power of professional learning communities. *Education as Change, 17*(1), 5–18.

Brown, B., & Schafer, M. (2006). Teacher education mathematical literacy: A modelling approach. *Pythagoras, 64,* 45–51.

Brown, B. A. (2010). Teachers' accounts of the usefulness of multigrade teaching in promoting sustainable human-development related outcomes in rural South Africa. *Journal of Southern African Studies, 36*(1), 189–207.

Departments of Basic Education & Higher Education and Training. (2011). *Integrated strategic planning framework for teacher education and development in South Africa 2011—2015: Technical report.* Pretoria: Departments of Basic Education & Higher Education and Training.

Dicker, A. (2015). Teaching mathematics in foundation phase multilingual classrooms: Teachers' challenges and innovations. *International Journal of Educational Sciences, 8*(1-i), 65–73.

Fricke, I., Horak., Meyer, E., & van Lingen, L. (2008). Lessons from a mathematics and science intervention programme in Tshwane township schools. *South African Journal of Higher Education, 22*(1), 64–77.

Gierdien, F. (2008). Teacher learning about probabilistic reasoning in relation to teaching it in an Advanced Certificate in Education (ACE) programme. *South African Journal of Education, 28,* 19–38.

Gierdien, M. F. (2014). On the use of spreadsheets algebra programs in the professional development of teachers from selected township high schools. *African Journal of Research in Mathematics, Science and Technology Education, 18*(1), 87–99.

Graven, M., & Venkat, H. (2014). Primary teachers' experiences relating to the administration processes of high-stakes testing: The case of Mathematics Annual National Assessments. *African Journal of Research in Mathematics, Science and Technology Education, 18*(3), 299–310.

Jita, L. C. & Mokhele, L. M. (2012) Institutionalising teacher clusters in south Africa: dilemmas and contradictions. *Perspectives in Education, 30*(2), 1–11.

Jita, L. C., & Mokhele, L. M. (2014). When teacher clusters work: Selected experiences of South African teachers with the cluster approach to professional development. *South African Journal of Education, 3*(2), 1–15.

Julie, C. (2006). Teachers' preferred contexts for mathematical literacy as possible initiators for Mathematics for Action. *African Journal of Research in Mathematics, Science and Technology Education, 10*(2), 49–58.

Julie, C. (2009). Appropriate contexts for teachers in-service courses: Perspectives of practising teachers. *South African Journal of Higher Education, 23*(1), 113–126.

Julie, C., Holtman, L., & Mbekwa, M. (2011). Rash modelling of mathematics and science teachers' preferences of real-life situations to be used in mathematical literacy. *Pythagoras, 32* (1), 11–18.

Kaino, L. M., Dhlamini, J. J., Phoshoko, M., Jojo, Z. M. M., Paulsen, R., & Ngoepe, M. (2015). Trends in mathematics professional development programmes in post-apartheid South Africa. *International Journal of Educational Sciences, 8*(i-1ii), 155–163.

Kazima, M., & Adler, J. (2006). Mathematical knowledge for teaching: Adding to the description through a study of probability in practice. *Pythagoras, 63*, 45–59.

Kazima, M., Pillay, V., & Adler, J. (2008). Mathematics for teaching: Observations from two case studies. *South African Journal of Education, 28*, 283–299.

Khuzwayo, M., & Mashiya, N. (2015). Gridlocked in a lesson plan triangle: The perceptions of in-service student teachers. *International Journal of Educational Sciences, 9*(3), 375–381.

Lampen, E. (2015). Teacher narratives in making sense of statistical mean algorithm. *Pythagoras, 36*(1), 1–12.

Leendertz, V., Blignaut, A. S., Ellis, S., & Nieuwoudt, H. D. (2015). The development, validation and standardisation of a questionnaire for ICT professional development of mathematics teachers. *Pythagoras, 36*(2), 1–11.

Likwambe, B., & Christiansen, I. M. (2008). A case study of the development of in-service teachers' concept images of the derivative. *Pythagoras, 68*, 22–31.

Mbekwa, M. (2006). Teachers' views on mathematical literacy and on their experiences as students of the course. *Pythagoras, 63*, 22–29.

Mhlolo, M., & Schäfer, M. (2012). Towards empowering learners in a democratic mathematics classroom: To what extent are teachers' listening orientations conductive to and respectful of learners' thinking. *Pythagoras, 33*(2), 1–9.

Mhlolo, M. K., Venkat, H., & Schäfer, M. (2012). The nature and quality of the mathematical connections teachers' make. *Pythagoras, 33*(1), 1–9.

Mokhele, L. M., & Jita, L. C. (2012a). When professional development works: South African teachers' perspectives. *Anthropologist, 1*(6), 575–585.

Mokhele, L. M., & Jita, L. C. (2012b). Institutionalising teacher clusters in South Africa. *Perspectives in Education.*

Mokhele, M. (2013). Empowering teachers: An alternative model for professional development in South Africa. *Journal of Social Sciences, 34*(1), 73–81.

Molefe, N., & Brodie, K. (2010). Teaching mathematics in the context of curriculum change. *Pythagoras, 71*, 3–12.

Mudaly, V., & Moore-Russo, D. (2011). South African teachers' conceptualisations of gradient: A study of historically disadvantaged teachers in an advanced certificate in education programme. *Pythagoras, 32*(1), 1–8.

Mwakapenda, W., & Dhlamini, J. (2009). Integrating mathematics and other learning areas: Emerging tensions from a study involving four classroom teachers. *Pythagoras, 71*, 22–29.

Ndlovu, M., & Mji, A. (2012). Pedagogical implications of students' misconceptions about deductive geometric proof. *Acta Academica, 44*(3), 176–205.

Ndlovu, M. (2014). The effectiveness of a teacher professional learning programme: The perceptions and performance of mathematics teachers. *Pythagoras, 35*(2), 1–10.

Ndlovu, M. C. (2011a). Re-envisioning the scholarship of engagement: Lessons from a university-school partnership project for mathematics and science teaching. *South African Journal of Higher Education, 25*(7), 1397–1415.

Ndlovu, M. C. (2011b). Students' perceptions of the effectiveness of their in-service training for the advanced certificate in education programme. *South African Journal of Higher Education, 25*(3), 523–541.

Ndlovu, M. C. (2011c). The pedagogy of hope at IMSTUS: Interpretation and manifestation. *South African Journal of Higher Education, 25*(1), 41–55.

Ndlovu, M. C. (2011d). University-school partnerships for social justice in mathematics and science education: The case of the SMILES project at IMSTUS. *South African Journal of Education, 31*, 419–433.

Nel, B. (2012). Transformation of teacher identity through a Mathematical Literacy re-skilling programme. *South African Journal of Education, 32,* 144–154.

Ono, Y., & Ferreira, J. (2010). A case study of continuing teacher professional development through lesson study in South Africa. *South African Journal of Education, 30,* 59–74.

Owusu-Mensah, J. (2014). The value of mentoring for mathematical literacy teachers in the South African school system. *International Journal of Educational Sciences, 7*(3), 509–515.

Paulsen, R. (2015). Professional development as a process of change: Some reflections on mathematics teacher development. *International Journal of Educational Sciences, 8*(1-ii), 215–221.

Pausigere, P., & Graven, M. (2014). Learning metaphors and learning stories of teachers participating in an in-service numeracy community of practice. *Education as Change, 18*(1), 33–46.

Phoshoko, M. (2015). Experiences of role players in the implementation of mathematics teacher continuous professional development in South Africa. *International Journal of Educational Sciences, 8*(1-ii), 241–248.

Plotz, M., Froneman, S., & Nieuwoudt, H. (2012). A model for the development and transformation of teachers' mathematical content knowledge. *African Journal of Research in Mathematics, Science and Technology Education, 16*(1), 69–81.

Posthuma, B. (2012). Mathematics teachers' reflective practice within the context of adapted lesson study. *Pythagoras, 33*(3), 1–9.

Pournara, C., Hodgen, J., Adler, J., & Pillay, V. (2015). Can improving teachers' knowledge of mathematics lead to gains in learners' attainment in mathematics? *South African Journal of Education, 35*(3), 1–10.

Rakumako, A., & Laugksch, R. (2010) Demographic profile and perceived INSET needs of secondary mathematics teachers in Limpopo province. *South African Journal of Educaiton, 30,* 139–152.

Setati, M. (2008). Access to mathematics versus access to the language of power: The struggle in multilingual mathematics classrooms. *South African Journal of Education, 28,* 103–116.

Shalem, Y., Sapire, I., & Huntley, B. (2013). Mapping onto the mathematics curriculum—an opportunity for teachers to learn. *Pythagoras, 34*(4), 1–10.

Shalem, Y., Sapire, I., & Sorto, M. A. (2014). Teachers' explanations of learners' errors in standardised mathematics assessments. *Pythagoras, 35*(1), 1–11.

Singh, S. K. (2011). The role of staff development in the professional development of teachers: Implications for in-service training. *South African Journal of Higher Education, 25*(8), 1626–1638.

Stols, G., & Kriek, J. (2011). Why don't all maths teachers use dynamic geometry software in their classrooms? *Australian Journal of Educational Technology, 27*(1), 137–151.

Stols, G., Ferreira, R., Pelser, A., Olivier, W. A., van der Merwe, A., De Viliers, S., et al. (2015a). Perceptions and needs of South African mathematics teachers concerning their use of technology for instruction. *South African Journal of Education, 35*(4), 1–13.

Stols, G., Mji, A., & Wessels, D. (2008). The potential of teacher development with geometer's sketchpad. *Pythagoras, 68,* 15–21.

Stols, G., Ono, Y., & Rogan, J. (2015b). What constitutes effective mathematics teaching? Perceptions of teachers. *African Journal of Research in Mathematics, Science and Technology Education, 19*(3), 225–236.

Tosavainen, T., Attorps, I., & Väisänen, P. (2013). Some South African mathematics teachers' concept images of the equation concept. *African Journal of Research in Mathematics, Science and Technology Education, 16*(3), 376–389.

Van der Merwe, T. M., & van der Merwe, A. J. (2008). Online professional development: Tensions impacting on the reflective use of a mathematics friendly forum environment. *South African Computer Journal, 42,* 59–67.

Van Laren, L., & Moore-Russo, D. (2012). The most important aspects of algebra: Responses from practising South African teachers. *African Journal of Research in Mathematics, Science and Technology Education, 16*(1), 45–57.

Van Staten, C. J., & Van der Westhuizen, D. (2013). Learn 2.0 technologies and the continuing professional development of secondary school mathematics teachers. *Journal for New Generation Sciences, 2*(2), 141–157.

Venkat, H., & Adler, J. (2012). Coherence and connections in teachers' mathematical discourses in instruction. *Pythagoras, 33*(3), 1–8.

Webb, L. (2015). The use of cartoons as a tool to support teacher ownership of mathematics curriculum change. *African Journal of Research in Mathematics, Science and Technology Education, 19*(1), 57–68.

Wessels, H., & Nieuwoudt, H. (2011). Teachers' professional development needs in data handling and probability. *Pythagoras, 32*(1), 1–9.

Wessels, H., & Nieuwoudt, H. (2013). Teachers' reasoning in a repeated sampling context. *Pythagoras, 34*(2), 1–11.

Chapter 23
Pedagogies of Emergent Learning

Ricardo Nemirovsky

Abstract We distinguish emergent learning from "teleological" learning, which is learning for the sake of passing pre-defined tests and goals. While teleological learning may succeed or fail, emergent learning is always going on in ways that move pass disciplinary boundaries and anticipated results. To advance a perspective on pedagogies of emergent learning we analyze selected episodes from a program for children who volunteered to enroll. The sessions alternated between the after school club they attended and an art museum. The program engaged the children in basket weaving, in the analysis of baskets exhibited at the museum, and with ways in which flat materials can be shaped in 3D space along distinct surface curvatures. These experiences have inspired us to outline two streams of pedagogical ideas that seem to nurture and go along with the unforeseeable paths of emergent learning.

Keywords Informal mathematics learning · Emergent learning
Pedagogy · Museum learning · Crafts and mathematics

23.1 Introduction

We contrast emergent learning with "teleological" learning, which is learning for the sake of passing pre-defined tests and goals. To grasp the nature of emergent learning and how it differs from teleological learning, we review one of the best known and most cited papers in mathematics education: "The case of Benny" (Erlwanger 1973). In sixth grade Benny was regarded as one of the best students in

---

This paper is the result of collaborative work with Cierra Rawlings and Bohdan Rhodehamel at San Diego State University, and Molly Kelton at Washington State University. Due to the ICME rules for invited lectures, they have not been included as co-authors. I want to thank them for their contribution and for their agreement to the latter.

R. Nemirovsky (✉)
Manchester Metropolitan University, Manchester, England
e-mail: bnemirovsky@mail.sdsu.edu

401

G. Kaiser et al. (eds.), *Invited Lectures from the 13th International Congress on Mathematical Education*, ICME-13 Monographs,
https://doi.org/10.1007/978-3-319-72170-5_23

his mathematics class. Since second-grade Benny had been using "Individually Prescribed Instruction" (IPI): a structured sequence of exercises punctuated by multiple-choice tests, such that 80% of the answers were required to be correct in order to advance in the sequence. Benny "was making much better than average progress through the IPI program" (p. 7), which indicated that his test responses had largely been correct according to the key provided by the IPI program. However, as Erlwanger interviewed him, he noticed that Benny was computing answers to problems by applying a multitude of self-generated rules many of which were incorrect, even though in particular cases they would lead to answers consistent with the key (e.g. the result of 0.7×0.5 is 0.35 because, on the left side, "there's two points in front of each number" (p. 8), then Benny used the same rule to evaluate: $0.3 + 0.4 = 0.07$). In addition to many idiosyncratic rules—Benny estimated that "in fractions we have 100 different kind of rules" (p. 10)—his five years of experiences with IPI led him to adopt certain views about the nature of mathematics and mathematics learning. The rules, he thought,

> were invented "by a man or someone who was very smart." This was an enormous task because, "it must have took this guy a long time... about 50 years... because to get the rules he had to work all of the problems out" (p. 12)

Applying diverse rules Benny was able to obtain different answers to the same problem, all of which he deemed to be true ones, although the IPI key accepted only one of them. Erlwanger asked Benny why the teachers would mark as wrong all these other true answers: "They mark it wrong because they just go by the key. They don't go by if the answer is true or not" (p. 12). This mismatch between the variety of true answers and the single one chosen by the key, Benny remarked, "is why nowadays we kids get the fractions wrong" (p. 11).

The practices involved in the use of IPI hinged on whether the students obtained adequate scores on its tests. We call this kind of learning "teleological:" learning for the sake of passing pre-defined tests and goals. At the same time, Benny learned many skills, ways of thinking, and forms of social awareness that were not pre-specified or even intended by the program, the school, and the participating teachers, such as the distinction between truth and key selection, his own confidence as a prolific maker of mathematical rules, or mathematics as an invention of a very smart and hard-working man. We will refer to this learning as "emergent." The concept of emergence is currently used in a range of disciplines, from complexity theory and thermodynamics of far-from-equilibrium systems, to system dynamics and organizational theory (Goldstein 1999; Kreps 2015). Characteristics of emergence include that it is largely unpredictable, not reducible to internal components and variables, self-organizing, and creative. Teleological Learning may succeed or fail; in the Case of Benny success was achieved to a certain degree, as he made more than average progress through the IPI program. Other researchers (Jacobson and Kapur 2012; Jacobson et al. 2016; Jacobson and Wilensky 2006) have elaborated on an approach to emergent learning that differs from ours because they base their work on an a computational model of learning.

In contrast to its etymological roots (Young 1987), the word "pedagogy" is nowadays strongly associated with formal teaching and schooling (Hamilton 1999).

This association elicits a paradoxical sense to the phrase "pedagogy of emergent learning" because school teaching is commonly seen as inherently teleological, as if, without explicit behavioral goals, teaching were to dissolve into a mass of incoherent and random interactions devoid of purpose. Emergent learning, being elusive to anticipated aims and predicted outcomes, appears, for the most part, to be an unintended byproduct of schooling practices that are bound to the achievement of testable results. Thus, clarifying what we mean by pedagogies of emergent learning is a critical matter.

We conceive of pedagogy of emergent learning as one that drifts and moves along unanticipated flows of emergent learning traversing educands and educators, one in which spontaneous memories, speculations, and projects of the participants may take center stage regardless of whether they accord with pre-conceived end-points. While a pedagogy of emergent learning seeks to instigate collective improvisation, it does preserve the asymmetry between educators and educands, although treating the axis of this asymmetry as, in the words of Rancière (1991 p. 13), will to will and not intelligence to intelligence. "Will to will" entails that educators plan, facilitate, and orchestrate the activities the group engages in; "not intelligence to intelligence" implicates that participants share, make sense and pursue these activities in their own ways nurtured by their desires, histories and contexts of life. In other words, while there is an inequality educand/educator in that the latter regulates and sets up the stage for their joint work, there is a primordial equality educand/educator in their autonomy for expression, recollection, concep-tualization, initiative, and insight.

Pedagogy of the Oppressed (Freire 1970) and *The Ignorant Schoolmaster* (Rancière 1991) seem to us inspiring for the development of a pedagogy of emergent learning. Freire saw the dialogue between educators and educands as necessitating humility and the sense that they are all learners: "At the point of encounter there are neither utter ignoramuses nor perfect sages; there are only people who are attempting, together, to learn more than they now know" (Freire 1970). Educators are also learners who struggle against their own assumptions and expectations, pursuing to "learn more than they now know." While it is said that the main goal of the pedagogy of the oppressed is "conscientização" (i.e. approxi-mately, to become aware), we think it is more accurate to say that its goal is to elucidate, to some extent, what "conscientização" amounts to in the context of the circumstances of the educands and educators, as well as the histories of their lives. What makes the pedagogy of the oppressed non-teleological is not the absence of goals, but coping with the ongoing persistent challenge of what the goals are, as well as the openness to their being constantly transformed into new, unanticipated, and often surprising and provisional ends. In other words, the goals themselves are emergent, which entails that they are diverse, shifting, ephemeral, situated, and co-generated.

We suggest that case studies are main sources to elaborate on pedagogies of emergent learning. An important example is the case of "SeanNumbers-Ofala" (2010, Online) focused on interactions in a third grade classroom taught by Deborah L. Ball. This paper is a case study based on a program entitled "Basket

Weaving and Curvature" that was conducted during the fall of 2015. The program was part of the InforMath project, one of whose main goals is to investigate/design informal learning environments amenable to the creation of new social images of mathematics—images that are more inclusive and inspiring than the prevalent ones in our society. Note that this is a goal without finish line and goalposts, not unlike the one of "conscientização." It is a goal irremediably recursive, the pursuit of which entails an ongoing questioning, hopefully insightful, of what it is about and where it comes from.

23.2 Basket Weaving and Curvature

This program is one of several that have been designed and conducted in the context of the InforMath project, which is a collaborative initiative including museum educators from three museums located in Balboa Park, San Diego, as well as faculty members and graduate students from San Diego State University. The children, all members of an after school program at the Boys and Girls Club of Southeast San Diego, volunteered to participate. A recruitment session was held at the Chula Vista clubhouse, where all students in grades 5–8 attended a brief presentation about Mingei museum and had the opportunity to engage with a weaving activity composed of yarn and a cardboard loom. The program consisted of six sessions that took place every other week and alternated locations between the museum and the Chula Vista clubhouse. Eleven students signed up, they were 9–12 years old, with four girls and seven boys, of which eight completed the program. To record each session, two stationary video cameras were used as well as head cameras worn by several of the kids. The Basket Weaving and Curvature program was designed and conducted by two museum educators, Lucera and Johanna, and two math educators, Ricardo and Cierra. Lucera led the activities during the sessions themselves. We will refer to the four of them as the "educators." The educators met in between sessions to design the ensuing ones and to prepare materials accordingly.

The springboard of the program was an exhibition hosted by the Mingei International Museum called "Made in America," which included outstanding craft products from each of the 50 US states. *Made In America* was in the process of installation when we began to envision the program. Lucera and Johanna had produced educational materials to accompany the exhibition. Apprehending the upcoming exhibition as a suitable arena for a program intermingling mathematics and crafts, to be attended by children from Southeast San Diego, were the initial issues we worked on. While the collection encompassed a wide variety of techniques and materials, during our preliminary visits we were particularly lured by several handcrafted baskets (see Fig. 23.1), as well as encouraged by Lucera's past workshop experiences, to engage children in basket weaving. We held several preparatory sessions. In one of them Lucera taught Ricardo and Cierra to create round baskets using two different techniques: coiling and weaving. In parallel to

Fig. 23.1 Some of the baskets included in the *Made in America* exhibition

this preparatory work, Ricardo and Cierra were participating in an online seminar with a mathematician, John McCleary, on topics of differential geometry. This seminar was one of the professional development initiatives held by the InforMath project. At the time, topics discussed in this seminar included surface curvature and geodesics. This overlap of activities evoked the idea of basket weaving as a set of techniques to create, out of flat materials, a shape in three-dimension space. Since 3D shapes can be characterized by the local curvature for each point of the surface, basket weaving appeared to be a suitable maker's context to encounter and use ideas about surface curvature.

23.3   Episode 1

The first session took place at the Mingei. After a warm up activity, the director of the museum and curator of *Made in America*, Rob Sidner, led a visit to the gallery floor explaining the history of the exhibition and conversing with the children about their impressions and questions. Afterwards the group gathered at the museum's workroom. Lucera initiated a discussion about the differences between straight and curved lines. She introduced the children to a tool we refer to as a "curvature instrument."

The curvature instrument had been designed by the educators over a three-week period before the beginning of the program. After trying out different designs, the final version consisted of a "cross" made out of cardstock with pipe cleaners in between; the two pieces of cardstock were stapled along the edges to keep the pipe cleaners in between. The curvature instrument is used by placing it over an object or certain shape of interest aligning one non-adjacent pair of arms along the orientation of maximal curvature, and contouring the remaining pair of arms to the shape of the object (see Fig. 23.2). The pipe cleaners help maintain the shape of the surface after the tool is detached from the object The idea of the curvature instrument arose from trying to figure out ways to support children to develop an intuitive sense of

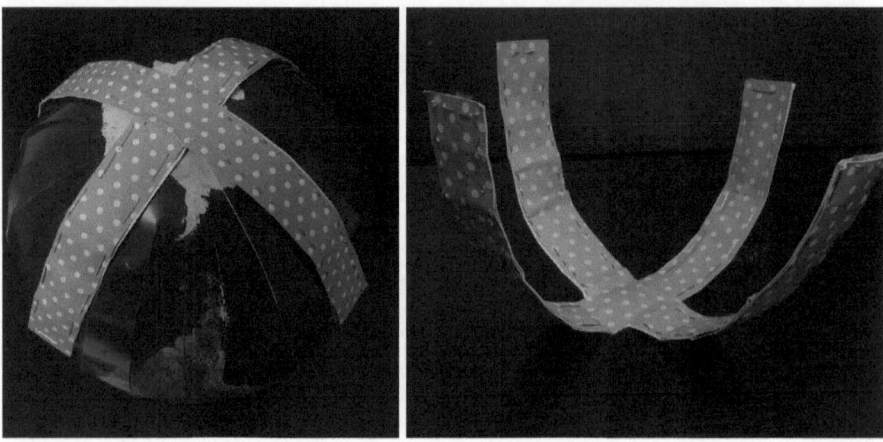

Fig. 23.2 Curvature instrument on a sphere (left) and removed (right)

Gaussian curvature at a given point on the surface. Gaussian curvature is obtained by multiplying the maximum and minimum linear curvature around the point of interest. Euler proved in 1760 that on smooth surfaces the maximum and minimum linear curvatures are perpendicular to each other, which necessitated the perpendicularity of the arms of the curvature instrument.

Lucera explained that when the two non-adjacent pairs are bent in the same direction it is said that the curvature is "positive," whereas if each of the two pairs are bent in opposite directions it is "negative"; if one or both non-adjacent pairs are flat the curvature is zero. She showed how the top of her head had a positive curvature whereas the inner side of a bent elbow or knee, has a negative one. The children then used the curvature instrument to ascertain different types of curvature on their bodies. During the last segment of the session Lucera showed how to weave a basket with pipe cleaners and yarn, and then the children selected materials and started to make their first basket.

The second session took place at the Chula Vista clubhouse. They reviewed the activities of the first session, watched and discussed a video showing craftsmen creating blown glass pieces, and continued work on their baskets. During the third session, at the Mingei, the group talked about their baskets and compared techniques (e.g. looping the yarn around each spoke or just alternating inside/outside each spoke). Afterwards they went to the gallery floor to observe and discuss different pieces, particularly woven baskets. Students were encouraged to speculate about what materials and processes went into creating the art pieces. Episode 1 took place during this visit to the gallery floor.

Fig. 23.3 **a** Outline of one of the many vertical reed spokes. **b** Slicing the basket horizontally along equator. **c** Tracing the lower half going outwards. **d** Tracing the upper half going inwards

- Annotated Transcript

1 Lucera: So, this basket ((see Fig. 23.3a)) also uses spokes. So, it has a bunch of reeds
2   ((which are the spokes, see white outline in Fig. 23.3a)) going up the sides
3   but I thought this basket was interesting, um, because it started out– if you
4   just imagine, like, slice it in half ((makes horizontal slicing motion with flat
5   hand, see Fig. 23.3b)) and the bottom is just like a regular bowl going out
6   ((makes upward swinging gesture following contour of lower half of
7   basket, see Fig. 23.3c)) but then it started going back in ((uses hand to trace
8   the contour of the upper half of basket going in, see Fig. 23.3d)). So, how do
9   you think they did it on this one? Yeah?

Commentary

Lucera started [1–3] by highlighting the vertical reed spokes traversing the basket. Then she imaginarily sliced the basket in half, to mark a difference between the bottom part (i.e. "going out" [5]) and the upper part (i.e. "going back in" [7]). She asked how the basket weaver managed to produce this difference outwards-inwards [8]. Lucera's question: "So, how do you think they did it on this one?" [8], was an invitation to conjecture the making of a difference. Generally speaking, woven baskets obtain a shape outwards by increasing the distance between spokes or widening the spokes themselves, and likewise turn inwards by decreasing the distance or spoke width—a relationship Lucera was familiar with. Lucera wished to discern, after calling their attention to the vertical spokes, how the children perceived the roles of the spokes in shaping up the basket. It is unlikely that any of the educators knew how to "explain" the relationship between variation of spoke width and curvature. This type of relationship is something that we literally grasp in the context of making, rather than the one of talking.

10 Ryan: So, like, what I found right here ((points at bottom of basket)) is like it's
11   going out ((curves hand to mirror contour of lower basket)) and then like
12   on this part ((points near equatorial region where handle reeds depart
13   from wall of basket)) it's going that way ((motions upward with hand
14   following angle of handle reeds, see Fig. 23.4a)) so, like, they can, they can
15   carry it ((points at handle))…
16 Lucera: Mm-hmm.
17 Ryan: …like a handle
18 Lucera: So, okay, so over here ((near equator of basket)) it's like making a turn
19   ((referencing contour of upper basket that curves back in)).
20 Ryan: Yeah.
21 Lucera: Can everyone see where he's pointing? Can you point where you're…
22 Ryan: Like, this point where it goes like that ((uses pointer finger to trace the angle
23   of the handle reeds, see Fig. 23.4b))…
24 Lucera: Okay, so he's noticed it's making a turn ((curves hand to model curvature of
25   upper basket))

Commentary

From his side, Ryan noticed something different around the equator line: the appearance of a spoke going upwards to hold the handle in position. This spoke is unique because while it appears to be an ordinary spoke on the bottom half, then it breaks free and becomes handle support. Lucera understood his "then like on this

(a)   **(b)**

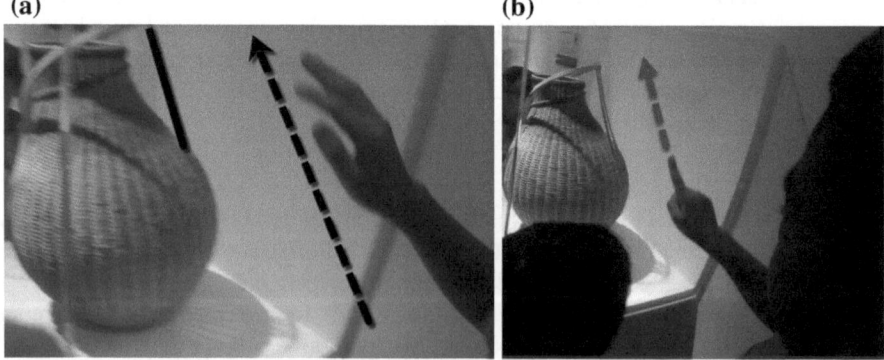

Fig. 23.4 **a** Ryan traces the spoke coming out of the woven reeds to support the handle. **b** Ryan traces the salient spoke again

part ((middle level)) it's going that way ((slanted upwards))" [12–13] as corroborating her "then it started going back in" [7], so that she re-described Ryan's words as "it's like making a turn" [16–17], from the lower to the upper half of the basket. Such mutual unawareness of differences between their accounts is a natural byproduct of the inherent ambiguity of utterances, exacerbated in this case by the inability to touch the basket, as well as the fact that we tend to be primed to perceive what we expect. Unless ambiguity turns out to be minimal, it is only through the insistence of a difference that we face it.

26 Lucera: Anybody else have ideas how this might've been made? Omar? Did you have
27   something?
28 Omar: I think they, they used, um, really, really, um, thin wood so it's easier for them
29   to, um, bend the ((points toward basket)) wood.
30 Lucera: Oh, okay. Okay. So, it's just a matter of using very thin wood so that they can
31   bend it ((makes upward sweeping, semicircular motion)). So you also
32   agree that there is some bending going on.
33 Omar: Yeah.
34 Lucera: Yeah?

Commentary

Omar thought that the basket maker had generated the outward/inward difference by bending the spokes, which required them to be made of an easy-to-bend material, such as thin wood. This might have resonated with his recent experiences weaving yarn around pipe cleaners that can be bent effortlessly. While isolated spokes show to contribute to the shape of the basket by their bent, the weaving reeds, as they bring the spokes into mutual relationships, make them bear and sustain the particular shape of the basket. This does not invalidate Omar's remark: had the spokes been made of rigid material, they would have refused to comply with the hands of the weaver as they interlaced the horizontal reeds. The main point we elicit in this commentary is the ongoing merging of perception and imagination, both materializing from memory: as Omar envisioned the (imaginary) making of this (perceived) basket, the salient feature that came to the present surface of memory—a memory that included the making of pipe cleaner baskets and much else—was the bending of the spokes.

35 Alexa: I think there is a little bit of bending going on, but like right there ((points
36   near equator of basket)), you can see that they're ((the spokes)) bigger and
37   then as they go up ((traces circles with finger as she raises arm)), they
38   become really small ((makes repeated pinching motions with index finger
39   and thumb)).
40 Lucera: Oh, okay. So let's look over here on this side ((see Fig. 23.5a)). So, she's
41   saying like kind of in the middle the ((indicates greater width with index
42   finger and thumb, see Fig. 23.5b)), the spokes, they get bigger, they get
43   thicker and then as it goes up ((raises arm, indicates lesser width with
44   index finger and thumb)), it gets smaller.

Fig. 23.5 a Other side of
basket. Note changing width
of reed spoke. **b** Lucera's
hand highlights greater width
of the spokes in the middle
height of the basket

Commentary

Alexa acknowledged that there is "bending going on" [35] but she foregrounded another variation: from wide to narrow width. She was referring to the spoke thickness, decreasing from the middle up. Note that Alexa gestured such movement up by tracing circles with her finger as she raised her arm [37]: Alexa imagined the making of this basket in terms of weaving reeds going around and gradually up. Just as in the making of a coiled basket, each circle of threaded reeds has a shorter and shorter perimeter in order to go inwards; the narrowing of the spokes generates such perimeter shortening. As in our previous commentary on Omar's remark, Alexa's utterance merges perception and imagination such that a particular feature came to the present from the depth of vast memories, recently stirred by her work with pipe cleaner basketry: the perimeter's shortening as circles move upwards. Lucera's requested the group to watch the basket from another side (see Fig. 23.5b), probably motivated by that one being the side that Alexa was observing, and perhaps also by that side of the basket being color-uniform (compare Fig. 23.5b and a), allowing for a more focused appreciation of the spoke width. From that side, Lucera highlighted the narrowing of the spokes [42–44].

45 Jake: Mmm.
46 Lucera: Does anyone else see that? Do you agree with that? Or are they the same
47   size all the way from the bottom to the top?
48 Omar: I think they're the same size.
49 Lucera: You think they're the same size? If you look at this little piece up here
50   ((makes measuring gesture with thumb and index finger near top of basket,
51   see Fig. 23.6a))– I wish I could touch it– and then this piece ((makes
52   measuring gesture near bottom)) and the middle ((makes measuring
53   gesture near equator, see Fig. 23.6b)) and down there ((makes measuring
54   gesture near bottom again, see Fig. 23.6c))...it's the same size? Who thinks
55   it's the same size? Raise your hand. ((four kids raise their hands)) ... Who
56   thinks it's a different size? ((four kids raise their hands, including one girl
57   who voted again))
58 Jake: I think it's just like an optical illusion.
59 Lucera: I think it's– Maybe it's an optical illusion? Okay, then that—yeah, that's

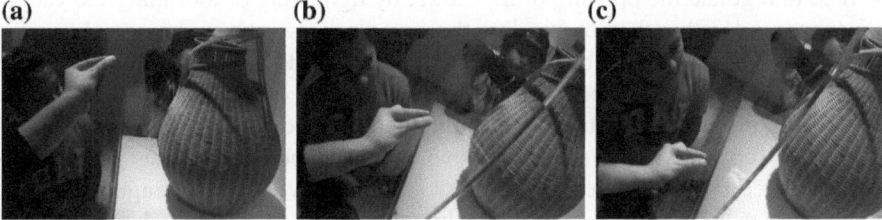

(a)   **(b)**   **(c)**

Fig. 23.6 a Lucera highlights spoke width at the top of the basket. **b** Lucera highlights spoke width in the middle of the basket. **c** Lucera highlights spoke width at the bottom of the basket

60   making me doubt myself but I think it's slightly different size. I think it, it
61   starts out, um, medium on the bottom ((uses fingers to indicate width near
62   bottom)) and then it gets a little bit thicker ((moves fingers up towards
63   equator to indicate width)) and then it gets thin at the top ((moves fingers
64   near top to indicate width)) Just slightly.
65 Ryan: I wish I could touch it.
66 Jake: Slightly

Commentary

Sensing that some of the children were unconvinced by Alexa's observation [45 and 48], Lucera responded by wanting to show the decrease in width. Limited by her inability to touch the basket [51], she marked the spoke thickness by the separation between the tips of her thumb and index fingers, as they slid vertically over the glass surface. However, as she was enacting the spoke thickness with her fingers, she started to hesitate, to the point of bringing into question the observation itself ("it's the same size?" [54]). Lucera turned to the children asking for a poll of opinions. Following the mixed polled opinions, Jakes remarked that it was "just" an optical illusion [58]. That perception is infused with the imaginary does not mean that the distinction between them vanishes: the question "do we see it or imagine it?", which corresponds to "is it there or is it an illusion?", still makes sense, and emerges with full force when the difference in question is feeble. Lucera accounted for her own doubts by deeming the width difference to be "slight" [60], and yet, she still thought that it was there [60–64]. While the seeing was ambiguous, "thought" brought to her a sense that, in all likelihood, the spoke width varied. Tenuous differences create possibilities for thinking and seeing to reach different conclusions.

23.4 Transition

Woven baskets obtain a shape outwards by increasing the distance between the edges of each vertical spoke and inwards by decreasing it, which can occur with or without a change in spoke width. However, this relationship had not been salient in the practice of weaving yarn with pipe cleaners because, we think, the children worked to regulate the opening of the basket by tightening or loosening the yarn as it went around, rather than by bending the pipe cleaners—the spokes—and keeping their shape and position stable while weaving; this made the tensioning of the yarn the primary method for regulating whether the wall of the basket would curve outward or inward. This observation prompted us to explore alternative crafts in which the separation between successive pairs of spokes becomes the primary manual/material difference engendering shape. While we were seeking alternatives, it happened that a colleague at SDSU, who is a quilter, mentioned fabric bowls and lent us a book about it. This serendipitous event launched us into investigating the

Fig. 23.7 Fabric cutout
ready to sew into a bowl

manufacture of fabric bowls and experimenting with different materials and techniques. The kind of fabric bowl we envisioned would be created by sewing the edges of a flat piece of fabric cut with a shape similar to the one shown in Fig. 23.7.

After a lengthy process of repeated experimentation, we ended up using cotton fabric ironed on both sides of a thick stabilizer (see Fig. 23.8). This material was then cut with a laser cutter according to templates generated in Geometer's Sketchpad (see Fig. 23.9). The two control points can be moved to change the radius of curvature of the two arcs of a circle enclosing each petal. The petals can be seen as equivalent to the spokes in a woven basket; the shape of the petals determines the separation between two successive in/out thread shifts at different heights, and regulating accordingly the overall shape of the bowl (Fig. 23.10).

23.5 Episode 2

During the 4th session, at the Chula Vista clubhouse, the children were asked to wrap large balls in paper and discuss the origin of the wrinkles appearing on the wrapping paper. After this initial experience transforming a flat surface into a curved 3D shape, Lucera introduced the materials for the fabric bowls. Each child chose the fabric and the template they wanted to use. Most of the 5th session at the Mingei was spent sewing the fabric bowls. Then they went to the gallery floor with their bowls to discuss and identify ways in which the shape of baskets exhibited in *Made in America* were similar or different than the shape of their fabric bowls. Episode 2 took place during this visit.

Fig. 23.8 A sewn fabric bowl

Fig. 23.9 Template
generated in GSP with ten
"petals"

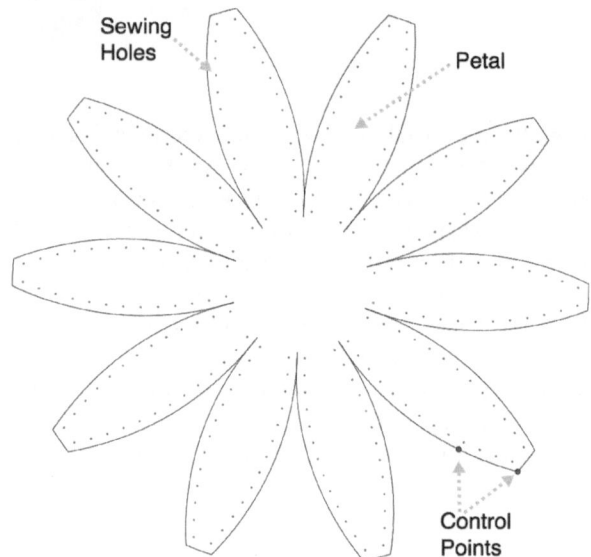

Fig. 23.10 A child sewing
his fabric bowl

- Annotated Transcript

1 Johanna: How about other people's baskets? Did you have a chance? Which one did
2   you notice?   [That looks like yours?
3 Gabriella: [Umm (.) Well, I… I noticed that THAT ((Gabriela points
4 at a basket on the opposite side of the room, see Fig. 23.11)) one over there…
5 …
6 Johanna: Oh, so you want to go all the way over here. Let's take a look.
7 Gabriella: ((while the group is walking towards the basket)) Yeah. If–um, Alexa
8 ((Gabriella looks towards Alexa)) sewed hers on up like, um, a basket, it would look
9 like this ((the one she had pointed at)).
10 Johanna: Oh, yeah. Okay, so you guys– 'cause yours ((Alexa's bowl)) isn't complete
11 yet ((See Fig. 23.12, Alexa's bowl is not completely sewn)) but we're
12 thinking that if that was sewn all of the way up that it would look really
13 similar to this ((basket selected by Gabriela))?
14 Allison: [Yeah, 'cause it's like she could pro'ly like bend it a little ((Allison
15   points at Alexa's bowl, on the upper side, see Fig. 23.13))
16 Gabriella: [Yeah, 'cause it's– it's small ((Gabriella shows how the slices become
17   narrower on the upper side, see right side of Fig. 23.13))
18 Allison: just to make it like the final thing to look like that ((like the basket they are
19   looking at)).

Fig. 23.11 Gabriella points
at a basket she had noticed

Fig. 23.12 Alexa holds her bowl, not yet completely sewn, whose overall shape will be similar to the exhibited basket

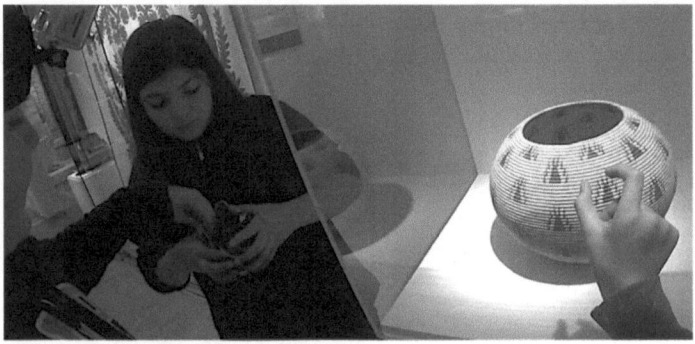

Fig. 23.13 Allison points at the upper side of Alexa's bowl while Gabriella, on the right side, shows that the upper side of the basket is "small"

20   Gabriella: 'Cause it goes small and then it gets fat ((Gabriella traces with her thumb
21   and her index finger a width that starts small, gets "fat", and then gets
22   smaller again, see Fig. 23.14))

Commentary

Alexa had chosen her template to be such that the edges of the petals fit an arc with a small radius of curvature—the type shown in Fig. 23.9. Most of the other children's templates were of the kind shown in Fig. 23.7. Her choice made of her bowl one that went conspicuously inwards over the upper half. This is the common feature that Allison indicated by touching Alexa's unfinished bowl and pointing at

Fig. 23.14 Gabriella traces the shape of the "petals" imaginarily forming the basket

the basket. Gabrielle traced on the glass panel the shape of a petal that would generate the contour of the basket see Fig. 23.14. Because this is a coiled basket that has no spokes or woven reeds, there is no physical indications on it of petal-like units; nevertheless, she found compelling, and others plausible, to imagine it split in petals with a certain outline, as if she were overlaying Alexa's bowl on the basket.

Returning to our theme of the imaginary infusing perception, both of them surging from memory, this episode is an occasion to elaborate further on the nature of memory and the significance of making and crafting for the genesis of memory. Memory is much ampler than individual recollections, both in the sense that what Allison and Gabrielle pointed out originated from group activity with a range of materials and tools, and that there was no single event from a past moment reenacted in their analysis of the basket. Furthermore, memory is deeply-rooted in materiality: the basket and Alexa's bowl remember countless versions of their own making, which they open to the grasp of others and things. Joining shaped petals is a "version" of the making of the basket—one that is markedly different from the processes that had been followed by its basket maker, but that was, in non-trivial ways, implicitly conveyed by them. The materiality that grounds memory allows for making and crafting being rich and nuanced sources: they bestow tangibility, texture, and bodily skill upon things, even when those things are out of touch or beyond the creative abilities of the perceivers. Neither Johanna nor the children could touch the basket, and yet, Alexa's bowl, unfinished and made out of other materials and techniques, passed onto the basket features graspable by skin, muscles, and sight. Because of this, making and crafting can be, among genetic sources of memory, exceptionally generous, just as the engagement with playing instruments can give birth to entirely new forms of music appreciation. Each craft gives in its own ways: basket weaving with yarn and pipe cleaners conferred to the basket shown in Fig. 23.3a attributes different from the ones sewing fabric bowls did to the basket shown in Fig. 23.14, such as the latter one being a composition of petals.

23.6   Towards a Pedagogy of Emergent Learning

Far from trying to demonstrate "best practices," we have shared some of our experiences with the Basket Weaving and Curvature program for the sake of investigating *a kind* of practice, whose main orientation is a quest for pedagogies of emergent learning. For this kind of pedagogy there are no best practices because no concrete attempt can be isolated from the circumstances of its development, the contingencies pervading its daily events, and the life history of the participant individuals and institutions. At most, given historic and contextual aspects, one can discriminate promising or rather-to-be-avoided ways of doing things. Ultimately, the character of a pedagogy of emergent learning is to be expressed by the ongoing outline of an ethics, which is a never-completed moving outline. Freire, for instance, emphasized the importance of "humility" on the part of educators and participants ("there are only people who are attempting, together, to learn more than they now know" Freire 1970). Like any other ethical rule, this is not a talisman. Not only because an act that seems humble to someone may strike as arrogant to someone else, but also because, aside from extreme cases, every judgment of this sort is necessarily bounded by invisible biases and more or less partial grasp of the circumstances. Ethical rules can be pursued not as maxims guiding behavior, but as efforts to keep certain questions alive (e.g. What does it mean, here and now, or there and yesterday, to be humble?). Part of this aliveness is the shared sense that there are no ideal or perfect actions and that, in retrospect, one can always imagine what appears to be more desirable ways of doing things, even though uncertainty about them cannot be dispelled (Nemirovsky et al. 2005). A pedagogy of emergent learning is distinct, we think as of now, by its openness to unanticipated courses of action, freedom from predefined testable outcomes, mostly voluntary participation, and, yes, humility. For the most part, these features make such pedagogies difficult to pursue, other than marginally, in formal education, but they can be central, we propose, in informal mathematics education (Nemirovsky et al. 2016), of which the Basket Weaving and Curvature program is an instance. We will delineate two streams of pedagogical ideas as recently inspired by our experiences in the program: (1) Explorations at the Edge; and, (2) Opening Avenues of Expression.

- Explorations at the Edge

The "edge" that we have in mind is one that adjoins or brings into contact two territories, like the edge of a sea bordering both, an expanse of water and a strip of country land. One side of the edge is a territory that appears firm and amenable to walk through, the other side is to be navigated with caution and wonder, without straying too far from the edge, as it is outpouring with questions and barely seen possibilities extending up to a remote horizon. Husserl thought that every object is located at an edge of that sort, demarcating, as it were, its sides directly perceived and the indefinite anticipations of the unseen sides: "…every object is not a thing isolated in itself but is always already an object in its horizon of typical familiarity and precognizance." (Husserl 1975, p. 122). He then wrote a crucial idea:

> But this horizon is constantly in motion; with every new step of intuitive apprehension, new delineations of the object result, more precise determinations and corrections of what was anticipated. (Husserl 1975, p. 122)

An exploration at the edge, we suggest, is an activity in which horizons are set in motion. An example of such exploration, as it took place in the Basket Weaving and Curvature program, was the work with fabric bowls. While neither the educators, nor the children, had ever sewn a fabric bowl, we were all acquainted with bowls and fabric. Fabric bowls and the techniques of their making were at an edge separating familiarities with various bowls and types of fabric from expanses circumscribed by a horizon of barely seen possibilities: What shapes can they have? Do they keep their shapes stably? How firmly can they hold content? What are suitable materials allowing for easy sewing? How do template shapes correlate with bowl shapes? and so forth. Certain skills that were for some participants on one side of the edge, were for others on the other side, such as sewing: the children told us that they had never sewn, and that, with few exceptions, they had never seen anyone sewing (a couple of grandmothers were the exception). Stemming from their concurrent participation in a geometry seminar, for Ricardo and Cierra the horizon of fabric bowls encompassed also the creation of flat maps for the rounded earth, as well as the distribution of Gaussian curvature on a 3D surface.

Star and Griesemer (1989) introduced the notion of "boundary object," which are objects, such as architectural drawings or soil samples, that are used and conceived differently by different disciplines and practitioners, while serving to coordinate their collaborative work. Similarly, exploring objects and techniques at the edge can nurture and mesh the diverse horizons of the explorers; an example of which, we think, took place in Line 20 of Episode 2, when Giselle traced on the glass enclosing a coiled basket, the shape of a petal.

The Basket Weaving and Curvature program included other explorations at the edge, some of which reached only an embryonary stage, such as the ones involving the curvature instrument and the paper wrapping of balls. The program has inspired us to propose that explorations at the edge, particularly when they are at the edge for all participants, including the educators, are very significant for pedagogies of emergent learning. Ultimately, it seems fair to say, emergent learning is the collective mobilizing of horizons.

- Opening Avenues of Expression

There is an important difference, particularly in the context of mathematics education, between representing and expressing (Whitacre et al. 2009). Instead of presenting-again, in a different format, what had been present before, an expression is an explosion of meaning without clear boundaries, subject to never-ending interpretations. It matters greatly whether we see a gesture, a diagram, a drawing, or an utterance as a representation or an expression. During the final session in which the children shared their work with parents and other adult attendants, a boy explained that the inside and outside colors he had chosen for his fabric bowl—dark on in the inside, light on the outside—were like some people he knew who looked nice from outside but were bad inside. In one of the individual interviews that Cierra conducted with the children, a girl said that she saw herself as an "art person," and then she pointed, as a mode of evidence, to her fabric bowl held on her other hand. Letting baskets, woven yarn around pipe cleaners, fabric bowls, and craftwork exhibited in *Made in America*, be expressions traversing disciplinary, institutional, and historical boundaries we customarily take for granted, amounts to opening avenues of expression. We propose that this is a major quality for the kind of pedagogy we try to understand. It is through expression that the emergent finds itself, for a gaze seeking a set representation, such as a certain definition or graph, is blind to emergent learning.

It is complex but possible to discern aspects of what has been learned in a program infused with qualities such as Explorations at the Edge and Opening Avenues of Expression. The analysis of videotaped episodes and interviews moves us to reckon that participants in the Basket Weaving and Curvature program learned, with various degrees of subtlety, that there is a relationship between the shape of petals and of sewn bowls, or that an art museum can be a fascinating place. Along the same lines, the authors of this paper sense the burgeoning appearance of seed-ideas about the roles of craftwork in mathematics learning. Had the exhibit been another one, or many of the contingent events populating the program been absent, a different learning would have emerged.

Acknowledgements Writing of this paper was supported by the InforMath project, funded by the National Science Foundation through grant DRL-1323587. All opinions and analysis expressed herein are those of the author and do not necessarily represent the position or policies of the Foundation. This paper is the result of collaborative work with Cierra Rawlings and Bohdan Rhodehamel at San Diego State University, and Molly Kelton at Washington State University. Due to the ICME rules for invited lectures, they have not been included as co-authors. I want to thank them for their contribution and for their agreement to the latter. I wish to thank the Mingei International Museum, the Boys and Girls Club of Chula Vista, and the Math Technology Lab at San Diego State University. In addition, I want to convey my gratitude for the invaluable work of Lucera Gallegos and Johanna Benson that made the Basket Weaving and Curvature program possible, and for the unforgettable eloquence and friendliness of the children who volunteered to participate.

References

Ball, D. (2010). SeanNumbers-Ofala. *Mathematics Teaching and Learning to Teach*, from https://deepblue.lib.umich.edu/handle/2027.42/65013.

Erlwanger, S. H. (1973). Benny's conception of rules and answers in IPI mathematics. *Journal of Children's Mathematical Behavior, 1*, 7–26.

Freire, P. (1970). *Pedagogy of the oppressed*. New York: Bloomsbury.

Goldstein, J. (1999). Emergence as a construct: History and issues. *Emergence: Complexity and Organization, 1*(1), 49–72.

Hamilton, D. (1999). The pedagogic paradox (or why no didactics in England?). *Pedagogy, Culture & Society, 7*(1), 135–152.

Husserl, E. (1975). *Experience and judgement*. Evanston: Northwestern University Press.

Jacobson, M. J., & Wilensky, U. (2006). Complex systems in education: Scientific and educational importance and implications for the learning sciences. *Journal of the Learning Sciences, 15*(1), 11–34.

Jacobson, M. J., & Kapur, M. (2012). Learning environments as emergent phenomena: Theoretical and methodological implications of complexity. In D. Jonasse & S. Land (Eds.), *Theoretical foundations of learning environments*. New York, NY: Routledge.

Jacobson, M. J., Kapur, M., & Reimann, P. (2016). Conceptualizing debates in learning and educational research: Toward a complex systems conceptual framework of learning. *Educational Psychologist, 51*(2), 210–218. http://doi.org/10.1080/00461520.2016.1166963.

Kreps, D. (2015). *Bergson, complexity, and creative emergence*. Hampshire, UK: Palgrave Macmillan.

Nemirovsky, R., Kelton, M. L., & Civil, M. (2016). Toward a vibrant and socially significant informal mathematics education. In J. Cai (Ed.), *Compendium for research in mathematics education*. Reston, VA: National Council of Teachers of Mathematics.

Nemirovsky, R., Lara-Meloy, T., DiMattia, C., & Ribeiro, B. T. (2005). Talking about teaching episodes. *Journal of Mathematics Teacher Education, 8*, 363–392.

Rancière, J. (1991). *The ignorant schoolmaster: Five lessons in intellectual emancipation*. Palo Alto, CA: Stanford University Press.

Star, S. L., & Griesemer, J. R. (1989). Institutional ecology, 'Translations' and bounday objects: Amateurs and professionals in Berkeley's museum of vertebrate zoology, 1907–39. *Social Studies of Science, 19*, 387–420.

Whitacre, I., Hohensee, C., & Nemirovsky, R. (2009). Expresiveness and mathematics learning. In W.-M. Roth (Ed.), *Mathematical representation at the interface of the body and culture* (pp. 275–308). Charlotte, NC: Information Age Publishing.

Young, N. H. (1987). Paidagogos: The social setting of a Pauline metaphor. *Novum Testamentum, 29*(2), 150–176.

Chapter 24
Connecting Mathematics, Community, Culture and Place: Promise, Possibilities, and Problems

Cynthia Nicol

Abstract In this essay I explore a critical pedagogy of place for mathematics education. Greenwood's (2013) theoretical framework of a critical pedagogy of place is used alongside frameworks for critical mathematics education to present an approach for connecting mathematics, community, culture and place. Drawing upon literature from both Indigenous and non-Indigenous scholars, theories of place-based education are examined. I introduce theories of mathematics education that advocate what Freire (1970/2000) calls 'problem-posing' practices to read (understand) and write (transform) the world with mathematics Gutstein (2006). Two place-based problems are presented, inspired and used by secondary/middle school teachers in a rural community. These problems provide examples and critiques of connecting mathematics, community, culture, and place. The essay concludes with reflections on the challenges and possibilities of a critical pedagogy of place for mathematics education in a world with increasing complex global issues.

Keywords Critical mathematics education · Place-based education Decolonization · Social justice

24.1 Introduction

Place-based or community-based education is receiving increased attention as an approach to education that connects school curriculum and local context for better understanding of complex global issues (Cajete 1994; Greenwood 2013; Smith and Sobel 2010; Sobel 1998). Advocating a holistic mindset, place-based education, begins at the local level to inspire student interest, engagement and participation in local community decision making and problem-solving (Smith 2002; Smith and

C. Nicol (✉)
University of British Columbia, 2125 Main Mall, Vancouver, BC V6T 1Z4, Canada
e-mail: cynthia.nicol@ubc.ca
URL: http://edcp.educ.ubc.ca/faculty-staff/cynthia-nicol/

© The Author(s) 2018   423
G. Kaiser et al. (eds.), *Invited Lectures from the 13th International Congress on Mathematical Education*, ICME-13 Monographs,
https://doi.org/10.1007/978-3-319-72170-5_24

Sobel 2010). Thus a curriculum that is grounded in activities and issues of the local place rather than abstractions of the environment, "can help children and youth begin to see themselves as actors and creators rather than observers and consumers" (Smith and Sobel 2010, p. viii). In a world with increasingly complex global issues including economic and social inequities, climate change, poverty, sustainability, resource depletion and mass displacement of people due to war, famine or environmental changes a deep understanding of place is required for understanding "the nature of our relationship with each other and the world" (Gruenewald 2003a, p. 622).

Smith (2002) articulated five different approaches to place-based education: (1) cultural studies involving students in studying local cultural or historical events; (2) nature studies by investigating the physical world; (3) real-world problem solving by locating, reflecting on and developing solutions to local community or school issues; (4) entrepreneurial opportunities by investigating relationships between vocation and place; and (5) participation in community decision-making. In recognizing that conceptions of place-based education ignored critical perspectives on culture, ecology, and schooling, while critical theory ignored attention to place, Gruenewald's (2003b) work adds a sixth approach framed as a "critical pedagogy of place."

In this chapter, I explore the nature of a critical pedagogy of place in the context of mathematics education by asking the question: What would it look like to connect mathematics, community, culture, and place? To answer this question I draw upon research with the Indigenous community of Haida Gwaii in Canada's Pacific northwest to provide a rural example of a critical pedagogy of place. To begin I first explore the nature of critical mathematics education. While some mathematics education scholars theorize and research mathematics education from a critical perspective (e.g., Frankenstein 1987; Gutstein 2012; Kumashiro 2015; Skovsmose 1994), few have explicitly considered the connection and role of place in mathematics education. Following a discussion of critical approaches to mathematics education, I examine theories of place and mathematics education. Next I discuss Greenwood's (2013) critical pedagogy of place drawing upon the processes of decolonization and reinhabitation, and use this framework to critique the case of connecting mathematics, community, culture, and place on Haida Gwaii. I conclude with a discussion of the possibilities and challenges of a critical pedagogy of place in mathematics education.

24.2 Critical Approaches to Mathematics Education

More than 30 years ago, Frankenstein, in the United States, writing in English, and Skovsmose, in Denmark, writing in Danish, developed ideas for critical approaches to mathematics education. Frankenstein, who was motivated by the "continuing injustices and the connections among those injustices to deeply entrenched institutional structures" (2012, p. 59), drew upon Freire's (1970/2000) approach to developing critical consciousness. She was one of the first to bring Freire's ideas of liberatory problem-posing education into the mathematics classroom (Frankenstein 1987). As critical thinkers, students could "develop their power to perceive critically *the way they exist* in the world *with which* and *in which* they find themselves" (Freire 1970/2000, p. 83). A problem-posing education recognizes people as historical beings with the human desire to move forward, drawing upon the past to improve the future, and accepting incompleteness as the process of "becoming more fully human" (p. 84). Thus a critical education affirmed teaching and learning in a way that both educators and students became teachers and learners, and where the generative themes of students' own historicity became starting places for education.

Frankenstein (1987, 2012), working with Freire's theory, designed statistics lessons for her adult education courses. Students explored problems developed from issues that concerned them, with a focus on unpacking the more hidden purposes and interests in various approaches to statistical analysis along with the contexts of those problems. Frankenstein (1987) provides sample problems for her adult students that were designed to teach mathematics, while raising political and social consciousness. Examples included: examine data on United States military spending or federal subsidies to nuclear power industries and calculate the total spent or total subsidized; examine data on United States food manufacturing to calculate the percentage of firms making the top percentage of net profits; examine decisions made around interest loan payments; examine survey report results; and analyze their own, or other students', mathematical error patterns. Statistics, argues Frankenstein, along with statistical and probability theories, provide rich opportunities for students to critically solve and pose problems related to issues in the public sphere, as well as to question underlying assumptions made based on statistical data and how they are used.

Like Frankenstein, Skovsmose, although working in Denmark, developed a critical approach to mathematics education that is "an open and uncertain concept" (Skovsmose 2012, p. 42) in "an open conceptual landscape" (Skovsmose 2016, p. 2). Critical mathematics education for Skovsmose is a broad field that is best characterized by the issues that drive it. One issue is the performative aspect of mathematics, where mathematics is *in action* in many different applications and practices including technological design construction, hypothetical reasoning and mathematical modeling, justification or legitimation, realization, and what Skovsmose terms the "dissolution of responsibility" (2011, p. 68). For example, a place of concern for mathematics–based action is mathematical modeling that could be designed to provide legitimacy for a decision that is already made, such as an oil company providing justification to increase its production while limiting environmental impacts. For all forms of mathematics, Skovsmose (2014) argues that "mathematics in action is in need of being carefully criticized" (p. 117).

This is particularly important when mathematics is considered a performative language where the tools of mathematics (its grammar or structure) format what innovations can be developed, how they are used, and the intentions of their use. Using examples from the airline industry Skovsmose (1994) illustrates how mathematics used in technologies for schedules and flight routines becomes "not only also descriptive but also prescriptive" (p. 55). Flight routines format or structure our social reality, for instance, in terms of how we might organize or plan our own daily activities. Thus, a critical mathematics education involves students in reflective knowing to become aware of the biases and blind spots, in order to make more apparent this formatting power of mathematics. Skovsmose provides possibilities for the design and implementation of critical mathematics education through using themes or problem-based learning contexts that draw upon students' interests and community experiences. One example of such a project is Economic Relationships in the World of a Child, a series of 12 units challenging young students to engage with mathematics and social issues related to spending pocket money, the child benefit allowance, and what to buy for a youth club. Skovsmose (2011) cautions, however, that although mathematics is in action in many different contexts, such as particular cultural settings, a critical mathematics curriculum needs to consider not only what is familiar to students (i.e., their backgrounds) but also their foregrounds (i.e., their possibilities and obstructions). Finally, at a meta-level, Skovsmose encourages an ongoing critical stance for the idea of critical mathematics education, as mathematics, its purposes, and formatting are neither fixed nor predetermined but continuously changing, and thus "always in need of critique" (2014, p. 119). What role do considerations of place/land have in critical mathematics education?

Gutstein (2006), also inspired by Freire's "problem-posing" curriculum, studied his own teaching of Grade 9 students in an urban American classroom. He extended Freire (1970/2000) ideas of literacy as involving both reading (making sense of the word in the world) and writing (using this sense-making to change the world), to mathematics (Gutstein 2006). Providing examples of projects co-constructed with students, Gutstein writes of challenging his students to read the world mathematically (do and understand mathematics) and write the world mathematically (use that understanding to change the world). One such project focused on using mathematics to understand arguments for recent home foreclosures in the students' community (Gutstein 2012). The mathematics of bank loans, subprime mortgage loans, profits and foreclosures as well as resulting neighbourhood displacement helped students better understand issues they were currently experiencing, and as Gutstein (2012) argues become more engaged in both mathematics and the world. Gutstein suggests that such problems drawing from students' lived experiences or their community knowledge, provide opportunities to learn classical or academic/ school mathematics, as well as engage students in critical inquiry from various perspectives.

Critical approaches to mathematics education developed by Frankenstein, Gutstein and Skovsmose take seriously approaches that draw on students' lived experiences—both their backgrounds and foregrounds—so that students can participate in understanding (reading) and intervening (transforming) the world. For each of these scholars importance is placed on participatory teaching practices that bring student and teacher in reciprocal relationships of co-learning. This means, as Freire (2005) argues, both teachers and students are co-creating, each is at the same time both teaching and learning. In considering place/land in shaping lived experiences, Cajete (1994, 2012), an Indigenous Tewa scholar from the United States, refers to this co-learning as an ecology of relationships connecting humans and place in which humans and place shape each other. "People make place" argues Cajete, "as much as place makes them" (1994, p. 84).

24.3 Theories of Place and Education

In considering the very idea of place, philosopher Edward Casey (1996), writes that "place is not a mere patch of ground, a bare stretch of earth, a sedentary set of stones" (p. 26). Instead, for Casey, "places not only *are*, they *happen*" (p. 27), they are "generative and regenerative," and from place "experiences are born and to it human beings (and other organisms) return for empowerment" (p. 26). Places are living and are lived, they "gather experiences and histories, even languages and thoughts" (p. 24) and places hold them in a kind of gathering action. A place bears on, or structures, the experiences of those (animate and inanimate) within place, while those occupying place organize living with place in a symbiotic co-creation.

Cajete (1994), writing from an Indigenous perspective, suggests that with place we have a dialogical relationship in which we learn more about ourselves, our relationships with each other, and our relationships with the more-than-human world. Place figures prominently in the discourse and life of Indigenous peoples for whom ancestral memories and stories are intimately connected to land and landscapes. Like Casey (1996) who describes place as event, Cajete (1999) conveys the animated nature of place, stating "place is ever evolving and transforming through the life and relationship of all its participants" (p. 193). Place, for Cajete (1999), is "not only a physical place", but also "a spiritual place, a place of being and understanding" where "interactions with places give rise to and define cultures and community" (p. 193).

Similarly, Indigenous scholar Michell et al. (2008) describes place with five dimensions: multidimensionality (more than physical and also emotional); relationality (epistemologically everything is in relation); experientially (experiences in place, on land and in relation to the human and non-human worlds ground meaningful learning); locality (places are specific and general at the same time) and where living place over time brings peoples' "landscapes [to] become reflections of their very souls" (Cajete 2000, p. 183); and land-based (place is land, and relationship between land and people is key). Thus place is described as local, experiential, land-based, and within a holistic perspective that "entails, physical, emotional, and spiritual characteristics" (Michell et al. 2008, p. 27). Bringing place and education together for Indigenous scholars such as Cajete (1999) and Michell (2013) involves considering place/land as teacher in a relational education that recognizes the interdependence of human, other-than-human, and more-than-human worlds toward sustainability, and "reinforces natural connections to land and community" (Cajete 1999, p. 201).

Place conscious education is the name Greenwood (2013) gives to this meeting of place and education. For Greenwood, place conscious education is a theoretical and philosophical stance that differs from the articulations of place-based education focusing on pedagogical strategies designed to improve student achievement through connections to the community (Smith 2002; Smith and Sobel 2010). Instead, Greenwood (2013) theorizes place conscious education as a critical pedagogy of place consisting of two goals: decolonization and reinhabitation. Greenwood uses decolonization to critique the cultural practices related to place, and reinhabitation to imagine new possibilities of consciousness between people and place. In Greenwood's (2013) words:

> Cultural decolonization involves learning to recognize disruption and injury in person-place relationships, and learning to address their causes. ... [D]ecolonization refers also to the educational process of identifying and unlearning patterned and familiar ways of experiencing and knowing to make room for practices that are unfamiliar. (p. 96)

Decolonization then involves an "awareness of potential settler impositions, and the desire to reveal and challenge these impositions" (Kerr 2014, p. 86). Coloniality scholars such as Quijano (2007), Fanon (1952/2008), and Memi (1965), examine Euro-centered colonialism as a formal system of domination, social and political, where colonizers' beliefs, knowledge and practices are considered superior to those first occupying place. European colonization, fuelled by the need for resources and materials in the Industrial Era, was one of domination over people, nature, land, and resources. This included political colonization, systematic repression, cultural colonization, and imposed patterns of meaning making. For example, a colonial view of land severs the relationships between humans and place/land, viewing land as something to be tamed, dominated, or conquered. Greenwood's (2013) decolonization then is a personal process of working toward "transforming or resisting oppressive relationships that limit people's ability to control their own life circumstances" (p. 96).

Reinhabitation is the second goal of a critical pedagogy of place. Reinhabitation, writes Greenwood (2013), "involves learning to live well socially and ecologically in a place, and learning to live in a way that does not harm other people and places" (p. 96). Decolonization involves recognition or unlearning colonial practices of dominance and oppression, conscious or unconscious, that could limit renewed relationships. Reinhabitation, on the other hand, involves moving from unlearning to relearning practices of being or inhabiting place that "involves taking a new stance toward one's own being and knowing" (p. 96).

24.4 Critical Pedagogy of Place and Mathematics Education

I am inspired by these views of place and education and argue for their inclusion within the broad field of critical mathematics education. Greenwood (2013) provides a series of questions related to place that I suggest are helpful in considering place and mathematics education as an approach to critical mathematics education. The questions provoke criticality toward the historical, socioecological, and ethical aspects of relations to place that can engage us in the inter-related practices of decolonization and reinhabitation (Greenwood 2013). The dual goals of decolonization and reinhabitation heed Tuck and Yang (2012)'s warning that decolonization as consciousness raising should not be the end goal but instead also require consideration of future actions. Greenwood's (2013, p. 97) critical questions include:

1. What happened here? (historical)
2. What is happening here now and in what direction is the place headed? (socioecological)
3. What should happen here? (ethical)

I bring place and mathematics education together in considering these questions through cases of logging practices and food growing on Haida Gwaii in Canada's Pacific northwest coast.

24.4.1 Land-Use on Haida Gwaii

As an example of a critical approach to mathematics education considering place/land relations, let me turn to an ongoing project located in Canada's Pacific northwest coast—Haida Gwaii, People of the Islands (Nicol et al. 2013; Nicol and Yovanovich 2011, in press). Haida Gwaii is a unique archipelago of over 150 islands located in northern British Columbia's Pacific Ocean, where all places intimately connect people to land and ocean. In fact, a Haida worldview is "everything is connected to everything else"—human, non-human and more-than-human worlds. Before European contact, tens of thousands of people (some Haida Elders say it is many more) lived on and with these islands. Monumental cedar trees, and seafood such as salmon, were resources managed in a sophisticated system of family governance. With European contact also came disease that decimated the Haida population to less than 1000 and required, out of necessity of survival, congregation into two main villages. In 1853 the British claimed Haida Gwaii as British land, and about 25 years later the Canadian government's Indian Act of 1876 declared all "Indians" or First Peoples as under the responsibility of the government, making illegal cultural activities such as the potlatch, used by coastal First Nations in cultural, social, and economic governance practices. With ancestral communities displaced or decimated by disease across the Islands the settler colonizer saw most of Haida Gwaii as unsettled and open for resource removal. Whaling stations, open-pit mining, clam canneries, sawmills and logging camps occurred over the years, much without the consultation, permission or decision-making of the Haida people. However, since 1985, the Haida have taken a stand first to logging companies then to the federal government, to reclaim governing rights over their lands. This conflict led to creating the Gwaii Haanas National Park Reserve and an historical agreement where the lands are co-managed by the Government of Canada and the Haida Nation.

Today Haida Gwaii's population is less than 5000, with just less than half the population identifying as having Indigenous ancestry. There is one public school district serving about 500 students, with approximately 70% identifying as being of Haida or First Nations ancestry. For the past 10 years, I have been working with the school district to explore the nature of creating mathematics education learning environments that are more responsive to the place and community of Haida Gwaii.

As a group that includes community members, Elders, artists, administrators and educators, we are exploring responsive mathematics education. We have co-created a number of lessons that bring mathematics, community, culture, and place together. I discuss two lessons related to land use, one focused on logging, the other on food growing.

24.4.1.1   Logging Practices

Land-use practices on Haida Gwaii can provide a context to critically examine mathematics and connections to place/land. The forests of Haida Gwaii contain an abundance of coniferous trees including red and yellow cedar, hemlock, and sitka spruce, some more than 600 years old. For generations, the Haida have used the bark, wood, and roots of cedar and spruce trees. Cedar was harvested for building magnificent longhouses and ocean-faring canoes; its bark and roots woven into blankets, clothing, baskets, and hats; and its wood carved into house poles and masks that hold ancestral stories. Monumental cedar trees, those over 140 years old and measuring more than 120 cm in diameter, were carefully selected for harvesting (Council of the Haida Nation 2016). If fallen, the entire tree was used; if left standing, only the bark was stripped in selected sections, preserving the tree's life for future harvests. Cedar trees, like all animate forms, together with the inanimate and spiritual, were considered part of the same world (Stewart 1984). Being part of the forest emanated a life-force. Before European contact the forests of Haida Gwaii and its monumental cedars formed an integral part of Haida life.

Need for aircraft construction grade wood during the early 1900s brought logging practices to Haida Gwaii. Spruce trees on the Islands were superior to elsewhere. With high strength to weight ratio, and tight, straight, uniform grain, Haida Gwaii spruce was perfect for aircraft frames. Heavy industrial clear-cut logging began in the 1950s and increased steadily for the next 40 years (Gowgaia Institute 2007). A mapping animation documenting the logging history from 1901 to 2004 provides a spatial map of logged areas and a visual representation of the harvesting rate (Gowgaia Institute 2004a).

Since the 1990s, the volume rate of harvest per year has decreased: approximately 2.4 million cubic metres were logged in 1984, 1.8 million cubic metres in 1994, 1.1 million cubic metres in 2004, and 840,000 m^3 in 2014 (Council of the Haida Nation 2016). Log barges the size of football fields carry logs off the Islands to Canadian and foreign markets, some carrying close to 30,000 m^3 of logs.

Middle school students examine and graph the change in volume harvested over the years. They engage in quantitative reasoning as they explore the relative size of log volumes, searching for comparable visualizations for 1 million cubic metres. What would hold 1,000,000 m^3? What else is 1,000,000 m^3 in volume? About how many cedar trees make 1,000,000 m^3? How many log barges were used each year to carry these logs to the mainland? Doing calculations and making referents for large numbers engages students in quantitative reasoning, but it is not sufficient for a critical pedagogy of place.

Asking Greenwood's (2013) historical, socioecological, and ethical questions from a mathematical perspective can lead to critically examining logging practices on Haida Gwaii in terms of who had the decision-making power for which areas would be harvested, how they would harvested, and at what rate. Such a study could call into question the underlying epistemologies of settlers who brought clear-cut logging practices to Haida Gwaii, practices that relegated land as subordinate to settler colonizer needs. How did it come to pass that trees were once logged at a rate of 2.4 million cubic metres per year? In this example, students examine what this rate means. Data documenting logging outputs, in terms of volume of wood harvested and land area logged, could be analyzed to determine how and in what ways rates of harvesting could be sustained or not, as well as who benefits, who does not and in what ways.

Furthermore, the mathematics used in designing the mapping tools used to calculate forest cut-rates could be examined (Gowgaia Institute 2004b). As maps are not neutral-free, students could examine the ways data were collected, represented, analyzed, and communicated, for creating animating maps with consideration of whose perspective is represented. Images of log barges carrying 15,000 tonnes of logs leaving the Islands, clear-cut mountains, and economic benefits can be used to prompt investigation of what is going on here, what is happening now, and what should be happening. Extensions include field trips to the forest scaling yard to learn the process of tree valuation and volume determination. Investigations of questions such as these can engage students in a critical mathematics education, where mathematics is used to make sense of the historical context of logging, current practices, and more sustainable future practices.

24.4.1.2 Food Growing

Over the past 10 years, sustainable food growing on Haida Gwaii has received increased attention. As an Island system separated from the mainland by large ocean bodies, it is a challenge for food to arrive on Haida Gwaii still fresh and affordable. A two-day drive plus seven to eight hour ferry ride make the price of shipped food to the Islands more than three times that found in mainland cities. For thousands of years, food harvesting of the Haida involved feasting off the range of marine wildlife including ocean fish such as salmon and halibut, shellfish such as clams and crabs, and kelp gathered from vast ocean kelp forests. As more settlers arrived on Haida Gwaii attracted by resource extraction industries such as logging, mining, and fishing, food-harvesting practices changed. Settlers claimed land, but harsh Island climates and short growing seasons challenged attempts to establish farms dedicated to food growing. Many Islanders then became dependent less on local food and more on purchasing food such as meat, fruit, and vegetables imported from the mainland.

With efforts of teachers in northern Haida Gwaii's only high school, and with collaboration and support of the community, a school greenhouse was built on school grounds in 2011. The greenhouse is now one of seven on the Islands and part of a Food-to-Cafeteria system that includes schools and the local hospital. The high school renovated its school kitchen to accommodate the integration of the Food-to-School program using greenhouse food for school cafeteria lunches, and integrating greenhouse activities across the curriculum. As the math and science teacher states:

> [D]epending on the course and the potential curricular links to gardening, I have expanded my activities to include full courses specifically dedicated to maintaining school gardens or short visits to the greenhouse for a quick extension linking a particular topic to a hands-on, garden related activity. [high school teacher's written reflection, 2012]

In linking school gardening activities with mathematics, the high school mathematics and science teacher involved students in planning, mapping and designing the shape and size of the raised soil bed boxes to optimize varying natural light intensity, sun angle, and greenhouse temperatures throughout the year. Students mapped garden layout designs, deciding where to seed various plants based on their mature height, light needs, and growth rate. They have studied soil composition and organic composting. Upon food harvesting, students created their own recipes, providing opportunities to study concentration ratios, scaling, volume, and proportion.

Returning to Greenwood's (2013) guiding questions offers an occasion to examine the kinds of activities needed for a critical mathematics education that includes place. Asking 'what happened here?', students and teachers could examine the conditions that led to the need to ship food to Haida Gwaii from the mainland. What forms of traditional marine knowledge or Indigenous knowledge of marine harvesting were practiced? What were the underlying values of these practices, and in what ways were they sustainable? And, what are the underlying values of food cultivation, farming, and agriculture? Next, students and teachers could examine socioecological issues that critique the need and use of greenhouse gardens, and the relationship of such food practices to traditional Haida food harvesting. Here students could examine large data sets focused on fish stocks and harvesting rates over various years. Studying patterns of marine harvesting and population growth on the Islands could provide contexts for discussion of socioecological issues of sustainability. Students could also engage in data collection and analysis of interviews and surveys to learn more about who takes advantage of the products of the greenhouse, who does not, and the patterns of food consumption over the year. Finally questions of what should happen could include consideration of further data collected from community members, to learn more about the benefits and possibilities of food-to-school programs, how Indigenous knowledge and harvesting practices are considered and included in current food-work practices, and what an Indigenous 'garden' might actually look like.

The Food-to-School program has gained momentum in Haida Gwaii. School food learning circles were created in 2013–2014, bringing teachers, farmers, Haida Elders, and chefs together to consider possibilities and future goals for the program. It began with settler initiative that is now including the visions and voices of Haida community members.

24.5 Conclusion: Challenges and Possibilities

To the conversation of critical mathematics education, I argue for a focus on place, and explore theories of education informed by place. I argue for a broad conceptualization of place, and draw upon historical, socioecological, and ethical questions posed by Greenwood (2013) to consider a critical mathematics education with place in mind. This focus on place differs from place-based education that tends to advocate connections to place in order to motivate and improve student achievement, as well as increase students' connection to places in order to better care for particular places (Smith and Sobel 2010). Certainly student outcomes are important, while pedagogies that re-connect students to places are crucial for rebuilding ecological relationships more globally. However, Indigenous scholars interested in place/land pedagogies push for theorizing land relations that pay attention to settler colonialism as an ongoing and incomplete project. As Cajete (2000) reminds us, places shape ways of being in the world. As first teachers, places represent learning environments that intimately connect human, non-human and more-than-human relational worlds. This is quite different from viewing place as the context for human activity, or background for human privileged use (e.g., resource extractions), or as a material object, or in terms of right of ownership. Such conceptions are what Bang et al. (2014) refer to as "conceptions of place in the service of settler colonial legitimacy" (p. 41). Instead, I am inspired by views of place/land that provide opportunities to challenge such colonial conceptions of land. Greenwood's historical, socioecological, and ethical questions provide one pathway toward a critical mathematics education that includes opportunities for increasing awareness of human/land relations, as well as the historical and ongoing colonizing of place.

In conclusion, I discuss the challenges and possibilities of a critical pedagogy of place in mathematics education through the following issues: social action, relevance, the role of mathematics, and urban versus rural places.

24.5.1 Social Action

Critical pedagogy conceptualized by Freire, Gutstein, and Frankenstein involves both reading the world with mathematics (using mathematics to understand the world), and writing the world with mathematics (using mathematics to change the world). The land-use lessons provide contexts to engage students in reading and writing the world with mathematics. Mathematics is used to make sense of past and current logging practices, analyze data represented in visual maps (volume of wood/ year) of logging practices, and use of mathematics to examine decisions about sustainability. These are examples of *reading* the world with mathematics. It is more difficult designing tasks that engage students in *writing* the world with mathematics. Few examples offered by Gutstein and Peterson (2013) actually involved students in moving from discussions of social action to enacting social action. Currently students on Haida Gwaii are neither participating in challenges to current logging practices, nor are they leading the food-to-school food movement. Both examples stop short of physical action engagement. Nonetheless, mathematics helped students understand the issues associated with each example and provided an occasion for further exploration. One could argue that discussing possibilities for social action could be the first step toward action engagement.

A further challenge related to social action involves considering which action is appropriate. How do teachers and students decide how to act or respond to issues presented? As Esmond (2014) found in her research, it is possible for students to engage in critical mathematics education with a social justice goal that strengthens rather than challenges unjust biases. It is also possible for students to engage in social action that may not change the situation. For example, in the case of logging practices on Haida Gwaii, it is possible that some students whose families are employed in the industry may deepen beliefs of land entitlement that denies Indigenous land claims. It is therefore important for educators to think carefully about problem contexts, be aware of the complexities of a critical pedagogy of place, and be prepared for an open kind of teaching with issues that may not be easily resolved.

24.5.2 Relevance

The land-use lessons were attempts to engage students both in developing mathematics competence and in what Freire (1970/2000) termed critical consciousness. Analyzing land-use and even marine-use practices on Haida Gwaii is a political issue affecting all families on Haida Gwaii in some way. A reason for students' lack of enacted social action in either lesson could rest on the degree of relevance of the issues for students. Freire argues for developing problems based on students' interests—generative themes that are part of students' culture and community. Although both logging and food growing are familiar to students, teachers chose both examples for students. In addition, it could be argued that land use, for example, on Haida Gwaii is not necessarily a topic of interest or of relevance to elementary or secondary school students. While all students on Haida Gwaii are familiar with logging trucks, have friends or family employed in the logging industry, and are witness to the sites of logged areas, most students lack personal experience or expertise in logging. Yet land-use and Indigenous claims to land are of high interest to communities on Haida Gwaii.

In conversations with youth, many spoke about a future need to balance logging, with maintaining old-growth forests, and with sustaining future yields. Nonetheless, the question of relevance is important. Enyedy et al. (2011) researched how, in a culturally relevant mathematics curriculum where students choose an issue to investigate, "different forms of relevance permeate and mediate students' sustained engagement" (p. 275). In a community mapping project these researchers found that relevance for students was negotiated throughout the project. Relevance could focus on mathematics content or context, on authentic purpose such as development of critical consciousness, or on familiar instructional practices. The land-use lessons were guided by relevance focused on content and purpose, where students' local experience and knowledge were considered in the design of the lessons. However, relevance of practices was not considered, and yet could be an important aspect for determining the extent to which students chose to engage or chose not to move toward social action. A critical pedagogy of place in mathematics education is not only about what places are relevant for learning to read and write the world with mathematics, but also about who decides on such relevance

24.5.3 The Place of Mathematics

Are problems focused on place somewhat removed from mathematics? Skovsmose (1994) asked himself this question when developing the idea of a critical mathematics education. Teachers discussing the land-use investigations also struggled about the role and place of school mathematics. Teachers often questioned whether or not students' activities were seen as mathematical, and the extent to which the activity met their required curricular outcomes. These concerns are similar to those raised by American teachers in Showalter's (2013) study, who questioned whether place-based mathematics education "compromises mathematical rigor" (p. 1). Showalter suggests that maintaining the level of mathematical sophistication appears easier in contexts where problems engage students in statistical analysis. In fact, many of the mathematics lessons offered by Frankenstein (1987, 2012), and by Gutstein and Peterson (2013), involve data analysis. This can also be seen in the food growing lessons. Students could analyze data sets on marine life and on harvesting practices as forms of traditional ecological knowledge in contrast to food growing. However, unlike the secondary school teachers in Showalter's study who found it difficult to connect meaningfully with place as a context for mathematical inspiration, the Haida Gwaii teachers could, in the particular cases of land-use, bring the context close to students, partly due to their own experiences with these activities. Nonetheless, I suggest both lessons provide opportunities for teachers and students to engage in mathematical work and to use mathematics to deepen understanding of local and global issues.

24.5.4   Place as Urban and Rural

The Haida Gwaii land-use examples were developed in a rural context. Is such work possible in other contexts, for example urban settings? Much of the place-based education research, such as that articulated by Smith and Sobel (2010), has focused on rural contexts, where land and ecological experiences are perhaps more easily accessed than in urban contexts. However, Rubel et al. (2016) argue that working with large data sets on questions of place provides opportunities to examine cities as places of economic and social inequities, providing opportunity for some, and disadvantage for others. Like Enyedy et al. (2011), Rubel et al. (2016) also use participatory mapping as a tool to reveal social injustices through spatial perspectives. Rubel et al. (2016) provide examples of possible investigations of reading and writing the urban context with mathematics. They suggest that the "urban setting is particularly conducive for TMSpJ [Teaching Mathematics for Spatial Justice] and participatory mapping because it is so densely populated and highly wired with cellular and data networks and services" (p. 561). Results from this work indicate that urban students are motivated to engage in critical analysis of place through data analysis and map making. Although there were unexpected outcomes and challenges of this work, students were able to use mathematics to identity issues and question injustices. This work provides evidence that a critical pedagogy of place in mathematics education is possible in not only rural settings such as Haida Gwaii but also urban contexts.

24.5.5   Concluding Remarks

A critical pedagogy of place for mathematics education involves attending to the generative themes of students' experiences to engage students critically with the world through mathematics, and to reason about the world with mathematics. Greenwood's (2013) goals of decolonization and reinhabitation, and his three guiding historical, socioecological, and ethical questions can provide a critical framework for teachers in connecting mathematics, community, culture, and place.

In addition to the challenges noted above, there can be resistance from educators, students, and parents, who question the purpose and place of a mathematics curriculum that has a goal of raising students' critical consciousness. For example, a Canadian news magazine, *Maclean's*, published an article by Reynolds (2012) titled "Why Are Schools Brainwashing Our Children?", arguing that teachers who are committed to social justice education, mathematics included, fall into the trap of imposing their own biases on students, and are therefore "brainwashing" students in the name of social justice. Reynolds further argues that teaching for social justice deters teachers from their main task of teaching children "properly." Reynolds (2012, para. 17) asks: "[D]oes too much time devoted to social justice divert attention from academic achievement and ironically promote a gross social injustice: students ill-prepared to contend with a complicated and competitive world?" Such a view conveys the message that what is important in mathematics education is successful completion of school mathematics, not politics or culture or learning about, and working for, social justice using mathematics. For some educators and parents, the mathematics classroom should remain pure and focused on school mathematics; for others, the classroom provides opportunities to write the world. The land-use tasks presented provide a counter narrative to Reynolds' claims, as neither teachers nor students came to think in one homogenous perspective. This is not to say, however, that all such tasks provide opportunities for deep conversations of justice that are transformative rather than supportive of further biases.

As we learn more about ways of connecting mathematics, community, culture, and place for a critical pedagogy of place, our understanding of what is possible in terms of teaching for justice will grow. Critical mathematics education with place in mind, I argue, could bring us closer to a vision of "teaching mathematics in a way where it can help us live in harmony with values that protect life and enhance understanding" (Fasheh 2012, p. 103). Or as Gloria Ladson-Billings (2015) states, pursuing "not social justice" but "just justice."

References

Bang, M., Curley, L., Kessel, A., Marin, A., Suzukovich, E., & Strack, G. (2014). Muskrat theories, tobacco in the streets, and living in Chicago as Indigenous land. *Environmental Education Research, 1*, 37–55.

Cajete, G. (1994). *Look to the mountain: An ecology of Indigenous education.* Skyland, NC: Kivaki Press.

Cajete, G. (1999). Reclaiming biophilia: Lessons from Indigenous peoples. In G. Smith & D. Williams (Eds.), *Ecological education in action: On weaving education, culture, and the environment* (pp. 189–206). Albany, New York: State University of New York Press.

Cajete, G. (2000). *Native science: Natural laws of interdependence.* Santa Fe, New Mexico: Clearlight Publishers.

Cajete, G. (2012). Contemporary Indigenous education: Thoughts for American Indian education in a 21st-century world. In S. Mukhopadhyay & W. M Roth, (Eds.), *Alternative forms of knowing (in) mathematics: Celebrations of diversity of mathematical practices* (pp. 33–51). Rotterdam, The Netherlands: Sense.

Casey, E. (1996). How to get from space to place. In S. Feld & K. Basso (Eds.), *Senses of place* (pp. 13–52). Sante Fe, New Mexico: School of American Research Press.

Council of the Haida Nation. (2016). *Forest form 2016.* http://www.haidanation.ca/. Accessed February 10, 2017.

Enyedy, N., Joshua, D., & Fields, D. (2011). Negotiating the 'relevant' in culturally relevant mathematics. *Canadian Journal of Science, Mathematics and Technology Education, 11*(3), 273–291.

Esmond, I. (2014). 'Nobody's rich and nobody's poor…it sounds good, but it's actually not': Affluent students learning mathematics and social justice. *The Journal of the Learning Sciences, 23*, 348–391.

Fanon, F. (1952/2008). *Black skin, white masks.* New York, NY: Grove Press.

Fasheh, M. (2012). The role of mathematics in the destruction of communities, and what we can do to reverse this process, including using mathematics. In O. Skovsmose & B. Greer (Eds.), *Opening the cage: Critique and politics of mathematics education* (pp. 93–106). Rotterdam, The Netherlands: Sense.

Frankenstein. M. (1987). Critical mathematics education: An application of Paulo Freire's epistemology. In I. Shor (Ed.), *Freire for the classroom: A sourcebook for liberatory teaching* (pp. 180–210). Portsmouth, NH: Heinemann.

Frankenstein, M. (2012). Beyond math content and process: Proposals for underlying aspects of social justice education. In A. Wager & D. Stinson (Eds.), *Teaching mathematics for social justice: Conversations with educators* (pp. 49–62). Reston, VA: National Council of Teachers of Mathematics.

Freire, P. (1970/2000). *Pedagogy of the oppressed.* New York, NY: Bloomsbury.

Freire, P. (2005). *Teachers as cultural workers: Letters to those who dare teach.* Boulder, CO: Westview Press.

Gowgaia Institute. (2004a). Logging Haida Gwaii 1901–2004. *SpruceRoots.* http://www.spruceroots.org/LogVideo/LogVid.html. Accessed February 10, 2017.

Gowgaia Institute. (2004b). Logging Haida Gwaii. *SpruceRoots.* http://www.spruceroots.org/Maps/Logging.html. Accessed February 10, 2017.

Gowgaia Institute. (2007). Old growth forest condition and distribution. *SpruceRoots.* http://www.spruceroots.org/Booklets/Booklet.html. Accessed February 10, 2017.

Greenwood, D. (2013). A critical theory of place-conscious education. In R. Stevenson, M. Brody, J. Dillon, & A. Wals (Eds.), *International handbook of research on environmental education* (pp. 93–100). New York: Routledge. https://doi.org/10.4324/9780203813331.

Gruenewald, D. (2003a). Foundations of place: A multidisciplinary framework for place-conscious education. *American Educational Research Journal, 40*(3), 619–654.

Gruenewald, D. (2003b). The best of both worlds: A critical pedagogy of place. *Educational Researcher, 32*(4), 3–12.

Gutstein, E. (2006). *Reading and writing the world with mathematics: Toward a pedagogy for social justice.* New York, NY: Routledge.

Gutstein, E. (2012). Connecting community, critical and classical knowledge in teaching mathematics for social justice. In S. Mukhopadhyay & W. M Roth, (Eds.), *Alternative forms of knowing (in) mathematics: Celebrations of diversity of mathematical practices* (pp. 299–312). Rotterdam, The Netherlands: Sense.

Gutstein, E., & Peterson, B. (2013). *Rethinking mathematics: Teaching social justice by the numbers.* Milwaukee, Wisconsin: Rethinking Schools.

Kerr, J. (2014). Western epistemic dominance and colonial structures: Considerations for thought and practice in programs of teacher education. *Decolonization: Indigeneity, Education & Society, 3*, 83–104.

Kumashiro, K. K. (2015). *Against common sense: Teaching and learning toward social justice.* New York, NY: Routledge.

Ladson-Billings, G. (2015). Social justice in education award lecture. *American Educational Research Association.* https://www.youtube.com/watch?v=ofB_t1oTYhI. Accessed June 15, 2017.

Memi, A. (1965). *The colonizer and the colonized.* Boston, MA: Beacon Press.

Michell, H. (2013). *Cree ways of knowing and school science.* Vernon, BC, Canada: J Charlton Publishing.

Michell, H., Vizina, Y., Augustus, C., & Sawyer, J. (2008). *Learning Indigenous science from place: Learning Indigenous science from place research study examining indigenous-based science perspectives in Saskatchewan first nations and métis community contexts.* University of Saskatchewan Aboriginal Education Research Centre, desLibris—Documents, & Canadian Electronic Library. Aboriginal Education Research Centre.

Nicol, C., Archibald, J., & Baker, J. (2013). Designing a model of culturally responsive mathematics education: Place, relationships and storywork. *Mathematics Education Research Journal, 25*(1), 73–89.

Nicol, C., & Yovanovich, J. (2011). *Tluuwaay 'Waadluxan: Mathematical adventures.* Skidegate, BC: Haida Gwaii School District 50.

Nicol, C., & Yovanovich, J. (in press). Sustaining living and learning culturally responsive pedagogy. In J. Archibald & J. Hare (Eds.), *Learning, knowing, sharing: Celebrating successes in K-12 Aboriginal education in British Columbia.* British Columbia Principals' & Vice-Principals' Association.

Quijano, A. (2007). Coloniality and modernity/rationality. *Cultural Studies, 21*, 168–178.

Reynolds, C. (2012). Why are schools brainwashing our children? *Maclean's, 125*(43). http://www.macleans.ca/news/canada/why-are-schools-brainwashing-our-children/. Accessed June 15, 2017.

Rubel, L., Hall-Wieckert, M., & Lim, V. (2016). Teaching mathematics for spatial justice: Beyond a victory narrative. *Harvard Educational Review, 86*(4), 556–579.

Showalter, D. (2013). Place-based mathematics education: A conflated pedagogy? *Journal of Research in Rural Education, 28*(6), 1–13.

Skovsmose, O. (1994). *Towards a philosophy of critical mathematics education.* Dordrecht, The Netherlands: Kluwer Academics Publishers.

Skovsmose, O. (2011). *An invitation to critical mathematics education.* Rotterdam, The Netherlands: Sense.

Skovsmose, O. (2012). Critical mathematics education: A dialogical journey. In A. Wager & D. Stinson (Eds.), *Teaching mathematics for social justice: Conversations with educators* (pp. 35–47). Reston, VA: National Council of Teachers of Mathematics.

Skovsmose, O. (2014). Critical mathematics education. In S. Lerman (Ed.), *Encyclopedia of mathematics education* (pp. 116–120). Dordrecht: Springer Science+Business Media. https://doi.org/10.1007/978-94-007-4978-8.

Skovsmose, O. (2016). What could critical mathematics education mean for different groups of students? *For the Learning of Mathematics, 36*(1), 2–7.

Smith, G., & Sobel, D. (2010). *Place-and community-based education in schools.* New York, NY: Routledge.

Smith, G. A. (2002). Place-based education: Learning to be where we are. *Phi Delta Kappan, 83,* 584–594.

Sobel, D. (1998). *Mapmaking with children.* Portsmouth, NH: Heinemann.

Steward, H. (1984). *Cedar.* Vancouver, BC: Douglas & McIntyre.

Tuck, E., & Yang, K. (2012). Decolonization is not a metaphor. *Decolonization: Indigeneity, Education & Society, 1,* 1–40.

Chapter 25
Relevance of Learning Logical Analysis of Mathematical Statements

Judith Njomgang Ngansop

Abstract Our work focuses on logic and language at a university in Cameroon. The mathematical discourse, carried by the language, generates ambiguities. At the university level, symbolism is introduced to clarify it. Because it is not taught in secondary school, it becomes a source of difficulties for students. Our thesis is as follows: "The determination of the logical structure of mathematical statements is necessary in order to properly use them in mathematics." We conducted our study in the predicate calculus theory. In the first part of the paper, a summary of the theory is presented, followed by a logical analysis of two complex mathematical statements. The second part is a report of two sequences of an experiment that was conducted with first-year students that shows that knowledge of the logical structure of a statement enables students to clarify the ambiguities raised by language.

Keywords Logic and language · Symbolism · Logical structure of statement Didactics

25.1 Introduction

The mathematical discourse is carried by language. As such, linguistic ambiguities are unavoidable. We can quote as examples the phrases "two by two" and "all . . . are not," which may have different[1] meanings according to the context. Besides, interpretation of statements whose quantification is implicit is problematic for a number of students.

The logico-mathematical symbolism introduced in mathematics courses in order to sort out these ambiguities is far from being shared by learners and even represents an obstacle in their understanding of statements. The handling of symbols is

---

[1] See Durand-Guerrier (2013) and Fuchs (1996), respectively.

J. N. Ngansop (✉)
University of Yaounde 1, PO Box 792, Yaounde, Cameroon
e-mail: judithnjomg@yahoo.fr

G. Kaiser et al. (eds.), *Invited Lectures from the 13th International Congress on Mathematical Education*, ICME-13 Monographs,
https://doi.org/10.1007/978-3-319-72170-5_25

441

learned neither in secondary school nor by university students. Switching from one language to another, namely, from a statement in a natural language to one written exclusively with mathematical variables or other relationship or operation symbols, constitutes for many students a difficult obstacle to overcome (Duval 1988).

Regarding the construction of a proof, Selden and Selden (1995) argue that when students cannot easily make the structure of a logical informal[2] statement explicit, they cannot easily determine the structure of the proof of these statements. Indeed, the logical structure of statements provides indications of how the proof can be undertaken.

The results of the studies that we listed above and a number of others that we are going to present further lead us to propose the following thesis:

The identification of the logical structure of mathematical statements is necessary for the good use of these statements in the learning of mathematics.

A research question that emerges is:

To what extent will conducting a logical analysis enable us to anticipate and analyze the difficulties students face in handling of mathematical statements?

We are carrying out our research in the framework of predicate calculus, which, according to Durand-Guerrier (2003), is the theory of reference for the analysis of mathematical discourse.

In the first part of this paper, we present some elements of predicate calculus that we used as tools to analyze statements. In the second part, we will illustrate with two examples the relevance of logical analysis of mathematical statements as a tool to anticipate students' difficulties. Indeed, the logical analysis of statements can help to anticipate the difficulties a priori in the determination of the structure of a sentence. In these analyses, we lay emphasis on *logical structure and proof* and *logical structure and language switching.*

In the third part, we present the result of an experiment involving first-year mathematics students. We conclude with the perspectives of the research.

25.2 Predicate Calculus as a Tool for Didactic Analyzing of Mathematic Statements

According to Cori and Lascar (2003), predicate calculus is somehow the first step into formalizing the mathematical activity.

---

[2] A statement that deviates from a version in the language of predicate calculus, i.e., it does not use such expressions as "for every," "there exists," "and," "or," "if . . . then," or "if and only if" with their variants (Selden and Selden 1995).

25.2.1 Some Elements of Predicate Calculus

In predicate calculus, the formal language consists of letters for variables and predicates, symbols for logical connectors, and both existential and universal quantifiers. The fundamental elements are atomic formulas, which are built with predicate letters and variables together, and terms. From atomic formulas, logical connectors, and quantifiers, complex statements can be built. But determining the truth value of these statements no longer obeys in most cases the principle of verifunctionality, as was the case in proposition calculus because in predicate calculus, "the complex propositions are not the aggregates of simpler propositions" (Tarski 1936/1972). Indeed, many complex statements are made of intertwined statements. The notion of satisfaction of the propositional function by an element of the discourse universe initiated by Tarski (1944/1972) allows giving a semantic definition of truth as an extension of propositional proposition calculus.

In mathematics, sentences such as:

Some integers are even. (1)
All integers are even numbers. (2)

are respectively true and false statements. They contain, in the first, the existential quantifier "some" and, in the second, the universal quantifier "all." These quantifiers are not part of the alphabet of propositional calculus. These sentences are considered in this system as entities and formalized by a letter, which is a propositional variable.

Let consider the mathematical negation of sentence (2):

There is at least one integer that is not even number. (3)

This can be formalized as $\neg p$, where p is interpreted using statement (2).

This formalization does not allow us to notice the change of quantifier from statement (2) to statement (3) and, *a fortiori*, the structure of the two sentences. Therefore, we cannot analyze them.

25.2.2 Quantification

In the standard language of predicate calculus, there exist two quantifiers: the noted universal quantifier $\forall$, whose meaning in natural language is "all," and the existential quantifier $\exists$, which means in spoken language "there exists at least one."

Given an interpreting domain, the universal quantifier changes an open statement into a true proposition when all the elements in the discourse universe satisfy the open statement;[3] if not, the proposition is false. The formalization of a universally

---

[3] An open statement is a statement containing a free variable, i.e., a variable that is not in the scope of a quantifier. For instance, "x is an even number."

quantified statement is "$\forall x, P(x)$," where x is a variable and P is a propositional function.

The existential quantifier transforms an open sentence into a true proposition if at least one element of the discourse universe satisfies the open sentence. In a case where no object satisfies the open sentence, the proposition is false. A formalization of an existential statement is "$\exists x, P(x)$," where x and P are as previously defined.

It is worth noting that in common language, the existential quantifier is not always explicit. It is the case of the following statement:

The set A has an upper bound.

To convert a given common language statement into predicate calculus language, we have to clarify its meaning, as we will see later. Let us consider that the implicit quantification of statements can have a major influence on the construction of the negation of such statements.

25.2.3   *Implication*

A formula of the type $P(x) \Rightarrow Q(x)$, where P and Q are predicates, is interpreted with an open statement. For any element a in the discourse universe, $P(a) \Rightarrow Q(a)$ is a material implication. It is false only if $P(a)$ is true and if $Q(a)$ is false. In the other cases, it is true. We will say in these cases that a satisfies the formula $P(x) \Rightarrow Q(x)$. Therefore, the connector $\Rightarrow$ in predicate calculus is defined from the material implication and is called *open implication*. As in proposition calculus, the contrapositive of the open implication $P(x) \Rightarrow Q(x)$ is the formula $\neg Q(x) \Rightarrow \neg P(x)$. It is an open implication equivalent to the preceding formula.

The formula $P(x) \Rightarrow Q(x)$ is interpreted in a structure by an open statement; it can be closed with a universal or existential quantifier.

The universal enclosure of the previous statement is $\forall x, P(x) \Rightarrow Q(x)$, which is called *formal implication* (Russell 1910/1989) or *conditional cluster* (Quine 1950). This proposition is true when in every instance of x the derived material implication is true. Therefore, it is obvious that to define the formal implication $\forall x, P(x) \Rightarrow Q(x)$, one should introduce each material implication $P(a) \Rightarrow Q(a)$, defined for a given series of objects.

Formal implication will generate two fundamental rules of deduction:

1. If $\forall x(P(x) \Rightarrow Q(x))$ and $P(a)$, then $Q(a)$.
2. If $\forall x(P(x) \Rightarrow Q(x))$ and $\neg Q(a)$, then $\neg P(a)$.

A formal implication being true can be inferred only in two cases:

When for an instance a of $x, P(a)$ is true, or when $\neg Q(a)$ is true.

For the rest, it is not possible to decide without further information.

It is worth noting that mathematical theorems are generally given in the form of a formal implication, but very often the quantifier is omitted. This expert practice does not always enable students to draw the distinction between an open statement and its universal enclosure: This can generate errors in the use of those statements.

25.2.4 Conclusion

We have made a short presentation of some elements of predicate calculus that make it possible to specify the vocabulary that we will use thereafter. Furthermore, unlike proposition calculus, where the sentence is considered as an entity, predicate calculus takes into account quantification and the status of the letters. It provides tools for analyzing complex statements.

The concepts encountered in this framework present a certain complexity in their use (Ben Kilani 2005; Durand-Guerrier 2003; Epp 1999; Njomgang Ngansop 2013; Durand-Guerrier et al. 2014). One finds them in statements whose logical level of complexity is high, because of the structure of these statements and in the way in which concepts are interwoven with them. We intend to highlight the complexity of two mathematical statements based on their logical analysis. This research shall be based on the logical elements presented above; they shall equally enable us do a priori and a posteriori analysis.

25.3 Examples of Logical Analysis of Mathematical Statements

As the students progress in their curriculum, they face mathematical statements that have increasing complexity. This is the case at the university level with the definition of the continuity of a numerical function of a real variable at point x_0. In secondary school, this definition is introduced with the notion of limit, while at university, it is the mixed or formal language that is used (Bloch and Ghedamsi 2005), but it is not always within the students' reach. This linguistic and mathematical complexity reinforces difficulties in the treatment of statements, but we are not going to linger on it.

In accordance with Quine (1950), we hold that the formalization of mathematical statements contributes to conceptual clarification. This is what guides the logical analyses of the two statements that we propose to examine.

The first statement that we suggest for our analyses is in elementary number theory, and the second one is in calculus.

25.3.1 The First Conjecture of Goldbach

(Pb1) An even integer greater or equal to 4 is the sum of two prime numbers.
We specify that the universe in consideration is the set of natural numbers.
We chose this statement for the following reasons:

- It is stated in common language and is apparently simple and understandable by the reader.
- It is a universally quantified conditional statement whose quantification is implicit. As pointed out in Sect. 25.1.2., this practice is a source of difficulties for students.
- Its initial form hides what must be done to prove this conjecture, while its logical structure shows it.

The formalization of this statement requires removing the implicit aspects inherent to natural language.

The proposed statement is in the form "Every A is B," where A stands for "even integer greater or equal to 4" and B stands for "the sum of two prime numbers." According to Epp (1999), this form can be changed to "for all x, if $A(x)$, then $B(x)$," which is formalized as:

$$\forall x, A(x) \Rightarrow B(x).$$

We are going to paraphrase statement (Pb1) in view of determining the logical structure.

Making explicit the conditional
Suppressing the bounded quantification and introducing a variable

(P1) "For every integer n, if n is even and greater or equal to 4 then, n is the sum of two prime numbers."

The variable n takes its values from the set of integers.

We have a formal implication where the universal quantification depends on the variable n. The antecedent is "n is even and greater than 4" and the consequent is "n can be written as the sum of two prime numbers."

Up to that point, the formulation of the consequent is not explicit; it concerns formalizing this property: "to be the sum of two prime numbers" by introducing two letters of variable.

Making explicit the existential quantifier
To say "the integer n is the sum of two prime numbers" implies that "one can find two prime numbers of which n is the sum," or, still, that "there are two prime numbers p and q such that their sum is equal to n." This clarification thus highlights the underlying existential quantifier.

This conjecture is stated thus:

For every integer n, if n is even and greater than 4, then there are two prime numbers p and q such as their sum is equal to n.

We note:

P: the property interpreted as "be even"
Q: the property interpreted as "be greater than or equal to 4"
S: the ternary relation interpreted as "be the sum of . . . and . . ."

We obtain the formal writing:

$$(P2)\forall n((\mathcal{P}(n) \wedge Q(n)) \Rightarrow (\exists p, \exists q, (P(p) \wedge P(q) \wedge S(n,p,q))$$

This highlights the statement form, which is a universally quantified conditional. Its antecedent is the conjunction of two atomic formulas, and its consequent is an existential statement.

In the clarification of the conditional which is done above, the limited quantification can be maintained within the set of even integers. The formulation obtained is:

(P3) For every even integer n, if it is greater or equal to 4, then there are two prime numbers whose sum is n.

After the clarification of the existential quantifier in the consequent it becomes:

(P4) For every even integer n, if it is greater or equal to 4, then, there exists two prime numbers whose n is the sum.

We can still delete the limited quantification; this brings about the appearance of a new implication:

(P5) For every integer n, (if n is even, then (if n is greater or equal to 4, then there exist two prime numbers whose sum is n)).

Written formally:

$$(P6)\forall n, [\mathcal{P}(n) \Rightarrow (Q(n) \Rightarrow (\exists p, \exists q, P(p) \wedge (q) \wedge S(n,p,q)))]$$

The (P6) written form is equivalent to (P2), for we have the logical equivalence:

$$[p \Rightarrow (q \Rightarrow r)] \equiv [(p \wedge q) \Rightarrow r]$$

The only variable that appears in the writing of (P1) is n, yet in (P2), we need three variables $(n, p,$ and $q)$ defined in $\mathbb{N} \times P \times P$. It is possible to have more variables by raising the formalization level of the statement: That is the case if we have to clarify that "p and q are prime numbers" and "n is even."

The paraphrase and the formalization helped us highlight:

1. The logical structure of the statement,
2. The pertinent variables for its treatment, and
3. The hidden existential quantifier and implicit universal quantifier at the beginning of the statement.

The logical structure of this conjecture gives us guidelines on what to do to make sure a given integer satisfies the implication.

25.3.2 A Fixed-Point Theorem

(u_n) designates a series defined by recurrence with the form "$u_{n+1} = f(u_n)$", where f is a continuous function in $\mathbb{R}$. We therefore have the following result:

(Pb2) If the series (u_n) is convergent, then its limit is the solution to the equation $f(x) = x$.

The logical reasons for the choice of this statement stem from the fact that:

- we are dealing with a theorem that is stated in combined language and simple at first sight;
- it contains non-explicit quantifiers, which makes its formalization complex; and
- the construction of its contrapositive in common language raises a problem caused by the presence of the anaphora.

Let us start with the study of the logical structure of statement (Pb2).

The study relies on the analyses of Durand-Guerrier (1996, pp. 151–153).

The initial formulation is not the same, but the changes bring it to the same formulation as ours.

We specify that the stated general theorem is well known in Terminal class. It is found in the *Terminale C* mathematics book in the syllabus in Cameroon[4] and also in the first-year university calculus course.

We are in the presence of a conditional statement whose structure is complex. It involves three distinct mathematical objects: the series (u_n), an equation, and the numerical function f which links the first two objects. The limit, which is mentioned in the consequent, is implicit in the antecedent. Indeed, to say a series converges means it admits a limit.

(Pb2) can boil down to the minimal statement where the equation is no longer explicit:

If the series (u_n) converges, then its limit is a fixed point of the function.

---

[4]In the collection CIAM manual *Terminal S*, it is in Chap. 13 (numerical series), paragraph 3, (complements on series), p. 286.

According to Durand-Guerrier (1996),

La simplicité apparente de cet énoncé cache en fait une structure complexe qui apparait lorsqu'on cherche à le formaliser, même partiellement. L'énoncé donné est d'ailleurs un intermédiaire nécessaire; en effet, pour formaliser l'énoncé, la présence d'un pronom nous oblige à introduire l'objet "limite." (p. 151)[5]

In fact, saying that the function converges, means admitting the existence of a real number l such that $\lim_{n \to +\infty} u_n = l$. In order to formalize this, the author uses as a discourse universe the reunion of the following sets: the set $\mathbb{R}$ of real numbers, the set of defined and continuous functions in $\mathbb{R}$ and with values in $\mathbb{R}$, and the set of numerical series.

She also chooses:

- a symbol for a two-place relation, R, that states that a series converges towards a given real; $R(u, l)$ is interpreted by "the series $u = (u_n)$ converges towards the real l;
- a predicate with two places denoted as S that expresses the relation between a series and the associated function: $S(u, f)$ is interpreted as "$u_{n+1} = f(u_n)$"; and
- a two-place predicate T that expressing the relation between a function and a fixed point: $T(f, l)$, which is interpreted as "l is a fixed point in function f."

When we are in the discourse universe, the theorem is formalized as:

$$\forall u, \forall f, \forall l, S(u, f) \land R(u, l) \Rightarrow T(f, l) \tag{a}$$

Given that $S(u, f)$ is true, the statement (Pb2) is going to be written:

$$R(u, l) \Rightarrow T(f, l) \tag{b}$$

Which is interpreted as "If the series (u_n) converges towards l, then l is a fixed point of f." This statement is actually quantified. It is written as:

$$\forall l, R(u, l) \Rightarrow T(f, l) \tag{c}$$

And it is interpreted as:

$$\forall l, \lim_{n \to +\infty} u_n = l \Rightarrow f(l) = l \tag{d}$$

The real l is an intermediary object, necessary in the treatment of this situation, where the objects at stake are the series (u_n) and the equation $f(x) = x$.

---

[5]The apparent simplicity of this statement hides in fact a complex structure that appears in the formalization process, even partially. The given statement is moreover a necessary intermediary; indeed, to formalize the statement, the use of a pronoun makes us present the object "limit." [our translation].

Contrary to what one might think, the letter l is connected to the universal quantifier, which the whole statement is on the scope. If that letter is introduced with the existential quantifier, which refers to the antecedent to convey the convergence of the series, we obtain the following open statement in l:

$$\forall f, \forall (u_n), (\exists l, \lim_{n \to +\infty} u_n = l) \Rightarrow f(l) = l \tag{e}$$

This is in contradiction with the fact that a theorem is a closed statement. Besides, this formulation produces a contrapositive which no longer bears its original meaning, namely, "if the equation $f(x) = x$ does not have a solution, then the series (u_n) does not converge."

25.3.3   Conclusion

We have analyzed two mathematical statements and highlighted their logical structures.

In the first statement, we move from a sentence with a linear structure (subject/copula/attribute) to a universally quantified conditional whose consequent is an existential statement. The analysis reveals complex logical structure. We will see in Sect. 25.4.2. that many students do not succeed recognizing the logical structure of this statement.

The second statement, given in mixed language, contains non-explicit quantifiers. To make them appear and to determine their real scope is fundamental for the use of the statement, mainly for the construction of its contrapositive, as we will see later.

25.4   An Experiment with Mathematics Undergraduate Students

In December 2010, we administered a questionnaire to 68 mathematics undergraduate students from the Higher Teachers Training College of Yaounde. After administering it, in January 2011 we organized a follow-up module with eight voluntary students who had previously answered the questionnaire. The aim consisted of identifying the representations that these students had when using logical concepts on the one hand and in teaching situations to clarify such concepts on the other hand. In this paper, we are interested in the justifications that the students gave.

The findings that we give stem from:

- For Problem 1, answers to the aforementioned questionnaire by 68 students, referred to as S1 to S68.
- For Problem 2, a task with eight voluntary students who had answered the questionnaire. We divided them into two groups of four people each. The work

was first carried out by each group, and then we put together the results of the two groups. The following results concern the students from the second group, who will be referred to as E5 to E8.

There was a week gap between the two experiments. Results from Problem 1 are from the questionnaire, while those from Problem 2 come from the workshop that followed the questionnaire administration.

25.4.1 Problem 1

Students were asked to write in formal language the following statement:

(Pb1) Every even integer n greater or equal to 4 is the sum of two prime numbers.

While administering the questionnaire, we specified to students that the scope was the set of integers.

Let us recall the formal writing of that statement

$$(P6) \forall n, [\mathcal{P}(n) \Rightarrow (Q(n) \Rightarrow (\exists p, \exists q, P(p) \wedge P(q) \wedge S(n,p,q)))]$$

where:

- Property P is interpreted as "be prime,"
- Property $\mathcal{P}$ is interpreted as "be even,"
- Property Q is interpreted as "be greater or equal to 4," and
- The ternary relation S is interpreted as "be the sum of ... and ... "

25.4.1.1 Results Analysis Grid

The clarification of both the antecedent and consequent underlines several levels of possible formalizations of this statement and brings us to consider possible answers according to two axes:

1. There is global structure of the sentence and explicit domain of quantification at the beginning of the sentence or not. We distinguish:

    a. There are universally quantified conditional statements where the antecedent and the consequent are respectively the correct expression or not in the formalized language in "n is an even integer greater than 4," and "n can be written as the sum of two prime numbers"

    b. An equivalence

    c. Formulations that are not conditional statements and that we have called "linear." It is a series of conjunctions or statements separated with a comma.

2. Translation of properties and introduction of variables (with or without quantifier).

We will adopt the coding below.
Following the first axis:

UQS: universally quantified conditional statement
NoQCS: non-quantified conditional statement
EquQ: universally quantified equivalence
EqunonQ: non-quantified equivalence
LQS: linear universally quantified statement
LnonQS: linear non-quantified statement

Following the second axis:

FrV: free variable

We do not signal bound variables because all should be bound given that we deal with a closed statement.
Examples of classification:

B designates the set of prime numbers
$$\forall n \in \mathbb{N}, (((\exists k \in \mathbb{N}, n = 2k) \, et \, n \geq 4) \Rightarrow (\exists(p, q) \in \mathbb{N} \times \mathbb{N}, n = p + q)), \text{ with } p \text{ and } q \text{ prime} \tag{25.1}$$

$$\forall n \in \mathbb{N}, n = 2k, \, et \, n \geq 4 \Rightarrow n = p + q, p \text{ and } q \text{ prime} \tag{25.2}$$

$$\forall n \in A, (\exists(p, q) \in B \times B, n = p + q) \text{ where } A = \{n \in \mathbb{N}/n \geq 4 \, and \, n \, even\} \tag{25.3}$$

$$\forall n \in 2\mathbb{N}, \text{ and } n \geq 4, \exists(p, q) \in B^2, n = p + q \tag{25.4}$$

Statement	Structure	Be even and greater than 4	Be the sum of two prime numbers
(1)	Universally quantified conditional in $\mathbb{N}$ UQS	Correctly stated	Prime numbers are introduced by the universal quantifier (UQ) but the property "to be prime" is stated at the end
(2)	A priori universally quantified conditional in $\mathbb{N}$ UQS	Is stated with a free variable to express that n is even k: FrV	The prime numbers are designated with letters of free variables and the property is stated at the end of the consequent FrV: p, q

(continued)

(continued)

Statement	Structure	Be even and greater than 4	Be the sum of two prime numbers
(3)	Linear universally quantified on the set of even numbers greater than 4 LQS	It characterizes the set A, and is correctly stated	Prime numbers are introduced at the beginning of the consequent by the UQ. The formulation is correct
(4)	Linear universally quantified statement in 2ℕ	The domain is made up of even numbers, and it is stated that the property be greater than 4	The prime numbers are introduced by the UQ. The formulation is correct

In consideration of what precedes, beyond a small number of possible global structures, we can expect to come across a wide range of formulations for this statement; this is all the more so as the mathematical uses are not homogenous from this viewpoint.

25.4.1.2 A Posteriori Analysis of the Results

Among the 68 students who took the test, only 25 proposed a formally written version of the item.

As we might have expected, productions are different from one to another in general, but we all the same find similar structures. We will present the answers of the students according to the first axis of our a priori analysis. In this table we delete the EqunonQ and LnonQS because all the students who answered the questionnaire proposed the universally quantified statements (Table 25.1).

Analysis according to the first axis
There is global structure of the sentence and explicit domain of quantification at the beginning of the sentence or not.

We have come across four types of formulations:

- Universally quantified conditional statements
- Non-quantified conditional statement
- Equivalence
- Linear quantified statements.

About the quantity of the sentence, except for the answers of two students, the scope of the universal quantifier binding the integer n was not specified, thus making the status of the letter n ambiguous: One cannot say with certainty whether the variable is free or is a generic element. But in the formal point of view, the variable is considered free.

We can attribute relative imprecisions about the quantifiers to school habits where the use of parentheses to mark the scope of the quantifier at the beginning of the sentence is not very common. It is when the quantifier is "internal" to the

Table 25.1 Distribution of answers according to the structure

UQS	NoQCS	EquQ	LQS	No answer
10	1	1	13	43

statement that we specify the scope. Therefore, we can devise a hypothesis for those answers: that the quantifier covers the whole sentence.

The conditional statements
Of the 10 students who produced universally quantified conditional statements, for eight of whom the quantification domain is $\mathbb{N}$. Among these eight answers, the clarification of the conjunction of "n is even" and "≥ 4" is present only once. In the antecedents of other answers, these two statements are separated with a comma:

$$S15 : \forall n \in \mathbb{N}, (n \geq 4), \exists k \in \mathbb{N}^*, n = 2k \Rightarrow \exists p_1, p_2 \in P/n = p_1 + p_2$$
$$S30 : \forall n \in \mathbb{N}, \ n \text{ even}, n \geq 4 \Rightarrow \exists p, q \in P, n = p + q$$
$$S16 : \forall n \in \mathbb{N}, \ n = 2k, k \in \mathbb{N} \ n \geq 4 \Rightarrow \exists p_1 \text{ and } p_2 \in \mathbb{N}, \text{prime}/n = p_1 + p_2$$

We make the hypothesis that it is the literal version of "every even integer n, greater or equal to 4."

Five consequents of the conditional statements are existential statements as in the three examples above; the others are not.

Linear statements
As is the case with conditional statements, we get the literal version "Every even integer n greater than or equal to 4" in some linear statements:

$$S27 : \forall n, n = 2k (k \in \mathbb{Z}^2_+) n > 4, \exists n_1 \text{and} \ n_2 \text{ prime as } n = n_1 + n_2$$
$$S31 : \forall n \in 2\mathbb{N}, \exists (p, q) \in P^2 \text{ such as } \geq 4 \wedge n = p + q, \text{ with } P \text{ a set of prime numbers}$$

S31's statement has incorrect syntax, and this leads to a modification in the meaning of the initial sentence. We paraphrase that answer thus:

For every even integer n, there is a couple of prime numbers (p, q) such that n is greater than 4 and is the sum of these two prime numbers.

The formulation is unsuitable because 2 is a counter-example to the associated open statement. Besides, we shall note the disappearance of the implication.

Analysis according to the second axis
Translation of properties and introduction of variables
Subsequently, we analyze the answers according to the second axis, that is to say, according to the manner in which the properties are expressed and the variables introduced. According to the clarification of the structure of the statement, except n, which is given in the initial sentence, subsidiary variables are introduced to define the two prime numbers and eventually the parity of an integer. The difficulty at this level could come from introduction of these hidden variables.

Table 25.2 Structure of statements and status of variables

	k FrV	p,q FrV	k,p,q FrV	n FrV	Closed statements	Total
The conditional statements	1	3	2	1	4	11
The linear statements	3	2	3	0	5	13
The equivalence	1					1

In Table 25.2, we have classified the students' answers according to the structure and the different free variables that they contain.

Let us give the name k to the variable which permits us to define the parity and p and q the variables that designate the two prime numbers.

According to the variable n

In S29's answer, if because n is a free variable in the consequent and is bound to the antecedent, then we have an open statement:

$$S29 : (\forall n \in \mathbb{N}, \exists p \in N - \{0, 1\}, n = 2p) \Rightarrow (\exists p, q \in \mathbb{N}, p \text{ and } q \text{ prime}/n = p + q)$$

This is due to an error in writing the parentheses; the universal quantifier binding n only marks the antecedent.

According to the variable k

We recall that this variable is used to algebraically define the parity of the integer n. Fourteen students chose to explicate the parity as shown below.

Twelve students produced statements where k is a free variable (type (k FrV and k, p, q FrV)). Among their answers, incorrect syntaxes are found:

$$S16 : \forall n \in \mathbb{N}, n = 2k, k \in \mathbb{N} \, n \geq 4 \Rightarrow \exists p_1 \text{ and } p_2 \in \mathbb{N} \text{ and prime}/n = p_1 + p_2$$
$$S40 : \forall n \in \{2k\}, k \in \mathbb{N}, n \geq 4, n = p_1 + p_2 \text{ with } p_1, p_2 \text{prime}$$

In S16's answer, one may think the two variables $p_1 et p_2$ have been bound, but the syntax is incorrect.

In S40's answer, the property "be prime" is at the end, whereas it ought to appear before the writing of n, and the prime numbers ought to have been bound with the existential quantifier.

Three students (S15, S29, and S42) used the existential quantifier to introduce it; k is a bound variable in their answer.

The other students did not make use of it as they have used sets in which the parity of the elements is a characteristic.

According to the variables p *and* q

The variables p and q should appear in the formalization of the statement in the writing of n as the sum of the two prime integers. They are introduced by the existential quantifier. A student (S02) has introduced them with the universal quantifier, which changes the significance of the statement, becoming: "every even

integer greater or equal to 4 is the sum of two prime numbers whatsoever." The
variables p and q appear as free variables in 10 statements produced by the students
(eventually with another name). For example:

S26 : $\forall n \in 2\mathbb{Z}, n \geq 4n = a + b$ with a, b of whole prime numbers

S67 : $\forall n = 2p, p \geq 2, p \in \mathbb{N} \Rightarrow n = T_1 + T_2$ and $D(T_1) = \{1, T_1\}; D(T_2) = \{1, T_2\}$

Among the 25 students who responded to this item, 16 produced open statements. Among the latter, four students (S40, S41, S44, and S67) neither introduced
the letters referring to prime numbers nor the variable k; three students (S26, S39,
and S45) did not introduce the letters which refer to prime numbers; three students
(S16, S42, S68) did not introduce k. Formally, the letters in their answer are free
variables.

In the 16 responses aforementioned, the students specified the variables' domain
after writing them:

$$= 2k, k \in \mathbb{N}$$

$n = p + q, p, q \, \text{prime}$

$n = p + q$, where p and q are prime integers...

We can hypothesize that there are generic elements for those students that they
introduced in some way.

25.4.1.3   Conclusion

This exercise has enabled us to account for the difficulties faced by a number of
students to identify the implicits in the formulations in common language on the
one hand and the management the students made of the variables on the other hand.
We can draw the following conclusions:

- The transformation of a statement in the form "all A is B": into a statement in the
  form "$\forall x, (A(x) \Rightarrow B(x))$" is not obvious: The students' answers are close the
  congruent statements of the initial statement, mostly regarding the antecedent.
  When the domain is $\mathbb{N}$, the formalization of the expression of "n even and
  greater than 4" is not made in the form of a conjunction.
- None of the conditional statements which have been suggested is correct.
- The syntax in the use of symbols is approximate and the phenomenon of imitation that has been seen with students (Gueudet 2008) is clearly there. Before,
  some denotations that the teachers had used are found in their work.
- The status of variables is not always taken into consideration. Some students
  gave as a symbolic formulation open statements where several free or generic

variables often appear. The nature of these variables is specified, but in an incorrect syntax from a logic point of view.

- The absence of the existential quantifier produces statements congruent to the initial statement that do not express this statement.
- We find again in the students' productions the same phenomena spotted by Selden and Selden (1995), namely, the poor capacity of students in making explicit the logical structure of informal statements.

In testing our results, we questioned 25 second-year students studying mathematics at the Higher Teachers Training College of Yaounde in 2015 who have given responses similar to the those that undergraduate students gave: None were correct.

In general, the results of the test a priori show that the academic standard does not have any major influence on the students' ability to satisfactorily perform the language shift. We come across practically the same formulations as those of first-year students, whereas the practice of formalism for at least an academic year let us assume that they would be more capable of handling this issue.

25.4.2 Problem 2

This problem is about inference rules: We are interested in the issue of identifying situations that permit or do not permit making deductions. Let us recall the statement:

In what follows, (u_n) is a series defined by recurrence as "$u_{n+1} = f(u_n)$," where f is a continuous function on $\mathbb{R}$. We then have the following result:

(Pb2) If the series (u_n) is convergent, then its limit is the solution to the equation f $(x) = x$.

The question: What can we say about the convergence of the series (u_n) if equation "$f(x) = x$" does not have a solution?

The table below presents the repartition of responses to this item in the questionnaire.

25.4.2.1 A Problematic Construction of the Contrapositive of (Pb2) by Students

We present and analyze in this part a sequence that happened between students of a group.

Answering the question asked, the students unanimously said that the series does not converge.

1 E6: If the equation $f(x) = x$ does not have a solution, we immediately deduce that f is not convergent. Because if we try to have a look at the . . . the reciprocal . . . the contrapositive of if f. is convergent, its limit is the solution of $f(x) = x$, heuuu, no, wait a moment.

5 E5: $f(x) = x$. It is false. You see a little moment, so that seems a bit clear in my head if, that the position . . .

6.1,[6] so, if $f(x) = x$ does not have a solution, then u_n is not convergent. To me, it looks crystal clear.

In their attempts to justify their answer, finally, the students decide to use the contrapositive of the statement that they formulate in common language as follows:

25 E5: And the contrapositive is very clear! The contrapositive says "if the limit of the series u_n" . . .

26 E7: . . . is not a solution to the equation $f(x) = x$

The two interventions can be summarized thus:

$$\text{If } \lim_{n \to +\infty} u_n = l \text{ and } f(l) \neq l \text{ then, } \lim_{n \to +\infty} u_n \neq l \qquad \text{(f)}$$

The construction of the contrapositive helps to discover the difficulties related to the implicit quantification. In common language, the literal expression of the contrapositive underlines a contradiction between the negation of the consequent "if the limit of the series does not satisfy the equation" and the negation of the antecedent "the series is not convergent," which means that the series does not have a finite limit. This is due to the phenomenon of anaphora.

The formal writing permits clarification of the implicit quantification on the object *limit* in expressing the convergence of the series (u_n); this object might have been introduced by the universal quantifier. This writing permits building the contrapositive because it dispels the ambiguity on the status of limit. The difficulties due to passing to the contraposition are dealt with in the debates below.

Regarding the contradiction stated in 25 and 26:

34 E5: I don't think the word limit can be in . . .

35 E7: what will be the contrapositive?

36 E6: Because the limit must first of all exist. Because if they say limit . . ., if you say now that . . .

37 E5: Hum, if the limit . . ., if the limit of the series (u_n) is the solution of . . ., that is to say, . . .

38 E6: if the limit . . ., it already exists, you see, don't you? It already exists . . .

39 E5: . . . and to say after that the series (u_n) is not convergent, this doesn't have meaning. You are following me, so for me I now say that as contrapositive we must say that if the equation $f(x) = x$ does not admit a solution in $\mathbb{R}$, then the series is not convergent. To me, I think that it is the contrapositive. Because as soon as they put

---

[6]"6.1" is the number of the question in the questionnaire given to the students.

the word *limit*, it creates a sort of misunderstanding; we no longer understand anything.

Having identified the contradiction in the first proposition of the contrapositive, the students use pragmatic arguments to construct it (37, 38, and 39).

These exchanges underline the difficulties related to the relations between the different objects introduced. E5 must substitute (Line 39) the word *limit* for the transformation of the sentence in order to be able to state the contrapositive. He obtains the correct contrapositive that corresponds to the one we have proposed further to the formalization of the initial statement in universally quantifying the letter of the variable that designates the limit. Indeed, the contrapositive of (c) is:

$$\forall l, \text{non } T(f, l) \Rightarrow \text{non } R(u, l) \tag{g}$$

Which is interpreted as:

$$\forall l, f(l) \neq l \Rightarrow \lim_{n \to +\infty} u_n \neq l \tag{h}$$

That is to say, if the equation $f(x) = x$ does not have solution, then, the series (u_n) does not converge. This is E5's formulation. The latter highlights the difficulty created by the presence of *limit* (39).

25.4.2.2 Conclusion

Table 25.3 shows that, of the 47 students who answered to this question in the questionnaire that has been proposed, 83% answered correctly. This result can be explained by the fact that this theorem is well known to students.

Exchanges above highlight difficulties students feel in justifying their answers through the construction of contrapositive of (Pb2) because of:

- The presence of anaphora and
- The non-explicit quantifiers.

The strategy of the students will be to eliminate the word *limit* (Line 34) in the antecedent of the contrapositive.

Moreover, the status of the series and the variable x remain ambiguous in their proposition of the contrapositive; they seem to be generic elements.

(1) From these exchanges, we can infer that the statement (Pb2) is appropriate to work, on one hand, on the choice of quantifiers in the formalization activities, and, on the other, on the importance of making quantifiers explicit in order to build the contrapositive of a statement. This problem could also permit making explicit some inference rules that will contribute to lighten certain reasoning. This problem is quite appropriate to work on the choice of quantifiers, firstly in formalization activities and secondly on the importance of clarification of quantifiers in constructing the contrapositive of a statement. This exercise could

Table 25.3 Students' responses

	The series does not converge	Nothing can be said	Other answer	No answer
Size	39	3	5	21

also clarify certain rules of inference that would contribute to lighten certain reasoning.

Conclusion and perspectives

We have shown the logical analysis of two mathematical statements:

- The first statement is the first of Goldbach's conjecturesThe hose clarification of the it logical structure that displays in the consequent, the existential quantifier that the formulation in common language was hiding;
- The second is an analysis theorem on $\mathbb{R}$ whose knowledge of the contrapositive is necessary in solving a problem.

These analyses highlight the complexity of those statements. This therefore urges us to question the capability of the students to effectively determine their own structure in their usage of proof-making activities.

Problem 1 shows that the transformation of a statement in the form "All A is B" into a statement in the form "$\forall x, (A(x) \Rightarrow B(x))$" is not obvious: The students' answers are close to congruent statements to the initial statement. The syntax in the usage of symbols is approximate and the phenomenon of imitation seen in students by Gueudet (2008) is quite visible.

About Problem 2, the sequences of exchanges show that the knowledge of logical structure on one hand helps dispel the language ambiguities: An alternative among the possible interpretations of a given statement in common language has to be made. On the other, savings in the cognitive point of view can be achieved by students when they know the form of the statements they are working out.

Making the logical structure of mathematical statements explicit is an activity that, given its importance, should be regularly practiced, with an emphasis on the semantic aspect. As a matter of fact, the teacher lecturer has to led the students to give a meaning to the symbols that they use. A perspective of this work is to elaborate situations that will enable students to become familiar with this type of exercise based on statements used in mathematics classes. Another prospect would be to develop the reverse activity, which consists of moving from formal language to common language. Indeed, the switch from formal to common language permits a good understanding of a statement. We believe that such an activity can help develop linguistic and language competences of the subject being learned.

References

Ben Kilani, I. (2005). *Les effets didactiques des différences de fonctionnement de la négation dans la langue arabe, la langue française et le langage mathématique* (Doctoral dissertation). Thèse en cotutelle entre l'Université de Tunis et l'Université Lyon 1.

Bloch, I., & Ghedamsi, I. (2005). Comment le cursus secondaire prépare-t-il les élèves aux études universitaires? Le cas de l'enseignement de l'analyse en Tunisie. *Petit X, 69,* 7–30.

Cori, R., & Lascar, D. (2003). *Cours de logique.* Paris: DUNOD

Durand-Guerrier, V. (1996). *Logique et raisonnement mathématique: Défense et illustration de la pertinence du calcul des prédicats pour une approche didactique des difficultés liées à l'implication* (Doctoral dissertation). Université Claude Bernard Lyon 1, Lyon.

Durand-Guerrier, V. (2003). Which notion of implication is the right one? Formal logical considerations to a didactic perspective. *Educational Studies in Mathematics, 53,* 5–34.

Durand-Guerrier, V. (2013). Quelques apports de l'analyse logique du langage pour les recherches en didactique des mathématiques 2011. Dans *Questions vives en didactique des mathématiques: problèmes de la profession d'enseignat, rôle du langage: actes de l'école d'été de didactique de mathématiques* (Vol. 1, pp. 233–265). Carcassonne: La Pensée Sauvage.

Durand-Guerrier, V., Kazima, M., Libbrecht, P., Njomgang Ngansop, J., Salekhova, L., Tuktamyshov, N., et al. (2014). Challenges and opportunities for second language learners in undergratuate mathematics. In *Mathematics education and language diversity* (pp. 85–101). Berlin: Springer.

Duval, R. (1988). Ecarts sémantiques et cohérence mathématique: Introduction aux problèmes de congruence. *Annales de Didactiques et de Sciences Cognitives, 1,* 7–23.

Epp, S. (1999). The language of quantification in mathematics instruction. In L. V. Dans & F. R. Curcio (Eds.), *Developing mathematical reasoning in grades K-12.* Reston: NCTM.

Fuchs, C. (1996). *Les ambiguïtés du français. Collection l'essentiel du français.* Paris: Orphys.

Gueudet, G. (2008). Investigating the secondary-tertiairy transition. *Educational Studies in Mathematics, 67,* 237–254.

Njomgang Ngansop, J. (2013). *Enseigner les concepts de logique dans l'espace mathématique francophone: aspect épistémologique, didactique et langagier. Une étude de cas au Cameroun.* (Doctoral dissertation). Thèse en cotutelle Université de Yaoundé et Université Lyon 1 Claude Bernard.

Quine, W. V. (1950). *Methods of logic.* New York: Holy, Rinehart, Winston.

Russell, B. (1989). *Principia Mathematica traduction française in RUSSELL, Ecrits de logique philosophique.* Paris: PUF (Original work published 1910).

Selden, J., & Selden, A. (1995). Unpacking the logic of mathematical statements. *Educational Studies in Mathematics, 29,* 123–151.

Tarski, A. (1972). La conception sémantique de la vérité et les fondements de la sémantique (G. Gilles-Gaston, et al., Trans.). Dans *Logique, sémantique, métamathématique* (Vol. 2, pp. 267–405). Paris: Armand Colin (Original work published 1944).

Tarski, A. (1972). Le concept de vérité dans les langages formalisés (G. Gilles-Gaston, et al., Trans.). Dans *Logique, sémantique et métamathématique* (Vol. 1, pp. 157–269). Paris: Armand Colin. (Original work published 1936).

Chapter 26
Understanding and Visualizing Linear Transformations

Asuman Oktaç

Abstract The aim of this chapter is to give an overview of the research that we have been conducting in our research group in Mexico about the linear transformation concept, focusing on difficulties associated with its learning, intuitive mental models that students may develop in relation with it, an outline of a genetic decomposition that describes a possible way in which this concept can be constructed, problems that students may experience with regard to registers of representation, and the role that dynamic geometry environments might play in interpreting its effects. Preliminary results from an ongoing study about what it means to visualize the process of a linear transformation are reported. A literature review that directly relates to the content of this chapter as well as directions for future research and didactical suggestions are provided.

Keywords Linear transformation · Visualization · Representation
Dynamic geometry · Linear algebra

26.1 Introduction

Linear transformation is one of the more abstract concepts studied in linear algebra. It is also one of the concepts with which students experience considerable difficulties (Sierpinska 2000; Sierpinska et al. 1999). Some of these difficulties may be related to their previously constructed function conceptions, since a linear transformation is a special kind of function between vector spaces. Trigueros and Bianchini (2016) observed that in the context of a modelling problem this relationship becomes clearer for students. Uicab and Oktaç (2006) observed that some students required an explicit formula for a transformation involved in a problem even in situations where it is not needed, as also mentioned in Sierpinska (2000).

A. Oktaç (✉)
Cinvestav-IPN, Mexico City, Mexico
e-mail: oktac@cinvestav.mx

G. Kaiser et al. (eds.), *Invited Lectures from the 13th International Congress on Mathematical Education*, ICME-13 Monographs,
https://doi.org/10.1007/978-3-319-72170-5_26

Karrer and Jahn (2008) report other types of difficulties such as conversion between registers, especially from the graphic register to others; belief that linear transformations can only be applied to polygonal objects; and thinking that a transformation that conserves straight lines is necessarily linear. These authors suggest the use of a dynamic geometry environment in which students can observe a linear transformation in three registers (graphic, algebraic, and matrix) simultaneously as well as the effect of making a change in one register on the others, in order to overcome these difficulties.

In our research group, we have been studying the linear transformation concept from different angles, including how it is constructed, associated difficulties, conceptions that students might develop, and representations. In this chapter the intention is to bring to the attention of an international audience selected work that has been conducted in Spanish about the learning of this notion. The findings reported here form part of a larger ongoing project about the understanding of Linear Algebra concepts. Although the data reported comes from studies conducted in Mexico and Chile, the observed phenomena might shed light on difficulties that students experience in other parts of the world as well.

26.2 Linear Transformations and Intuitive Models

One of our early interests in starting to study the understanding of linear transformations was to determine the kinds of intuitive models, in Fischbein's (1989) sense, that students develop in relation with this concept. Our first study in this direction (Molina and Oktaç 2007) placed emphasis on geometric contexts, since these are favored less in linear algebra courses. Indeed, after analyzing some linear algebra textbooks, Karrer and Jahn (2008) concluded that the graphic register is the least used. In our study, five master's students in mathematics education were given pairs of figures such as the one shown in Fig. 26.1 that showed a region or some vectors in the plane and were asked if there could exist a linear transformation that mapped the configuration in the figure on the left to the figure on the right. When

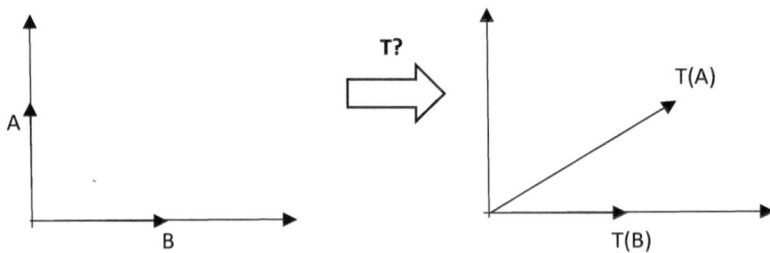

Fig. 26.1 Possible linear transformation given by its effect on two vectors (adapted from Molina and Oktaç 2007)

confronted with the problem in Fig. 26.1, which corresponds to a shear transformation, all the students responded in the same manner, saying that there was no such linear transformation; they explained their answer by reasoning that a linear transformation cannot leave one vector fixed and change the other one. Later we observed this phenomenon with graduate students in scientific fields in the case of a reflection about the y-axis as well (Ramírez Sandoval and Oktaç 2012).

Hermes, one of the interviewed students, thought for a long while when he saw the next problem, shown in Fig. 26.2. He then suddenly grabbed the previous question (Fig. 26.1) and said that it was possible to have such a linear transformation and that his previous answer was wrong. We wondered what had happened. What made him change his mind and how were these two questions related? When we asked him about it he said that he was considering only special transformations such as rotation and dilation (Molina and Oktaç 2007). Actually, it seems that there are two things that led him to change his mind: First, he assumed that in the second problem such a linear transformation exists, probably because of textbook illustrations and classroom examples showing that rectangular regions are mapped to rectangular regions under linear transformations. Second, he saw that in the image figure one side of the rectangle shrank and the other side expanded, and this made him realize that his argument that a linear transformation should do "the same thing" to both vectors does not hold in general. In other questions that showed the effect of general linear transformations where no immediate geometric interpretation was observable, the same difficulty was observed, since the students were looking for prototype transformations.

The tendency to think in prototype transformations such as rotation, reflection, and dilation was present in all the students we interviewed. A similar phenomenon was reported in Sierpinska (2000). Some students were able to think in terms of compositions of these known transformations, but not beyond. However this is not to say that rotation is a simpler transformation than shear; it all depends on the context. For example, Trigueros and Bianchini (2016) observed that students had a more difficult time finding the formula of a rotation (since it contains trigonometric functions) when working on a modelling problem than with the formula of a shear transformation; in this context rotation is considered more complex.

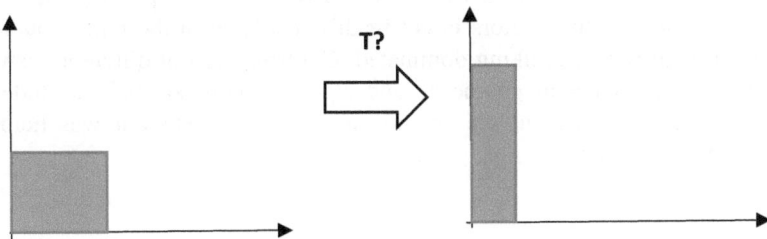

Fig. 26.2 Possible linear transformation given by its effect on a region (adapted from Molina and Oktaç 2007)

As Hillel (2000) comments, not all linear transformations have simple geometric interpretations, even in two and three dimensional spaces. However, for students these prototypes that are associated with certain movements in the plane replace the definition, giving rise to intuitive models that act as substitutes for mathematical theory (Fischbein 1989). These models impose certain "results" in the substitute theory; for example, "a linear transformation does the same thing to all vectors." Since this statement is a simplified version of what happens to the plane geometrically under a linear transformation, "doing the same thing" is also interpreted in a simplistic way visually. According to Fischbein (1989) "the intuitive model manipulates from *behind the scenes*, the meaning, the use, the properties of the formally established concept. The intuitive model seems to be stronger than the formal concept" (p. 10).

Another conception that we came across in this study is the one that associates a linear transformation to each vector instead of the whole plane. A similar phenomenon was also observed in Dreyfus et al. (1998). Textbook illustrations that show one vector and its rotated image, in order to exemplify a rotation, for example, might contribute to students developing this viewpoint. One of the conclusions at which we arrived is that some students focus on the objects involved, such as vectors, and not on the processes that are transforming them (Molina and Oktaç 2007).

After this first study, we wanted to know whether these intuitive models prevailed only in geometric contexts or they were also present in algebraic contexts and, if so, in what way. This time we designed an interview (Ramírez Sandoval and Oktaç 2012; Ramírez Sandoval 2008) that consisted of two parts. In the first part we included questions similar to the ones presented in Molina and Oktaç (2007), and the second part consisted of equivalent problems presented algebraically, asking whether a linear transformation could exist that maps a given pair of vectors to another. The intention was, if students' answers to algebraic problems differed from geometric ones, to confront them and see how they reconciled the conflicting responses. The interview was applied to five master's students who were specializing in different science subjects.

Geometrically, we observed the same kinds of conceptions and intuitive models that students had developed as in the previous investigation. Algebraically, though, this was not the case. Algebraic symbols, as opposed to images of geometric objects, are not "given directly to the mind" (Sierpinska 2000, p. 233); the effect of the transformation on the vectors is not readily available in the tuple notation. In this context, algorithmic thinking dominated. Confrontation of different answers to the equivalent problems in geometric and algebraic contexts helped students to recognize their mistakes, but we are not sure to what extent it was helpful in constructing the concept.

26.3 How Is the Linear Transformation Concept Constructed?

Subsequently we wanted to understand how the linear transformation concept can be constructed in the mind of an individual. In order to research this we adopted APOS (Action–Process–Object–Schema) theory as a framework and made a genetic decomposition that consisted of descriptions of mental structures and mechanisms through which students might come to comprehend the topic in question (Roa-Fuentes and Oktaç 2010). According to Arnon et al. (2014) these structures and mechanisms "involve a spiral approach where new structures are built by acting on existing structures" (p. 26). We contemplated that the construction of the linear transformation concept can start in one of two ways: Either the student constructs a general transformation concept first and then the linear transformation is constructed as a special case, or the linearity properties can be constructed without reference to a transformation concept; in either case the function schema assimilates the vector space object so that new kinds of functions with domain and range as vector spaces can be considered (Roa-Fuentes and Oktaç 2010). We did not find empirical evidence for the version that starts with the construction of the transformation concept (Roa-Fuentes and Oktaç 2012), probably because the instruction that the student participants followed was not based on that approach.

According to our validated genetic decomposition, students with an Action conception can verify the linearity conditions for specific vectors and linear transformations but have difficulty in imagining the verification of the conditions for all the vectors of the domain and thinking about the concept of linear transformation in a general way. When these actions get interiorized, they give rise to Processes related to the two conditions of linearity, which are then coordinated to construct a new Process that can be called the Process of linear transformation. In this way, the sum and multiplication by scalar properties can be combined in a linear combination version, which unites them. The importance of this coordination was evidenced with the observation of a student who was able to cite the properties, but when it came to verifying whether a given transformation was linear, relied only on one of them (Roa-Fuentes and Oktaç 2012). This also shows the difference between the mathematical definition and the cognitive construction of a concept. When there is need to apply actions on this Process, it is encapsulated into an Object that can be modified; these actions can consist of composing linear transformations or asking questions about their general properties, such as the conditions for a linear transformation to be invertible (Roa-Fuentes and Oktaç 2012). Due to lack of previous research from an APOS perspective on the topic, in this study we took into consideration only the algebraic representation as a starting point.

26.4 Role of Registers of Representation

Conscious of the role that different registers of representation (Duval 1993) and their interplay might have on the understanding of the linear transformation concept, we undertook a study (Ramírez-Sandoval et al. 2014) in which we identified, on one hand, successful cases of coordination of different registers and, on the other, different ways in which this attempt proves unsuccessful; we also discussed a phenomenon called *mixing of registers*. We defined this notion as the simultaneous use of more than one register without coordination: "The mixing of registers consists in the use of representations without respecting the rules of formation of the register that they supposedly belong to, mixing rules of formation of two or more registers" (p. 244).

In a study that made use of representations, Wawro (2009) set up a teaching experiment in order to determine the connections that students might make between the matrix representation of a linear transformation defined on R^2 and its geometric effect on the plane, as well as to provide a context for feeling the need for a change of basis and exploring this notion. This way students were able to realize how different components of a matrix contributed to the geometric effect of a linear transformation, especially in the case of stretch and shear. Now, in order to illustrate the notion of mixing of registers, we give an example from our study in which during an interview the student Franco was working on the following problem adapted from Wawro (2009).

In Fig. 26.3 the letter M appears first in 12-point normal font, and then in 16-point italics font. Can you find a matrix that transforms the M on the left to the one on the right?

Franco's strategy consisted of taking a pair of vectors from the image on the left and from the image on the right, converting these graphical representations to algebraic ones and forming a system of equations whose solution would give the entries of the matrix. His strategy was reasonable, but when he started working on the graphical register he began mixing the rules of the synthetic graphical register and the Cartesian graphical register. After he drew the shape in Fig. 26.4, he assigned the coordinates (0, 3) to the vertical vector and (1, −2) to the diagonal one.

Fig. 26.3 Looking for a matrix that transforms the *M* on the left to the one on the right (adapted from Wawro 2009)

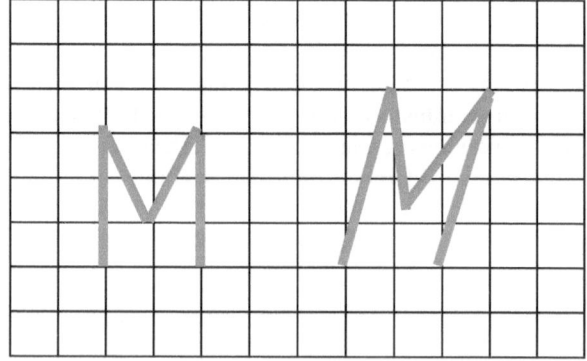

Fig. 26.4 Franco's
M (adapted from
Ramírez-Sandoval et al.
2014)

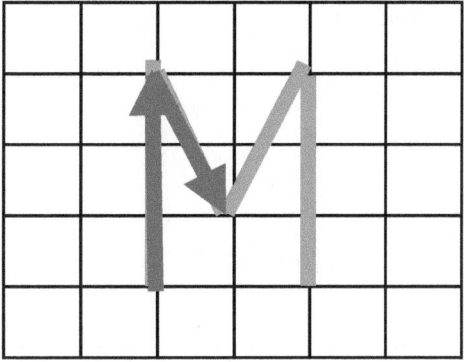

This means that he was reading the coordinates as in the Cartesian plane, but placing a vector to start with the end point of another one as in the synthetic graphic register, leading to an existence of two distinct origins for the two vectors (Ramírez-Sandoval et al. 2014); this prevented him from solving the problem successfully. Wawro et al. (2012) also reported that some students placed the starting point of a vector on the tip of another, giving rise to a "floating origin".

As a result of this study we note that in the literature there is a lack of consensus on the names and characteristics of registers of representation in linear algebra. We offer a characterization of two graphical registers (synthetic and Cartesian) as well as algebraic and matrix registers (Ramírez-Sandoval et al. 2014).

26.5   Integrating Dynamic Geometry Software

Dreyfus et al. (1998) designed a course using a dynamic geometry environment so that students could have a coordinate-free geometric entry into linear algebra in which they could experiment with linear as well as non-linear transformations; in this approach the linearity conditions were interpreted geometrically. They sustain that a dynamic geometry environment

> gives far more visibility to transformations than a paper and pencil environment. In fact, a variable vector and its variable image under the transformation can be placed on the screen simultaneously. In this situation, the effect of a transformation is directly observable, thus indirectly lending some visibility to the transformation itself. (pp. 218–219)

They also warn us against the pitfalls of computer environments and related designs, since students tend to develop conceptions of a linear transformation as the image of a vector as opposed to a function transforming the plane.

In a study that we conducted (Romero Félix and Oktaç 2015a, b), university students went through instruction on the topic of linear transformations using GeoGebra applets, after which they were given a questionnaire and then interviewed with the aim of characterizing their mental constructions in the presence of

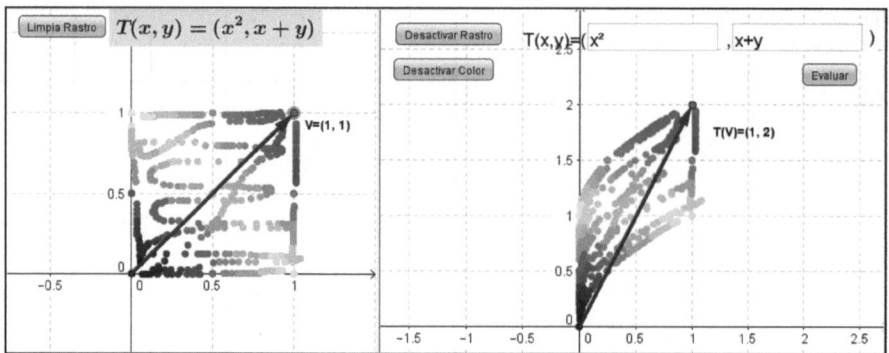

Fig. 26.5 Image of a region under a non-linear transformation (Romero Félix 2016, p. 93)

Fig. 26.6 Static
representations with the
influence of dynamic ones
(adapted from Romero Félix
2016)

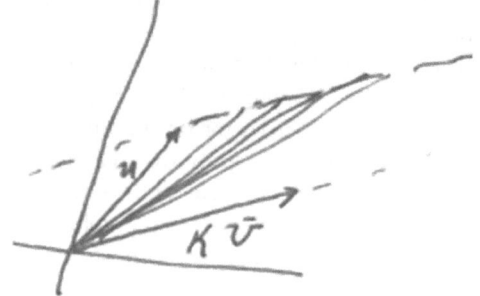

static and dynamic representations. The activities that were used during instruction were specially designed to motivate the construction of mental structures antici- pated by the genetic decomposition. For example, students worked with problems of the type shown in Fig. 26.5, where they could trace the image of a particular region under a transformation and observe the differences between the effect that corresponds to a linear transformation and the one that corresponds to a non-linear one; the aim of this kind of activity was to aid in the interiorization of actions into processes where students can start thinking about the image of a region instead of images of particular vectors. After having worked with these kinds of activities, even when asked to produce graphical representations in paper and pencil envi- ronments, students generated images that showed traces of vectors, obviously with the influence of this dynamic environment (Fig. 26.6).

In this study it was found that the dynamic geometry environment together with the instructional design that was directed towards helping students construct the conceptions expected by the genetic decomposition aided in the interiorization of Actions. About constructing a Process conception, Romero Félix (2016) points out the following:

> Interiorization toward Processes requires analyzing a *sufficient* quantity of repetitions in order to achieve an internal version of them; the sufficiency is reached when a significant reflection that permits taking control of the steps of the Actions is made. Dynamic representations make a greater quantity of information available for students, practically in an immediate and continuous manner, through intuitive manipulations of representations. (p. 167)

> We propose that dynamic graphical representations work for the students as catalysts of the mechanism of interiorization. (p. 168)

In general, the Object conception is hard to reach and even after completing undergraduate courses, students do not show evidence of being able to apply actions on processes (Arnon et al. 2014). Care was taken during instruction to provide students with activities that were intended to aid in the encapsulation of processes. For example, after they had to come up with a linear transformation that sends a square region to itself, they had to modify it so that this time it would send the same square region to a bigger square or a general rectangle. Students in this study in general were able to develop an Object conception for linear transformations in the plane, but not outside of this context. This points out the importance of providing students with the opportunity to work with transformations in different vector spaces. According to Romero Félix (2016) "because of the analog nature of graphical registers, treatments in these without reference to the represented Objects would be difficult.... [I]n this way formation or interpretation of graphical representations refers to the properties of Objects more directly" (p. 167).

To the best of my knowledge, this was the first time dynamic geometry software was used in relation with research from the viewpoint of APOS theory. As a result of this study, a genetic decomposition of the linear transformation concept that includes representations was produced (Romero Félix 2016), as well as an articulation between the theoretical approaches of APOS (Arnon et al. 2014) and registers of representation (Duval 1993).

26.6 Visualization of Linear Transformations

Visualization of linear transformations has applications in computer graphics and robotics and in general where a study of geometrical representation of objects and their motion and transformation are involved. There have been different didactic suggestions in the literature as to how to visualize linear transformations. Monagan (2002) proposes both the use of CAS to display the images of different objects, such as circles, instead of the square regions that are normally used to illustrate the effect of a linear transformation and the use of animations to get an idea of the effect on the whole plane. Triantafyllou and Timcenko (2013) recommend displaying the matrix of a linear transformation and a geometric object on the screen so that when changes are introduced in the matrix the effect on the object and on the whole grid covering the screen can be observed. Hern and Long (1991) argue that three dimensions are necessary in order to study the effect of a linear transformation visually, making use of shapes such as a cube, a sphere, or a tetrahedron.

Dubinsky (1997) makes a distinction between visualization of objects and visualization of processes:

One observation I would make about Harel's approach is that it focuses on the *objects* of linear algebra that is, vectors and their various relationships such as membership in a subspace, geometric and coordinate representations, etc. All of these objects can be visualized geometrically—at least in the lower dimensions. It is quite a different matter to visualize the linear algebra *processes* which transform these objects, that is, linear transformations. According to Piaget, visual perception (which is the main tool used by Harel) is possible for static objects, but not for dynamic processes. To visualize the latter, he argues, it is necessary to perceive a set of static phenomena and to reason about them in making mental constructions of dynamic processes. (p. 91)

Indeed Piaget and Inhelder (1969), in an experiment about anticipatory images, asked children aged 5 through 9 to draw the steps of the process through which an arc becomes a straight line. The children's drawings in which the intermediate states lacked the important characteristics of the transformation gave evidence of their static view. These authors contemplate the following:

No matter how adequately we try to visualize the transformation of an arc into a straight line or vice versa, our images proceed in jumps and do no more ... than to take instantaneous 'snapshots' amid the continuum, instead of attaining it as a transformation.... Thus, images cannot exhaust the operation.... The results are: inability to anticipate in imagination, inadequate intermediate images and ... evaluation based only on the starting and finishing points. At the operational level, on the other hand, there appears a new type of image based on symbolic imitation of these operations which succeeds in multiplying 'snapshots' to stimulate a continuum and in anticipating approximately the continuation of the sequence thus evoked. (p. 119)

About the visualization of function transformations, Eisenberg and Dreyfus (1994) make a difference between a *static view* which consists in "*moving a graph from an initial state to a final one* with the graph having moved and changed throughout a transformation" (p. 58) and a *dynamic view* that consists in viewing "it as a mapping which is moving every point in the plane to a new location" (p. 58). In the first approach, the emphasis is on the two static states, and in the latter approach the focus is on the dynamic movement, which starts with the initial state and ends in the final one. Cognitively speaking, they imply the involvement of different structures or conceptions. In their experiment Eisenberg and Dreyfus observed that for the majority of the students "there was no view of the process, no view of a transformation performing some change" (p. 59).

When classifying student concept images about functions and linear transformations, Zandieh et al. (2012) discuss *morphing*, which

involves a beginning state of an entity that changes or is morphed into an ending state of the same entity. There must be a clear sense that the beginning entity did not simply move to the new location (ending entity), nor was it replaced by the new output (ending entity), but that there was actually a metamorphosis of the beginning entity into the ending entity. (p. 527)

This has to do with imagining the process of linear transformation dynamically.

Considering the problem described in this section, we decided to investigate the perception of processes. Beyond the visibility, which might refer to objects such as vectors and their images under a linear transformation, we wanted to explore how the process of a linear transformation might be conceived (Camacho Espinoza and Oktaç 2017). We decided to interview a linear algebra instructor in order not to deal with difficulties concerning knowledge of the content. Luis, the linear algebra instructor whom we interviewed, was asked the following question:

Can you describe how the linear transformation associated to the matrix $\begin{bmatrix} 1 & -1 \\ -1 & 1 \end{bmatrix}$ transforms the plane geometrically?

He had to his disposition the GeoGebra software, which he made use of, as we will see. We wanted to find out how he was thinking about the linear transformation and if he made use of arguments about a continuum in his discussion of it. We will now go through his reasoning as he works on this problem.

To determine the image of the whole plane under this transformation, Luis first talks about sweeping the plane with vertical lines, as can be seen from his drawing in Fig. 26.7. Subsequently considering a generic $x = a$ line and after doing some algebraic calculations, he identifies its image as the line $y = -x$; surprised by the result and gesturing with his hands, he says "But what do you know? Not only this line, but all the vertical lines will be transformed into only one. It is very interesting that all this will be only one" (Fig. 26.7). Not satisfied with the result that he

Fig. 26.7 Sweeping the plane with vertical lines and finding their image

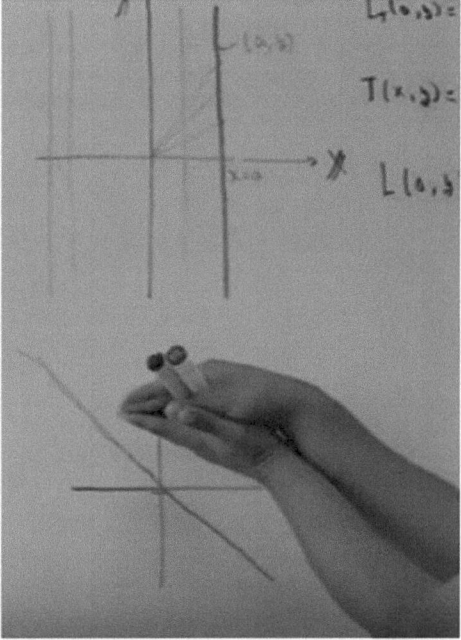

Fig. 26.8 Sweeping the
plane with horizontal lines

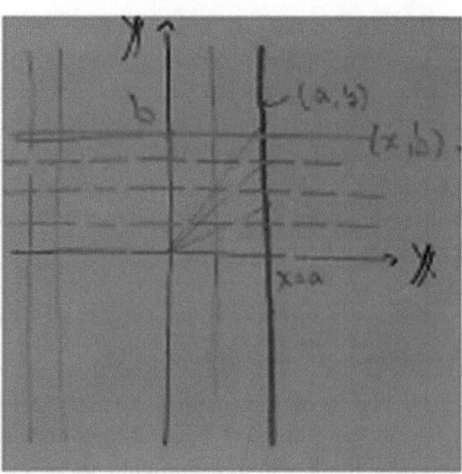

obtained, he then decides to sweep the plane with horizontal lines and when he
arrives at the same image, still surprised, he says: "It is interesting that all the
horizontal lines will go to one line. Obviously, in this way you are sweeping the
whole plane. So don't tell me the whole plane will collapse into the minus identity"
(Fig. 26.8).

Still not content, this time he wonders what the image of a square would be
(Fig. 26.9) and calculates it (Fig. 26.10).

"It will go to a line segment. I am squashing it, but something tells me that it's
not true," he says.

> What would happen, instead of straight lines, ... because maybe I was wrong about the
> geometrical object ... because this is how you sweep the plane, right? We will leave the lines
> and we will use something more interesting. We will sweep the plane with concentric circles.

Fig. 26.9 A square whose
image is to be calculated

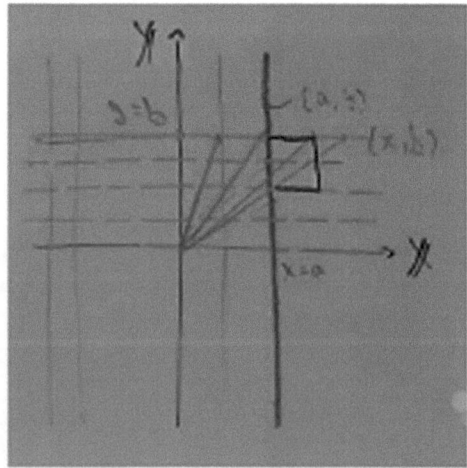

Fig. 26.10 Image of the
square in Fig. 26.9

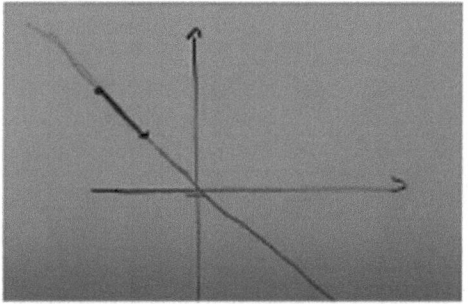

Shortly he realizes that the calculations would be quite messy and gives up on it.

It turns out that from the beginning he was thinking that the transformation was a rotation. In other words, the image that he was obtaining as a result of his calculations and the image that he had in mind did not coincide. He gets convinced of the image being a line only after realizing that the rank of the transformation is 1. At that point he decides to explore with the dynamic geometry software and works with an applet previously designed for this linear transformation that shows a vector and its image simultaneously on the screen (Fig. 26.11). The applet made possible to move a vector on the screen while observing the effect on the image vector; it was also possible to activate a box that allowed visualization of a chosen line and its image. Manipulation of this applet helped Luis identify the kernel of the linear transformation as well as determine the images of specific regions of the plane.

The interviewer tries to find out how he is thinking about the process of this particular linear transformation.

I: With what you have observed so far, can you describe how the transformation deforms the plane?

Luis: It maps it into a line.

I: OK, the image of the plane is a line under this transformation. But can you describe *how* it does it?

Fig. 26.11 Screen showing a
vector and its image

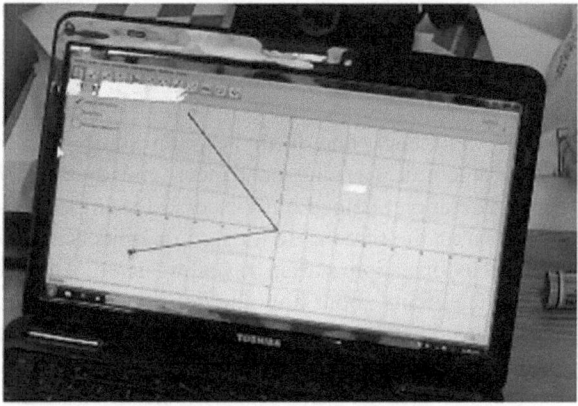

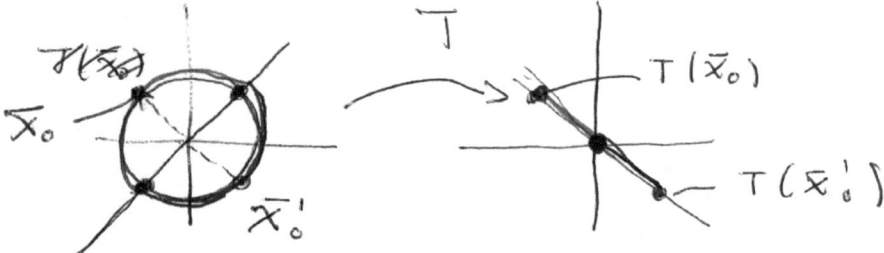

Fig. 26.12 The image of a circle (Camacho Espinoza and Oktaç 2017)

Luis: All the vectors that are above the line $y = x$ are mapped to the second quadrant, and all the vectors below the line $y = x$ are mapped to the fourth quadrant.

Luis: When you start approaching the identity line you start decreasing the norm, because the identity line is the kernel.

By his own initiative, Luis also determines the image of a circle with center at the origin as the line segment lying on the line $y = x$ whose middle point is the origin. He also identifies the images of the two semicircles determined by the line $y = x$ as the two segments of the line $y = -x$, lying above and below the origin (Fig. 26.12).

When the interviewer insists on finding out how he thinks about the process of the linear transformation, Luis gets uncomfortable:

I: Can you describe in which way the vectors of the plane are being transformed?

Luis: Wow! That's a very intimate and strong question.

I: We can determine the images of different objects. But can we describe how those images are being obtained? What is happening to the plane?

Luis: No, actually, I wouldn't be able to find the elements to do that.

Luis: I understand what you mean. You want a geometric argument explaining how this happens, but I don't think so....

Luis: The only precise manner in which I showed how it happens is when I swept the plane with lines, point by point. I didn't get to do it with the circles, but anyway we would have reached the same conclusion.

Luis: You take little pieces: All you have to do is to segment it into little pieces, that's the simplest way.

Luis went back and forth between static-geometric (he was drawing on the board), algebraic (writing on paper), and dynamic-geometric (software) environments. The applet helped him discover the importance of the line $y = x$, which was not shown on the screen. He focused on objects and their images, through which the linear transformation started revealing itself. However he was reluctant to make further statements about how the deformation of the plane took place.

The visualization of the process may not be too difficult in the case of prototype transformations such as rotations and projections, but even with a slight increase in complexity, such as when two transformations are composed, it becomes considerably harder, as in this case where a projection and a dilation are involved.

Dreyfus et al. (1998) mention the following when discussing the effect of a dynamic environment on the conceptions that students may develop of a linear transformation:

> A variable vector has an unstable existence. Only while being dragged does it exist as such: a variable vector. When dragging stops, only a very partial record remains on the screen: An arrow with given, potentially variable length and direction. The variability remains only potential, in the eye (or mind) of the beholder. If the student looking at the screen is not aware of this variability, the variable vector has ceased to exist as such. (p. 218)

According to Piaget:

> Given sufficient practice, geometrical intuition can enable one to "see in space" the transformations themselves, even at times the most complex and the furthest removed from common physical experience. This is because the image rests on a spatialized imitation of operations which are themselves spatial. (Piaget and Inhelder 1969, p. 137)

Zazkis et al.'s (1996) definition of visualization is in line with this point of view:

> Visualization is an act in which an individual establishes a strong connection between an internal construct and something to which access is gained through the senses. Such a connection can be made in either of two directions. An act of visualization may consist of any mental construction of objects or processes that an individual associates with objects or events perceived by her or him as external. Alternatively, an act of visualization may consist of the construction, on some external medium such as paper, chalkboard or computer screen, of objects or events that the individual identifies with object(s) or process(es) in her or his mind. (p. 441)

26.7 Didactical Suggestions and Future Direction for Research

The combined use of different techniques may help in understanding and interpreting the different aspects of the process of a linear transformation. As discussed earlier in this chapter, the use of dynamic geometry software and specially designed applets may form part of these strategies. Working on particular transformations as well as studying the effects of changing the entries of a matrix representation on the image obtained (of geometric objects and of the whole plane) may help in this direction.

Another visual representation of linear transformations of the plane consists in generating a vector field;[1] for that, vectors (in the sense of physics) showing the

---

[1] I thank Franz Pauer for bringing it to my attention after my talk at ICME-13.

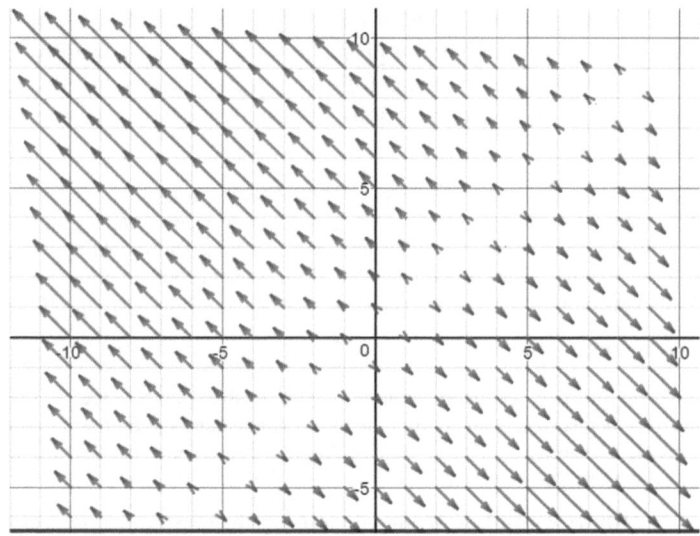

Fig. 26.13 Vector field associated with the linear transformation $T(x,y) = (x - y, -x + y)$ (generated by the vector field generator Desmos at https://www.desmos.com/)

images of selected points under the transformation are placed in the plane, as if they were translated from the origin to those points, and then usually those vectors are scaled down proportionally so that they do not overlap. For example the transformation represented by the matrix $\begin{bmatrix} 1 & -1 \\ -1 & 1 \end{bmatrix}$ has as its formula $T(x,y) = (x - y, -x + y)$. The associated vector field is shown in Fig. 26.13.

This approach might be worth exploring, taking into account that giving meaning to this kind of representation would require another kind of interpretation. For example, for the vector field in Fig. 26.13, the $y = x$ line should be interpreted as the kernel, since there are no vectors lying on it. The fact that all the vectors have the same direction (but not the same orientation) should be interpreted as the image being a line whose equation can be calculated using the slope of the vectors; the magnitudes of the vectors also provide clues about where the images lie. As mentioned before, combined with other strategies, vector fields might prove useful in visualizing the effect of the linear transformation on the whole plane.

In order to help students develop intuitive models compatible with mathematical theory, we suggest working with linear transformations in the two dimensional plane as well as outside of it, where domain and range vector spaces are different. Making use of different registers of representation to help in their coordination is recommendable for establishing relationships between these representations.

Currently we are working on incorporating characteristic values and vectors into our research with the aim of assisting the understanding and visualization of linear transformations. Our hope is to characterize this process, which would in turn lead to recommendations for overcoming difficulties as well as for constructing the concept.

References

Arnon, I., Cottrill, J., Dubinsky, E., Oktaç, A., Roa Fuentes, S., Trigueros, M., et al. (2014). *APOS theory—A framework for research and curriculum development in mathematics education.* New York: Springer.

Camacho Espinoza, G., & Oktaç, A. (2017). Exploración de una transformación lineal de R^2 en R^2. El uso de geometría dinámica para ampliar o adecuar construcciones mentales. In I. M. Gómez-Chacón, et al. (Eds.), *Proceedings of the Fifth Mathematical Working Space Symposium* (pp. 253–266). Greece: University of Western Macedonia.

Dreyfus, T., Hillel, J., & Sierpinska, A. (1998). Cabri-based linear algebra: transformations. In *Proceedings of CERME-1 (First Conference on European Research in Mathematics Education),* Osnabrück. http://www.fmd.uni-osnabrueck.de/ebooks/erme/cerme1-proceedings/papers/g2-dreyfus-et-al.pdf. Accessed December 13, 2016.

Dubinsky, E. (1997). Some thoughts on a first course in linear algebra at the college level. In D. Carlson, C. R. Johnson, D. C. Lay, R. D. Porter, & A. Watkins (Eds.), *Resources for teaching linear algebra* (Vol. 42, pp. 85–105). USA: MAA Notes.

Duval, R. (1993). Registros de representación semiótica y funcionamiento cognitivo del pensamiento. In F. Hitt (Ed.), *Investigaciones en Matemática Educativa II* (pp. 173–201). Mexico: Grupo editorial Iberoamérica.

Eisenberg, T., & Dreyfus, T. (1994). On understanding how students learn to visualize function transformations. In E. Dubinsky, A. Schoenfeld, & J. Kaput (Eds.), *Research in collegiate mathematics education* (Vol. 1, pp. 45–68). Providence, RI: American Mathematical Society.

Fischbein, E. (1989). Tacit models and mathematical reasoning. *For the Learning of Mathematics, 9,* 9–14.

Hern, T., & Long, C. (1991). Viewing some concepts and applications in linear algebra. In W. Zimmermann & S. Cunningham (Eds.), *Visualization in teaching and learning mathematics* (pp. 173–190). USA: Mathematical Association of America.

Hillel, J. (2000). Modes of description and the problem of representation in linear algebra. In J.-L. Dorier (Ed.), *On the teaching of linear algebra* (pp. 191–207). Dordrecht: Kluwer Academic Publishers.

Karrer, M., & Jahn, A. P. (2008). Studying plane linear transformations on a dynamic geometry environment: Analysis of tasks emphasizing the graphic register. *ICME-11, TSG 22.* http://tsg.icme11.org/document/get/237. Accessed December 27, 2016.

Molina, G., & Oktaç, A. (2007). Concepciones de la Transformación Lineal en Contexto Geométrico. *Revista Latinoamericana de Investigación en Matemática Educativa, 10*(2), 241–273.

Monagan, M. B. (2002). 2D and 3D graphical routines for teaching linear algebra. In *2002 maple summer workshop.* Waterloo, ON, Canada: Waterloo Maple Inc. http://www.cecm.sfu.ca/CAG/papers/monVis4LA.pdf. Accessed December 29, 2016.

Piaget, J., & Inhelder, B. (1969). Mental images. In P. Oléron, J. Piaget, B. Inhelder, & P. Greco (Eds.), *VII intelligence. Experimental psychology: Its scope and method* (pp. 85–143). London: Psychology Press.

Ramírez Sandoval, O. (2008). *Modelos intuitivos que tienen algunos estudiantes de matemáticas sobre el concepto de transformación lineal* (Unpublished masters' thesis). Cinvestav-IPN, Mexico.

Ramírez Sandoval, O., & Oktaç, A. (2012). Modelos intuitivos sobre el concepto de transformación lineal. *Actes du Coloque Hommage à Michele Artigue, Université Paris Diderot- Paris 7, Paris, France.* http://www.uqat.ca/telechargements/info_entites/Didactiques %20des%20math%C3%A9matiques.%20Approches%20et%20enjeux_Atelier7.pdf. Accessed December 13, 2016.

Ramírez-Sandoval, O., Romero-Félix, C. F., & Oktaç, A. (2014). Coordinación de registros de representación semiótica en el uso de transformaciones lineales en el plano. *Annales de Didactique et de Sciences Cognitives, 19*, 225–250.

Romero Félix, C. F. (2016). *Aprendizaje de transformaciones lineales mediante la coordinación de representaciones estáticas y dinámicas* (Unpublished doctoral thesis). Cinvestav-IPN, Mexico.

Romero Félix, C. F., & Oktaç, A. (2015a). Coordinación de registros y construcciones mentales en un ambiente dinámico para el aprendizaje de transformaciones lineales. In I. M. Gómez-Chacón, et al. (Eds.), *Actas Cuarto Simposio Internacional ETM* (pp. 387–400). Madrid, España.

Romero Félix, C. F., & Oktaç, A. (2015b). Representaciones dinámicas como apoyo para la interiorización del concepto de transformación lineal. *Anales de XIV CIAEMIACME,* (pp. 511–522). Chiapas, Mexico.

Roa-Fuentes, S., & Oktaç, A. (2010). Construcción de una descomposición genética: análisis teórico del concepto transformación lineal. *Revista Latinoamericana de Investigación en Matemática Educativa, 13*(1), 89–112.

Roa-Fuentes, S., & Oktaç, A. (2012). Validación de una descomposición genética de transformación lineal: un análisis refinado por la aplicación del ciclo de investigación de APOE. *Revista Latinoamericana de Investigación en Matemática Educativa, 15*(2), 199–232.

Sierpinska, A. (2000). On some aspects of students' thinking in linear algebra. In J.-L. Dorier (Ed.), *On the teaching of linear algebra* (pp. 209–246). Dordrecht: Kluwer Academic Publishers.

Sierpinska, A., Dreyfus, T., & Hillel, J. (1999). Evaluation of a teaching design in linear algebra: The case of linear transformations. *Recherches en Didactique des Mathématiques, 19*(1), 7–40.

Triantafyllou, E., & Timcenko, O. (2013). Developing digital technologies for undergraduate university mathematics: Challenges, issues and perspectives. In L. H. Wong, C.-C. Liu, T. Hirashima, P. Sumedi, & M. Lukman (Eds.), *Proceedings of the 21st International Conference on Computers in Education* (pp. 971–976). Uhamka Press.

Trigueros, M., & Bianchini, B. (2016). Learning linear transformations using models. *First Conference of International Network for Didactic Research in University Mathematics.* Montpellier, France. https://hal.archives-ouvertes.fr/hal-01337884/document. Accessed December 27, 2016.

Uicab, R., & Oktaç, A. (2006). Transformaciones lineales en un ambiente de geometría dinámica. *Revista Latinoamericana de Investigación en Matemática Educativa, 9*(3), 459–490.

Wawro, M. (2009). Task design: Towards promoting a geometric conceptualization of linear transformation and change of basis. *Twelfth Conference on Research in Undergraduate Mathematics Education*, Raleigh, NC. http://iola.math.vt.edu/media/pubs/Wawro_LONG.pdf. Accessed December 15, 2016.

Wawro, M., Larson, C., Zandieh, M., & Rasmussen, C. (2012). A hypothetical collective progression for conceptualizing matrices as linear transformations. *Fifteenth Conference on Research in Undergraduate Mathematics Education*, Portland, OR. http://iola.math.vt.edu/media/pubs/Wawro-et-al-2012-Italicizing-N-paper-small.pdf. Accessed December 29, 2016.

Zandieh, M., Ellis, J., & Rasmussen, C. (2012). Student concept images of function and linear transformation. In S. Brown, S. Larsen, K. Marrongelle, & M. Oehrtman (Eds.), *Proceedings of the 15th Annual Conference on Research in Undergraduate Mathematics Education* (pp. 320–328). Portland, OR: SIGMAA-RUME.

Zazkis, R., Dubinsky, E. & Dautermann, J. (1996). Coordinating visual and analytic strategies: A study of students' understanding of the group D_4. *Journal for Research in Mathematics Education, 27*(4), 435–457.

Chapter 27
Mapping the Relationship Between Written and Enacted Curriculum: Examining Teachers' Decision Making

Janine Remillard

Abstract I offer an approach to representing and examining the relationship between curriculum resources and the performance of teaching, for the purpose of analyzing teachers' design work. The approach builds on the assumptions that teaching is a design activity, that curriculum resources are tools that convey complex instructional ideas, and that, in using these tools, teachers interact with them and selectively leverage resources to design and enact instruction. I introduce the instructional design arc as a unit of analysis, referring to an episode in a lesson, prompted by the teacher, and that require the teacher to make instructional design decisions in the moment. When compiled into a lesson map, these design arcs model the episodic and emerging contours of the enacted lesson, representing teachers' planned and in-the-moment decisions. Using data from 3rd to 5th grade mathematics classrooms in the USA, I analyze instructional design arcs within mathematics lessons, focusing on teachers' design work.

Keywords Written curriculum · Enacted curriculum · Teaching
Mathematics instruction

27.1 The Relationship Between Written and Enacted Curriculum

In his exploration of the "teacher-tool relationship," Brown (2009) uses sheet music and different artists' renditions of the same classic jazz song, *Take the A Train*,[1] to illustrate the ways the same song, as written, can be performed in substantially

---

This lecture was based on analysis reported in Remillard et al. 2015.

---

[1]Written by Billy Strayhorn.

J. Remillard (✉)
University of Pennsylvania, Philadelphia, USA
e-mail: janiner@upenn.edu

© The Author(s) 2018   483
G. Kaiser et al. (eds.), *Invited Lectures from the 13th International Congress on Mathematical Education*, ICME-13 Monographs,
https://doi.org/10.1007/978-3-319-72170-5_27

different ways. Both performances, Brown points out, have "essential similarities," yet they "sound distinctly different" (p. 17). Curriculum materials, Brown (2009) goes on to argue, have similarities to sheet music, both in terms of their form and role in guiding performance. They are both "static representations" of intended activity and the "means of transmitting and producing" it, but they are not the activity itself.

Curriculum materials and sheet music are also similar in that "they are intended to convey rich ideas and dynamic practices," but they do so "through succinct shorthand that relies heavily on interpretation" (Brown 2009, p. 21). At the same time, curriculum materials are often designed to "influence common practice by introducing innovative approaches and ideas." Most critically, and often overlooked, curriculum materials "require craft in their use; they are inert objects that come alive only through interpretation and use by a practitioner" (p. 22).

The relationship between sheet music and musical performance is also apt as a metaphor for teaching and the curriculum-teacher relationship because, in many ways, teaching is a live performance. Like different performances of the same musical score, different enactments of the same written curriculum will not only vary in style, pace, and emphasis, they are likely to also vary in quality. Some teaching performances come closer than others to meeting the mathematical and pedagogical goals specified in the curriculum or intended by the teacher. This variation in quality may be attributed to a teacher's grasp of the mathematics, ability to connect to learners, or manage a classroom. For those using curriculum materials, I suggest, the quality of instruction is related to the teachers' ability to interpret, make decisions about, and leverage the resources in this tool as she designs and enacts instruction.

27.2   Research Questions and Analytical Focus

The analysis presented in this paper is particularly concerned with examining the relationship between curriculum resources and the performance of teaching. My approach draws on Brown's (2009) and others' idea that teaching is a design activity; that curriculum materials are tools that convey complex instructional ideas and that, in using them, teachers interact with these tools and selectively leverage available resources to design and enact instruction. Using video recordings of elementary teachers' mathematics lessons, together with interviews and artifacts detailing their reading of the teacher's guide, I consider the following conceptual question:

> How can enacted lessons be conceptualized and represented for the purpose of analyzing the design work teachers do during enactment?

This question is conceptual, requiring us to build a model of the work of teaching that can be used to analyze teaching performance and its relationship between the written curriculum guides. As described later in the results section,

I introduce the concept of *instructional design arcs* to model instructional episodes in a lesson that are prompted by the teacher and that require the teacher to make instructional design decisions in the moment. These arcs not only model the episodic and emerging contours of the enacted lesson, but serve as units of analysis in my effort to examine teachers' design work during the enacted curriculum and its relationship to the written or planned curriculum. My aim in this analysis is to build a tool to examine empirical questions about the relationship between the enacted and written curriculum, including how to understand variation across teachers and different types of curriculum resources.

27.3 Theoretical and Conceptual Perspectives

This study builds on three overlapping, framing perspectives: an adaptive view of curriculum (Stein et al. 2007; Remillard and Heck 2014), teaching as design work (Brown 2009), and a participatory view of teachers' use of curriculum materials (Remillard 2005).

27.3.1 Curriculum Enactment as an Adaptive Process

An adaptive perspective asserts that instantiations of curriculum unfold and develop over several temporal phases, from the ideal indicated by official policy, to the written, the teacher intended, and the enacted (Remillard and Heck 2014; Stein et al. 2007). Valverde et al. (2002) described textbooks and instructional materials as mediators between the intended and implemented curriculum. They are written to work on the behalf of teachers and students "as the links between the ideas presented in the intended curriculum and the very different world of the classroom" (p. 55). Rather than focusing on fidelity of implementation, research from this perspective examines how curriculum artifacts are transformed by policy makers, textbook authors, teachers, and students from one phase to the next, such as written to enacted, and considers factors that influence these transformations.

27.3.2 Teaching as Design and Curriculum Use as Participatory

Teachers, then, are critical decision makers within this adaptive framework. I draw on Brown's (2009) assertion that teaching involves design work, even when using curriculum materials. Teachers make design decisions when they read curriculum materials and moblize them to plan a lesson. They also make design decisions while enacting lessons (Brown 2009).

Further, I view teachers' use of curriculum materials as a participatory process, involving interactions between the teacher and the curriculum resource. This view emphasizes the interactive and transactional nature of this work, framing curriculum use as a dynamic and ongoing relationship between teachers and resources, a relationship shaped by both the teacher and characteristics of the resource (Remillard 2005).

A fundamental theoretical thread running through these perspectives is Vygotsky's (1978) notion of practice as inseparable from tools, both employed and produced through the process and as deeply rooted in the particular context. Whereas some views of teaching and, most notably, curriculum use emphasize implementation or brokering of existing, fully formulated resources, this perspective frames teaching as a design activity and curricular resources as contributing partners in the generative work (Brown 2009; Remillard 2005).

27.3.3 Curriculum Fidelity

Many studies of teachers' use of curriculum materials focus on fidelity or the closeness of the enacted curriculum to that specified in the teacher's guide (O'Donnell 2008). Some scholars have suggested that fidelity is a complex and underspecified concept, often used in problematic ways. At the same time, Remillard (2005) notes, "It would be inaccurate and irresponsible to conclude that all interpretations of a written curriculum are equally valid." As a result, the field is "in need of ways to characterize reasonable and unreasonable variations or instantiations of a particular curriculum that are tied to features most central to its design" (pp. 239–240). Brown et al. (2009) offer an approach to studying curriculum use that differentiates fidelity to the written curriculum from fidelity to the authors' intended opportunities to learn. My analysis aligns with efforts by these researchers to conceptualize fidelity between the written and enacted curriculum as alignment to the intended opportunities to learn.

The conceptual and methodological work described in this paper is aimed at representing the curriculum enactment process in relation to the curriculum resources being used. I am interested, to some extent, in the alignment between intended opportunities to learn suggested in the curriculum guides and those opportunities made available in the classroom. More importantly, however, I am interested in understanding and describing the processes through which the enacted curriculum is designed by teachers' decisions, both when planning lessons and when enacting them.

27.4 Methods

My data were drawn from the corpus of qualitative data collected for the ICUBiT[2] study. A principal goal of research project was to identify key components of the capacity required for elementary teachers to use curriculum resources productively in their mathematics teaching.

27.4.1 Data Sources

Data collection relied on a *teaching set* methodology (Cobb et al. 2009; Simon and Tzur 1999), which involves collecting video records of multiple lessons along with associated artifacts and then using specific events or practices observed in the data as a basis for teacher interviews. The ICUBiT study collected teaching sets for 25 teachers. For the analysis in this paper, I selected four teachers, two using *Investigations in Numbers, Data, and Space* (TERC 2008) and two using *Math in Focus* (Kheong 2010). See Table 27.1 for details.

Two teaching sets were collected for each teacher, one in the fall and one in the spring. The teaching set included 3 video recorded lesson observations, a completed curriculum reading log (CRL) for the lessons taught during the week of observation, and a follow-up interview. Prior to the fall teaching set, each teacher completed an introductory interview, during which the teachers provided information about professional background and curriculum use. The CRLs consisted of a copy of the relevant lesson in the teacher's guide on which teachers used colored highlighters to indicate which parts of the guide they read for various reasons. During the follow-up interview, the interviewers asked teachers to respond to questions about the observed lessons and the CRL.

27.4.2 Data Analysis

Each lesson was divided into episodes (arcs), identified by a distinct mathematical purpose held by the teacher. Each episode began with a prompt by the teacher. Based on my analysis of the lesson, the CRL, and the follow-up interview, I coded each prompt as written, adapted, inserted, or improvised, to indicate its relationship to the lesson as described in the guide. These terms overlap with those introduced by Brown (2009) to characterize how teachers use curriculum resources, offload,

---

[2]The Improving Curriculum Use for Better Teaching is directed by Janine Remillard and Ok-Kyeong Kim.

Table 27.1 Participating teachers

ID	Name[a]	Grade	Curriculum	Full years teaching	Years using Curriculum
008	Maya Fiero	4	*Math in focus*	9	2
009	Meredith Frankl	5	*Math in focus*	1	1
061	Ingrid Navarra	5	*Investigations*	14	13
063	Irma Nelson	4	*Investigations*	25	12

[a]All names are pseudonyms

adapt, and improvise, as follows. A *written* prompt was drawn from the curriculum guide and aligns closely with the mathematical topic and objectives of the text, even if the teacher makes minor modifications. Brown used the term offload to refer to episodes during a lesson that a teacher relied on fully the written lesson. Similar to Brown, I labeled *Adapted* prompts as instances when the teacher used the curriculum guide as a resource, but made modifications in structure, approach, or objective. Brown used the term improvised to refer to all instances in which the teacher replaced elements of the written lesson with different activities or approaches. My codes differed, depending on whether the revision was planned or unplanned. I coded prompts as *inserted* if they were not resourced from the teacher's guide, but were planned in advance. I used *Improvised* to refer to prompts that were not in the teacher's guide, but appeared to be developed in the moment, in response to classroom situations. I also identified *omissions* from the curriculum guides and coded as significant those that I deemed to be fundamental to the designed lesson plans, with respect to accomplishing the learning objectives of the lesson. The coded lessons were then used to build a lesson map, as described in the section that follows.

27.5   Conceptualizing and Representing the Enacted Curriculum

In the section below I describe my approach to analyzing and representing the enacted curriculum in relation to teachers' design decisions. In the process, I provide a brief introduction to each of the four teachers analyzed for this paper. I also discuss how this analytical approach might be used to examine teachers' design decisions, their pedagogical design capacity, and the relationship between the written and enacted curriculum.

27.5.1 Two Types of Design Decisions

My analysis of video recorded lessons in relation to teachers' planned curriculum (measured through CRLs), supplemented by follow-up interviews revealed the range of design decisions teachers make when enacting instruction. For this analysis, I distinguish between two types of decisions: (a) *planned decisions*, made in advance of the lesson, and (b) *in-the-moment-design decisions* (*IMDDs*), made during the enacted lesson. Planned decisions refer to those made in advance. They include identifying the goals and tasks of the lesson. Often, they involve designing what Castro et al. (2007) refer to as *instructional moves*, instructions, questions, guidelines, or other types of prompts offered by the teacher during a lesson. I think of them as *initiating prompts* because they are designed to prompt student engagement or participation in some sort of mathematical work.

When using a curriculum-use lens, planned decisions also reflect the teacher's decisions in relation to the designed, specified, or suggested moves described in the teacher's guide. That is, they involve determining which parts of the written lesson to use and how to use them. In my analysis of planned decisions, I considered (a) which elements in the teacher's guide teachers chose to use or omit, (b) whether teachers used the elements as written or adapted them in some way, and (c) the types of insertions teachers made to the written curriculum.

Not all decisions teachers make during a lesson can be pre-planned. During any lesson, a teacher must make instructional-design decisions in the moment, in response to how students respond to her initial prompt. I call these decisions *in-the-moment design decisions* (IMDDs). Unlike pre-planned initiating prompts, these teacher moves are all responsive in nature. Consider the following illustration.

On December 7th, Irma Nelson, introduced her 4th grade students to the first activity in the lesson entitled *Strategies for Multiplication*. The lesson, the third session in a set of four on doubling and halving, was near the end of Book 3, *Multiple Towers and Division Stories*, in the *Investigations* program.

To introduce the activity, the teacher's guide instructs the teacher to write two expressions on the board, taken from a workbook page the students had completed two days prior: 16×3 and 16×6.

The teacher's guide offers the following prompt to begin the session:

Let's look again at these two problems. How are they related to each other?

How are the two answers related?

The guide then suggests the teacher collect a few responses about why the product is doubled when one factor is doubled. Then ask students to share story contexts or representations from their previously completed work that show why this is true. Finally, the guide suggests that if no student suggests using an array, the teacher should introduce it herself by drawing two open arrays on the board representing 16×3 and 16×6.

Ms. Nelson wrote these expressions on the overhead projector and began by asking: "Ok, um, let's take a look at, um, a problem 16 times 3 and another problem that's similar is 16 times 6, ok? What's the first thing that you notice about these problems?"

Although Ms. Nelson did not begin with the question worded exactly as it was in the guide, she initiated the task as designed by the authors. Then, as suggested, she called on a student and the following exchange transpired, which included several opportunities for IMDDs:

Ana: 3 is double 6.
Teacher: 3's double 6? Or…is 3 twice as big as 6? Ok, what you said makes me think that 3 is twice as large as 6.
Ana: 3 is, uh, two put together is 6.
Teacher: Ok so if you double up 3 you're gonna get 6. Carl, what do you notice?
Carl: I know that the 16's are still the same.
Teacher: The 16's are still the same.
Lana: Since the 3 is half of the 6, when you get your answer for 16 times 3, um, you'll just have to double that.
Teacher: Only doubling it then, all right so that will make it easier to solve the second problem. Kent?
Kent: She got it.
Teacher: She got yours? All right. Um, how could you show this if you wanted to do that in an array? How might you show that?

In the example above, Ms. Nelson displayed the suggested expressions and asked an open question that prompted students to make observations about similarities between them. As the guide suggested, she collected students' responses about the relationship between the factors in and product of the two expressions. I assess that she decided to use this part of the guide as *written*.

What transpired in the minute after offering the initiating prompt appears to reflect the intent of the curriculum authors; students made observations about the similarities between the two expressions. The particulars of the intended exchange are not specified in the guide, which advises teachers to collect students' responses about why the product is doubled using story contexts or representations. The teacher must enact these decisions in the moment. In this short example, Ms. Nelson made IMDDs in order to respond to several students' observations about the two expressions. She questioned Ana's partially correct response, accepted Carl's observation without further probing, and restated Lana's suggestion, pointing out that her observation could help them solve the second expression.

My analysis revealed that, regardless of the detail of planned decisions, made in advance of the lesson, teachers experience multiple opportunities to make additional design decisions while the lesson was being enacted. Simply put, any time a teacher asks a question or prompts students to respond in some way, she must then navigate their responses in relation to her intended objective. Teachers, for example, make decisions about which students to call on and how to respond to what they say, be their responses correct, partially formulated, incorrect, or unrelated. They must decide when their specific objective underlying the prompt has been met and it is appropriate to move on or whether an additional instructional move needs to be improvised.

The need for teachers to make IMDDs is not necessarily a reflection of poor planning or underdeveloped resources. Rather, they reflect the substantive distinction between the written, planned, and enacted curriculum (Remillard and Heck 2014; Stein et al. 2007). Like the distinction between sheet music and a musical performance (Brown 2009), the enacted curriculum is richer, more detailed and varied than the succinct representation of a lesson in a teacher's guide. Even more significant is the fact that enacted lessons are co-constructed with students; they are shaped by students' actions and teachers' moves in relation to them. In a study of professional development sessions, Remillard and Geist (2002) used the term *openings in the curriculum* to describe similar instances, which required the facilitator to make "on-the-spot decisions, about how to guide the discourse" (p. 13).

27.5.2 Instructional Design Arcs

Through my analysis of teachers' decisions during mathematics lessons, I observed that these enacted lessons were comprised of a series of instructional episodes that typically begin with a planned instructional prompt and follow with a segment of time during which the teacher guides classroom interactions toward a particular mathematical purpose. An instructional episode ends when the teacher initiates a new prompt, usually, but not always, because the purpose has been met. I refer to these episodes as *instructional design arcs*.

I see instructional design arcs as the basic building block of an enacted lesson. They are not unanticipated, but they cannot be fully planned. Navigating these arcs is at the heart of the work of teaching. To varying degrees, curriculum authors anticipate these arcs and provide guidance to help teachers navigate them. Regardless, I posit, understanding how teachers make IMDDs in order to guide instructional arcs, including how they mobilize curriculum resources in the process, is critical to understanding the work of teaching. In the following sections, I describe my use of lesson maps to represent teachers' instructional prompts, the instructional design arcs that result, and their relationship to elements and supports in the teacher's guide.

27.5.3 Using Lesson Maps to Model the Enacted Curriculum

I use the following four examples of lessons, one from each of the four teachers in this analysis to illustrate the different analytical features of the lesson maps. These examples were also selected to introduce the four teachers in my analysis.

27.5.3.1 Example 1: Ms. Nelson's Enacted Lesson

The lesson map for Ms. Nelson's 12.05.2012 lesson, which took place two days before the lesson introduced earlier (comparing 16×3 and 16×6), is shown in Fig. 27.1. The timeline along the bottom of the map represents the time of the lesson in minutes. The solid circles represent instructional prompts, introduced by the teacher, and defined by an identifiable mathematical purpose. The arcs that follow represent the instructional design arcs that resulted. An arc ends when a new prompt was offered, which usually occurred when the teacher's mathematical purpose was met or the teacher deemed a new prompt was appropriate.

In this lesson, all prompts fall on the lower horizontal line, which indicates that they were drawn from the teacher's planned lesson. The color of circles indicates the source of the arcs in relationship to the curriculum guide. The first arc began with an inserted prompt, marked by the red circle, which represents a timed multiplication exercise that the teacher inserted into the lesson from outside of the curriculum guide. Then, at minute 5:48, Ms. Nelson began the lesson based on the teacher's guide. She directed students to a story problem in their workbook: Ms. Santos has 168 apples. She wants to pack them into boxes of 28. How many boxes does she need? She asked, "Is this a multiplication problem or a division problem?" This arc aligned with the curriculum guide and is therefore coded as written, indicated by a black circle. After the class discussed the problem and various ways to solve it, at minute 10:56, Ms. Nelson directed the students to solve the division problem using a strategy suggested by one of the students. This arc was coded as adapted because the teacher significantly changed the mathematical purpose of the task, indicated by the gray circle.

The remainder of the lesson, represented by the lesson map, was comprised of several more arcs sourced from the teacher's guide as written. The long arc, initiated at time 39:52 min involved students working in small groups to complete a page of practice problems included in the lesson, while the teacher circulated. Typically, segments involving small group or individual student work were coded as a single arc, based on a single objective, even though the teacher's interactions with individual students or groups might have been initiated by distinct prompts.[3] Occasionally, although not in this case, I found that teachers interrupted

---

[3]Project video records do not provide us with a reliable record of these exchanges during small-group work periods.

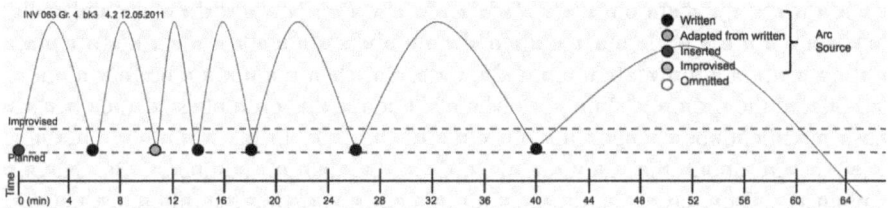

Fig. 27.1 Lesson map represents prompts and design arcs in Irma Nelson's 12.05.2011 lesson

small-group work periods to engage the entire class in an exchange; these were coded as new prompts. The lesson ended after 64 min, at the end of the long arc.

27.5.3.2 Example 2: Ms. Navarra's Improvised Arcs

Not all prompts initiating design arcs came from the teacher's planned curriculum. It was not unusual for a teacher to initiate a new arc with an *improvised* prompt, based on an IMDD. These prompts are marked with a yellow circle, placed on the improvised line on the map. As described above, within each instructional arc, teachers make many IMDDs. An IMDD becomes a prompt, initiating a new arc when it has a distinct mathematical objective. I considered an arc improvised when there was evidence that it had occurred spontaneously, often in response to an event that occurred during the previous episode. I also consulted the follow-up interview transcript and the teacher's CRL to determine whether a prompt had been preplanned.

Ingrid Navarra's lesson on decimals on 02.06.2012 illustrates an improvised arc, along with inserted and adapted arcs (Fig. 27.2). Ms. Navarra, a 5th grade teacher also using *Investigations,* taught a lesson drawn from Book 6, Session 1.3: *Decimals on the Number line.* Like Ms. Nelson's, her lesson map shows she began the lesson with an inserted prompt. Ms. Navarra's inserted prompt was in the form of a review. She asked students to tell her some things they knew about decimals.

At 2:30 min, Ms. Navarra moved into Activity 2 in the teacher's guide, called *Introducing Decimals on a Number Line.* The teacher's guide recommends drawing a number line on the board that begins at 0 and ends at 2, and having students place

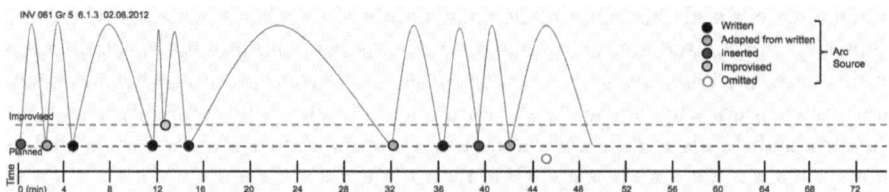

Fig. 27.2 Lesson map of Ingrid Navarra's 02.06.2012 lesson

the following numbers on it: 0.3, 0.5, 1.25, 1.8. Ms. Navarra drew a number line that went from 0 to 1 and modified the numbers students were asked to place, writing them as follows: 0.3, 0.5, 1.25, 0.05. She then pointed to the four numbers and offered the prompt: "Who goes where?" I coded this prompt as adapted because these modifications changed the objective of the task. In the follow-up interview, when asked about the addition of 0.05 to the list of numbers, Ms. Navarra explained, "They have a hard time with this book knowing the difference between what 0.05 is and 0.5 or 0.2 and 0.02. So, I'm always trying to throw those in there."

She initiated an improvised arc at 13:19 min. Students were working on the activity in the lesson labeled *Ordering Tenths and Hundredths*. Using a set of decimals cards, which included decimals to the hundredths place, the students were to place the cards in order from smallest to largest. The guide also suggests the teacher check students' decimal ordering after they finish their work. Ms. Navarra, following the mathematical objective of the activity in the text, which is "ordering decimals and justifying their ordering through reasoning about decimals representations, equivalents, and relationships," asked students to explain the strategy they used to order the Decimal Card set. A student explained that his group counted by 5's. At that point, Ms. Navarra initiated an improvised arc by asking how much 0.05 (pointed at the number) would be if it were money. As the excerpt below illustrates, she pointed to several decimal numbers and asked students to give the value in money.

Teacher: You counted by 5's. All right, if this were money, how much would this be?
Student: 5 cents.
Student: A nickel.
Teacher: How much would this be? [Points to other numbers on the overhead]
Student: 50 cents, a half dollar.
Student: 2 quarters.
Teacher: 2 quarters a half dollar, 50.
Student: Three quarters...
Student: One quarter.
Student: A dime.
Student: Uh, 90 cents.
Student: 9 dimes.
Teacher: Ok, now. It's gonna get a little harder...

This improvised arc lasted approximately two minutes and was aimed at pushing students to connect the value of decimal numbers to related amounts of money. Typically, improvised arcs are initiated by what is happening in the enacted lesson. In this case, the *counting by* 5's strategy used by a group prompted her to be sure they understood what those "5's" represented. In the follow-up interview, Ms. Navarra said: "This is a nickel versus two quarters. And I always take it back to money, 'cause they like money" (Follow-up Interview, Spring 2012).

Ms. Navarra's lesson map also includes one significant omission (open red circle) in minute 45. I discuss omissions in the following section.

27.5.3.3 Example 3: Ms. Fiero's Significant Omissions

I use Ms. Fiero's lesson on multiplying 2-digit numbers by multiples of tens to illustrate my identification of significant omissions. All five of the curriculum programs in the study include more options than can possibly be used in a single lesson. They also include tasks and suggestions that are designed to be optional, allowing the teacher to tailor the curriculum to the particular students (Remillard and Reinke 2012). It is my expectation that teachers will make adaptations to the written curriculum based on their assessments of students' need and their preferences. Thus, I anticipate that teachers will omit certain suggestions and I do not attempt to capture all of them in the lesson maps. I do, however, identify omissions that I assess to be fundamental to accomplishing the learning objectives of the designed lesson plans.

Maya Fiero, a 4th grade teacher using *Math in Focus,* made what I consider significant omissions during her 12.05.2011 lesson. One of the lesson objectives listed in the teacher's guide was for students to "Multiply by 2-digit numbers with or without regrouping." In the first part of the lesson students are introduced to the multi-step approach to multiplying numbers by multiples of 10 based on the associative property of multiplication, shown on the student page in Fig. 27.3.

The teaching notes that accompany this introductory page state:

Students learn to multiply by 2-digit numbers in the form of tens.

- Help students recall the strategy for multiplying a number by tens by working through the examples in the Student Book.
- In the first example, express 10 as 1 ten. So $4 \times 10 = 4 \times 1$ ten = 4 tens = 40.
- Using the strategy, work through the second example with students.
- First, express 20 as 2 tens. So $3 \times 20 = 3 \times 2$ tens = 6 tens = 60.
- For students who cannot visualize multiplying by tens, use a place-value chart to show the connection.

The teacher's guide also suggests beginning the lesson with the following 5-minunte warm up:

- Have students work in pairs. Each partner takes turns giving and solving a multiplication problem to multiply 1-digit numbers by tens and hundreds mentally, for example, 6×100.
- Repeat with 2-digit numbers. Encourage students to identify the pattern.
- This activity helps recapitulate the previous lesson and provides a warm-up for this lesson. (p. 86)

Ms. Fiero's lesson map is shown in Fig. 27.4. The lesson began, as is typical for Ms. Fiero, with a warm up drawn from another source, indicated with a red

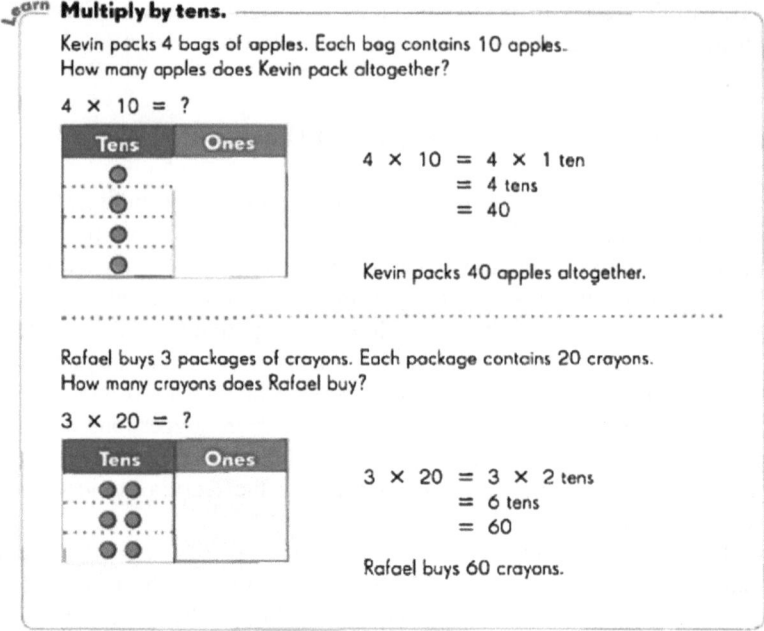

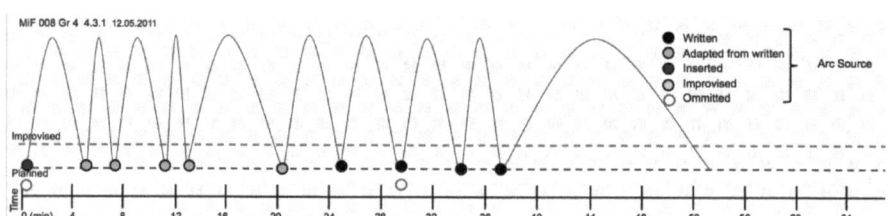

Fig. 27.4 Lesson map of May Fiero's 12.05.2011 lesson

"inserted circle; this one involved practicing 2- by 1-digit multiplication. As a result, she omitted the warm up, which involved students practicing multiplying single-digit numbers by 10 and 100. This omission is marked with an open circle because facility with multiplying numbers by 10 is anticipated in the introduction to multiplying numbers by multiples of 10.

Before moving onto this introduction page of the lesson, Ms. Fiero guided students through a review of the previous day's multiplication work on 1- and 2-digit multiplication. Because these activities involved material drawn from the previous days' lesson, I coded each prompt as adapted. At 25:52 min, Ms. Fiero began the introduction shown in Fig. 27.4. They began with a short discussion of title, objective, and key vocabulary listed on the first page of the student workbook. With the next prompt, at 29:41, she moved the class to considering how to multiply numbers by 10. The guidance in the teacher's guide states: "Help students recall the strategy for multiplying a number by tens by working through the examples in the Student Book." Rather than using the examples in the student book, Ms. Fiero, reminded the students that they had multiplied numbers by 10 previously. She wrote 81 × 10 on the white board to illustrate. For the next several minutes, the students struggled to provide an answer. Even when she moved the class onto working with the two examples on the page (Fig. 27.3), students had difficulty recognizing that 4 × 10 was equivalent to 4 tens or 40. The teacher's guide included the following suggestion: "For students who cannot visualize multiplying by tens, use a place-value chart to show the connection." Even though the place value charts were on the student page, Ms. Fiero did not refer students to these models once. I marked this as another significant omission.

My review of the lesson description in the curriculum, in light of the difficulties students were having multiplying numbers by 10, 20, or other multiples of 10, suggests that the omissions described above left out critical steps of the designed learning sequence. For this reason, I coded them as significant omissions.

It is important to note that identifying omissions of lessons closely tied to the curriculum is easier than lessons that have little relationship to the curriculum. The following example illustrates a lesson comprised fully of inserted arcs, from which omissions are not indicated.

27.5.3.4 Example 4: Ms. Frankl's Inserted Arcs

Meredith Frankl was a 5th grade teacher using the *Math in Focus* program. Her CRL, which provides details of her planned lesson, indicates that she used the mathematical topic of the lesson, the relationship between area and perimeter, to guide the design of the lesson. As the lesson map (Fig. 27.5) indicates, the lesson was comprised of 9 instructional design arcs, all inserted by the teacher. Although the mathematical topic of the lesson overlapped with the lesson in the written curriculum, the enacted lesson did not reflect the written curriculum in objective, structure, tasks, or approach. For this reason, it was not possible to identify omitted tasks.

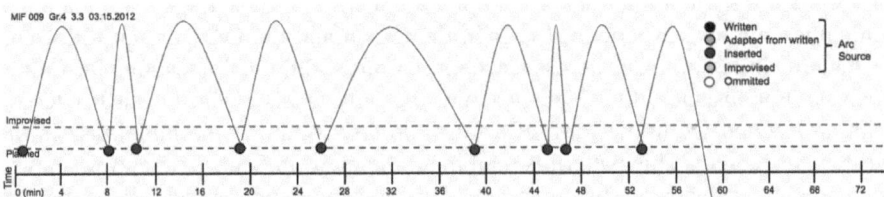

Fig. 27.5 Lesson map of Meredith Frankl 03.15.2012 lesson

Ms. Frankl's lesson map illustrates that regardless of the role the written curriculum plays in shaping it, the enacted lesson can be represented in terms of mathematical episodes initiated by the teacher and through which the teacher must steer the students. The coding system also allows researchers to represent the relationship between these episodes and the written curriculum. In the following section, I discuss some empirical possibilities and next steps based on this approach.

27.6 Conclusion and Next Steps: Using Lesson Maps to Examine the Enacted Curriculum and Factors that Influence It

In addition to representing the contours of the enacted lesson, I see lesson maps as analytical tools to examine the relationship between the written and enacted curriculum, factors that influence this relationship including teacher and curriculum capacity, and implications for lesson quality. I conclude by discussing some possible uses of lesson maps to pursue initial questions in this area.

27.6.1 Examining Patterns in Teachers' Enacted Curriculum

Lesson maps allow us to explore patterns and themes in an individual teacher's lesson and make comparisons to other teachers. Examining several maps of one teacher's lessons along side those of another allows us to explore possible patterns and contrasts. I have found a great deal of internal consistency among each teacher's lesson maps, but have, thus far, observed little similarity between the maps of different teachers using the same curriculum program (Remillard et al. 2015). Lesson maps, and patterns across them, provide a starting place to probe individual teachers' design decisions more deeply. For instance, patterns in the placement of inserted or adapted prompts, or significant omissions, can point to a teacher's interpretation and goals that merit further study. Further, particular types of arcs might be associated with teachers' steering moves during other lesson episodes. Ms. Fiero's decision to omit the warm-up (see Sect. 27.5.3.3) might be related to difficulties students experienced later in the lesson and the IMDDs that followed.

Patterns in lesson maps across teachers can raise additional questions for further analysis. For instance, Ms. Fiero was not the only teacher who began her lesson with an inserted arc in the form of a warm-up task, omitting the initial task in the written curriculum. This pattern is seen in all the maps in Sect. 27.5 and was noted in many other maps in the data set. This tendency may reveal an important phenomenon in teachers' design decisions that merits further exploration. Understanding these decisions can provide insight into factors that influence the quality of the enacted lessons.

27.6.2 Understanding in-the-Moment-Design Decisions

The lesson maps provide a skeletal representation of the lesson in terms of structure and source, but offer limited detail about the interactions that take place within instructional design arcs. Further analysis is needed to examine the interactions within the arcs in order to understand how the enacted lesson unfolds, including how teachers' decisions during instruction influence this unfolding. I introduce the term IMDD to refer to the decisions teachers make during these instructional episodes. More work is needed to understand IMDDs in context and their consequences for the quality of the enacted lesson. I see promise in uncovering patterns in the types of IMDDs teachers make and the factors that influence them, including tracing possible influences that the teacher's guide has on IMDD's. I hypothesize that a teachers' ability to make high quality IMDDs is a critical component of pedagogical design capacity (Brown 2009).

Acknowledgements I wish to thank my collaborators in this analysis: Gina A. Cappelletti and Elena A. Dominguez. I also thank Rowan Machalow for her editorial assistance and substantive feedback. The work presented in this paper is part of the ICUBiT project, a multi-year study of elementary teachers' interactions with curriculum materials and the capacity required to use these resources productively. The ICUBiT study was supported by the National Science Foundation (Grants No. 0918141 and 0918126). Any opinions, findings, or recommendations expressed in this paper are those of the authors and do not necessarily reflect the views of the National Science Foundation. Research team members who have contributed to the data collection and analysis include: Ok-Kyeong Kim, Napthalin Atanga, Luke Reinke, Dustin Smith, Joshua Taton, and Hendrik Van Steenbrugge.

References

Brown, M. W. (2009). The teacher-tool relationship: Theorizing the design and use of curriculum materials. In J. T. Remillard, B. A. Herbel-Eisenmann, & G. M. Lloyd (Eds.), *Mathematics teachers at work: Connecting curriculum materials and classroom instruction* (pp. 17–36). New York: Routledge.

Brown, S. A., Pitvorec, K., Ditto, C., & Kelso, C. (2009). Reconceiving fidelity of implementation: An investigation of elementary whole-number lessons. *Journal for Research in Mathematics Education, 40*(4), 363–395.

Castro, A., Brown, S. A., Pitvorec, K., & Ditto, C. (2007). Fidelity of implementation: Characterizing teachers' instructional moves in the context of a standards-based curriculum. In T. Lamberg & L. R. Wiest (Eds.), *Proceedings of the 29th Annual Meeting of the North American Chapter of the International Group for the Psychology of Mathematics Education* (Vol. 2, pp. 513–520). Stateline (Lake Tahoe), NV: University of Nevada, Reno.

Cobb, P., Zhao, Q., & Dean, C. (2009). Conducting design experiments to support teachers' learning: A reflection from the field. *Journal of the Learning Sciences, 18*(2), 165–199.

Kheong, F. H., Sharpe, P., Soon, G. K., Ramakrishnan, C., Wah, B. L. P., & Choo, M. (2010). *Math in focus: The Singapore approach by Marshall Cavendish.* Boston: Houghton Mifflin Harcourt.

O'Donnell, C. A. (2008). Defining, conceptualizing, and measuring fidelity of implementation and its relationship to outcomes in K–12 curriculum intervention research. *Journal for Research in Mathematics Education, 78*(1), 33–84.

Remillard, J. T. (2005). Examining key concepts of research on teachers' use of mathematics curricula. *Review of Educational Research, 75*(2), 211–246.

Remillard, J. T., Cappelletti, G., & Dominguez de Diclo, E. (2015, April). *Mapping the relationship between the written and enacted curriculum: Examining teachers' curriculum decision making.* Paper presented at the annual meeting of the American Educational Research Association, Chicago, IL.

Remillard, J. T., & Geist, P. (2002). Supporting teachers' professional learning though navigating openings in the curriculum. *Journal of Mathematics Teacher Education, 5*(1), 7–34.

Remillard, J. T., & Heck, D. (2014). Conceptualizing the curriculum enactment process in mathematics education. *ZDM The International Journal on Mathematics Education, 46*(5), 705–718.

Remillard, J. T., & Reinke, L. T. (2012, April). *Complicating scripted curriculum: Can scripts be educative for teachers?* Paper presented at the annual meeting of the American Educational Research Association, Vancouver, BC, CA.

Simon, M., & Tzur, R. (1999). Explicating the teachers' perspective from the researcher' perspectives: Generating accounts of mathematics teachers' practice. *Journal for Research in Mathematics Education, 30*(3), 252–264.

Stein, M. K., Remillard, J. T., & Smith, M. S. (2007). How curriculum influences student learning. In F. K. Lester (Ed.), *Second handbook of research on mathematics teaching and learning* (pp. 319–369). Greenwich, CT: Information Age Publishing.

TERC. (2008). *Investigations in number, data, and space* (2nd ed.). Glenview, IL: Pearson.

Valverde, G. A., Bianchi, L. J., Wolfe, R. G., Schmidt, W. H., & Houang, R. T. (2002). *According to the book: Using TIMSS to investigate the translation of policy into practice through the world of textbooks.* Dordrecht, The Netherlands: Kluwer.

Vygotsky, L. S. (1978). *Mind in society.* Cambridge, MA: Harvard University Press.

Chapter 28
Building Bridges Between the Math Education and the Engineering Education Communities: A Dialogue Through Modelling and Simulation

Ruth Rodriguez Gallegos

Abstract This chapter shows the importance of building communication bridges between two apparently disconnected academic communities: the mathematicians' and the engineers'. The starting point is an overview of an approach to teach mathematics through modeling and simulation of real problems at a private university in the northeast of Mexico that mainly focuses on the training of future engineers. The need to build communication bridges between the mathematics and the engineering education communities seems to be fundamental in order to rethink mathematics education's goals of being prepared to face the challenges posed by today's increasingly changing environment. The results and experience of mathematics professors teaching engineering students show some of the advantages of incorporating new ways of visualizing and understanding phenomena. Furthermore, these new ways allow students to have a new vision of mathematics and a deeper understanding of several math concepts.

Keywords Engineering · Mathematics · Modelling · Simulation
Holistic

28.1 Introduction

The objective of this chapter is to show the importance of building communication bridges between two apparently disconnected academic communities: the mathematicians' and the engineers'. The main goal is to show the importance of introducing a new register of a concept in a mathematics course in order to improve the students' understanding and learning. This new register is the result of the interaction between mathematics education and engineering education. This idea sheds

R. Rodriguez Gallegos (✉)
Tecnológico de Monterrey, Monterrey, México
e-mail: ruthrdz@itesm.mx

light on how to teach mathematics based on a specific engineering point of view and context.

28.2 Literature Review

From an international perspective, studies such as the report of the Program for International Student Assessment (PISA; OECD 2009) state the need to train people in developing the skills of mathematical literacy to solve problems in the future. PISA defines mathematical literacy as

> the capacity to identify, to understand, and to engage in mathematics, and to make well-founded judgments about the role that mathematics plays, as needed for an individual's current and future private life, occupational life, social life with peers and relatives, and this individual's life as a constructive, concerned, and reflective citizen. (p. 17)

This idea of applying mathematics in the context of the students leads us to visit the theme of teaching and learning of mathematics, which has been treated for more than 40 years by the mathematics education community, mainly from the perspective of modeling (Blum et al. 2007) and simulation of real phenomena. A particular example of this is a study that focused on the training of future engineers (Rodríguez 2015) and even using technology (Rodríguez and Quiroz 2015) at a private university in the northeast of Mexico. Previous studies (such as Rodríguez 2015; Rodríguez and Quiroz 2015) allow us to highlight the richness of a work where students can improve their understanding of the mathematical notions learned in class when faced with the idea that they can describe several phenomena.

In Sect. 28.1.2, we show some theoretical background on the teaching and learning of mathematics through an overview of modeling.

28.2.1 The Field of Engineering Education

In the context of engineering education, studies (Bourn and Neal 2008) focusing on a very specific population of future engineers have asserted the prevailing need that an individual's basic education should develop generic skills to complement and reinforce disciplinary skills. The generic skills (Bourn and Neal 2008, p. 12) listed below aim to develop the global dimension in shaping the future engineer and stress the need and importance of these skills in several areas different from mathematics.

Generic skills

1. *Holistic thinking*, critical enquiry, and analysis and reflection
2. Active learning and practical application
3. Self-awareness and empathy

4. Strong communication and listening skills

Its first place in the list shows the importance of developing students' holistic thinking, as they will also play a vital role as engineers and citizens of the 21st century. The authors suggest that the key to understand global skills is to recognize the complex nature of the world we live in and that the future is uncertain and there is not necessarily a series of easily identified solutions.

Hence, the need to develop holistic thinking as an important skill for students and future citizens of the 21st century is made explicit.

Zeroing in on these assertions has led us to rethink the teaching of mathematics. Unlike traditional methods that have a prescribed order of contents and structure, developing competences for the 21st century entails introducing complexity, changes, uncertainty, interconnectedness, multiple meanings, and interpretations in an unstructured universe. We believe that holistic thinking can contribute to enriching the modeling-based teaching of mathematics to meet the unknown needs of the 21st century. Bourn and Neal (2008) state that holistic thinking "requires understanding not only the complexities within the engineering systems but also the relationship between engineering systems and their context" (p. 8). Previously, Jowitt (2004; cited in Bourn and Neal 2008) stated that "a more holistic/systems view of the world is now required—one in which engineers need to be more fully aware of the economic, social and environmental dimensions of their activities and more skilled in meeting their objectives (p. 8)."

In essence, systems thinking is the ability to see a problem or situation holistically from multiple perspectives and understand the relationships, interconnections, and complexities between the different parts that make up the whole (Meadows 2008).

Based on the request to train students of basic education in this area, we decided to explore the importance of developing holistic thinking in future engineers. Since holistic thinking is also related to systems thinking (ST), our proposal is to think how ST skills can be included in math education. Bourn and Neal (2008) report the work done by Senge (2006) in this regard—learning about organizations from a business approach—and it became the trigger to show the advantages and benefits of incorporating systems thinking in a math class. Therefore, we want to emphasize that the idea of introducing ST in a math course is not new, but it has been studied little in recent research, at least in the math education community. In this section, we want to revisit some important works in this direction.

Several authors (Bourguet and Pérez 2003; Bourguet 2005; Fisher 2001, 2011a, b; Caron 2014) in the literature consider that introducing concepts of system dynamics (SD) is very natural in mathematics. SD models explore possible futures and ask "what if" questions. SD is a specific technique of ST where the use of simulators such as iThink or STELLA helps to show a representation of the model and operate with it. Bourguet and Pérez (2003) establish in their work that the SD models and differential equations are two effective representations to express the changes of things over time. SD uses symbolic and graphical representations as

Table 28.1 A language of systems thinking in mathematics courses (based on Fisher 2001, 2011a)

Concept	Mathematical form	Vensim form	Vensim button
Stock This represents the main amount that is to be accumulated. The values increase or decrease over time, and this shows the way things are	Current amount = $A(t)$ (This is an unknown or dependent variable; time or t is usually the independent variable)	Current Amount	T Box Variable
Flow This represents actions or activities that cause the stock value to increase or decline over time	Flow = $\frac{dA}{dt}$ (This is a rate of change or derivative)	flow	Rate
Arrow/connector This serves as an information or action wire showing the relations between the unknown (variable A (t)) and its derivative(s)	(These are the assumptions or hypotheses that we make or the physical laws that govern the phenomenon.)		Arrow
Variable/converter This is used to represent additional and important logic to the model. (It is often a modifier for the flow.)	(This is a parameter of the equation, usually called k)	variable	A Variable

well as computer simulation models to represent and understand the dynamics of a situation (Table 28.1).

Some reflections made from a particular theoretical perspective are presented to undertake more comprehensive studies in engineering education for the 21st century. It is our belief that math colleagues could require several generic skills to expand their vision of the first approach to modeling and simulation of complex phenomena as well as those of a social nature (Rodríguez 2015; Rodríguez and Bourguet 2014).

28.2.2 Teaching and Learning in Math Education: A Differential Equations Course

All over the world, and specifically in Mexican universities, the teaching of differential equations (DE) predominantly focuses on analytical methods rather than on qualitative and numerical methods. In spite of the wealth of knowledge of both approaches in the teaching of DE (Artigue 1995; Arslan et al. 2004), this and other developments have been evidenced in the community of mathematics education for over 20 years (Blanchard 1994; Blanchard et al. 2006).

In contrast, some changes have been reported in daily classroom activities. While successful innovative proposals for teaching DE (especially internationally) have been documented over the past few years (Blanchard 1994; Kallaher 1999; Blanchard et al. 2006; Nathan and Klingbeil 2014) and some other research on the subject has been published, few changes can be observed in classrooms and academic programs at various universities nationwide, particularly in the area of engineering.

This proposal aims to acknowledge the importance of the changes in three registers (algebraic, numeric, and graphic: "rule of three"; Douady 1986; Janvier 1987; Duval 1988, 1995; Artigue 1992; Fisher 1997, 2001, 2011a, b), the modeling approach (Blum et al. 2007; Rodríguez 2015), and the effective use of technology in the teaching/learning process of DE (Rodríguez and Quiroz 2015). We also incorporate a fourth register, the verbal (word problems: "rule of four"; Fisher 1997), and in the last years, we have recognized the importance of dealing with "reality" through physical experiments (such as building an electric circuit; Rodríguez 2015) and the effective use of technology in the teaching/learning process of DE (Rodríguez and Quiroz 2015) or a simulated "reality" through the use of simulators. In this study, we propose the necessity of integrating a new register (the fifth? "rule of five"? Senge 2006; Fisher 1997, 2001; Caron 2014) to work in a math course. We consider that the introduction of the fifth register in a DE course could help to promote another kind of perspective and way of regarding the problems and contexts studied in this course. In the next section, we want to further explain this specific technique from the system dynamics viewpoint. It is our hope to shed some light on the wealth of integrating the two seemingly separate disciplines, systems thinking and mathematics.

Figure 28.1 illustrates different ways in which technology allows "bringing reality" into the classroom in several ways:

a. Experimenting with physics and using sensors to "see" what happens to the magnitudes under study. Some of the sensors are portable technology. Texas Instruments calculators; voltage sensors; and temperature, motion, and various graphical interfaces are also used (Rodríguez and Quiroz 2015; Rodríguez 2015; Quezada-Espinoza and Zavala 2014; Wang et al. 2014).
b. Understanding the phenomenon through remote labs (e-labs), where a real phenomenon may occur. Students manipulate and study at a distance (Ramírez and Macías 2013).
c. Studying a phenomenon that is videotaped and studied through video analysis software such as Tracker (Olmos 2012).
d. Simulation of an experiment that failed when conducted in the classroom is possible as the University of Colorado (United States) proposed in PhET (Quezada-Espinoza et al. 2015; Rehn et al. 2013).
e. Building simulations with open/free simulators. At a slightly more advanced level, students are invited to build simulations using Vensim (Rodríguez and

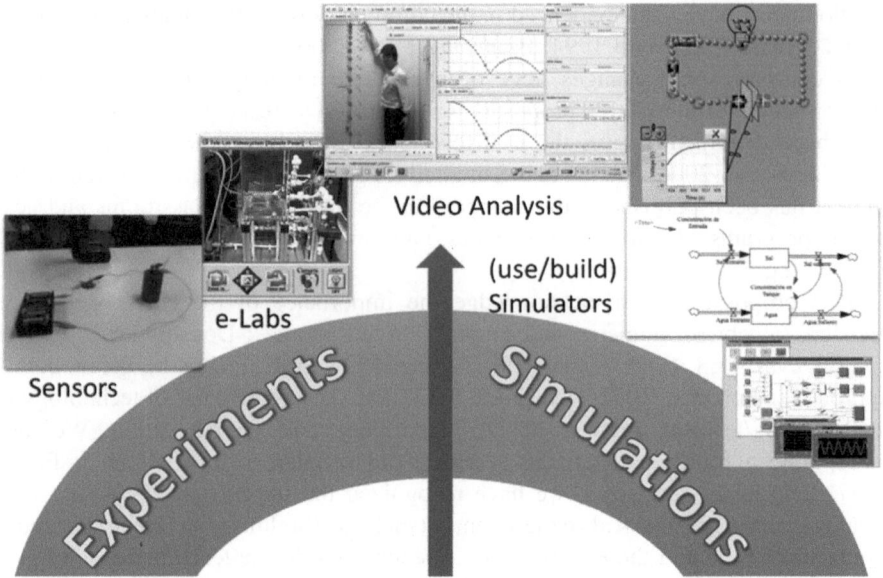

Fig. 28.1 Different uses of technology to portray "reality"

Bourguet 2014) and/or more sophisticated and specific software such as MatLab/Simulink (Smith and Campbell 2011) for the particular case of control theory.

28.2.2.1 Some Theoretical Background About Representations in Math Education

This section is devoted to the analysis of a theoretical framework regarding researchers in didactics of mathematics on the idea of representations.

Let us analyze a theoretical framework related to researchers in didactics of mathematics on the idea of representations. In 1987, Janvier presents the importance of the processes of conversion between representations. Duval (1988, 1995) independently emphasizes the importance of articulating representational registers and the importance of analyzing representations by identifying visual variables (for example, in a graphical representation of a function) that allow them to be linked to significant symbolic units. Duval (1995) provides an approach in the learning of mathematics by studying cognitive problems in depth when performing the processes of conversion between representations. Table 28.2 is a concrete example about the DE notion as explained by Rodríguez (2015).

In Douady's (1986) previous work, used by Artigue (1995) to discuss a problem in differential equations, she proposes the use of frames of representation, alluding to the importance of a student being able to transit, convert, and recognize a

Table 28.2 Different representations of a DE notion in an electrical context

Graphical representation	Numerical representation	Analytical representation	Word problem (verbal representation) or physical representation
		DE: $$R\frac{dq}{dt} + \frac{1}{C}q(t) = E(t),$$ $$q(0) = qo$$ Solution of the DE: $$q(t) = \frac{E}{C}\left(1 - e^{\frac{t}{RC}}\right)$$	

mathematical concept in its different classical representations as an object (numerical, graphical, and algebraic; "rule of three"; Fisher 1997; also verbal representation). In this work, Artigue basically explains how the graphic and qualitative records are left aside for a long time due to the increasing use of technology in the classroom, which puts the qualitative analysis at stake even though it is a fundamental part of the solution in a DE course. It is precisely at that point that we concluded that a fourth valuable representation is not only the verbal (word problem), but the real one in the sense that the student recognizes the real context that the DE models or simulates (Rodríguez 2015; the so-called "real" representation or simulated using an experiment). Finally, the purpose is to investigate whether a fifth representation—diagrams of stocks and flows using SD modeling— would help students make better sense of it considering that they are learning how to model real phenomena for the first time.

For this chapter, and in the following sections, we want to show a specific development of an educational research study about ways to improve the teaching of mathematics using modeling and simulations built by the students themselves (Item e in the list above). In particular, this is exemplified by the introduction of holistic and/or systemic thinking in the training of sophomore students of engineering in a specific course on differential equations. Through the introduction of a new language and vision of the phenomena, qualitative studies can give feedback that allows the introduction of a new vision, a new approach, and a new language for modeling (Fisher 2011a, b). The goal is then to study how the introduction of a new representation in a math course (DE course) in a university in Mexico helps the students better understand the topic.

28.3 The Research Question

Our major interest in our previous project was to explore and identify the most important uses of the mathematical objects engineers frequently utilize. This study is part of a bigger project whose purpose is to give some examples of the use of mathematics in other specialty subjects. In this case, for this chapter, we focus our attention on a specific kind of engineering. In Rodríguez and Bourguet (2014), we presented and justified our interest in a particular community of industrial engineering students with a minor in systems engineering, focusing in particular on the uses and meaning related to the DE math object. Considering the above, we identified some useful information on how math professors can obtain some important ideas of the DS approach (see Rodríguez and Bourguet 2014) to promote generic skills in addition to the mathematical ones to better educate the new global engineer of the 21st century, in particular, those skills related to the importance of holistic thinking.

I decided that some important DS ideas would be designed and implemented in a DE course. Based on Fisher's work (1997, 2001, 2011a, b) and on interaction with an expert in system engineering who teaches a system dynamics course, the series of activities devised and used is shown in Sect. 28.4.1. Finally, the research question for this chapter is:

> How does the introduction of DS-based activities using a new representation in a DE course affect our students, the future engineers?

The answer to this question will be the result of describing the methodology used when introducing a new fifth representation. This representation helped the students better understand the DE notion as an object or as a model of different situations and contexts.

28.4 Methodological Approach

The methodology for this exploratory and descriptive study was mainly a qualitative analysis (Creswell 2013) that had the purpose of leading us to have a more comprehensive outlook on the importance of how the student received this new representation.

It is very important to highlight that the design of the activities was the result of two years of collegiate work between a math professor and an industrial engineering professor. During this time, we had two-hour weekly meetings (32 weeks per year, 64 total) to discuss the elements in common between both disciplines. Each professor analyzed the course, the textbooks, the technology used, and their counterpart's evaluation. (Both professors attended each other's 16-week course). Finally, each professor established the advantages and disadvantages of each other's approach. This work was the basis to establish two phases. Specifically, the design phase (Phase 1), which inspired this study, was fully documented in Rodríguez and Bourguet (2014).

The methodological part of this study was divided into two phases, as follows:

Phase 1: The qualitative analysis from October 2013 to January 2014. We
          designed a series of two activities to be implemented in two sessions of

1.5 h each. We presented four exercises (see Sects. 28.4.2.1–28.4.2.4), with the last one (28.4.2.4) was specially built and analyzed by the students themselves. We focused on the part of using software to build a new representation of the exercise proposed, what the students' perception of this new language was, and how students would solve real problems in the DE course. This last exercise represented a kind of a real problem that is not usually studied in a traditional DE course, because the context is more on the social side, related to an ecological issue. The kind of DE proposed was not easy to solve by hand; thus, the use of specific technology was important and so was the mathematical difficulty of the DE, which was emphasized since one of the models was a non-linear DE.

Phase 2: The qualitative analysis from November to December 2014. We decided to use the results of an institutional survey related to three specific indicators to know more about the student's perceptions regarding this kind of modelling in a DE course (including the use of software to introduce and use a new representation of a DE).

In this chapter, we analyze the design of the activity eventually implemented and the results of the students' perceptions about the introduction of a new representation (issued from a system dynamics approach) in a math/DE course.

28.4.1 Sample

A total of 123 students completed an activity (during two sessions) and subsequently answered a survey at the end of the course. These students belonged to 24 different engineering majors. Of the four analyzed groups, two were honors courses. An honors course has a maximum number of 25 students. These students have a grade point average of between 90 and 100 points, and they speak at least one foreign language (usually English). Overall, the academic profile of these students is higher from the rest.

28.4.2 The Design Process: Instruments for Phase 1

28.4.2.1 Test Over the 2014 Fall Term: Session 1, Part A. Total Time: 90 Min; Time for Part A: 45 Min

During the first 45 min of the session, the Vensim software and the philosophies of systemic thinking and system dynamics were introduced. During this time, a mathematical model previously developed in class (Week 5) involving the filling of a water tank was discussed. How to model the system of a water tank being filled

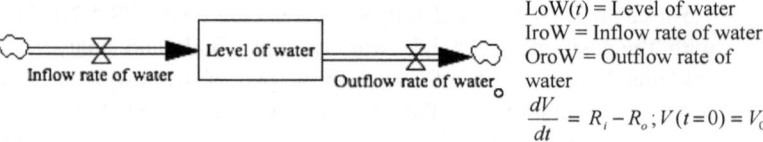

$$\mathrm{LoW}(t) = \text{Level of water}$$
$$\mathrm{IroW} = \text{Inflow rate of water}$$
$$\mathrm{OroW} = \text{Outflow rate of water}$$
$$\frac{dV}{dt} = R_i - R_o ; V(t=0) = V_0$$

Fig. 28.2 Example 1 of Vensim diagram

and emptied as a function of its incoming and outgoing flows of water was shown (see Fig. 28.2). This was interesting for the students because of the use of software and was a way to help the students become familiar with the graphic language used by Vensim.

28.4.2.2 Session 1, Part B. Total Time: 90 Min; Time for Part B: 45 Min. Week 14 of 16

In the second part of the session, during the last 45 min, the students were asked to observe the software for a second time so that they could adapt it to the studied situation in class. They first studied the system of two tanks, but then salt was added to the incoming flow. At the time, what we were concerned about was the variable amount of salt in the tank $S(t)$. Figure 28.3 shows an example of a tank of water mixed with salt.

28.4.2.3 Session 2, Part C. Total Time: 90 Min; Time for Part C: 45 Min

The philosophy of system dynamics was presented again in a new problem during the first 45 min of the session. Over that time, a mathematical model previously developed in class (Week 3) involving the infection of a virus was discussed (see Fig. 28.4).

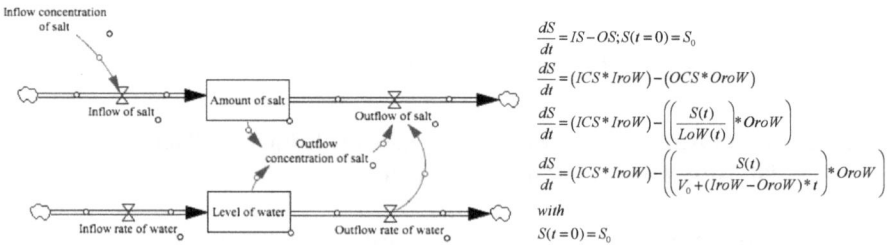

$$\frac{dS}{dt} = IS - OS; S(t=0) = S_0$$
$$\frac{dS}{dt} = (ICS * IroW) - (OCS * OroW)$$
$$\frac{dS}{dt} = (ICS * IroW) - \left(\left(\frac{S(t)}{LoW(t)}\right) * OroW\right)$$
$$\frac{dS}{dt} = (ICS * IroW) - \left(\left(\frac{S(t)}{V_0 + (IroW - OroW) * t}\right) * OroW\right)$$
with
$$S(t=0) = S_0$$

Fig. 28.3 Example 3 of a diagram in Vensim for a tank with water mixed with salt

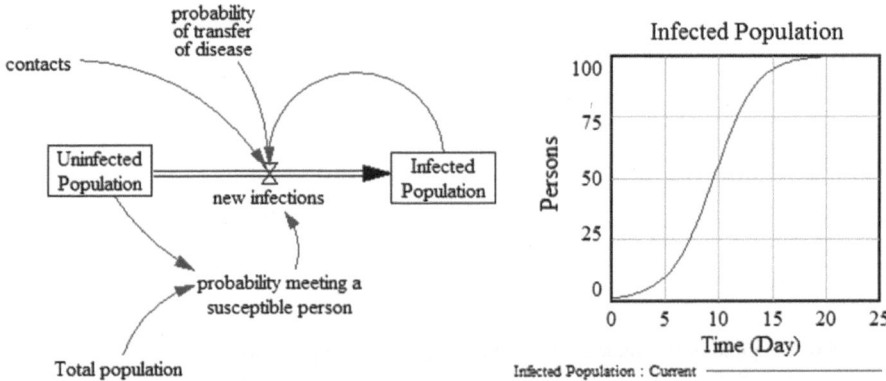

Fig. 28.4 Example of a diagram in Vensim (stock and flow diagram) and graphical representation

---

15. Suppose a species of fish in a particular lake has a population that is modeled by the logistic population model, with a growth rate k, carrying capacity N, and time t measured in years. Adjust the model to account for each of the following situations.
a) 100 fish are harvested each year.
b) One third of the fish population is harvested annually.
c) The number of fish harvested each year is proportional to the square root of the number of fish in the lake.
(From Exercise 16). Suppose that the growth rate parameter is $k = 0.3$ and the carrying capacity is $N = 2{,}500$ in the logistic population model of Exercise 15. Suppose $P(0) = 2{,}500$.
d) Which situation, a, b, or c, is the most threatening to the environment? Support your answer.

Fig. 28.5 Exercise 15 and 16 in Blanchard et al. (2006, p. 18)

28.4.2.4   Session 2, Part D. Total Time: 90 Min; Time for Part D: 45 Min

Later on, the students were asked to solve another problem that was designed based on Blanchard et al. (2006, p. 18, exercise 15 and 16, see Fig. 28.5) but rewritten in the case format.

Question d was included in the problem, and it was the professors' decision to address it since it was not originally posed in the problem. During the 2014 Fall Term, the statement of this exercise was modified to introduce it as a case study. The way it was written changed it considerably, giving the student the freedom to make decisions according to the situation described. This time, the problem led the students to think of a scenario in which they had to make a decision based on three possible ways of fishing. They also had to support their answers based on the simulator according to the statement of the problem in the book (see Fig. 28.6).

It is important to emphasize that the 3 ODE are not linear, hence the difficulty in obtaining the analytical solution for the analytical methods. Thus, the use of the simulator Vensim was well suited to solve this case and to allow each team of students take a position on the last question asked:

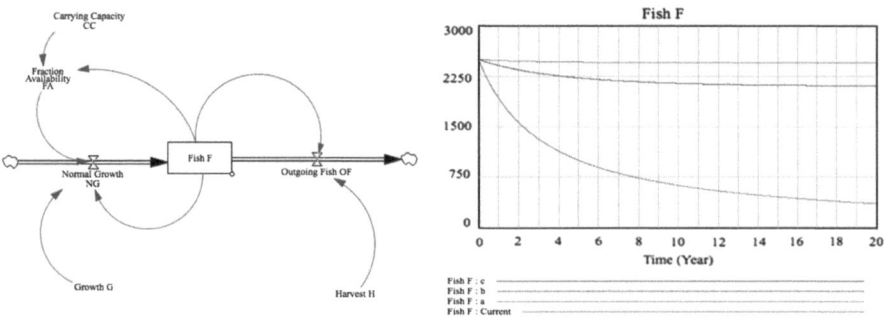

Fig. 28.6 Example of the student representation for the case proposed

Which situation, a, b, or c, is the most threatening to the environment? Support your answer

In the next section, we show a student's analytical response obtained using the simulation software Vensim.

28.4.2.5 Findings: A Student's Abridged Analytical Response in Phase 1

It is important to note that this answer was finally obtained with the simulator by the students themselves during their experience. We may notice that Option a causes greater damage to the environment while the other two remain constant. Furthermore, Option b is less harmful than c.

In the rest of this section, we want to reflect about the comments and decisions made by the students in studying this case in a differential equations course. In both groups, the students concluded that Option b was less harmful.

However, the focus in this paper is on the richness of the discussion among the students in each team after analyzing the question and making their decision using the results shown by the simulator.

For example, over the class discussion, we aimed at having the fishing company behave in a responsible manner. From the financial point of view, the company would make greater profit by catching more fish. However, if they wanted the company to behave in an environmentally friendly manner, Option b was the right one. Finally, many students agreed on an "intermediate" option, Option c. Another important matter discussed was whether to seek more benefit or more stability in the short to medium term (8 years) or in the long term. (It is worthwhile noticing how the time domain was simulated over a period of 100 years). Something interesting to note was that in one group, a couple of teams proposed alternative solutions of the first 3 proposals. A student proposed that instead of catching 100 fish per year to consider 150 (a variant of b). Prior to this proposal, they had to modify the value of the parameters using Vensim, an issue that we wanted to promote in using this software in this case.

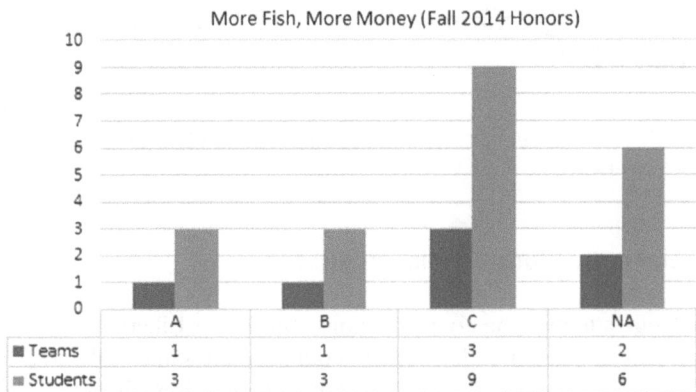

Fig. 28.7   Students' answers to the exercise

In the second part of Session 2, the logistic model was presented from the view of system dynamics. The answers of 21 students were analyzed and are presented in Fig. 28.7.

The problem presented was not an easy task to solve in class. This caused the different teams to rethink their answers, and, although the problem seemed easy, it was not possible to give a unique answer. Furthermore, the students had difficulties in establishing a conclusion without the help of the software. We had assumed that incorporating a systemic viewpoint in order to understand the complete setting of the problem would be helpful. Some matters of social, ethical, and sustainable development interest appeared within the arguments the students gave to the problem. This kind of modeling practice observed in most of the cases of engineering fields is seldom addressed in a traditional math class.

To conclude this part, we would like to comment that the qualitative results obtained in the answers for the activity/case led us to think that it is advisable to include software such as Vensim in the DE course. This allows a new representation of the DE object to be given from the system dynamics viewpoint.

Regarding our research question, we have established that the most relevant aspects of introducing the systems perspective in a math course (or DE course) and using this software are related to the students' learning of how to make adjustments in the parameters of the DE, understanding behaviors and setting relationships between variables in the equation, and understanding the meaning of the numerical and graphical solutions of a DE: All of these are beyond the analytical methods and solutions. We consider that this evidence observed in the answers of the students in this exercise gave us some useful information to show the richness of introducing a systemic approach in the teaching of a mathematics course such as the DE course.

28.4.3   Phase 2: Qualitative Analysis (2 Semesters: Fall 2014)

In this last phase, we analyze some results of an institutional survey from a descriptive, statistical point of view.

28.4.3.1   About the Student Opinion Survey

We focused our attention on four questions. Since the Likert scale is used institutionally, it was decided that this survey would use the same scale. We considered that these questions could help us have a general idea of how the new exercises in a math class using simulation with systems thinking viewpoint would allow the student to have better understanding of how mathematics is relevant in real and work life.

Question TR: Theory and Reality (TR). "The professor implemented learning activities that allowed students to understand the relationship of the content of the course with reality." It is important to emphasize that the indicator theory and reality is usually lower for math classes, especially because of the traditional and near-sighted perspective that stops students from realizing the importance of math and its applications in their everyday life. It is also worthwhile noticing that this question is related to Generic Skill 2, active learning and practical application.

Question CC: Comprehension of Concepts (CC). "The professor facilitated the understanding of the content through clear explanations." It is mentioned in the mathematical section of this paper that the introduction of the systems perspective and the use of a simulator such as Vensim is to allow the students to have another representation (the fifth representation: stock-and-flow diagrams) of the DE for them to have better understanding of this math concept.

Question RDL: Research Documents in Library (RDL). "The professor promoted the use of the query library materials (books, magazines, digital library, and databases) to support learning activities." This question is about how to deal with complex problems, beyond those traditionally seen in a math class; therefore, there is the need to search information in other sources in addition to the one found in the textbook of the math class.

Question IC: Intellectual Challenge (IC). "The professor always demanded your best while maintaining a high intellectual challenge to favor your learning." This question refers to how to deal with complex problems beyond those traditionally seen in a math class. We also included space for the students to express a general opinion about the professor's overall performance during the semester. We called this indicator GOP, General Opinion about the Professor (GOP). We considered it a signal of the complete design of the differential equations course. It is very important to remember that this course is based on active learning and modeling and simulation practices.

We compared two different honors groups and in Fall 2013, this group (F13H, see Fig. 28.8) did not consider an SD modelling approach including the new representation.

Figure 28.8 shows charts of the 2013 and 2014 fall semesters.

Figure 28.8 shows that there are three important differences concerning the introduction of systems thinking in the four analyzed indicators in these two years:

(a) The CC indicator decreases from 1.44 to 1.18 (−0.26 points).
(b) The TR indicator decreases from 1.25 to 1.06 (−0.19 points).
(c) The IC indicator decreases from 1.24 to 1.06 (−0.18 points).
(d) The RDL indicator decreases from 1.18 to 1.06 (−0.12 points).

28.4.3.2 Findings: A Brief Analysis of Students' Responses in Phase 2

As a result of the changes implemented in the course, we can conclude that students perceived that the professor facilitated the understanding of the content through clear explanations (CC indicator). We could infer that the design and incorporation of these activities in the DE course helped to better understand what a DE is by studying the different representations of this object (analytical, numerical, graphical, stocks and flows diagrams, and real situation). The students acknowledged the importance of the DE object since it was helpful to model other real situations (like social/ecological issues, shown in the TR indicator) in addition to those related to physical phenomena (traditionally studied in a math course). The students also highlighted the use of the intellectual challenge (IC indicator) that the course imposed for the activities carried out (including this one) as well as the necessity to search for additional materials different from the textbooks (RDL indicator).

With respect to our research question, we consider that the results of this institutional survey allowed us to give elements that visualize the effects of introducing SD activities such as those shown in Sect. 28.3.2. We consider it important to comment that in the institution where the activity was developed, the TR and CC indicators are difficult to improve since mathematics traditionally has little relation to real problems. We believe that an approach like the one shown above could help

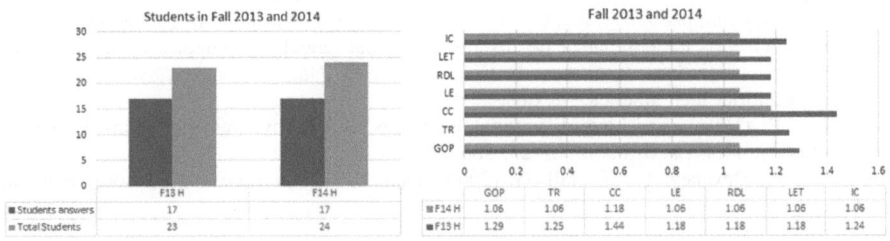

Fig. 28.8 Student sample and results in fall 2013 and 2014 (Honors)

the students make sense of the math concepts by linking them with real contexts, especially one that interests them such as concern for the environment.

28.5 Conclusions

The intention of the chapter arises from the idea of improving the teaching and learning of mathematics at all educational levels by incorporating basic ideas from an engineering field. The first part showed our interest in improving the understanding the future engineers have of a math concept. As an example, we identified that the differential equation tool is of great value for professionals at all levels in many disciplines. We identified the uses of differential equations that industrial engineers with a minor in systems engineering use in a specific course: systems dynamics. We also mentioned that the math professors could obtain some important ideas of the SD approach to promote generic skills such as holistic thinking/systems thinking, active learning, and practical application. We think that these ideas can be obtained from engineering tools from other discipline areas. Based on the results obtained in the methodological part with the students' surveys, we observed that they perceived that the introduction of a new representation such as that offered by SD modeling in terms of a dynamic programming with stocks-flows allowed the students to better understand the notion of DE. This new representation helped students to give another meaning to the components involved in a specific context and their connections, but above all, we found that the richness of having this new way of representing a DE could give students a control tool to use to model more complex problems in addition to the numerical methods already available. Hence, we consider that future studies in this direction would be valuable.

References

Arslan, S., Chaachoua, H., & Laborde, C. (2004). Reflections on the teaching of differential equations. What effects of the teaching of algebraic dominance? *Memorias del X Congreso Internacional de Matemática Educativa (ICME XI)*. Dinamarca.

Artigue, M. (1992). Functions from an algebraic and graphic point of view: Cognitive difficulties and teaching practices. In E. Dubinsky & G. Harel (Eds.), *The concept of function: Aspects of epistemology and pedagogy, MAA notes 25* (pp. 109–132). Washington, DC: MAA.

Artigue, M. (1995). La enseñanza de los principios del cálculo: problemas epistemológicos, cognitivos y didácticos. In Gómez, P. (Ed.), *Ingeniería didáctica en educación matemática*. Grupo Editorial Iberoamérica. México.

Blanchard, P. (1994). Teaching differential equations with a dynamical systems viewpoint. *The College Mathematics Journal, 25*, 385–393.

Blanchard, P., Devaney, R., & Hall, G. (2006). *Differential equations* (3a edición). Belmont: Cengage.

Blum, W., Galbraith, P. L., Henn, H.-W., & Niss, M. (2007). Introduction. In W. Blum, P. L. Galbraith, H.-W. Henn, & M. Niss (Eds.), *Modelling and applications in mathematics education* (pp. 45–56). The 14th ICMI-study 14. Berlin: Springer.

Bourguet, R. E. (2005). *Desarrollo de Pensamiento Sistémico usando ecuaciones diferenciales y dinámica de sistemas. En Reunión de Intercambio de Experiencias en Estudios sobre Educación del Tecnológico de Monterrey (RIE)*. Monterrey. Recuperado en: http://www.mty. itesm.mx/rectoria/dda/rieee/.

Bourguet, R. E., & Pérez, G. (2003). On mathematical structures of systems archetypes. In *Proceedings of the 21st International System Dynamics Conference*. New York, USA: System Dynamics Society.

Bourn, D., & Neal, I. (2008). *The global engineer. Incorporating global skills within UK higher education of engineers*. Engineers against Poverty. Leading Education and Social Research. Institute of Education. University of London.

Caron F., Lidstone D. & Lovric M. (2014). Complex dynamical systems. In S. Oesterle & D. Allan (Eds.), *Actes du Groupe canadien d'étude en didactique des mathématiques* (pp. 137–148). Alberta, 30 mai–3 juin 2014.

Creswell, J. W., & Creswell, J. W. (2013). *Qualitative inquiry and research design: Choosing among five approaches*. Los Angeles: SAGE Publications.

Douady R. (1986). Jeux de cadre et dialectique outil-objet. *Recherches en didactique des mathématiques*, 7.2, 5–31.

Duval, R. (1988). Graphiques et équations: l'Articulation de deux registres. *Annales de Didactique et de Sciences Cognitives*, *1*, 235–253.

Duval, R. (1995). *Sémiosis et pensée humaine: Registres sémiotiques et apprentissage intellectuels*. Neuchâtel: Peter Lang.

Fisher, D. (1997). *Seamless integration of system dynamics into high school mathematics: Algebra, calculus, modeling courses*. Recuperado en: ftp://www.clexchange.org/documents/ implementation/IM1997-07IntegrationSDMath.pdf.

Fisher, D. (2001). *Lessons in mathematics: A dynamic approach*. Australia: Stella Software.

Fisher, D. (2011a). *Modeling dynamics systems: Lessons for a first course* (3rd ed.). Australia: Stella Software.

Fisher, D. M. (2011b). "Everybody thinking differently": K-12 is a leverage point. *System Dynamics Review*, *27*, 394–411. https://doi.org/10.1002/sdr.473.

Janvier, C. (Ed.). (1987). *Problems of representation in the teaching and learning of mathematics*. London: Lawrence Erlbaum Associates.

Jhori, A. (2009). Preparing engineers for a global world: Identifying and teaching strategies for sensemaking and creating new practices. In *Proceedings of the 39th ASEE/IEEE Frontiers in Education Conference*.

Jowitt, P. (2004). *Systems and sustainability: Sustainable development, civil engineering and the formation of the civil engineer*.

Kallaher, M. (1999). *Revolutions in differential equations, exploring ODES with modern technology*. MAA Notes 50, Washington, DC: MAA.

Meadows, D. (2008). *Thinking in systems. A primer*. USA: Chelsea Green Publishing.

Nathan, K., & Klingbeil, N. (2014). *Introductory math to engineering applications*. New Jersey: Wiley.

Niss, M., Blum, W., & Galbraith, P. (2007). *Introduction. ICMI Study 14: Applications and modelling in mathematics education* (pp. 3–32). New York: Springer.

Olmos, O. (2012). Laboratorio para la creación de recursos didácticos para física y matemáticas a través del video análisis: Video-Learning Lab. *Memorias del 7o. Congreso en Innovación y Tecnología Educativa (CITE)*. Monterrey, N.L.

Organization for Economic Cooperation and Development [OCDE]. (2009). *PISA 2009 results. What students know and can do: Students perfomance in reading, mathematics and science.* Retrieved for http://www.oecd.org/pisa/keyfindings/pisa2009keyfindings.htm.

Quezada-Espinoza, M., del Campo, V., & Zavala, G. (2015). Technology and research-based strategies: Learning and alternative conceptions. In *2015 Physics Education Research Conference Proceedings* (pp. 271–274). American Association of Physics Teachers. http://doi.org/10.1119/perc.2015.pr.063.

Quezada-Espinoza, M., & Zavala, G. (2014). El uso de calculadoras con sensores en el aprendizaje de circuitos eléctricos. *Latin American Journal of Physics Education, 8*(4), 1–10.

Ramírez, D., & Macías, M. (2013). Solving material balance problems at unsteady state using a remote laboratory in the classroom. *American Society of Engineering Education (ASEE) International Forum Proceedings.* Atlanta, Estados Unidos. Recuperado en: http://www.asee.org/public/conferences/20/papers/8178/viewsthash.rUAqjad8.dpuf.

Rehn, D. A., Moore, E. B., Podolefsky, N. S., & Finkelstein, N. D. (2013). Tools for high-tech tool use: A framework and heuristics for using interactive simulations. *Journal of Teaching and Learning with Technology, 2*(1), 31–55.

Rodríguez, R. (2015). A differential equations course for engineers through modelling and technology. In G. Stillman, W. Blum, & M. S. Biembengut (Eds.), *Mathematical modelling in education, research and practice. Cultural, social and cognitive influences* (pp. 545–555). New York: Springer. Print ISBN: 978-3-319-18271-1, Electronic ISBN: 978-3-319-18272-8. http://www.springer.com/us/book/9783319182711.

Rodríguez, R., & Bourguet, R. (2014). *Diseño interdisciplinario de Modelación Dinámica usando Ecuaciones Diferenciales y Simulación. Latin American and Caribbean Consortium of Engineering Education (LACCEI 2014).* Guayaquil, Ecuador. http://www.laccei.org/LACCEI2014-Guayaquil/index.htm.

Rodríguez, R., & Bourguet, R. (2015). Building bridges between mathematics and engineering: Modeling practices identified through differential equations and simulation. *American Society of Engineering Education (ASEE) Annual Conference and Exposition, Conference Proceedings.* Atlanta, Estados Unidos. https://www.asee.org/public/conferences/56/papers/13153/view.

Rodríguez, R., & Quiroz, S. (2015). El papel de la tecnología en el proceso de educación matemática para la enseñanza de las Ecuaciones Diferenciales. *Revista Latinoamericana de Investigación en Matemática Educativa, 19*(1). https://doi.org/10.12802/relime.13.1914; ISSN: 2007–6819.

Romo-Vázquez, A. (2014). La modelización matemática en la formación de ingenieros. *Educación Matemática,* 314–338. Recuperado de http://www.redalyc.org/pdf/405/40540854016.pdf.

Senge, P. (2006, 1990). *The 5th discipline: The art and practice of the learning organization.* New York: Doubleday/Currency.

Smith, C., & Campbell, S. (2011). *A first course in differential equations, modeling, and simulation.* Boca Ratón: CRC Press.

Société Européenne pour la Formation des Ingénieurs. (2016, 23 de julio de). Obtenido de http://www.sefi.be/.

Sterman, J. D. (2000). Learning in and about complex systems. *In Business dynamics: System thinking and modeling for a complex world* (Cap. 1, 5–10). Estados Unidos de América: Irwin/McGraw-Hill.

Tecnologico de Monterrey. (2015). *Challenge based learning.* Recuperated in: http://observatory.itesm.mx/edu-trends.

Universidad de Alcalá. Fundación General. (2012, 26 de 08 de). *Programas de la UE. Obtenido de Proyecto "USo+I: Universidad, Sociedad e Innovación. Mejora de la pertinencia de la educación en las ingenierías de Latinoamérica"*: http://areadecooperacion.fgua.es/2012/03/proyecto-alfa-usoi-universidad-sociedad.html.

Vensim www.vensim.com.

Wang, C.-Y., Wu, H.-K., Lee, S. W.-Y., Hwang, F.-K., Chang, H.-Y., Wu, Y.-T., et al. (2014). A review of research on technology-assisted school science laboratories. *Educational Technology & Society, 17*(2), 307–320.

Chapter 29
Constructing Dynamic Geometry: Insights from a Study of Teaching Practices in English Schools

Kenneth Ruthven

Abstract Any technology retains a degree of fluidity in its conception, shaped not just by its designers but by its subsequent users. This chapter applies this perspective to one form of software which has attracted particular attention in mathematics teaching: dynamic geometry. Drawing on a study conducted in professionally well-regarded mathematics departments in English secondary schools, the chapter sketches the wider curricular context, provides an overview of each of three contrasting cases of teaching practices making use of dynamic geometry, and presents cross-cutting themes through which these contrasts can be characterised. Critical variables include the degree to which teachers see student use of the software as promoting mathematically-disciplined interaction, analysis of apparent mathematical anomalies as supporting learning, and dragging as a means of focusing attention on continuous variation. The chapter concludes by discussing how teaching practices might productively be developed, and how such development might be supported by further research.

Keywords Case study · Dynamic geometry · School mathematics
Teaching practices · Technology integration

29.1 Overview

In recent years, dynamic geometry has attracted attention in mathematics education. However, like any mathematical software, it leaves the user considerable scope to interpret how it might serve as a curricular and pedagogical resource. After sketching the wider context, this chapter outlines three illustrative cases of teaching practices featuring use of dynamic geometry in lower-secondary mathematics lessons in English schools (drawn from Ruthven et al. 2008). Initially, the focus is on

K. Ruthven (✉)
Faculty of Education, University of Cambridge, 184 Hills Road,
Cambridge CB2 8PQ, UK
e-mail: kr18@cam.ac.uk

© The Author(s) 2018   521
G. Kaiser et al. (eds.), *Invited Lectures from the 13th International Congress
on Mathematical Education*, ICME-13 Monographs,
https://doi.org/10.1007/978-3-319-72170-5_29

the way in which the teacher within each case characterises their teaching practice and offers a supporting rationale for it. Next, taking a researcher perspective informed by relevant literatures, these cases are compared in terms of cross-cutting themes intended to capture commonalities and contrasts between them. The chapter concludes by discussing how teaching practices might productively be developed further, and how such development might be supported.

29.2   The Evolving Design of Dynamic Geometry and Framing of Its Use

In the field of social studies of technology, the idea of 'interpretative flexibility' (Kline and Pinch 1999) acknowledges that conception of a technology remains fluid beyond the initial stage of the design of a product, continuing into the subsequent stage in which it is taken up by users. The ways in which a technology is employed become aligned with user concerns and adapted to the situations in which use takes place. This opens the way to variation in modalities of use between different user groups and between different settings for use, and to change in these modalities over time. Similarly, in research on the diffusion of educational resources, it has come to be recognised that teachers act as interpreters and mediators of curriculum materials (Remillard 2005). Teachers typically select from and adapt curriculum materials, and they necessarily incorporate these materials into wider systems of classroom practice (Ball and Cohen 1996). Both these traditions, then, highlight the scope for interpretation of an innovative resource such as dynamic geometry

Two particular packages served to define the new class of software which came to be known as 'dynamic geometry'. *Geometer's Sketchpad* was originally conceived simply as a program "to draw accurate, static figures from Euclidean geometry" (Goldenberg et al. 2008, p. 58). In the course of its development, however, the idea of creating a dynamic—rather than static—figure was borrowed from contemporary drawing software, so that points and segments of a figure could be dragged while preserving the properties defining it. Inspired by earlier software for visualizing discrete graphs, dragging was designed into *Cabri Geometry* from the start, although views on its significance differed considerably (Laborde and Laborde 2008). Relatively rapidly, however, dragging became accepted as the key defining feature of dynamic geometry software.

Although dynamic geometry systems were developed with educational purposes in view, neither *Cabri* nor *Sketchpad* was initially devised with a particular pedagogical approach in mind (Goldenberg et al. 2008). However, pioneering work by mathematics educators associated dynamic geometry with a pedagogical orientation in which such software served "to create experimental environments where collaborative learning and student exploration are encouraged" (Chazan and Yerushalmy 1995, p. 8). Nevertheless, a substantial national survey conducted in the United States (Becker et al. 1999) conveyed a rather different picture. Although

Sketchpad was chosen by more than one in five high school mathematics teachers who nominated a 'most valuable' software title, further evidence indicated an association between teachers nominating *Sketchpad* and reporting skill-development as their main objective for computer use (p. 46). This suggests, then, that modes of use of dynamic geometry in mainstream classrooms may differ markedly from the exploratory orientation advocated by many proponents.

Much of the pioneering development of dynamic geometry systems took place in countries—notably France and the United States—which have retained a more strongly Euclidean spirit within their school geometry curriculum (Hoyles et al. 2001). This Euclidean lineage of dynamic geometry might be expected to fit poorly with a national curriculum which refers—as did the English one framing the practice to be studied here—not to *Geometry* but to *Shape, Space and Measures*. However, the scope to employ the software as a means of supporting observation and measurement resonates with a longstanding orientation of English school mathematics. In her comparative ethnography, Kaiser (2002) noted the predominance in English mathematics classrooms of example-based checking as a means of validating results, not just employed by students, but also encouraged by their teachers. Exemplifying this trend, she cites a lesson in which one of the 'circle theorems' was established by the teacher setting each student to draw and measure three diagrams to test the result, and then arguing for its acceptance on the strength of these accumulated checks.

At the time of the study to be discussed, some use of dynamic geometry at lower-secondary level had been given official endorsement and more detailed expression in a government-sponsored elaboration of the English national curriculum (Department for Education and Employment 2001). The emphasis of many of the illustrations was on empirical exploration, but some examples presented observation of a dynamic figure as a precursor to proving with static diagrams. While this official guidance explicitly recognised the knowledge required to use classical manual tools for the construction and measurement of geometric figures, it overlooked equivalent aspects of using dynamic software. For instance, while the knowledge required to make use of a protractor to measure angles by hand was carefully specified, no attention was given to the distinctive knowledge required to measure angles with dynamic software.

29.3 Study Design

The objective of this study was to understand what mathematics teachers at lower-secondary level in England regarded as successful use of dynamic geometry. To address this objective, a multiple case-study design was chosen, employing methods which aimed at characterising teachers' thinking and practice, first in their own terms, and then in terms of broader constructs informed by the research literature. The cases which I discuss here were identified through a process intended to elicit professionally well-regarded practice in using digital tools in mathematics teaching. Relatively few schools nominated on this basis reported making use of dynamic geometry: cases were then chosen for further study so as to capture the range of approaches reported and to follow up teachers who had been particularly informative. Data portfolios for each case were assembled through a procedure involving classroom observation followed up by teacher interview, including the copying of associated curricular resources. These portfolios were analysed thematically in two stages. The first stage of analysis was within-case, adopting an emic approach intended to capture and distil into a case narrative the terms in which the teacher responsible characterised and explained the case and offered a supporting rationale for it. The second stage was cross-case, adopting an etic approach, taking a researcher perspective informed by relevant literature, and aimed at identifying important commonalities and contrasts across cases. Full details of the design of the study and methods employed can be found in the original report (Ruthven et al. 2008). Equally, the individual case narratives and the cross-case analysis are reported there in full.

29.4 Case Outlines

These case outlines (adapted from Ruthven et al. 2008; Ruthven 2012) summarise each case in terms of the main emic themes identified in the analysis (as indicated by the titles of subsections). They characterise the practice associated with each case as elaborated by the teacher responsible. Each case is summarised in terms of four main themes, encapsulated in the subsection titles and then expressed in the teacher's own words.

29.4.1 Case N

This example, focusing on angle properties in the circle (see Fig. 29.1), was nominated in these terms by the teacher concerned:

> The one that I do like to do is the one with the circle theorem that says the angle at the centre is twice the angle at the circumference, because that covers the same theorem as the angles in the same segment, and the angle at the semi-circle is 90 degrees, and you can cover a lot of different circle theorems by doing that one demonstration. And the students find it very, very difficult to believe if they don't see it on the computer... And yet when you're dragging it round this circle using Cabri... it just gets that they start to believe a lot more and they are more convinced of its truth.

A corresponding 40-minute lesson started with 25 minutes of activity led by the teacher, during which he constructed and manipulated a dynamic figure, projected from his computer onto an ordinary whiteboard. This was followed by individual student activity on a (pencil and paper) textbook exercise.

Fig. 29.1 Figure used in
Case N lesson

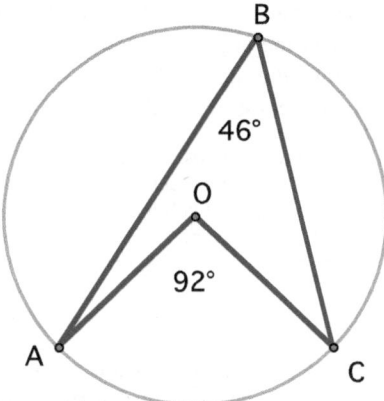

29.4.1.1   Maintaining Students' Attention and Lesson Progression Through Dynamic Presentation and Tactical Questioning

The teacher valued dynamic geometry presentation for holding the attention of students:

> I think that holds their attention more. The fact that they can see that you can pick up and drag these shapes around, and then the angles change as well automatically, so all the numbers are changing... That sort of movement, that dynamism, helps to keep their attention.

He used tactical questioning of carefully chosen students with a view to supporting the progression of the whole-class segment of the lesson towards a target result:

> [I] pick on students that I think may have a problem with it... If students that I think are going to have a problem with it understand it, then I can be fairly confident that the others understand it as well... If [I] ask a student a question and they don't know the answer, I won't give up on them, I'll carry on asking or trying to get the correct response out of them... Not only is it benefiting the student you've asked, everybody else is trying to come to the correct solution as well. And so I do that to help reinforce what's being presented.

29.4.1.2   Making Properties Apprehensible and Convincing to Students Through Purposive Dragging

The teacher saw purposive dragging of the figure as a powerful means through which he could make properties—such as the unchanging measure of the angle at the circumference—apprehensible and convincing to students:

> If you do it on the board and you drag the thing round, then they tend to be much more convinced by what they see. So, I think the technology helps because they can actually see it getting dragged round, they see the angle doesn't change and they are much more convinced.

To make underlying relationships between changing measures—such as between the angle at the circumference and the angle at the centre—discernible by students, the teacher led a process of dragging the figure to generate pairs of values from which students themselves could identify the target pattern:

> Trying to get across the point that the angle at the centre is twice the angle at the outside… Because the angle automatically changes as you drag the point round, you can write up pairs of values, [and] the students can deduce that themselves… So the technology helps a great deal in that respect because you're not just telling them a fact, you're allowing them to sort of deduce it and interact with what's going on.

29.4.1.3   Making It Easy for Students to Identify Properties by Pre-empting Possible Confusions

The teacher took great care to anticipate and pre-empt situations which might confuse students:

> [I] keep things running through the lesson in my own head, and looking for possibilities where students may become confused, or things that might cloud the issue, so that I can do something about that before it becomes an issue in the classroom.

In particular, he designed the dynamic figures that he used and manipulated them in ways intended to make target properties as readily discernible by students as possible:

> I obviously pre-prepared the circle with the lines and angles already marked in. Also for this group, I made sure the angles were always integer values… That way you don't have half angles to deal with. So the angle at the centre was always an even number of degrees because that way the angle at the outside can be halved quite successfully… So I did that to help make it a little bit easier for them to spot the rule.

For example, the teacher avoided dragging figures into positions where an angle —such as the angle at the centre of the circle—would become reflex, resulting in the measure of its smaller counterpart angle being shown by the software:

> When you move things around, if the three points you are measuring swap over somehow, then it starts measuring a different angle.

29.4.1.4   Avoiding the Disadvantages of Software Use by Students Through Teacher Presentation

The teacher limited software use to his giving a presentation so as to avoid demands and difficulties of students having to use it:

> If I wanted the students to do it, it, it would take a long time in order for them to master the package and I think the cost benefit doesn't pay there… And there's huge scope for them making mistakes and errors, especially at this level of student.

He was not persuaded that the educational returns from students learning to use the software were sufficient given a curriculum which was more factual than investigative, and with relatively few topics which would benefit from dynamic geometry treatment:

> It's a difficult program for the students to master... The return from the time investment ... would be fairly small... And the content of geometry at foundation and intermediate level just doesn't require that degree of investigation. So they need to learn certain facts... but most of those facts can be learned quite well enough without Cabri.

29.4.2 Case P

This example, focusing on angle properties in various configurations, was nominated in these terms by the teacher concerned:

> All of our angle work at [lower secondary] is done [with Cabri]... Most of the tasks are... designed with what we want to achieve from it in mind. So if we want them to see that the angles on a straight line add to 180 it's designed exactly for that purpose... So they should come to the right conclusion but... they feel that they've done it on their own and they've explained it.

A corresponding 50-minute lesson started with 15 minutes of activity led by the teacher, using an interactive whiteboard. This was followed by the students working in trios at a computer, guided by a teacher-devised worksheet, to investigate the angle sums of dynamic polygons that they themselves constructed (see Fig. 29.2).

29.4.2.1 Developing Students' Broad Understanding of Space and Shape Through Exploring Dynamic Figures

The emphasis of the investigation in this lesson was more on promoting students' broad understanding of shape and space than their knowledge of specific results:

Fig. 29.2 Figure used in Case P lesson

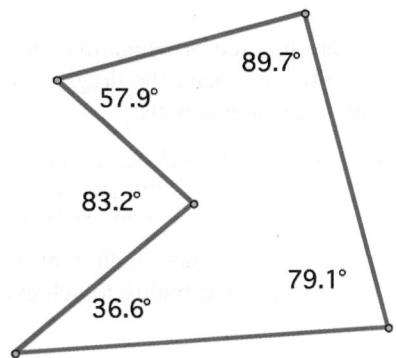

The work on angles in polygons is not so important; that is extension work, and we do it again, that topic, later in the year… So it's sort of a lead up for that as well. But in terms of key concepts, it's not really the curriculum topic that's important, but the understanding of space and shape in geometry, and how that works.

While other investigations which aimed at establishing specific results were more structured, this lesson was concerned with developing broader spatial awareness, and so was more fluid:

[This] lesson was more about them getting to know the software, them having an awareness of space, and thinking how shapes grow and what happens to some of the corners as they grow, and whether that's related to the shape. And that was much more fluid… I was a bit more adding on to the end of a topic, and you can be a bit wilder there, and not have to follow the curriculum as such.

29.4.2.2 Giving Students Experience of Geometrically-Principled Interaction with the Software

The teacher saw an important feature of the software as being the way in which its design around geometrical principles shaped student interaction with it:

The package is geometry-based, and it is from-first-principles geometry… One of the main parts of this lesson was that they could learn the software, and have some idea of how shapes and points relate to each other, and to see that the software works geometrically.

In particular, the teacher could build on students' experience of making use of the software to draw out the way in which they had been enacting geometrical principles:

When they were trying to measure the angle, that really brought out the idea of what is an angle… Just the action of doing it really made a fuss about that for them, and they really understood that angles, these three points that are on two lines, and what it means.

29.4.2.3 Focusing Students' Attention on Mathematical Essentials Through Structured Software Use

In more focused investigations, the teacher preferred to structure students' use of the software around the dragging of simple prepared figures in order to focus on mathematical essentials:

It does add complications, because it's quite a difficult piece of software. So that's why we structure the work so they just have to move points. So they don't have to be complicated by that, they really can just focus on what's happening mathematically.

She saw the ease with which figures could be dragged to create multiple examples as contributing to achieving this focus:

As they move the shapes around, they can see what's actually happening, as if they were drawing it on a page, and doing different drawings. But it obviously removes the need for them to have to redraw things. So it's easier for the kids who find it hard to draw well. So it really helps them, and they can focus on the learning of how the angles match, and what they add up to.

29.4.2.4 Supporting Students in Questioning Unexpected Results and Learning from Them

The teacher sought to promote students' learning through supporting them in identifying and analysing the sometimes confusing way in which the software measured angles:

I wanted to draw attention to... how the software measures the smaller angle, thus reinforcing that there are two angles at a point and they needed to work out the other... Because a lot of them had found that they'd got the wrong answers, and [that] it measured the obtuse angle rather than the reflex angle, so I highlighted that, because that was important in terms of understanding the software. Next lesson, we'll talk a lot more about what we learned from it.

She valued anomalous situations of this type for developing students' critical mathematical thinking about results produced by the software:

For me, success is when the kids produce something and then say "This can't be right because it's not what I expect"... Because they're going to make mistakes. But if they look at it... they can sense that there's something wrong... So we talked about how we'd overcome that... I think that happened in slightly different ways around the room, but it was one of the key things that the kids learned, that you can't assume that what you've got in front of you is actually what you want, and you have to look at it... and question it, which is very powerful.

This reflected her wider emphasis on supporting students in thinking through apparently conflicting states of affairs to a coherent resolution:

Where there is a conflict like that and the child's not understood something or finds there's something not right, I question them about what's not right and why it's not right, and therefore what do they think it should be?... All the time [I'm] subconsciously thinking, what will challenge this child, what will open the door for them to take this step through.

29.4.3 Case Q

This example, focusing on the idea of the 'centre' of a triangle (see Fig. 29.3), was nominated in these terms by the teacher concerned:

We'd done some very rough work on constructions with compasses and bisecting triangles. And then I extended that to Geometer's Sketchpad on the interactive whiteboard... And we... bisected the sides of a triangle. And [the pupils] noted that [the perpendicular

Fig. 29.3 Figure used in
Case Q lesson

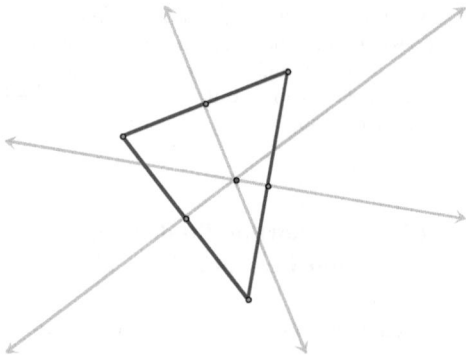

bisectors] all met at a point… And we moved it around and it wasn't the centre of the triangle. Sometimes it was inside the triangle and sometimes outside… So we had… the whole lesson, just discussing what's the centre of a triangle.

The corresponding lesson extended over two 45-minute sessions on consecutive days. Both sessions started with 10–20 minutes of teacher-led activity, using an interactive whiteboard. This was followed by student activity at individual computers, guided by a teacher-devised worksheet. In the first session, students themselves constructed a dynamic triangle and the perpendicular bisectors of its edges, in order to investigate the properties of the bisectors. In the second session, they investigated how changing the shape of the triangle affected the position of the point of concurrence.

29.4.3.1 Giving Students Experience of Finding Rules and Patterns Within Abstract Geometry

The teacher reported that the lesson followed on from work on construction by hand, extending this beyond the official curriculum:

Geometry's so vague in school maths at the moment… I mean the… national curriculum way would be to do the construction, and then on loci and stuff like that… [But] I went on more of the geometry way.

This involved learning to explore the properties of familiar shapes and conceptualise them in more abstract geometric terms:

The main thing is the idea that you can look at a shape that you're fairly familiar with and do things to it, and find new ideas. It's this idea of coming across abstract geometry and finding rules and patterns within it, which is what geometry's all about really.

The teacher's focus was more on students learning about geometrical exploration than on their mastering particular content:

It's less about learning the actual facts, than about the ideas of exploring an abstract geometrical idea, and finding that there's lots of different rules… They will have manipulated shapes, they'll have got used to… the idea of trying to describe things. There's a lot of maths there that isn't directly learning facts, and I think that's a really important part of doing that sort of exercise.

29.4.3.2   Emphasising Mathematical Rules Through Clarifying Student Instructions to the Computer

The teacher saw the way in which the software required clear instructions to be given in mathematical terms as a key characteristic:

I always introduce Geometer's Sketchpad by saying "It's a very specific, you've got to tell it. It's not just drawing, it's drawing using mathematical rules."… They're quite happy with that notion of… the computer only following certain clear instructions.

The teacher helped students to identify mistakes in their constructions and analyse them mathematically:

[Named student] had a mid-point of one line selected and… a perpendicular line to another, and he didn't actually notice… When I was going round to individuals; they were saying "Oh, something's wrong"; so I was [saying] "Which line is perpendicular to that one?"

He saw such difficulties as helping him to draw out the mathematical ideas at stake:

A few people… drew random lines… because one of the awkward things about it is the selection tool… [But] quite a few discussions I had with them emphasised which line is perpendicular to that edge… So sometimes the mistakes actually helped.

29.4.3.3   Making Mathematical Properties Stand Out for Students Through Prompting Dragging

The teacher noted the crucial part he played in making key mathematical properties stand out for students by prompting them to drag vertices of the dynamic triangle—for example, so as to bring out the concurrence of the perpendicular bisectors:

They didn't spot that they all met at a point as easily. I think it just doesn't strike them as being particularly unusual… I don't think anybody got that without some sort of prompting. It's not that they didn't notice it, but they didn't see it as a significant thing to look for.

Through similar prompting of dragging he helped students to appreciate how the position of the point of concurrence—inside or outside the triangle—could be related to the size of the largest angle of the triangle—acute or obtuse:

It was nice to see the way that the point, the central point, went from inside to outside. They were able to move that around and look how the angle was changing. And what sort of rules... I led them very closely on that.

29.4.3.4 Making Learning Less Vague Through Getting Students to Write a Rule Clearly

Getting students to formulate their findings in explicit mathematical terms was an important issue for the teacher:

They've got to actually write down what they think they've learned. Because at the moment, I suspect they've got vague notions of what they've learnt but nothing concrete in their heads.

Accordingly, he sought to sharpen the precision with which students expressed their conclusions:

I was focusing on getting them to write a rule clearly. I mean there were a lot writing "They all meet" or even, someone said "They all have a centre"... So we were trying to discuss what 'all' meant, and a girl at the back had "The perpendicular bisectors meet", but I think she'd heard me say that to someone else, and changed it herself; "Meet at a point".

This refining of mathematical expression was assisted by the provisionality of the text box in which students entered accompanying sentences alongside their figures on the screen:

The fact that they had a text-box... and they could change it and edit it; they could actually then think about what they were writing, how they describe. I could have those discussions. With handwritten, if someone writes a whole sentence next to a neat diagram and you say..."Can you add this in?", you've just ruined their work. But with technology you can just change it, highlight it and add on an extra bit, and they don't mind.

29.5 Cross-Cutting Themes

The second stage of analysis sought to identify and conceptualise salient issues across cases, in broader analytic terms. The resulting themes identify important dimensions of practice, comparing the cases so as to characterise commonalities and contrasts in each dimension. Each subsection heading encapsulates a theme and extracts from quotations from the case outlines are included in the subsequent text to signal the evidential base for the theme.

29.5.1 Employing Dynamic Geometry to Support Guided Discovery

In all of the cases, teachers appealed to some notion of learning through guided discovery, although the practical expressions of this idea differed. In one case, discovery by the whole class was closely guided by the teacher who found that "the technology helped a great deal" by making it easier for students to "spot the rule", and thus for the teacher "to get the correct response out of them", so that "you're not just telling them a fact, you're allowing them to sort of deduce it and interact with what's going on" [N]. In other cases, the classroom approaches involved more delegation to students working individually or in pairs, through tackling 'investigations' intended to develop their general understanding of shape and skills of inquiry. Ultimately, however, this activity was structured with the intention that students "should come to the right conclusion but... feel that they've done it on their own" [P]. In this respect, teachers acknowledged their important role in "drawing attention to" [P] and "prompting" [Q] target results.

29.5.2 Evaluating the Costs and Benefits of Student Software Use

In all of the cases teachers alluded to benefits of dynamic geometry—compared to classical manual tools—in facilitating extensive work with figures. However, there were important differences of perspective on whether the software should be used by students, leading to approaches based variously on *avoiding, minimising,* or *exploiting* its demands. In some cases, the teachers viewed dynamic geometry as "a difficult program for the students to master" with "huge scope for... mistakes and errors" [N]; as "quite a difficult piece of software" which "does add complications" [P]. In one of these cases, then, the software was used only for teacher presentation on the grounds that "it would take a long time... for [students] to master the package" and "the return from the time investment... would be fairly small", so that "the cost benefit doesn't pay" [N]. In the other case where concerns about the accessibility of the software to students were expressed, unless the required figure was straightforward for students to construct for themselves, the normal pattern was "to structure the work, so [that students]... don't have to be complicated by that [and]...can just focus on what's happening mathematically" through providing them with a prepared figure [P]. Such concerns were not expressed in the final case [Q], perhaps because the teacher treated the construction of dynamic figures by students, and the development of their proficiency with the software, as a vehicle for developing and disciplining their geometrical thinking and expression.

29.5.3 Handling Apparent Mathematical Anomalies of Software Operation

A further challenge for teachers was in handling apparent mathematical anomalies in the way in which the software operated, such as those arising when figures were dragged to positions where an angle becomes reflex (with the software displaying the measure of the associated acute or obtuse angle), or where an arithmetical relationship between angle measures was obscured by values being rounded. In one case, the teacher took great care to avoid exposing students to such anomalies, through vigilant dragging [N]. The other case tackling the same type of topic (and indeed an identical topic in one lesson) provided a striking contrast. The teacher actually sought "to draw attention to... how the software measures the smaller angle" so that the apparent anomaly of measurement for reflex angles could be resolved through mathematisation, "thus reinforcing that there are two angles at a point and [that students] needed to work out the other" [P]. Moreover, in this case, the teacher considered that "one of the key things that the kids learned" was "that you can't assume that what you've got in front of you is actually what you want, and you have to look at it... and question it" [P]. These strategies differed, then, between—in the first case—*concealing* anomalies of software operation and—in the second case—*capitalising* on them for purposes of mathematical knowledge building.

29.5.4 Supporting Learning Through Analysis of Mathematical Discrepancies

Capitalising on such anomalies formed part of a wider teaching strategy in which "where there is a conflict like that and the child's not understood something or finds there's something not right", the teacher's preferred response was to "question them about what's not right and why it's not right" [P]. The further case in which students worked with the software also emphasised the value of exploiting errors and anomalies so as to promote mathematical thinking and knowledge building. This teacher suggested that, where students encountered difficulties in constructing figures with the software, "sometimes the mistakes actually helped", by leading to discussions which drew out the mathematical ideas at stake [Q]. In the final—and contrasting—case, one of the reasons that the teacher offered for not having students themselves work with the software was the "huge scope for them making mistakes and errors". As already noted, this teacher sought to avoid exposing students to such anomalies by himself manipulating the software; and, while he reported that through questioning he would "pick on students that... may have a problem" and "carry on... trying to get the correct response out of them", this seems to have been more a means of managing lesson pace and progression than of supporting rethinking [N].

29.5.5 Promoting Mathematically-Disciplined Interaction Through the Software

In both the cases where students did construct figures using the software, teachers stressed the mathematical discipline involved. Experiencing "from-first-principles geometry" was intended to help students "see that the software works geometrically", and to give them "some idea of how shapes and points relate to each other" [P]. Likewise, the ideas of "the computer only following certain clear instructions" and of the software "not just drawing [but] drawing using mathematical rules" were made explicit to students [Q]. In addition, in this case, getting students to formulate their conclusions as a mathematically precise sentence was helped by using the software "text-box [to] change it and edit it… [so as to] actually then think about… how [to] describe" [Q].

29.5.6 Privileging a Mathematical Register for Framing Figural Properties

One case [Q] was distinctive in going beyond the official curriculum to a more classical emphasis on use of a *geometrical register* to frame and analyse figural properties. A concern that students should have experience of "finding rules and patterns" in "abstract geometry" led to them being asked to "write a rule clearly" as a means of "trying to describe things" in directly geometrical terms [Q]. By contrast, in the other cases, figural properties were inferred less directly from arithmetical patterns in the numerical measures generated by a dynamic figure. Because "you can pick up and drag these shapes around… so all the numbers are changing", students could "see [that] the angle doesn't change", or use "pairs of values… [to] deduce [for] themselves" the relationship between two angles [N]. By using a dynamic figure emphasising this more familiar *arithmetical register*—privileged by an official curriculum placing emphasis on the development of numeracy—students could "focus on the learning of how the angles match, and what they add up to" [P].

29.5.7 Incorporating Dynamic Manipulation into Mathematical Discourse

Finally, the case records show how mathematical discourse was developing to incorporate dynamic manipulation. Quite often, dragging was presented simply in terms of moving or changing a figure to generate discrete static examples: "move the triangle around and try different triangles within seconds" [Q]; "change it and you've got then an unlimited number of shapes" [P]. However, dragging was sometimes framed in more dynamic terms. Working in the directly geometrical register, for example, attention was drawn to "the way that the point, the central point, went from inside to outside [the triangle]" so that students "were able to move that around and look how the angle was changing" [Q]. Working in the more prevalent arithmetic register, attention was focused more sharply on the constancy, variation and covariation of measures; so that, for example, students "can actually see it getting dragged round, they see the angle doesn't change" or by "dragg[ing] these shapes around… so all the numbers are changing" [N].

29.6 Discussion and Conclusion

In line with the idea of interpretative flexibility, this study found important differences in teachers' 'constructions' of dynamic geometry, even if their teaching practices all appealed to some idea of employing the software to support guided discovery. Teacher assessments of the costs and benefits of software use by students were influenced by the extent to which such use was seen as promoting mathematically-disciplined interaction. Approaches to handling apparent mathematical anomalies of software operation depended on more fundamental pedagogical orientations towards analysis of mathematical discrepancies as a means of supporting learning. Dynamic manipulation entered mathematical discourse when dragging was used to focus attention on continuous variation rather than being treated as an efficient means of generating multiple static figures. In summary, then, not only is there a considerable gap between the aspirations of advocates for the educational potential of dynamic geometry and actual patterns of use in mainstream teaching but patterns of use vary markedly within that mainstream practice.

These teaching practices observed in England did share some basic characteristics that the ICMI Study on Mathematics Education and Technology identified in projects aiming to implement use of digital technologies at a national scale (Sinclair et al. 2010). In particular—in line with the ICMI findings—dynamic geometry was treated more as providing pedagogical support than as provoking curriculum change, and patterns of classroom (inter)activity tended to be less open than the accompanying pedagogical rhetoric suggested. In England, these characteristics reflected a statutory curriculum privileging the paper-and-pencil medium and classical instruments of hand construction; references to dynamic geometry in curricular guidance that were sporadic, optional and superficial; and high-stakes external assessment that excluded any use of dynamic geometry.

Indeed, recent developments in England have exacerbated these conditioning factors with a new curriculum in which the main substantive guidance on use of digital tools is now simply that "teachers should use their judgement about when ICT tools should be used" (Department for Education 2013, p. 2), with one subsequent fleeting reference to the possibility of using such tools to "derive and illustrate properties of… plane figures… using appropriate language and technologies" (p. 8), against a default assumption that classical tools should be privileged, for example, for "standard ruler and compass constructions" (p. 8).

Consequently, recent graduate students that I have supervised have had great difficulty in tracking down examples of dynamic geometry use in English schools, with those found largely following the patterns exhibited in Cases N and P. Nevertheless, recently we have been able to describe one example of more developed practice in greater detail (Bozkurt and Ruthven 2016).

Under such circumstances, it is easier to envisage success for developments in teaching practice that treat dynamic geometry as a pedagogical aid in enriching the existing curriculum. In the English context, for example, such lines of development might encourage use of tasks which incorporate a richer system of related arithmetic patterns, and of figures which provide better visual support for directly geometric reasoning (for illustrations, see Ruthven 2005). However, a stronger aspiration for development would see dynamic geometry treated (for both curriculum and assessment) as a standard mathematical tool rather than as a marginal and occasional pedagogical aid. In the discussion after the lecture on which this chapter is based, there were suggestions that curriculum schemes (notably textbooks) in some educational systems (specifically California USA, Denmark and France) were now integrating use of dynamic geometry in a more systematic way. It would indeed be interesting to have analyses of such developments reported at future ICME conferences, examining the manner in which dynamic geometry is interpreted and its use developed, both in the curriculum materials themselves and in the teaching practice associated with their use.

In effect, the field of mathematics education is itself still 'constructing' dynamic geometry. With the many varieties of dynamic geometry software, convergence towards mature common standards remains limited (Mackrell 2011). However a major European project, Intergeo, has sought to facilitate exchange and use of dynamic geometry resources by teachers through creating an online repository with the potential to enrich the readily available range of tried and tested dynamic tasks (Kortenkamp and Laborde 2011). While these resources were developed by individual members of the community, the project sought to enhance their collective usability in three particular ways:

- by specifying a common file format based on open standards, so enabling teachers to employ their software of choice;
- by tagging each resource with metadata, so helping teachers to search for relevant resources;
- by developing a standard quality questionnaire to be completed by users of a resource, so providing prospective users with reviews.

Although such developments are of considerable value, they have also highlighted continuing obstacles. Following the introduction of the quality questionnaire, for example, the overwhelming majority of responses were made without the reviewer actually having trialled the resource in the classroom (Trgalova et al. 2011).

Indeed, it is arguable that the field of mathematics education has only scratched the surface in generating the knowledge needed to support a more thorough integration of dynamic geometry. Some areas where productive development could take place are clear. Arzarello et al. (2002) have identified a rich range of dragging techniques, each associated with particular patterns of geometrical reasoning. However, in the teaching practices observed in English schools, only a small subset of these techniques were used, largely tacitly: Dragging defining points of a figure to generate multiple examples; Dragging defining points of a figure to explore patterns of dynamic (co)variation. Development of this aspect would be supported if the field were to build a tighter mathematical theorisation of dragging, establish a more widely accepted register for discussing it, and create well-tested curricular sequences (with supporting didactical analyses) which make effective use of a range of dragging techniques and provide widely accepted norms for these. Developing a robust practical framework of this type would provide a firmer basis for developing more reflective use of a richer repertoire of dragging techniques by teachers and students.

Likewise, Laborde (2001) has identified an important gradation in curricular scenarios which make use of dynamic software, increasingly challenging for teachers to conceive and implement:

- Facilitates material aspects of familiar task: e.g. constructing a diagram showing a triangle and its perpendicular bisectors.
- Assists mathematical analysis of familiar task: e.g. through dragging vertices of a dynamic triangle to identify concurrence of the perpendicular bisectors as an invariant property.
- Substantively modifies a familiar task: e.g. through dragging vertices of a dynamic triangle to identify conditions under which its circumcentre is positioned internally or externally.
- Creates task which could not be posed without dynamic software: e.g. through constructing circles sharing a common free centre but each passing through a different vertex of the triangle, then dragging that free centre to identify positions where the circles become concurrent (see Fig. 29.4).

The challenge here is not just to devise dynamic geometry scenarios at those latter two levels which go beyond the classical curriculum. Rather it is to move beyond creating relatively isolated scenarios to establishing well-articulated sequences of scenarios which provide a sound basis for developing the mathematical and technical (i.e. instrumental) knowledge of students in a coordinated and progressive manner. (For example, one can imagine a curricular progression in which the task shown in Fig. 29.4 occurs earlier in a sequence leading to the tasks used in Case Q). More thoroughgoing integration of dynamic geometry into the curriculum as a mathematical tool depends on the development of progressive sequences of this type, and of a principled basis for their design (see, for example, Coutat et al. 2016).

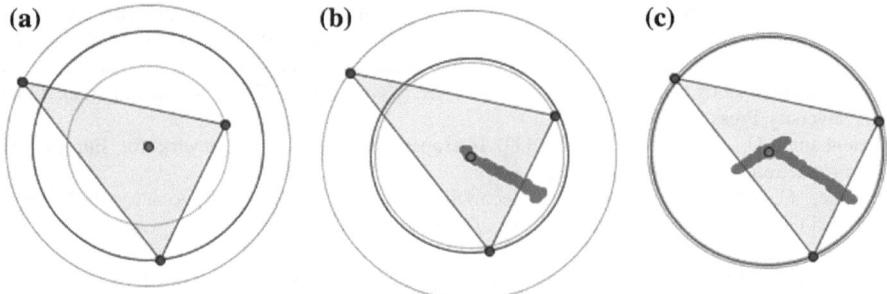

Fig. 29.4   a Construction of concentric circles through individual triangle vertices; **b** dragging of free centre to identify locus for concurrence of two circles; and then **c** further dragging to identify position for concurrence of all three circles (and thus centre of the circumcircle of the triangle)

Acknowledgements   Funding for the study discussed in this chapter came from the UK Economic and Social Research Council (R000239823). The research was carried out in collaboration with Sara Hennessy and Rosemary Deaney. Thanks are also due to the participating teachers. This chapter draws on and adapts material from previously published work (Ruthven et al. 2008; Ruthven 2012). Finally, thanks go to the ICME editor and reviewers for their helpful suggestions.

References

Arzarello, F., Olivero, F., Paola, D., & Robutti, O. (2002). A cognitive analysis of dragging practises in Cabri environments. *ZDM, 34*(3), 66–72.

Ball, D. L., & Cohen, D. K. (1996). Reform by the book: What is—Or might be—The role of curriculum materials in teacher learning and instructional reform. *Educational Researcher, 25* (9), 6–8 & 14.

Becker, H., Ravitz, J., & Wong, Y. (1999). *Teacher and teacher-directed student use of computers, teaching, learning, and computing: 1998 National Survey Report #3*. Centre for Research on Information Technology and Organizations, University of California Irvine.

Bozkurt, G., & Ruthven, K. (2016). Classroom-based professional expertise: A mathematics teacher's practice with technology. *Educational Studies in Mathematics, 94*(3), 309–328.

Chazan, D., & Yerushalmy, M. (1995). Charting a course for secondary geometry. In R. Lehrer & D. Chazan (Eds.), *New directions in the teaching and learning of geometry* (pp. 67–90). Hillsdale, NJ: Lawrence Erlbaum.

Coutat, S., Laborde, C., & Richard, P. R. (2016). L'apprentissage instrumenté de propriétés en géométrie: propédeutique à l'acquisition d'une compétence de démonstration. *Educational Studies in Mathematics, 93*(2), 195–221.

Department for Education and Employment [DfEE]. (2001). *Key stage 3 national strategy: Framework for teaching mathematics*. London: DfEE.

Department for Education [DfE]. (2013). *Mathematics programmes of study: Key stage 3*. London: DfE.

Goldenberg, E., Scher, D., & Feurzeig, N. (2008). What lies behind dynamic interactive geometry software? In G. Blume & M. Heid (Eds.), *Research on technology in the learning and teaching of mathematics: Cases and perspectives* (pp. 53–87). Greenwich, CT: Information Age.

Hoyles, C., Foxman, D., & Küchemann, D. (2001). *A comparative study of geometry curricula*. London: Institute of Education.

Kaiser, G. (2002). Educational philosophies and their influence on mathematics education—An ethnographic study in English and German mathematics classrooms. *ZDM, 34*(6), 241–257.

Kline, R., & Pinch, T. (1999). The social construction of technology. In D. MacKenzie & J. Wajcman (Eds.), *The social shaping of technology* (pp. 113–115). Buckingham: Open University Press.

Kortenkamp, U., & Laborde, C. (2011). Interoperable interactive geometry for Europe: An introduction. *ZDM, 43*(3), 321–323.

Laborde, C. (2001). Integration of technology in the design of geometry tasks with Cabri-Geometry. *International Journal of Computers for Mathematical Learning, 6*(3), 283–317.

Laborde, C., & Laborde, J.-M. (2008). The development of a dynamical geometry environment: Cabri-géomètre. In G. Blume & M. Heid (Eds.), *Research on technology in the learning and teaching of mathematics: Cases and perspectives* (pp. 31–52). Greenwich, CT: Information Age.

Mackrell, K. (2011). Design decisions in interactive geometry software. *ZDM, 43*(3), 373–387.

Remillard, J. (2005). Examining key concepts in research on teachers' use of mathematics curricula. *Review of Educational Research, 75*(2), 211–246.

Ruthven, K. (2005). Expanding current practice in using dynamic geometry to teach about angle properties. *Micromath, 21*(2), 26–30.

Ruthven, K. (2012). The didactical tetrahedron as a heuristic for analysing the incorporation of digital technologies into classroom practice in support of investigative approaches to teaching mathematics. *ZDM, 44*(2), 627–640.

Ruthven, K., Hennessy, S., & Deaney, R. (2008). Constructions of dynamic geometry: A study of the interpretative flexibility of educational software in classroom practice. *Computers and Education, 51*(1), 297–317.

Sinclair, N., Arzarello, F., Gaisman, M. T., Lozano, M. D., Dagiene, V., Behrooz, E., et al. (2010). Implementing digital technologies at a national scale. In C. Hoyles & J.-B. Lagrange (Eds.), *Mathematics education and technology-rethinking the terrain* (pp. 61–78). New York, NY: Springer.

Trgalova, J., Soury-Lavergne, S., & Jahn, A. P. (2011). Quality assessment process for dynamic geometry resources in Intergeo project. *ZDM, 43*(3), 337–351.

Chapter 30
Exploring the Contribution of Gestures to Mathematical Argumentation Processes from a Semiotic Perspective

Cristina Sabena

Abstract A multimodal perspective on mathematics thinking processes is addressed through the semiotic bundle lens and considering a wide notion of sign drawing from Vygotsky's works. Within this frame, the paper focuses on the role of gestures in their interaction with the other signs (speech, in particular) and investigates the support they can provide to mathematical argumentation processes. A case study in primary school in the context of strategic interaction games provides data to show that gestures can support students in developing argumentations that depart from empirical stances and shift to a hypothetical plane in which generality is addressed. In this regard, by combining synchronic and diachronic analysis of the semiotic bundle, specific features of gestures are pointed out and discussed: the semiotic contraction, the condensing character of gestures, and the use of gesture space in a metaphorical sense.

Keywords Argumentation · Gestures · Multimodality · Semiotic contraction
Semiotic bundle

30.1 Introduction

At the turn of the millennium, in 2000, the provocative essay *Where Mathematics Comes From* by George Lakoff and Rafael Núñez pointed out the crucial role of perceptual and bodily aspects on the formation of abstract concepts, including mathematical concepts (Lakoff and Núñez 2000). The new stance emphasized sensory and motor functions, as well as their importance for successful interaction with the environment. Criticizing the platonic idealism and the Cartesian mind–body dualism, Lakoff and Núñez advocated that all kinds of ideas, including the most sophisticated mathematical ideas, are founded on our bodily experiences and develop through cognitive metaphorical mechanisms.

C. Sabena (✉)
Universita' di Torino, Turin, Italy
e-mail: cristina.sabena@unito.it

© The Author(s) 2018
G. Kaiser et al. (eds.), *Invited Lectures from the 13th International Congress on Mathematical Education*, ICME-13 Monographs,
https://doi.org/10.1007/978-3-319-72170-5_30

The book aroused a great interest in mathematics education and prompted many research studies highlighting the role of bodily and kinesthetic experiences in mathematical learning (Arzarello and Robutti 2008; de Freitas and Sinclair 2014; Edwards 2009; Ferrara 2014; Nemirovsky 2003; Radford 2014; Roth 2009; for an overview, see Gerofsky 2015).

More recently, embodied stances seem to receive a certain confirmation by neuroscientific results on "mirror neurons" and "multimodal neurons," which are neurons firing when subjects performs actions, when they observe somebody else doing the same action, and when they imagine it (Gallese and Lakoff 2005). On the basis of these results, Gallese and Lakoff (2005) provide a new theoretical account on how the brain works, according to which "action and perception are integrated at the level of the sensory-motor system and not via higher association areas" (p. 459). In particular, such an integration would appear to be crucial not only for motor control, but also for *planning actions*, an activity typical of what is generally understood as "thinking."

The terms *multimodal* and *multimodality* come therefore to indicate a feature of human cognition opposed to "modularity." On the other hand, in the communication field the term *multimodal* is used with reference to multiple modalities that we have to communicate and express meanings to our interlocutors: words, sounds, images, and so on (Kress 2004). These communicative affordances have been acquiring increasing attention due the diffusion of new technological affordances, which are constantly developing new possibilities of interaction with them through our body.

In this paper, in line with Radford et al. (2009), multimodality refers to the importance and mutual co-existence of a variety of cognitive, material and perceptive modalities or resources in the mathematics teaching-learning processes, and more in general in the formation of mathematical meanings: "These resources or modalities include both oral and written symbolic communication as well as drawing, gesture, the manipulation of physical and electronic artifacts, and various kinds of bodily motion" (pp. 91–92). Including the embodied aspects in the analysis of mathematical thinking and learning brought to the fore the study of gestures as an important cognitive and communicative manifestation.

On the other hand, the attention to embodied and multimodal aspects needs to come to terms with the consideration of the social, historical, and cultural aspects in the genesis of mathematical concepts (Schiralli and Sinclair 2003; Radford et al. 2005). Mathematics is indeed "inseparable from the symbolic instruments" and the act of knowing is a "culturally shaped" phenomenon (Sfard and McClain 2002, p. 156) in which use of tools and signs play an important role.

This paper takes a semiotic stance to analyze gestures not as isolated variables, but rather as part of the multimodal resources at the students' disposal in order to bridge the gap between everyday experience and formal mathematics. The multimodal resources will be considered as signs entering in meaning-making processes, and will be analyzed through the *semiotic bundle* lens (Arzarello 2006). Previous research from this perspective has suggested that gestures can contribute not only to the semantic content of mathematical ideas but also to the logical structure that organizes them in mathematical arguments (Arzarello and Sabena 2014). In this line of research and adopting a case study methodology, the following research question will be addressed:

> What specific contribution can gestures, when they are considered as signs in semiotic bundles, provide to students' argumentation processes?

In the next sections, the theoretical framework for the research is presented: It is constituted by theoretical elements and results from gesture studies in psychology and by a semiotic perspective for multimodality grounded on Vygotsky's account of signs and on the semiotic bundle notion. Afterwards, selected data from a case study in primary school in the context of strategic interaction games will be analyzed and discussed according to an analytical generalization stance.

30.2 Gestures as Multimodal Resources

Gestures accompanying discourses are a widespread phenomenon (not only Italian!), as the pioneering work by Kendon has documented since the 80s (Kendon 1980). Since then, psychological and psycholinguistic studies have been stressing that speech and gestures are closely linked and that gesturing is relevant in communication and thinking processes (McNeill 1992, 2005; Goldin-Meadow 2003).

McNeill (1992) found speech and gestures to be closely linked in many respects: They are temporally synchronous in phonological (the central phase of the gesture coinciding with the peak of the phonological phrase), semantic (at the meaning level), and pragmatic (their function in the discourse) aspects. Also, in child development, gesture and speech proceed together. At the cognitive level, some scholars have identified their function being important in lightening the working memory, offering the possibility for cognitive resources to do their best to reorganize (Goldin-Meadow et al. 2001).

These cognitive interpretations provide elements that can explain, for example, why we gesticulate in telephone conversations (de Ruiter 1995), why when we are prevented gesturing our discourse becomes less fluid, or why even blind from birth use gestures while speaking. These phenomena cannot be explained only in terms of interpersonal communicative dimension, and so gestures are claimed to have a constitutive role also in thinking processes.

Vygotsky (1934/1986) had already stressed the constitutive role of language in thinking by saying that "thought is not merely expressed in words; it comes into existence with them" (p. 218); since then psychological studies on gestures have pushed in the direction of extending this constitutive role of language to the speech-gesture unity. Quoting McNeill (1992), we can say that "gestures do not just reflect thought but have an impact on thought. Gestures, together with language, *help constitute thought* [emphasis original]" (p. 242). It is within this Vygotskian hypothesis that I frame the role of gestures in mathematical activities.

Gesture studies have provided other tools of analysis, such as categories for their classification. McNeill (1992) has classified gestures as:

- *Iconic*: if they bear a relation or resemblance to the semantic content of discourse (for example, inclining two hands to indicate a roof);
- *Metaphoric*: similar to iconic gestures, but with the pictorial content presenting an abstract idea that has no physical form (a classical example is the hand in the act of holding an object, when referring to idea of "a certain topic" in the discourse);
- *Deictic*: if they indicate objects, events, or locations in the concrete world.
- *Beats*: if they contribute to stress some parts of the discourse.

Deictic gestures are usually performed with the extended forefinger (sometimes with hand-held objects, such as a pen) and are also called *pointings*. Apparently simple, pointing is indeed a complex act. Besides concrete pointings (such as indicating a book on the table), research has also identified *abstract pointings*, when the hand or fingers are extended in the space as to indicate something, but the space it actually empty. In McNeill's (1992) interpretation, "the speaker appears to be pointing at empty space, but in fact the space is not empty; it is full of conceptual significance. Such abstract deixis implies a metaphoric use of space in which concepts are given spatial forms" (p. 173). Such a classification is not based on the physical features of gesture, but by considering the relationships with contextual information: this entails that the interpretative process needs to take into account the broader context in which a gesture is performed. A second remark concerns the fact that the same gesture may belong to more than one category; therefore, the categories have to be considered dimensions along which a gesture can be featured, more than in a classificatory view.

Furthermore, gestures are sometimes characterized by repetition: distinct features of a gesture recur over the length of a discourse (although not necessarily in consecutive gestures), and the recurrence can be signaled by the form of the hand shape, its location, orientation, motion, rhythm, and so on (McNeill et al. 2001). This phenomenon is called *catchment* and may be related to discourse cohesion:

> By discovering the catchments created by a given speaker, we can see what this speaker is combining into larger discourse units – what meanings are being regarded as similar or related and grouped together, and what meanings are being put into different catchments or are being isolated, and thus are seen by the speaker as having distinct or less related meanings. (McNeill et al. 2001, p. 10)

Catchments may therefore be of great importance because they can give information on the underlying meanings in speech and dynamics. In a classroom setting, studying the catchments could provide clues on the evolution of meanings in students. In addition, catchments can contribute to the organization of an argument at a logical level, as discussed in Arzarello and Sabena (2014).

30.3 A Semiotic Approach to Multimodality

The choice of adopting a semiotic approach to study the role of multimodality and gestures in mathematical activities stems basically from two considerations. The first is epistemological, dealing with the assumption that mathematical objects are not directly perceivable using our senses and need by their nature to be mediated by signs, such as the graph of a function for the function concept. Indeed, signs and transformations between them are at the heart of the mathematical activities:

> The significance of semiosis for mathematics education lies in the use of signs; this use is ubiquitous in every branch of mathematics. It could not be otherwise: The objects of mathematics are ideal, general in nature, and to represent them—to others and to oneself— and to work with them, it is necessary to employ sign vehicles, which are not the mathematical objects themselves but stand for them in some way. (Presmeg et al. 2016, pp. 1–2)

The second consideration is psychological, concerning how meanings are formed and evolve. In Vygotsky's account of human cognitive development (or *cultural development*), signs play a crucial role (Vygotsky 1931/1978). By virtue of their social meaning, signs serve individuals as a way to exert voluntary control on their behavior, in a way similar to the way that road signs signal events to individuals to regulate their conduct. Analogous with tools in labor activities, signs work, on the individual psychological level, as "stimuli-means" standing for some characteristic or aspect of the socially shared experience and steering one's own mental processes:

> The invention and use of signs as auxiliary means of solving a given psychological problem (to remember, compare something, report, choose, and so on) is analogous to the invention of tools in one psychological respect. The signs act as instruments of psychological activity in a manner analogous to the role of a tool in labor. (Vygotsky 1931/1978, p. 52)

From this perspective, signs are considered in their functional role as psychological tools that allow the subjects to reflect and plan actions and act as cultural mediators (Radford and Sabena 2015). This is a very general idea of a sign that does not assign prescriptions on what can be a sign and which specific features it should have: A gesture can also be considered a sign, as Vygotsky himself highlighted in his famous example of the pointing gesture to illustrate the internalization process starting from the meaning assigned by the mother to the child's hand movement (Vygotsky 1931/1978).

In order to include gestures as well as other more classical registers, Arzarello developed the semiotic bundle construct (Arzarello 2006; Arzarello et al. 2009) as a system made of different signs (or semiotic resources) and their mutual relationships that are produced by students and possibly the teacher during mathematics activities: words (spoken or written), written diagrams, gestures, tools, and so on. Similarly to Radford's idea of "semiotic system" (Radford 2002), the semiotic bundle includes both the classical registers, with precise and codifiable rules of productions and transformation (Duval 2006), and the embodied ones, allowing us to provide a semiotic account of the multimodal processes occurring while learning and teaching mathematics. An example can be constituted by students' words, gestures, and drawn figures while solving a geometrical problem.

The semiotic bundle is characterized by two key features:

- A systemic character, revealed by a synchronic analysis of the relationships between the different kinds of signs at a certain moment (like a sort of "semiotic picture")
- A dynamic nature revealed by a diachronic analysis focusing on the evolutions of signs and of their transformations over time (a sort of "semiotic movie")

Synchronic and diachronic analysis—which are distinguished only for the sake of analysis—are performed by considering closely the video recordings from classroom activities students are engaged in, together with their multimodal transcripts (i.e., transcripts that include not only words but also record gestures and other kinds of signs). It is interesting to remark how the possibilities offered by new technologies for the study of the interaction between students and the teacher in the classroom gave a boost to considering multimodal aspects in mathematics learning. Although in the early 90s attention to the "classroom discourse" in the teaching learning of mathematics had already emerged, it was the use of video recordings that opened up the possibility to observing phenomena that had hitherto been unnoticed due to its undetectability. Gestures and other embodied resources have thus begun to be considered among the resources through which communication and conceptualization are realized. In other words, the strong push towards new theories that has come from methodological aspects has required new analytical tools, such as the semiotic bundle.

In the following, the semiotic bundle lens is applied to empirical data on mathematical argumentation processes carried out by primary students. The analysis will use a fine-grain focus on video-recording data with the aim of identifying and theorizing key gestural phenomena that play a role in such processes.

30.4   A Case Study: The Race to 20

The case study is based on a strategic interaction game called "Race to 20," which was used by Brousseau to illustrate the theory of didactical situations (Brousseau 1997). The case study is part of a design-based research project on the use of strategy games for fostering problem-solving and argumentation processes from primary to secondary school.

The game is played by two players who struggle to reach the number 20 by adding alternatively small numbers. Specifically, the first player chooses a number between 1 and 2, then the second player must add 1 or 2 to the previous number and say the result, then the first player adds 1 or 2, and so on. The player saying 20 wins the game. In game theory, it is a perfect information game with complete information, based on sequential decision-making. As the reader may know or can check, the winning strategy consists of starting as the first player and in following the number sequence 2-5-8-11-14-17-20. In this kind of game, the player needs to determine, for any move by the opponent player, the right move in order to win the game. These processes may be related to the logical scheme of coordinating a universal qualifier with an existential one, as it is the case in many mathematics theorems.

I will refer to data from a classroom discussion in Grade 4. The discussion follows some lessons in which the students played the game in pairs by writing the added numbers on the top of arrows from left to right and the results in a line from left to right (similar to Fig. 30.1).

This semiotic template was introduced by the teacher in order to allow the students to keep record of both the winning numbers and the added ones. This record was meant to support them in determining regularities that are at the base of the winning strategy.

The discussion is engineered right after the students have finished a classroom team tournament, in which representatives of each team have played the matches at the blackboard (the last match is shown in Fig. 30.1). The teacher initiates the discussion by making explicit the goal of providing a strategy to win the game and justifying it.

We will focus in detail on the specific contribution of some children—Giulio, Eliana, and Elisa—but first let us give some contextual information, in order to make the analysis understandable.

As the discussion starts, numbers 14 and 17 are soon identified as winning numbers: In some cases justifications are based on the possible moves of the two players, as Marta states:

> Marta You have to get first to 14 and then to 17. Because if you do 14 plus 1 and you get 15 and then you do plus 2 and you get to 17, then . . . you do plus 1 to arrive to 18 and the other does 2 and gets to 20. Whereas if from 14 you do plus 2, you arrive at 16 and the other one does plus 1 to arrive at 17, the other 19 if he does plus 2, you do plus 1, and you get 20. So however, from 14 to 17 you arrive anyway to 20.

In other cases, they rely on the empirical observation of what did actually happen during the tournament. Through backward induction, the number 11 begins to be identified as a winning number and related to 14 and 17:

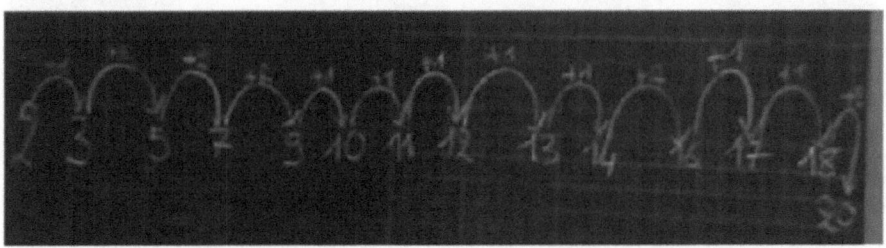

Fig. 30.1 Semiotic template introduced by the teacher to play the Race to 20

Diego: 11 maybe is an important number, because maybe my team adds 2 and it is 13, the other team adds 1 and arrives at 14, I add 1, 15, they add 2 and it is 17.

At this point, after about 20 min of classroom discussion, Giulio proposes a general rule to identify all winning numbers:

Giulio: I think that for the winning numbers you always remove 3: from 20 you remove 3 and you arrive at 17; from 17 you remove 3 and you arrive at 14, I think that another winning number could be 11, could be . . . 8, could be . . . 5, could be . . . 2.

Giulio's strategy identifies all winning numbers starting from the winning result, 20, and moving backwards through repeated subtractions in a process of backward induction, as described in game theory.

The strategy is expressed verbally in general terms, without simulations of moves, and it is accompanied by several gestures. Table 30.1 reports the transcript of the initial part of Giulio's utterance, enriched with the gesture component and the Italian original words. The underlined words indicate that they are co-timed with the shown gesture and the same convention will be used in the following tables.

When saying "you always remove 3," Giulio moves his hand from right to left (from the child's perspective): This movement can be interpreted as indicating subtraction, with reference to the number line, which is often used in the Italian curriculum. With this interpretation, the gesture can be classified as a metaphoric gesture indicating subtraction. I remark that in the original Italian version, Giulio uses the term *togli*, which is used both in everyday contexts to say "remove, take away" and in mathematical context in primary school to indicate "subtract."

Table 30.1 Multimodal transcript of the first part of Giulio's strategy

I thik that for the winning numbers you always remove 3	From 20	you remove 3	and you arrive at 17
Secondo me dato che i numeri vincenti si toglie sempre 3	*Da 20*	*togli 3*	*e arrivi a 17*
Open hand moving from right to left	*Three extended fingers*	*Three extended moving from right to left*	*Abstract pointings downwards*

The hand movement from right to left is repeated when the rule is applied in order to identify which are the winning numbers. From Table 30.1 (Pictures b and c) it can be noticed that when Giulio is saying "from 20 you remove 3," his hand is moving leftwards. Now, three fingers are pointed out, and overall we find two metaphorical references condensed in a single gesture:

- right-to left movement → subtraction
- three fingers → number 3

When completing the sentence and uttering the winning numbers, Giulio is performing abstract pointing gestures downwards that are co-timed with the uttered numbers (in Table 30.1, the case of 17 is reported in the last column).

If we consider the whole sequence, we see that the subtraction of 3 is repeated in order to obtain 14; this repetition is also expressed by the repetition of the same gesture configuration of the three fingers extended (pictures are not reported for reasons of space). The same metaphorical gesture is hence repeated in a catchment expressing that number 3 is *always* subtracted, in order to get *all* the winning numbers.

Afterwards ("I think that another . . ."), we can notice a type of *semiotic contraction* occurring within the semiotic bundle: Speech reduces, mentioning only the winning numbers, and shifts to a hypothetical level ("could be . . ."); also, gestures appear to reduce in their movements, ending up with quick abstract pointing left and downwards co-timed with the utterance of the winning numbers (8, 5, and 2).

At this point, the teacher asks Giulio to explain his idea. Here it is the verbal transcription of Giulio's argument:

Teacher: Explain well this idea.
Giulio: Because . . . that is I don't know, if I arrive at 2 . . . I don't know, I begin, I make 1, no I make 2, he arrives and makes 1, I put 2 and I arrived at 5, which I think is another winning number . . . yes, arrived at 5 . . . it is a winning number, I think. Then . . . he adds 2, say, I add 1 and I arrived at 8, which is another winning number. She adds 1, I add 2 and I arrive at . . . 11, which is a winning number. He adds 2, I add 1, and I arrive at 14, which is another winning number, he adds 1 I add 2, we arrive at 17 which is a winning number, he adds 1 or 2, I add 1 or 2, and I win.

The subtraction turns now into an onward movement that starts from the very first move (number 2) of an imagined match between himself and another player. This movement is produced by means of a repetition of the same linguistic structure: "He adds . . . I add . . . and I arrive at . . . , which is a winning number." This repetition is not just a mere repetition of words, but is performed with a *rhythmical structure* in sound, which is preserved along the entire sentence and contributes to convey the general character of the found rule (similar to what was discussed in algebraic context in Radford et al. 2007).

Table 30.2 Multimodal transcript of the first part of Giulio's argument

I begin . . . , I make 2	he arrives and puts 1	I put 2	and I arrived at 5
Io inizio . . . , faccio 2	*lui arriva lì e mette 1*	*Io metto 2*	*e sono arrivato a 5*
Two fingers pointed out	*Hand held open upwards as containing something*	*Two fingers pointed out*	*Hand held open upwards as containing something*

Gestures are constantly present, starting from the first simulated move to the last (winning) one. Table 30.2 reports some pictures of gestures accompanying the very first part of Giulio's sentence.

While uttering the moves of an imagined match, Giulio is performing again two kinds of metaphoric gestures. When indicating his own moves, the gesture indicates the uttered numbers by pointing out the correspondent number of fingers, i.e., two fingers when saying 2 (Pictures a and c in Table 30.2). When referring to the other players' moves or to the obtained result, the hand is held open upwards as containing something (Pictures b and d in Table 30.2): In this case the metaphoric reference is made to underline a certain kind of generality of the uttered numbers.

At a certain point, when mentioning an opponent move, Giulio uses a linguistic expression that in English may be translated as "say" or "for example" (in Italian it is *tipo*) and that can be interpreted as expressing the germs of the concept of "any number." When uttering "say," the student performs a gesture consisting of an open hand quickly turned around (Table 30.3, Picture a).

The gesture is again metaphoric and the semiotic bundle of words and gestures underlines that the number 2, chosen to indicate the opponent's move, is to be considered as one possibility among others (all the possible moves): It is one generic move. This is indeed a very delicate logical relationship to manage: the articulation between a universal qualifier (for any move from my opponent player) with an existential one (the move that I choose after him). We see that the gesture-speech combination allows the students to successfully manage it.

From this moment on, when uttering the imagined moves by the two players, abstract pointing gestures are enacted with left-right *spatial alternation*, which indicates visually the alternation between the two players in the game (Table 30.3, Pictures b and c). This spatial alternation can be interpreted as helping the student to keep control of the argument at the local level, that is to say, to control the choice of

Table 30.3 Multimodal genericity conveyed within the semiotic bundle

Then . . . he adds 2, say	. . . She adds 1,	I add 2
Poi...lui aggiunge 2, tipo,	*... lei aggiunge 1,*	*Io aggiungo 2*
Open hand quickly turned around	*Abstract pointing at the left*	*Abstract pointing at the right*

the moves and counter-moves in the imagined sequence. The gesture spatial alternation is repeated several times (catchment) for any couples of moves and counter-moves, realizing the same rhythm of the accompanying words. As indicated by McNeill (2005), gesture catchments provide the discourse with cohesion. In this case, it contributes in structuring the entire argument at a global level.

The written template through which the game has been played has possibly helped Giulio in developing his strategy, working as an interiorized tool. As a matter of fact, when saying the general rule, he is looking at the blackboard (Table 30.1, Picture a), where the record of last match is still written (see Fig. 30.1). We remark that this match had not been played according to Giulio's strategy and that after this initial moment, we do not find any explicit reference to performed matches, for instance, with pointing gestures to the blackboard or to his notebook: This could be another index of the general level reached by Giulio in his argument.

The discussion focuses then on Giulio's strategy. Some students immediately agree with Giulio and produce their own argumentations, such as Eliana:

Eliana: I agree with Giulio because practically any time you have to reach a lucky number you must add 3, because first you add 1 and then you add 2 or first you add 2 and then add 1.

Eliana makes explicit that winning or "lucky" numbers can be reached by *adding* 3 (while Giulio mentioned subtraction by 3) and produces an argument for this by referring to the two numbers 1 and 2 that can be played in the game. When she says "you must add 3," she accompanies her speech with a gesture with the right hand turning from left to right (Table 30.4, Picture a), which may be referring to the addition on the number line.

Eliana explains also where this number 3 comes from, i.e., the combination of the possible subsequent moves of the two players. When uttering the numbers added to compose 3, she moves first the left hand from left to right (Picture b in Table 30.4), then the right hand with the same movement (Picture c in Table 30.4);

Table 30.4 Eliana's argument on the strategy of adding 3

you must add 3	because first_you add 1	and then you add 2	or first you add 2	and then add 1
devi aggiungere 3	*perché prima aggiungi 1*	*e poi aggiungi 2*	*o prima aggiungi 2*	*e poi aggiungi 1*
Right hand turning from left to right	*Left hand moving from left to right*	*Right hand moving from left to right*	*Left hand moving from left to right*	*Right hand moving from left to right*

she mentions the two possible combinations (1 + 2 or 2 + 1) and repeats the gestural combination in a catchment (Pictures c and d in Table 30.4).

As in the case of Giulio, we can notice the spatial alternation left-right as a metaphoric reference for the alternation of the two players, and again we can notice a catchment. But differently from the case of Giulio, now the fact that there are two alternating players is expressed by Eliana *only* through her gesture, because in her speech she uses always the pronoun "you," possibly in impersonal sense.

Again, we may identify two metaphoric components condensed in a single gesture:

- left-to-right movement → addition
- spatial alternation → players alternation

Right after Eliana, Elisa intervenes:

Elisa: So overall . . . if you play . . . you add 3 every time, and so if you can arrive at the numbers that there are 3 [*pointing out index and thumb; Picture a in* Table 30.5], that is if . . . Overall it is 3 [*shifting the pointed fingers from left to right; Picture b in* Table 30.5], because if you add 1 [*placing the pointed fingers at her left; Picture c in* Table 30.5] and the other adds 2 [*placing the pointed fingers at her right, Picture d in* Table 30.5], if you add 2 and the other adds 1 [*repeating the sequence with pointed fingers at her left and then right as in Pictures c and d in* Table 30.5], overall it is 3 [*shifting again the pointed fingers as in Picture b in* Table 30.5] and so you must be able to pick the numbers that are . . .

Teacher: At a distance . . .

Elisa: . . .of 3.

Elisa accompanies her speech with a gesture performed with two fingers pointed as if they were holding a little stick. This gesture is performed for the first time

Table 30.5 Elisa's condensing gesture

and so if you can arrive at the numbers so that there are 3	that is if . . . overall it is 3	because if you add 1	and the other adds 2
e quindi se tu riesci ad arrivare ai numeri in cui...in cui ci sono 3	*cioè se...in tutto fa 3*	*perché se tu aggiungi 1*	*e l'altro aggiunge 2*
	Holding-stick gesture shifted from left to right	*Holding-stick gesture placed at left*	*Holding-stick gesture placed at right*
Gesture performed with two pointed fingers as holding a little stick			

when she says "there are 3" (Picture a in Table 30.5) and is kept until the end of the sentence. It indicates metaphorically a fixed distance, similar to what has been described in an early calculus context in previous studies (Arzarello et al. 2009; Sabena 2007, 2008). The word *distance* is never uttered (it will be uttered immediately later, after a prompt from the teacher): The gesture is complementing her words and providing further meaning to her multimodal discourse.

When saying "overall it is 3," the holding-stick gesture is shifted from left to right (Picture b in Table 30.5): The left-to-right movement indicates that the fixed distance (of 3) allows one to pass from a winning number to the following one in the sequence. The girl is keeping her eyes towards the blackboard, where the last match played is still written, according to the template chosen by the teacher (Fig. 30.1). In this template, the subsequent moves are written one after the other, in a horizontal way. The horizontal movement of Elisa's gesture may be interpreted against this background, suggesting that the semiotic choice of the teacher has been useful for developing the students' argument. At the same time, the gesture may be referring metaphorically to an addition on the interiorized number line. We find another example of a gesture condensing different meanings through its metaphoric references and its dynamism. Through the condensing character of the gestures and in synergy with speech, the different meanings come to be connected to build an important part of the argument.

Through a gesture repetition or catchment, the condensing gesture is then combined with spatial alternation referring again to the two players (Pictures c and d in Table 30.5): The catchment provides support in shifting from a relationship between moves to a relationship between numbers that can justify the moves and indicates that such a distance does not vary over the entire winning sequence in the game.

30.5 Discussion

This paper adopted a multimodal perspective on mathematics teaching and learning processes and chose a semiotic tool in order to address it: the semiotic bundle, with its wide notion of sign drawing from Vygotsky's works, and its systemic and dynamic features. It focused in particular on the role of gestures in interaction with the other semiotic resources used in the classroom—speech first of all, but also written signs—and addressed primary students' mathematical argumentation processes in the context of strategic interaction games.

Through a case study and qualitative-interpretative analysis, it has been shown that gestures may contribute to carrying out argumentations that depart from empirical stances and shift to a hypothetical plane in which generality is addressed. From this case study, we can get also some insights on *how* gestures can do this. In particular, specific features have been identified: semiotic contraction, the condensing character of gestures, and the use of gesture space in a metaphorical sense combined with catchments. They will be briefly discussed, referring to the data analysis reported above.

When Giulio identifies and/or expresses (we do not have sufficient data to determine) the general rule of "always removing 3," we see his sentences becoming shorter and shorter and at the end just expressing the winning numbers, accompanied by abstract pointing gestures (Table 30.1, Fig. d). This is a type of semiotic contraction that has also been found in other contexts and at different ages, such as pattern generalization and function graphs (Sabena et al. 2005; Sabena 2007). From an epistemological point of view, semiotic contraction characterizes modern mathematical symbolism, and from a cognitive point of view it is a precious mechanism. Radford (2008) relates contraction to focusing attention to the elements that are relevant for a certain situation and to a deeper level of consciousness: "Contraction is the mechanism for reducing attention to those aspects that appear to be relevant. This is why, in general, contraction and objectification entail forgetting. We need to forget to be able to focus" (p. 94).

Semiotic contraction can be found also in what Vygotsky (1934/1986) calls "inner speech," which is described discussing language as a paradigmatic signs system. Inner speech is described at a structural level by syntactic reduction and phasic reduction and at semantic level by agglutination. *Syntactic reduction* is a specific form of abbreviation that curtails the subjects of sentences and leaves pure predication. Syntactic articulation results are therefore minimized to the pure juxtaposition of predicates. *Phasic reduction* consists of minimizing the phonetic aspects of speech, namely curtailing the words themselves (for example, writing "u" instead of "you"). *Agglutination* consists in combining words, gluing different meanings (concepts) into one expression (for example, "highway," formed by "high" and "way"). Nowadays, instant messaging communication systems on our smartphones extensively exploits semiotic contractions (combined with additional iconic features, such as the emoticons), typically in informal communication by people sharing most of the contextual information.

With syntax being reduced, Vygotsky (1934/1986) claims that semantics undertakes a contrary movement, with meaning coming to the fore: "With syntax and sound reduced to a minimum, meaning is more than ever in the forefront. Inner speech works with semantics, not phonetics" (p. 244).

We may observe that gestures, because of their spatial and kinesthetic nature, do not need processes of agglutination to combine meanings, as some languages do: It is their enactment itself that may produce the same result of combining meanings as agglutination does. A specific form of semiotic contraction characterizing gestures is in fact what I call *blending* or *condensing gestures*, which are gestures expressing (at least) two different meanings. We have seen two examples above:

- Giulio, with the right-to-left gesture with three fingers pointed out, indicates both the number 3 and subtraction (Table 30.1, Pictures b and c).
- Elisa, with two fingers pointed as if she were holding a little stick, shifted from left to right, which may interpreted as indicating both a fixed distance and the fact that this distance allows one to pass from one winning number to the next one (Table 30.5, Pictures b and c). The co-timed speech specifies that this distance is 3, obtained as the sum of 1 and 2.

In both examples, gestures condensed or blended two meanings by combining a dynamic component with the hand shape: This dynamic feature has been observed also by Calbris (2011) in what she calls "polysign gestures." In previous studies in the mathematical domain (Sabena 2007, 2008, 2010), the condensing or blending character of gestures has been identified in functions and graphs contexts and associated with iconic features of gestures. In the reported study, this feature is shown in an arithmetic domain and associated with the metaphoric feature of gestures, as McNeill (1992) classified them. Condensing two different meanings, each of these gestures establishes two different kinds of metaphorical references, one of which calls into play the number line, a didactical tool suggested in the Italian curriculum. By exploiting space in order to reason about numbers, the number line itself has a metaphorical nature. A double or even multiple blending process seems therefore to be activated by some metaphorical gestures typical of the mathematical domain. This theme requires deepening the reflection of what "metaphorizing" means at a cognitive and at a semiotic level, and further research is needed (for preliminary results using cognitive metaphors and blended spaces, see Sabena et al. 2016).

Metaphoricity appears to be related also to the use of gesture space with spatial alternation, as we have seen in Giulio and in Eliana. Giulio moves his hand left and right when mentioning the moves and countermoves in his imagined match (Table 30.3, Pictures b and c), while Eliana alternates her left and right hand for the same purpose (Table 30.4, Pictures b and c and Pictures d and e). Through gesture spatial alternation, empty spaces acquire meanings, which in the data appear related to a sort of local level, either in the game (imagined subsequent moves, in the case of Giulio), or in the mathematical argument (numbers to add to compose 3, in the case of Eliana).

As seen in the analysis, such a spatial alternation is repeated many times, realizing what in gesture studies is called a catchment and is interpreted as providing the discourse with cohesion (McNeill 2005). In this case study, gesture catchment is interpreted as supporting the students in structuring the entire argument at a global level. Previous results about how opposite spatial locations are exploited gesturally to indicate mutually excluding cases seem to confirm this interpretation (Arzarello and Sabena 2014).

For the sake of analysis, the different gestural features contributing to providing general meaning and structure to the argumentation process have been discussed here one after the other. However, as can be observed going back to the data analysis, many of these features intertwine; furthermore, the analysis of gestures needs to take into account the entire semiotic bundle. For example, only a systemic analysis of words and gestures can show how, even if it is describing a certain hypothetical match between himself and another player, Giulio's argument contains essential aspects conveying generality: the rhythmical repetition of the same linguistic structure, accompanied with a corresponding catchment (Tables 30.2 and 30.3); the use of generic words accompanied by a generic gesture (Table 30.3, Picture a); and the use of abstract pointings while uttering the possible moves (Table 30.3, Pictures b and c). In the case of Eliana's argument, it is striking to observe how gestures and words complete each other in a synchronic way.

If we analyze the children's contribution in a diachronic way, further observations may be drawn that provide elements to describe the classroom discussion evolution in a multimodal perspective. To give an example, it is interesting to see how Giulio's spatial alternation with his right hand evolves in Eliana's alternation of the two hands one after the other (see Table 30.4); in this latter case, the subject in her speech is not changing (it is always "you"), showing a tension towards the arithmetical relationships rather than on the strategic game interaction. This paves the way to the following Elisa's intervention about the "numbers so that there are 3."

A final consideration is reserved for the didactical implications of such a fine-grained analysis. In this paper, little attention has been devoted to the didactical variables of the situation. Of course, the teacher's choices are never neutral with respect to the use of any semiotic resource in the classroom, gestures included. An example in the data is the semiotic template through which the Race to 20 has been presented to the students and through which they play the game (Fig. 30.1). We have seen that this choice—which resonates with the didactic tool of the number line—has provided an essential tool for the students not only to play the game, but also for developing argumentations about how to win it. In particular, Elisa's multimodal argument about the "distance of 3" between winning numbers shows a relation to the semiotic written template through which the game was played.

The results of this analysis appear therefore to offer elements for validating the choice of the teacher. It is beyond the scope of the analysis, however, to discuss why for Elisa (and for some students) it did work, whereas for others, further reflection was needed. Classroom discussion appears indeed to be a suitable means for allowing the development of multimodal argumentations such as the one described, in which the students may exploit gestures as semiotic resources. This requires, of course, that gestures are considered legitimate in the classroom (as it happens in the analyzed case: The teacher supports Elisa in her multimodal argument by considering the contribution of her gesture and offering her the missing word). Even more, the teacher can contribute to classroom mathematical activity through her gestures in order to make the mathematical discourse evolve towards culturally established mathematics forms (see the "semiotic game" in Arzarello et al. 2009). Ongoing research indicates that the teacher can have an important role in the evolution of signs within the semiotic bundles and in building "multimodal semiotic chains" that make mathematical meaning progress through argumentation processes (Maffia and Sabena 2015, 2016). Further extensive research is still needed in order to unveil and exploit fully the potentiality of gestures as didactical means in the classroom.

References

Arzarello, F. (2006). Semiosis as a multimodal process. In L. Radford & B. D'Amore (Guest Eds.), *Revista Latinoamericana de Investigación en Matemática Educativa, Special issue on semiotics, culture, and mathematical thinking* (pp. 267–299).

Arzarello, F., Paola, D. Robutti, O., & Sabena, C. (2009). Gestures as semiotic resources in the mathematics classroom. *Educational Studies in Mathematics, 70*(2), 97–109.

Arzarello, F., & Robutti, O. (2008). Framing the embodied mind approach within a multimodal paradigm. In L. English, M. Bartolini Bussi, G. Jones, R. Lesh, & D. Tirosh (Eds.), *Handbook of international research in mathematics education* (2nd ed., pp. 720–749). Mahwah, NJ: Erlbaum.

Arzarello, F., & Sabena, C. (2014). Analytic-structural functions of gestures in mathematical argumentation processes. In L. D. Edwards, F. Ferrara, & D. Moore-Russo (Eds.), *Emerging perspectives on gesture and embodiment* (pp. 75–103). Charlotte, NC: Information Age Publishing, Inc.

Brousseau, G. (1997). *Theory of didactical situations in mathematics*. Dordrecht: Kluwer.

Calbris, G. (2011). *Elements of meaning in gesture*. Amsterdam: John Benjamins Publishing Company.

de Freitas, E., & Sinclair, N. (2014). *Mathematics and the body*. Cambridge, UK: Cambridge University Press.

de Ruiter, J. P. (1995). Why do people gesture at the telephone? In M. Biemans & M. Woutersen (Eds.), *Proceedings of the center for language studies opening academic year '95–96* (pp. 49–56). Nijmegen: Center for Language Studies.

Duval, R. (2006). A cognitive analysis of problems of comprehension in a learning of mathematics. *Educational Studies in Mathematics, 61*, 103–131.

Edwards, L. D. (2009). Gestures and conceptual integration in mathematical talk. *Educational Studies in Mathematics, 70*(2), 127–141.

Ferrara, F. (2014). How multimodality works in mathematical activity: Young children graphing motion. *International Journal of Science and Mathematics Education, 12*(4), 917–939.

Gallese, V., & Lakoff, G. (2005). The brain's concepts: The role of the sensory-motor system in conceptual knowledge. *Cognitive Neuropsychology, 22*, 455–479.

Gerofsky, S. (2015). Approaches to embodied learning in mathematics. In L. D. English & D. Kirshner (Eds.), *Handbook of international research in mathematics education* (3rd ed.). New York: Routledge.

Goldin-Meadow, S. (2003). *Hearing gesture. How our hands help us think*. Cambridge, Massachusetts, and London, England: The Belknap Press of Harvard University Press.

Goldin-Meadow, S., Nusbaum, H., Kelly, S. D., & Wagner, S. (2001). Explaining math: Gesturing lightens the load. *Psychological Science, 12*, 516–522.

Kendon, A. (1980). Gesticulation and speech: Two aspects of the process of utterance. In M. R. Key (Ed.), *The relation between verbal and nonverbal communication* (pp. 207–227). The Hague: Mouton.

Kress, G. (2004). Reading images: Multimodality, representation and new media. *Information Design Journal, 12*(2), 110–119.

Lakoff, G., & Nùñez, R. (2000). *Where mathematics comes from: How the embodied mind brings mathematics into being*. New York: Basic Books.

Maffia, A., & Sabena, C. (2015). Networking of theories as resource for classroom activities analysis: The emergence of multimodal semiotic chains. In C. Sabena & B. Di Paola (Eds.), *Teaching and learning mathematics: Resources and obstacles, Proceedings of the CIEAEM 67, Quaderni di ricerca didattica*, 25-2 (pp. 405–417). Aosta, July 20–24, 2015.

Maffia, A., & Sabena C. (2016). Teacher gestures as pivot signs in semiotic chains. In C. Csikos, A. Rausch, & J. Szitànyi (Eds.), *Proceedings of 40th Conference of the International Group for the Psychology of Mathematics Education* (Vol. 3, pp. 235–242). Szeged, Hungary: PME.

McNeill, D. (1992). *Hand and mind: What gestures reveal about thought*. Chicago: University of Chicago Press.

McNeill, D. (2005). *Gesture and thought*. Chicago: University of Chicago Press.

McNeill, D., Quek, F., McCullough, K.-E., Duncan, S., Furuyama, N., Bryll, R., ... Ansari, R. (2001). Catchments, prosody and discourse. *Gesture, 1*(1), 9–33.

Nemirovsky, R. (2003). Three conjectures concerning the relationship between body activity and understanding mathematics. In N. A. Pateman, B. J. Dougherty, & J. T. Zillox (Eds.), *Proceedings of the 27th Conference of the International Group for the Psychology of Mathematics Education* (Vol. 1, pp. 105–109). Honolulu, HI: PME.

Presmeg, N., Radford, L., Roth, W.-M., & Kadunz, G. (2016). *Semiotics in mathematics education*. ICME-13 Topical Surveys. https://doi.org/10.1007/978-3-319-31370-2_1.

Radford, L. (2002). The seen, the spoken and the written. A semiotic approach to the problem of objectification of mathematical knowledge. *For the Learning of Mathematics, 22*(2), 14–23.

Radford, L. (2008). Iconicity and contraction: A semiotic investigation of forms of algebraic generalizations of patterns in different contexts. *ZDM, 40*(1), 83–96.

Radford, L. (2014). Towards an embodied, cultural, and material conception of mathematics cognition. *ZDM—The International Journal on Mathematics Education, 46*, 349–361.

Radford, L., Bardini, C., & Sabena, C. (2007). Perceiving the general: The semiotic symphony of students' algebraic activities. *Journal for Research in Mathematics Education, 38*(5), 507–530.

Radford, L., Bardini, C., Sabena, C., Diallo, P., & Simbagoye, A. (2005). On embodiment, artifacts, and signs: A semiotic-cultural perspective on mathematical thinking. In H. L. Chick & J. L. Vincent (Eds.), *Proceedings of the 29th Conference of the International Group for the Psychology of Mathematics Education* (Vol. 4, pp. 113–120). Melbourne: University of Melbourne, PME.

Radford, L., Edwards, L., & Arzarello, F. (2009). Beyond words. *Educational Studies in Mathematics, 70*(3), 91–95.

Radford, L., & Sabena, C. (2015). The question of method in a Vygotskian semiotic approach. In A. Bikner-Ahsbahs, C. Knipping, & N. Presmeg (Eds.), *Approaches to qualitative research in mathematics education* (pp. 157–182). New York: Springer.

Roth, W. M. (Ed.). (2009). *Mathematical representation at the interface of body and culture.* Charlotte, NC: Information Age Publishing.

Sabena, C. (2007). *Body and signs: A multimodal semiotic approach to teaching-learning processes in early Calculus* (Ph.D. dissertation). University of Torino, Italy.

Sabena, C. (2008). On the semiotics of gestures. In L. Radford, G. Schumbring, & F. Seeger (Eds.), *Semiotics in mathematics education: Epistemology, history, classroom, and culture* (pp. 19–38). Rotterdam, The Netherlands: Sense Publishers.

Sabena, C. (2010). Are we talking about graphs or tracks? Potentials and limits of 'blending signs'. In M. M. F. Pinto & T. F. Kawasaki (Eds.), *Proceedings of the 34th Conference of the International Group for the Psychology of Mathematics Education* (Vol. 4, pp. 105–112). Belo Horizonte, Brazil: PME.

Sabena, C., Krause, C., & Maffia, A. (2016). L'analisi semiotica in ottica multimodale: dalla costruzione di un quadro teorico al networking con altre teorie. *Relazione al XXXIII Seminario Nazionale di ricerca in didattica della matematica Giovanni Prodi*, Rimini 28–30 Gennaio 2016. http://www.airdm.org/sem_naz_2016_25.html.

Sabena, C., Radford, L., & Bardini, C. (2005). Synchronizing gestures, words and actions in pattern generalizations. In *Proceedings of the 29th Conference of the International Group for the Psychology of Mathematics Education* (Vol. 4, pp. 129–136). Melbourne, Australia: University of Melbourne, PME.

Schiralli, M., & Sinclair, N. (2003). A constructive response to 'Where mathematics comes from'. *Educational Studies in Mathematics, 52*(1), 79–91.

Sfard, A., & McClain, K. (2002). Analyzing tools: Perspectives on the role of designed artifacts in mathematics learning. *Journal of the Learning Sciences, 11*(2&3), 153–161.

Vygotsky, L. S. (1931/1978). *Mind in society. The development of higher psychological processes.* In M. Cole, V. John-Steiner, S. Scribner, & E. Souberman (Eds.). Cambridge, MA: Harvard University Press.

Vygotsky, L. S. (1934/1986). *Thought and language* (Revised edition, A. Kozulin, Ed., Trans.). Cambridge, MA: MIT Press. (Original work published in 1934).

Chapter 31
Improving Mathematics Pedagogy Through Student/Teacher Valuing: Lessons from Five Continents

Wee Tiong Seah

Abstract This chapter focuses on the construct of values/valuing, using the findings of the large-scale, 'What I Find Important (in mathematics learning)' [WIFI] study to explore how values/valuing promotes effective (mathematics) pedagogy. The analysis of some 16,000 questionnaires collected from 19 economies reveals the absence of any relationship between values and specific actions, suggesting that the actions that reflect what are being valued are culturally-dependent. Students in economies which perform well in the PISA assessments were also found to value connections, understanding, communication, and recall in their mathematics learning, whereas their peers at the other end of the league table appeared to value relevance and practice more. The notion of intrinsic and extrinsic valuing will be discussed. In acknowledging the presence of value differences and conflicts that arise from inter-personal interactions in mathematics lessons, teachers' capacity to engage with values alignment is highlighted.

Keywords Values · Intrinsic/extrinsic valuing · Values alignment
Volition · Conation

31.1 Introduction

Hattie's (2015) more than 1200 meta-analyses of some 65,000 studies involving about 250 million students has identified factors associated with students' academic success at school. Amongst the 195 factors, 89 of these displayed effect sizes of 0.4 or more, which Hattie considered to be the hinge point above which the factors are worth employing to advance student learning. Six key findings were summarized from these 89 interventions that mattered, suggesting that the key to effective teaching lies in the valuing of just a couple of main ideas. One of these key findings

W. T. Seah (✉)
Melbourne Graduate School of Education, The University of Melbourne,
Melbourne, Australia
e-mail: wt.seah@unimelb.edu.au

G. Kaiser et al. (eds.), *Invited Lectures from the 13th International Congress on Mathematical Education*, ICME-13 Monographs,
https://doi.org/10.1007/978-3-319-72170-5_31

refers to heightened impact on student learning "when teachers base their teaching on students' prior learning" (Hattie 2015, p. 81), emphasizing the valuing of *prior learning*. Another key finding identifies teachers setting "appropriate levels of challenge" (p. 81), suggesting the importance and valuing of *challenge*. The message here is that effective teaching practices can be described in ways which are generic, focusing just on the essence that is of value. Thus, for example, for the intervention 'appropriate levels of challenge', *challenge* is being valued and is the focus; what constitute appropriate levels and indeed what they look like seem to be flexible and able to be defined in context.

This focus on values and valuing to account for effective or successful learning has also been evidenced in individual studies, some of which would have been analyzed by John Hattie in his meta-analysis exercise. For instance, a Nuffield Foundation-commissioned review asserted that

> high attainment may be much more closely linked to cultural values than to specific mathematics teaching practices. This may be a bitter pill for those of us in mathematics education who like to think that how the subject is taught is the key to high attainment. But study after study shows that countries ranked highly on international studies – Finland, Flemish Belgium, Singapore, Korea – do not have particularly innovative teaching approaches. (Askew et al. 2010, p. 12)

This chapter focuses on these culturally-situated values, using the findings of a large-scale research study to explore the sorts of values/valuing that are associated with effective (mathematics) pedagogy. It will begin with a review of research that had been conducted on values and valuing in the context of mathematics education. This review would highlight the process of valuing as involving both cognition and affect, how its evaluation is complicated by the fact that values are invisible, implicit, and not always activated, and how it provides one with the want to embrace it. This will be achieved through reflecting on the findings of the large-scale, 'What I Find Important (in mathematics learning)' Study. The discussion will be presented next, emphasizing that constituent actions of values are culturally-dependent rather than absolute. The notion of intrinsic and extrinsic valuing will be discussed, with a tentative proposal of how these might be related to mathematical performance. The absence of correlation between values and constituent actions will be discussed. Lastly, the inevitable and prevalent instances of values alignment will be highlighted in the context of the data analysed.

31.2   The Nature of Values and Valuing in Mathematics Education

Although the concept of values and valuing in school education is not new (e.g. moral education programs), the acknowledgement of its role in the teaching and learning of individual school subjects is a relatively recent research activity. Values in mathematics and in mathematics education were first proposed by Bishop

(1988a, 1996) respectively. For the former, "the three value components of culture - White's sentimental, ideological and sociological components - appear ... to have pairs of complementary values associated with mathematics" (Bishop 1988b, p. 185), namely, *rationalism* and *objectism*, *progress* and *control*, *mystery* and *openness*.

While Seah and Andersson's (2015) conception are rather similar, it is also more explicit in highlighting two aspects of values and valuing in mathematics education. One aspect acknowledges that the values that are being espoused in mathematics education need not stem from mathematics lessons alone, but from the wider sociocultural context as well. The other aspect that is more explicitly stated is that the valuing that are inculcated through mathematics education goes beyond being in students' memories, and they in fact 'swing back' to affect the quality of mathematics learning. For them, values and valuing reflect

the convictions which an individual has internalised as being the things of importance and worth. What an individual values defines for her/him a window through which s/he views the world around her/him. Valuing provides the individual with the will and determination to maintain any course of action chosen in the learning and teaching of mathematics. They regulate the ways in which a learner's/teacher's cognitive skills and emotional dispositions are aligned to learning/teaching in any given educational context. (p. 169)

31.2.1 Values and Valuing as Involving Both Cognition and Affect

Seah and Andersson's (2015) definition above implies that values are neither cognitive nor affective constructs per se. Instead, valuing is regarded as being both cognitive and affective in nature (see also, Hartman, n.d.; Huitt 2004).

Rather than being an affective construct as it was generally known (e.g. Bishop 1996; Krathwohl et al. 1964), the process and act of valuing invariably involve reasoning and thinking. Even though Krathwohl et al.'s (1964) taxonomy of educational objectives might refer to the affective domain, the 'organization' phase involves the individual relating the values s/he subscribes to amongst themselves such that these values co-exist, which is a task that involves thinking and reasoning.

Similarly, Raths et al. (1987) conception regards successful attainment of a value as involving all seven criteria, namely, choosing freely, choosing from alternatives, choosing after thoughtful consideration of the consequences of each element, prizing and cherishing, prizing through affirming to others, acting with the choice, and acting repeatedly in some pattern of life. Clearly, the choosing components involve reasoning, whilst the prizing components involve affect.

31.2.2 Values and Valuing as Being Socio-cultural

Values and valuing is also socio-cultural in nature. What we value reflect years of learning and influence from our historical experiences and social interactions as members of the cultures we belong. Indeed, the notion of cultures has been regarded as "an organised system of values which are transmitted to its members both formally and informally" (McConatha and Schnell 1995, p. 81).

The discussion thus far has signaled a perspective to learning where the learner's objectivised actions are culturally and symbolically mediated by values, and which can be examined through activity theory. Activity theory provides a useful theoretical framework also in that it explains how the mediation gets internalised within cultures, giving the learner a particular identity that characterises him or her in culturally unique ways. In particular, the Cultural Historical Activity Theory [CHAT] embodies the construct of values very well. CHAT represents the third generation of the activity theory approach to understanding learning and education. While the first and second generations were associated with Vygotsky's sociocultural theory of teaching and learning, in which students participate in negotiation and co-construction of knowledge (Haenen et al. 2003), and Leontiev's activity theory, the set of specific notions, claims, and arguments that consider the relationship between a subject (typically an individual human) and the object (Kaptelinin and Nardi 2012). In this third-generation interpretation of the activity theory, Engeström's activity system model extended Leontiev's original concept of subject-object interaction to become a three-way interaction between 'subject', 'object', and 'community'. The new theory went beyond a focus on activity systems to emphasise the interactions between and amongst activity systems, so that learning is meaningful through a process of multi-voicedness, difference, and conflict negotiation. Gummesson (2006) had argued that the main outcome of this process is value co-creation. In the classroom, for example, pedagogical activities take place through the interaction of what students, teachers, and indirectly, the wider community value. The interactions have brought together the different things that teachers and their students value similarly and differently, and the co-creation of values can be perceived as the agreed-upon, aligned values that facilitate the continued functioning of the activity systems in interaction. Importantly, while the first generation of the activity theory focuses on the individual learner, and the second generation directs the attention to the community within which learning takes place, CHAT considers as the unit of analysis joint activity amongst individuals in the learning environment. In relation to values, thus, we can imagine values not only as being acquired over time, but that the negotiations of values between and amongst activity systems would also lead to values being challenged and refined on an ongoing basis, depending on the opportunities for one's values to come into contact with values from other activity systems.

31.2.3 Values and Valuing Driving Performance

Research evidence has also supported the belief that mathematics performance is related to students' valuing. In addition to the Nuffield Foundation-commissioned report (Askew et al. 2010) mentioned above which highlighted the role of cultural values, there is also more recent research by Jerrim (2014), who sub-divided the Australian dataset for PISA 2012 by broad student ethnicity, specifically, high-performing East Asian, low-performing East Asian, Indian, British, and native Australian. Given that the students in the sample were second-generation immigrants experiencing an Australian mathematics education with their native Australian peers, it can be assumed that the factors underlying the differences existed beyond the school level, with their different emphases and valuing on different aspects of school education.

Schukajlow's (2017) study with 192 Years 9/10 students in Germany demonstrated a similar relationship between student valuing and mathematics performance. Differences were found, however, between performance on problems related to real-life scenarios and problems which were not. Schukajlow had flagged this for further investigations, and it represents existing research effort into understanding how values might be used to further enhance the mathematics learning experience of young children.

Such an association between valuing and mathematics performance is important, and even more so given that what are being valued also affect the cognitive processes and affective states that in turn influence the quality of mathematics learning. As such,

> the extent to which the educational aspirations of students and parents are the result of cultural values or determinants of these, and how such aspirations interact with education policies and practices is an important subject that merits further study. (OECD 2014, p. 20)

In responding to this call, the guiding assumption is that students' possession or acquisition of relevant valuing allows each of them to apply appropriate cognitive skills and to develop positive affective states which promote desirable outcomes in mathematics learning, whether these be related to measurable performance or to relational understanding.

In addition to being culturally-referenced, what is being valued is also invisible and implicit. Due to the inevitable presence of competing and overriding values (Seah 2005), what one values is not articulated in all situations. Indeed, Takuya Baba had likened values and valuing to the underground roots of a tree, which are not only invisible and implicit, but also crucial to supporting and nurturing the healthy growth of what is visible of the tree above the ground, such as student results.

Herein lies one of the most important aspect of values and valuing, that is, how it supports the development of cognitive functioning and nurturing of affective states. It is as if attending to the cognitive and affective development of mathematics learners alone is not sufficient to bring about meaningful learning. The learner should want to engage, to understand, to learn, and perhaps to achieve as well in the first place. As the saying goes, 'you can bring a horse to the water, but you can't make it drink'. That is, facilitating the valuing of relevant attributes in mathematics learning by the learners themselves is a crucial—and often forgotten—component of mathematics pedagogy, for this in turn supports the development of cognitive functioning and nurturing of affective states that would more directly impact on the quality of learning.

31.3 The 'What I Find Important (in Mathematics Learning)' [WIFI] Study

In fact, the 'What I Find Important (in mathematics learning)' [WIFI] study was conceptualised with this guiding assumption in mind. The objective of the WIFI study has been to find out what students in the last two years of primary schooling and what 15-year-old students value in their mathematics learning experiences.

The desire to facilitate a 'mapping of the scene' has necessitated a large-scale study, which also highlighted the need for a methodology that allowed for the assessment of student values in time-efficient ways. Thus, instead of adopting the qualitative approaches such as in Chin and Lin (2000) or in Clarkson et al. (2000), the WIFI study made use of the questionnaire method. This way, a large number of students could be surveyed for the attributes of mathematics education which they personally find important, and also so that the data collected could be interpreted efficiently using the SPSS software.

The WIFI questionnaire is divided into four sections. Section A is made up of 64 items, each of which being a mathematics classroom activity (e.g 'outdoor mathematics activities', 'explaining my solutions to the class') or a pedagogical norm (e.g 'shortcuts to solving a problem'). Student respondents were expected to rate on the 5-point Likert scale the extent to which an activity or norm was important to each of them. Section B is made up of 10 continuum dimension items, in which opposing values are located on both ends of each continuum dimension (e.g 'how the answer to a problem is obtained' vs. 'what the answer to a problem is') and student respondents needed to indicate where s/he stood in relation to the two opposing values. Section C is an open-ended, scenario-stimulated responses section, providing for another means of identifying what students valued in their mathematics learning. Students' demographic and personal information were collected in Section D. Examples of the questionnaire items and of the layout can be seen in Seah et al. (2016). The WIFI questionnaire can also be accessed online at: https://www.surveymonkey.com/r/WIFI_maths.

The questionnaire was administered in class, that is, student participants filled in the questionnaire individually in their own classroom setting, with the exercise facilitated by their mathematics teachers. The questionnaire is available in hardcopy and online versions, and any participating school will select one of the two possible formats for all its student participants.

To date, some 20,000 questionnaires have been completed (with more than 16,000 analysed) across the 19 different economies. Only the findings from the analysis of responses to Section A items are reported in this chapter. These items are listed in the Appendix.

Each participating economy is represented by a team of local researchers. Each team was responsible for administering the questionnaire in its own context. The quantitative analysis of the questionnaire data was conducted centrally by the Australian research team, however. The results generated by SPSSwin® were then returned to the respective research teams, the intention being that the sense-making could be done in a culturally-meaningful manner by their own researchers.

Upon receipt of the raw data from each participating economy, initial data screening was carried out to test for univariate normality, multivariate outliers (using Mahalanobis' distance criterion), and homogeneity of variance-covariance matrices (using Box's M tests). A Principal Component Analysis (PCA) with Varimax rotation was used to examine the questionnaire items. The significance level was set at 0.05, while a cut-off criterion for component loadings of at least 0.45 was used in interpreting the solution. Items that did not meet the criteria were eliminated. For each economy's data, the Kaiser-Meyer-Olkin (KMO) measure of sampling adequacy was noted, and Bartlett's test of sphericity (BTS) (Bartlett 1950) was also checked for significance at the 0.001 level, so that factorability of the correlation matrix could be assumed, which demonstrated that the identity matrix instrument was reliable and confirmed the usefulness of the PCA. According to the cut-off criterion, the number of items that were removed from the original 64 was understandably different between economies.

The research team in each participating economy then interpreted these components, assigning a value label to each. This is a distinguishing feature of this study, in that the cultural-situatedness of valuing has meant that the researchers from each participating economy analysed and interpreted their own PCA components, and no attempt was made for each group's criteria for interpretations to be shared or made consistent across all participating economies.

31.4 What the Top Performers Value

The key research question guiding the conduct of the WIFI Study is: What do students value in their respective mathematics learning? As we saw above, it is expected that the students' valuing is shaped in context, that is, influenced by societal, ethnic, religious, family, school and other institutional cultures.

This chapter, however, focuses on what students in the top performing PISA2012 economies valued in mathematics learning. Given the relationship between student valuing and mathematics performance (see above), and given the interest in many countries across the world to understand how the top performing economies consistently lead the pack in different ranking exercises, it is hoped that the findings reported in this chapter can inform researchers on the valuing that

Table 31.1 What students in top performing economies valued

HKG [3/65]	TWN [4/65]	KOR [5/65]	MAC [6/65]
Understanding	Connections	Understanding	Achievement
Control	Recall	Connections	Humanism
Effort	Effort	Fun	Practice
Ideas	Exploration	Fluency	Technology
Recall	Openness	Accuracy	Communication
ICT	Communication	Collaborative reflection	Mathematical development
Feedback		Efficiency	
Connections		Communication	
Learning approach		Mystery	

HKG Hong Kong, *TWN* Taiwan, *KOR* Korea, *MAC* Macau

might account for students' mathematics achievement at the national level, thereby deepening what we understand about excellence in mathematics learning. Table 31.1 shows these valuing for Hong Kong, Taiwan, Korea and Macau, which were placed third, fourth, fifth and sixth by student performance in PISA2012. These economies continue to lead the world in subsequent PISA tests. For example, in PISA2015, they were ranked second, fourth, seventh, and third respectively (Thomson et al. 2016).

The number of attributes listed for each economy is different from that of another economy, since these are associated with the number of components that were elicited from the respective PCA. The attributes had been named independently by the respective research teams, based on each team's cultural interpretation of the questionnaire items which had loaded onto the components. The order of listing of the attributes in Table 31.1 reflects, for each economy, the order of the components that were derived from the PCA exercises. As shown in Table 31.1, it may be said that generally, the top performing economies have students which valued *connections*, *understanding*, *communication*, *recall*, and *ICT*.

It is also necessary to check if students in the economies which did not perform as well might have been valuing the same aspects in mathematics education. Accordingly, the students' valuing for Turkey and Thailand—ranked 44th and 50th respectively amongst the 65 surveyed economies—were referred to, as shown in Table 31.2.

Given that students in Turkey and Thailand valued *ICT* as well, it is unlikely that information and communication technology in general would have contributed to student achievement in mathematics. It might well be that certain aspects within ICT use would enhance or promote student learning and achievement, whereas other aspects of ICT use might have the opposite effect, such that its valuing was nominated by different groups of students. While this may be a possibility (which will be briefly explored below)—and indeed, this signals further research about how different groups of students might value different aspects of ICT—it is also reasonable to remove it from the list of attributes associated with top performing

Table 31.2 What students in Turkey and Thailand valued

TUR [44/65]	THA [50/65]
Relevance	Process
Practice	Fact
ICT	Practice
Collectivism	Relevance
Objectism	ICT
	Learning from others

TUR Turkey, *THA* Thailand

students' valuing. In other words, it can be deduced that students from top per-forming countries in the PISA2012 assessment valued *connections, understanding, communication,* and *recall.*

It is reassuring that many of these values are being explicitly promoted in many current-day mathematics curriculum documents. For example, amongst the 5 pro-cess standards identified for the NCTM 'Principles and Standards for School Mathematics' (2000) are *connections* and *communication.* Incidentally, the other two values—that is, *understanding* and *recall*—are being reflected in the current Australian Curriculum (ACARA 2016) and Victorian Curriculum (VCAA 2017) for Mathematics, if we associate *recall* with being an aspect of *fluency.*

These attributes are observed to be different in nature from those which were valued by countries like Turkey and Thailand (see Table 31.2), which did not perform as well relative to the other participating countries. The valuing of *con-nections, understanding, communication,* and *recall* was concerned with paying attention to the attributes of the nature or structure of the mathematics discipline. These values might thus be considered to be intrinsic in nature. On the other hand, the valuing of *relevance* and *practice* by students in Turkey and Thailand highlights the importance given to what can be done with mathematical knowledge and skills, or what can be done externally to the discipline itself to acquire the knowledge and skills. These values can thus be considered to be more extrinsic. It appears that students' mathematical performance might be related to not just valuing of indi-vidual attributes, but also, to the extent to which the valuing is related to intrinsic characteristics.

31.5 Valuing and Constituent Actions

As suggested by the questionnaire format, the invisible nature of valuing is com-pensated for in this study by focussing on the observable actions that are expressed by the valuing associated to it. Given the cultural nature of valuing, it was assumed that the same valuing can take different forms in different settings. This was investigated by examining the questionnaire items that loaded onto the same

Table 31.3 How *ICT* was valued across three economies

HKG	JPN	GHA
Using the calculator to check the answer	Learning mathematics with the computer	Using the calculator to check the answer
Learning mathematics with the computer	Learning mathematics with the internet	Using the calculator to calculate
Using the calculator to calculate	Using the calculator to check the answer	
Learning mathematics with the internet	Mathematics games	

HKG Hong Kong, *JPN* Japan, *GHA* Ghana

valuing across different economies. Table 31.3 shows the constituent actions corresponding to the valuing of *ICT* in Hong Kong, Japan and Ghana, for example.

Although it may initially look as if the valuing of *ICT* can be described by a common set of classroom actions, it is important to note that no one action can be found across all three economies. The interplay between valuing and its constituent actions is indeed complicated. While the valuing of *ICT* is associated with the learning of mathematics content with the computer and with internet in Hong Kong and Japan, these actions could not be found amongst the Ghana data, even though Ghanaian students also valued *ICT*. At the same time, Japanese students' valuing of *ICT* was uniquely associated with mathematics games.

It needs to be noted that the similarity of activities emphasised in different education systems does not necessarily increase the chance of these cultures valuing the same attribute. In the example above, the classroom activities embraced by the Japanese students were similar to what their peers in Korea seemed to be preferring too. Yet, the cultural interpretations of these classroom activities in Japan and in Korea were such that the sets of activities were seen to reflect different valuing; while they referred to a valuing of *ICT* in Japan, it was a valuing of *fun* in Korea.

It is worthy to consider the impact on the effectiveness of mathematics pedagogy in different education systems when the same valuing is emphasised differently across the institutions. We can see this in the three economies' valuing of *ICT* above, where it was speculated that an economy's infrastructure might be a contributing factor to this difference. On the other hand, the differences might come about through culturally different conceptions of pedagogy and of education. Here we may consider Macao and Ghanaian students' valuing of *achievement* through their preferences for 15 and 16 classroom activities respectively. Although 6 (e.g. 'understanding concepts/processes', 'working out the mathematics by myself') of these preferred activities were common between the two economies, there were still up to 10 activities which were regarded by students to be important in their respective education systems. In Macao, the students' valuing of *achievement* through understanding and working out the mathematics individually was supported by activities such as 'shortcuts to solving a problem' and 'practising how to use mathematics formulae', which both pointed to means of achieving in

Table 31.4 How students in Korea and Hong Kong valued *connections*

KOR	HKG
Relating mathematics to other subjects in schools	Relating mathematics to other subjects in schools
Appreciating the beauty of mathematics	Appreciating the beauty of mathematics
Connecting mathematics to real life	Connecting mathematics to real life
Stories about recent developments in maths Stories about mathematics Explaining my solutions to the class	Learning the proofs
Students posing maths problems Looking out for maths in real life Making up my own maths questions Mathematics puzzles Mathematics debates	
Looking for different possible answers	
Investigations	

KOR Korea, *HKG* Hong Kong

mathematics through efficient and fluent working-out. Students in Ghana also aimed to achieve in mathematics through efficient and fluent practices, though these appeared to take on different forms. There, the preferred means were 'teacher asking us questions' and 'remembering the work we have done'.

Even amongst the top performers which are generally perceived to be of similar culture (i.e. East Asian), the same attribute being valued is portrayed through different classroom actions. Table 31.4 provides an example for Korea and Hong Kong's valuing of *connections*. Although there are three actions which were in common across the two economies, that is, 'relating mathematics to other subjects in schools', 'connecting mathematics to real life', and 'appreciating the beauty of mathematics', in each economy the students were also demonstrating their valuing of *connections* through other actions. For example, students in Korea appeared to regard their explanations of their solutions 'in public' in class as a means of valuing *connections*, possibly through the need for the presenting students to be able to establish how concepts and knowledge are interconnected in their respective solutions. However, this classroom action was not identified with the valuing of *connections* by their peers in Hong Kong classrooms, even though such a classroom activity is also commonly found there.

31.6 Determining Valuing from Particular Actions

In the same way that any attribute of mathematics learning and teaching can be valued through different classroom activities, any activity needs not point to any one particular valuing. Rather, the implementation of any activity in a mathematics classroom can point to the valuing of one or more of several possible valuing. What this means for the assessment and identification of valuing is that some kind of triangulation is needed through the observation of multiple supporting activities.

This other aspect of the absence of a one-to-one correspondence between valuing and classroom activities was observed earlier on in the study, when the different research teams were aligning individual questionnaire items of classroom activities with the valuing that was assumed to being reflected. For example, the emphasis given by students in Turkey for small-group discussions would reflect the valuing of one or more of the following attributes of mathematics education: *collaboration, communication, efficiency, fun, humanism, openness, question posing, practice,* and *representation*. Thus, whatever a teacher's intention or valuing is when small-group discussions is part of his/her professional practice in the mathematics classroom, students may not be able to understand what teacher valuing is being espoused through it. However, to the extent that the teacher is able to express whatever is being valued through a variety of classroom activities, students will be able to triangulate these to understand this valuing. There are implications here for research designs involving data collection through lesson observations: Multiple observations might be needed for the teachers (and students) to display a range of actions and activities, so that the underlying intentions, philosophies and valuing can be 'sieved out' from amongst the possible attributes valued. If repeated observations is not possible, then post-lesson interviews or discussions with the participants involved would be necessary to clarify these underlying valuing.

Elsewhere, Clarke's (2004) documentation of the Japanese classroom practice of teacher between-desk instruction—which the Japanese educators called 'kikan-shido'—might lead the Western academic community to associate it with teacher elicitation of student difficulties and subsequent teacher individualised explanation. However, this classroom practice has been noticed in Shanghai (Lopez-Real et al. 2004) and German mathematics lessons too. Of importance is how this similar act of teacher between-desk instruction actually expresses different valuing amongst the three economies. In Shanghai, teachers made use of their monitoring of student work to encourage students to think further. In Germany, however, the monitoring and correction of student work seemed to be absent, where the teachers appeared to be using the opportunities to ask questions for the purpose of stimulating students' mathematical thinking. Thus, even though kikan-shido might be observed in German, Japanese and Shanghai mathematics lessons, the teachers across these three cultures were portraying different valuing with regards to mathematics pedagogy.

Similarly, analysed data from the TIMSS Video Study (Hiebert et al. 2003) have suggested that even though the classroom activity of problem-solving may be embraced in many mathematics education systems, this same form should not be taken to imply that the same attributes of mathematics pedagogy are being valued. Indeed, it is instructive to note that the high performing mathematics education systems emphasise *connections* that are facilitated through the problem-solving tasks, whereas many of the other mathematics education systems emphasise *procedure*. Thus, doing what effective mathematics education systems do does not imply that the same benefits will be gained. Rather, culturally-appropriate classroom activities are means through which the features of mathematics learning that matter are valued, expressed and operationalised.

31.7 Implications for Teacher Practice

The analysed data from the top performing PISA2012 economies suggest that student performance in mathematics was related to students' valuing of *connections, understanding, communication,* and *recall*. Given that PISA items assess students' ability to apply their mathematical knowledge in novel problems—which would require students to demonstrate knowledge, skill and application—the four attributes being valued do cover the various aspects of being able to excel in the assessment. Many of Hattie's (2015) top classroom interventions (which refer to school education generally and which also include background variables such as 'home environment' and 'ethnicity') are related to these four attributes, such as classroom discussion (effect size = 0.82), feedback (effect size = 0.73), formative evaluation (effect size = 0.68), concept mapping (0.64), and mastery learning (effect size = 0.57). Significantly, none of the last 50 interventions in the list of 195 appeared to relate to these four valuing.

This student valuing of *connections, understanding, communication,* and *recall* reflect intrinsic valuing, as opposed to extrinsic valuing which would emphasise such valuing as *application* and *relevance*. There are implications here for professional practice in the mathematics classroom, even though curriculum documents might emphasise both these categories of attributes. This is important, not least because the inculcation of extrinsic valuing can be more appealing to students and can also be easier to convey to them. On the other hand, it is likely that teachers' efforts to prompt students' appreciation and subsequent valuing of intrinsic valuing can actively be derailed by students routinely asking questions such as, "when are we ever going to use this?" It thus appears that students need to appreciate the utilitarian aspects embedded within intrinsic valuing.

Mathematics pedagogical approaches or strategies can be defined by what they value with regards to the teaching and learning of mathematics. At the level of the intended curriculum, the valuing that is embedded in these pedagogical approaches or strategies may not be explicitly linked to the specific approaches or strategies, but are rather merely mentioned in the introductory or rationale sections only. As a result, too, these valuing are not explicitly stated in the text or by the teacher. To the extent that this valuing can be considered the heart and soul of the particular pedagogical approach or strategy, it is important that preservice or in-service teachers who are being introduced to it is made aware of the underlying valuing.

The data collected from different economies have indicated that merely 'transplanting' a new pedagogical approach or strategy in one's classroom might not make clear to students the underlying valuing that is being advocated or taught. This is why expensive projects which attempted to introduce Japanese classrooms in the USA and Chinese classes in the UK have largely failed to achieve their respective objectives. Teaching students with the new approach or strategy alone is likely not able to realise its intended benefits to mathematics learning. The professional discourse might need to change from 'we are learning skill ABC or technique DEF' or similar, to one of 'through this skill ABC or technique DEF, we are learning to value attribute XYZ' or similar.

In this way, it adds another dimension to how values and valuing play a key role in (mathematics) lesson planning. Not only is a focus on valuing in lesson planning expected to promote students' cognitive and affective engagement, it also allows teachers to adopt/adapt and reap maximal potential out of teaching approaches or strategies they are introduced to.

31.8   Shaping Student's Valuing

The discussion above assumes that students' valuing can be and are being shaped in the mathematics education process. After all, that values are internalised and stable variables does not imply that they cannot be modified. Furthermore, modification and (re-)shaping may be easier when the individual is still young.

From a practical perspective, the data collected from around the world have also provided empirical evidence that value change takes place between the primary and secondary school years. In Japan, primary school students surveyed valued *process*, but their secondary school peers were valuing *product*, which can be viewed as the opposing attribute to *process* when considering the engagement with and completion of mathematical tasks. On the other hand, in Hong Kong, it was found that when students progressed from primary to secondary schools, they experienced a drop in their valuing of *understanding*, *recall* and *control*. Of course, given Hong Kong's excellent performance in international assessments and this relationship to the students' valuing of *understanding* and *recall*, it is also reasonable to assert that the reduced valuing was still significant enough to be highly regarded by 15-year old Hong Kong students who aspire to work excellently.

Teachers thus play the role of value agents even as they engage in mathematics teaching. In many ways, one can argue that this has always been a role that is played out by teachers everywhere. Teacher agency in shaping and modifying students' valuing is very real indeed, although it can be more explicit than it normally is. In fact, this teacher role is also often advocated in curriculum statements.

Teacher teaching, shaping and modification of student valuing can take on different forms. One of the more innovative forms is the introduction of role-play activities in mathematics lessons, either through students taking on roles which correspond to particular nominated valuing (e.g. being a student who values *progress*), or taking on the role of teacher or peer tutor, which would necessitate student evaluation of the valuing that underlies effective teaching/tutoring.

31.9 Values Alignment

Interactions between teachers and students, as well as teachers' pedagogical tasks and activities in class, both bring to the fore what teachers and their students valued similarly and differently. Teacher effectiveness can depend on the extent to which teachers are able to negotiate these inevitable value differences, so as to bring about a learning environment in which everyone's valuing are aligned and inter-personal relationships are in harmony. After all, "all relationships ... are claimed to be strengthened by aligned values" (Branson 2008, p 381). Value alignment thus involves teachers making in-the-moment decisions, acknowledging that teacher practice is situated in a socially co-constructed setting. Several teacher strategies of value alignment have been reported by Seah and Andersson (2015), and further research is being conducted in this area to empower teachers to recognise, align and shape the valuing that underlies cognitive and affective functionings of mathematics teaching and learning.

Thus, values alignment should be regarded as already being part of day-to-day interactions. When individuals come together with their own value systems, they will always need to negotiate about different preferences and intentions to ensure that the interaction is successful. This calls for values alignment to take place, and it does not mean that one party needs necessarily to impose his/her/their values to the rest. There can always be middle-path compromises, for example. At the same time, it is important to note that any consideration of values being aligned (or not) is mutual between and amongst the individuals involved, both teachers and their students. In this manner, student agency is acknowledged.

31.10 Summarising Ideas

Valuing refers to an individual's embrace of convictions which are considered to be of importance and worth. It provides the individual with the will and grit to maintain any 'I want to' mindset in the learning and teaching of mathematics. In the process, this conative variable shapes the manner in which the individual's reasoning, emotions and actions relating to mathematics pedagogy develop and establish. The argument in this chapter is that more effective teaching and learning can only take place by paying attention to what are being valued by teachers and students respectively, and through teachers' purposeful shaping of students' valuing and alignment of the diverse values that are enacted upon by these teachers and their respective students.

Data collected and analysed from the various economies participating in the WIFI Study have demonstrated that any particular valuing is manifested in one of many possible classroom practices. Similarly, any one classroom practice is not reflective of any one valuing only.

The quantitative WIFI Study has allowed us to identify what students valued in mathematics learning across 19 different economies. It was found that students in 4 top performing mathematics education systems in PISA2012 (i.e. ranked third to sixth) generally valued *connections, understanding, communication,* and *recall.* On the other hand, students in two of the education systems which did not perform well were valuing *relevance* and *practice.* A distinction between intrinsic and extrinsic valuing is proposed; further studies are recommended to explore the extent to which intrinsic valuing fosters greater mathematical performance. In addition, *ICT* was valued across both types of mathematics education systems. Thus, there is a need to further examine the effectiveness in student valuing of *relevance, practice* and *ICT.*

School-aged students are in the process of defining and internalising what they each value in life and in mathematics learning. Teachers' awareness of what they themselves value, and purposeful and explicit portrayal of these valuing, are expected to facilitate students' development of what they value in mathematics and in mathematics learning. Additionally, given the inevitable opportunities for differences in what teachers and their students value, teachers assume an important task of aligning the different and potentially conflicting valuing, such that meaningful mathematics learning is facilitated.

Appendix 1: WIFI Questionnaire (Section A Only)

Note that the questionnaire layout has been altered here, to suit the publication guidelines.

The Third Wave Project

Study 3: What I Find Important (in Maths Learning)

Student Questionnaire

Section A
For each of the items below, tick a box to tell us how important it is to you when you learn mathematics.

	Absolutely important	Important	Neither important nor unimportant	Unimportant	Absolutely unimportant
1. Investigations					
2. Problem-solving					
3. Small-group discussions					
4. Using the calculator to calculate					
5. Explaining by the teacher					
6. Working step-by-step					
7. Whole-class discussions					
8. Learning the proofs					
9. Mathematics debates					
10. Relating mathematics to other subjects in school					
11. Appreciating the beauty of maths					
12. Connecting maths to real life					
13. Practising how to use maths formulae					
14. Memorising facts (e.g. Area of a rectangle = length X breadth)					
15. Looking for different ways to find the answer					
16. Looking for different possible answers					
17. Stories about mathematics					
18. Stories about recent developments in mathematics					
19. Explaining my solutions to the class					
20. Mathematics puzzles					
21. Students posing maths problems					
22. Using the calculator to check the answer					
23. Learning maths with the computer					
24. Learning maths with the internet					
25. Mathematics games					
26. Relationships between maths concepts					
27. Being lucky at getting the correct answer					
28. Knowing the times tables					
29. Making up my own maths questions					
30. Alternative solutions					
31. Verifying theorems/hypotheses					
32. Using mathematical words (e.g. angle)					
33. Writing the solutions step-by-step					
34. Outdoor mathematics activities					
35. Teacher asking us questions					
36. Practising with lots of questions					
37. Doing a lot of mathematics work					
38. Given a formula to use					
39. Looking out for maths in real life					
40. Explaining where rules/formulae came from					
41. Teacher helping me individually					
42. Working out the maths by myself					
43. Mathematics tests/examinations					
44. Feedback from my teacher					
45. Feedback from my friends					
46. Me asking questions					

(continued)

(continued)

	Absolutely important	Important	Neither important nor unimportant	Unimportant	Absolutely unimportant
47. Using diagrams to understand maths					
48. Using concrete materials to understand mathematics					
49. Examples to help me understand					
50. Getting the right answer					
51. Learning through mistakes					
52. Hands-on activities					
53. Teacher use of keywords (e.g. 'share' to signal division; contrasting 'solve' and 'simplify')					
54. Understanding concepts/processes					
55. Shortcuts to solving a problem					
56. Knowing the steps of the solution					
57. Mathematics homework					
58. Knowing which formula to use					
59. Knowing the theoretical aspects of mathematics (e.g. proof, definitions of triangles)					
60. Mystery of maths (example: 111 111 111 × 111 111 111 = 12 345 678 987 654 321)					
61. Stories about mathematicians					
62. Completing mathematics work					
63. Understanding why my solution is incorrect or correct					
64. Remembering the work we have done					
65. Comments (if any)					

References

Askew, M., Hodgen, J., Hossain, S., & Bretscher, N. (2010). *Values and variables: Mathematics education in high-performing countries*. Retrieved from: http://www.nuffieldfoundation.org/sites/default/files/Values_and_Variables_Nuffield_Foundation_v__web_FINAL.pdf.

Australian Curriculum Assessment and Reporting Authority (ACARA). (2016). *The Australian curriculum: F-10 mathematics*. Retrieved from Sydney, Australia.

Bartlett, M. S. (1950). Tests of significance in factor analysis. *British Journal of Psychology, 3*, 77–85.

Bishop, A. J. (1996, June 3–7). *How should mathematics teaching in modern societies relate to cultural values—Some preliminary questions*. Paper presented at the Seventh Southeast Asian Conference on Mathematics Education, Hanoi, Vietnam.

Bishop, A. J. (1988a). *Mathematical enculturation: A cultural perspective on mathematics education*. Dordrecht, The Netherlands: Kluwer Academic Publishers.

Bishop, A. J. (1988b). Mathematics education in its cultural context. *Educational Studies in Mathematics, 19*, 179–191.

Branson, C. M. (2008). Achieving organisational change through values alignment. *Journal of Educational Administration, 46*(3), 376–395.

Chin, C., & Lin, F.-L. (2000). A case study of a mathematics teacher's pedagogical values: Use of a methodological framework of interpretation and reflection. *Proceedings of the National Science Council Part D: Mathematics, Science, and Technology Education, 10*(2), 90–101.

Clarke, D. (2004). *Kikan-shido: Between desks instruction.* Paper presented at the 85th Annual Meeting of the American Educational Research Association, San Diego, CA.

Clarkson, P., Bishop, A. J., FitzSimons, G. E., & Seah, W. T. (2000). Challenges and constraints in researching values. In J. Bana & A. Chapman (Eds.), *Mathematics Education Beyond 2000: Proceedings of the Twenty-Third Annual Conference of the Mathematics Education Research Group of Australasia Incorporated held at Fremantle, Western Australia,* July 5–9, 2000. (Vol. 1, pp. 188–195). Perth, Australia: Mathematics Education Research Group of Australasia Incorporated.

Gummesson, E. (2006). Many-to-many marketing as grand theory. In R. Lusch & S. Vargo (Eds.), *The service dominant logic of marketing: Dialogue, debate and directions* (pp. 339–353). Armonk, NY: M. E. Sharpe Inc.

Haenen, J., Schrijnemakers, H., & Stufkens, J. (2003). Sociocultural theory and the practice of teaching historical concepts. In A. Kozulin, B. Gindis, V. S. Ageyev, & S. M. Miller (Eds.), *Vygotsky's educational theory in cultural context* (pp. 246–266). Cambridge, UK: Cambridge University Press.

Hartman, R. S. (n.d.). *Self-knowledge, values, and valuations.* Retrieved from https://www.hartmaninstitute.org/self-knowledge-values-valuations/.

Hattie, J. (2015). The applicability of visible learning to higher education. *Scholarship of Teaching and Learning in Psychology, 1*(1), 79–91.

Hiebert, J., Gallimore, R., Garnier, H., Givvin, K. B., Hollingsworth, H., Jacobs, J., et al. (2003). *Teaching mathematics in seven countries: Results from the TIMSS 1999 video study.* Washington, DC.

Huitt, W. (2004). Values. *Educational psychology interactive.* Valdosta, GA: Valdosta State University.

Jerrim, J. (2014). *Why do East Asian children perform so well in PISA? An investigation of Western-born children of East Asian descent.* London: University of London.

Kaptelinin, V., & Nardi, B. (2012). *Activity theory in HCI: Fundamentals and reflections.* Williston: VT Morgan & Claypool Publishers.

Krathwohl, D. R., Bloom, B. S., & Masia, B. B. (1964). *Taxonomy of educational objectives: The classification of educational goals (Handbook II: Affective domain).* New York: David McKay.

Lopez-Real, R., Mok, A. C., Leung, K. S., & Marton, F. (2004). Identifying a pattern of teaching: An analysis of a Shanghai teacher's lessons. In L. Fan, N.-Y. Wong, J. Cai, & S. Li (Eds.), *How Chinese learn mathematics: Perspectives from insiders* (pp. 282–412). Singapore: World Scientific.

McConatha, J. T., & Schnell, F. (1995). The confluence of values: Implications for educational research and policy. *Educational Practice and Theory, 17*(2), 79–83.

National Council of Teachers of Mathematics (NCTM). (2000). *Principles and standards for school mathematics.* Reston, VA: NCTM.

OECD. (2014). *PISA 2012 results: Ready to learn: Students' engagement, drive and self-beliefs.* Retrieved from Paris, France.

O'Keefe, C. A., Xu, L. H., & Clarke, D. J. (2006). Kikan-shido: Through the lens of guiding student activity. In J. Novotná, H. Moraová, M. Krátká, & N. Stehlíková (Eds.), *Proceedings of the 33rd Conference of the International Group for the Psychology of Mathematics Education* (Vol. 4, pp. 265–272). Prague: PME.

Raths, L. E., Harmin, M., & Simon, S. B. (1987). Selections from 'values and teaching'. In J. P.F. Carbone (Ed.), *Value theory and education* (pp. 198–214). Malabar, FL: Robert E. Krieger.

Schukajlow, S. (2017). Are values related to students' performance? In B. Kaur, W. K. Ho, T. L. Toh, & B. H. Choy (Eds.), *Proceedings of the 41st Conference of the International Group for the Psychology of Mathematics Education* (Vol. 4, pp. 161–168). Singapore: PME.

Seah, W. T. (2005). Immigrant mathematics teachers' negotiation of differences in norms: The role of values. In M. Goos, C. Kanes, & R. Brown (Eds.), *Mathematics Education and Society: Proceedings of the 4th International Mathematics Education and Society Conference* (pp. 279–289). Queensland, Australia: Griffith University.

Seah, W. T., & Andersson, A. (2015). Valuing diversity in mathematics pedagogy through the volitional nature and alignment of values. In A. Bishop, H. Tan, & T. Barkatsas (Eds.), *Diversity in mathematics education: Towards inclusive practices* (pp. 167–183). Switzerland: Springer.

Seah, W. T., Andersson, A., Bishop, A., & Clarkson, P. (2016). What would the mathematics curriculum look like if values were the focus? *For the Learning of Mathematics, 36*(1), 14–20.

Thomson, S., Bortoli, L. D., & Underwood, C. (2016). *PISA 2015: A first look at Australia's results*. Melbourne, Australia: Australian Council for Educational Research.

Victorian Curriculum and Assessment Authority (VCAA). (2017). *The Victorian curriculum F-10: Mathematics*. Victoria, Australia: Victorian Curriculum and Assessment Authority.

Chapter 32
About Collaborative Work: Exploring the Functional World in a Computer-Enriched Environment

Carmen Sessa

Abstract The purpose of this paper is to address two main concerns in mathematics education. The first is finding ways of bridging the gap between the worldviews of a university research team and secondary school mathematics teachers. The second is meaningful and implementable ways of introducing technological tools in regular classrooms in order to teach and explore functional relationships. Whereas these two issues have been discussed in the literature, this contribution blends these two issues in the context of Argentina while proposing general insights for the mathematics education community at large. This paper outlines and describes the different stages of the formation and functioning of a collaborative team of researchers and teachers and discusses some didactical complexities encountered.

Keywords Collaborative group · Integration of ICT · Design proposal

C. Sessa (✉)
Universidad Pedagógica Nacional, Buenos Aires, Argentina
e-mail: sessacarmen@gmail.com

C. Sessa
Universidad de Buenos Aires, Buenos Aires, Argentina

© The Author(s) 2018
G. Kaiser et al. (eds.), *Invited Lectures from the 13th International Congress on Mathematical Education*, ICME-13 Monographs,
https://doi.org/10.1007/978-3-319-72170-5_32

32.1   The Journey that Led Me to Work in a Collaborative Group[1]

I have been working in the area of didactics in mathematics for the last 25 years. I originally trained in mathematics, and my shift towards didactics was accompanied by an important change in the conception of teaching that I held at the time. I went from considering the problem of teaching as a question of organizing a discourse (logically organized and attractive for the listener/student) to considering it as a double interaction process: the interaction between the students and a problem or task and the interaction between teachers and their students regarding their productions.[2]

Looking at teaching in this way obliges one to attend to various matters: first, the need for powerful tasks that promote varied and rich productions from the students and, second, a teacher's management of classroom activities that fosters the students' work and, based on their production, is able to build relations between this work and the knowledge wished to be consolidated in the classroom.

The beginning in didactics of mathematics. The first years of research into the didactics of mathematics were the result of teamwork with many colleagues[3] and were centered on understanding how the teaching system works at the beginning of the learning of algebra. This led us to get in touch with teachers and students and to analyze and interpret different programs and curricular documents. This was a very rich period during which the findings of research in this field provided us with conceptual tools for the analysis.

The didactic engineering[4] stage. Not long after starting, we needed to test and study how the situations we created with the objective of provoking specific work from the students would function in a classroom situation. Briefly, the work process was the following: First, we produced a teaching situation or sequence, then we contacted various teachers who were willing to teach it in their classrooms. Next, we undertook to communicate to the teachers the greater and smaller objectives of the overall sequence and of its individual parts. (This instance was usually very delicate. How to convey the subtlest objectives of each part of the sequence, without imposing the teacher as a script to follow? Then there was the fact that the

---

[1]These are the teachers who now form the collaborative group in which I work: Marina Andrés, María Brunand, Marité Coronel, Rosa Escayola, Claudia Kerlakian, Sabrina Maffei, Esteban Romañuk, Débora Sanguinetti, Marina Torresi, and Martín Tornay. Gema Fioriti, from the National University of General San Martin, has participated in this group since its very beginnings and in all these years; the interaction with her has enriched my work. I must mention too that most of the ideas on this paper came about in interaction with the members of the research team at the Pedagogical University: Betina Duarte, Enrique di Rico, Mara Cedrón, Valeria Borsani, Juan Pablo Luna, and Rosa Cicala.

[2]These are the central issues of Guy Brousseau's Theory of Situations. See, for example, Brousseau (1997).

[3]Among them, I want to mention Patricia Sadovsky and Mabel Panizza.

[4]See, for example, Artigue (1998).

teacher was time for joint meetings). After this, we went to the classroom to take notes on the set up of the situation, and finally we analyzed the work produced with a small group of "specialists."

This research model proved to be unsatisfactory. We observed it locked inside our academic university environment. Although the work spoke of the classroom and of teaching, it was far away from it. We were not satisfied with the kind of relation we had with the teachers: It was always respectful but distant.

The emergence of a collaborative group: Associate teachers with our research. In 2006, we could form a group made up of secondary school teachers, academics specializing in mathematical didactics, and students in pre-service teacher training, with the objective of thinking together about the teaching of mathematics. The group's work was based on the design and analysis of teaching situations that the group's teachers then implemented in their classrooms. The practical realization of these situations was then once again analyzed by the group. Our work could be considered a kind of collaborative didactics engineering (Sensevy 2011). I have been working in this collaborative group ever since, and it has undergone different periods and continues to undergo changes in its conformation.

32.2   Different Stages in the Consolidation of the Collaborative Group

We can identify four stages in the development of the collaborative group's work.

Stage 1. This stage corresponds to a period in which the usual work environment in the classrooms of Argentina involved pencil and paper and blackboard. During this stage, we developed a teaching proposal centered on quadratic functions. The proposal and its findings were set out in a curricular document: Mathematics, Quadratic Functions, Parabolas, and Second-Degree Equations. Contributions to Teaching. Middle School. the document was completed in 2009 and was published in 2014 (Sessa et al. 2014).

It was the founding moment for associating teachers with our research. In the development of didactic engineering, the university team provided important experiences and didactic proposals. The necessary symmetry[5] was in the process of building.

Stage 2. Beginning in 2009 and for 2 years, the group worked in one of its most autonomous periods and decided to call itself the "Grupo de los lunes" ("Monday Group") because of the day it used to meet on. At the same time, the nationwide distribution of netbooks in secondary schools begins to take place, with the idea of

---

[5]We take from Sensevy (2011) the central idea that to install collaboration, it is necessary to build a symmetry between researchers and teachers. This is based on an equalization of legitimacy in relation to the work carried out, rather than in the denial of differences.

providing "one for each student" and with the manifest intention of "closing the technological gap" among different sectors of society. Computers started to reach schools, and the Monday Group wanted to study how to incorporate them into the proposals it was making. Considering this challenge, we elaborated a didactic proposal about an introduction to polynomial functions (Fioriti and Sessa 2015).

With different formations and concerns, the problem we faced was new for all; achieving real progress in building the symmetry required for the group work.

Each person's contribution was merged in the production and analysis of the elaborated proposal. The dynamism of the GeoGebra program and the possibility of obtaining multiple and linked representations on the screen were the two potentialities of this program that we "learned" during this work. On the other hand, we were able to identify in this stage new concerns, including how to manage the collective spaces of work in the classroom and how the students would keep a record of their work on the computers.

In 2012 the collaborative group found a place within the Universidad Pedagógica, the institution where I have been working for the last 5 years. Two more stages took place in this period.

Stage 3. Once in the institutional framework of the Universidad Pedagógica, the group reconsidered the proposal it had elaborated in Stage 1, which was originally designed to teach the quadratic function in a paper and pencil context, and worked on adapting it so that the computer could be incorporated into the students' work. The modification of the original proposal was thought of not only in terms of new tasks designed to work with GeoGebra files, but fundamentally in relation to the new aspects that had to be taken into account for the students' and teachers' work.

A fundamental concern was to preserve the didactic intentions of the original proposal or to eventually enrich them, but, while teachers felt secure and in control of the proposal that had been elaborated at Stage 1 with paper, pencil, and blackboards, the adaptation that we developed was done in a constant backdrop of uncertainty, a sensation provoked by the process of migrating towards work being done using the computer.

Many questions surfaced at this stage: How will mathematical knowledge be transformed? How can the work and interaction with the software be done independently by the students and be linked and integrated to the moments of collective dialogue and discussion? How can we solve unforeseen situations that will surely arise in the students' work in their interaction with the software, about which we had no previous repertoire? Some of these questions were anticipated while others arose from the classroom work with this new presence. In the classroom, we could see the need for a teaching action to organize and sustain the students' work, both individually and collectively. The notion of instrumental orchestration (Trouche 2004a; Drijvers and Trouche 2008) gave us tools to conceptualize this space of teachers' decisions.

Stage 4. This stage is currently underway. From new additions in Monday's Group, we decided to think of situations for the introduction to working with functions that involved some aspect of modeling and were directed towards students in the initial years of secondary school. The research of Arcavi and Hadas (2000) and Arcavi (2008) was discussed in the collaborative group and served as inspiration for the proposals we made.

We want to pay special attention to the students' mathematical work; in particular, we will try to identify the existence of knowledge more closely related to technological contexts and more anchored in mathematics. In relation to the teachers' work and considering our earlier stages, we hope to design a possible orchestration that will take into account teachers' room for movement in the management of their classes. In particular, we want to develop didactic techniques that will allow teachers to recover their students' productions made with GeoGebra.

Some remarks about the Monday Group and the incorporation of the GeoGebra

Looking at all the work in retrospect and before commenting on some specific examples, we will add some general questions regarding the Monday Group's work.

Our shared vision of a math class. The members of the Monday Group share some principles regarding mathematical work in the classroom. Our objective is to involve students in the real activity of producing knowledge. To do so it will be necessary to propose challenging problems to the students and generate an environment in the classroom that will encourage them to explore, produce different solutions, and contribute ideas. Attempts, solutions, and ideas are the raw material with which a teacher organizes classroom interaction. The collective space for discussion is appropriate for studying the validity of reasoning processes and procedures, advancing in terms of precision, presenting new problems, speculating, and making conjectures and studying them. In this space, students can get involved in the elaboration of mathematical theory.

Where do these preoccupations lead us when we consider the incorporation of a program like GeoGebra into classroom work?

Changes to take into account. The inclusion of work that uses educational software in the teaching and learning processes establishes the need to take into account changes in relation to the students' mathematical work and the mathematical-didactic work of the teachers.

When referring to the students' activity, changes appear in both the problems and tasks that can be proposed and in the possible techniques that are constituted. Tasks will be created that are unthinkable without the computer.

The instrumental approach, which recognizes the complexity of the teaching of mathematics mediated by technology, gives us theoretical elements with which to think about our work. In this approach, the use of a technological tool implies a process of instrumental genesis in which the object or artifact becomes an instrument. This instrument is a psychological construct, combining the artifact and schemas (in the sense of Vergnaud 1990) that the user develops to use for specific types of tasks (Drijver et al. 2010).

The construction of the instrument must be understood in a double movement: a movement directed towards the artifact, where users take the artifact in their hands and adapt it to their work habits (*instrumentalization*), and a user-oriented movement in which both the limitations imposed by the device and the possibilities offered by it contribute to structuring user activity (*instrumentation*; Trouche 2004b).

In terms of the teachers' work, new spaces requiring decision taking have appeared in collective planning and other more personal spaces have come into play in the management of each teacher's classroom. We found the idea of *instrumental orchestration* (Drijvers and Trouche 2008) relevant in order to pay attention to these teacher decision-taking spaces when working with the inclusion of computers. This notion includes both those spaces related to the tasks and the ways of solving them (which includes the methods and techniques that the students are expected to develop) as well as those related to the instruments and their organization for individual and group work.

In terms of the way the group works, I would like to highlight the fact that the production of classroom activities is developed in interaction with the teachers' work. The presence of teachers-in-activity as part of the research group makes it possible to constantly question the feasibility of what is being proposed.

I have borrowed words and concepts from Fernanda Delprato, a young Argentine woman researcher at the University of Cordoba, when we talk about the search for the (re)signification and reciprocity of different knowledges and meanings that are made possible by the mutual recognition of different visions arising from the spaces occupied by each member of the group (be they teachers or researchers (Delprato 2013).

These are ideas that we find once again in authors such as Gerard Sensevy, in discussions about cooperative engineering. Sensevy criticizes a position of a certain duality that exists in the world of education regarding how teachers and researchers are considered. "According to this duality, teachers are seen as 'practical agents' trapped in a practical relation with their work, while investigators uphold a theoretical position. In this division of work, educational research must be an applied research (in which the practical agents must apply the 'scientific results' to their practice)" (Sensevy et al. 2013, p. 1032).

The paradigm of design-based research—which the author places in cooperative engineering—positions itself in contrast to this duality, proposing a different form of relationship between teachers and researchers.

Using the words of these authors, I would say that, for the Monday Group, the idea of the cooperative elaboration of a proposal and the posterior analysis of how it was developed in certain classrooms supposes eliminating the classic duality about persons who "think" and persons who "act," because all the participants get involved in the conceptual work and, at the same time, think about its concrete realization.

Looking in perspective, we can identify three dimensions in the production of the collaborative work:

- The collaborative production of a teaching sequence for a particular curriculum subject, with the incorporation of the computer into the mathematical work of students in the classroom.
- The study of didactical phenomena associated to computer mathematical work in the secondary school classroom.
- The reflection on the collaborative working device itself, a device that is modified in the search for the genuine conformation of a collaborative group in which we are included.

Based on the last two dimensions, necessarily imbricated, I have tried so far to point out different questions and challenges that the group has to face in the different stages and the ways they faced them. Although both themes have been studied in the literature, the expected contribution of this paper is to reflect on the convergence of both and the synergy and problems that occur when the two perspectives are merged for a common production.

In the next three examples, I will try to show some didactical complexities encountered in the work of the Monday Group.

32.3   Three Examples Which Illustrate Products and Processes in the Working Group

In the rest of this paper, we will present three examples, each one chosen from each one of the three stages during which the Monday Group has worked with computers. We will illustrate some areas of both products and processes through the work of the Monday group.

First example: New tasks for students involving new didactic questions that teaching must consider

This example corresponds to our first encounter with computers being used in the mathematical work of the students. The teaching proposal we developed to introduce students to working with polynomials and polynomial functions focused on:

- The production of higher degree functions as a result of the product of two lower degree functions. Basically, we created higher degree functions as a result of linear and quadratic functions.
- The strong presence of graphs, to the point that both the students and the teachers/researchers talked of "multiplying" straight lines and parabolas.
- The possibility of working with high degree polynomials with roots of multiplicity 2, 3, or more, their formulas and their graphs.

Within this context and after many days of work, the students were expected to come up with the formula and the graph of a sextic function without zeroes or be able to justify that the required function did not exist.[6] Although the students had no problem coming up with functions like this by multiplying three parabolas without zeroes, they were not always able to produce a screen in which the factor parabolas and the resulting sixth-degree function could be seen simultaneously. Let's see three different students' answers to this.

–   The first two students explained:

A sixth-degree function can, in fact, not cross the x axis because it contains three parabolas that can never have a zero. An example is shown in Fig. 32.1.

The sextic function has values for "y" that are so negative that they can't be seen by the naked eye.

–   A second pair of students was able to zoom in sufficiently so as to be able to see the sextic equation on the screen (Fig. 32.2).
–   A third pair of students explained that they modified the coefficients of the parabolas so that they became a little flatter, thus making it possible to see the sextic equation on the screen (also by zooming in a little; Fig. 32.3).

I bring this first example to identify a new task to be solved in the classroom, inherent in the work carried out by the computer: to make something look good on the screen. The previous examples illustrate a variety of positions that can be taken when faced with this task:

–   Not dealing with it.
–   Solve it using the program's tools.
–   Solve it using mathematical knowledge.

Although in this case the work was done in written form and included individual feedback, the example shows the new questions that teachers will have to deal with in the classroom: handling the interaction among student's solutions when they were elaborated from such different positions.

A more general reflection as a result of this specific example. Because of our work throughout the years and on different teaching proposals, we have run into very different and often unexpected answers and ways of solving problems proposed by students with computers. They range, as in this example, from answers based on computer skills and program-provided tools to answers based on mathematical knowledge specified by the students. And in between there are a wide range of gray areas and variations. It is then up to the teacher to find ways of working in the classroom to relate these varied solutions. Regarding this, we find ourselves confronted with the following questions: How can we make a mathematical question about a student's answer that focuses mainly on the program's

---

[6]This represents a new task in the high school classroom. Teachers stated they had never worked on high degree functions, even less so if they were factorized and were presented with their graphs.

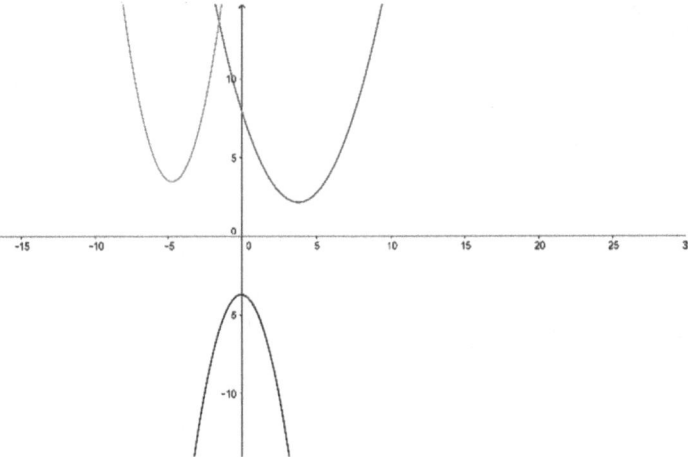

Fig. 32.1 The image presented by the first pair of students

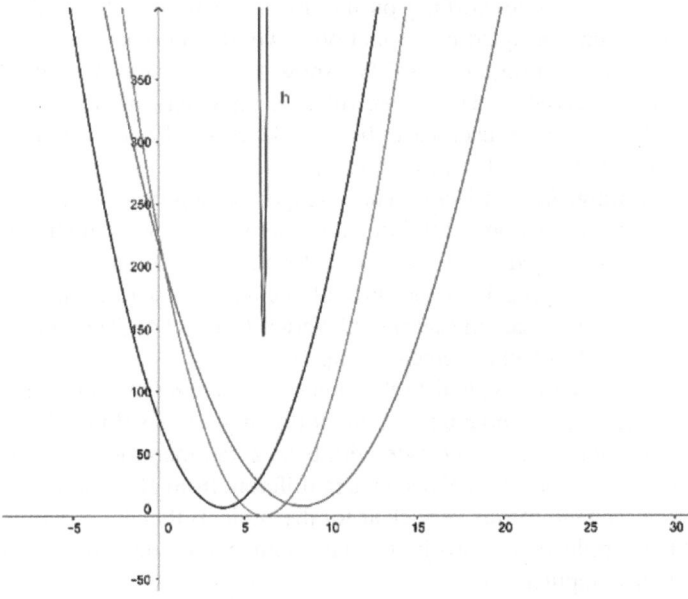

Fig. 32.2 Screen the second pair of students got

tools and is described in terms of actions on the computer? How can we move the most non-mathematical actions and discourses towards others that incorporate mathematical relations?

We will return to this question in our third example.

Fig. 32.3 Screen the third
pair of students got

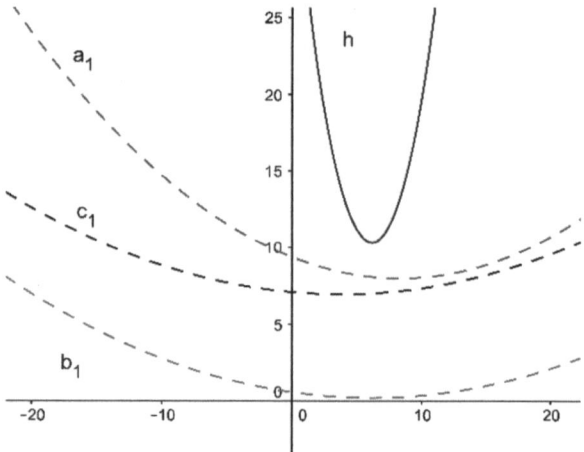

Second example: The rounding and the autonomy of the students

In the next stage, while working on the transformation of the proposal we had already thought up for quadratic functions, we ran into problems due to the "rounding off" that the program does. We show two episodes where the group had to make decisions about it. They were different, not only because of the mathematical activity required in each case, but also because of our growing confidence in mathematical work with the program.

Second example, first episode. The unexpected appearance of the "rounding off" took place in a classroom while students were working with the first problem of the proposal. That obliged us to take some "controlling measures."

Students were studying how the area of rectangles inscribed in a right-angle isosceles triangle with sides measuring 11 varied (Fig. 32.4). They had a dynamic model of this on a GeoGebra screen to explore.

At one point, they must calculate the area of a rectangle having "base 2." Using GeoGebra, some of them came up with the following screens (Figs. 32.5 and 32.6):

In both cases, they saw a rectangle with base 2 and therefore with a height of 9 and, nevertheless, the program gives an area different from 18. Since the calculation was easy to do by hand, it was very clear for the students that the program had made a mistake. This produced a chaos in the classroom! Some students even went so far as to shut their computers off.

In many cases, the area values that the program displayed when the different rectangles were dragged in did not match with the base values that it showed.

Fig. 32.4 Drawing of one of
the rectangles of the family

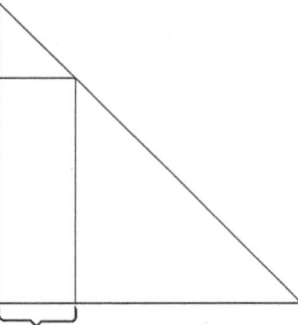

Fig. 32.5 Screen with a
rectangle of base 2 and area
different from 18

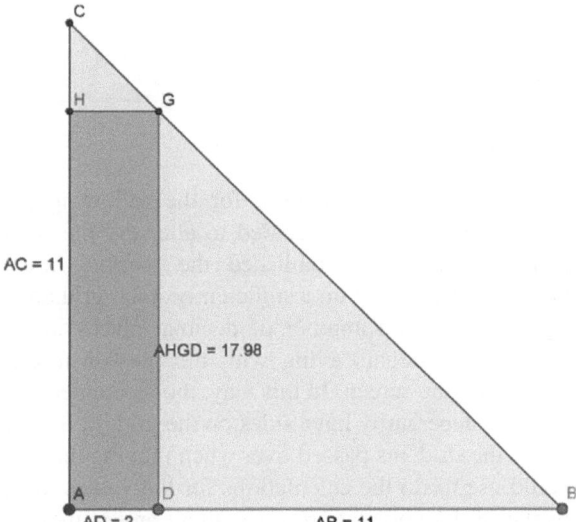

The machine used many more decimal digits to calculate than it showed on the
screen and, in fact, the students did not reach the value of 2 but rather some value
sufficiently close to 2.[7]

As the whole purpose of the activity was to study the variation of the rectangle's
area as a function of the length of one of its sides, measurements played an
important role. When thinking about students being introduced to the study of the
variation of sizes via dynamic models, these difficulties can contribute to the fact
that they will not use the model to answer the questions they are asked, as occurred
in the classes we just mentioned.

---

[7]The issue was that the problem could not be fixed by asking the program to display more
rounding off decimal digits. Even if we allow more decimal digits for the base values, the program
would calculate the area using more digits than that anyway. So it would once again display a
result for the area that would not be the right one.

Fig. 32.6 Other screen with
a rectangle of base 2 and area
different from 18

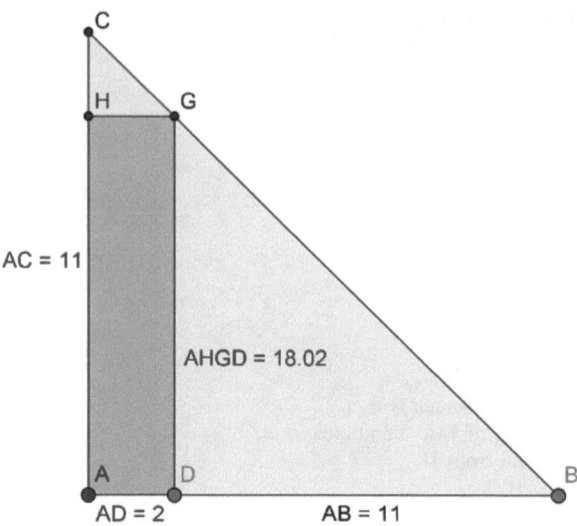

Faced with this situation, for the following presentation of this problem in another classroom, we decided to alter the file so that everything would turn out happily exact. We established the working area: We regulated the mouse's movement by setting up a sufficiently small grid and we included attractors in it and then we fixed the number of decimal places it would display. Additionally, we offered the students a file with the pre-constructed triangle to insure a certain position on the screen. In this way, the rectangles—that they built to begin the work —would necessarily have sides on the grid. In this way, we were able to control the values the students passed over when moving the mouse and the values the program would use to do the calculations for the values displayed on the screen. Finally, it displayed the complete result without any rounding off. We decided to share with the students some superficial information about this file.[8]

Second example, second episode. One of the problems we designed to work with GeoGebra was assigned at an exploratory place, but, as a way of making some of the mathematical relations in play more explicit, we introduced some numeric values into the problem so that the final answer could not be completed by the program and would require some paper and pencil work. Below is the specific problem from the proposal.

---

[8]For more detail see Sessa et al. (2015).

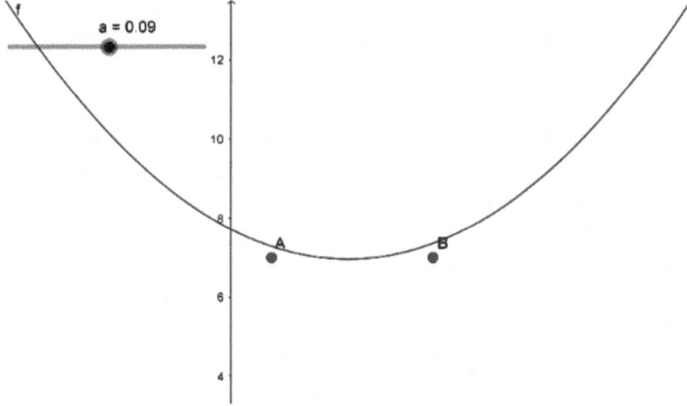

Fig. 32.7 An approximation to the requested parabola

Problem 5 (to be done with GeoGebra)

- Enter the points (1; 7) and (5; 7).
- Enter the parameters a and c and the function $f(x) = a(x - 2.9)^2 + c$
- Modify the values for a and c so that, if possible, the graph of $f(x)$ passes through the given points.

We wanted the students to explore this on a GeoGebra screen by moving parameters. In Fig. 32.7, you can see an attempt that gives an approximate answer.

They might be able to "visualize" a solution in the graphic view (Fig. 32.8).

They would then go to the algebraic view to verify whether the function really had a value of 7 at 1 and at 5. We expected that the algebraic view would show that

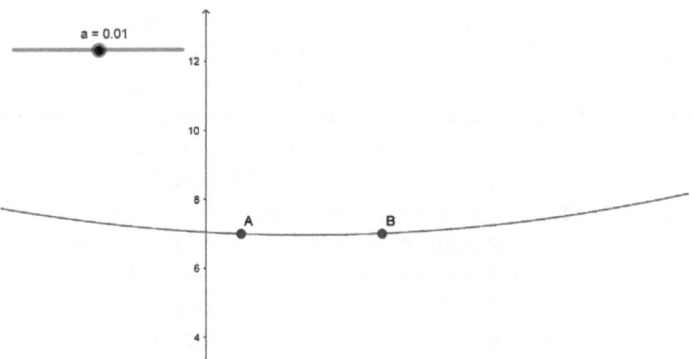

Fig. 32.8 Screen with an apparent solution

it did not happen. So the idea was that they should leave the program and find the reasons why it could not be done. However, when we asked GeoGebra for the evaluation of the function obtained from $f(x) = 0.01 \ (x - 2.9)^2 + 6.96$ at 1 and 5, the program answered 7 for both values. We were now once again facing the problem of incompatibility of information between the graphic view and the algebraic one, and even within the algebraic view. The function, of course, is not a solution, and if we zoomed in more, we would that in fact its graph does not pass through the points in question.

This time, we thought we could control this problem if we could anticipate the number of digits the calculation would have. To do this, we decided to set the file with a parameter of a 2-decimal digit and "rounding off" using 4 decimal places. In this way, we would be able to make it display the complete result. Doing that in the previous function, we get $f(1) = 6.991$ and $f(5) = 7.0041$.

This time, unlike the episode above, we thought that we could share these decisions with the students. We feel that the fact that they can estimate the number of figures in the result has a formative value both in mathematics as well as in the mathematical work done with the program. This would allow the students to understand more about how the machine works in order to prepare it to respond to the working necessities in each future problem. The decision to share this with the students was also the result of the more solid position that the group had in referring to the mathematical work with the program.

We as teachers also go through a process of *instrumental genesis* (Trouche 2004b). We believe we have advanced a little, but there is much left to do:

- Creating situations in which students can work in spite of coming up with "rounded off" results.
- Working in classrooms in which each student will have a different configuration of their working area.

All of this will occur if we maintain the objective of building greater student autonomy in working with GeoGebra.

Third example. From the actions on the computer to mathematical question

This is an example of our current work. On one hand, there was a change in the subject: We went back into the curriculum. On the other hand, we have incorporated new teachers while others had to discontinue their participation in the group. As the group settled again, some discussions reappeared. But it was never the same discussion because the people were not the same. Some of us already had history in the Monday Group and our ideas were different than they had been in the beginning. Among the new members, there were some teachers who were already using computers, so they brought new experiences to the discussion. Also, there were new members who had not used the computer in their classes yet and came to the Monday Group to see what it would be like. I will refer briefly to a question that arose in the shared planning stage, which is what we are going through at the moment. We are designing a small teaching sequence around a problem that we

Fig. 32.9 One of the triangles of the dynamic model

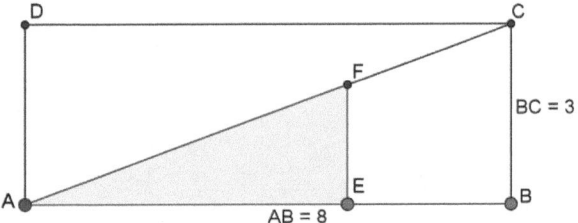

read in an article by Arcavi (2008). In a GeoGebra file, a dynamic construct is presented (Fig. 32.9).

Students can move point E to obtain other triangles. They must study the function that relates the base of the triangle with its area. They are 13-year-old eighth graders who do not know how to calculate the area of one of these triangles knowing only the base. But they can calculate the area when the base is 4 or 2. We ask them to calculate those values and then invite them to draw those points in the second graphical view by entering the ordered pair in the entry bar. To continue the graph of the function, we decided to present the "dynamic point" tool. This tool allows us to get a representation coordinated with the dynamic figure of the triangle.

The decision to introduce the "dynamic point" tool makes us take into account the complexity of this kind of representation of a function:

- It is linked to a dynamic situation in which two variables have been chosen. Each state of this situation, i.e., each specific triangle, determines both the first and the second coordinate of point P.
- As a representation of the function, it always requires some time to become a representation. It is not a representation at any given moment. Moreover, the representation requires our movement. Therefore, it is a representation in action.

In our sequence, we invite students to move E in Graphical View 1 and observe the path of the point P in Graphical View 2, especially noticing that P passes over the points already marked (Figs. 32.10 and 32.11).

At first, we decided not to activate the trace of P. We propose the following task:

(5) Explore the path of point P in the second graphical view and answer the questions.
(a) What is the area of the triangle when the base $AE = 3$?
(b) What will be the base of a triangle that has an area equal to 6?

In each case, explain how you reached those answers.

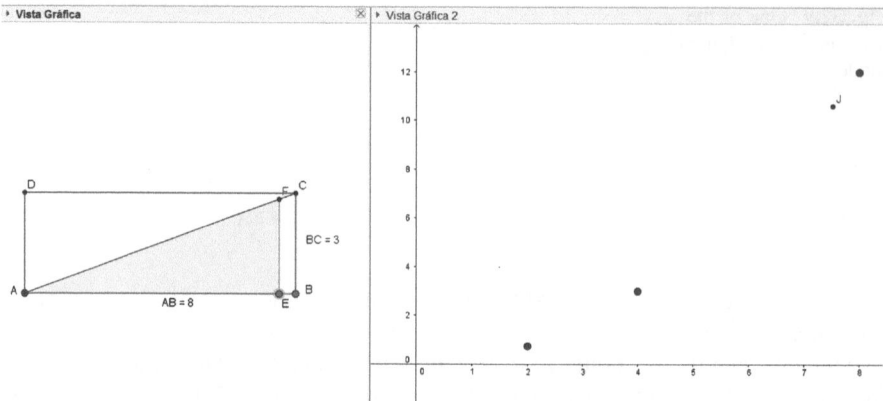

Fig. 32.10 Screen with a triangle and the associated point P

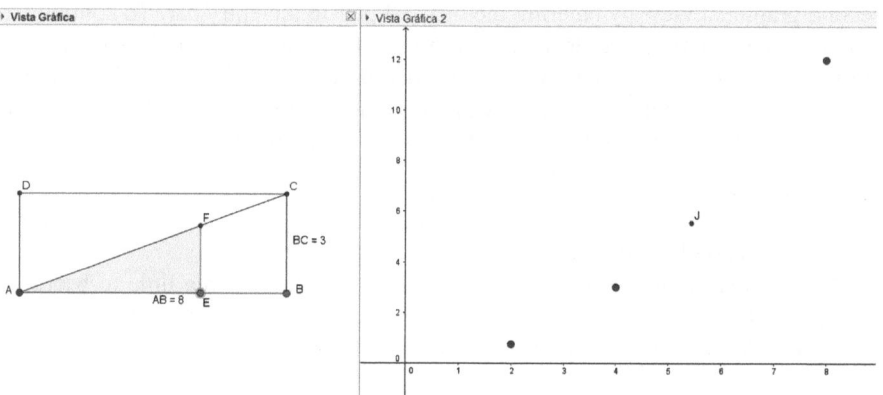

Fig. 32.11 Another triangle and its corresponding point P

The lack of computer presence in middle and high school math classes and our group's work path led to a shortage in our repertoire of student responses to working with the computer. Moreover, in the face of certain tasks we do not have experience that allows us to anticipate how they could be carried out. These two issues reconfigure the planning task, forcing us to pass ourselves through the experience of producing answers by working with the computer. In tune with this, at a meeting of the Monday Group we all set out to explore the task, working in pairs.

The work focused on appealing to tools of the program that allowed visualization and greater precision (zoom, changes in the configuration, "stretching" the axes, putting on a grid, etc.). It was work that took a long time and in which we did not put into play many mathematical relationships, and it left the pairs with a an unpleasant feeling about the task. In the final discussion of Monday Group it seemed that the balance was inclined to the side of giving little epistemic value to both the anticipated gestures of the students and the possibility of discussing them in the classroom.

The university team resumed this discussion at their weekly meeting, trying to identify what could be discussed in the classroom as a result of Problem 5 and the more accurate search work. Along these lines we had separate two issues:

1. On the one hand, in the search for precision, you first have to move the point E in Graphical View 1 to achieve a triangle of the family that fulfills the requested conditions: in Part a, the measurement of the base, in Part b, the measurement of the area. These measurements are read in the Cartesian plane that appears in Graphical View 2 from the location of P that is determined. Once point P is located, the problem involves obtaining information about the other magnitude as accurately as possible. This necessary coordination between a triangle and a point on the Cartesian plane is an opportunity to speak again of the link between the Cartesian graph of the relation and the geometric situation that allows us to define the relation.

2. On the other hand, once we locate point P, because we, perhaps visually, have ensured the value of one of the coordinates and we want to know the value of the other, tools such as scaling, the use of zoom, the "stretching" of the axes, and the addition of a grid all bring the appearance of new values on the screen for the coordinates of point P. These actions do not modify point P (because point E was not moved in Graphical View 1); however, some students may consider that they have obtained, for example, different area values for base 3. This last fact would be an opportunity to talk about the necessary uniqueness of the area value for a single triangle (which shows in Graphical View 1). Making explicit these questions in the classroom can contribute to the understanding that the base-area relationship is a function and to the understanding of the concept of a function itself.

These two questions refer to Lagrange (2000), who states, about the actions of a user and the responses produced by software, that it is necessary to distinguish which actions produce changes in the mathematical object and which produce changes in what can be seen from the representation that the program makes of that mathematical object.

We take from semiotic mediation theory (Mariotti 2009) the idea of pivot sign and extend it to the idea of *question pivot*. We think that a teaching intervention promoting a reflection on what is realized by and about the response given by the software allows construction of mathematical meanings in relation to the signs of the artifact. A question from the teacher such as "When we zoom, will point P change?" would allow reflection on the most artefactual actions in terms of mathematical objects.

This event refers us to what was formulated in first example:

– How can we make a mathematical question about a student's answer that focuses mainly on the program's tools and that is described in terms of actions on the computer?

– How can we move the most non-mathematical actions and discourses towards others that incorporate mathematical relations?

32.4   Coda

In this paper, we wanted to show some aspects of the didactic complexity that involve the incorporation of the computer in the mathematical work of the students. We did this by showing the close work of a collaborative group of school teachers and university team who faced this problem. We think that the convergence of views and approaches and the diversity of experience that characterize our collaborative group create good conditions for thinking about teaching and learning in key transformations for real and proper integration of TIC.

As part of a process, we raise some issues that also mark a way forward in our study:

– Gaining greater confidence in our work as teachers and greater autonomy in the work of students.
– Thinking about gestures and teacher discourse that allow us to weave bridges in the collective space of the classroom between work that is more focused on the tools of the program and the mathematical knowledge to which it is pointed.
– Advancing in the co-construction of working devices in the collaborative group.[9]

References

Arcavi, A. (2008). Modelling with graphical representation. *For the Learning of Mathematics, 28* (2), 2–10.

Arcavi, A., & Hadas, N. (2000). Computer mediated learning: An example of an approach. *International Journal of Computers for Mathematical Learning, 5*, 15–25.

Artigue, M. (1998). Ingeniería didáctica. In M. Artigue, R. Douady, L. Moreno, & P. Gómez (Eds.), *Ingeniería didáctica en educación matemática*. Colombia: Una empresa docente.

Brousseau, G. (1997). *Theory of didactical situations in mathematics*. Dordrecht: Kluwer Academic Publisher.

Delprato, M. F. (2013). *Condiciones para la enseñanza matemática a adultos de baja escolaridad* (Ph.D. Tesis - Ciencias de la Educación). Universidad Nacional de Córdoba, Córdoba, Argentina.

Drijvers, P., Doorman, M., Boon, P., Reed, H., & Gravemeijer, K. (2010). The teacher and the tool: instrumental orchestrations in the technology-rich mathematics classroom. *Educational Studies in Mathematics, 75*, 213–234.

---

[9]I want to especially thank Abraham Arcavi who very generously gave me relevant advice that helped make the writing of this paper clearer.

Drijvers, P., & Trouche, L. (2008). From artefacts to instruments: A theoretical framework behind the orchestra metaphor. In G. W. Blume & M. K. Heid (Eds.), *Research on technology and the teaching and learning of mathematics: Vol. 2. Case and perspectives (pp. 363–392).* Charlotte: Information Age Publishing.

Fioriti, G., & Sessa, C. (2015). *Introducción al trabajo con polinomios y funciones polinómicas. Incorporación del programa GeoGebra al trabajo matemático en el aula.* Editorial Unipe. Universidad pedagógica. http://editorial.unipe.edu.ar/wp-content/uploads/2015/10/Libro-Matemática-UNIPE.pdf.

Lagrange, J. B. (2000). L'intégration d'instruments informatiques dans l'enseignement: une aproche par les techniques. *Educational Studies in Mathematics, 43,* 1–30.

Mariotti, M. (2009). Artifacts and signs after a Vygotskian perspective: The role of the teacher. *ZDM Mathematics Education, 41,* 427–440.

Sensevy, G. (2011). *Le sens du savoir. Éléments pour une théorie de l'action conjointe en didactique.* Bruselas: De Boeck.

Sensevy, G., et al. (2013). Cooperative engineering as a specific design-based research. *ZDM Mathematics Education, 45,* 1031. https://doi.org/10.1007/s11858-013-0532-4.

Sessa, C., et al. (2014). *Matemática, función cuadrática parábola y ecuación de segundo grado. Aportes para la enseñanza. Nivel Medio.* Ministerio Educación Gobierno a Ciudad de Buenos Aires. http://estatico.buenosaires.gov.ar/areas/educacion/curricula/media/matematica/matematica-cuadratica.pdf.

Sessa, C., et al. (2015). La transformación del trabajo matemático en el aula del secundario a partir de la integración de las computadoras. In A. Pereyra (Ed.), *Prácticas pedagógicas y Políticas Educativas. Investigaciones en el territorio bonaerense.* Buenos Aires: Editorial UNIPE.

Trouche, L. (2004a). Managing the complexity of human/machine interactions in computerized learning environments: Guiding students' command process through instrumental orchestrations. *International Journal of Computers for Mathematical Learning, 9,* 281–307.

Trouche, L. (2004b). Environnements informatisés et mathématiques: quels usages pour quels apprentissages. *Educational Studies in Mathematics, 55,* 181–197.

Vergnaud, G. (1990). La théorie des champs conceptuels. *Récherches en Didactique des Mathématiques, 10*(23), 133–170.

Chapter 33
Re-centring the Individual in Participatory Accounts of Professional Identity

Jeppe Skott

Abstract Studies of professional identity are generally conducted using participatory frameworks and from the perspective of a particular development initiative. They provide understandings of teachers' move towards more comprehensive participation in the practices the initiative promotes. Studies in line with this main trend, however, leave questions of teacher identity unanswered when teachers are not enrolled in long-term development programmes. I argue that to address such questions a different framework is needed, one that maintains the participatory stance, but focuses on the individual teacher rather than a development initiative. It is the intention of the Patterns-of-Participation framework (PoP) that I introduce to re-centre the individual in this sense. To make my point, I discuss how research frameworks may be conceptualized and compared and use the resulting "frameworks framework" to contrast studies of the main trend with the intentions of PoP.

Keywords Professional identity · Teacher development · Research frameworks
Social practice theory · Patterns of Participation (PoP)

33.1 Introduction

The notion of professional identity has attracted increasing attention in research on and with teachers over the last decade, both in mathematics education and beyond. The construct of identity is generally conceived in processual and participatory terms, and often the intention is to understand how teachers' experiences with programmes for educational development inform and transform their tales of themselves as professionals as well as their contributions to the practices that evolve at their schools and in their classrooms. In this sense the research interest in identity may be seen as a supplement and to some extent as a challenge to most research on

J. Skott (✉)
Linnaeus University, Vaxjo, Sweden
e-mail: jeppe.skott@lnu.se

G. Kaiser et al. (eds.), *Invited Lectures from the 13th International Congress on Mathematical Education*, ICME-13 Monographs,
https://doi.org/10.1007/978-3-319-72170-5_33

teachers' knowledge and beliefs, which to a greater extent relies on constructivist interpretations of human learning.

In what follows I build on this relatively recent, participatory approach. However, I argue that there is also a need for frameworks that re-centre the individual in studies of teacher identity. I present one such framework called Patterns of Participation, PoP, and ask how it differs from other participatory approaches to research on and with teachers, including what it has to offer in terms of interpretive potential of teachers' professional identities. The moral of the story is that both approaches are needed, and that PoP is helpful for understanding teacher development in the majority of cases in which they are not involved in comprehensive programmes for teacher development (TD programmes).

To make my argument, I first discuss the notion of a research framework before outlining perspectives on identity in key references beyond mathematics education. These two sections form the backdrop for a discussion of a general trend in identity studies in mathematics education (Sect. 33.4). I then discuss identity as conceived in symbolic interactionism, a further inspiration for the PoP-approach, which I present in Sect. 33.6. I round off by comparing and contrasting the general trend and PoP studies of identity.

33.2   Theories and Frameworks

Mathematics education research is characterised by a mixture of what Steiner (1984) calls theoretical imports and home-grown theories. Consequently the role of and mutual relationships between different theoretical approaches is a recurrent theme in the field. Phrasing this discussion in terms of theory networking, Bikner-Ahsbahs and Prediger (2010) place propensities to engage with a variety of theoretical perspectives on a continuum ranging from ignoring other theories to integrating them globally. Comparing and contrasting different theories are placed approximately in the middle of the continuum.

It is not obvious what it takes for an approach or a framework to qualify as theory and what role theories may have in mathematics education research. Apparently with inspiration from mathematics, Niss argues that a theory is a hierarchically ordered network of concepts and a related set of claims about some field of investigation (Niss 2007). The claims are, according to Niss, either "fundamental" in the sense of being beyond justification within the theory itself, or derived from some such set of claims by formal or experimental/experiential means. Romberg (1998) talks about research conducted by the *National Center for Research in Mathematical Sciences Education* and says that by theory he and his colleagues mean "a set of statements about the causal relationships between and among a number of variables used to describe features of classroom communication in mathematics" (p. 387).

Adopting a somewhat broader perspective, Radford (2008) argues that a theory may be thought of as an ordered triple of basic *Principles*, *Methodologies*, and paradigmatic *Q*uestions. The *P*rinciples are the system of views and statements that serves to "delineate the frontier of what will be the universe of the discourse and the adopted research perspective" (p. 320). As an example Radford mentions how cognition is considered important, but is interpreted differently depending on what system of principles (e.g. constructivist or socio-cultural) is adopted. The *M*ethodology is the set of methods used, but also the reasons for using them, and the arguments for turning pieces of data material into data, that is, making them worthy of analysis. The paradigmatic *Q*uestions are the ones that were addressed initially in the field and that continue to orient it with regard to what and how questions are asked. Like Bikner-Ahsbahs and Prediger (2010), Radford is interested in theory networking and suggests that (*P*, *M*, *Q*)-triples may be helpful for instance for considering how to link the *P*rinciples of one theory with the *M*ethodology of another.

My intention at present is not to suggest ways of combining certain elements of one theory with different elements from another. Instead I discuss a compare-and-contrast approach to networking in the particular case of frameworks currently used in the study of professional identity. In particular I discuss differences and similarities between what I see as a major trend in mathematics education research on identity on the one hand and PoP on the other. To do so, I begin by discussing the notion of a research framework. I have previously used this "frameworks framework" to consider the relationship between main-stream belief research and PoP (Skott 2015a). The differences are in this case more obvious than the ones I am pointing to in the present context. However, I suggest that the approach is also useful when comparing frameworks currently used to study professional identities.

The notion of a framework carries different metaphorical connotations irrespectively of whether it is—or in the particular case functions as—a theoretical, a conceptual, or a practical one (cf. Eisenhart 1991). It may limit what one can see or focus on (cf. a picture frame); it may be a support structure that upholds the interpretations made and ensures that they are sound and do not "collapse" (cf. construction frameworks for buildings); or it may allow one to go to places from where one can see well-known landscapes from new vantage points or even to travel to new territories and experience things never before imagined (cf. a bicycle frame). Irrespectively of whether any or all of these metaphors carry weight in a particular case, the framework constitutes an argument for why the approach adopted makes sense for the purposes of the particular study. A framework is "a basic structure of the ideas (i.e., abstractions and relationships) that serve as the basis for a phenomenon that is to be investigated" (Lester 2010, p. 69).

In this section I draw on and extend Radford's (2008) discussion of the (*P*, *M*, *Q*)-triple, even though the frameworks that I compare, those used in participatory studies of professional identity, may not qualify as theory in the sense of Niss or Romberg (cf. above). The rationale is that it is also helpful to be able to compare and contrast these more loosely structured approaches to empirical study.

I use the notion of a framework in a narrow and in a broader sense (see Fig. 33.1). In the narrow sense it includes

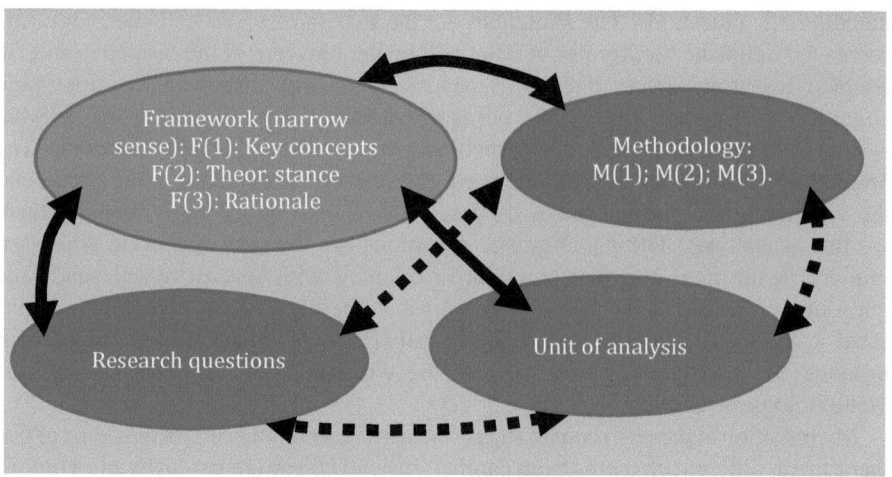

Fig. 33.1 The "frameworks-framework"

F(1)   a set of key concepts: in research on and with teachers the set may include mathematical knowledge in or for teaching; beliefs; participation; practice; interaction; identity;

F(2)   a theoretical stance: used to interpret the meaning, relative significance of, and relationships among the concepts in F(1). For instance, primarily acquisitionist and primarily participatory approaches to understanding the acts of teaching may be used to develop different interpretations of concepts that are nominally the same or similar;

F(3)   a rationale: the reasons for engaging in a line of study, such as developing novel understandings of issues under investigation, supporting educational development, or some combination of the two.

In a wider sense a research framework includes also the unit of analysis, the paradigmatic research questions, and the methodology. In line with Radford (2008), I consider the methodology a triad of M(1): the methods used; M(2): the reasons for using them; and M(3): how issues and relationships in data material (e.g. transcripts from a video recorded classroom observation) become data, that is, worthy of attention in the subsequent analysis.

As indicated by the arrows in Fig. 33.1, the elements of the frameworks-framework are considered reflexively related. It is not assumed that the choice of framework (in the narrow sense) or the methods follow linearly from the research questions. There is in this interpretation no unidirectional and almost causal relationship between the questions and the other elements of the framework (in the broader sense). The elements of the frameworks-framework are considered mutually dependent, as for instance the meaning of the research questions is elusive, if other elements are not considered.

33.3 Researching Identity—Inspiration from Outside Mathematics Education

The participatory and situative orientation is apparent in many studies of identity (e.g. Beauchamp and Thomas 2009; Beijard et al. 2004; Brown and McNamara 2011; Cobb and Gresalfi 2011; Diversity in Mathematics Education Center for Learning and Teaching 2007; Hodgen and Askew 2007; Horn et al. 2008; Teacher Education Quarterly 2008). Key references and main sources of inspiration come from discourse analysis and social practice theory and include people like Gee (2000–2001), Holland et al. (1998), Lave (1988, 1996), and Wenger (1998). Far from considering identity a stable personality trait, these scholars view it as ever-evolving and constantly renegotiated in social practice.

Gee (2000–2001) describes identity as "being recognized as a certain kind of person in a given context" (p. 99). Such recognition is based on "a combination" of for instance ways of speaking, acting, dressing, and feeling that actively invites or at least leaves one open for interpretations of being a particular kind of person in that particular situation. Identities may be viewed (and have been viewed) as primarily the result of forces of *N*ature (e.g. being an identical twin); as related to one's position in an *I*nstitution (e.g. being a university professor); as *D*iscursively established (e.g. being a charismatic person); and based on one's allegiance to and involvement with a set of practices in an *A*ffinity group. (e.g. being a Star Trek fan). These *N*-, *I*-, *D*-, and *A*-identities interrelate in a myriad of ways. For instance, the *N*-identity of being a child with ADHD may be Institutionally sanctioned when the child gets a diagnosis, and there are Affinity groups that arrange support activities for ADHD children. Gee asks two questions about how identities are established and sustained. A macro-level question concerns how institutions and discourses function so as to make recognition of some "combination" possible in a particular context? A micro-level counterpart is how "a combination" becomes recognised, contested, or renegotiated in particular face-to-face interactions?

Social practice theory views individual identities as contextually embedded and dependent, and therefore multiple, fluctuating, and always in-the-making. Holland et al. (1998) define identity as "the imaginings of self in worlds of action", and as identities are "lived in and through activity [they] must be conceptualized as they develop in social practice" (p. 5). There are two interrelated aspects to this, one of which concerns how people position themselves and each other in everyday interactions. The other aspect is related to figured worlds, that is to "socially and culturally constructed realm[s] of interpretation, in which particular characters and actors are recognised, significance is assigned to certain acts, and particular outcomes are valued over others" (Holland et al. 1998, p. 52). Such imagined and collective as-if worlds and their related discourses and practices orient action and sense-making and serve in open-ended ways to "figure the self" (p. 28). Figured identities, then, concern narrativised versions "that make the world a cultural world" (p. 127).

Also working in social practice theory, Wenger (1998) argues that identity may be viewed as negotiated "ways of being a person in [a] context" (p. 149). He elaborates on this for instance by saying that "identity [...] is a layering of events of participation and reification" connected to a practice (p. 151). Practice is viewed as a process and outcome of collective learning. Engaging with one another in the pursuit of a joint enterprise, people use or jointly develop a repertoire of modes of participation in the practice, negotiating its meaning and the character of their own membership in the community in question in the process.

Although Wenger's distinction between participation and reification does not parallel the one from Holland et al. between positional and figurative identities, both pairs of concepts point to the emerging interplay between immediate social encounters and social markers or identifiers (e.g. being recognised as a qualified teacher) as sources of identity. Empirical studies informed by and contributing to social practice theory have focused on a range of different social constellations, from Alcoholics Anonymous in the US to tailors in Liberia and emphasised how individuals gradually come to participate more profoundly in the practices involved (cf. Wenger 1998) or orient themselves towards the figured worlds in question (cf. Holland et al. 1998).

One difference between the concepts of practice and figured world is that the latter may gain a social existence for the individual without her or his involvement in the renegotiation of the meaning of the broader enterprise. It has been argued, for instance, that the current reform movement in mathematics education (*the reform*) may qualify as a figured world for a teacher (e.g. Ma and Singer-Gabella 2011). Within the reform discourse there are certainly certain characters that are recognised and certain acts and outcomes that valued over others (cf. the definition of figured world above). In spite of that, *the reform* hardly qualifies as a practice in which the teacher contributes to the renegotiation of its broader social meaning, although (s)he may of course renegotiate its role and meaning as it is dealt with in a particular development initiative or among her colleagues and as it relates to his/her own contributions to the practices of the classroom. In spite of the difference, however, both practices and figured worlds may significantly orient contextually dependent identities as they evolve in classrooms and at schools.

33.4   Mathematics Education Research on Identity

Significant parts of the growing scholarship on mathematics teachers' professional identities draw on the broader scholarship on identity outlined above (Cobb and Gresalfi 2011; Cobb et al. 2009; Hodgen 2011; Hodgen and Askew 2007; Horn et al. 2008; Ma and Singer-Gabella 2011; Sfard and Prusak 2005). In what follows I do not claim to do justice to the field as a whole, but refer to four studies that I consider representative for a reasonably significant part of the field. Or rather, the four studies are non-representative in the sense that they are more comprehensive and better documented than most, but at least somewhat representative for my present purposes, that is, as far as their frameworks are concerned. In this sense I suggest that they are representative of a somewhat general trend in mathematics education research on professional identity.

Hodgen and Askew (2007) refer to Holland et al. (1998) in their report on a case study of a primary teacher, Ursula, and analyse how her emotional relationships with mathematics develop over the 3½ years of her participation in a TD initiative. In their analysis, Hodgen and Askew relate emotion to the teachers' "figured" and "positional" identities (cf. above), that is, aspects of identity that are either relatively stable in the sense of cutting across teachers' participation in different communities of practice or more local and position them in relation to one particular context. They argue that a focus on identity helps understand why professional change is difficult to achieve, not least in mathematics, and use the construct to interpret how Ursula develops from distancing herself from mathematics to challenging dominant norms for school mathematics and "constructing a strong and powerful image of a different mathematics teaching" (p. 482).

Cobb and Gresalfi (2011) seek to document how teachers relate to and come "to identify with the vision of high quality mathematics instruction" promoted by the TD initiative in which they take part (p. 271). Cobb and Gresalfi argue that it is useful to view identity as "a set of practices and expectations that shape participation in particular contexts" (pp. 273–274), and they focus on the extent to which teachers in the study identify with "others' expectations for competent teaching" (p. 275). They draw on Gee's (2000–2001) distinction between institutional and affinity identity and investigate how middle school teachers involved in the comprehensive TD programme react to tensions and conflicts between the "normative affinity identities for teaching" established within the TD programme and the normative institutional identities of the schools at which the teachers work.

A somewhat similar approach has been taken to prospective teachers. Both Horn et al. (2008) and Ma and Singer-Gabella (2011) draw on Holland et al.'s notion of figured world, and investigate how prospective teachers' identities develop. Horn et al. define identity as "the way a person understands and views himself, and is often viewed by others, at least in certain situations—a perception of self that can be fairly constantly achieved" (p. 62). Working with secondary teachers, they investigate the reflexive relationships between such identities and two different contexts, the university-based teacher education programme (TEP-world) and the prospective teachers' experiences from their school placements (Field-world). They focus on how research participants' descriptions of what a good teacher is and of themselves as teachers-to-be relate to their engagement in these two different "worlds" of their pre-service education, while acknowledging that "the figurative RealWorlds of the interns' own past experience" (p. 63) also play a part. Ma and Singer-Gabella work with prospective elementary teachers and—like the other authors mentioned above—distinguish between the reform and the traditions of school mathematics. They argue that teacher education programmes that value the reform must move beyond teaching relevant content and make students familiar with and participants in the figured world [of the reform], both reshaping their various models of identity of children and teachers and introducing or promoting an appropriate construction of mathematics. (p. 10)

All the studies above emphasise changes in professional identity as teachers become involved in two distinct sets of practices, one of which represents the traditions of school mathematics, while the other introduces current reform recommendations for the subject as taught in school. The delineation of these two sets of practices constitutes a normative dimension in these studies. Further, the ambition is generally to understand and support teachers' gradual engagement with the latter of the two sets of practices. The analytical focal point, then, is on teachers' identity trajectory as they are introduced to TD-practices promoting the figured world of the reform. It is generally implied that identity change is a long-term endeavour that it takes more than yet another course in mathematics for teachers to accomplish. It requires teachers to become "a 'different' teacher and a 'different' person" (Hodgen and Askew, p. 474). The extent to which programmes are successful in this constitutes the empirical and analytical dimension of the studies.

I suggest that the two dimensions of the studies referred to above, the normative and the empirical/analytical, indicate a trend in studies of mathematics teachers' professional identities. In general, reform-oriented practices become the centre of attention. Phrased differently the trend is to prioritise a particular set of practices, those related to the PD or teacher education programme, and a related figured world, the reform. In passing it should be noted that this is in line with the theoretical references for this line of research, as most studies in social practice theory in a somewhat similar sense foreground a particular practice or figured world, be it claims processing at an insurance company (Wenger 1998), girl scouts selling cookies (Rogoff 1995), or romance at a university campus (Holland et al. 1998). Centring the reform does not entail a disregard for how individual teachers engage with the practices that the reform promotes. Indeed, these and other studies of professional identity explicitly analyse how individuals or groups of teachers construct new narratives about themselves in relation to school mathematics (Hodgen and Askew 2007); come to identify with practices promoted by the PD initiative (Cobb and Gresalfi 2011); use practices and meaning systems of teacher education contexts as "resources for them to understand their own emerging sense of themselves as teachers" (Horn et al. 2008, p. 67); or focus on "relationships among prospective teachers' identities as learners, doers, and teachers of mathematics and the contexts and practices in which they are situated" (Ma and Singer-Gabella 2011, p. 9). The overall intention may be phrased metaphorically in the terminology of Lave and Wenger (1991), as attempts to support prospective or practising teachers' movement from peripheral to more comprehensive or substantial modes of participation in practices envisaged by the figured world of the reform (cf. Fig. 33.2).

Mathematics education research on identity, as exemplified above, has contributed with significant understandings of the potentials of different approaches to professional development, including the challenges that may arise if the intentions

Fig. 33.2 Moving towards
more comprehensive
participation in practices
promoted by the reform

of PD initiatives are in conflict with dominant traditions of school mathematics. However, this line of research leaves questions related to professional identity and teacher learning unanswered in the majority of cases in which teachers are not engaged in long-term development programmes. I suggest that there is a need to supplement this line a study and adopt a somewhat different approach, if the intention is to address such questions.

Such an approach should not prioritise, but also does not disregard, current reform efforts, and studies need to comply neither with the normative nor with the empirical/analytical dimension of other identity studies as outlined above. Rather than foregrounding and centring current recommendations for reform, I suggest addressing the latter of the two questions that Gee asked (cf. Sect. 33.3), the one concerned with how micro-interactions develop and sustain identities, and to re-centre the individual by focusing on the experiences of the teacher in those interactions. In fact the suggestion is to define professional identity as teachers' experiences of being, becoming, and belonging as school and classroom interactions unfold. In order to develop such a perspective, I seek inspiration from symbolic interactionism.

33.5   Interactionist Approaches to Identity

A part of symbolic interactionist writing on professional identity has focused on socially constructed, individual meaning making and pays attention to "the experience of work from the point of view of those who engage in it" (Shaffir and Pawluch 2003, p. 894). This includes how individuals become part of an occupation by engaging in the practices involved in ways that are deemed appropriate and legitimate within the community. This line of research concerns the key concept of self as developed by Mead (1913, 1934) and the inherent I-me duality. According to Mead, the me is the result of a reflective approach to oneself, that is, of viewing oneself from the perspective of others. One takes the attitude of others to oneself, interpreting their actual or expected actions symbolically, including their possible reactions to one's own behaviour. In this terminology "'the me' [is] that group of organized attitudes to which the individual responds as an 'I'" (Mead 1934, p. 186).

There are different theoretical and methodological approaches to identity in symbolic interactionism. Structural symbolic interactionists view self as a set of multiple identities, each a result of internalising role expectations stemming from particular positions within social structures. In this tradition identities are defined as "the meanings that persons attach to the multiple roles they typically play in highly differentiated contemporary societies" (Stryker and Burke 2000, p. 286). Although identities are multiple as they relate to a variety of different organisations, groups, or institutions, they are considered relatively stable in terms of time and space. They do vary, though, in terms of salience, that is, in terms of the probability that they are invoked in social interaction. From this perspective the self is viewed as a salience hierarchy of identities from different, but somewhat durable social constellations (Stryker 2008; Stryker and Burke 2000). Smith-Lovin (2007) builds on Stryker's work, but suggests that in highly segregated institutional settings people rarely encounter situations in which a multitude of high-salience identities are activated at the same time. According to her, people may have multiple identities and complex selves; but they rarely enact more than one significant identity in the same context.

Other symbolic interactionists, working more in Blumer's tradition (Blumer 1966, 1969), adopt a less structural and more dynamic and situated perspective. According to Blumer "Mead saw the self as a process" (1966, p. 535). In this interpretation the I acts, but the individual instantaneously views her- or himself through the eyes of individuals, specific groups, or generalised others and adjusts her/his actions accordingly. Identity may then be seen as related to the shifting versions of the self that emerge in social interaction. This perspective has methodological implications, and Blumer suggests what he calls a naturalistic approach to empirical research. In contrast to the structural perspective that seeks to develop generalizable hypotheses that may be refuted by quantitative means, Blumer's perspective is qualitative and interpretive. He describes the methods in terms of exploration and inspection (Blumer 1969).

Exploration is the primarily descriptive phase of "getting close to social life" by using a range of qualitative approaches, that is, "any ethically allowable procedure that offers a likely possibility of getting a clearer picture of what is going on in the area of social life" (Blumer 1969, p. 41). Such procedures may include for instance observations, interviews, text analyses, and discussions among people closely connected to the focus of the study. Inspection is, in Blumer's terminology, the phase of developing theoretical accounts of what he refers to as the analytical element of the study. The analytical element may for instance be processes, modes of organization, networks, and relations among networks. As an example he mentions the assimilation of girls in organized prostitution and argues that inspection should consist of the "careful scrutiny of [empirical] instances with an eye to disengaging the generic nature of such assimilation" (Blumer 1969, p. 44).

33.6   PoP and Teacher Identity: Re-centring the Individual

Elsewhere I have categorised different approaches to research on teachers' beliefs, two of the categories being belief enactment and belief activation (Skott 2015b). They share the view of teachers' mathematics related beliefs as relatively stable mental constructs that are at least potentially important for teachers' contributions to classroom practice; the main difference between them is the extent to which it is expected that teachers' beliefs impact practice. To some extent the view of beliefs as mental constructs embedded in the individual parallels the perspective on identity in structural symbolic interactionism as outlined above, although the belief-activation approach to beliefs is less focussed on structure and more on immediate interaction. The resemblance is that structural symbolic interactionists (Stryker 2008; Stryker and Burke 2000) conceive of identities as affectively laden and relatively stable cognitive schemas that are activated to different degrees in different situations. They are not, then, situated, but variably salient and therefore differentially enacted in different situations.

The PoP framework grew out dissatisfaction with how belief research tends to ignore the field's conceptual and methodological problems. As a result PoP challenges—among others—both the belief-enactment and the belief-activation approaches (Skott 2013, 2015b). Conceptually the argument is that there is little to gain by assuming the existence and behavioural impact of reified mental constructs in the form of temporally and contextually stable beliefs, and that a more dynamic perspective may be helpful when seeking to understand teachers' acts and meaning-making. From this perspective teaching is not seen as enactment or activation of temporally and contextually stable beliefs, but as constant negotiation of the teacher's own contributions to the practices that unfold at the instant in view of her/his prior engagement in other practices and figured worlds. This perspective on teachers' acts and meaning-making is based on empirically grounded analyses of teaching-learning processes in novice teachers' classrooms. My argument in the present context is that it provides a useful approach also to understanding teacher identity.

For a very brief illustration of the use of PoP in identity studies I refer to a study of Anna, a young, Danish novice teacher of mathematics for middle and lower secondary school. I followed Anna in four two-week periods over the first three years of her teaching career at Northgate Primary and Lower Secondary School, a well-functioning school with relatively few social problems in a well-to-do area of a major city in Denmark. The data for the study are from interviews and informal conversations with Anna, including some using stimulated recall; from classroom observations; from observations of team meetings and short teacher development initiatives; and from interviews with the leadership and with the three other teachers in Anna's team. The team is a group of teachers, who teach almost all subjects to the three classes in a year group. In the first year of the study Anna's team teaches year 7 and the team follows the same three classes until they leave the school after grade 9.

At the time of her graduation from college, Anna is highly committed to her profession. She says that the four years of her pre-service education were tremendously important to her, and that she developed from seeing teaching as something she wanted to study to seeing it as "something that has become part of you" (interview 1). Part of her commitment is to the reform and Anna consistently emphasizes student communication, reasoning, and modelling as important parts of school mathematics. She also distances herself from the other mathematics teachers at Northgate, convinced that they think and act differently when teaching the subject. However, her team is important to Anna and she explicitly wants to learn from older team members. Talking about Ian, who has 25 years of teaching experience, she says that she wants to "maybe copy a little of what [he] does" (second interview). One reason for this is that she finds the other team-members' interest in building close relations with the students much in line with her own. In the first half of the study Anna consistently talks about her positive relationship with the students and half-jokingly describes it as being "somewhere between a mother and a friend".

Over the three years of the study, Anna's allegiance shifts. She has, she says, been recognized as a good mathematician by the other teachers of mathematics. Also, she has developed a trusting relationship with the leadership, and they now often ask for her advice on administrative and educational matters. Besides, Anna is increasingly aware that everybody at Northgate considers it a privilege to work at the school, but one that comes with some obligations. Both her team and the leadership talk about the high level of ambition among everybody employed at the school and say that you are expected to be committed and take your own initiatives. While all of this indicates a stronger allegiance to her colleagues and to the school in general, Anna's commitment to the reform has weakened. Also, she now considers herself so much older than the students that the significance and character of her relation to them has changed.

Neither Anna's position among her colleagues nor her awareness of the high expectations connected to teaching at Northgate outline in detail what it entails to teach mathematics to a class of lower secondary students. However, these developments coincide with and are probably related to Anna's fading commitment to the reform and to a change in her initial interpretation of what it means to have good relations with the students. Taken together this changes Anna's experiences of being a teacher; of when she is doing a good job and what it means to become better; of what it takes to be recognised as competent colleague; and of what it means to belong at Northgate.

This brief outline of the study of Anna indicates that in line with the main trend in other studies of identity, the PoP framework draws the concepts of practice and figured worlds as developed by Wenger (1998) and Holland et al. (1998) respectively. However, and somewhat in contrast to that trend, PoP does not focus on a set of practices in a particular development initiative and on the figured world(s) it promotes (e.g. the reform). Drawing on symbolic interactionism, PoP re-centres the individual teacher and foregrounds a situated view of identity (Fig. 33.3).

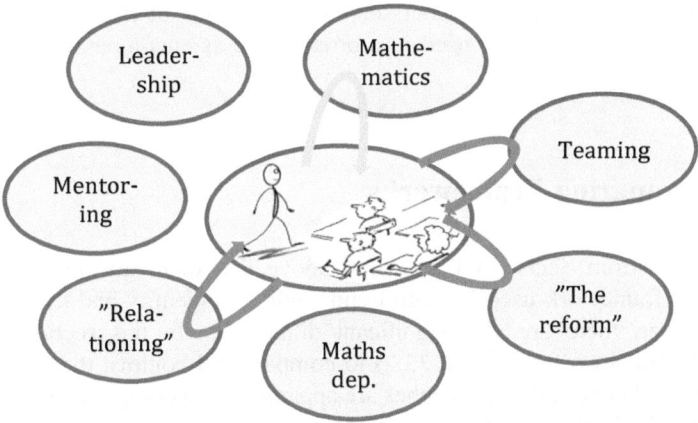

Fig. 33.3 Recentring the individual

Interpreting or envisaging students' reactions to her own behaviour, the teacher may take the attitude to herself of the students in question and for instance seek to overcome their frustration with a problem solving task, promote or sustain her own mathematical or professional authority, support the students' self-confidence, solve an evolving disciplinary problem, and many more. But doing so, she may also draw on practices and figured worlds beyond the classroom, such as those stemming from her teacher education programme or a recent development initiative; from more or less systematic collaboration with her colleagues; from a dominant discourse about the school as promoted by the leadership; from recurrent discussions with parents at PTA meetings or with others with an interest in the running of the school; and even from practices and figured worlds that are less immediately related to school life. Such practices and figured worlds may function as generalized others for the teacher in question as she interacts with her students. And conversely, as teachers work in teams, communicate with the leadership and the parents, individually prepare for tomorrow's teaching, or engage in other professional activities, they may draw on and discursively reengage with each of the other practices as well as with previous classroom interaction. If and how the teacher experiences herself as a good mathematician, a close ally of the students, a valued colleague, a trusted professional, a promoter of the reform, or as any of the opposites to these or other positive characteristics of being a teacher depends on the attitudes she takes to herself at the instant. And as the significance of and relationships among practices and figured worlds beyond the classroom change, so does the experience of being in it.

From this perspective and in the terminology of symbolic interactionism, professional identity relates to the shifting versions of the me that evolve in interaction. More specifically, it may be defined as the experiences of being, of becoming, and of belonging that emerge as teachers engage in the acts of teaching. In this definition the term of teaching is used in a broad sense as encompassing all forms of engagement in the profession.

33.7 Comparing Frameworks

It is apparent from Sects. 33.4 and 33.6 above that there are many similarities between the framework used in main-trend studies of identity and the one used in PoP. However, there are also significant differences. In this section I use the frameworks framework (cf. Sect. 33.2) to compare and contrast the two.

Similarities between the approaches are apparent when comparing two aspects of the frameworks in the narrow sense, the key concepts and the theoretical stance. Beyond the construct of identity, key concepts of both include the ones of interaction, practice, participation, and figured world. Also, and as reflected in these key concepts, the theoretical stance is similar to the extent that both approaches are inspired by social practice theory and/or discourse analysis. This is non-trivial, as other studies of identity draw on for instance personality theory or psycho-social theory (Skorikov and Vondracek 2011). Also, it means that the concept of identity is in both cases considered dependent on individual engagement with particular social practices.

There are also similarities between the frameworks in the broader sense. In both cases the unit of analysis is concerned with the individual-practice(s) interface. In terms of methodology, both use a multitude of qualitative techniques (observations, interviews, text analyses) in longitudinal studies in order to analyse changes in the relation between research participants and significant social practices. Also, both lines of study tend to analyse data without a ready-made set of codes and categories, and often they make explicit reference to coding procedures inspired by grounded theory.

There are, however, also significant differences between the frameworks. In the case of the main trend, the rationale of supporting teachers to move towards comprehensive participation in practices promoted by the reform has implications for other parts of the framework. The research questions concern "the changes that teachers go through as they determine whether it is worthwhile to attempt to change their teaching practice" (Cobb and Gresalfi 2011, p. 270); "how teachers can become engaged with professional development (PD) in primary mathematics", despite the emotionally problematic relationship they often have with mathematics (Hodgen and Askew 2007, p. 470); and understanding "what, exactly, are the contributions of teacher education to teachers' eventual practice?" (Horn et al. 2008, p. 61). The paradigmatic question that orients this field seems to be if and how educational innovation may support identity change so that teachers may relate more productively to the intentions of the reform.

In contrast studies using the PoP framework have no element of intervention at present, and the ambition is to develop understandings of teacher learning in the vast majority of cases in which teachers are not enrolled in long-term development initiatives.

The limited emphasis on interventions and normative issues in PoP leads to differences also in other parts of the framework when compared to the general trend. The key concept of identity is also in PoP a social construct, but it is not linked to a set of practices that are prioritised in advance. It follows that two key questions are what and how prior practices and figured worlds play a part for a teacher's experiences of being, becoming, and belonging as she engages with others (students, colleagues, the leadership, short-term development opportunities, etc.) as part of her profession? Another question is what changes there are in the significance of and mutual relationships among these prior practices and figured worlds over time? The unit of analysis may still be phrased as person-in-practice(s), but in PoP the expectation is that there are multiple practices involved, rather than merely the ones of the reform and the tradition, and it is an open question what practices and figured worlds turn out to be important.

Finally there are differences in methodology although the methods used are mainly the same. Studies in line with the general trend use methodological triangulation to provide different modes of access to the same unit of analysis (the research participants' engagement with traditional and reform-oriented practices). In PoP multiple methods are used for a different reason, as it is not assumed that classroom observations, interviews, and other methods shed light on the same key construct. In line with the theoretical stance adopted, the experiences of being a teacher, of becoming increasingly recognised as one, and of belonging in a particular professional context is expected to be decidedly different in a research interview, in a team meeting with colleagues, and in a classroom interaction. The purpose of using different methods is exactly that they provide some access to the teacher's participation in and experiences from *different* practices and figured worlds. Such access may be helpful for understanding her contributions to the practices that evolve in the classroom. Further, the parts of the data material that are turned into data are not only those that relate fairly immediately to reform or tradition in school mathematics. Rather, any data material that point to any practice or figured world that appears to orient the teacher's action or meaning making as they relate to the profession becomes data.

As mentioned before, there are obvious advantages to researching how teachers engage with the practices promoted by TE- or PD initiatives, and how aspects of the reform become a figured world they can attend to when teaching. However, I do suggest that a different approach is needed, if the intention is also to understand how teachers' identities relate to instruction and to their participation in school life in general when they are not engaged in long-term PD initiatives. The suggestion is that there is a need to re-centre the individual, while maintaining the participatory approach of other studies of identity. PoP is one way to do that. Studies of for instance Anna (cf. Sect. 33.6) indicate that there are issues much beyond those related to the teaching and learning of mathematics that orient teachers' acts and meaning-making, issues that are elusive, if we limit the focus to the traditions and reform of the school subject. Phrased in more positive terms, PoP and other

approaches that re-centre the individual may shed light on how unexpected practices and figured worlds come to play a role for teachers' contributions to their
interactions with others at their schools. Referring to the metaphorical description of
a framework as a bicycle frame (cf. Sect. 33.2), I suggest that PoP allows one to go
to unexpected places and see things never imagined before. And although the initial
emphasis in PoP is not on interventions, such new experiences may even allow us
to look at potentials and problems of educational innovation in new ways. Maybe
they are needed for productive educational development.

References

Beauchamp, C., & Thomas, L. (2009). Understanding teacher identity: An overview of issues in
    the literature and implications for teacher education. *Cambridge Journal of Education, 39*(2),
    175–189.
Beijard, D., Meijer, C. M., & Verloop, N. (2004). Reconsidering research on teachers' professional
    identity. *Teaching and Teacher Education, 20*, 107–128.
Bikner-Ahsbahs, A., & Prediger, S. (2010). Networking of theories—An approach to exploiting
    the diversity of theoretical approaches. In B. Sriraman & L. English (Eds.), *Theories of
    mathematics education. Seeking new frontiers* (pp. 483–506). Berlin: Springer.
Blumer, H. (1966). Sociological implications of the thought of George Herbert Mead. *American
    Journal of Sociology, 71*(5), 535–544.
Blumer, H. (1969). *Symbolic interactionism. Perspective and method*. Berkeley: University of Los
    Angeles Press.
Brown, T., & McNamara, O. (2011). *Becoming a mathematics teacher. Identity and identifica-
    tions*. Dordrecht: Springer.
Cobb, P., & Gresalfi, M. S. (2011). Negotiating identities for mathematics teaching in the context
    of professional development. *Journal for Research in Mathematics Education, 42*(3), 270–304.
Cobb, P., Gresalfi, M. S., & Hodge, L. L. (2009). An interpretive scheme for analyzing the
    identities that students develop in mathematics classrooms. *Journal for Research in
    Mathematics Education, 40*(1), 40–68.
Diversity in Mathematics Education Center for Learning and Teaching. (2007). Culture, race,
    power and mathematics education. In F. K. Lester (Ed.), *Second handbook of research on
    mathematics teaching and learning* (Vol. 1, pp. 405–433). Charlotte, NC: NCTM/Information
    Age Publishing.
Eisenhart, M. (1991). *Conceptual frameworks for research circa 1991: Ideas from a cultural
    anthropologist; implications for mathematics education researchers*. Paper presented at the
    13th Annual Meeting of the North American Chapter of the International Group for the
    Psychology of Mathematics Education, Blacksburg, VA.
Gee, J. P. (2000–2001). Identity as an analytic lens for research in education. *Review of Research
    in Education, 25*, 99–125.
Hodgen, J. (2011). Knowing and identity: A situated theory of mathematics knowledge in
    teaching. In T. Rowland & K. Ruthven (Eds.), *Mathematical knowledge in teaching* (pp. 27–
    42). Dordrecht: Springer.
Hodgen, J., & Askew, M. (2007). Emotion, identity and teacher learning: Becoming a primary
    mathematics teacher. *Oxford Review of Education, 33*(4), 369–487.
Holland, D., Skinner, D., Lachicotte, W., Jr., & Cain, C. (1998). *Identity and agency in cultural
    worlds*. Cambridge, MA: Harvard University Press.

Soto-Andrade 2015; Sfard 2008, 2009; Soto-Andrade 2006, 2007, 2015; and many others).[1]

In a broader perspective, increasing agreement has arisen in cognitive science that metaphorising (looking at something and seeing something else) serves as the often unknowing foundation for human thought (Gibbs 2008). As suggested by Johnson and Lakoff (2003), our ordinary conceptual system, in terms of which we think and act, is fundamentally metaphorical in nature. Lakoff and Núñez (2000) highlight the intensive use we make of conceptual metaphors that appear— metaphorically—as inference-preserving mappings from a more concrete 'source domain' into a more abstract 'target domain', enabling us to fathom the latter in terms of the former.

Elementary examples of (conceptual) metaphors in mathematics education are the two foremost metaphors for multiplication, to wit, the 'area metaphor' and the 'grafting metaphor' (Soto-Andrade 2014), illustrated in Fig. 34.1 for the case of 2 times 3 and 3 times 2.

Notice that the area metaphor allows us to see commutativity of multiplication as invariance of area under rotation. We 'see' that $2 \times 3 = 3 \times 2$, without counting and knowing that it is 6. On the contrary, the grafting (or concatenated branching) metaphor does not allow us to 'see' at a glance the commutativity of multiplication. As realised by Lakoff and Núñez (Núñez, personal communication, December 2012) this fact suggests that multiplication is not really commutative. In more precise terms, there might be 'multiplications', inspired by the grafting metaphor, that are not commutative! Indeed, think of *composition* of permutations, of matrices and of operators. We have here then two different metaphors for the 'same' mathematical object, each with a different scope. The first one, the area metaphor, which is quite close to East Asian crossing metaphor for multiplication, where you count the number of crossings of, in this case, 2 lines and 3 lines, is quite friendly and lets us see immediately the commutativity of multiplication. The second one (multiplication is concatenation), points in a different direction, does not exhibit commutativity (Mac Lane 1998) as an obvious property and is in fact more profound: It reshapes our understanding of multiplication and it unfolds into category theory in contemporary mathematics. A case of a felicitous metaphor opening up the way to deep and far-reaching generalisations of a seemingly innocent elementary concept (see Manin 2007)!

Notice that, as argued by Sfard (1997), metaphorising appears here as a circular autopoietic process (Maturana and Varela 1980) rather than as a unidirectional mapping. So a more appropriate metaphor than the arrow metaphor to describe metaphorising would be the ouroboros (the snake eating its own tail), an outstanding metaphor of circularity, self-reference and organisational closure in living systems (see Soto-Andrade et al. 2011). The ouroboros indeed plays an important role in Maturana and Varela's theory of autopoietic systems (Maturana and Varela 1980), appearing even on the cover of the first Spanish edition of their work

---

[1]For a recent survey of the role of metaphor in mathematics education, see Soto-Andrade (2014).

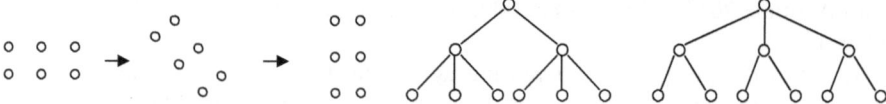

Fig. 34.1 Area metaphor (left) and grafting metaphor (right) for multiplication

(Maturana and Varela 1973; see also Soto-Andrade et al. 2011, Fig. 34.1), and in cognition as enaction (Varela et al. 1991). Recall that the latter was also metaphorised by Varela by the famous *Drawing Hands* lithograph by Escher, where each hand draws the other into existence.

Our approach to the learning of mathematics emphasises the poietic (from the Greek *poiesis* = creation, production) role of metaphorising, which brings concepts into existence. For instance, we bring the concept of probability into existence when, while studying a symmetric random walk on the integers, we look at the walker (a frog, say) and we see it *splitting* into two equal halves that go right and left instead of jumping equally likely right or left (Soto-Andrade 2007, 2014, 2015). This 'metaphoric sleight of hand' that turns a random process into a deterministic one allows us to reduce probabilistic calculations to deterministic ones where we just need to keep track of the walker's splitting into pieces: The probability of finding the walker at a given location after n jumps is just the portion of the walker landing there after n splittings.

We remark that a different metaphoric way of bringing mathematical notions into existence, called *reification* by Sfard (2008), where a *process* is seen as an *object*, is exemplified by the case of fractions: Splitting a whole into 3 equal parts and keeping 2 of them becomes the number 2/3. Of course, splitting the whole into 6 equal parts and keeping 4 is a different (but equivalent) process whose reification is the same number, 4/6 = 2/3. Saying that 4/6 and 2/3 are just *equivalent* fractions instead of *equal* fractions is here a sign of incomplete reification.

Although in the literature metaphor and representation are often used as synonyms, we draw here a distinction: we *metaphorise* to construct concepts (as in the above example) and we *represent* to explain concepts. Typically, metaphors are arrows going *upwards*, from a down-to-earth domain to a more abstract one, and representations are arrows going *downwards*, i.e., the other way around. In this connection, it is pertinent to recall that in the German school of didactics of mathematics, originally mostly concerned with primary mathematics education and going back to Pestalozzi (Herbart 1804; vom Hofe 1995), representation and metaphor were quite present: as *Darstellung*—representation aiming at explaining something to others—and *Vorstellung*—a personal way to figure out or fathom something, operationally equivalent to metaphor (Soto-Andrade and Reyes-Santander 2011). So metaphorising was already recognised and appreciated at the beginning of the 19th century in German didactics of mathematics, well before its irruption from cognitive psychology and linguistics into mathematics education (Lakoff and Núñez 2000).

The ubiquity of metaphor in mathematics education should not be underestimated: Besides bringing into existence mathematical concepts or objects or helping learners to fathom them, unconscious metaphorising often dramatically shapes the way teachers teach, for instance. A foremost example is afforded by the metaphor 'teaching is transmitting knowledge'. Indeed, when confronted with it, many teachers reply: This is not a metaphor, teaching *is* transmitting knowledge! What else could it be? Unperceived here is the 'acquisition metaphor' (Sfard 2009; Soto-Andrade 2007) for learning, dominant in mathematics education, that sees learning as acquiring an accumulated commodity. The alternative, complementary metaphor is the 'Participation Metaphor': learning as participation (Sfard 2009). This dichotomy is well expressed in Plutarch's metaphor: 'A mind is a fire to be kindled, not a vessel to be filled' (Sfard 2009, p. 41).

Paraphrasing Bachelard (1938), who advocated epistemological vigilance, we suggest nowadays to practise metaphorical vigilance, i.e., the art of noticing (Mason 2002) our unconscious or implicit metaphors, that shape our way of interacting with the world and particularly our approach to teaching and learning.

Last but not least, metaphorising plays also a key epistemological role: We have claimed elsewhere (Díaz-Rojas and Soto-Andrade 2015) that—metaphorically—a theory is in fact just the 'unfolding' of a metaphor (the involved unfolding process, however, may be laborious and technical).

A paradigmatic example is the 'tree of life' metaphor in Darwin's theory of evolution. Also, Brousseau's theory of didactical situations (Brousseau and Warfield 2014) may be seen as an unfolding of the 'emergence metaphor' that sees mathematical concepts emerging in a situation instead of being parachuted from Olympus as in traditional and abusive teaching. The 'grafting' metaphor above for multiplication (Soto-Andrade 2014) unfolds into category theory in mathematics Mac Lane (1998). We use in fact the metaphorical approach as a meta-theory to describe other theories relevant to us in terms of their generating metaphors, something more helpful to fathoming how they arise than just describing them a posteriori. We exemplify this below in the case of Varela's enaction.

34.2.2 Enactivism in Mathematics Education

An unfolding metaphor for enaction is Antonio Machado's famous poem (Machado 1988, p. 142; Thompson 2007; Malkemus 2012): 'Caminante, son tus huellas el camino, y nada más; caminante, no hay camino, se hace camino al andar' ['Wanderer, your footsteps are the path, nothing else; there is no path, you lay down a path in walking.'], cited by Varela (1987, p. 63) himself when he introduced what he called the *enactive approach* in cognitive science (Varela et al. 1991). In his own words: 'The world is not something that is given to us but something we engage in by moving, touching, breathing, and eating. This is what I call cognition as enaction since enaction connotes this bringing forth by concrete handling' (Varela 1999, p. 8).

Notice *en passant* to what extent the 'laying a path in walking' metaphor is transversal to the traditional one for learning as following a well-marked path given in advance.

Before proceeding any further, however, to avoid confusion given the somewhat polysemic current status of the terms *enactivism, enactivist, enaction, enact, enacting* and *enactive*, we will adhere to the following usage.

The now prevalent terms *enactivism* and *enactivist* will always refer to Varela's anti-representationalist 'enactive program' (Varela et al. 1991, p. xx), which sees cognition as embodied action, more precisely, cognition as enaction, as metaphorised by Machado's verse. Key aspects of enaction are: perceptually guided action, embodiment and structural coupling through recurrent sensorimotor patterns (Varela et al. 1991; Reid and Mgombelo 2015). In an aphorism: 'All doing is knowing, and all knowing is doing' (Maturana and Varela 1992, p. 26). We will speak then of an 'enactivist approach' to problem solving or to mathematics education. We will also use the term 'enaction' exclusively in Varela's sense (Maturana and Varela 1992).

On the other hand, unless otherwise explicitly stated, 'enact', 'enacting' and 'enactive' are to be understood in the sense of everyday language and also in the sense of Dewey (1997) and Bruner (1966), i.e., as synonyms of 'acting out' or 'acted out', in an embodied way. So 'enacting a metaphor' just means 'to act it out', with your body (see Example 34.4.1). This fully coincides with the use of 'enactive' in Gallagher and Lindgren (2015), where they refer to 'enactive metaphors' (metaphors in action, that we act out bodily) as opposed to what they call 'sitting metaphors'. We use 'enactive metaphorising' below in this sense.

As mentioned above, in mathematics education the term *enaction* may be traced back to Bruner (1966), who was following the traces of Dewey's (1997) 'learning by doing'. Bruner's enaction, which means essentially acting out or doing, is however far less radical than Varela's, in that it does not challenge the notion of a given reality 'out there' that we perceive or represent more or less successfully. Dewey, however, already emphasised the role of sensorimotor coordination in perception, acknowledging that movement is primary and sensation is secondary (Dewey 1896; Gallagher and Lindgren 2015).

In particular, the enactivist notions of structural determinism and structural coupling (Maturana and Varela 1992; Varela 1999; Varela et al. 1991) have provided new insights on learning, problem solving and problem-posing processes: Learning is not determined by a didactical environment but arises from the interaction of the learner's structure and environment, which plays at most the role of a 'trigger'. Traditionally, however, problem solving entails problems given beforehand, lying 'out there' in the world, waiting to be solved, independently of us as cognitive agents. In the enactivist perspective, because of our structural coupling with the world (Varela 1996; Varela et al. 1991), we bring forth emergent problematic situations instead. This is what Varela calls problem posing. This diverges from the usual gas fitter metaphor for problem solving, where solvers look into their toolboxes of predefined strategies and choose the appropriate one for solving the problem at hand (Soto-Andrade 2007; Proulx 2008). In the enactivist perspective, mathematical strategies emerge continually in the interaction of solver and problematic situation (Proulx 2013; Thom et al. 2009).

At present, an enactivist didactics of mathematics unfolds where the teacher is an enactivist practitioner acting in situation and learning appears as an emergent, situated and embodied process (Brown 2015; Brown and Coles 2012; Proulx 2008, 2013; Proulx and Simmt 2013). For a recent survey of enactivist theories, see Goodchild (2014).

According to Varela, we are always 'enacting' a world, most of the time unconsciously. So we cannot choose to enact or not to enact (in Varela's sense); enaction is just the way we cognise as living beings. We may nevertheless entertain the 'representationalist illusion' (a privilege of humankind!) that we are perceiving and representing an objective reality 'out there'. Also, we can choose to enact (in the everyday sense of the word of bodily acting out) a given metaphor or situation or not, for instance. Paradoxically, we are definitely able to teach in a way that ignores enaction (in Varela's sense) and does not allow for enacting (as bodily acting out): a non-enactivist stance that paves the way for cognitive bullying. Our enactivist approach to education, distilled in the 'lying down a path in walking' metaphor for cognition and learning, leads us on the contrary to foster metaphor enacting among the learners.

34.2.3 Research Questions

Along the lines of our stated research aim in the Introduction, we intend to address here the following research questions:

- When and how does metaphorising arise from learners in a problem-solving situation, particularly idiosyncratic metaphorising?
- How does metaphorising correlate with the emergence of new ideas or insights to tackle challenging situations?
- Is metaphorising enactive most of the time? How relevant for learning are action-based enactive metaphors?
- What is the influence of learners' non-metaphoric enacting in mathematical problem-solving situations?

34.3 Methodology

Our methodology adheres to the enactivist perspective, where we focus on the learners doing and knowledge is not metaphorised—by the researcher—as an object to be captured or held by a learner (Sfard 2008, 2009).

According to this and our research objectives, we suggest and propose various challenging situations to the learners and observe how they tackle them. We pay attention to the whole spectrum of emerging strategies, to whether they metaphorise or enact and how the emergence and the quality of their ideas and insights correlate with their metaphorising and enacting. We do not focus on trying to measure their knowledge over time but on monitoring their being mathematical as a means to tackle a challenge co-emergent with their doing.

Our experimental methodology relies on qualitative approaches and field observation, especially multiple case study (Yin 2003); participant observation techniques and ethnographic methods (Spradley 1980; Brewer and Firmin 2006). Our experimental background includes a broad spectrum of learners (seven cohorts) with whom we carried out didactical experiences based on a metaphor-intensive enactivist approach in 2015 and 2016 that included the following:

A. Fifty students in a one-semester first-year mathematics course in the social sciences and humanities option of the Baccalaureate Programme of the University of Chile (two cohorts: 2015 and 2016).
B. Thirty-five prospective secondary school physics and mathematics teachers in a one-semester course in probability and statistics at the same university (two cohorts: 2015 and 2016).
C. Twenty (5 graduate and 15 undergraduate) students majoring in mathematics in an optional course on random walks at the University of Chile (2015).
D. Fifty participants in a two-session workshop, each session consisting of 1.5 h, on enactive metaphoric approaches to mathematical problem solving, held at the annual meeting of the Chilean Mathematics Education Society, in Valparaiso (2016). Participants included in-service secondary and primary school teachers, prospective secondary teachers, post-graduate students in mathematics education, researchers in mathematics education and some undergraduate and graduate math students.
E. Twenty in-service primary school teachers engaged in a 15-month professional development programme (mathematics option) at the University of Chile at Santiago (2016).

These cohorts were chosen because they constituted a rather broad spectrum of learners with whom our overarching approach could be tested while performing our usual teaching duties at the university and facilitating invited workshops elsewhere.

Regarding data recollection, learners, working most of the time in random groups of three to four, were observed by the teacher or facilitator and an assistant, the latter assuming the role of participant observer or ethnographer. Field notes and transcripts of the generated dynamics were taken (especially of critical moments of the work sessions, such as emergence of metaphors, horizontal confrontations between the students, and didactic tension build up), snapshots of their written output (on paper or whiteboard) in problem-solving activities were taken, short videos of their enacting moments were recorded. We recall that all these data are also acts of interpretation, where a researcher learns in co-emergence with a research situation (Reid 1996).

Regarding data analysis, categories involved in the initial phase of the observation included:

- Learners' participation and engagement (level estimates).
- Questions and answers (from teacher and learners, frequency, relative weight, spaces for pondering).
- Horizontal (peer) interaction (level estimate).
- Metaphors, especially idiosyncratic ones (emergence, spontaneously or under prompting, variety).
- Arising of gestural language of learners and teacher.
- Expression and explicit acknowledgement of affective reactions from the learners.
- Enacting (acting out) of metaphors and situations by the learners.

Recall that in an enactivist methodological framework the initial categorical grid evolves according to the flow of activity in the classroom and the reactions of learners and teachers in an autopoietic way (Reid 1996; Maheux and Proulx 2015).

To address our research objective we kept track of ideas and actions emerging after either spontaneous or prompted metaphorising (see examples below).

From data analysis, we compared cognitive and affective reactions of the different cohorts and inferred the profile and strength of the prevailing didactical contract (Brousseau et al. 2014), usually installed during secondary math education for most learners.

Moreover, learning in Cohorts A, B and E was assessed through monthly tests (where students had to solve contextual problems in a limited time), compulsory and optional exercises and challenges as homework (Cohorts A, B, C and E). Process assessment was also done by observing their acting and behaviour and recording their production during individual and group work sessions and lectures (all cohorts).

34.4   Illustrative Examples of Enactive Metaphorising

We report and discuss below two types of examples of enactive metaphorising in challenging mathematical situations that we experimented with in the above cohorts.

34.4.1 The Sum of the Exterior Angles of a Polygon and the Inner Acute Angles of a Star (Cohorts D and E)

34.4.1.1 Problem: Everybody Knows About the Inner Angles of a Triangle and Their Sum. But What About the Exterior Angles? And Their Sum? Also What About the Same for a General Polygon?

As suggested by our enactivist perspective, we prompt the learners to notice and to voice their reactions, cognitive and affective, to this 45-min challenge. We intend in this way to facilitate a circular interaction between the problem and the learners that could trigger a reshaping of the challenge. Most of them, however, have trouble in acknowledging a negative reaction. After renewed prompting some dare to ask: Why should we be interested in the exterior angles of a polygon and their sum? A few (in-service and prospective teachers alike) complain about the prescriptive way this sort of geometry is usually taught. After a while, most of them agree that one needs to re-signify exterior angles: What are they good for?

After some polling, we found that a majority of learners (students as well as teachers) prefer inner angles to exterior angles and wonder about the meaning, usefulness or relevance of exterior angles.

We observed that to tackle this problem almost every in-service and prospective secondary mathematics school teacher in our country calculates first the sum of the inner angles (usually by triangulating), finding that it depends on the number n of sides of the polygon, as $(n-2)$ times $180°$. Then, replacing each interior angle by $180°$ minus the corresponding exterior angle and calculating, they wind up discovering that the sum of all exterior angles of a (convex) polygon is $360°$, independently of its number of sides! This is surprising for most of them! At this point some students (more often girls than boys, in our courses) ask: Isn't there a simpler way to get this? Others feel frustrated because they have 'calculated blindly' and without insight.

This is the usual way in which teachers and students 'get into' the proposed task (Proulx 2013). Unfortunately, this is *the only* approach found in almost every textbook, where the mathematical content 'exterior angles of a polygon' is then checked as having been covered.

Our metaphoric approach suggests, however, prompting the students to metaphorise a polygon first (not just recite its definition) to help them to get into the problem in more transparent ways. Their metaphorising will depend, of course, on their previous history and experiences. We observed the rather slow emergence of two main competing metaphors:

- A polygon is an enclosure between crossing sticks (most popular among in-service primary teachers).

- A polygon is a closed path made out of straight segments. Interestingly, some learners *say* that a polygon is a closed plane figure, while *drawing* a circuit in the air with their index!

Notice that the first metaphor carries the viewpoint of the eagle (who sees from above) and the second one, the viewpoint of the ant (who crawls down to earth). A high-speed version of the second one is quite familiar to children nowadays in video games.

Among primary school teachers (Cohorts D and E), enacting the first metaphor triggers the idea of manipulating the sticks, translating them in convenient ways, so as to make clearly visible the exterior angles first, and then shifting them parallel to themselves to shrink the polygon to a minimum, preserving its shape. So in fact they zoom out the polygon! In this way they *see* that the sum of all exterior angles is clearly 360° instead of getting this value by blind calculation.

Enacting the second metaphor also allows the learners to *see* that the sum of exterior angles is a whole turn, when 'laying down a polygonal path in walking'.

Indeed, we observed that the in-service primary teachers in Cohort E, working in groups, had one of them (who had trouble *seeing* in this way that the sum of exterior angles was a whole turn) 'lay down a polygon in walking', following the instructions of his peers: Walk 5 steps, stop, turn 45° to your left, walk 7 steps, and so on. In this way they realized that exterior angles rather than inner angles were the necessary and convenient data to inflect or bend the path of the walker as desired. Addition of all exterior angles occurred when the walker made a complete circuit and came back to his starting point with his nose pointing in the same initial direction. Learners also noticed that this metaphor suggests a natural generalisation, involving a signed sum, for the case of a non-convex polygon!

Recently a third metaphor was suggested by one of our former mathematics students:

- A polygon is a wheel of the Flintstones' car.

Learners realised quickly that when the Flintstones' car runs, its wheels turn, and when they complete a whole turn, their exterior angles (arising as the successive angles between the wheel's sides and the ground) add up to a whole turn!

We see in this example that metaphorising and enacting can make a dramatic difference in understanding that is within the reach of 'everybody', as opposed to the 'blind' unappealing calculation found everywhere. Our appraisal of inner and exterior angles also changes: Exterior angles appear now to be more natural and friendlier than inner angles: a dissident view indeed! In particular, learners realised that it is smarter to figure out first the sum of all exterior angles of the polygon and then deduce the value of the sum of the inner angles, which is contrary to the usual procedure and a valuable idea for the next challenge.

34.4.1.2   The Five-Pointed Japanese Star Problem, The Enactive Way

A typical challenge in Hosomizu's little red book (Hosomizu 2008) is to calculate the sum of the inner acute angles of a five-pointed star (eventually non-regular). Several clever approaches are discussed there, although none of them are enactive or metaphoric. We posed this problem to Cohorts D and E. When posed from scratch, before Problem 34.4.1.1, everybody tackled it in a geometric-algebraic way: Some learners in Cohort D even wrote down a whole system of equations, taking advantage of the inscribed pentagon in the star. Most took the star to be regular and computed dutifully the value of each inner acute angle. Then they conjectured that the total sum would remain constant if the star were deformed. So they more or less converged to the approaches illustrated in Hosomizu (2008), although less clever on the average. Nobody thought of 'laying down a star in walking' (following the circuit usually used to *draw* the star) to instantly see that the sum of all exterior angles at the points of the star equals 2 whole turns and from there get the sum of all inner acute angles (as 5 half turns minus 4 half turns = 1 half turn). Several in-service secondary teachers avowed nevertheless that they preferred the algebraic approach that they felt more at home with.

Learners in Cohort E, who worked on this problem after having worked out Problem 34.4.1.1, wondered for a while which closed path to walk to lay down the star before settling for the one they use to *draw* the star. After learners in Cohort D solved this problem using 'angular yoga', as in Hosomizu (2008), we proposed to them Problem 34.4.1.1, which they discussed and finally solved metaphorically and enactively. They went then back to the five-pointed star to find a friendly circuit to walk and solve the problem. Learners in Cohorts D and E noticed that this enactive metaphoric approach worked equally well for irregularly drawn seven-pointed stars and more generally for stars with an odd number of points.

34.4.2   Probabilistic Enacting

34.4.2.1   Falk's Urn and Fischbein's Test

The following challenging question (Falk and Konold 1992; Fischbein and Schnarch 1997) was proposed to learners in Cohorts A, B and C.

> John and Mary each receive a box containing 2 black marbles and 2 white marbles.
>
> John extracts a marble from his box and finds out that it is white. Without replacing this marble, he extracts a second marble. Is the likelihood that this second marble is also white smaller than, equal to or greater than the likelihood that it is black?
>
> Mary extracts a marble from her box that she puts in her pocket without looking at it. Now she extracts a second marble that turns out to be white. Is the likelihood that the marble in her pocket is white smaller than, equal to or greater than the likelihood that it is black?

Learners had no trouble with John's drawings, but roughly 60% of learners in Cohort A and 40% of learners in Cohorts B and C thought that the fact that Mary's second marble be white had no effect whatsoever on the likelihood of the first one being black. The remaining learners thought intuitively that since the second one was white it was more likely that it was drawn from a box with more white than black marbles, so it was more likely that the first marble was black. Just a few learners in Cohorts A and B had the idea of simulating many times to figure out what would be more likely. Others (learners in Cohort C included) metaphorised the whole process as a two-step random walk on a binary tree, or better on a grid, and computed diligently the non-required probabilities (the question was qualitative and couched in everyday language, not in probabilistic language). They found correctly that the probability of the marble being black in both cases is 2/3. They realised, however, that they did not really *see* why probabilities were the same and why 'there was no time arrow'.

Following our enactivist perspective, we prompted the students to enact (act out) the experiment. Extracting the marbles, they ended up with two marbles by the box, in the first case the first one being white and the second one being hidden under a cap, in the second case, the first one being hidden and the second one being white. They realised then that they had just extracted two marbles from the box and hidden one of them, the other one being white!

A variant of this enactment that we suggested to the students, inspired from a remark by M. Borovcnik (personal communication at ICME 13, July 27, 2016), starts by grabbing a marble from the box with one hand and then another one with the other hand, keeping both fists closed. They realise then by themselves that it is just a matter of opening first one fist or the other and that they could have also grabbed the two marbles simultaneously.

34.4.2.2 Drawing Balls from an Urn Without Replacement: Metaphorising as a Random Walk and Enacting (Proposed to Learners in Cohorts B and C)

Problem: From an urn containing 3 red balls and 5 blue balls, 5 balls are drawn one after another at random without replacement. How likely is that the 5th ball drawn is red?

We have discussed this problem, proposed to learners in Cohort C, in detail elsewhere (Soto-Andrade et al. 2016) in the simpler case of the 3rd ball drawn from a (2, 3) urn instead of a (3, 5) urn. We comment here on further experimentation with learners in Cohort B in 2016 and new ways of enacting it (acting it out).

We observed that most students in Cohort B, when exposed to the problem in a test, dutifully calculated the requested probability with the help of a lush possibility tree with probabilities assigned stepwise using a hydraulic metaphor (that sees probabilities as portions of one litre of water that drained downwards from the 'root' of the tree). Nevertheless most of them did not realise that the probability of a red ball at any drawing was always 3/8, because they did not even notice that for the second drawing it was 21/56 = 3/8! One or two intuited that order did not matter, but most of them were quite surprised in a subsequent stage, when working in groups in the classroom they finally found that the probability of the nth ball drawn being red was the same for all n up to 8. We then prompted them to enact the process by actually drawing marbles from a (3, 5) bag. Some were a bit reluctant to do so. A good performing student said bluntly:

> I do not see how enacting can help me to solve the problem. What else do I get from enacting that I do not get from thinking? I just need to think about it!

Nevertheless, afterwards he gave the following intuitive argument to see that all probabilities were the same. Keeping the first 4 drawn marbles in his closed left hand, he said:

> Now I have to choose a 5th one from the 4 marbles remaining in the bag. But it is clear that this is equivalent to choosing one of the hidden 4 marbles in my left hand! So it amounts to choosing 1 marble from the whole bunch of 8 marbles!

All other students put the drawn balls carefully in a line, one by one (they did not throw them away!). This helped several of them to see the invariance of the probability of drawing a red marble. No one put them insightfully in a circle, as an undergraduate female student[2] in Cohort C did for the (2, 3) bag in 2015, but they really appreciated the idea when told.

Now, a new enaction, suggested by M. Borovcnik (personal communication at ICME13, July 27, 2016) is to grab sequentially first, five marbles from the bag, with five hands (of three students) keeping the five fists closed, and subsequently opening them in the same sequence, or in another one, e.g. the fifth fist first. Eventually the grabbing could be also simultaneous! Enacting in this way all students *saw* the invariance of the probability of 'red', not just a few clever eidetic students.

To get a better grasp of the drawing process, learners also metaphorised it as a 2D random walk—from the source (3, 5) to the sink (0, 0) or from the source (8, 3) to the sink (0,0)—that in turn they metaphorised as a splitting process, whose transition probabilities they calculated with the help on a hydraulic metaphor. They realized then that the associated (deterministic) 'barycentric walk' provides a friendly metaphor for the 'expected walk' of the walker. They intuitively guessed that the barycentric walk should proceed geodesically along a line whose slope corresponds exactly to the probability of red at any drawing in the case where they represent the initial state of the urn by (8, 3).

---

[2]Notwithstanding that Chile's boys-girls PISA math performance gap is extreme among OECD countries (OECD 2016, p. 198).

34.5 Discussion and Conclusions

Motivated by an enactivist perspective, we have shown by way of illustrative examples how metaphorising and enacting (acting out) mathematical objects, processes and situations can make a significant difference in the ideas and insights that may emerge from learners tackling a mathematical challenge. In the cases considered (34.4.1 and 34.4.2) concerning geometry and probability, we observed notably that there was a dramatic contrast between blind calculation before metaphorising and sudden insight when metaphorising or enacting. We also saw how different insights were triggered by different metaphors or enactings. In Problem 34.4.1.1, for instance, we collected in all one blind calculation and three different insights leading to the answer triggered by three different metaphors for a polygon with different levels of enactivity (bodily engagement), the foremost one being 'laying down a polygon in walking'.

Very concretely, we observed that when they enact, learners have to make up their minds: What do I do with the balls I draw from the urn? Throw them away? Keep them in my hand or in my closed fists? Put them carefully in a row or a circle on the table? Each way of enacting—determined primarily by the solvers' structures and histories—suggests various different ideas and insights that do not emerge so easily when they just think about a problem. Our learners working on the challenges in Cases 34.4.1 and 34.4.2 indeed discovered unforeseen mathematical relations or facts in their bodily actions (see Abrahamson and Trninic 2015).

We nevertheless found that, surprisingly, metaphorising and enacting were quite difficult for most of the observed learners. Persistent prompting and plenty of time was often needed to elicit both among them. Notice that learners in Cohort A, for instance, came straight from secondary school (where cognitive bullying prevails). Even so, students in Cohort A and in-service primary school teachers were more prone to metaphorise than prospective or in-service secondary school math teachers.

Particularly, we noticed that metaphorising a polygon, for instance, was a very unusual challenge for students, prospective and in-service teachers alike: a violation of the prevailing didactical contract. But once they felt they were allowed to, even prompted to, metaphors began to arise among them, shyly and slowly at first. They came later to gradually appreciate the operational virtues of metaphorising.

In fact, we observed chains of metaphors emerging that completely transformed a given problem (e.g., Sects. 34.4.1.1 and 34.4.2.2) and allowed learners to better fathom what was going on. From an enactivist perspective, they were not just reacting to a problem out there or looking for a solving strategy that had been stocked beforehand in their personal toolkit but rather shaping and transforming the problem, eventually because they did not like it (see Proulx 2013). Moreover, metaphorising a mathematical object, such as a polygon, may show them the way to guess and discover meaningful or significant properties amidst the huge set of properties entailed by its formal definition.

Interestingly, no more than one student out of 20 on the average tried spontaneously to enact (act out) an opaque problematic situation. Some prospective teachers even voiced their disbelief regarding the usefulness of enacting, because math problems are just a matter of thought!

Recalling the well-documented strong negative feelings of school children towards mathematical content taught the traditional way, it seemed to us a bit paradoxical to observe widespread 'emotional anaesthesia' in most of our learners, who had trouble in acknowledging and expressing their emotional reactions, especially negative ones, towards mathematical content. Only primary school teachers and students in Cohort A escaped this condition to some extent. We interpret this syndrome as a consequence of the didactic contract (Brousseau et al. 2014) associated to the prevailing cognitive abuse in our culture, where students are expected to understand a mathematical content or not, but not to like or dislike it. Expression of affect is then ignored and repressed.

This phenomenon seemed important to us because we observed that metaphorisation, for instance, may be often triggered by disliking of a proposed problem: The learner tries to metaphorise it in order to see it otherwise, wearing friendlier attire. So in fact negative emotions may foster creativity!

We noticed a remarkable convergence of our claims and experimental findings regarding the positive incidence of enacting in the arising of new insights in problem-solving situations with very recent research in cognitive science (e.g., Glenberg 2015; Vallée-Tourangeau et al. 2016; Abrahamson and Trninic 2015).

We may conclude that metaphorising and enacting (in the sense of bodily acting out) play indeed a key role in the learning of mathematics, especially for non-mathematically inclined learners who have been cognitively abused by traditional learning. Since cognitive bullying is to a great extent institutionalised by the prevailing unspoken didactic contract that is functional in thwarting metaphorising, enacting and affect in teaching and learning contexts, it seems urgent to reshape this contract to allow for and foster these processes. This endeavour deserves further research, taking into account the relevance of collaborative group work and learners' horizontal interaction and participation.

As an open end, we would like to extend longitudinally our study to involve the pupils of in-service and pre-service teachers we have worked with, to further investigate the incidence of metaphorising and enacting in their learning and their role as an antidote to cognitive bullying.

Acknowledgements Funding from PIA-CONICYT Basal Funds for Centres of Excellence Project BF0003 and from University of Chile Domeyko Fund (Interactive Learning Networks Project) is gratefully acknowledged.

References

Abrahamson, D., & Trninic, D. (2015). Bringing forth mathematical concepts: Signifying sensorimotor enactment in fields of promoted action. *ZDM Mathematics Education, 47*(2), 295–306. https://doi.org/10.1007/s11858-014-0620-0.

Bachelard, G. (1938). *La Formation de l'esprit scientifique*, Paris: Librairie philosophique Vrin.

Brewer, P., & Firmin, M. (Eds.). (2006). *Ethnographic and qualitative research in education.* New Castle: Cambridge Scholars Press.

Brousseau, G., Sarrazy, B., & Novotna, J. (2014). Didactic contract in mathematics education. In S. Lerman (Ed.), *Encyclopedia of mathematics education* (pp. 153–159). Berlin: Springer.

Brousseau, G., & Warfield, V. (2014). Didactic situations in mathematics education. In S. Lerman (Ed.), *Encyclopedia of mathematics education* (pp. 163–170). Berlin: Springer.

Brown, L. (2015). Researching as an enactivist mathematics education researcher. *ZDM Mathematics Education, 47*, 185–196.

Brown, L., & Coles, A. (2012). Developing "deliberate analysis" for learning mathematics and for mathematics teacher education: How the enactive approach to cognition frames reflection. *Educational Studies in Mathematics, 80*, 217–231.

Bruner, J. (1966). *Toward a theory of instruction.* Harvard, MA: Harvard University Press.

Cantoral, R. (2013). *Teoría Socioepistemológica de la Matemática Educativa. Estudios sobre construcción social del conocimiento* (1a ed.). Barcelona: Editorial Gedisa.

Chiu, M. M. (2000). Metaphorical reasoning: Origins, uses, development and interactions in mathematics. *Educational Journal, 28*(1), 13–46.

Dewey, J. (1896). The reflex arc concept in psychology. *Psychological Review, 3*(4), 357–370.

Dewey, J. (1997). *How we think.* Mineola, NY: Dover. (Original work published 1910).

Díaz-Rojas, D., & Soto-Andrade, J. (2015). Enactive metaphoric approaches to randomness. In K. Krainer & N. Vondrová (Eds.), *Proceedings CERME9* (pp. 629–636). Prague: Charles University & ERME.

English, L. (Ed.). (1997). *Mathematical reasoning: Analogies, metaphors, and images.* London: Lawrence Erlbaum Associates.

Falk, R., & Konold, C. (1992). The psychology of learning probability. In F. Sheldon & G. Sheldon (Eds.), *Statistics for the twenty-first century* (pp. 151–164). Washington, DC: MAA.

Fischbein, E., & Schnarch, D. (1997). The evolution with age of probabilistic, intuitively based misconceptions. *Journal for Research in Mathematics Education, 1*(28), 96–105.

Freire, P. (1970/1972). *Pedagogía del Oprimido* (4a ed.). Buenos Aires: Siglo XXI Editores.

Freire, P., & Faúndez, A. (2014). *Por una pedagogía de la pregunta: Crítica a una educación basada en respuestas a preguntas inexistentes* (2a ed.). Buenos Aires: Siglo XXI Editores.

Gallagher, S., & Lindgren, R. (2015). Enactive metaphors: Learning through full body engagement. *Educational Psychology Review, 27*, 391–404.

Gattegno, C. (1971). *What we owe children: The subordination of teaching to learning.* London: Routledge Kegan.

Gibbs, R. W. (Ed.). (2008). *The Cambridge handbook of metaphor and thought.* Cambridge: Cambridge University Press.

Glenberg, A. M. (2015). Few believe the world is flat: How embodiment is changing the scientific understanding of cognition. *Canadian Journal of Experimental Psychology, 69*, 165–171.

Goodchild, S. (2014). Enactivist theories. In S. Lerman (Ed.), *Encyclopedia of mathematics education* (pp. 209–214). Berlin: Springer.

Hall, E. T. (1959). *The silent language.* Greenwich, CT: Fawcett.

Herbart, J. F. (1804). *Pestalozzi's Idee eines ABC der Anschauung als ein Cyklus von Vorübungen im Auffassen der Gestalten.* Göttingen: Röwer. Retrieved from http://digi.ub.uni-heidelberg.de/diglit/herbart1804.

Hosomizu, Y. (2008). *Entrenando el pensamiento matemático. Edición Roja.* Tsukuba: Tsukuba Incubation Lab.

Johnson, M., & Lakoff, G. (2003). *Metaphors we live by.* New York, NY: The University of Chicago Press.

Johnston-Wilder, S., & Lee, C. (2010). Developing mathematical resilience. In *BERA Annual Conference 2010*, September 1–4, 2010, University of Warwick.

Lakoff, G., & Núñez, R. (2000). *Where mathematics comes from*. New York, NY: Basic Books.

Libedinsky, N., & Soto-Andrade, J. (2015). On the role of corporeality, affect and metaphoring in Problem solving. In P. Felmer, J. Kilpatrick, & E. Pehkonen (Eds.), *Posing and solving mathematical problems: Advances and new perspectives* (pp. 53–67). Berlin: Springer.

Mac Lane, S. (1998). *Categories for the working mathematician*. Berlin: Springer.

Machado, A. (1988). *Selected poems*. Cambridge, MA: Harvard University Press.

Maheux, J.-F., & Proulx, J. (2015). Doing|mathematics: Analysing data with/in an enactivist-inspired approach. *ZDM Mathematics Education, 47*, 211–221. https://doi.org/10.1007/s11858-014-0642-7.

Malkemus, S. A. (2012). Towards a general theory of enaction. *The Journal of Transpersonal Psychology, 44*(2), 201–223.

Manin, Y. (2007). *Mathematics as metaphor*. Providence, RI: American Mathematical Society.

Mason, J. (2002). *Researching your own practice: The discipline of noticing*. London: Routledge.

Maturana, H., & Varela, F. J. (1973). *De Máquinas y Seres Vivos*. Santiago, Chile: Editorial Universitaria.

Maturana, H., & Varela, F. J. (1980). *Autopoiesis and cognition: The realization of the living*. Dordrecht: Reidel.

Maturana, H. R., & Varela, F. J. (1992). *The tree of knowledge: The biological roots of human understanding*. Boston, MA: Shambhala.

OECD. (2016). *PISA 2015 results (Volume I): Excellence and equity in education*. Paris: PISA, OECD Publishing. http://dx.doi.org/10.1787/9789264266490-en.

Proulx, J. (2008). Structural determinism as hindrance to teachers' learning: Implications for teacher education. In O. Figueras & A. Sepúlveda (Eds.), *Proceedings of the Joint Meeting of the 32nd Conference of the International Group for the Psychology of Mathematics Education and the XX North American Chapter* (Vol. 4, pp. 145–152). Morelia, Michoacán, Mexico: PME.

Proulx, J. (2013). Mental mathematics emergence of strategies, and the enactivist theory of cognition. *Educational Studies in Mathematics, 84*(3), 309–328.

Proulx, J., & Simmt, E. (2013). Enactivism in mathematics education: Moving toward a re-conceptualization of learning and knowledge. *Education Sciences & Society, 4*(1), 59–79.

Reid, D. (1996). Enactivism as a methodology. In L. Puig & A. Gutierrez (Eds.), *Proceedings of the Twentieth Annual Conference of the International Group for the Psychology of Mathematics Education* (Vol. 4, pp. 203–210). Valencia: PME.

Reid, D. A., & Mgombelo, J. (2015). Survey of key concepts in enactivist theory and methodology. *ZDM Mathematics Education, 47*(2), 171–183.

Sfard, A. (1997). Commentary: On metaphorical roots of conceptual growth. In L. English (Ed.), *Mathematical reasoning: Analogies, metaphors, and images* (pp. 339–371). London: Erlbaum.

Sfard, A. (2008). *Thinking as communicating*. Cambridge: Cambridge University Press.

Sfard, A. (2009). Metaphors in education. In H. Daniels, H. Lauder, & J. Porter (Eds.) *Educational theories, cultures and learning: A critical perspective* (pp. 39–50). New York NY: Routledge.

Shulman, L. (1999). Taking learning seriously. *Change, 31*(4), 10–17.

Soto-Andrade, J. (2006). Un monde dans un grain de sable: Métaphores et analogies dans l'apprentissage des maths. *Annales de Didactique et de Sciences Cognitives, 11*, 123–147.

Soto-Andrade, J. (2007). Metaphors and cognitive styles in the teaching-learning of mathematics. In D. Pitta-Pantazi & J. Philippou (Eds.), *Proceedings CERME 5* (pp. 191–200). Larnaca: University of Cyprus. Retrieved from http://www.mathematik.uni-dortmund.de/~erme/CERME5b/.

Soto-Andrade, J. (2014). Metaphors in mathematics education. In: S. Lerman (Ed.), *Encyclopedia of mathematics education* (pp. 447–453). Berlin: Springer.

Soto-Andrade, J. (2015). Une voie royale vers la pensée stochastique: les marches aléatoires comme pousses d'apprentissage. *Statistique et Enseignement, 6*(2), 3–24.

Soto-Andrade, J., Diaz-Rojas, D., & Reyes-Santander, P. (2016) Random walks as learning sprouts in the learning of probability, communication to TSG 14. *ICME 13*. Retrieved from http://iase-web.org/documents/papers/icme13/ICME13_S15_Soto-Andrade.pdf.

Soto-Andrade, J., Jaramillo, S., Gutiérrez, C., & Letelier, J. C. (2011). Ouroboros avatars: A mathematical exploration of self-reference and metabolic closure. In T. Lenaerts, M. Giacobini, H. Bersini, P. Bourgine, M. Dorigo, & R. Doursat (Eds.), *Advances in Artificial Life ECAL 2011: Proceedings of the Eleventh European Conference on the Synthesis and Simulation of Living Systems* (pp. 763–770). Cambridge, MA: The MIT Press.

Soto-Andrade, J., & Reyes-Santander, P. (2011). Conceptual metaphors and "Grundvorstellungen": A case of convergence? In M. Pytlak, T. Rowland, & E. Swoboda (Eds.), *Proceedings of CERME 7* (pp. 735–744). Rzészow, Poland: University of Rzészow. http://www.mathematik.uni-dortmund.de/~erme/doc/cerme7/CERME7.pdf

Spradley, J. P. (1980). *Participant observation*. New York, NY: Holt, Rinehart & Winston.

Thom, J. S., Namukasa, I. K., Ibrahim-Didi, K., & McGarvey, L. M. (2009). Perceptually guided action: Invoking knowing as enaction. In M. Tzekaki, M. Kaldrimidou, & H. Sakonidis (Eds.), *Proceedings of the 33rd conference of the International Group for the Psychology of Mathematics Education* (Vol. 1, pp. 249–278). Thessaloniki: PME.

Thompson, E. (2007). *Mind in life: Biology, phenomenology, and the sciences of mind*. Cambridge, MA: The Belknap Press of Harvard University.

Vallée-Tourangeau, F., Steffensen, S. V., Vallée-Tourangeau, G., & Sirota, M. (2016). Insight with hands and things. *Acta Psychologica, 170*, 195–205. https://doi.org/10.1016/j.actpsy.2016.08.006.

Varela, F. J. (1987). Lying down a path in walking. In W. I. Thompson (Ed.), *Gaia: A way of knowing* (pp. 48–64), Hudson, NY: Lindisfarne Press.

Varela, F. J. (1996). The early days of autopoiesis: Heinz von Foerster and Chile. *Systems Research, 13*, 407–417.

Varela, F. J. (1999). *Ethical know-how: Action, wisdom, and cognition*, Stanford: Stanford University Press.

Varela, F. J., Thompson, E., & Rosch, E. (1991). *The embodied mind: Cognitive science and human experience*. Cambridge, MA: The MIT Press.

vom Hofe, R. (1995). *Grundvorstellungen mathematischer Inhalte*. Heidelberg: Spektrum Akademischer Verlag.

Watson, A. (2008). Adolescent learning and secondary mathematics. In P. Liljedahl, S. Oesterle, & C. Bernèche. (Eds.) *Proceedings of the 2008 Annual Meeting of the CMESG* (pp. 21–32). Canadian Mathematics Education Study Group, Université de Sherbrooke.

Weissglass, J. (1979). *Exploring elementary mathematics*. New York, NY: W.H. Freeman.

Yin, R. K. (2003). *Case study research: Design and methods* (3rd ed.). Thousand Oaks, CA: Sage.

Chapter 35
Number Sense in Elementary School Children: The Uses and Meanings Given to Numbers in Different Investigative Situations

Alina Galvão Spinillo

Abstract This research investigated number sense in second grade Brazilian children (7–8 years old) from different social backgrounds. Study 1 (interview) aimed to identify the general uses given to numbers by children in everyday life situations. Study 2 (multiple choice tasks) examined how children assign meaning to numbers by asking the participants to make judgments about numerical situations involving both numbers and measurement and to provide justifications for their responses. The uses given to numbers in Study 1 were classified into different types: school uses, outside school uses, intellectual abilities and professional uses. The data in Study 2 were analysed according to correct responses and the types of justifications given. Both studies showed that there are some differences between children from different social backgrounds. On the whole, the children presented number sense that needs to be taken into account in the school setting.

Keywords Number sense · Children · Social backgrounds · Uses and meaning of numbers

Numbers and quantities, in a broad sense, are part of our daily life, from childhood to adulthood, in the most diverse contexts: at home, on the streets, at school and at work. They are part of the activities we carry out and the plans and the decisions we make. We are surrounded by numbers, and in order to function properly and efficiently in this environment we need to be numerate.

Being numerate involves familiarity with the world of numbers: to be able to think quantitatively in a variety of situations, to be able to employ efficient systems of representation and to understand the logical rules that govern the mathematical concepts inserted in these situations (Nunes and Bryant 1996). Being numerate is related to what in the literature is called number sense.

A. G. Spinillo (✉)
Federal University of Pernambuco, Recife, Brazil
e-mail: alinaspinillo@hotmail.com

G. Kaiser et al. (eds.), *Invited Lectures from the 13th International Congress on Mathematical Education*, ICME-13 Monographs,
https://doi.org/10.1007/978-3-319-72170-5_35

One may say that number sense is a cognitive ability that allows individuals to interact successfully with the various resources that the environment provides so that they can generate appropriate solutions to deal with daily activities that involve the use of mathematics. According to several authors (Cebola 2007; Greenes et al. 1993; Greeno 1991; Reys 1989), number sense can be defined as good intuition about numbers, their uses, meanings and relationships, which allows the individual to handle, in an efficient and flexible manner, situations which involve numbers and quantities. It is a skill that develops gradually from knowledge about the properties and meanings of numbers in varied contexts and from the construction of relationships that are not restricted to the use of algorithms.

Reys (1989) defines number sense as an understanding of numbers and operations that allows the application of appropriate resolution strategies and the processing of information, interpreting and communicating it accordingly. Similarly, Greenes et al. (1993) stress that number sense is the capacity for understanding the mathematical relationships involved in problem situations. Godino et al. (2009) and McIntosh et al. (1992) stress that this understanding needs to be flexible, given that the same strategies do not apply to all situations.

Faced with such a broad definition, it is important to be able to identify, both from a psychological and an educational point of view, what the indicators of number sense would be.

Spinillo (2006), based on an analysis of the literature on the subject (e.g., Greeno 1991; Yang et al. 2004), points out some indicators of numerical sense with the objective of contributing to a conceptual understanding of this topic and the creation of educational alternatives that will effectively make students numerate. Without intending to exhaust all possibilities of manifestation of a numerical sense, the author presents and exemplifies several indicators: estimating; performing flexible numerical computation; making quantitative judgments; establishing inferences; using anchors; recognising the plausibility of a result; recognising the absolute and relative magnitude of numbers; understanding the effect of operations on numbers; being able to use and recognise that one instrument or representation medium may be more useful or appropriate than another; and being able to recognise uses, meanings and functions of numbers in different situations.

Flexible numerical computation, mental calculations and estimates, assessed using oral problem-solving tasks, have been the most frequently investigated indicators (Yang 2003; Yang et al. 2004). Spinillo and colleagues, by means of judgment tasks, have investigated the effect of operations on numbers (Spinillo 2011); situations involving measurements (Spinillo and Batista 2009); and the meanings, uses and functions of numbers in different situations (Ribeiro and Spinillo 2006).

The understanding of the meanings and uses numbers can have in everyday life has been highlighted in curricular proposals (NCTM 1989) and by researchers. Cebola (2007), for example, states that children gradually discover what numbers are for and gradually begin to understand that numbers are what allow us to count, order or name something. Children realise that it is through numbers that one can (i) indicate the number of elements in a set (number as cardinal), (ii) say in which position an athlete has arrived in relation to other athletes (number as ordinal) and (iii) refer to a car number plate (number as nominal). According to the author, when thinking about numbers and when using them, children broaden their numerical sense. The present paper discusses, based on data obtained in two different studies, the meanings and uses that children assign to numbers in their daily life.

Like any other mathematical knowledge, number sense has its origin in everyday activities performed in the most different social contexts: at home (Blevins-Knabe and Musun-Miller 1996; Clements and Sarama 2007; Siegler 2009), at school (Brocardo and Serrazina 2008; Cebola 2007; Jordan et al. 2009) and on the streets and in the workplace (Nunes et al. 1993; Gainsburg 2005). In terms of socio-economic levels, Clements and Sarama (2007) observed that middle-class parents engage more frequently in mathematical activities with their children than low-income parents. According to Siegler (2009), children from low-income families have experiences which are less favourable to the development of number sense, and begin school with a limited knowledge of mathematics. For Nunes et al. (1993), who have studied low-income children and adolescents who carry out selling and buying activities on the streets, these children present a well-developed mathematical knowledge, although they adopt ways of reasoning different from those valued at school.

On the whole, these studies show that children from different social classes have different mathematical experiences in their daily lives that generate distinct types of mathematical knowledge (not necessarily better or worse). Thus, it seems relevant to examine whether children from different social classes would differ in their ability to intuitively understand numbers and operations. This is the issue addressed in this paper.

Given the large variety of number-sense indicators, the present study focuses specifically on one of these indicators: the uses and meanings assigned to numbers. The purpose of this paper is to discuss the results derived from two different investigative situations. In both studies, the participants are low-income and middle-class Brazilian children (7–8 years old), attending the second grade of elementary school in the city of Recife, Pernambuco, Brazil. None of the participants was engaged in any type of informal commercial activity.[1]

---

[1]It is important to mention that children from low-income families whose parents work as street vendors usually help them with their informal commercial transactions, such as selling snack foods (popcorn, popsicles, sweets, cupcakes etc.) and seasonal fruits. For more details about these informal commercial activities see Carraher et al. (1985) and Saxe (1991). The study conducted by Saxe also describes the four-phase cyclical structure of this practice.

35.1   Study 1

Study 1 consisted of clinical interviews whose objective was to identify the general uses given to numbers by children in everyday life situations. Three key questions were presented to each child: (1) What are numbers for? (2) What are sums for? and (3) Why do we measure things? These questions are associated with three fields of mathematical knowledge considered relevant in the national curricular proposal for elementary school in Brazil: numbers, arithmetic operations and measures (MEC/SEF 1997).

Forty children, 20 from a low-income background and 20 from a middle-class background, took part in the interviews. Each interview was recorded and the responses given were analysed and classified into different types according to the use given to numbers, to operations and to the activity of measuring. The responses were classified by two independent judges between whom the reliability of coding assessment was 82.5%. The cases of disagreement were evaluated by a third independent judge and the final classification was determined by agreement between two of the three judges. The types of responses are described and exemplified below:

Type 1 (school uses): Children provide answers in which they give school-related uses to numbers, operations and measurements.[2]

What are numbers for?

> 'To do the homework'.
> 'To have good grades at school'.

What are sums for?

> 'To learn what is on the board, in the book, in the notebook'.

Why do we measure things?

> 'So that we know the size of something if the teacher asks'.

Type 2 (outside-school uses): Children provide answers in which they relate numbers, operations and measurements to everyday situations:

What are numbers for?

> 'So that we know how many biscuits there are in a packet, for example'.

What are sums for?

> 'To know the total, to count money, to pay the electricity and the water bill'.
> 'So that we always get the right change'.

---

[2]None of the answers given have included more than one type of usage.

Why do we measure things?

> 'To know how tall someone is. I am taller than my sister'.
> 'To know how big a wardrobe is and see if it can go through the door'.

Type 3 (professional uses): Children provide answers in which they relate numbers, operations and measurements to professional activities:

What are numbers for?

'We need to know how to add, subtract. If we want to work, we need to study a lot'.

What are sums for?

> 'To find a job when we grow up'.
> 'To have a good salary in the future'.

Why do we measure things?

> 'Because I want to be a dressmaker when I grow up, and I'll have to measure things'.
> 'To be a good engineer. To build something like a house'.

Type 4 (intellectual abilities): Children provide answers in which they associate numbers, operations and measurement activities with intellectual success/achievements:

What are numbers for?

> 'To be clever'.

What are sums for?

> 'Because if we don't know how to add, to subtract, we are stupid'.

Why do we measure things?

> 'To learn things'.
> 'To know more'.

As it can be seen in Table 35.1, and revealed by the Mann-Whitney U test, middle-class children assigned outside-school uses to numbers more often than low-income children did ($U = 8$, $p = 0.0290$), while low-income children assigned

Table 35.1 Percentage of types of responses in each group

	Low-income ($n = 60$)	Middle-class ($n = 60$)
School uses	46.7	40
Outside school uses	30	43.4
Professional uses	13.3	5
Intellectual abilities	10	11.6

professional uses to numbers more often than middle-class children did ($U = 106$, $p = 0.0078$).

For low-income children, school uses were more frequent than other uses (Friedman test: $X = 15.529$, $p = 0.001$), whereas for middle-class children, school uses and outside-school uses were approximately the same and more frequent than the other uses (Friedman Test: $X = 11.571$, $p = 0.003$). It is important to mention that future uses related to professional activities were rarely observed among middle-class children (only 5%).

Relating these uses to the three fields of mathematical knowledge investigated (numbers, operations and measures), it was observed that, according to the participants:

(1) Numbers were assigned school uses primarily associated with performing arithmetic operations ('Numbers serve to solve the problems in the book, don't they?'). There were no uses or meanings related to the ordering or naming of things, which, according to Cebola (2007), are relevant aspects. When assigned outside-school uses, numbers were associated with counting situations ('They serve to count the things we have. To know how many things we have.').
(2) Arithmetic operations were essentially associated with school uses ('To find the answer to the math problem.').
(3) Measuring activities were mostly associated to outside-school uses ('To know the height of the person, to see if they have already grown.').

On the whole, children in both groups tended to give school-related uses to numbers and to sums in particular, whereas they tended to give outside-school uses to measurement.

To explain this result, it seems necessary to look at the mathematical activities the participants perform both in and out of school contexts, particularly in their living context. It is possible that the children in this study consider the activity of counting as a school-context activity. This is because in Brazil, mathematics textbooks directed to the first years of elementary school tend to favour the solution of arithmetic operations, either alone or in the solution of word problems (Mandarino 2009). On the other hand, measuring activities are not often proposed at school, so they were not associated with school uses. The opposite, however, seems to occur in the family context, as observed by Spinillo and Cruz (2016). Through natural observations in two different living contexts—home and orphanage—the authors verified that situations requiring measurements were the most frequent ones, especially among children living with their families. The children often had to measure the length of objects (e.g., the height of a wardrobe) or the distance between them (e.g., between the wardrobe and the bed), whereas arithmetic operations were performed mainly when doing homework. This was the case in both the family home and orphanage contexts. This possible explanation needs to be investigated further in future research through systematic observations of the school and living contexts of children from different social backgrounds.

35.2   Study 2

The aim of Study 2 was to examine how children assign meaning to numbers. The participants were 40 children, 20 from a low-income background and 20 from a middle-class background. Each child was shown a number on a card, and asked to say what that number meant, choosing one of three alternatives presented orally. The multiple choice task consisted of 12 trials, and the alternatives referred to meanings assigned to numbers in everyday life, for example: 'Do you think this number (child is shown a card with the number 6) is (a) the age of a person, (b) a car number plate or (c) the number of books in a library?' and 'Do you think this number (child is shown a card with the number 401) is (a) the number of pills someone took in one day, (b) the number of a flat or (c) someone's age?'

In order to understand better the child's way of thinking, when the correct answer was chosen,[3] the examiner asked the child why that number could not correspond to one of the other two alternatives (that is, the incorrect ones). The answers were audio-recorded and classified as vague or as precise, as exemplified below:

Do you think this number (shows a card with the number 82) is (a) the number of cars in someone's garage, (b) someone's age or (c) someone's telephone number?

Vague justification:

Child: It is someone's age.
Examiner: And why can it not be, for instance, someone's phone number?
Child: Because it can't.
Examiner: But why not?
Child: It just can't.
Examiner: Then explain to me why it cannot be.
Child: Because it's impossible.

Precise justification:

Child: It is someone's age.
Examiner: And why can it not be, for instance, someone's phone number?
Child: Because a phone number is a very long number. It has got many numbers. 82 only has two numbers, it is too short to be someone's phone number.

Do you think this number (shows card with the number 5900) is (a) the total number of pages in a comic book, (b) the amount of money someone has in the bank or (c) the number of floors in a building?

---

[3]The order in which the correct alternative was presented was randomized, so that in the first four trials the correct answer was presented as the first alternative, in the following four trials as a second alternative and in next four trials as the third.

Table 35.2 Percentage of the types of justifications given to the correct answers per group

	Low income ($n = 157$)	Middle class ($n = 163$)
Vague justification	62.4	28.2
Precise justification	37.6	71.8

Vague justification:

Child: An amount of money.
Examiner: And why could it not be the number of pages in a comic book?
Child: Because it cannot. This is money. I know it.
Examiner: And how do you know this? Can you explain it to me?
Child: I just know it.

Precise justification:

Child: The amount of money in a bank.
Examiner: And why could it not be the number of pages in a comic book?
Child: No. Comic books are not like this. It would be too many pages and no child would read a comic book with so many pages.

The percentage of correct responses did not vary between the two groups (Low income: 65.4% and Middle class: 67.9%), but the types of justifications varied (see Table 35.2).

As shown in Table 35.2, the results differ between groups due to the percentage of vague justifications being significantly higher among low-income children than the percentage of precise justifications (Friedman test: $X = 17.316$, $p = 0.026$). The opposite is observed among middle-class children (Friedman test: $X = 29.746$, $p = 0.000$). Also, vague justifications occurred more often among low-income (Mann-Whitney U test: $U = -2.919$, $p = 0.004$) than among middle-class children, whereas precise justifications were more frequently observed among the middle-class children (Mann-Whitney U test: $U = -3.163$, $p = 0.002$).

Thus, it is possible to say that children from different social backgrounds are able to successfully identify the different meanings assigned to numbers by society. The meanings assigned to numbers that were explored in this study were essentially number as a quantity of elements (number as cardinal, e.g., the number of books in a library) and as an identity (number as a nominal, e.g., a car number plate or a phone number). It would be interesting to investigate whether the two groups of children would differ in relation to the other different meanings that numbers may have in daily life beyond those examined in this study. For example, number as an ordinal (an athlete's position in relation to other athletes in a competition) and number as a measure (different dimensions such as length, volume, etc.).

However, the groups of children who participated in this study differ in their ability to express the foundations that underlie their judgments, since middle-class children can provide more precise justifications than lower-income children. A possible explanation for this result is given below.

35.3 General Discussion and Conclusions

Taken together, the results from both studies show that there are some similarities and some differences between children from different social backgrounds with regard to the uses and meanings attributed to numbers in everyday life.

The first similarity is that children in both groups were able to identify the different meanings attributed to numbers in society. Regardless of their social background, they tended to attribute school uses to numbers and sums in particular, and to give outside-school uses to measurement. Also, although somewhat infrequent, there were children in both groups who associated mathematical knowledge with intellectual gains.

However, the groups also differ in some respects. Whereas low-income children tended to assign school uses to numbers, operations and the activity of measuring, middle-class children tended to attribute both school and extra-curricular uses to numbers. This suggests that middle-class children perceive a greater diversity of uses for numbers than low-income children, for whom numbers are mainly associated with the school context. It is important to mention that the uses related to professional activities are more frequently given by low-income children, who associated mathematical knowledge with work and subsistence.

Another aspect to be stressed is a child's ability to explain the bases that guide their judgments when they attribute meaning to numbers. The most remarkable difference between the groups is that middle-class children provided more accurate justifications than low-income children. This result can be explained in the light of Vergnaud's theory (1983, 1997), specifically in relation to what he calls 'theorems in action', which can be briefly defined as a non-explicit knowledge. Therefore, one can assume that low-income children have knowledge in action that allows them to appropriately assign different meanings to numbers. They are not, however, able to verbally state the basis of their judgments. On the other hand, middle-class children have a propositional knowledge that allows them to assign different meanings to numbers, as well as to verbally express the way they think, that is, explain the bases of their judgments. A possible explanation for this is that middle-class children are more used to giving explanations about the bases of their judgments than low-income children. However, it is necessary to be cautious when interpreting such data since, while insightful, the results derived from the reported studies are not definitive and other explanatory alternative hypotheses need to be considered in future research.

While the concept of number seems to be related to logical development, following a similar path independent of social environment (see Piaget's (1965) ideas about conservation of quantity and class inclusion, for instance), number sense seems to be a type of knowledge subject to greater variability, being dependent on the social experiences that individuals have with numbers in their everyday life. One may say that number sense is not the same for all: In other words, it is not equally distributed in society. This possibility needs to be further explored in future research with regard to number sense. However, based on results obtained in previous studies, such as those conducted by Nunes et al. (1993) and Saxe (1991), it is possible to assume that children acquire their mathematical knowledge from the interplay of activities they perform at home, at school and on the streets.

Teachers and researchers need to be fully aware of this fact. Teachers need to take it into account in school settings, especially during the early years of elementary school. Researchers should consider the relevance of using a variety of methodological recourses and investigative contexts when examining number sense in children. Different investigative contexts allow us to explore different aspects of number sense and to go beyond. For instance, future research could investigate, at least partially, where number sense comes from and use natural observations to look at the mathematical activities that children perform at home. Comparisons between children from different social backgrounds and also from different home environments, such as family and orphanage, could be of great importance, as the partial results of a recent exploratory study indicate (Spinillo and Cruz 2016). Such studies can help to identify and describe the mathematical activities performed by children in their home context, a context that has been little investigated with relation to mathematical knowledge.

References

Blevins-Knabe, B., & Musun-Miller, L. (1996). Number use at home by children and their parents and its relationship to early mathematical performance. *Early Development and Parenting, 5*, 35–45.

Brocardo, J., & Serrazina, L. (2008). O sentido do número no currículo de matemática. In J. Brocardo, L. Serrazina, & I. Rocha (Eds.), *O sentido do número: Reflexões que entrecruzam teoria e prática* (pp. 97–115). Lisboa: Escolar Editora.

Carraher, T. N., Carraher, D. W., & Schliemann, A. D. (1985). Mathematics in the streets and in schools. *British Journal of Developmental Psychology, 3*, 21–29.

Cebola, G. (2007). Do número ao sentido de número. In J. P. Ponte, C. Costa, A. I. Rosendo, E. Maia, N. Figueiredo, & A. F. Dionísio (Eds.), *Atividades de investigação na aprendizagem da matemática e na formação de professores* (pp. 223–239). Lisboa: Secção de Educação Matemática da Sociedade Portuguesa de Ciências da Educação.

Clements, D. H., & Sarama, J. (2007). Effects of a preschool mathematics curriculum: Summative research on the building blocks project. *Journal for Research in Mathematics Education, 38*, 136–163.

Gainsburg, J. (2005). School mathematics in work and life: What we know and how we can learn more. *Technology in Society, 27*, 1–22.

Godino, J. D., Font, V., Konic, P., & Wilhem, M. R. (2009). El sentido numérico como articulatión flexible de los significados parciales de los números. In J. M. Cardeñoso & M. Peñas (Eds.), *Investigación en el aula de matemáticas. Sentido numérico* (pp. 117–184). Granada: SAEM.

Greenes, C., Schulman, L., & Spungin, R. (1993). Developing sense about numbers. *Arithmetic Teacher, 40*(5), 279–284.

Greeno, J. G. (1991). Number sense as situated knowing in a conceptual domain. *Journal of Research in Mathematics Education, 23*(3), 170–218.

Jordan, N. C., Glutting, J., & Ramineni, C. (2009). The importance of number sense to mathematics achievement in first and third grades. *Learning and Individual Differences, 20*(2), 82–88.

Mandarino, M. C. F. (2009). Que conteúdos da matemática escolar professores dos anos iniciais do ensino fundamental priorizam? In G. Guimarães & R. Borba (Eds.), *Reflexões sobre o Ensino de Matemática nos Anos Iniciais de Escolarização* (pp. 101–118). Recife: SBEM.

McIntosh, A., Reys, B. J., & Reys, R. E. (1992). A proposed framework for examining basic number sense. *For the Learning of Mathematics, 12*(3), 2–8.

MEC/SEF. (1997). *Parâmetros Curriculares Nacionais, Matemática* (Vol. 3). Brasília, Brasil: Secretaria de Educação Fundamental.

National Council of Teachers of Mathematics. (1989). *The principles and standards for school mathematics*. Reston, VA: NCTM.

Nunes, T., & Bryant, P. (1996). *Children doing mathematics*. Oxford: Wiley-Blackwell.

Nunes, T., Schliemann, A. D., & Carraher, D. W. (1993). *Street mathematics and school mathematics*. Cambridge: Cambridge University Press.

Piaget, J. (1965). *The child's conception of number*. New York: Norton.

Reys, B. J. (1989). Conference on number sense: Reflexions. In J. T. Sowder & B. P. Schapelle (Eds.), *Establishing foundations for research on number sense and related topics: Report of a conference* (pp. 70–73). San Diego: Diego State University Center for Research in Mathematics and Science Education.

Ribeiro, L. M., & Spinillo, A. G. (2006). Preschool children's number sense. In J. Novotná, H. Moraová, M. Krátiká, & N. Stehliková (Eds.), *Proceedings of the 30th Conference of the International Group for the Psychology of Mathematics Education* (Vol. 1, p. 417). Prague, Czech Republic: PME.

Saxe, G. B. (1991). *Culture and cognitive development: Studies in mathematical understanding*. New Jersey: Lawrence Erlbaum Associates.

Siegler, R. S. (2009). Improving the numerical understanding of children from low-income families. *Child Development Perspectives, 3*, 118–124.

Spinillo, A. G. (2006). O sentido de número e sua importância na educação matemática. In M. R. F. de Brito (Ed.), *Soluções de problemas e a matemática escolar* (pp. 83–111). Campinas: Alínea.

Spinillo, A. G. (2011). Number sense in children: Understanding number as an operator when adding and subtracting. In B. Ubuz (Ed.), *Proceedings of the 35th Conference of the International Group for the Psychology of Mathematics Education* (Vol. 4, pp. 201–209). Ankara, Turkey: PME.

Spinillo, A. G., & Batista, R. M. F. (2009). A sense of measurement: What do children know about the invariant principles of different types of measurement? In M. Tzekaki, M. Kaldrimidou, & H. Sakonidis (Eds.), *Proceedings of the 33rd Conference of the International Group for the Psychology of Mathematics Education* (Vol. 5, pp. 161–168). Thessaloniki, Greece: PME.

Spinillo, A. G., & Cruz, M. S. S. (2016). Number sense in children from different living contexts. In C. Csikos, A. Rausch, & J. Szitànyi (Eds.), *Proceedings of the 40th Conference of the International Group for the Psychology of Mathematics Education* (Vol. 1, p. 240). Szeged, Hungary: PME.

Vergnaud, G. (1983). Multiplicative structures. In R. Lesh & M. Landau (Eds.), *Acquisition of mathematics: Concepts and processes* (pp. 127–174). London: Academic Press.

Vergnaud, G. (1997). The nature of mathematics concepts. In T. Nunes & P. Bryant (Eds.), *Learning and teaching mathematics: An international perspective* (pp. 5–28). Sussex: Psychology Press.

Yang, D.-C. (2003). Teaching and learning number sense—An intervention study of fifth grade students in Taiwan. *International Journal of Science and Mathematics Education, 1*, 115–134.

Yang, D.-C., Hsu, C.-J., & Huang, M.-C. (2004). A study of teaching and learning number sense for sixth grade students in Taiwan. *International Journal of Science and Mathematics Education, 2*, 407–430.

Chapter 36
Uncovering Chinese Pedagogy: Spiral Variation—The Unspoken Principle of Algebra Thinking Used to Develop Chinese Curriculum and Instruction of the "Two Basics"

Xuhua Sun

Abstract Many international research studies are conducted in the Western deductive tradition strongly influenced by a geometric perspective. During the past decades, the missing paradigm from an algebraic tradition has rarely been explored. I intend to present the algebraic perspective that structures inductive tradition in an effort to understand Chinese curriculum and instruction of the "Two Basics" and its unspoken principle, spiral variation. This study can deepen our understanding how the inductive reasoning that underpins early Chinese algebra provides a foundational cultural perspective for interpreting "indigenous" principles and their application. This discussion can enlighten our understanding of the Chinese tradition of mathematics education, which can in turn shed light on the research into algebra education from the perspective of problem variation.

Keywords Chinese curriculum · Spiral variation · Algebra history
Chinese mathematics history

36.1 Introduction

Algebra is one of the most daunting branches of school mathematics (Radford 2015), yet it is generally considered an essential worldwide language for any study of advanced mathematics, science, or engineering and also for such applications as medicine and economics. Cross-national studies have provided insight into the cultural and educational factors that may influence the learning of mathematics (e.g., Cai and Wang 2006). A range of studies on the differences in the mathematical thinking of students have found that Chinese students prefer to use

X. Sun (✉)
University of Macau, Zhuhai, China
e-mail: xhsun@umac.mo

© The Author(s) 2018   651
G. Kaiser et al. (eds.), *Invited Lectures from the 13th International Congress
on Mathematical Education*, ICME-13 Monographs,
https://doi.org/10.1007/978-3-319-72170-5_36

symbol-based strategies and algebraic solutions and U.S. students prefer concrete, pictorial-based strategies in problem solving (e.g., Cai 2000). A corresponding difference in the approach of teachers is that U.S. teachers put more emphasis on the use of concrete examples to aid student understanding, while Chinese teachers tend to emphasize the abstract reasoning beyond the concrete after presenting concrete examples (e.g., Cai and Wang 2006). Some corresponding studies have made an effort to document and analyze how the Chinese curriculum and instructional practice supports the development of algebraic thinking in students (e.g., Cai and Knuth 2005). This historic-cultural aspect of algebraic development may allow us to examine the deeper educational roots beyond the current curricula and instruction, which have been insufficiently explored. This study will examine the legacy of ancient China's algebraic development (China in this paper denotes mainland China exclusively from a historical perspective). I intend to discuss how the inductive reasoning that underpins early Chinese algebra provides a foundational cultural perspective for interpreting "local" principles and their application. I will begin with an introduction to the Chinese tradition of mathematical education from an algebraic perspective, where has been unknown in the West.

36.2   The Legacy of Ancient China: Generalization of a Solution Method, an Algebraic Development Framework

The detailed Chinese tradition of algebra has rarely been reported in Western historical literature. For example, Chinese history is omitted from the classic mathematical literature edited by Kline (1972) and the history of algebra's development (Sfard 1995). As Wu (1995) points out: "there are two core thoughts/paths through the mathematical history of the world. One is axiomatic thought from the Greek Euclidean system. Another is mechanistic thought which originated in China and influenced India and the whole world" (cited in Guo 2010). For example, the Chinese remainder theorem, the solution of modular equations, was discovered in the fifth century CE by the Chinese mathematician Sunzi and described by Aryabhata in the sixth century. Special cases of the Chinese remainder theorem were also known to Brahmagupta in the seventh century and appeared in Fibonacci's *Liber Abaci* in 1202 (Pisano and Sigler 2002; Li 2005). The axiomatic method is renowned for its influence on the development of geometry and non-Euclidean geometry, the foundation of real analysis, and Cantor's set theory, which stands for rigor, clarity, and absolute truth (Guo 2010). However, mechanistic thought, also called the algorithmic method, which aims to find invariant strategies by performing calculations, processing data, and automating reasoning, has received little attention, despite being the representative system of traditional Chinese arithmetic and algebra from which most of the classic works of ancient Chinese mathematics originated. The most brilliant example of the application of the algorithmic method is the arithmetic algebra system known as *The Nine Chapters on the Mathematical Art* (*JZSS*;1000 BC–200 AD). Using this logical tool, Chinese mathematicians attempted to convert geometric problems into algebraic problems (Guo 2010), in contrast to the Greek approach of converting algebraic problems into geometric problems. This directly influenced Asian countries such as Korea, Japan, Mongolia, Tibet, and Vietnam (Martzloff 1997, pp. 105–110). Although Chinese algebraic development was limited by the nature of its language which lacks letters (characters rather than letters were and continue to be used in China), a flourishing series of advanced classic algebraic works were developed. In contrast to Greek geometry, various algorithms for solving equations were the main focus, from high-degree polynomial equations to linear equations. Even indefinite equations were created from applied mathematics (Guo 2010; Li 2005), for example, by providing algorithms to calculate the extraction of square/cube roots and the rules for calculating positive and negative numbers as a foundation for solving equations and equation systems (irrational numbers and negative numbers were first identified in ancient China (Guo 2010)). Eighteen problem-solving methods for systems of linear equations with 2, 3, 4, and 5 unknowns were presented in *JZSS*. Gaussian elimination (19th century), was first introduced about 2000 years earlier (Shen et al. 1999). Compared with the approaches to find two numbers, known as syncopated algebra from the *Arithmetica* of Diophantus (250 AD), this was considerably earlier and more systematic, presenting the first systematic use of irrational and negative numbers. In fact, the concept of variable, called *tian yuan shu* (天元术), was in systematic use in China long before that of Francois Viete (1540–1603). *Tian yuan shu* denotes a strategy of the heavenly unknown, which played an important role in the Chinese algebraic approach to solving polynomial equations in the 13th century. It first became known through the writing of Li Ye in his work *Ceyuan Haijing* (测圆海镜) in 1248. Meanwhile, *tian yuan shu* spread to Japan, where it was called *tengen jutsu* in *Suanxue Qimeng* (算學啓蒙), authored by Zhu Shijie, and played important role in the development of Japanese mathematics (*wasan*) in the 17th and 18th centuries (Mikami 1913). In fact, the general root of high-degree equations to solve the numerical solution of the program *zheng fu kai fangshu* (正負开方术), the mechanical algorithm in *Shushu Jiuzhang* (数书九章), was written by Qin Jiushao. The algorithm for eliminating and solving polynomial equations with four unknowns, *Si Yüan Yü Jian* (四元玉鑒; *The Jade Mirror of the Four Unknowns*, with the four elements, heaven, earth, man, and matter, representing the four unknown quantities) was written by Zhu Shijie in 1303 AD. This deals with simultaneous equations and with equations of degrees as high as 14, marking the peak in the development of Chinese algebra (Guo 2010).

Shu (术), a term broadly used in problem solving, played an important role in the development of the ancient Chinese mathematical system, which stemmed from the spirit of "general methods" in the problem-oriented tradition of Asian mathematics "to produce new methods from real problems, promote them to the level of a general method, generalize them into *shu*, and deploy these *shu* to solve various similar problems which are more complicated, more important, and more abstruse" (Wu 1995, p. 46). In some of Liu Hui's commentary on *JZSS*, *du shu* (都术, "the basic algorithm") was highlighted to describe basic algorithms that are much more generalized than specific algorithms for a specific class, and can thus be applied to broader classes of problems (Guo 2010). *Jinyou shu* (今有术) is one of these (Guo 2010). Although algebra (e.g., equations and systems of equations) existed in several ancient civilizations, including the Babylonian, Greek, Egyptian, Indian, Chinese, Arab, and European, the clear framework for a more generalized solution appeared only in the Chinese literature in terms of the generalization *pu shixing* (普適性) and *du shu* (都术, "the basic algorithm"), an algebraic framework beyond a question-answer algorithm, *wen-da-shu* (問答术), which is in the form of a statement of a concrete problem followed by a statement of the solution and an explanation of the procedure that led to the solution. In contrast to the axiomatic approach—a strategy for deducing propositions from an initial set of axioms in the geometric tradition of Egypt and Greece that has dominated the intellectual world since the time of Greek philosophers such as Thales, Anaximander, and Aristotle—the inductive approach was always the more dominant tool of abstraction in ancient China (Wu 1995). Other Chinese treatises that contain structures similar to those in the *JZSS* usually emphasize the algebraic framework too include *Haidao Suanjing* (海島算經), *Zhang Qiujian Suanjing* (張丘建算經), *Wuchao Suanjing* (五曹算經), *Wujing Suanshu* (五經算術), *Figu Suanjing* (緝古算經), *Shushu Jiyi* (數術記遺), and *Xiahou Yang Suanjing* (夏侯陽算經).

The inductive reasoning used within the algebraic framework, as opposed to deductive reasoning, is in fact frequently used today in science, philosophy, and the humanities because it can lead to unknown predictions and new knowledge, which deductive reasoning cannot. Its application has been questionable, however, due to uncertain conclusions drawn from relatively limited cases or experiences. However, algebraic thinking is to some extent born of the inductive reasoning system rather than deductive reasoning. It is worth noting that although algebra was developed in the West from ancient Babylonian mathematics (Høyrup 2002), it does not use the clear algebraic framework described above, but rather the concrete-problem and concrete-solution method [e.g., the tablet AO8862 1800/1600 BC (Spagnolo and Di Paola 2010, p. 52)]. The classic early algebra work of mathematician Mohammed ibn Musa al-Khowarizmi, the author of Aljabr w'al muqabala, which provided the modern word *algebra* also failed to emphasize general solutions beyond the concrete in the way the Chinese mathematical literature did (Guo 2010). Chemla (2009) showed that some of the algorithms in *JZSS* were built not just to solve a specific problem but rather the general class of problems they represented. The whole structure of *JZSS* seems to call for this general procedure and encourages the search for general formulations in algebraic rhetoric, as pointed out by Spagnolo and Di Paola (2010).

However, it is interesting to note that the ancient Chinese developed algebra only, not geometry (Euclid's *Elements* was introduced into China in the 17th century). This encourages us to enquire whether there was a specific ecology in China that was conducive to the development of algebra. The historical, social, and cultural foundations of the development of algebra have been neglected from the international perspective. In this paper, we attempt to fill in the gap of lack of recognition of the historical beginnings of algebra in China and, in particular, provide an argument that the inductive reasoning that underpins early Chinese algebra provides a foundational cultural perspective for interpreting pedagogical approaches.

36.3   What Are the Key Features of Ancient Chinese Mathematics?

Jiu Zhang Suan Shu (九章算术), the most classic work of Chinese mathematics, used 246 word problems categorized into nine categories to spread mathematical knowledge, which also reflects ancient China's pedagogical approach. The structure of *Jiu Zhang Suan Shu* (as pointed out by Liu Hui) emphasizes *lu* (率; Sun and Sun 2012), *jin you shu* ("ratio equation"), and the *qitong* theorem as the core ideas (Sun and Sun 2012), and its mathematical problems are arranged into nine categories by the idea of categorization (Guo 2010).

The ideas of categorization stressing the above invariance-variation concept appeared in the preface below as the central guiding spirit in Liu Hui's commentary in the 2000-year-old Chinese textbook, *JZSS*, which has played a similar role in Asian countries to that of Euclid's Elements: "Although they (knowledge tree) are diverse, their branches grow from the same root" ("故枝條雖分而同本榦者,知發其一端而已"; Guo 2010, p. 178).

The invariance-variation relationship is represented by the idea of categorization in the *JZSS*, described as the ideology of "categorizing to unite categories (以類合類)" in ancient China (Guo 2010, p. 76). The concept of categorization was illustrated by classifying the 246 variant problems into the nine categories (歸類) below.

1. *Fangtian* (方田): rectangular fields
2. *Sumi* (粟米): millet and rice, the exchange of commodities at different rates, pricing
3. *Cuifen* (衰分): proportional distribution, the distribution of commodities and money at proportional rates
4. *Shaoguang* (少广): the lesser breadth, division by mixed numbers
5. *Shanggong* (商功): consultations on works, volumes of solids of various shapes
6. *Junshu* (均输): equitable taxation
7. *Yingbuzu* (盈不足): excess and deficit, linear problems solved using the principle known later in the West as the rule of false position
8. *Fangcheng* (方程): the rectangular array, systems of linear equations
9. *Gougu* (勾股): base and altitude, problems involving the principle known in the West as the Pythagorean theorem.

After the emergence of the *JZSS*, the concept of categorizing became the model for mathematical task design in traditional applied mathematics, which has played a role as an associated pedagogy of the *JZSS*.

Almost all problems in ancient China were placed into categories in the classic mathematics texts by Wucao (五曹算經) and Xiahouyang (夏侯陽算经; Wang 1996). Before the Western system was imposed on the Chinese curriculum, the categorizing model was the unspoken task design framework. For example, mathematical problems grouped into the following categories were typical of the Chinese curriculum (Wang 1996):

1. Difference/sum category
2. Speed category
3. Tree-planting category
4. Age category
5. Availing category
6. Engineering category
7. Profit category.

Through the traditional logic of the Greeks, the axiomatic approach has remained the cornerstone of mathematics in the West. Accordingly, a definition/theorem-based model stressing content knowledge gradually formed the fundamental idea of mathematical task design in the West. This has played an important role in the history of Western mathematical education, where word problems, labeled application problems (应用题), play a role in knowledge application. In contrast, the problem-solving approach and applied mathematics in *JZSS* mainly remained the cornerstone of mathematics in the East. Its associated categorization model in *JZSS* gradually formed the unspoken but fundamental framework of mathematics task organization/design in China (Sun 2013). It is interesting to note that this model stresses the category-based inductive tradition rather than the definition/theorem-based deductive tradition of the West, where word problems with variations play a role in relation-oriented knowledge introduction (Bartolini Bussi et al. 2013). In short, the idea of categorization reflects the ancient curriculum practice using the variant–invariant (from concreteness to abstract logic) spirit above.

36.4   The Key Features of Chinese Pedagogy in Current Teaching Practice: From a Single Problem to a Class of Problems with Variation

The tradition of categorizing was not implemented after the Western mathematics curriculum was imported into China in 1878 (Wang 1996). However, Chinese curriculum developers emphasized the Two Basics and, after 1878, developed an associated pedagogy with variation problems stressing the categorization process, from the variant concreteness to the invariant abstract application. This pedagogy of problem design centered on the idea of expanding a single problem to a class of problems with variation problems. It also aimed to establish the necessary and sufficient conditions to determine each category of problem set using two similar and important parameters of mathematical structure, the dimensions of possible variation, and the associated range of permissible change, as pointed out by Watson and Mason (2005, 2006). This practice is called *bianshi* (變式) in Chinese, where *bian* stands for changing and *shi* means form. Although it has spread into a wide range and variety of forms in China (Sun 2007), "indigenous" variation practice in mathematics refers to the "routine" daily practice commonly accepted by Chinese teachers, the local experience used broadly in the design of examples or exercises to extend the original examples, known widely as "one problem, multiple changes" (OPMC,—題多變, "varying conditions and conclusions"), "one problem, multiple solutions" (OPMS,—題多解, "varying solutions"), and "multiple problems, one solution" (MPOS, 多題一解, "varying presentations"; Sun 2007, 2011a, 2016).

According to Kieran (2004, 2011), the global meta-level algebra activities essential to the other generational, transformational activities of algebra include studying change, generalizing, analyzing relationships, and noticing structure for which algebra is used as a tool. The "routine" activities of varying conditions and conclusions, varying solutions, and varying presentations above play the role of concept connections, solution connections, and presentation connections (Sun 2011a, b, 2016). Systematically, they provide a platform to support analyzing relationships and noticing structure and, therefore, can support meta-level algebra development.

This practice, rarely apparent in the West, is a typical daily routine in the local curriculum and regarded as a natural strategy for deepening understanding, which perhaps makes this practice distinctive. This strategy, easily found in school teaching materials (such as textbooks or teaching plans) and any piece of learning material (such as student exercises or worksheets) is followed after school in China. As mentioned before, Chinese arithmetic development, textbooks, textbook reference books, and particular variation practices provide useful clues for understanding the Chinese mathematics education system rarely known outside of the Chinese community.

In contrast to variation problems, contextualization problems are prioritized as the general curricular trend in the West (Clarke 2006). However, contextualization problems to facilitate engagement mainly provide examples of the same concept and solution method, missing the chance to make timely connections between concepts and methods. In this light, variation problems suggest a way in which Western counterparts can learn from the content-oriented curricula in China. Compared to contextualization problems, variation problems are clearly a double-edged sword that can increase the learning challenge because they require the use of multiple concepts, solutions, and conceptual development.

36.5 Why Is the Task Design Principle Important?

In seeking a basic algorithm as a demonstration tool, the problem variations described earlier aim to avoid heuristic trial and error (such as arithmetic) by eliciting reasoning, using variation as scaffold for discerning the invariant, a kind of pre-algebraic thinking. It could be a helpful transition from arithmetic thinking to algebra. For example, here is a prototype example of an OPMC variation in which the concept of subtraction is always introduced in the Chinese textbook below as: $1 + 2 = 3$, $2 + 1 = 3$, and $3 - 1 = 2$. Within the problem set, there are two concepts of addition and subtraction behind three similar problems made with 3, 2, and 1. Clearly, this OPMC provides a setting in which learners can reflect and generalize between the concepts of addition and subtraction in order to concentrate on the relationships, a kind of pre-algebraic thinking, involved. In contrast, the concept of subtraction is introduced in some U.S. textbooks using a problem set such as this: $4 - 1 = 3$, $5 - 3 = 2$, and $3 - 2 = 1$. Within the problem set, instead of embedding two concepts as done above, only the concept of subtraction is included. Clearly, the variation of solutions, conditions/conclusions, and presentations can be used to emphasize the invariant elements as a possible way of generalization, providing the transition from arithmetic thinking to algebra. In contrast, the "one-thing-at-the-time" design based on the notion of consolidating one topic or skill before moving on to another that is broadly used in most textbook development in Europe and throughout the world would clearly provide fewer opportunities for "making connections" (Sun 2011a, b) compared to those of contemporaneous variation approaches (e.g., Rowland 2008).

The variations described earlier elicit the idea that variability is at the heart of algebra, and aim to provide a platform to transit from arithmetic thinking to the relational algebraic thinking. Variation plays the role of meta-logic to access algebra. From the perspective of the Chinese philosophy (Hua 1999) and language, it is an important framework for algebraic thought development from the arithmetic stage rooted in Chinese cultural logic (Sun 2016). In contrast to the deductive Western cultural thinking derived from Euclid's *Elements* and Aristotle's logic, variation can support another kind of inductive reasoning to discern invariance, which does not rely on logic to refer to the type of divalent but rather on extensive use of the idea of variability as the initial form of expression. This serves as a bridge or a schema for relational thinking to transition from arithmetic to algebra, which indicates the process of generalization: a new logic for algebraic development using the idea of variability (Sun 2016).

Based on this perspective, it is easy to note the difference when we compare Chinese and Western curricula (Sun et al. 2013). For example, different task design features for addition and subtraction are found in Chinese versus Portuguese textbooks, featuring invariant versus variant concept/solution methods embedded in their examples. In Chinese textbooks, addition and subtraction are almost always connected using the OPMC transformation principle rather than separated into different chapters as in a Portuguese textbook. Although Chinese textbook authors appear to use multiple concepts for every example, the underlying invariant concept is about part-part-whole relations and the invariant knowledge is about relations between numbers. In contrast, the addition examples in the Portuguese textbook use multiple underlying concepts, such as counting, combining, and adding. The subtraction examples also use multiple concepts such as subtracting, comparing, and identifying inverse operations, but do not connect these simultaneously to additional concepts.

Although Chinese textbook authors use multiple solution methods in every example, the particular methods in the Portuguese textbook that depend on counting and doubling are rarely introduced. Only one specific solution method, "make 10," is explicitly addressed among all the addition/subtraction examples in the first six chapters (Sun et al. 2013). In contrast, the additional examples in the Portuguese textbook suggest multiple solution methods, such as "doubles," "doubles plus 1," "compensation," (e.g., $6 + 8 = 7 + 7 = 14$), and "reference number" (e.g., $6 + 7 = 5 + 1 + 5 + 2 = 10 + 3 = 13$). The subtraction examples use multiple solution methods such as "counting back," "tables for addition to subtraction," and "identifying the inverse operation of subtraction as addition." Thus the learner might get a temporary sense of these methods from being offered a variety of suitable examples without getting an overall understanding of the whole additive relation. The underlying Portuguese design principle is not made explicit, but we can infer that it is about learning "one thing at a time" and is hence more fragmented and less dependent on laying down basic foundational principles for future work (Sun et al. 2013).

36.6   The "Indigenous Principles" in Mathematics: Spiral Variation

In addition to the Chinese philosophy and language conductive to algebraic thinking mentioned above (Sun 2016), it is not surprising to note that the Two Basics is not only regarded as the explicit principle of local curriculum design (Sun et al. 2013) but also the central aspect of the unified teaching framework of the Ministry of Education (1963, 2001). The Two Basics, i.e., basic knowledge and basic skills, is a Chinese term stressing the basic facts, basic concept, basic principles, invariant aspects behind the variant–invariant (from concreteness to abstract logic) idea or inductive, algebraic thinking mentioned above. They are described as "indigenous" principles for designing educational tasks (Zhang 2006). However, some research indicates that variation plays a more important role in the Chinese curriculum. For example, as Marton (2008) argued:

> Chinese students do very well when compared to students from other cultures. Teachers spend much more time on planning and reflecting than teachers in other countries, and they develop their professional capabilities by the teaching, in which patterns of variation and invariance, necessary for learning (discerning) certain things, are usually brought about by juxtaposing problems and examples, such as illustrations that have certain things in

common while resembling each other in other respects. By such careful composition, the learner's attention is drawn to certain critical features ... instead of just going through problems that are supposed to be examples of the same method of solution. ... [There] is a very powerful pedagogical tradition in the Chinese culture. (p. 1)

In fact, the statement above is consistent with the two most repeated local terms in the Chinese mathematical pedagogy, the Two Basics and "variation teaching" (變式 教学), which indicates that, on one hand, the invariant aspect should be stressed for curriculum development and, on the other, the variation aspect should be stressed for instruction as its tool. They indicate the original local notions of teaching practice, stressing the invariant and variant elements, respectively, which may be related to the categorization approach of Chinese language, "grasping ways beyond categories" (*yifa tongli*, 以法通類) and "categorizing to unite categories" (*yilei xiang-cong*,以類相從), discussed above. Here, a distinct instructional feature of the problems is to develop the ability to identify the category to which a problem (识类) belongs and to discern the different categories (归类), in other words, to discern the invariant from the variant elements of different problems and recognize the category each problem belongs to (Sun 2011a, 2016). The process of discerning the invariant from the variant elements can provide the chance to generalize the common feature and see the deep structure behind different problems, which is needed in the process of algebraization (Mason 1996, 2011). Obviously, both Two Basics in the traditional Chinese curriculum and instruction and the model of the spiral variation focus on the variant–invariant (from concreteness to abstract logic) idea. The Two Basics stress the aspect of knowing the invariant aspect: the cognitive product. The model of spiral variation stresses the aspect of the cognitive tool: the variation process. To elucidate the "hidden" principles of task design, we use spiral variation to illustrate the structural aspects, emphasizing the core ("month" is the central idea for month naming in Chinese) and line variation (the naming stresses the expression of order in a linear way). This directly reflects the meta-rule of grouping by category (以類合 類) of the Chinese language and philosophy (Sun 2016). In our past and present research, we follow previous studies (e.g., Gu et al. 2004; Marton 2008; Sun 2007) in seeking a theoretical model for designing a mathematical curriculum based on China's local language (Marton et al. 2010), philosophy, and practice. We thus propose the spiral variation curriculum model: an invariant, relation-oriented model based on the practice of variation (Fig. 36.1).

The spiral variation theory of learning emphasizes perception of the underlying invariant as a necessary condition for learners to be able to discern the old aspects of an object of algebra learning. Thus, spiral variation theory spells out the conditions of inductive learning and explains algebra learning failures in a specific way: When learners do not learn what was intended, they have not discerned the necessary invariant aspects. So, the very core idea of spiral variation theory is that perception of the underlying invariant aspects is a necessary condition of algebra learning: What aspects we attend to or discern are of decisive significance for how we understand or experience the object of algebra learning. Algebra learning cannot happen without the learner having perceived the underlying invariant, local term of "Two Basics." The spiral variation model for curriculum design denotes how a

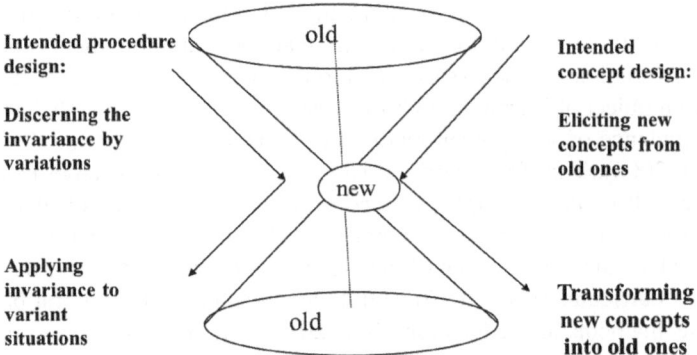

Fig. 36.1 Spiral variation model for curriculum design based on Chinese practice (Sun 2016, p. 22)

relationship-oriented model aims to achieve the "Two Basics" through systematic problem variation. The model is situated in the context of a content-focused and an exam-driven, textbook-centered system. It emphasizes three important aspects of variation in task design to develop the hypothetical learning trajectory (Sun et al. 2006; Sun 2007, 2013).

1. Variation (the vertical aspect of task design) is key to developing learning in a new light and provides a chance to link new concepts with old ones. The issue of variation in problem sets directly reflects the old Chinese proverb, "no clarification, no comparison" (沒有比較就沒有鑒別), rather than "consolidating one topic or skill before moving on to another," and highlights invariance through variation (變中发现不變) and the application of invariance to variant situations (以不變應萬變). It also reflects "grasping ways beyond categories" and "categorizing to unite categories," namely, the dynamic categorization approach above.
2. Emphasizing of the underlying invariance (a lesson's key points [重点], difficult points [难点], and critical points [关键点] as the central aspects of Chinese lesson plans (Yang and Ricks 2012); "key" pieces or "concept knots" as the central aspect of Chinese knowledge organization (Ma 1999); "Two Basics" as the central aspect of the Chinese curriculum goal) is necessary condition for developing algebra learning (making learning stable and coherent).
3. The horizontal, vertical, and central aspects combine to form a spiral structure (a similar principle in physics states that spiral movement can be decomposed into horizontal, vertical, and centripetal movements).

Many international research studies are conducted in the Western deductive tradition based on a geometric perspective. During the past decades, the missing paradigm from algebraic tradition has rarely been explored. The rationale for the model to articulate the framework is more aligned to the cultural roots of algebra. In addition, it is in line with the argument that the variation theory of learning emphasizes variation as a necessary condition for learners to be able to discern new aspects of an object of learning and how variation can be used to enhance students' learning, evidence of which been reported (e.g., Huang and Yeping 2017; Watson and Mason 2006). This framework enriches the specific perspective from algebra development. It could be helpful to reconsider the significance of algebra development through the widespread daily practice of variation (Sun 2011a, b; Sun et al. 2007), its curriculum significance (Sun et al. 2006), its significance for the reform of the Chinese curriculum (Sun 2016), and its relationship to the cultural background of Chinese mathematical education (Sun 2011b). It has also been piloted with efficient results in non-mainland China as a proposed practical design framework for curriculum development (Sun 2007, 2016; Wong et al. 2009). The explicit discussion on the curriculum framework of design and variation practice in task design in China for development of algebraic thinking could be helpful in reflecting China's own hidden tradition. For example, the current reform in China that completely follows the Western strand model may not be wise for development of the algebraic thinking (Ma 2013; Sun 2016). Due to space limitations, we will not elaborate further.

36.7 Application in Italy and Hong Kong: Transposition of Problems with Variation in Italy and in Hong Kong

Bartolini Bussi et al. (2013) reported two cases of transposition of problems with variation in Italy and Hong Kong. To try to find "cues" in the problem text and link addition to subtraction, the system of the nine problems in Table 36.1 has been used as a prompt in teacher education and development and tested by practicing teachers in several classrooms from second grade onwards to foster this approach to algebraic reasoning as early as possible. Teacher-researchers who collaborated in the pilot study did not implement the same Chinese task but redesigned it to tailor it to the Italian tradition (see Table 36.1) and to their individual teaching styles and belief systems.

This is a system of nine problems involving addition and subtraction where the organization in rows refers to the already mentioned combination, change, or comparison categorization and the organization in columns refers to the same arithmetic operation (either addition or subtraction; see MPOS above). In each row there is a problem (in the shaded cell) and two variations (see OPMC above). The "routine" activities of varying the conditions above play the role of concept

Table 36.1 A summary system of problems with variation in second grade from Bartolini Bussi et al. (2013, p. 558)

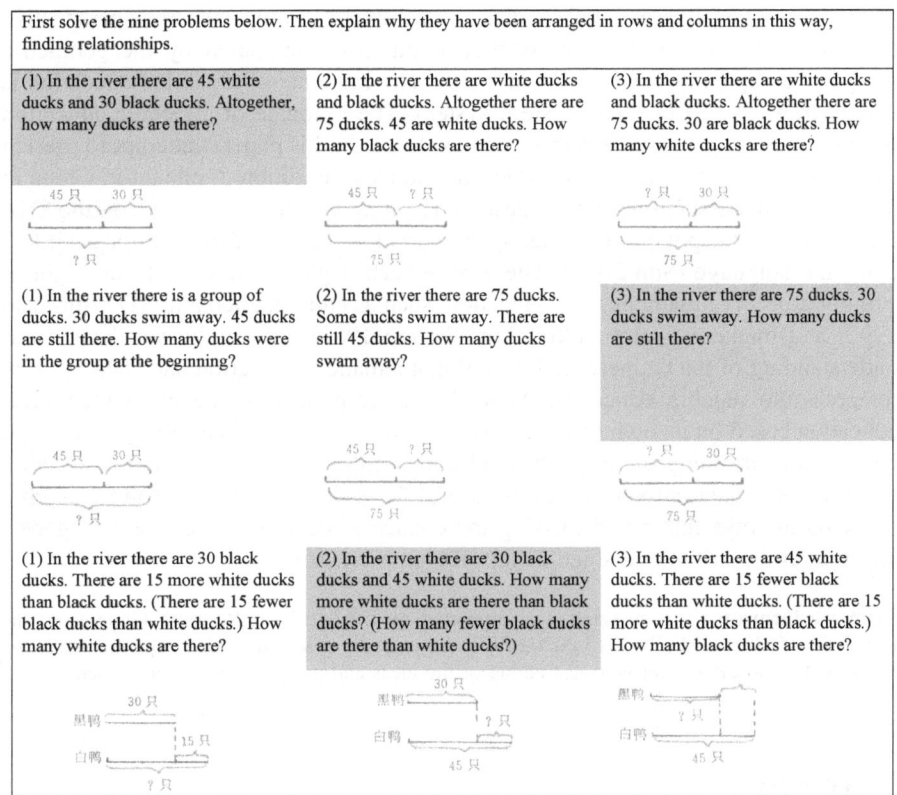

connections that provide a platform to support analyzing relationships and noticing structure. Therefore, they can support the meta-level algebra development.

In 2006, an experiment was carried out and tested in three schools on a treatment group where a textbook was developed that heavily emphasized relationships with problem variations in the division of fractions. In the control group in another three schools, the traditional Hong Kong textbook was used, which was heavily influenced by English principles and placed only light emphasis on relationships. The experimental treatment group achieved a better conceptual understanding of fractions, division, and multiplication compared with the control group (Sun 2007). Similar experiments in other content areas (ratio, volume, and columns) confirmed these findings (e.g., Wong et al. 2009).

36.8 Conclusion

Obviously, many international studies are conducted from a Western perspective under the influence of Western deductive tradition dominated by the geometric perspective (Spagnolo and Di Paola 2010). The rationale of algebraic education is mainly transferred from the Western system. The rationale of algebraic education from the Eastern system has rarely been explored. In this paper, I attempt to present the algebraic perspective derived from the inductive tradition dominant in China in the hope of understanding the Chinese curriculum with its instructions on the Two Basics and its unspoken principle, spiral variation, derived from the local philosophy and language (Sun 2016). These have been neglected to date. In the light of cultural aspects of mathematical education, such as ethnomathematics (D'Ambrosio 1992) and mathematical enculturation (Bishop 1988), this study can deepen our understanding of the Chinese tradition of mathematical education and shed light on research into algebra education. Specially, a Chinese rationale of mathematical education based on its own historical tradition rather than the Western system could be far more meaningful for both local and non-local curriculum and instruction development. This rationale could be useful for task design in developing algebra curricula in ways that avoid missing the chance to develop the concept of generalization at the arithmetic stage using problem variation.

Acknowledgements This research was supported by a Multi-Year Research Grant from the University of Macau (MYRG2015-00203-FED). I am greatly indebted to Maria G. Bartolini Bussi for critical feedback and for help in framing of the ideas and perspectives presented here.

References

Bartolini Bussi, G., Sun, X. H., & Ramploud, A. (2013). A dialogue between cultures about task design for primary school. In C. Margolinas (Ed.), *Task design in mathematics education. Proceedings of ICMI Study 22* (pp. 409–418). Oxford, UK. Retrieved February 10, 2015 from https://hal.archives-ouvertes.fr/hal-00834054v3.

Bishop, A. J. (1988). *Mathematical enculturation: A cultural perspective on mathematics.* Dordrecht: Kluwer.

Cai, J. (2000). Mathematical thinking involved in U.S. and Chinese students' solving process-constrained and process-open problems. *Mathematical Thinking and Learning: An International Journal, 2*, 309–340.

Cai, J., & Knuth, E. (Ed.). (2005). Developing algebraic thinking: Multiple perspectives. *ZDM Mathematics Education, 37*(1), 68–76.

Cai, J., & Wang, T. (2006). U.S. and Chinese teachers' conceptions and constructions of representations: A case of teaching ratio concept. *International Journal of Mathematics and Science Education, 4*, 145–186.

Chemla, K. (2009). On mathematical problems as historically determined artifacts: Reflections inspired by sources from ancient China. *Historia Mathematica, 36*(3), 213–246.

Clarke, D. J. (2006). Using international comparative research to contest prevalent oppositional dichotomies. *ZDM Mathematics Education, 38*(5), 376–387.

D'Ambrosio, U. (1992). *Ethnomathematics: A research program on the history and philosophy education.* Dordrecht: Kluwer.

Gu, L., Huang, R., & Marton, F. (2004). Teaching with variation: A Chinese way of promoting effective mathematics learning. In L. Fan, N. Y. Wong, J. Cai, & S. Li (Eds.), *How Chinese learn mathematics: Perspectives from insiders* (pp. 309–347). Singapore: World Scientific.

Guo, S. (2010). *Chinese history of science and technology.* Beijing: Science Press (in Chinese).

Høyrup, J. (2002). *Lengths, widths, surfaces: A portrait of Old Babylonian algebra and its kin. Studies and sources in the history of mathematics and physical sciences.* Berlin & London: Springer.

Hua, C. (1999). *I Ching: The book of changes and the unchanging truth.* Los Angeles: Seven Star Communications.

Huang, R., & Yeping, L. (Eds.). (2017). *Teaching and learning mathematics through variation. Confusian heritage meets western theories.* Boston, MA: Sense.

Kieran, C. (2004). The core of algebra: Reflections on its main activities. In K. Stacey, H. Chick, & M. Kendal (Eds.), *The future of the teaching and learning of algebra: The 12th ICMI study* (pp. 21–34). Dordrecht, The Netherlands: Kluwer.

Kieran, C. (2011). Overall commentary on early algebraization: Perspectives for research and teaching. In J. Cai & E. Knuth (Eds.), *Early algebraization: A global dialogue from multiple perspectives* (pp. 579–593). New York: Springer.

Kline, M. (1972). *Mathematical thought from ancient to modern times.* Oxford, UK: Oxford University Press.

Lee, L. (Eds.). *Approaches to algebra* (pp. 65–86). Dordrecht: Kluwer.

Li, W. (2005). The development and influence of Chinese classic mathematics. *Journal of Chinese Science Academy, 1,* 31–36.

Ma, L. (1999). *Knowing and teaching elementary mathematics: Teachers' understanding of fundamental mathematics in China and the United States.* Mahwah, NJ: Lawrence Erlbaum Associates, Inc.

Ma, L. (2013). A critique of the structure of US elementary school mathematics. *Notices of the American Mathematical Society, 60*(10), 1282–1296.

Marton, F. (2008). *Building your strength: Chinese pedagogy and necessary conditions of learning.* Retrieved February 28, 2009 from http://www.ied.edu.hk/wals/conference08/program1.htm.

Marton, F., Tse, S. K., & Cheung, W. M. (2010). *On the learning of Chinese.* Rotterdam: Sense.

Martzloff, J. C. (1997). *A history of Chinese mathematics.* Berlin, Heidelberg: Springer.

Mason, J. (1996). Expressing generality and roots of algebra. In N. Bednarz & C. Kieran (Eds.), Of mathematics with pedagogical implications. *Notices of the American Mathematical Society, 39* (10), 1183–1184.

Mason, J. (2011). Envisioning what is possible in the teaching and learning of algebra. In J. Cai & E. Knuth (Eds.), *Early algebraization: A global dialogue from multiple perspectives. Advances in mathematics education* (pp. 557–578). Berlin: Springer.

Mikami, Y. (1913). *The development of mathematics in China and Japan.* Leipzig: Teubner.

Ministry of Education. (1963). Retrieved February 24, 2013 from: http://202.109.208.240/database/b1009/ziyuanku/material/dagang/dg1963.htm.

Ministry of Education. (2001). Retrieved February 24, 2013 from: http://math.cersp.com/CourseStandard/CEDU/200509/406.html.

Pisano, L., & Sigler, L. E. (2002). *Fibonacci's Liber Abaci.* Berlin: Springer.

Radford, L. (2015). Early algebraic thinking: Epistemological, semiotic, and developmental issues. In S. J. Cho (Ed.), *The Proceedings of the 12th International Congress on Mathematical Education* (pp. 209–227). Cham: Springer.

Rowland, T. (2008). The purpose, design, and use of examples in the teaching of elementary mathematics. *Educational Studies in Mathematics, 69*, 149–163.

Sfard, A. (1995). The development of algebra: Confronting historical and psychological perspectives. *Journal of Mathematical Behavior, 14*, 15–39.

Shen, K., John, C., & Anthony, W. C. L. (1999). *The nine chapters on the mathematical art: Companion and commentary* (p. 358). Oxford, UK: Oxford University Press.

Spagnolo, F., & Di Paola, B. (Eds.). (2010). *European and Chinese cognitive styles*. Berlin: Springer.

Sun, X. (2007). *Spiral variation (Bianshi) curricula design in mathematics: Theory and practice* (Unpublished doctoral dissertation). The Chinese University of Hong Kong, Hong Kong (in Chinese).

Sun, X. (2011a). Variation problems and their roles in the topic of fraction division in Chinese mathematics textbook examples. *Educational Studies in Mathematics, 76*(1), 65–85.

Sun, X. (2011b). An insider's perspective: "Variation Problems" and their cultural grounds in Chinese curriculum practice. *Journal of Mathematical Education, 4*(1), 101–114.

Sun, X. (2013). *The fundamental idea of mathematical tasks design in China: The origin and development*. Paper presented in ICMI STUDY 22: Task Design in Mathematics Education, University of Oxford, UK, July 22–26, 2013.

Sun, X. (2016). *Spiral variation: A hidden theory to interpret the logic to design Chinese mathematics curriculum and instruction in mainland China*. Singapore: World Scientific (in Chinese).

Sun, X., Neto, T. B., & Ordóñez, L. E. (2013). Different features of task design associated with goals and pedagogies in Chinese and Portuguese textbooks: The case of addition and subtraction. In C. Margolinas (Ed.), *Task design in mathematics education. Proceedings of ICMI Study 22* (pp. 409–418). Oxford, UK. Retrieved February 10, from https://hal.archives-ouvertes.fr/hal-00834054v3.

Sun, X., & Sun, Y. (2012). *The systematic model LÜ (率) of Jiu Zhang Suan Shu and its educational implication in fractional computation*. Paper presented at the 12th International Congress on Mathematics Education, Seoul, Korea July 08–15, 2012.

Sun, X., Wong, N., & Lam, C. (2006). The unification of structure and function in mathematics problems' variation. *Curriculum, Textbook and Pedagogy, 5* (in Chinese).

Sun, X., Wong, N., & Lam, C. (2007). The mathematics perspective guided by variation (Bianshi). *Mathematics Teaching, 10* (in Chinese).

Wang, Q. (1996). *History of mathematics education in China's elementary schools*. Jinan, China: Shandong Education Press.

Watson, A., & Mason, J. (2005). *Mathematics as a constructive activity: Learners generating examples*. Mahwah: Lawrence Erlbaum Associates, Inc.

Watson, A., & Mason, J. (2006). Seeing an exercise as a single mathematical object: Using variation to structure sense-making. *Mathematical Thinking and Learning, 8*(2), 91–111.

Wong, N. Y., Lam, C. C., Sun, X., & Chan, A. M. Y. (2009). From "exploring the middle zone" to "constructing a bridge": Experimenting the spiral bianshi mathematics curriculum. *International Journal of Science and Mathematics Education, 7*(2), 363–382.

Wu, W. (1995). *Mathematics mechanization: Mechanical geometry theorem-proving, mechanical geometry problem-solving and polynomial equations-solving*. Beijing: Beijing Normal University Press (in Chinese).

Yang, Y., & Ricks, T. E. (2012). How crucial incidents analysis support Chinese lesson study. *International Journal for Lesson and Learning Studies, 1*, 41–48.

Zhang, D. Z. (2006). *The "two basics": Mathematics teaching in mainland China*. Shanghai: Shanghai Education Press.

Chapter 37
Digital Pedagogy in Mathematical Learning

Yahya Tabesh

Abstract Digital pedagogy is a learning paradigm that can allow learners to be active partners in discovering and developing their own mathematical knowledge. In this sense, Piaget's constructivist principles lay the foundation for developing digital pedagogy. In the paper that follows, we present a novel, intuitive, digital mathematical learning model. The model is focused on problem solving through computational thinking and is targeted to empower teenagers. More features and outcomes of this model will be discussed as well. As a foundation moving forward, the "use-modify-create" framework offers a helpful progression for developing computational thinking over time. It illustrates the benefits arising from engaging youth with progressively more complex tasks and giving them increasing ownership of their learning. The gained knowledge and skills of this cognitive learning both empower learners and enhance creativity. In its essence, we aim to develop the utopia of digital pedagogy in mathematical learning.

Keywords Digital pedagogy · Computational thinking · Gamification of education · Project-based learning · Problem solving

37.1 Introduction

We review an effective digital pedagogy for mathematical learning. We intend to present a way to create a cognitive-learning digital environment in today's ubiquitously connected world and align with the surrounding and dynamic cultural

---

In memory of **Seymour Papert** (1928–2016).

Digital pedagogy is a legacy of Seymour Papert. I developed this work based on his work "Mindstorms" in the last four years. I delivered a presentation on digital pedagogy on July 29, 2016 at ICME-13. He passed away on July 31, 2016.

Y. Tabesh (✉)
Sharif University of Technology, Tehran, Iran
e-mail: tabesh@sharif.ir

G. Kaiser et al. (eds.), *Invited Lectures from the 13th International Congress on Mathematical Education*, ICME-13 Monographs,
https://doi.org/10.1007/978-3-319-72170-5_37

669

trends of the mobile computational device. Smart computational devices compose a medium for digital pedagogy and affect in the way people think and learn.

To enhance creativity, connecting the developmental psychology of Piaget (1966) to the digital pedagogy in mathematics learning is key to developing an innovative learning model. In the modern approach towards teaching and learning mathematics, students should be partners and active agents in the learning process and problem solving (Boaler 2008). This modern approach in mathematics learning is more experimental and collaborative based on "learning by doing" (Dewey 1897), and students' learning gain is organic as they are partners in the learning process. They learn and develop their own knowledge step by step through innovative and creative thinking, experiences and discoveries, and collaboration and teamwork. Access to information and online resources has the potential to change the means of learning, allowing for a personalized and collaborative learning environment that is no longer restricted to schools and classes. The gained knowledge and skills of such cognitive learning empower students for everyday activities such as data analysis, reasoning, and problem solving. Digital pedagogy is a way to create such an environment for cognitive learning that requires rich toolkits rather than force-fed knowledge. Nevertheless, an interactive learning platform and digital resources are needed to develop the new paradigm for cognitive learning.

The growth mindset is also another important aspect of the digital pedagogy. In a growth mindset, students understand that their talents and abilities can be developed and expanded through effort, good learning, and persistence. In contrast, a fixed mindset proposes that students' basic abilities, intelligence, and talents are inherent characteristics and thus are not expandable: They have a certain amount of "smartness" and that's that. In the growth mindset, students do not necessarily believe everyone possesses the same intelligence, but they suppose anyone can get smarter if they work at it (Dweck 2006; Boaler 2016). Digital pedagogy is the ultimate paradigm to attract learners and support a growth mindset. Problem solving ability or "smartness" grows with experience on the digital pedagogic platform.

We intend to present a digital platform for mathematical learning through problem solving that enables creative engagement, develops mathematical skills, and supports a growth mindset. Briefly, we go through the background of digital pedagogy, introducing a theoretical model for the digital platform, and finally we discuss a case study and some experimental results.

37.2   Background

Digital pedagogy of mathematics learning is a legacy of Seymour Papert (Blikstein 2013); we summarize his work to touch on how his ideas have affected mathematics education. We take the opportunity to adapt his approach to the advancements of computational technology. We will go also describe the Piagetian learning path, but in the context of Papert, who connected it to digital pedagogy.

We proceed to see how Piagetian developmental psychology has been connected to digital learning through Papert's work, reflected in Mindstorms (Papert 1980), but we discuss technological advancements as well.

37.2.1 Mindstorms

Innovative learning models must reflect on what is happening in the surrounding culture and use dynamic cultural trends as media to carry educational interventions. It has become commonplace to say that today's culture is marked by a ubiquitous smart computational device. Smart computational devices can contribute to mental processes, not only as instruments we improve at using (e.g., touch screens), but also in more essential conceptual ways, influencing how we may think. Smart computational devices will enter the private worlds of learners everywhere and can create a new paradigm to form new relationships with knowledge in a personalized way. In this regard, the whole process of mathematics learning should be involved in a dialectical interaction between new technologies and new ways of learning mathematics.

Digital pedagogy is essential in reflection of the Piagetian learning path, and in this sense we should create an environment in which learners surf and interact with their environment to learn how to "talk mathematically" as a means to capture mathematical concepts and ideas implicitly. Smart computational devices are the best tools to create such a new paradigm for a learning environment that can help learners learn and develop their own mathematical knowledge organically. Computational devices can address personalized learning; they are unique in providing us with means to counteract what Piaget saw as an obstacle.

37.2.2 Piagetian Learning Path

To touch and understand how computational technology and a digital learning environment can be a medium for knowledge development, we should look at the Piagetian learning path.

Piaget is the leading theorist of learning without deliberate teaching. This, however, does not imply a spontaneous atmosphere that leaves the learner alone; rather, it means supporting learners to build their own intellectual structures. In this sense, we are looking for an environment where mathematics can become a natural vocabulary, a learning environment with the proper emotional and cultural support where learners can learn not only that they can excel at mathematics but also that they can share the joy of mathematical experiences. This concept shows how to use computational devices as vehicles to develop digital pedagogy in a new environment.

We are focused on the Piagetian learning path as the natural, spontaneous learning of people in interaction with their environment. Piaget's thoughts have been underplayed because they offered no possibilities for action in the world of traditional education. But in the learning environment of the digital pedagogy, enriched by smart computational devices and supported by artificial intelligence, Piaget's principles can come to fruition. For many years his ideas could not be expanded due to lack of means of implementation, but digital pedagogy is going to make it available.

37.3   A Model

Our dream is to create a digital learning environment in which the task is not to learn a set of formal rules but to develop sufficient insight for mathematical concepts. We look for a digital environment to grow learners' mathematical mindsets through experience and a flourishing joy of mathematics. We expect an empowering platform to enhance creativity and develop mathematical skills and naturally explore domains of mathematical ideas in the sense of Piagetian learning path. We would like to present a model for such a learning paradigm as an online interactive "learning-by-doing" environment. We consider problem solving and algorithmic thinking as the means of exploring mathematical concepts on a platform that can expose computational thinking.

37.3.1   Characteristics

We consider a gamified learning environment that utilizes the computational thinking process and computational mathematical skills. It should be a personalized and collaborative learning platform targeted towards teenagers.

Computational thinking concepts were envisioned by Papert (1980, 1996) and involve how to use computation to enhance thinking, create new knowledge, and change patterns of access to knowledge. More recently, however, Wing (2006) brought a different approach and new attention to computational thinking. She considered the topic a fundamental skill for everyone's analytical ability, along with reading, writing, and arithmetic, and as a process to formulate a complicated problem and algorithmically solve it. We consider computational thinking based on Papert's enhancement of thinking, but specifically with a problem-solving approach in the sense of algorithmic thinking. In brief, computational thinking combines critical thinking with the computing power as the foundation for innovating solutions to real-life problems.

We also consider the gamification of mathematical concepts as a framework. Games bring a new approach to pedagogy (Gee 2007; Devlin 2011) and possess the potential to create interaction and insert motivation; players are driven to their virtual goals and learn by doing. Allowing players to make mistakes through experimentation in a risk-free environment brings about learning by doing implicitly through mistakes: Players "feel" their way around games, and, by receiving instant feedback to their actions, they can adjust their problem-solving strategies accordingly. Put simply, games bridge the gap between formal knowledge and intuitive understanding. Another crucial aspect of games is the immense amount data generated by players that can be used as feedback for assessment of the learning process. The basic idea is to implicitly ease learners into the world of mathematics while they are enjoying themselves.

37.3.2 Playground

We consider computational thinking as the process for problem solving on the proposed platform; we also go for functional programming as a tool to formalize intuition about the problem-solving process.

Computational thinking is a four-stage problem-solving framework consisting of decomposition, pattern recognition, abstraction, and algorithm design, as shown in Fig. 37.1. We have enriched and connected the stages with a "playground" as a place for experimental problem solving.

In this model, the playground is an easily accessible place where learners can tackle problems through experimentation.

We also consider functional programing as a toolbox on the playground for problem solving. The functional programming paradigm explicitly supports a pure functional approach to problem solving, which involves composing the problem as a set of functions to be executed. Functional programming is a style that avoids changing state, so it is a powerful tool that can be used in a modular form for problem solving. Such modularity is key, as it specifically empowers learners to utilize what they have built in the past for future solutions.

Gained knowledge in this model empowers learners in reasoning, problem solving, and algorithmic thinking in a gamified fashion. We can gather users' data and analyze them through the design-based research method (Brown 1992), which

Fig. 37.1 Four steps around playground

should be embedded in the platform. Results of the analyzed data can be used to improve the platform and also bring recommendation and feedback to the learners.

37.4   A Case Study

Piaget (1966) showed how learners construct a world out of materials in their environment. Papert (1980) has also mentioned that experience with games is a bridge between formal knowledge and intuitive understanding. In this sense, we have developed and considered a gamified digital learning platform as a case study.

To develop a case study for the first approach, we went through digital mathematical puzzle games in an interactive fashion, but we found they only attracted students who showed a proclivity for mathematical thought rather than the general population. But, in the revised version we considered that the following objectives should be achieved by the platform:

- Enable creative engagement
- Develop mathematical skills
- Support a growth mathematical mindset
- Be collaborative and social.

We went through the next version (Polyup 2016) to look for the above objectives, and we received positive responses to the prototype from a variety of teenagers in focus groups. We will present the platform and have a brief look at the results of test cases.

37.4.1   Platform

The developed platform (Polyup 2016) for mobile computational devices is about problem solving on a functional programming platform through computational thinking. Functional programming is achieved with lambda calculus (Revesz 1998) and provides a theoretical framework for describing functions and their evaluation. Functional programming is a style of building structures and elements in a modular form that treats computation as the evaluation of mathematical functions and avoids changing-state and mutable data. To develop computation through functions, we are using a postfix, or Reverse Polish Notation, to avoid parentheses in expressions and computation (McCarthy 1960).

The platform is a user friendly environment where the user is equipped with numbers, operations, and basic functions. The user can do computation in a functional modular form; computation simply goes top to bottom with postfix. Users can drag and drop numbers and operations on stacks to script a program and

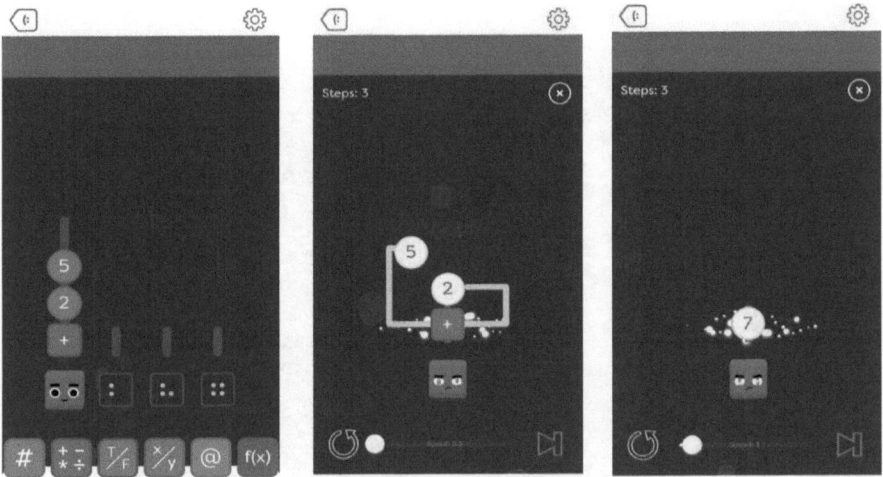

Fig. 37.2 Computational thinking playground

run it to calculate the output of the desired function. The platform and an example are shown in Fig. 37.2.

The platform is a collaborative and social playground for problem solving through personal experiences and also supports growth of a mathematical mindset. Many puzzles are preloaded in a step-by-step fashion for users to develop their own knowledge in a gamified interactive platform. To make the platform social, users can develop their own puzzles and share them with friends.

Despite its simplicity, the platform is Turing complete. It also features a chatbot as a sidekick, or a mentor, to help problem solvers. Users' data are gathered on the platform to be used through a smart system for analytics, which provides puzzle recommendations for users and is also used for advancement of the platform.

Advanced functional techniques such as recursion also are a central part of the learning environment and are shown visually; an example script to compute triangular numbers, as well as its visual running form, is shown in Fig. 37.3. Developed functions in a modular form can be reused to address more complex problems.

37.4.2   Feedback

We have tested the platform in a variety of classes and schools, from middle schools to high schools. A summary of students' feedback from seven different classes is shown in Table 37.1; the figures shown are the mean of students' responses on a scale of 1–10.

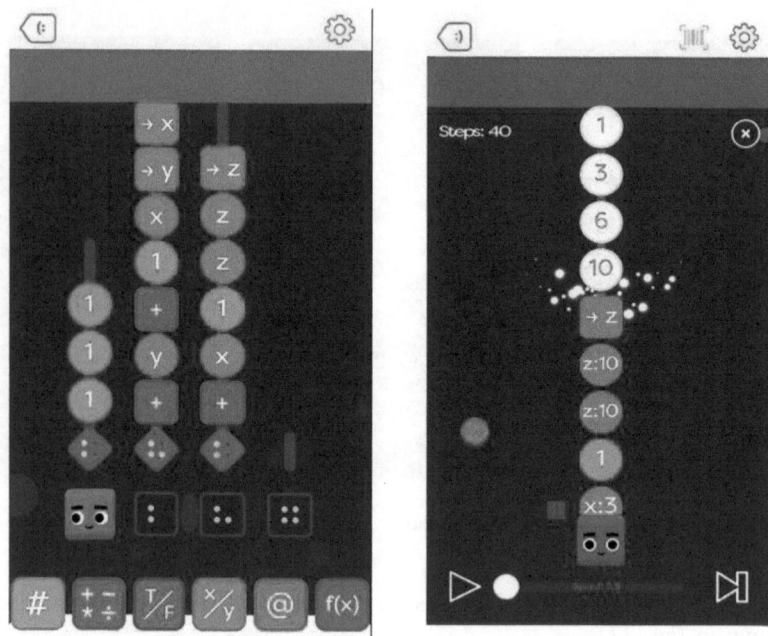

Fig. 37.3 Triangular numbers on the playground

Table 37.1 Feedback of students

Grade level	8	9	8	9	10	8	11
Number of students	17	12	19	16	13	18	18
How much did you like the platform?	6.94	7.33	7.40	6.93	7.38	7.22	8.13
How much do you like the script language?	6.76	6.08	7.05	6.5	6.46	7.11	6.88
How did you like the training puzzles?	6.82	6.42	6.00	5.85	6.85	6.50	6.61
How would you like ability to control digital art with the functions that you can develop on the platform?	7.18	7.50	8.10	5.62	8.23	8.17	7.50
How would you like the ability to control robots with the functions that you can develop on the platform?	6.19	7.17	8.10	6.18	7.77	7.83	7.63
How much do you like working with another player on a shared playground?	6.71	6.92	6.95	5.5	6.62	7.00	6.75
How likely would you be to recommend the platform to your friends?	6.71	7.17	7.30	6.62	6.62	6.78	7.38

Tests consisted of one-hour sessions, starting with introducing the platform to the students and then having them solve 10–15 selected puzzles while learning implicitly about functional programming. In the last 10 min, they had a chance to develop their own puzzles, in which significant achievements were observed.

Feedback was generally positive. As observed in the live sessions, students were very well engaged and would also recommend the platform to their friends, as supported by the first and last questions of the survey. Another important observation lies in how students got involved in the technicalities of scripts and functional programming, a learning path we observed to be very natural and organic.

Another key observation lies in the questions that asked if they were interested in controlling robots or developing digital arts through functional programming. Their very high levels of interest show how important it is to reconcile mathematical problem solving with applicable skills. With a connected functional programming environment, users can become creatively engaged with the software and hardware tools in their daily lives and change these objects' functionality to better suit their interests and needs.

37.5 Conclusion

We studied significant works of Seymour Papert as a pioneer in developing digital pedagogy, and these works provided the base for our adaptation to the recent advancement of mobile computational technology. We found the opportunity to develop a digital learning environment that can engage learners in an experiential and growth-mindset fashion such that they can develop their own knowledge.

We also received positive feedback from the users, which provides motivation for further development of digital pedagogy in mathematical learning. The interest of users to develop applicable skills reveals the deficiency that current platforms have in connecting problem-solving ability to real-life applications such as digital arts and robotics. With the modularity of functional programming and the creativity of computational thinking, modification of various objects is natural. Once these tools are available, the process of "use-modify-create" will bring opportunities for endless creativity among the youth.

The translation of problem solving to allow it to have a tangible impact outside the educational environment is a novel approach that will attract and motivate a greater general audience to become engaged in computational thought. Through digital experiences, learners will develop their own mathematical ability and ultimately spread the joy of mathematics.

Acknowledgements A sincere acknowledgement goes to the Department of Mathematics at Stanford University for hosting me over the last four years as I worked on the pedagogy of mathematical learning project. Also, special thanks to Jo Boaler, Professor of Mathematics Education at Stanford University, and Shima Salehi, a Ph.D. student in the Graduate School of Education at Stanford University, for their valuable advice.

References

Blikstein, P. (2013). Seymour Papert legacy: Thinking about learning, and learning about thinking. In *Proceedings of Interaction Design and Children Conference*. http://portalparts.acm.org/2490000/2485760/fm/frontmatter.pdf. Accessed June 25, 2016.

Boaler, J. (2008). *What's math got to do with it?* New York: Penguin.

Boaler, J. (2016). *Mathematical mindsets*. San Francisco: Jossey-Bass.

Brown, A. (1992). Design experiment: Theoretical and methodological challenges in creating complex interventions in classroom settings. *The Journal of the Learning Science, 2*(2), 141–178.

Devlin, K. (2011). *Mathematics education for a new era: Games as a medium for learning*. Boca Raton: CRC Press.

Dewey, J. (1897). My pedagogic creed. *School Journal, 54*, 77–80.

Dweck, C. (2006). *Mindset, the new psychology of success*. New York: Ballantine Books.

Gee, J. P. (2007). *What video games have to teach us about learning and literacy*. Basingstoke: Palgrave Macmillan.

McCarthy, J. (1960). Recursive functions of symbolic expressions and their computation by machine, part I. *Communication of the ACM, 3*(3), 184–195.

Papert, S. (1980). *MINDSTORMS, children, computers, and powerful ideas*. New York: Basic Books Inc.

Papert, S. (1996). An exploration in the space of mathematics educations. *International Journal of Computers for Mathematical Learning, 1*(1), 95–123.

Piaget, J. (1966). *La Psychologie de l'enfant*. Paris: Presses Universitaires de France.

Polyup Casual Modding Platform (2016). http://www.polyup.com. Accessed July 1, 2016.

Revesz, G. E. (1998). *Lambda-calculus, combinators, and functional programming*. New York: Cambridge University Press.

Wing, J. M. (2006). Computational thinking. *Communications of the ACM, 49*(3), 33–35.

Chapter 38
Activity Theory in French Didactic Research

Fabrice Vandebrouck

Abstract The theoretical and methodological tools provided by the first generation of activity theory have been expanded in recent decades by the French community of cognitive ergonomists, followed by a sub-community of researchers working in the didactics of mathematics. The main features are, first, the distinction between tasks and activity and, second, the dialectic between the subject of the activity and the situation within which this activity takes place. The core of the theory is the twofold regulatory loop that reflects both the codetermination of the activity by the subject and by the situation and the developmental dimension of the subject's activity. This individual and cognitive understanding of activity theory mixes aspects of Piaget's and Vygotsky's frameworks. In this paper, it is first explored in association with a methodology for analysing students' mathematical activities. We then present findings that help to understand the complexity of student mathematical activities when working with technology.

Keywords Mathematics · Tasks · Activity · Mediations · Technologies

38.1 Introduction

Activity theory is a cross-disciplinary theory that has been adopted to study various human activities, including teaching and learning in ordinary classrooms, where individual and social levels are interlinked. These activities are seen as developmental processes mediated by various contextual elements—here we consider the teacher, the pair and the artefact (Vandebrouck et al. 2012, p. 13). Activity is always motivated by an object: a characteristic that distinguishes one activity from another. Transforming the object into an outcome is another key feature of activity. Subjects and objects form a dialectic unit: subjects transform objects, and at the same time subjects are transformed, mainly in Vygotsky's sense of internalisation

F. Vandebrouck (✉)
Université Paris Diderot, LDAR, Paris, France
e-mail: vandebro@univ-paris-diderot.fr

© The Author(s) 2018
G. Kaiser et al. (eds.), *Invited Lectures from the 13th International Congress on Mathematical Education*, ICME-13 Monographs,
https://doi.org/10.1007/978-3-319-72170-5_38

(Vygotsky 1986). This framework can be adapted to describe the actions and interactions that emerge in the teaching/learning environment and that relate to the subjects, the objects, the artefacts and the outcomes of the activity (Wertsch 1981).

Activity theory was originally developed by Leontiev (1978), among others. A well-known extension is the systemic model proposed by Engeström et al. (1999), which is referred to as the third generation of activity theory. It expresses the complex relationships between the elements that mediate activity in an activity system. In this paper, we take a more cognitive and individual perspective. This school of thought has been expanded over the course of the past four decades by French researchers working in the domain of occupational psychology and cognitive ergonomics and has since been adapted to the didactics of mathematics. The focus is on the individual as a cognitive subject and an actor in the activity, rather than the overall system—even if individual activity is seen as embedded in a collective system and cannot be analysed outside the context in which it occurs.

An example of this adaptation has already been well established internationally. Specifically, it refers to the distinction between the artefact and the instrument, which is used to understand the complex integration of technologies into the classroom. The notion of instrumental genesis (or instrumental approach) was first introduced by Rabardel (1995) in the context of cognitive ergonomics, then extended to didactics of mathematics by Artigue (2002), and it is concerned with the subject-artefact dialectic of turning an artefact into an instrument. In this paper, we draw upon and try to encompass this instrumental approach.

First, we describe how activity theory has been developed in the French context. These developments are both general and focused on students' mathematical activity. Next, we present a general methodology for analysing students' mathematical activity when working with technology. We then develop an application example and describe our findings. Finally, we present some conclusions.

38.2 Activity Theory in the French Context

The first notable feature of activity theory in the French context is the distinction between *tasks* and *activity* (Rogalski 2013). Activity relates to subjects, while tasks relate to objects. Activity refers to what subjects engage into complete tasks: external actions but also inferences, hypotheses, thoughts and actions they decide to take or not. It also concerns elements that are specific to subjects, such as time management, workload, fatigue, stress, enjoyment and interactions with others. As for the task—as described by Leontiev (1978) and extended in cognitive ergonomics—this refer to the goal to be attained under certain conditions (Leplat 1997).

Activity theory draws upon two key concepts: the *subject* and the *situation*. The subject refers to an individual person, who has intentions and competencies (potential resources and constraints). The situation provides the task and the context for the task.

Fig. 38.1 Codetermination
of activity and twofold
regulatory loop

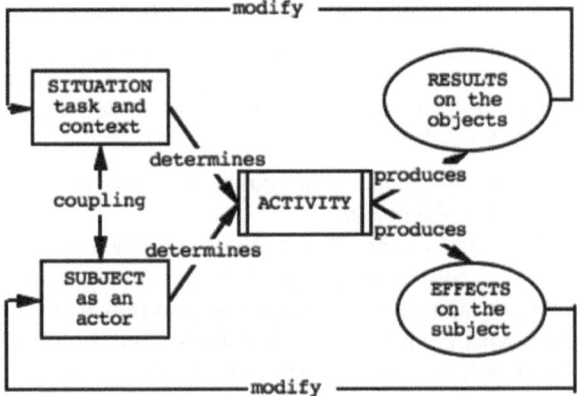

Together, situation (notably task demands) and subject codetermine activity. The dynamic of the activity produces feedback in the form of a twofold regulatory loop (Fig. 38.1) that reflects the developmental dimension of activity theory (Leplat 1997).

The concept of twofold regulation reflects the fact that the activity modifies both the situation and the subject. On the one hand (upper loop), the situation is modified, giving rise to new conditions for the activity (e.g., a new task). On the other hand (lower loop), the subject's own knowledge is modified (e.g., by the difference between expectations, acceptable outcomes and the results of actions).

More recently, the dialectic between the upper and lower regulatory loops (shown in Fig. 38.1) has been expanded through a distinction between the productive and constructive dimensions of activity (Pastré 1999; Samurcay and Rabardel 2004). Productive activity is object oriented (motivated by task completion), while constructive activity is subject oriented (subjects aim to develop their knowledge). In teaching/learning situations, especially those that involve technologies, the constructive dimension in the students' activity is key. The teacher aims for the students to develop constructive activity. However, especially with computers, students are mostly engaged in producing results, and the motivation of their activity can be only towards the productive dimension. The effects of their activity on students' knowledge—as it is stipulated by the dual regulatory loop—are then mostly indirect, with fewer or without any constructive aspects.

The last important point to note is the fact that French activity theory mixes the Piagetian approach of epistemological genetics with Vygotsky's socio-historical framework in order to specify the developmental dimension of activity. As Jaworski writes in Vandebrouck (2013, p. vii):

> The focus on the individual subject—as a person-subject rather than a didactic subject—is perhaps somewhat more surprising, especially since it leads the authors to consider a Piagetian approach of epistemological genetics alongside Vygotsky's sociohistorical framework.

Rogalski (Vandebrouck 2013, p. 20) responds:

The Piagetian theory looks from the student's side at epistemological analyses of mathematical objects in play while the Vygotskian theory takes into account the didactic intervention of the teacher, mediating between knowledge and student in support of the
students' activity.

The dual regulation of activity is consistent with the constructivist theories of
Piaget and Vygotsky.

The first author (Piaget 1985) provides tools to identify the links between
activities and development through epistemological analyses. Vergnaud (1982,
1990) expands the Piagetian theoretical framework regarding conceptualization and
conceptual fields by highlighting situation classes relative to a knowledge domain.
We therefore define the students' learning—and development—with reference to
Vergnaud's conceptualization.

On the other hand, Vygotsky (1986) stresses the importance of mediation within
a student's zone of proximal developmental (ZPD) for learning (scientific concepts). Here, we refine the notion of mediation by adding a distinction between
procedural and constructive mediations in the context of the dual regulation of
activity. Procedural mediations are object oriented (oriented towards the resolution
of the task), while constructive mediations are more subject oriented. This distinction can be seen as an extension to what Robert and Hache (2008) call teachers'
procedural and constructive aids. A more detailed exploration of the complemen-
tarity of Piaget and Vygotsky can be found in Cole and Wertsch (1996).

38.3 General Methodology for Analysing Students' Mathematical Activities

Following activity theory, we postulate that students' learning depends directly on
their activity, even though other elements can play a part—and even if activity is
partially inaccessible to us and differs from one student to another. Students'
activity is developed through the actions that are carried out to complete tasks.
Through their actions, subjects aim to achieve goals, and their actions are driven by
the motivation for the activity. Here, we draw upon the three levels originally
introduced by Leontiev (1978): activity associated with a motive, actions associated
with goals and operations associated with conditions. Activity takes place in a
specific situation, such as in the classroom, at home, or during a practical session.
The actions involved by the proposed precise tasks can be external (i.e., spoken,
written, or performed), or internal (e.g., hypotheses or decisions) and partially
converted in operations. As Galperine (1966) and Wells (1993) note, the three
levels are relative and, for instance, operations can be considered as actions that
have been routinised.

Here, we use the generic term mathematical activities (rather than activity) to
refer to students' activity on a specific mathematical task in a given context.

Mathematical activities refer to everything that surrounds actions and operations (also non-actions, for instance). They are a function of a number of factors (including task complexity, but extending to the characteristics of the context and all mediations that occur as tasks are performed) that contribute to regulation and intended development in terms of mathematical knowledge.

Two methodological levels can be adopted from the dynamic of activity within the twofold regulatory loop. First of all, regulations can be considered at a local level as short-term adjustments of activities to previous actions and as procedural learning (also called functional regulations; upper loop in Fig. 38.1). Secondly, at a global level, regulations are mostly constructive ones (also called structural regulations) and correspond to the long-term development of the subject (linked with conceptualization).

38.3.1   The Local Level

At the local level, the analysis focuses on students' activities in the situation, in the form of tasks, their context and their completion by students with or without direct help from the teacher. The initial step is an a priori analysis of the tasks given to students (by the teacher, the computer, etc.), which is closely linked to the situational context (e.g., the students' academic level and age). We use Robert (1998) categorization to characterise these tasks.

First, we identify the mathematical knowledge to be used for a given task: the representation(s) of a concept, theorem(s), definition(s), method(s), formula(s), types of proof, etc. The analysis aims to answer several crucial questions: Does the mathematical knowledge to be used already exist for students or is it new? Do students themselves have to find the knowledge to be used? Does the task only require the direct application of this knowledge without any adjustment (technical task) or does it require adaptations and/or carrying out subtasks? A list of such adaptations can be found in Robert and Horoks (2007): mix of knowledge, the use of intermediaries, change of register (Duval 1995), change of mathematical domain or setting (Douady 1986), introduction of steps or choices, use of different points of view, etc. Tasks that require the adaptation of knowledge are referred to as complex tasks and encourage conceptualization, as students become able to more readily and flexibly access the relevant knowledge, depending, however, on the implementation in the classroom.

The a priori analysis of tasks leads us to describe what we have called the *intended students' activities* associated with the tasks. Here we draw upon Galperine (1966) functions of operations and adapt them to mathematical activities. Galperine distinguishes three functions: orientation, execution and control. Next, we use three 'critical' mathematical activities that are characteristic of complex tasks (Robert and Vandebrouck 2014).

- First, *recognizing activities* refer mainly to orientation and control. They occur when students have to recognise mathematical concepts as objects or tools that can be used to solve the tasks they are given. Students may also be asked to recognise modalities of application or adaptation of these tools.
- Second, *organizing activities* refer mainly to orientation: Students have to identify the logical and temporal steps in their mathematical reasoning, together with any intermediaries.
- Third, *treatment activities* refer to all of the mathematical activities associated with execution on mathematical objects. Students may be asked to draw a figure, compute, substitute, transform expressions (with or without giving the steps), change registers, change mathematical domains, etc.

Following Vygotsky, we supplement our local analysis of intended students' activities by developing ways to analyse classroom teaching (a posteriori) and to approach effective students' activities as functions of the different mediations that occur. For this, we use videos and observations in the classroom. We also record students' discussions, teachers' discourses and writings and capture students' computer screens to identify observable activities. The data that is collected concerns how long students spend working on tasks, the format of their work (the whole class, in small groups, by pairs of students, etc.), its nature (copying, reading, calculation, investigation, written or oral, graded or not, etc.) and all elements of the context that may modify intended activities. This highlights, at least partially, the autonomy given to students, the nature of mediations and opportunities for students to show initiative in relation to the adaptation and availability of knowledge. Multiple aspects of mediations are analysed with respect to their assumed influence on student activities. Some relate to their format (interactions with students, between students, with teacher, with computers, etc.), while others concern the specific ways of taking into account the mathematical content (mathematical aids, assessment, reminders, explanations, corrections and evaluations, presentation of knowledge, direct mathematical content, etc.).

Two types of mediations have already been introduced: modifying intended activities or adding to activities (effective or when last observed). The first are object oriented; here we use the term *procedural mediations*. These mediations modify intended activities and correspond to instructions given by the teacher, the screen, or other students, directly or indirectly, before or during task completion. They are often seen in open-ended questions from the teacher such as 'What theorem can you use?' They can be given by the computer giving feedback that transforms the task to be performed or with some limitations in the provided tools that give indirect indications to students about the way to perform the task. These procedural mediations may lead to the subdivision of a complex task into subtasks. They usually change knowledge adaptations in complex tasks and simplify the intended activities in such a way that it becomes more like a technical task (for instance, students having to apply a contextualised method).

The second type of mediation is more subject oriented; here we use the term *constructive mediations*. They are designed to add something to the students' activities and the knowledge that can emerge from these activities. They can take the form of a simple summary of what has been developed by students, an explanation of choices, a partial decontextualisation or generalization, assessments and feedbacks, a discussion of results, etc. On some computers, the way students have achieved a geometrical figure can be replayed in order to help them recall the order in which the instructions have been given without any mistakes.

It should be noted here that our framework leads to the hypothesis that there is an internal transformation of the subject in the learning process: Constructive mediations aim to contribute to this process. However, the mediations can be constructive for some students and remain procedural for others. On the contrary, some procedural mediations can become constructive for some students, for instance, if they are able on their own to extract a generalization from a local indication. Moreover some constructive mediations—but also perhaps productive—can belong to some students' ZPD in Vygotsky's sense or they can remain out of the ZPD. When they belong to the ZPD, their identification can help to appreciate the explicit links between the expression of the general concepts to be learned and their precise applications, in contextualised tasks, based on the necessary dynamic between them. Distinguishing between the kinds of mediations and the way they do or do not belong to some students' ZPD can be very difficult.

38.3.2   The Global Level

The local level can be extended to a global level that takes into account the set of mathematical activities, the link with the intended conceptualization (long-term constructive loops) and teaching practices in the long term. We link students' mathematical activities to the intended conceptualization of the relevant mathematical notion, establishing a 'relief map' of this mathematical notion. This relief map is developed from an epistemological and mathematical analysis of the notion, the study of the curricula and didactical analyses (e.g., students' common difficulties). This global analysis focuses on the similarity between students' activities (intended, observed, or effective) and the set of activities that characterise the intended conceptualization of the relevant notion.

However, the didactical analysis of one teaching session is insufficient. It is necessary to take into account, on a day-to-day basis, all of the tasks students are asked to complete and teachers' interventions. We use the term *scenario* to describe a sequence of lessons and exercises on a given topic. The global scenario could be understood as a long-term 'cognitive road' (Robert and Rogalski 2005).

38.4 Example of Application: The 'Shop Sign' Situation

To illustrate the utilization of our activity theory, this section presents an example of a situation that aims to contribute to students' conceptualizations of the notion of function. Some limitations of the methodology at the global level are then outlined.

The example relates to a GeoGebra family of figures for learning functions. This family refers to mathematical situations that lie at the interface between two mathematical domains: geometry and functions. There are many possible examples. We call them "shop sign" situations because they share the idea that some coloured areas of the figures are the lit areas of shop signs (Artigue et al. 2011), which depends on some moving variables in the figure.

In Fig. 38.2, *ABCD* is a square, with *A* at the origin and *AB* = 4. *E* is a mobile point on the segment [*CD*]. We consider the sum of the areas of the square (*DFGE*) and the triangle (*ABG*). The task is to find the minimum of the sum of the areas as *E* moves.

The task is set for Grade 10 students (15 years old). One solution is to identify *DE* as an independent variable *x*. Then *f(x)*, the sum of the two areas, is equal to x^2 (for the square) plus $4(4 - x)/2$ (for the triangle): equivalent to $x^2 - 2x + 8$. In the French curriculum at Grade 10, the derivative is not known and students must compute and understand the canonical form $(x - 1)^2 + 7$ as a way to identify the minimum 7 for the distance *DE* = 1 (which is the actual position on the figure).

Students work in pairs on computers. They have already worked with functions in the traditional pencil-and-paper context, and they also have manipulated GeoGebra for geometrical tasks that do not refer to functions. In this new situation, GeoGebra helps them to begin the task by making conjectures about the minimum. Students can also trace the graph of the function, as shown in Fig. 38.6. Then, in the algebraic register, they can find the canonical form of the function *f(x)* and the characteristics of the minimum.

Fig. 38.2 Shop sign

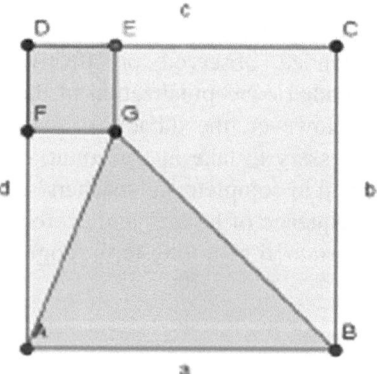

We first identify the relief map of the notion of function and the intended conceptualization. We then give the a priori analysis of the task and the intended students' activities. We finish with the observation of two pairs of students to identify observable and effective activities.

38.4.1 The Global Level: Relief Map of the Notion of Function and Intended Conceptualization

The function is a central concept in mathematics and links it to other scientific fields and real-life situations. It both formalises and unifies (Robert and Hache 2008) a diversity of objects and situations that students encounter in secondary school: proportionality, geometrical transformations, linear, polynomial growth, etc. A diversity of systems of representations (numerical, graphical, algebraic, formal, etc.) and a diversity of perspectives (pointwise, local and global) are frequently combined when working with them (Duval 1995; Maschietto 2008; Vandebrouck 2011). As it is summarised by Artigue et al. (2007), the processes of teaching and learning of function entail various intertwining difficulties that reinforce one another in complex ways.

Educational research (Tall 2006; Gueudet 2008; Hitt and Gonzalez-Martin 2016) shows that an efficient conceptualization of the notion requires a rich experience that illustrates the diversity illustrated above and the diversity of settings in which functions are used (Douady 1986). It also means that functions are available as tools for solving tasks and can be flexibly linked with other concepts. There must be a progression from embodied conceptualizations (where functions are highly dependent on physical experience) to perceptual conceptualizations (where they are considered dialectically and work both as processes and objects), paving the way for more formal conceptualizations (Tall 2004, 2006).

At Grade 10, the intended conceptualization can be characterised by a set of tasks in which functions are used as tools and objects. They can be combined and used to link different settings (including geometrical and functional); numerical, algebraic and graphical representations; and the dialectic between pointwise and global perspectives. The shop sign task is useful in this respect, as students have to engage in such mathematical activities. A priori optimization tasks in geometrical modelling help to build the intended functional experience and link geometrical and functional settings.

Technology provides a new support for physical experience, as the modelling process provides new systems of representation and helps to identify the dynamic connections between them. It also offers a new way to approach and connect pointwise and global perspectives on functional objects and supports the building of rich functional experiences. Arzarello and Robutti (2004) famous contribution uses sensors to introduce students to the functional domain. The framework is already an activity theoretical framework together with more semiotic approaches, but it is not in a context of dynamic geometry. There have been many experiences that involve learning functions through dynamic geometrical situations. For instance, Falcade et al. (2007) studied the potential of didactical engineering with Cabri-Géomètre. The authors take a Vygotskian perspective about semiotic mediations that is more precise than our adaptation of Vygotsky inside activity theory, but which is also more restrictive in the sense that they do not consider deep connections between given tasks and mathematical activities. Moreover, it does not concern ordinary classrooms. More recently, Minh and Lagrange (2016) analysed students' activities on functions using Casyopée. This software is directly built for the learning of functions and the authors adopted the model of mathematical working spaces (Kuzniak et al. 2016). They built on three important challenges for students in the learning of functions: to consider functional dependencies, to understand the idea of independent variable and to make sense of functional symbolism. The aims of the shop sign family is consistent with such a progression, which is close to Tall's introduced above (Tall 2006).

38.4.2   The Local Level: A Priori Analysis of the Task and Students' Intended Activities

The task is to identify the position of E on $[DC]$ in order that the sum of the areas $DFGE$ and AGB are minimal (Fig. 38.2). This requires actual knowledge about geometrical figures and functions. However, it assumes that the notion of function is available, i.e., students have to identify the need for a function by themselves.

In a traditional pencil-and-paper environment, students first draw a generic figure. They can try to estimate—using geometrical measurements—some values for the areas for different positions of E. They can draw a table of values, but this kind of procedure is usually not enough to obtain a good conjecture of the minimum value. Moreover such a procedure can reinforce the pointwise perspective because it does not bring the continuous aspects of the function at stake. Usually, the teacher quickly asks students to produce algebraic expressions of the areas. Students try themselves to introduce an algebraic variable ($DE = x$), or the teacher gives them procedural aids.

In the example given here, the teacher provided students with a sheet of paper showing a figure similar to the one given in Fig. 38.2 and the instructions as summarised in Fig. 38.3.

Figure 38.3 shows that the overall task is divided into three subtasks. Organizing activities are directed by procedural mediations (functional regulation), which is a way to ensure that most students can engage in productive activity.

First: construction of the figure (with GeoGebra)

Second: conjecture (experimentation and observations with GeoGebra)

Troisième partie : Démonstrations
  1. On pose DE = x. Exprimer en fonction de x la somme des aires du carré EDFG et du triangle GAB.

Third: algebraic proof

Fig. 38.3 Main instructions given to students

38.4.2.1   A Priori Analysis of the First Subtask: The Construction of the Figure

In the geometrical subtask, students have to identify the fixed points (A, B, C, D), the free point (E) on $[DC]$ and the dependent points $(F$ and $G)$. The order of construction is crucial to the robustness of the final figure, but is not important in the paper-and-pencil environment. Consequently, organizing activities—the order of instructions—are more important in the GeoGebra environment.

This subtask also requires students to make choices. It is possible to draw either G or F first, and the sequence of instructions is not the same. Moreover, there are other choices that have no equivalent in the paper-and-pencil environment: whether to define the polygons (the square and triangle) with the polygon instruction or by the length of their sides, whether to use analytic coordinates of fixed points or a geometrical construction, whether to use a cursor to define E, etc. These choices refer not just to mathematical knowledge but also to instrumental knowledge (following the instrumental genesis approach). This means that treatment activities include instrumental knowledge and are more complex than in the traditional environment. Once the construction is in place, students can verify its robustness—a treatment that is also specific to the dynamic environment.

38.4.2.2   A Priori Analysis of the Second Subtask: The Conjecture

There is no task really equivalent to this subtask in the paper-and-pencil environment. This again leads to specific treatment activities. These are engaged with the feedback provided by the software, which assigns numerical values of the areas $DFGE$ and AGB, according to the position of E. However, students are required to redefine $DFGE$ and AGB as polygons if they have not already used this instruction to complete the first subtask (Fig. 38.5). They also have to create in the GeoGebra environment a new numerical value that is the sum of the two areas in order to refine their conjecture. It is not clear to what extent these specific treatment activities refer to mathematical knowledge, and we will return to this point later.

38.4.2.3   A Priori Analysis of the Third Subtask: The Algebraic Proof

This subtask appears similar to its equivalent in the paper-and-pencil environment. However, as students already know the value of the minimum, the motivation for activity is different and only relates to the proof itself. The most important step is the introduction of x as a way to pass from the geometrical setting to the functional setting. This step brings recognizing activities (students must recognise that the functional setting is needed), which is triggered by a procedural mediation (the instructions given on the sheet).

Students have to determine the algebraic expression of the function. Existing knowledge about the area of polygons must be available. They also have to recognise a second-order polynomial function associated with specific treatments. The treatment activity that remains is obtaining the canonical form (as students have not been taught about derivatives, they must be helped in this by the teacher). Finally, the recognition of the canonical form as a way to obtain the minimum of the area and the position of E that corresponds to this minimum correlates with the importance of the dialectic between pointwise and global perspectives on functions.

38.4.3   A Posteriori Analysis: Observable and Effective Activities

Students worked in pairs. The teacher only intervened at the beginning of the session (to ensure that all students were working) and at the end (to summarise the session). Students mostly worked autonomously, although the teacher helped individual pairs of students. The following observations are based on two pairs of students: Aurélien and Arnaud, and Lolita and Farah.

38.4.3.1   Analysis of the First Pair of Students' Activities: Aurélien and Arnaud

This pair took a long time to construct their figure (more than 20 min). They began with A, B, C, D in sequence, using coordinates and then drawing lines between pairs of points. This approach is closest to the paper-and-pencil situation, and while it is time-consuming it is not crucial for global reasoning. They then introduced a cursor —a numerical variable j that took a value between 0 and 4—in order to position E on $[D, C]$. However, the positioning of F at $(0, 3)$ was achieved without the cursor, which led to a wrong square (Fig. 38.4). G was drawn correctly. After they had completed their construction, they moved the cursor in order to verify that their figure was robust; an operation which revealed that the figure was not (Fig. 38.4).

This mediation from the screen is supposed to be a constructive mediation: It does not change the nature of the task and is supposed to permit a constructive

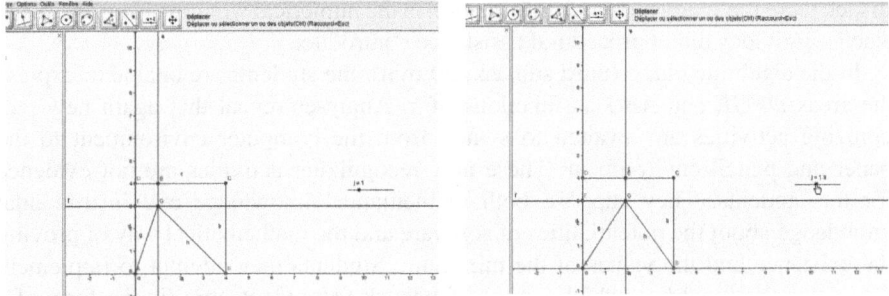

Fig. 38.4 Exploring the robustness of the shop sign

regulation of students' activities (the lower loop in Fig. 38.1). However, the mediation does not encounter the students' ZPD, and it is insufficient for them to regulate their activity by their own. In fact, the mediation supposes new recognizing activities specific to dynamic geometry on computers that these students are not able to develop.

In this case, the teacher makes a procedural mediation and helps the students to rebuild their figure ('You use the polygon instruction to make *DFGE*... then again to make the polygon *ABG*.'). Once the two polygons have been correctly drawn, the values of their areas appear in the numerical window of GeoGebra (called *poly1* and *poly2*, shown on the left-hand side of the screens presented in Fig. 38.5).

In the conjecture phase (second subtask, 8 min), the students made the conjecture that the sum is always 8 ('Look, it's always 8...'), by computing *poly1* + *poly2* in their mind. The numerical window of GeoGebra now shows 18 different pieces of information, including the areas of *DFGE* (*poly1*) and *ABG* (*poly2*). Students must introduce another numerical variable, *poly3*, that is equal to the sum of *poly1* + *poly2*. However, this requires new organizing activities that GeoGebra does not help with.

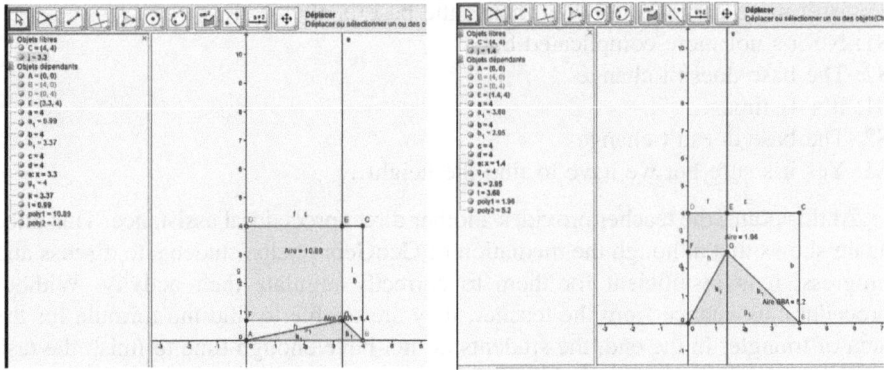

Fig. 38.5 Exploration of varying areas by moving the point *E* on [*DC*]

In fact, there is already too much information in the numerical window. Here again, the teacher provides direct procedural assistance ('introduce $poly3 = poly1 + poly2$').

In the algebraic phase (third subtask, 20 min), the students are unable to express the areas $DFGE$ and ABG as functions of x. Analyses reveal that again new recognizing activities are awaited to switch from the computer environment to the paper and pencil environment. These new recognizing activities are not evidence for the students. They suppose both mathematical knowledge and instrumental knowledge about the potentialities of software and the mathematical way of proving the existence and the values of the minimum. Students then attempt to implement $DE = x$ in the input bar, which leads to feedback from GeoGebra (in the form of a syntax error), which informs them that their procedure is wrong—but does not provide any guidance about what to do instead. It is difficult to know whether to categorise this kind of mediation as procedural or constructive as it does not add any mathematical knowledge.

The teacher asks the students to try to find a solution with pencil and paper (procedural assistance). However, the introduction of x, which is linked to the change of mathematical setting (adaptation of knowledge), seems very artificial. The students start working on their algebraic formula by looking at their static figure, with E positioned at $(1, 4)$. The base of the triangle measures 4 and its height is 3. One of the pair suggests that 'it depends on x', meaning that each algebraic expression ends in x, as the following dialogue between the two students shows:

Student 1: This is $4x$
Student 2: Base times height… so the base is 4
S1: $4x$… it's 4 times x, because it's a function of x
S2: Oh? The height?
S1: Yeah the height… $3x$
S2: No, it can change
S1: Yeah but in this case
S2: Look, (he moves the cursor) this isn't $3x$ here
S1: Humm… OK, listen…
S2: Before for the square we found it because x squared is always the area… this isn't more complicated than that… the base is always 4…
S1: No it's not more complicated but…
S2: The base doesn't change
S1: It's $4x$ times…
S2: The base doesn't change
S1: Yes it's sure but we have to find the height…

At this point, the teacher provides another direct procedural assistance. This once again shows that although the mediation of GeoGebra helps students to discuss and progress, it is insufficient for them to correctly regulate their activity. Without procedural assistance from the teacher, they are unable to find the formula for the area of triangle. In the end, the students do not have enough time to finish the task by themselves.

At end of the session, the teacher gives a procedural explanation to the whole class of how to find the canonical form (as '$x^2 - 2x + 8 = (x - \_)^2 + \_$'). Although Aurélien and Arnaud write it down, they do not make the link between it and their classroom work. Consequently, they do not understand the motivation for the activity and cannot make sense of the explanation of the canonical transformation given by the teacher.

Then the teacher gives a constructive explanation about the meaning of the coefficients in the canonical form and the way they give the minimum and the corresponding value of x. However, based on what we see in Aurélien and Arnaud's activities, it is too early, and they do not make the link with their numerical conjecture. In other words, the collective mediation of the teacher seems too far from the students' ZPD, and it is not at all constructive for this pair of students.

38.4.3.2 Analysis of the Second Pair of Students' Activities: Lolita and Farah

Lolita and Farah are better students and quickly draw their robust figure. Their numerical conjecture is correct and the teacher gives them another subtask: Find a graphical confirmation of their conjecture. The procedural instruction is to find a new point, M, whose abscissa is the same as E and ordinate is the value *poly1 + poly2* (Fig. 38.6).

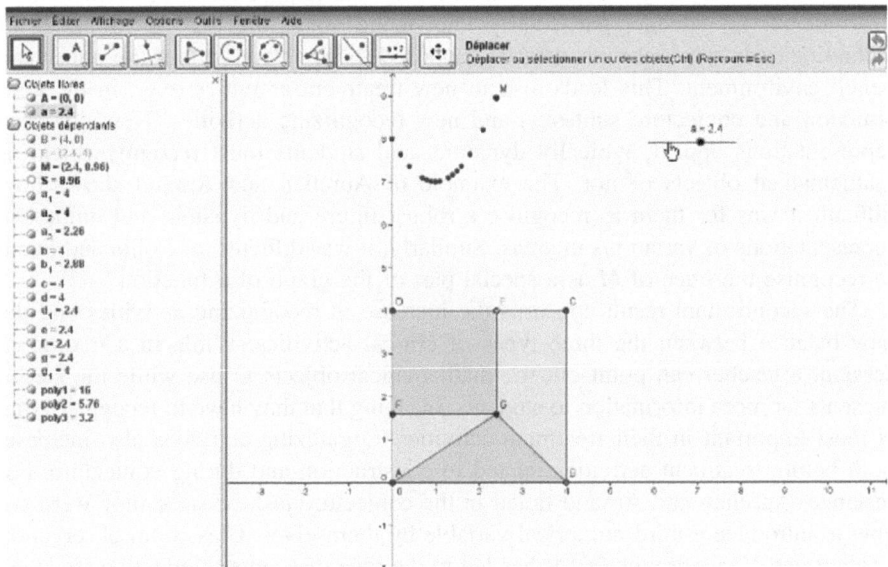

Fig. 38.6 The shop sign task showing part of the graph of the function

When moving the cursor, the trace of M can be interpreted as the graph of the function $f(x)$. However, Lolita and Farah do not recognise this. One says 'this is not a curve' and then 'the minima, we have seen this for functions but here...'

They only recognise the trace as a part of a parabola (geometrical setting) and associate its lowest point with the value of the minimum area.

The graphical observation confirms to Lolita and Farah that their numerical conjecture was correct. However, this is a proof for them and they do not understand the motivation of the third subtask, which does not make sense to them. Although they succeed in defining the algebraic expression of the function and they find the canonical expression, they do not make the link with their graphical observation.

Here again, the teacher's summary of how to obtain the canonical form of the function, the value of the minimum, and the corresponding value of x is not useful for this pair, as it is not the problem they encountered. A constructive intervention about the motivation for the third subtask and how the canonical form was linked to the conjecture would have been a mediation that was closer to their ZPD.

38.5 What Does This Tell Us About Students' Mathematical Activities?

The main result concerns complex activity involving technology: Here the complexity is introduced by mathematical activities that require either mathematical or instrumental knowledge, particularly knowledge about the real potentialities of technologies in contrast with what is supposed to be solved within the paper and pencil environment. This leads also to new treatment activities (e.g., in the construction and conjecture subtasks) and new recognizing activities. New onscreen representations appear, typically dynamic, and students must recognise them as mathematical objects or not. The example of Aurélien and Arnaud shows how difficult it was for them to recognise a robust figure and dynamic and numerical representations of variations in areas. Similarly, it was difficult for Lolita and Farah to recognise the trace of M as a special part of the graph of a function.

The second main result concerns the increase in recognizing activities and the new balance between the three types of critical activities. While in a traditional session, a teacher can point out the mathematical objects to use while the screen presents far more information to students, meaning that they have to recognise what is most important in their treatment activities. Organizing activities also increase, both before treatment activities related to construction and during conjecture. For instance, Aurélien and Arnaud failed in the conjecture task because they were not able to introduce a third numerical variable by themselves. Classroom observation (Robert and Vandebrouck 2014) has led to the idea that most of effective students' activities are treatment activities, as the teacher must make procedural interventions before most students can begin the task. Recognizing and organizing activities are mostly activities for the best students. These students often have an idea of how to begin the resolution of the task, are able to adapt quickly their knowledge and develop all three types of critical mathematical activities, whereas weaker students find it difficult to engage in the task, waiting for any procedural assistance from the teacher. In classroom sessions that use technology, students are confronted with all of these critical activities and have to deal with them by themselves, which may help to explain the difficulty of weaker students.

A further finding concerns mediations. In such sessions, a teacher's mediations are mostly procedural and clearly aim to foster productive activity. Onscreen mediation leads to specific new recognizing activities (dynamism) but is insufficient for all students—not only weaker students—to regulate their own activity. It appears that most of the time this mediation is not procedural or constructive enough, leading to more teacher intervention. Moreover, it seems that onscreen mediation is always associated with treatment activities and does not help students in their recognizing or organizing activities.

The last point concerns constructive mediation and the heterogeneity of the students' knowledge (and ZPD). Student activities in classroom sessions that use technology are difficult for teachers to evaluate. Even if they try to manage the best 'average' constructive mediations for all students, our examples show that this is very challenging. This raises the question of what the real impact of such sessions is with respect to the intended conceptualization. The availability and recognition of functions as tools to complete such tasks was not really investigated, in the sense that the independent variable x was given to students (on paper) and none of them returned to the geometrical setting as in the traditional modelling cycle as described by Kaiser and Blum (Maass 2006). Moreover, Aurélien and Arnaud did not explore the dynamic numerical-graphical-algebraic flexibility, which was one of the aims of the session, while Lolita and Farah did, but lacked the constructive mediations needed to complete the cycle.

38.6   Conclusion

We have presented activity theory in the context of French didactics, notably the dual regulation found in the activity model, which was first developed in ergonomic psychology and then adapted to didactics of mathematics in order to study students' activities. Other works, which we have not discussed here, have looked at teachers' practices in some different ways (Robert 2012; Robert and Rogalski 2005). An important component of this model is the impact of activity on subjects, which represents the developmental dimension of students' activity. This focus highlights the commonalities and complementarities of the constructivist theories of Piaget (extended to Vergnaud's conceptual fields) and Vygotsky. The connection between activity theory, the work of Piaget and Vygotsky and didactics of mathematics provides a theoretical foundation for a dual approach to students' activity from the viewpoint of mathematics (the didactical approach) and subjects (the cognitive approach).

Our analysis does not provide a model of students' activity (or teachers' practices). However, it leads to the identification of similarities and differences in terms of the relations between subtasks, students' ways of working, mediations, and mathematical activities and compares this complex task with the traditional paper-and-pencil environment. One of the specificities of our approach is the deep connection between the students' activities analysis and the a priori task analysis, including mathematical content. But we do not look for the teacher's own intention, unlike what is done in some English research (for instance, Jaworski and Potari 2009). Moreover, we do not attempt to raise the global dynamic between individual and collective interactions and learning. We should take now a threefold approach to the investigation of students' practices: didactical, cognitive and socio-cultural. As Radford (2016) argues, with respect to Mathematical Working Spaces (Kuzniak et al. 2016), the individual-collective dynamic remains difficult to understand in both our activity theory and Mathematical Working Spaces, which are discussed together. This represents a new opportunity to better investigate the socio-cultural dimension of activity theory—especially as developed by Engestrom—and integrate it into our didactical and cognitive approach.

References

Artigue, M. (2002). Learning mathematics in a case environment: The genesis of a reflection about instrumentation and the dialectics between technical and conceptual work. *International Journal of Computers for Mathematical Learning, 7*(3), 245–274.

Artigue, M., Batanero, C., & Kent, P. (2007). Mathematics thinking and learning at post-secondary level. In F. Lester (Ed.), *Second handbook of research on mathematics teaching and learning* (pp. 1011–1049). Greenwich, CT: Information Age Publishing, Inc.

Artigue, M., Cazes, C., Hérault, F., Marbeuf, G., & Vandebrouck, F. (2011). The challenge of developing a European course for supporting teachers' use ICT. In M. Pytlak, T. Rowland, & E. Swoboda (Eds.), *Proceedings of the 7th Congress of the European Society for Research in Mathematics Education* (pp. 2983–2984). Rzeszów: University of Rzeszów.

Arzarello F., & Robutti O. (2004). Approaching functions through motion experiments. *Educational Studies in Mathematics, 57*(3), Special issue CD Rom.

Cole, M., & Wertsch J. (1996). Beyond the individual-social antinomy in discussions of Piaget and Vygotski. *Human Development, 39,* 250–256.

Douady, R. (1986). Jeux de cadre et dialectique outil-objet. *Recherches en Didactique des Mathématiques, 7*(2), 5–31.

Duval, R. (1995). *Sémiosis et pensée humaine: Registres sémiotiques et apprentissages intellectuels.* Berne: Peter Lang.

Engeström, Y., Miettinen, R., & Punamaki, R. L. (Eds.). (1999). *Perspective on activity theory.* Cambridge, UK: Cambridge University Press.

Falcade R., Laborde C., & Mariotti M. A. (2007). Approaching functions: Cabri tools as instruments of semiotic mediation. *Educational Studies in Mathematics, 66,* 317–333.

Galperine, P. (1966). Essai sur la formation par étapes des actions et des concepts. In D. A. Leontiev, A. Luria, & A. Smirnov (Eds.), *Recherches psychologiques en URSS* (pp. 114–132). Moscou: Editions du progrès.

Gueudet, G. (2008). Investigating the secondary-tertiary transition. *Educational Studies in Mathematics, 67,* 237–254.

Hitt, F., & González-Martín, A. S. (2016). Generalization, covariation, functions, and calculus. In A. Gutiérrez, G. L. Leder, & P. Boero (Eds.), *The second handbook of research on the psychology of mathematics education. The journey continues* (pp. 3–38). Rotterdam: Sense Publishers.

Jaworski, B., & Potari, D. (2009). Bridging the macro- and micro-divide: Using an activity theory model to capture sociocultural complexity in mathematics teaching and its development. *Educational Studies in Mathematics, 72,* 219–236.

Kuzniak, A., Tanguay, D., & Elia, I. (2016). Mathematical working spaces in schooling. *ZDM Mathematics Education, 48*(6), 721–737.

Leontiev, A. (1978). *Activity, consciousness and personality.* Englewood Cliffs: Prentice Hall.

Leplat J. (1997). *Regards sur l'activité en situation de travail.* Paris: Presses Universitaires de France.

Maass, K. (2006). What are modelling competencies? *ZDM Mathematics Education, 38*(2), 113–142.

Maschietto, M. (2008). Graphic calculators and micro straightness: Analysis of a didactic engineering. *International Journal of Computer for Mathematics Learning, 13,* 207–230.

Minh, T.-K., Lagrange, J.-B. (2016). Connected functional working spaces: A framework for the teaching and learning of functions at upper secondary level. *ZDM Mathematics Education, 48* (6), 793–807.

Pastré P. (1999). Apprendre des situations. *Education permanente, 139.*

Piaget, J. (1985). *The equilibration of cognitive structures: The central problem of intellectual development* (T. Brown & K. J. Thampy Trans.). Chicago: University of Chicago Press.

Rabardel, P. (1995). *Les hommes et les technologies, approche cognitive des instruments contemporains.* Paris: Armand Colin.

Radford, L. (2016). The epistemic, the cognitive, the human: A commentary on the mathematical working space approach. *ZDM Mathematics Education, 48*(6), 925–933.

Robert, A. (1998). Outil d'analyse des contenus mathématiques à enseigner au lycée et à l'université. *Recherches en Didactique des Mathématiques, 18*(2), 139–190.

Robert, A. (2012). A didactical framework for studying students' and teachers' activities when learning and teaching mathematics. *International Journal for Technology in Mathematics Education, 19*(4), 153–158.

Robert, A., & Hache, C. (2008). Why and how to understand what is at stake in a mathematics class? In F. Vandebrouck (Ed.), *Mathematics classrooms: Students' activities and teachers' practices* (pp. 23–74). Rotterdam: Sense Publishers.

Robert, A., & Horoks, J. (2007). Tasks designed to highlight task-activity relationships. *Journal of Mathematics Teacher Education, 10*(4–6), 279–287.

Robert, A., & Rogalski, J. (2005). A cross-analysis of the mathematics teacher's activity. An example in a French 10th-grade class. *Educational Studies in Mathematics, 59,* 269–298.

Robert A., & Vandebrouck F. (2014). Proximités en acte mises en jeu en classe par les enseignants du secondaire et ZPD des élèves: Analyses de séances sur des tâches complexes. *Recherches en Didactique des Mathématiques, 34*(2/3), 239–285.

Rogalski, J. (2013). Theory of activity and developmental frameworks for an analysis of teachers' practices and students' learning. In F. Vandebrouck (Ed.), *Mathematics classrooms: Students' activities and teachers' practices* (pp. 3–22). Rotterdam: Sense Publishers.

Samurcay, R., & Rabardel, P. (2004). Modèles pour l'analyse de l'activité et des compétences: Propositions. In R. Samurcay & P. Pastré (Eds.), *Recherches en Didactique Professionnelle* (Chapitre 7). Toulouse: Octarès.

Tall, D. (2004). Thinking through three worlds of mathematics. In Dans (Ed.), *Actes de 28th Conference of the International Group for Psychology of Mathematics Education* (pp. 281–288). Bergen, Norway.

Tall, D. (2006). A theory of mathematical growth through embodiment, symbolism and proof. *Annales de Didactique et de Sciences Cognitives, 11,* 195–215.

Vandebrouck, F. (2011). Points de vue et domaines de travail en analyse. *Annales de Didactique et de Sciences Cognitives, 16,* 149–185.

Vandebrouck, F. (Ed.). (2013). *Mathematics classrooms: Students' activities and teachers' practices*. Rotterdam: Sense Publishers.

Vandebrouck, F., Chiappini, G., Jaworski, B., Lagrange, J.-B, Monaghan, J., & Psycharis, G. (Eds.). (2012/13). Activity theoretical approaches to mathematics classroom practices with the use of technology (Part 1 & Part 2). *International Journal for Technology in Mathematics Education*, *19*(4) & *20*(1).

Vergnaud, G. (1982). Cognitive and developmental psychology and research in mathematics education: Some theoretical and methodological issues. *For the Learning of Mathematics*, *3*(2), 31–41.

Vergnaud, G. (1990). La théorie des champs conceptuels. *Recherches en Didactique des Mathématiques*, *10*(2–3), 133–169.

Vygotsky, L. (1986). *Thought and language*. Cambridge, MA: MIT Press.

Wells, G. (1993). Reevaluating the IRF sequence: A proposal for the articulation of theories of activity and discourse for the analysis of teaching and learning in the classroom. *Linguistics and Education*, *5*(1), 1–37.

Wertsch, J. V. (1981). The concept of activity in soviet psychology: An introduction. In J. V. Wertscher (Ed.), *The concept of activity in soviet psychology*. Armonk, NY: M.E. Sharpe Inc.

Chapter 39
The Effect of a Video-Based Intervention on the Knowledge-Based Reasoning of Future Mathematics Teachers

Naďa Vondrová

Abstract The article focuses on the professional vision of pre-service mathematics teachers. Drawing on literature about its development in video-based interventions, the article focuses on the effect of a video-based intervention on pre-service teachers' (n = 32) knowledge-based reasoning as a component of professional vision. The intervention had features compatible with situated cognition learning theory. The participants' knowledge-based reasoning was tracked in participants' written reflections on mathematics lessons shown on video before and after the intervention. An important feature of the intervention lies in balancing the videos to avoid the learning effect and a possible influence of the video content. The study showed a decrease in subjective judgments and negative comments about the lessons; however, there was a decrease rather than increase in higher-level interpretations. Possible reasons for this are discussed against results of similar intervention studies. Implications for teacher education are given.

Keywords Professional vision · Noticing · Pre-service mathematics teachers
Knowledge-based reasoning · Evaluation · Interpretation

39.1 Introduction

Pre-service teachers (PSTs) must develop a range of skills, including a certain type of noticing skill, that is different from lay people's skills. When observing a mathematics lesson, either live or on video, during their mathematics education programme, they are expected to notice aspects of the lesson that are deemed important for the development of pupils' knowledge. Yet the complexity of a mathematics lesson is such that if they direct attention to something, they do so at the expense of something else. Much research has been undertaken in recent years

N. Vondrová (✉)
Faculty of Education, Charles University, Magdalény Rettigové 4,
116 39 Praha 1, Czech Republic
e-mail: nada.vondrova@pedf.cuni.cz

© The Author(s) 2018
G. Kaiser et al. (eds.), *Invited Lectures from the 13th International Congress on Mathematical Education*, ICME-13 Monographs,
https://doi.org/10.1007/978-3-319-72170-5_39

focusing on what it is that PSTs do and do not notice in a mathematics lesson and how they make sense of it.

Teacher education programmes contribute to the development of noticing in different ways. For example, in our previous research (Simpson et al. 2017), we found that a two-year master's programme for future mathematics teachers without a focus on noticing and with two short school practice placements does not significantly influence PSTs' patterns of attention when observing a videoed lesson. On the other hand, there is a body of research showing that video-based interventions do influence noticing in important ways (see Sects. 39.2.4 and 39.5). In the above research, we were interested to see in what way a quite short video intervention spanning three seminars would or would not influence future mathematics teachers' patterns of attention. In many ways, we confirmed the results found in existing literature. The PSTs increasingly focused on the mathematical aspect of the lesson and on students rather than on the teacher, and their comments were more specific than general. We also looked briefly at their knowledge-based reasoning and found that the PSTs increasingly described and evaluated less, but also interpreted less when reflecting on what they saw in the video. The aim of this article is to elaborate on the above research and to present some further findings about the nature of PSTs' knowledge-based reasoning and how it was affected by the video intervention.

39.2 Theoretical Background and a Review of Literature

39.2.1 Teacher Noticing

The concept of noticing is usually described as consisting of the processes of attending to particular events in the lesson and making sense of these events. Probably the most influential in the field is the characterisation by van Es and Sherin (2002; cited in Sherin and Star 2011), who propose three aspects of noticing:

> (a) identifying what is important or noteworthy about a classroom situation; (b) making connections between the specifics of classroom interactions and the broader principles of teaching and learning they represent; and (c) using what one knows about the context to reason about classroom events. (p. 573)

In this paper, 'pattern of attention' will be used to refer to the first process, while the other two processes will both be referred to as 'knowledge-based reasoning'. The concept of noticing is often conflated with professional vision, which is characterised as seeing phenomena in a scene from the area of expertise that are different from those arising from lay viewings of the same scene (Goodwin 1994). For example, Sherin et al. (2011) understand noticing as '*professional vision* in which teachers selectively attend to events that take place and then draw on existing knowledge to interpret these noticed events' (p. 80).

It is obvious, given the complexity of observing a lesson, that observers must split their attention between what they see as noteworthy and what they choose to neglect; it is not a passive process. Moreover, as Schoenfeld (2011) points out, the observer's knowledge, beliefs and orientations will have an impact on where attention is actually directed.

Much research has been aimed at the patterns of PSTs' attention in general. It mostly concludes that they pay more attention to the teacher and classroom management than to students or mathematical content and its implementation in the lesson (e.g., Santagata et al. 2007; Alsawaie and Alghazo 2010). Moreover, PSTs' comments are found to be rather evaluative. Generally, studies do not distinguish between 'more and less important' moments to be noticed. Some even say that before teachers are able to attend to important moments, they have to develop the ability to notice trivial classroom features (Star et al. 2011). However, they do add that it is not clear 'whether it is better to focus first on improving teachers' awareness of the full range of (trivial and important) events (as was done here [in their course]) or to focus explicitly on only important events from the outset' (Star et al. 2011, p. 132). In some studies, the authors speak about salient features of mathematical instruction to be noticed (e.g., Mitchell and Marin 2015).

Naturally, the ability to 'identify noteworthy events in a teaching situation depends on one's image of what is important in teaching' (Alsawaie and Alghazo 2010, p. 227). Further, what the authors of a video-based intervention see as important in teaching mathematics will also have an impact on such ability for the participants. For the presented study, the important moments are the ones that are generally accepted as playing the key role in pupils' learning of mathematics: the types of tasks used by teachers and the kinds of discourses that they orchestrate when implementing them (Hiebert et al. 2003). Moreover, in line with the constructivist view of learning, pupils' active role in developing their mathematics knowledge is emphasised. Thus, the concept of opportunity to learn is important; namely: '[the] circumstances that allow students to engage in and spend time on academic tasks such as working on problems, exploring situations and gathering data, listening to explanations, reading texts, or conjecturing and justifying' (Kilpatrick et al. 2001, p. 333).

It includes 'considerations of students' entry knowledge, the nature and purpose of the tasks and activities, the likelihood of engagement, and so on' (Hiebert and Grouws 2007, p. 379) and is seen as the single most important predictor of pupils' achievement.

In the review of literature, I will restrict myself to studies on video-based interventions with future mathematics teachers aimed at noticing.

39.2.2 Structure of Video-Based Interventions

The instructional strategy employed when embedding classroom videos into a course is informed by the learning goal and purpose at hand (Blomberg et al. 2014).

In literature, we can find courses based on situated cognition learning theory, which suggests 'that learning should be rooted in authentic activity; that learning occurs within a community of individuals engaged in inquiry and practice; that more knowledgeable "masters" guide or scaffold the learning of novices; and that expertise is often distributed across individuals' (Whitcomb 2003, p. 538). In such a case, 'video is used as a problem anchor to elicit learners' mental action. Video thus represents a complex example from which learners can collectively derive principles or rules' (Blomberg et al. 2014, p. 447). Another approach is based on cognitive learning theory, according to which learning involves the storage and access of knowledge in long-term memory; it is necessary to avoid overloading the learner's working memory, so prompts are used and explicit guidelines are given, etc. In such interventions, videos are used as illustrations of previously taught principles and rules (Blomberg et al. 2014).

The learning goal and purpose influence the way videos are embedded in tasks. Video-based interventions utilise various scaffolds to develop noticing. Their leaders provide participants with some framework that draws their attention to particular features of the lesson and that they can use to account for what they notice. An example is the Mathematical Quality of Instruction (MQI) analysis framework (Mitchell and Marin 2015), which utilises aspects such as teacher mathematical error or imprecision, use of mathematics with pupils, cognitive demand of task and student work with mathematics. Roth McDuffie et al. (2014) provided participants with 'four lenses of analysis of lessons' (namely, teaching, learning, task, and power and participation). The Lesson Analysis Framework draws attention to four aspects of the lesson and particularly to connections between them: the learning goal of the lesson, pupils' learning, specific and instructional activities and alternative strategies (Santagata et al. 2007; Santagata and Angelici 2010; Santagata and Guarino 2011; Santagata and Yeh 2014; Yeh and Santagata 2014). Participants in Star and Strickland (2008) study used a 'five observation categories framework' to observe a lesson, namely, classroom environment, classroom management, tasks, mathematical content and communication.

Video-based interventions vary in length, ranging from short interventions, such as four sessions within one month (Santagata and Angelici 2010), five sessions in 10 weeks (Mitchell and Marin 2015) or eight sessions in three months (Blomberg et al. 2014) to whole-semester courses (e.g., Star and Strickland 2008) and differ in the type and number of videos used. For example, Santagata and Angelici (2010) only used one video of the whole mathematics lesson, while Santagata et al. (2007) used three and Alsawaie and Alghazo (2010) used 10. Others (e.g., Blomberg et al. 2014; Roth McDuffie et al. 2014) used video clips only. Some also used videos of interviews with individual pupils (e.g., Santagata and Guarino 2011; Yeh and Santagata 2014). Some used videos of the participants' own teaching (e.g., Mitchell and Marin 2015), and some interventions were complemented with a field experience (Stockero 2008; Santagata and Guarino 2011).

39.2.3 Measuring Effects of Video-Based Interventions

Some video-based interventions use an 'experimental vs. control group' design (Alsawaie and Alghazo 2010; Blomberg et al. 2014; Santagata and Yeh 2014). Others investigate effects of two different types of scaffolds (Blomberg et al. 2014; Santagata and Angelici 2010). Still others do not have a control group and examine the effect of the intervention only (Santagata et al. 2007; Mitchell and Marin 2015; Roth McDuffie et al. 2014).

There are basically two types of measures used in video-based interventions. In the first (Stockero 2008; Roth McDuffie et al. 2014), the participants' responses are treated together. The development in noticing is usually studied in group discussions. The second measures in what way the individual responses differ before and after the intervention (e.g., Santagata et al. 2007; Star and Strickland 2008; Alsawaie and Alghazo 2010; Santagata and Angelici 2010; Santagata and Guarino 2011; Blomberg et al. 2014; Mitchell and Marin 2015). The tasks used for the individual pre- and post-tests are usually based on the analysis of videos of teaching of the whole lesson or its parts (an exception is the use of learning journals in Blomberg et al. 2014). In some studies, the same video is used in both tests (Santagata et al. 2007; Santagata and Angelici 2010; Santagata and Guarino 2011; Yeh and Santagata 2014; Mitchell and Marin 2015) while in others, videos of different lessons are analysed (Star and Strickland 2008; Stockero 2008; Alsawaie and Alghazo 2010; Santagata and Yeh 2014). Simpson et al. (2017) have noted problems with both. In the former, it is difficult to discount the learning effect (especially for short-term interventions): Is any change in the participants' pattern of attention the result of the intervention or the fact that they see the video for the second time? In the latter, it is not taken into account that different videos may provide participants with different stimuli—the post-test video may include more moments (and/or more visible moments) in which students appear in the foreground than in the pre-test video; thus, it is no wonder that more student-centred comments appear in the post-test. In the study presented in this text, this problem is dealt with by balancing videos (see Sect. 39.3.3).

For the analysis of comments, different frameworks are used. van Es and Sherin (2008, 2010) framework is widely used. It identifies four dimensions with several codes. The first is Actor, which splits into the focus of Teacher and Student (or students); Curriculum Developer (a comment referring to a textbook author, curriculum documents, etc.); Self (observers discuss themselves in relation to the video) and Other. The second dimension is Topic and it includes Classroom Management, Climate (the social environment), Mathematical Thinking, Pedagogy and Other. The third dimension is Stance, which overarches the pattern of attention and knowledge-based reasoning. It includes Describe (a recounting of what is seen), Evaluate (a judgment about what is seen) and Interpret (making inferences or links to what is seen which might help account for it or understand it). Finally, the dimension of Specificity captures whether the comment relates to a specific event in the lesson (Specific) or to some aspect of the whole class or whole lesson or makes a generalization beyond the class (General). In some studies (such as van Es and Sherin 2010), the authors also coded whether the comment was related to the video or not.

Other authors have developed the framework further. An example is Stockero (2008), who, using Manouchehri (2002) levels of reflection to better capture the quality of reflection, elaborated Stance into: Description, Explanation (connecting interrelated events and exploring cause and effect issues), Theorizing (adding support to an analysis by a reference to research or course reading or providing 'substantial evidence from transcripts and/or student written work as justification', p. 377), Confronting (considering alternate explanations for events and/or considering others' point of view) and Restructuring (showing evidence of Theorizing and Confronting by considering alternative instructional decisions and 'of re-examining his or her fundamental beliefs and assumptions about teaching and learning', p. 377). Similarly, Roth McDuffie et al. (2014) distinguished four quality levels, from descriptions with general impressions and evaluative comments at Level 1 to the analysis and interpretations of relationships between teaching strategies and students' thinking at Level 4.

39.2.4 Results of Video-Based Interventions and Our Previous Work

All the above studies report changes in the pattern of attention and knowledge-based reasoning after video-based interventions. PSTs increasingly focus on students rather than the teacher, and they are better observers of the mathematical content. They use fewer subjective evaluative comments. There is mixed evidence in terms of the development of the interpretation skill. To avoid repetition, results of related research will be further elaborated in Sect. 39.5.

Simpson et al. (2017) report on an intervention study of the pre- and post-test design whose aim was to find how pre-service mathematics teachers developed in regard to their pattern of attention following their participation in a video-based intervention.[1] The data were coded using Sherin and van Es' framework and quantitative methods were used to look for statistically significant differences in PSTs' comments in the pre- and post-tests.

The PSTs' written reflections were significantly longer after the intervention— on average more than 50% longer than those before it. The PSTs commented less on self in relation to the video and more on students in the video. There was an increase in the frequency of the mathematical thinking code after the intervention, i.e., the PSTs noticed mathematical aspects of the lesson more at the expense of Classroom Management and Pedagogy. Their comments became more descriptive

---

[1]The intervention and the methodology of the study whose results are reported here are given in Sect. 39.3.

and less evaluative, but at the same time also less interpretative. The responses were significantly more specific after the intervention. To sum up, the study suggested a markedly similar shift in attention to that seen in other studies (e.g., Santagata et al. 2007; Mitchel and Marin 2015) except for interpretation, which has been reported to increase in some studies (e.g., Stockero 2008; Alsawaie and Alghazo 2010; Roth McDuffie et al. 2014).

Our previous work has mostly been focused on the pattern of attention. In this article, I will look at the second process of professional vision, that is, knowledge-based reasoning. I will revisit the same data from the intervention study for further analysis to answer the research question:

> In what way is the PSTs' knowledge-based reasoning as demonstrated in the written analysis of a lesson on video affected by a video-based intervention?

Our previous research has also shown that there are differences in the pattern of attention which depend on the lesson observed. Thus, the second research question is:

> Are there any differences in PSTs' knowledge-based reasoning that depend on the lesson observed?

39.3   Methodology

39.3.1   Participants

The participants were Czech mathematics PSTs in the first semester of a four-semester master's programme. They had completed bachelor degrees in either Mathematics or Mathematics with a Focus on Education, but had had no formal education in teaching mathematics. Most were in their early or mid-20s. Five students were already qualified as teachers of other subjects and wanted to widen their qualification for mathematics. Six students had limited experience teaching mathematics. In total, 32 PSTs participated in the study. There was no selection made, all the PSTs in that year level participated.

39.3.2   Intervention

The intervention formed part of a mathematics education course that was taught by the author and was attended by all participants of the study. During the intervention, no school practice placement was assigned to the participants. In this course, prior to the intervention, PSTs were introduced to a theory of concept development process in mathematics, constructivist approaches to the teaching of mathematics and the division of mathematics tasks into procedural and making-connections types (taken from TIMSS). The theory was illustrated by either written cases or short video clips of mathematics lessons. Concrete prompts directing the PSTs'

Table 39.1 Description of the video-based intervention

Home study	Task 1: PSTs watched a Czech lesson CZ2 (TIMSS 1999 video study), answered questions and suggested an alternative to the core of the lesson
Session 1	Discussion about the PSTs' responses to Task 1 Two clips from two Czech Grade 8 mathematics lessons on tasks in geometry with very different approaches shown and discussed Short discussion about the Pythagorean theorem and its teaching
Home study	Tasks 2 and 3: PSTs watched two Czech lessons from Grade 8, both focussing on the Pythagorean theorem and its use but with different enactments of the objective. PSTs analysed them from the point of view of phenomena identified in Session 1 and chose a moment in the lessons where (a) a learning opportunity is lost, (b) the teacher reacts to a pupil and (c) a pupil's (mis)understanding is visible. After uploading their responses, the PSTs were assigned responses from two of their peers to comment upon (in Moodle module Workshop)
Session 2	Discussion about the PSTs' responses to Tasks 2 and 3
Home study	Tasks 4 and 5: The same as for Tasks 2 and 3. The two videos were Swiss and Czech lessons from Grade 8 with contrasting practices. PSTs described moments in which (a) the teacher reacts to pupils' mistakes and (b) pupils' (mis) understanding is visible and the strategies the teacher uses to make pupils more active. The Workshop module was used again
Session 3	Discussion about PSTs' responses to Tasks 4 and 5

attention were used and thus this video use was more aligned with the cognitive learning theory approach.

As stated above, the course's main aim was for the PSTs to develop their ability to notice features of the lesson salient to its success, and the tasks prepared for the intervention complied with it. The intervention was based on the situated cognition learning theory. There were online (through the Workshop module in Moodle, in a virtual learning environment [VLE]) and in-person (in sessions) opportunities for participants to work cooperatively, which follows the social cognition view approach.

The intervention spanned three guided-observation sessions (each about 120 min), with home study tasks, over a three-month period (see Table 39.1). The tasks consisted of watching videos of mathematics lessons and were set within the Moodle VLE. The videos were purposefully selected from a pool of videos used in preceding years with PSTs in which noteworthy events were visible and were motivating for PSTs to comment upon. In line with video-based courses in teacher education, the lessons were not used as examples of good practice. For example, Seago (2004) found that 'the most useful video clips were based on situations where there was some element of confusion (either the students' or the teachers') that typically arises in classrooms' (p. 267). This was confirmed by Sherin et al. (2009), who say that 'video clips should provide something for teachers to puzzle over or speculate about… it is through this process of inquiry that teacher learning will likely occur' (p. 215).

Some videos were of lessons from other countries as 'the exposure to alternative practices helps observers to become aware of their own cultural routines' (Santagata et al. 2007, p. 127). Recordings of whole lessons were used for home tasks and clips for sessions. In the whole-lesson video, all important elements to understand the lesson are present: 'goals for students' learning, instructional activities, strategies for monitoring students' thinking and assessing their learning, curriculum and pedagogy, and so on' (Santagata et al. 2007, p. 127).

During the sessions, the PSTs' responses to home tasks were discussed. The course teacher was drawing their attention to important moments (see Sect. 39.2.1) that they might not have noticed, e.g., by asking one PST to present his or her comments and inviting others to comment on them, or by showing the appropriate part of the video. To reduce PSTs' inclination towards criticism of the teacher in the video, the course teacher repeatedly reinforced the norms that 'included respecting others' ideas and providing evidence for claims' (Stockero 2008, p. 376).

39.3.3   The Pre- and Post-tasks

Two videos were selected for the pre- and post-intervention tasks: HK01 and HK04 (both from TIMSS 1999 Video Study). They capture Grade 8 lessons and are about half an hour in length. The topic of HK01 is square roots and HK04 is about linear identities. The piloting of the videos with an earlier group of PSTs had shown that they were lessons with which the participants could identify (Brophy 2004); despite the cultural differences between Hong Kong and the Czech Republic, the approach taken to teaching mathematics and to managing and organizing the class resembled a common approach taken in the Czech Republic.

The lessons were provided to the participants on a disk and they were accompanied by Czech subtitles. The PSTs were asked to write a reflection of the lesson; no prompts for the reflection were provided. There was no time or word limit and the PSTs were assured that they were not being assessed or judged on their responses. They were encouraged to write about what they found interesting and/or important.

To balance the videos and to avoid possible confusion caused by the use of the same or different videos for the pre- and post-tasks (see Sect. 39.2.3), the PSTs were randomly assigned to comment on one of the two videos before the first session and the complementary video after the last session. The videos were not discussed during the sessions.

39.3.4   Analysis

The PSTs' written responses were divided into units of analysis, each representing some articulated observation. They were usually whole sentences; however, in some cases they were a clause where a sentence appeared to contain a shift of focus (e.g., from the teacher to a student). Across the pre- and post-intervention responses, there were 1591 units of analysis.

39.3.4.1 Pattern of Attention

We used the framework developed by van Es, Sherin and their colleagues. The process of analysis is described in detail in (Simpson et al. 2017); however, as our new analysis is based on it, it should also be briefly mentioned here. Each unit of analysis was allocated codes based on the four dimensions given in Sect. 39.2.3. The descriptions of the categories in van Es and Sherin (2008, 2010) were used to create a coding manual and an inductive process of coding scripts and agreeing on meanings of codes was undertaken by two coders. 'Inter-rater reliability was assessed using Janson and Olsson (2001) multidimensional extension of Cohen's kappa, and once a good-to-excellent level of agreement ($\iota = 0.71$) was achieved, the coders were randomly assigned all remaining responses to code' (Simpson et al. 2017).

Examples of units of analysis and their codes are in Table 39.2.

39.3.4.2 Knowledge-Based Reasoning

Based on the study of literature and mainly on Stockero (2008),[2] a more refined framework to capture the nature of PSTs' reasoning about events was developed. It was used for the units of analysis which were coded as Evaluate and Interpret in the Stance category. Table 39.3 presents the framework and gives examples. For the sake of completeness, I also include Description, even though only statements that go beyond description and in which an observer engages with the information, makes judgements about it and/or interprets it are relevant in this text.

Statements from all the categories (except for Alteration I) were also given a value describing the way the PST saw their content. It could be rather negative ('The teacher does not prompt further to find out if the pupil understands what the mistake was.' Q2), positive ('I appreciate that the lesson was conducted through simple questions, clear for pupils.' Q2) or neutral ('The approach I offer seems to be oriented more to a concept: what I am learning, rather than process; this is the way to proceed.' Q4). Alteration I statements were rather negative in nature as the PSTs typically suggested an alternative when they did not like what had been done in the lesson. However, there were cases in which the PST did not openly criticise the

---

[2]Stockero (2008) used her five levels of reflection on the analysis of group discussion and her units of analysis were much broader than in the presented study. After trying it on our data, I found it impossible, for example, to apply Confronting and Restructuring to the very short units of analysis I had. Thus, I modified the framework.

Table 39.2 Examples of coding for the pattern of attention

Unit of analysis	Actor	Topic	Stance	Specificity
The teacher does not react to mistakes	Teacher	Pedagogy	Describe	General
However, the teacher says herself why the result cannot be −4	Teacher	Mathematical thinking	Evaluate	Specific
I find it interesting how often the class laughs at the pupil at the blackboard	Student	Climate	Evaluate	General

Table 39.3 Framework for knowledge-based reasoning

	Code	Description	Example
Q0	Description	Reproducing facts with no further elaboration	The teacher walked around and checked pupils' work
Q1	Subjective evaluation	Judgement of what is seen in the video	I find this way of teaching really useful and sense making
Q2	Explanation	Simple explanation of what is seen	She tried to make pupils use proper terms to develop mathematical vocabulary
Q2a	Alteration I	Suggestion of general or trivial alternative actions	The pupils should be given an opportunity to find it for themselves
Q3	Interpretation	Explanation of what is seen based on a theory	I find it important because thanks to the fact that the solving steps were not difficult, the pupils were attentive and did not lose motivation
Q4	Alteration II	Suggestion of elaborated alternative actions based on a theory	[A detailed proposal of teaching equations in a different way through concrete equations]
Q5	Prediction	Considering the effect of what is seen on pupils' knowledge in the future	The importance of a square root as inverse to squaring was shown, and something was hinted at that will later help pupils understand binomial equations

event but suggested an alternative action anyway. Thus, Alteration I statements were not given any value.

Without the units coded Describe in the first stage of analysis, 1046 units of analysis were left. The coding was done by the coders in a manner similar to that in the first stage.

39.3.4.3 Quantitative Methods

Finally, I used statistical methods to find differences between PSTs' knowledge-based reasoning in pre- and post-tests. Due to the small number of comments coded as Interpretation, Alteration II and Prediction, this was done only for Evaluation, Explanation and Alteration I.

First, the assumptions of normality and homogeneity of variance were assessed. The results of the Shapiro-Wilk test were significant for Q1 Evaluation ($W = 0.93$, $p = 0.035$). This suggests that the difference is unlikely to have been produced by a normal distribution. A Q–Q scatterplot was used to further evaluate the normality of data, which showed that normality cannot be assumed. Thus, the Related-Samples Wilcoxon Signed Rank test, which does not require normality, was used for Evaluation. For Q2 Explanation ($W = 0.97$, $p = 0.413$) and Q2a Alteration I ($W = 0.95$, $p = 0.138$), the Shapiro-Wilk test showed that normality can be assumed, thus a paired samples t-test was conducted in these cases. The same applies to the negative/positive/neutral nature of comments for which normality assumption and assumption of homogeneity of variances was met (both Shapiro-Wilk and Levene's tests were not significant).

As observed by Simpson et al. (2017), a difference was found in comments related to HK01 and to HK04 in the PSTs' pattern of attention. I was therefore interested to see whether any differences might also occur in the knowledge-based reasoning. The Shapiro-Wilk test showed that normality can be assumed and thus a paired samples t-test was used.

39.4 Results

39.4.1 Knowledge-Based Reasoning

Table 39.4 shows that prior to the intervention, the most statements on average were coded as Explanation, followed by subjective Evaluation. PSTs provided little interpretation of what they saw. Furthermore, they did not suggest elaborated alternative actions and make predictions. On the other hand, the task in the pre-test did not encourage them to do so explicitly.

It is also worth pointing out that one fifth of the comments are of an evaluative nature; that is, the PSTs make a judgement without providing a plausible explanation for it. Statements such as 'I like the structure of the lesson' or 'I did not like the lesson at all' were common.

Table 39.4 and Fig. 39.1 depict the development in the quality of comments beyond description between the pre- and post-tests. The only significant difference was found in Q1 Evaluation (the result of the Related Sample Wilcoxon Signed Rank Test was significant, with standardised test statistic of 2.724 and $p = 0.006$).

Table 39.4 Relative number of codes in the pre- and post-tests

	Description	Evaluation	Explanation	Alteration I	Interpretation	Alteration II	Prediction
Pre (%)	26	20	38	9	6	1	1
Post (%)	36	13	41	4	0	1	6

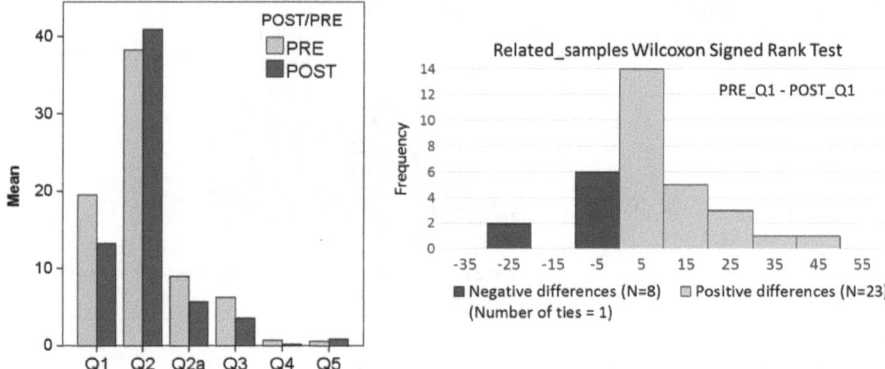

Fig. 39.1 Codes for knowledge-based reasoning and their distribution in the pre- and post-tests (left) and the change in Q1 Evaluation between pre- and post-tests (right). *Source* SPSS

This suggests that the PSTs had significantly more evaluative comments before the intervention. In the pre-test, the evaluative comments appeared on average in 20% of cases ($M = 0.20$, $SD = 0.13$), while in the post-test it was only in 13% ($M = 0.13$, $SD = 0.10$). For 72% of individual PSTs, the percentage of evaluative comments decreased in the post-test compared to the pre-test (Fig. 39.1 on the right: The black colour represents the PSTs with more evaluative comments in the post-test and the grey colour PSTs with more evaluation in the pre-test).

It should be noted that PSTs increasingly described after the intervention and evaluated less (Simpson et al. 2017), which was taken as a sign of learning. The lower level of Q1 Subjective evaluation observed here can be seen in a similar way. The PSTs gained new knowledge in the course and the intervention and were more cautious about jumping to evaluative conclusions than before the intervention. Rather, they included more descriptions in their reflections, showing that they noticed more events: These descriptions are more specific after the intervention.

There are small shifts in the other code values, but none of them are significant. There is a small increase in Explanation and Prediction, which goes in the direction of expert-like, knowledge-based reasoning. The decrease in Alteration I also does not have to be a negative result, as the PSTs suggested rather general or trivial alternative actions not based on theory.

In terms of the negative/positive/neutral nature of comments, the only difference that proved to be significant is in neutral comments ($t(31) = 4.36$, $p < 0.001$). After the intervention, the PSTs made significantly more neutral comments (on average 41%, $M = 0.41$, $SD = 0.14$) compared to the situation at the beginning when only 24% of all non-descriptive comments were neutral ($M = 0.24$, $SD = 0.17$). When looking at individual PSTs, for 26 of them (81%), the percentage of neutral comments increased in the post-test (Fig. 39.2).

Fig. 39.2 The change in neutral comments between pre- and post-tests

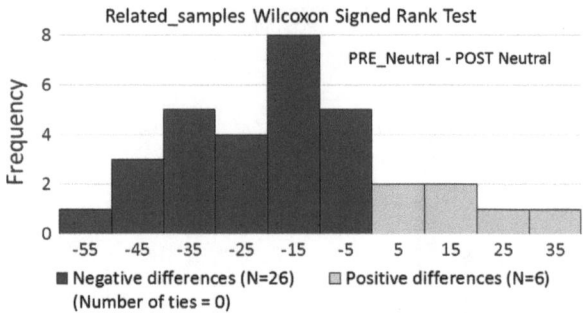

39.4.2 Differences for HK01 and HK04

In Simpson et al. (2017), we found differences in comments about the two lessons used in the pre- and post-tests. Namely, responses for HK04 focused less on Teacher and more on Curriculum Developer (Actor category) and less on Classroom Management and more on Mathematical Thinking (Topic). There was no difference for Stance and Specificity between the two lessons. In the re-analysis of data, no statistically significant difference in terms of knowledge-based reasoning (Q1–Q5) was found. The only significant difference appeared in comments with a negative tinge ($t(31) = 4.29$, $p = 0.000$). The mean of negative comments related to HK01 ($M = 0.37$, $SD = 0.23$) was significantly higher than the mean for HK04 ($M = 0.21$, $SD = 0.23$). For HK01, an average 37% of comments that went beyond description were critical in nature, while it was only 21% for HK04. This adds a further argument to being cautious when using different videos in the pre- and post-tests. The content of the lesson matters and might distort results.

39.5   Discussion and Conclusions

The study explored the influence of a video-based intervention on PSTs' knowledge-based reasoning. In previous research, it was established that this particular intervention led to changes in the pattern of attention that were markedly similar to changes seen in the literature. The exception was the category related to the levels of knowledge-based reasoning, i.e., Interpretation. In contrast to some other studies, the PSTs, rather than providing more interpretation after the intervention, provided less interpretation. With a more refined framework, I reached the conclusion that the PSTs did make progress in their knowledge-based reasoning, but only at a lower level. They provided fewer evaluative judgments and more explanation. This finding accords with, for example, Mitchell and Marin (2015) and Stockero (2008). Still, the participants in the presented study did not display more interpretation, elaborated alternatives or predictions.

The question is why statements coded as Interpretation disappeared after the intervention, considering that the participants underwent the intervention and the mathematics education course within which it was embedded and in which some theoretical notions were introduced. Why were there at least some attempts to interpret events before the intervention and after it none, considering that the task of the pre- and post-tests was the same? At least two explanations are plausible. First, the pre-test was done immediately after the part of the mathematics education course in which the theory mentioned was introduced (see Sect. 3.2). Thus, the PSTs had it fresh in their mind and used it in the pre-test. The post-test followed after three months and the theory may have been forgotten. This has an important implication for the course: The theory was probably not continually reinforced and the PSTs could not apply it. Another explanation is that the PSTs not only became more reluctant to make judgments but also grew reluctant to interpret things, as if the more they had learned, the more they had realised how complex the teaching-learning situations were and that there were no easy interpretations. In fact, the significance of increase in neutrality[3] of comments further confirms the above consideration. The PSTs after the intervention did not jump easily to conclusions and did not make as many critical comments. The increase in neutral comments at the expense of both negative and positive ones (the change was not statistically significant) may point to the PSTs' attempt to avoid evaluation and be impartial.

In Simpson et al. (2017), two patterns of results across relevant studies were noted:

> In the studies by Sherin and van Es (2009) and van Es and Sherin (2010), there is a very direct movement towards increased interpretation with roughly balanced decreases in description and evaluation. However, Mitchell and Marin (2015), Blomberg et al. (2014), and our [video-based intervention] all show increases in description, generally at the expense of evaluation.

These differences may be attributed to several factors, some of which concern the methodology of research. First, some researchers may have a different threshold for coding a comment as interpretation. However, in the presented study, there was no significant change even if the codes of Explanation and Interpretation are taken together. Second, Sherin and van Es (2009, 2010) used a group measure in their studies. The same applies for Roth McDuffie et al. (2014), who report that by the end of their course, PSTs regularly analysed and interpreted what they attended to. The data in their study were transcriptions of group discussions throughout the course. The group discussions may be deeper in that their participants picked up on one another's ideas and developed the interpretation further. For instance, if one person provides an interpretation, even the ones who would have not been able to come up with it themselves as individuals may grasp it. Such a discussion will likely produce more cases of interpretation.

---

[3]This is in contrast with studies that report a movement from positive comments to more critical ones and to proposing alternatives (e.g., Santagata et al. 2007; Santagata and Angelici 2010).

Third, if we look at another study that reports a significant change from no interpretation to interpretation supported by evidence and offering pedagogical alternatives (Alsawaie and Alghazo 2010), we can see that, unlike in the presented study, their task in the pre- and post-tests directly called for such knowledge-based reasoning ('Highlight and critique important events in the lesson. If you were the teacher, how would you handle things differently? If you were a student in this class, would you be able to learn what was taught in the lesson well?', p. 229). Thus, seeing more interpretation in their case might be at least in part due to the task itself.

Next, Simpson et al. (2017) suggest that one possible reason for an increased interpretation might lie in the way the video-based intervention was organised and in the tasks used. Indeed, studies comparing two different forms of video-based interventions, such as Santagata and Angelici (2010) and Blomberg et al. (2014), showed that there were important differences in knowledge-based reasoning between the participants of the two interventions. Thus, the tasks PSTs undertook in our intervention might not have motivated them to use theory, so they did not feel they needed to do so in the post-task. The same may help account for the increased interpretation reported in Mitchell and Marin (2015). Their intervention was similar to the presented study in that it aimed at salient features of mathematical instruction, but they used a very specific framework (MQI), which provided PSTs with guidance in the analysis of lessons during the intervention and which they could use when doing the post-test. Perhaps a more detailed look at the intervention tasks would make this conclusion more secure.

Interestingly, Blomberg et al. (2014) found that only the group of PSTs who took the situated strategy course were able to maintain a focus on engaging consistently in higher-level categories of evaluating, which in their case included detailed explanations, and integrating. This would mean that the intervention in the presented study did not sufficiently implement the situated strategy. However, there was also an important difference between our study and Blomberg et al. (2014). Our data consisted of a written analysis of one lesson at one point of time, while Blomberg and colleagues coded a learning journal that the participants were to write during the course. After each session, they were asked to write what they learned and while doing so, they were guided by eight questions (e.g., 'Provide examples… that confirm and/or contradict what you learned today.', p. 451). This may motivate the PSTs towards deeper reflection, adding support to the effects of the intervention itself. An implication may be to also include learning journals in video-based interventions. This does not explain, however, why the cognitive group in Blomberg et al. (2014) did not do so well as the situated learning group.

Note that in our previous research it has been found that the two-year master's programme that included two 4-week school practice placements did not lead to any changes in the pattern of attention (Simpson et al. 2017). Stockero (2008) showed that a video-based intervention followed immediately by field experience leads to significant gains in higher level reflection. The same applies to research by Santagata and Guarino (2011). An implication for teacher education would be to couple the video-based intervention with school practice placement.

For Sherin and van Es' research (2008, 2009, 2010), there is another possible reason for increased interpretation. Their participants were experienced teachers and, moreover, they learned from reflecting on the videos of their own teaching.[4]

To sum up, the question is whether it is reasonable to expect that PSTs at the beginning of their master's studies with very limited or no teaching experience are able to make high-level professional reasoning. In Simpson et al. (2017), we speculate that 'teacher education programs might need two distinct phases to develop noticing: the first concentrated on shifting attention and second on theorizing'. Thus, there is a question of whether we would get greater gains in knowledge-based reasoning if the video-based intervention was organised later in the master's programme when PSTs have more knowledge of mathematics education concepts.

The study presented above has its limitations. First, a one-to-one correspondence is presumed between what is written down and what is actually noticed. The PSTs were able to notice an event but for whatever reason chose not to record it. Second, with different measures, we might have reached different results, for example, if we used a group measure or if we asked more targeted questions in the pre- and post-tasks. Third, even though the study provided more insight into the nature of PSTs' knowledge-based reasoning, its quality was not investigated. For example, further exploration could be undertaken of whether the explanations or alternatives proposed by the PSTs are plausible and coincide with an expert's (e.g., experienced teachers or educators) view. The question also arises of what the results for interpretation would be if no theory was presented in the course prior to the intervention or if the intervention was organised later in the two-year master's programme.

Acknowledgements The research reported in this paper has been financially supported by project PROGRES Q17, "Teacher Preparation and the Teaching Profession in the Context of Science and Research".

References

Alsawaie, O., & Alghazo, I. (2010). The effect of video-based approach on prospective teachers' ability to analyze mathematics teaching. *Journal of Mathematics Teacher Education, 13*(3), 223–241.

Blomberg, G., Sherin, M. G., Renkl, A., Glogger, I., & Seidel, T. (2014). Understanding video as a tool for teacher education: Investigating instructional strategies to promote reflection. *Instructional Science, 42*(3), 443–463.

Brophy, J. (2004). *Using video in teacher education*. San Diego, CA: Elsevier.

Goodwin, C. (1994). Professional vision. *American Anthropologist, 96*(3), 606–633.

Hiebert, J., Gallimore, R., Garnier, H., Givvin, K. B., Hollingsworth, H., Jacobs, J., et al. (2003). Understanding and improving mathematics teaching: Highlights from the TIMSS 1999 video study. *Phi Delta Kappan, 84*(10), 768–775.

---

[4]The question is whether reflecting on one's own teaching might help PSTs notice more. This is something that I am exploring in a different study at the moment.

Hiebert, J., & Grouws, D. A. (2007). The effects of classroom mathematics teaching on students' learning. In F. K. Lester (Ed.), *Second handbook of research on mathematics teaching and learning* (pp. 371–404). Charlotte, NC, USA: Information Age Publishing.

Janson, H., & Olsson, U. (2001). A measure of agreement for interval or nominal multivariate observations. *Educational and Psychological Measurement, 61*(2), 277–289.

Kilpatrick, J., Swafford, J., & Findell, B. (Eds.) (2001). *Adding it up: Helping children learn mathematics*. Washington, DC, USA: National Research Council.

Manouchehri, A. (2002). Developing teaching knowledge through peer discourse. *Teaching and Teacher Education, 18*, 715–737.

Mitchell, R. N., & Marin, K. A. (2015). Examining the use of a structured analysis framework to support prospective teacher noticing. *Journal of Mathematics Teacher Education, 18*(6), 551–575.

Roth McDuffie, A., Foote, M. Q., Bolson, C., Turner, E. E., Aguirre, J. M., Bartell, et al. (2014). Using video analysis to support prospective K–8 teachers' noticing of students' multiple mathematical knowledge bases. *Journal of Mathematics Teacher Education, 17*(3), 245–270.

Santagata, R., & Angelici, G. (2010). Studying the impact of the lesson analysis framework on preservice teachers' abilities to reflect on videos of classroom teaching. *Journal of Teacher Education, 61*(4), 339–349.

Santagata, R., & Guarino, J. (2011). Using video to teach future teachers to learn from teaching. *ZDM Mathematics Education, 43*(1), 133–145.

Santagata, R., & Yeh, C. (2014). Learning to teach mathematics and to analyze teaching effectiveness: Evidence from a video- and practice-based approach. *Journal of Mathematics Teacher Education, 17*(6), 491–514.

Santagata, R., Zannoni, C., & Stigler, J. W. (2007). The role of lesson analysis in pre-service teacher education: An empirical investigation of teacher learning from a virtual video-based field experience. *Journal of Mathematics Teacher Education, 10*(2), 123–140.

Schoenfeld, A. H. (2011). Noticing matters. A lot. Now what? In M. G. Sherin, V. R. Jacobs, & R. A. Philipp (Eds.), *Mathematics teacher noticing: Seeing through teachers' eyes* (pp. 223–238). London: Taylor & Francis.

Seago, N. (2004). Using video as an object of inquiry for mathematics teaching and learning. In J. Brophy (Ed.), *Using video in teacher education* (pp. 259–286). San Diego, CA: Elsevier.

Sherin, B., & Star, J. R. (2011). Reflections on the study of teacher noticing. In M. G. Sherin, V. R. Jacobs, & R. A. Philipp (Eds.), *Mathematics teacher noticing: Seeing through teachers' eyes* (pp. 66–78). London: Taylor & Francis.

Sherin, M. G., Linsenmeier, K. A., & van Es, E. A. (2009). Selecting video clips to promote mathematics teachers' discussion of student thinking. *Journal of Teacher Education, 60*(3), 213–230. https://doi.org/10.1177/0022487109933696.

Sherin, M. G., Russ, R. S., & Colestock, A. A. (2011). Assessing mathematics teachers' in-the-moment noticing. In M. G. Sherin, V. R. Jacobs, & R. A. Philipp (Eds.), *Mathematics teacher noticing: Seeing through teachers' eyes* (pp. 79–94). London: Taylor & Francis.

Sherin, M. G., & van Es, E. A. (2009). Effects of video club participation on teachers' professional vision. *Journal of Teacher Education, 60*(1), 20–37.

Simpson, A., Vondrová, N., & Žalská, J. (2017). Sources of shifts in pre-service teachers' patterns of attention: The roles of teaching experience and of observational experience. *Journal of Mathematics Teacher Education*. Advance online publication. https://doi.org/10.1007/s10857-017-9370-6.

Star, J. R., Lynch, K., & Perova, N. (2011). Using video to improve preservice mathematics teachers' abilities to attend to classroom features: A replication study. In M. G. Sherin, V. R. Jacobs, & R. A. Philipp (Eds.), *Mathematics teacher noticing: Seeing through teachers' eyes* (pp. 117–133). London: Taylor & Francis.

Star, J., & Strickland, S. (2008). Learning to observe: Using video to improve preservice mathematics teachers' ability to notice. *Journal of Mathematics Teacher Education, 11*(2), 107–125.

Stockero, S. L. (2008). Using a video-based curriculum to develop a reflective stance in prospective mathematics teachers. *Journal of Mathematics Teacher Education, 11*(5), 373–394.

van Es, E. A., & Sherin, M. G. (2002). Learning to notice: Scaffolding new teachers' interpretations of classroom interactions. *Journal of Technology and Teacher Education, 10*(4), 571–596.

van Es, E. A., & Sherin, M. G. (2008). Mathematics teachers' "learning to notice" in the context of a video club. *Teaching & Teacher Education, 24*(2), 244–276.

van Es, E., & Sherin, M. G. (2010). The influence of video clubs on teachers' thinking and practice. *Journal of Mathematics Teacher Education, 13*(2), 155–176.

Whitcomb, J. A. (2003). Learning and pedagogy in initial teacher preparation. In I. B. Weiner, (Ed.), *Handbook of psychology* (Vol. 7, pp. 533–556). Hoboken, NJ: Wiley.

Yeh, C., & Santagata, R. (2014). Preservice teachers' learning to generate evidence-based hypotheses about the impact of mathematics teaching on learning. *Journal of Teacher Education, 66*(1), 21–34.

Chapter 40
Popularization of Probability Theory and Statistics in School Through Intellectual Competitions

Ivan R. Vysotskiy

Abstract Since 2004, in accordance with the Federal Educational Standards, probability theory and statistics has been included into teaching practice in Russian schools. This paper focuses on one form of this work: organization of intellectual competitions on probability theory and statistics for school students. Since 2008, the Moscow Center for Continuous Mathematical Education has conducted the Internet Olympiad for students in school years 6–11. In addition to the traditional problems, participants are offered a choice to write an essay on a proposed topic. This article attempts to classify those topics and highlight the most popular ones among the students. In addition, this paper makes a short overview of selected problems that from the organizers' point of view represent promising and prospective trends in the teaching of probability and statistics at school. The article is addressed to education specialists, teachers, and researchers who specialize in probability theory and statistics.

Keywords Probability · School math education · Olympiad on probability
Math intellectual competition · Moscow center for continuous mathematical education

40.1 Introduction

In 2004, elements of probability theory and statistics were introduced into the school mathematics curricula in Russia in accordance with the federal educational standards. Since 2012, problems on probability and data representation have been included in the Unified State Exam in mathematics. The Concept of Development of Mathematical Education in Russia was approved by the government in December 2013. It states that probability theory and statistics are important sections of school mathematics. In 2015, the Federal Exemplary Curricula were developed,

I. R. Vysotskiy (✉)
Moscow Center for Continuous Mathematical Education, Moscow, Russia
e-mail: i_r_vysotsky@hotmail.com

© The Author(s) 2018
G. Kaiser et al. (eds.), *Invited Lectures from the 13th International Congress on Mathematical Education*, ICME-13 Monographs,
https://doi.org/10.1007/978-3-319-72170-5_40

where probability and statistics appear as complete sections that determine content for each educational level.

Meanwhile, educational practices have caused serious difficulties that have been inescapable when forming a new school subject that is completely different from the traditional courses that are being taught in universities. Difficulties in the preparation of teachers have followed from these difficulties.

This paper focuses on only one of the popularization dimensions aimed at the formation of public inquiry into the field of mathematics: the methodology and practice of intellectual competitions for school children on probability theory and statistics. The example considered was the Olympiad that has been held by the Moscow Center for Continuous Mathematical Education (MCCME) since 2008.

The Olympiad Organizing Committee is ready to cooperate with colleagues from all countries. In particular, we can provide English versions of all Olympiad materials.

40.2   Background

We often hear from mathematicians that because probability theory and statistics are too difficult, they should not be taught in school. This opinion has grown out of complications that follow studying probability theory in universities, which traditionally is deductive and based on combinatorics and wide knowledge of calculus. A combinatorial approach to probability theory is typical for many generations and has grown from the Soviet period, when probability theory was torn out of statistics upon being announced as "a social science." In fact, combinatorics is not directly related to basic ideas of statistics and probability.[1] This is just a way to enumerate elements of vast probabilistic spaces and prove theorems. Experience and intuition are primary, and no combinatorial tricks are useful without them. One should meaningfully consider chances of events, especially unlikely events that play a significant role in daily life.

The problem is that events are less obvious than figures or numbers, while concepts of chance and volatility are not as intuitive as length, area, or volume. An event and its chances make special types of abstract objects and their formalization into mathematical notions is much more complex than a formalization of a picture (transition to geometry) or a quantity (to arithmetic or algebra).

The second problem is that for the majority of children, the concept of volatility and instability of events is alien to them until a certain age. At what age a child is ready to perceive changeable models and determine which models they should be has yet to be discovered. In the Soviet period, the science of the laws of cognitive

---

[1] In talking about the probability theory in school, we often omit "theory" to simplify the text and follow the tradition that has been formed in Russian educational terminology.

activity was destroyed (Petrovsky 1991). Its place was taken by the paradigm of "a clear sheet of paper."

However, if in early childhood the rejection of variability possibly serves as a defensive mechanism that simplifies adaptation to social and natural conditions and minimizes the number of necessary rules of behavior, then later an absence of general ideas of randomness and volatility starts to hinder efforts to understand abstract concepts of probability and statistics. Study of statistics in adulthood does not improve this situation (Kakihana and Watanabe 2013).

The first thing to take care of is to make basic concepts of probability and statistics clear and familiar to math teachers who have difficulty in the transition from teaching abstract mathematical facts to applying mathematical concepts and laws to the solution of practical problems.

In addition, one of the most important aspects of education is the popularization of knowledge. If algebra, geometry, and other sections of traditional school mathematics show no shortage of additional popular scientific and popular literature for adults and children of all ages, an analysis of the situation in the area of probability theory has shown a distinct lack of a popular literature and other forms of popularization (Bunimovich et al. 2009a, b). A number of popular books for school children were published in the Soviet Union, mainly in the 1950s through the 1980s (see, for example, Kolmogorov et al. 1982; Mosteller 1985). The number of new materials appearing in Russia is vanishingly small, even if we take into account translations of foreign publications on probability and statistics for students (e.g., Chjun and Ait-Sahlia 2007). At the same time, the amount of popularization literature on statistics and probability theory in the world has been increasing. This is partly due to the increasing importance of teaching probability and statistic at school and partly due to the growth of the role of stochastic methods in different sectors of the global economy (Bunimovich et al. 2009a). Popularization activities should be carried out in various forms. In addition to special literature and mathematics circles, Olympiads and other intellectual competitions of different levels for children and adults are of great value. Recent years have shown a spontaneous increase in number and quality of Internet sites dedicated to popular mathematics, in particular in the field of probability theory and statistics.

40.3 The Olympiad, Its Main Principles and Description

In 2008, the first Olympiad was administered by MCCME on the initiative of Yuri Tyurin, Alexei Makarov, and Ivan Yaschenko. In following years, many mathematicians and educators (E. Bunimovich, V. Bulychev, P. Semenov, et al.) participated in the selection and preparation of Olympiad problems. The rules of the Olympiad and its requirements for participation are simple. The Olympiad is open to everyone and is held during a calendar month. Olympiad problems are designed for students in Grades 6–11.

The Olympiad has two rounds. The first round, which is invitational, happens in schools. In 2015, the number of participants was 2450 and in 2016 it was 2890. In 2016, an intramural final round was added that consisted of two contests: first, a competition involving a statistical experiment and, second, problem solving.

The problems in the Olympiad are designed to be recommended to students starting from a certain grade level. For example, a task that can be solved using only intuitive ideas, finite enumerations, and classical probability definitions is recommended for students in Grade 6 and older. If a solution requires the use of simple transformations within the algebra of events, the task is recommended for students in Grade 7 and older. A problem involving the characteristics of random variables will be offered to students in Grade 8 and older. There is no upper limit to the 'age' of a problem Age differentiation occurs at the stage of grading solutions and awarding the winners.

Olympiad problems are placed on the website for free access for about a month. According to the rules of the Olympiad, participants can use any help, reference books, etc. Only "openly hiring adult labor" is considered non-sporting and is discouraged.

Materials from previous years are published on the Olympiad website in the Archive section at http://ptlab.mccme.ru/olympiad.

Olympiads from 2008 to 2017 have been published. Articles dedicated to this competition have been published in (Vysotskiy 2012; Vysotskiy et al. 2009).

40.4   Essay Tasks

A distinctive feature of the Olympiad is that it also includes essay tasks in addition to problems. The essays are evaluated separately, regardless of the age. The essays turned out to be important and most attractive part of the Olympiad.

Participants are required to analyze the offered situation and write a short essay on a given topic. These essays immerse students in uncertain situations that require imagination and activities involving estimation that take into account real limitations and the nature of the data. Actions in an uncertain situation play a crucial role in the formation of the statistical and mathematical culture of students, because instead of performing the steps of a known or studied algorithm, Olympiad participants become researchers who plan experiments themselves. Students have to determine important and unimportant factors of random experiments, interpret the results, invent a method to describe the data, and form a hypothesis. None of the proposed essays requires students to check the formulated hypothesis, as the mathematical tools available to school students are clearly insufficient for that.

In some situations, the creation of a hypothesis itself is a very complicated task that requires a high culture of regulatory activity from a student. Moreover, situations often arise where the number of possible plausible hypotheses is large. In such cases, the authors try to repose the situation in such way that either it contains a hypothesis formulated explicitly or the task implicitly directs the students' actions (see two examples below). The most important part of the essay task for a student is to search for data with the aim of finding a pattern or checking the data's correspondence with certain assumptions. Some tasks require students to collect data independently through short surveys. The others demand an independent search on the Internet.

Some essay tasks are designed to generate critical scientific thinking in students, immersing students in situations where the arguments provided deliberately contain an inaccuracy of some kind, an error, or an unreasonable conclusion. Students are invited to understand the shortcomings of the study and make research steps. In my experience, tasks of this type appeared to be the most attractive to the participants. Among the submitted essays there have been very bright and original works, which we placed on the Olympiad website in the Solutions section.

Below we provide several essay tasks from previous years.

All essays can be classified by their educational objectives and the methods used to perform them:

1. *Checking a statement* using either independently collected data or raw data provided in the essay task. In this case, the essay usually offers to an Olympiad participant a partially proposed algorithm of actions.
2. A *study that may require a survey* of the student's classmates, friends, and parents followed by data processing with the formulation of a plausible hypothesis or the refutation of an implausible one. A student's ability to organize collected data, present it in the most appropriate way, and produce hypotheses are important here.
3. A *search for a statistical method* for solving a problem (see Essays 1 and 2). This is aimed at developing constructive thinking and skills. An unbounded search appears very attractive to students. An essay about the estimation of a number of people in a crowd was the most frequently selected among participants of all age groups in 2013. None used a method that the authors of the problem considered the most natural; instead we received many original ideas.
4. A *search of an error* in a complex and extensive reasoning presented in the formulation of the task (see, e.g., Essay 6). Solving tasks of this type helps to develop a critically destructive way of thinking, which is an integral part of intellectual culture. Speaking specifically on the implementation of an essay about the relationship between air humidity and levels of snowfall, we note that many students succeeded in noticing flaws in the author's arguments, but the main problem—the unsuitability of the linear regression for the case—was not mentioned by any of them.

This shows that the basic ideas of statistics stay on the border between intuitive and conscious.

Examples of essay tasks

1. **The number of people in a crowd (2012).** The photo on Fig. 40.1 shows a crowd of people. How one can estimate (approximately calculate) the number of

Fig. 40.1 The number of people in a crowd

people in this crowd? Try to develop an appropriate method, use it and make such estimation. Describe your method in detail, explain why it works correctly and how to use it, and what you got using it. We are looking forward to your answer: How many people are there in the photo?

2. **Which way is faster (2014)**? Is it true that an aircraft takes a different amount of time when it flies east and west? Is it always so? Go to the site of a large airport or a major airline; it is even better if you consider several airlines or airports. Select flights from east to west and vice versa. Collect and process the necessary information. How different are the durations of flights one way and the other? Does it depend on the distance? You need to come up with a statistical measure that describes the difference. Is it stable? The difficulty is that a mere difference between the time there and the time back does not give us much: We have to take into account not only very long flights but also relatively short ones. If such a difference does exist, what causes it? Is it possible to estimate consistency and power of this amazing factor? Are east and west really guilty? Maybe a similar pattern can be observed with other flights, for example, from north to south and back? Many questions arise. Try to locate, describe, and analyze data and use your imagination.

3. **Haga's problem (2011)**. Professor Kazuo Haga from the University of Tsukuba is the inventor of origamics (geometrical origami). Once he posed an interesting question: A paper square is divided into light and dark parts by four semicircles (Fig. 40.2, left), giving a graceful ornament that resembles a flower.
   If we choose a point in a light part and then make folds so that all vertices meet this point, we get a pentagon (Fig. 40.2, right). So the union of light parts is a "domain

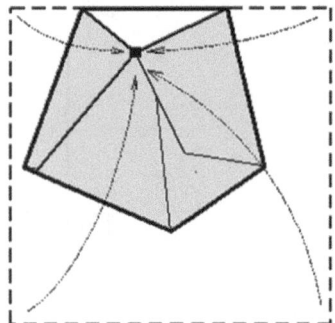

Fig. 40.2 Haga's problem

of pentagons" or simply a "5-domain." The dark parts make a "6-domain." The four vertices and the center of the square give quadrangles. Professor Haga writes: "I noticed that when asked to choose a random point, the majority of people mark a point leading to a hexagon. Pentagons are much rarer. Very few choose points that make quadrangles. The question is: Is the number of those who choose a point in a certain domain proportional to the domain's area?"

If sides of the square measure 1 unit, then each semicircle on the left figure has radius of 0.5 and its area is $\pi/8$. Therefore, the area of the petals (6-domain) equals $\pi/2 - 1 \approx 0.57$, while all the rest (5-domain) has the area of about 0.43. The difference is not too big. The ratio of those who choose points in the corresponding domains does not fit the ratio of the areas. What is the reason for this: Why are points outside the petals less attractive than points inside?

Conduct an experiment. You will need some a number of paper squares. Ask as many people as possible to mark a point on the clear square. Putting all points together on a new square gives the distribution of the points. Maybe some properties of the distribution will help to explain why "hexagon admirers" appear more often than Professor Haga could anticipate, having compared the domain areas.

4. **Height correlation (2009).** A teacher once decided to show her students that heights of boys and girls are independent random values. In order to do this, she did some research. In every class, she chose 10 girls and 10 boys at random, then composed random boy/girl pairs and wrote down their heights as x_k and y_k for every pair. When she was done and had put all the results on the scatter chart (Fig. 40.3), she found to her horror that all the points were grouped around a slanted line. This meant that there was an obvious correlation between heights of boys and heights of girls! How could this be?

   Write a short essay in which you try to explain whether the teacher made a mistake in her findings and, if so, what her misjudgment likely was.

5. **Insurance (2010).** The insurable value of a car depends on its age. Agents of insurance company ABC estimate the depreciation in a very simple way:

Fig. 40.3 Height correlation

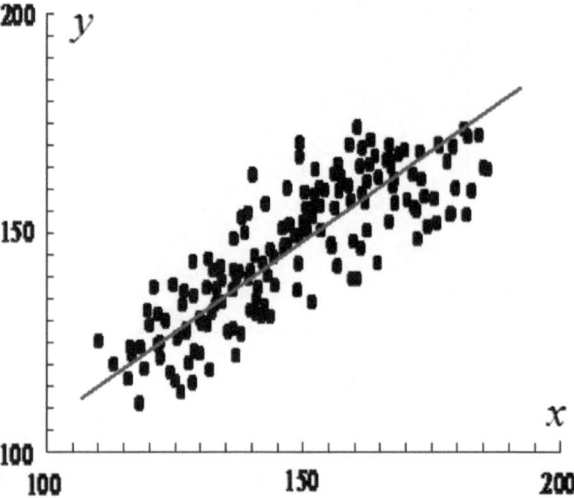

Cars older than two years old lose 10% of their value yearly. Using http://www.auto.ru and other available sources, conduct research on the topic of whether the price policy of the company ABC corresponds to actual practice in the used car market. In your research, take into account that among the cars there are some vehicles whose values do not meet the average for the cars of the same model and age.

40.5 Problems That Require a Solution

In addition to three essays, the Olympiad traditionally includes 16 problems in statistics and probability theory. Some problems are easy and admit a very simple solution by brute force or short reasoning. More complex problems require, in addition to looking for the key to the problem, students to have the ability to perform operations on events and some knowledge of probability properties. Finally, there are some very complex problems that surrender only to those who devote enough time to thought and attempts and who have studied the literature and problems from previous years. Authors deliberately include in the Olympiad special problems whose solution requires complex transformations. Such problems are few, but they must be in an online competition, as participation in this type of problem implies that the participant has done scientific research. A few examples are listed below. We wanted to give a sample of problems that show a wide range of methods and topics used in probability. The difficulty of each problem is indicated by its number of points and recommended grade.

Examples of problems

1. **Defective coins (sixth grade and older, 1 point).** For the anniversary of the Saint Petersburg mathematical Olympiad, the mint produced three commemorative coins. One coin was made correctly, the second coin had tails on both sides, and the third coin had heads on both sides. The Director of the Mint chose one of these coins without looking, tossed it, and it came up tails. Find the probability that the second side of this coin also is a tail.

2. **Three targets.** A shooter fires on three targets as many times as needed to hit all three. The probability to hit for one shot is p.

   (a) **(Seventh grade and older, 2 points).** Find the probability that the shooter will fire exactly five times.

   (b) **(Eighth grade and older, 2 points).** Find the expected value for the number of shots.

3. **Intersecting diagonals (ninth grade and older, 3 points).** In a convex polygon with an odd number of vertices equal to $2n+1$, two random diagonals are chosen independently. Find the probability that these diagonals intersect inside the polygon.

4. **Draws (ninth grade and older, 6 points).** Two hockey teams of equal strength have agreed that they will play until the total score reaches 10. Find the expected value for the number of times a draw will happen.

5. **Stunning news.** A conference is attended by 18 scientists, of whom 10 know some stunning news. During a coffee break, all scientists are randomly divided into pairs and in each pair a scientist who knows the news tells it to the other if the other did not know it yet.

   (a) **(Ninth grade and older, 1 points).** Find the probability that after the coffee break the number of scientists who know the news will be equal to 13.

   (b) **(10th grade and older, 4 points).** Find the probability that after the coffee break the numbers of scientists who know the news will be equal to 14.

   (c) **(Ninth grade and older, 3 points).** Denote by X the number of scientists who will know stunning news after the coffee break. Find the expected value of X.

6. **Mini-Tetris.** A tall rectangle of width 2 is open at the top, and randomly oriented L-shaped triminos fall into it. (A trimino is a piece that looks like a domino piece but consists of 3 squares. There are two types of trimino, straight and L-shaped.)

   (a) **(Ninth grade and older, 3 points).** Let k triminos fall into the rectangle. Find the expected value of the height of the resulting polygon.

   (b) **(10th grade and older, 6 points).** Let 7 triminos fall. Find the probability that the resulting figure will have a height of 12.

40.6   The Statistical Experiment Contest

In 2016, the Olympiad had three rounds for the first time. In the final (intramural) round, participants were offered a topic where they were to develop a statistical experiment. Organizers took a well-known scheme of taste tests as a base for it. All participants asked to design an experiment aimed at revealing the threshold of sensitivity to sweetness using a weak aqueous solution of sugar. All participants were asked to remember the following.

1. Hygiene. It is unacceptable to have more than one person drink water from the same cup.
2. Various effects that distort the result are possible. For example, a person may be less sensitive to a less saturated sugar solution after a stronger one. How can this effect be reduced?
3. There may be many tiny factors that affect taste. Should we regard them all as non-significant? Should some of them be taken into account?
4. How to process the collected data. We do not assume a deeply scientific approach, but we hope that a proposed procedure will be convincing.

Participants presented several plans for such an experiment. After a discussion, the best-justified plan was implemented. (On Fig. 40.4, left: the course of the experiment; right: the best plan author Amelia, fifth grade,[2] Republic of Bashkortostan.)

During the discussion about this form of work, the organization committee came to the following conclusions:

1. This experiment may become the most dynamic and exciting part of the Olympiad.
2. The topic of the experiment should be chosen carefully and designed so that learners should be able to replicate the experiment in the existing conditions within the announced time.
3. The form of the experiment should be chosen in such a way that all participants can take part together with their parents, accompanying persons, etc., regardless of age.
4. The experiment should be planned so that all participants will be busy at all times or will at least have a chance to busy themselves by performing tests, collecting statistics, processing collected data, etc. Optimally the experiment should be organized so that work is done in groups with different duties for all members in a group.

---

[2]Not a mistake. We have announced the Olympiad for 6–11 grades but if a fifth grader wants to take part—why not? She is a really smart girl.

Fig. 40.4 Left: the course of the experiment; right: the best plan author Amelia, fifth grade

40.7 Conclusion

The experience of the Olympiad on probability theory shows that despite the fact that this branch of mathematics has traditionally been considered difficult and unusual for school learning, interest in it is gradually increasing.

In 2015, the Olympiad school tour was held for the first time and was attended by 2465 students from 20 regions of Russia, and in the 2016 tour, 2859 students from 41 regions participated, as the Olympiad was joined by leading schools in many regions where such a competition had not previously been known.

Growth has also been indicated by the increasing number of requests from teachers and students to provide methodological support for teaching and learning. During the 2014–2015 study year, the methodological site for teaching of probability and statistics (http://ptlab.mccme.ru) received 876 queries. During the 2015–2016 school year (up to May 13) the number of such inquiries was 1764. The proportion of queries related to the Olympiad only (applications, rules, results, appeals, etc.) was 20% (downloads). These statistics show that the Olympiad plays an important role in the popularization of probability theory and statistics as a school subject.

Of course, the Olympiad should not be the only means of promoting this subject. In addition, there must be math circles (out-of-class activities), numerous publications on probability and statistics in teachers' and popular magazines (such as *Mathematics*, *Mathematics in School*, *Kvant*, and *Kvantik*). At the same time, the number and variety of probabilistic problems on the national exam for primary and high school courses has increased.

Unlike the problems from the regular school course, the Olympiad tasks are much more diverse in subject matter and level of difficulty. Taking advantage of this, the Olympiad developers are gradually expanding the range of tasks and inventing new forms, some of which later will be included in school courses and methods.

References

Bunimovich, E. A., Bulychev, V. A., Vysotskiy, I. R., et al. (2009a). About probability theory and statistics in the school course. *Mathematics in School, 7,* 3–13.

Bunimovich, E. A., Vysotskiy, I. R., et al. (2009b). Terminology, notations and agreements in a school course of probability theory and statistics. *Mathematics, 17,* 13–27.

Chjun, K. L., & Ait-Sahlia, F. (2007). *An elementary course of probability theory. Stochastic processes and financial mathematics.* Moscow: BINOM, Laboratoriya Znanii.

Kakihana, K., & Watanabe, S. (2013). Statistics education for lifelong learning. In *Proceedings of 6th East Asia Regional Conference on Mathematics Education (EARCOME6)* (Vol. 3, pp. 318–322), March 17–22, 2013, Phuket, Thailand.

Kolmogorov, A. N., Zhurbenko, I. G., & Prokhorov, A. V. (1982). *Introduction to probability theory.* Moscow: Nauka.

Mosteller F. (1985). *Fifty challenging problems in probability with solutions* (3rd ed.). Moscow: Nauka, FIZMATLIT.

Petrovsky, A. V. (1991). Ban on a comprehensive study of childhood. In *Repressed science* (pp. 126–135). Leningrad: Nauka.

Vysotskiy, I. R. (2012). Olympiad problems on PT. In *Mathematics* (January issue, pp. 61–62). Pervoe Sentyabrya Publishers.

Vysotskiy, I. R., Borodkina V. V., et al. (2009). Online olympiad on probability theory and statistics. In *Mathematics* (Vol. 15, pp. 17–25). Pervoe Sentyabrya Publishers.

Chapter 41
Noticing in Pre-service Teacher Education: Research Lessons as a Context for Reflection on Learners' Mathematical Reasoning and Sense-Making

Helena Wessels

Abstract Professional noticing of learners' mathematics reasoning is a crucial ingredient of a mathematics teacher's set of teaching competencies. Research lessons in the lesson study process, with its focus on learner reasoning, provide a structured environment for the building of mathematical knowledge as well as for reflection and the development of teacher professional noticing and sense-making. This paper reports on the depth and growth in noticing of three pre-service teachers during research lessons in their third and fourth years, using the Van Es noticing framework. The study showed that two of these teachers' noticing shifted to higher levels over the two years, with greater focus on learners' mathematical reasoning and sense-making than on teacher actions and teaching. Prospective teachers need well-structured and focused opportunities, individually as well as in groups, to learn to notice and make sense of learner thinking and reasoning.

Keywords Professional noticing · Lesson study · Research lessons
Mathematical reasoning · Reflective practice

41.1 Introduction

The preparation of pre-service teachers to teach mathematics for understanding is an important focus in mathematics education, and the development of reflective practice is one of the cornerstones in this process. Teaching for understanding goes hand in hand with a problem-centred approach to the teaching of mathematics as described in the work of Murray et al. (1998), Hiebert et al. (1997) and Stein et al. (2008). The problem-centred approach is based on the premise that learners build their own conceptual knowledge and mathematical understanding through solving

H. Wessels (✉)
Research Unit for Mathematics Education,
Stellenbosch University, Stellenbosch, South Africa
e-mail: hwessels@sun.ac.za

© The Author(s) 2018
G. Kaiser et al. (eds.), *Invited Lectures from the 13th International Congress on Mathematical Education*, ICME-13 Monographs,
https://doi.org/10.1007/978-3-319-72170-5_41

problems independently or through interaction with others. Preparing pre-service teachers to teach for understanding using a problem-centred approach is challenging as it is so different from the more traditional way of teaching that they have usually been exposed to during their own school years as well as during school practicums during their studies (Santagata 2011). Prospective teachers therefore need professional development and experience to scaffold their ability to consider children's mathematical responses through a different lens, interpret what they see, and to 'imagine themselves in the future acting responsively and freshly rather than habitually. The mark of improving research capacities for individuals lies in their being able to imagine themselves in the future acting (responding) more appropriately than before' (Mason 2011, p. 38). Research studies have shown that prospective teachers lack in this kind of noticing and lesson analysis skills (Barnhart and Van Es 2015; Santagata 2011).

Pre-service teachers also need repeated opportunities for reflection over time:

> [They] are not likely to develop such a complicated array of knowledge, skills, and dispositions simply by watching a video or demonstration lesson or reading standards or articles; they are likely to need repeated cycles of study, trial in the classroom, reflection, refinement, and trial again. (Takahashi et al. 2013, p. 243)

Designing teacher preparation programs that foster the appropriate knowledge, skills and dispositions for pre-service teachers has been an enduring challenge in mathematics education (Hiebert et al. 2003). At Stellenbosch University, lesson study has been implemented in their preparation program for prospective foundation phase teachers for the past five years to develop pre-service teachers' knowledge, skills and dispositions through reflective practice and to scaffold their awareness and interpretation of learners' mathematical reasoning, specifically during lessons (Paolucci and Wessels 2017). The focus of the research described in this paper is on determining the growth and depth of pre-service teachers' noticing of the mathematical reasoning of learners during research lessons over a two-year period in this preparation program.

41.2 Literature Review

41.2.1 Reflection and Professional Noticing

The linking of theory and practice is one of the crucial issues in teacher education and systematic, purposeful reflection is considered vital in this process (Dewey 1933; Korthagen et al. 2006).

Schön (1987) distinguishes between *reflection-in-action* and *reflection-on-action*. Reflection-in-action (in-the-moment decisions) entails teachers' ability during a lesson to reshape what they do while they are doing it, think on their feet and use knowledge and experience to interpret the situation and act on it. Thorough preparation and anticipating learner strategies and misconceptions provide points of reference for reflection-in-action and enable didactical flexibility. Reflection-on-action (looking back), on the other hand, entails the often-documented looking back after an experience and reflecting on what happened and why it happened. Both these notions play an important role in preparing pre-service teachers to bridge the gap between theory and practice and become reflective practitioners. However, pre-service teachers often do not know what to reflect on or how to reflect in lessons they observe: 'Without structured support and appropriate framing, pre-service teachers' analyses tend to focus on the actions and behaviors of the teacher rather than student thinking, learning and sense-making and tend to be judgmental and lack evidential support and coherence' (Barnhardt and Van Es 2015, p. 84).

Noticing can be described as one of the core practices in mathematics teacher education (Choy 2016; Grossman et al. 2009) and can provide appropriate support and framing for the development of productive reflection on learners' mathematical reasoning during lessons. A growing number of researchers regard teacher noticing as a useful framework for promoting adaptive and responsive teaching and for in-service and prospective teachers to learn from their teaching by focusing on learner mathematical reasoning (Sherin et al. 2011a). Galbraith (2015) describes noticing as 'an essential ability of a perceptive and effective mathematics teacher' (p. 151). It entails attending to student mathematical reasoning and making sense of this information to inform teaching decisions and teacher moves. Noticing is referred to in different ways in the literature: a discipline and an intentional systematic set of practices (Mason 2002), the sizing up of students' ideas and responding (Ball et al. 2001), a set of skills (Jacobs et al. 2010), two processes (Sherin et al. 2011a) and a goal-oriented decision-making process (Schoenfeld 2011). Noticing progresses through three interrelated phases: attending to learner mathematical reasoning in classroom interactions, the interpretation of learner reasoning in this setting and deciding what actions should be taken based on inferences from this analysis (Barnhart and Van Es 2015; Jacobs et al. 2010; Van Es 2011). The three phases can entail reflection-in-action of the teacher during the lesson as well as reflection-on-action during the post-lesson discussion by observers and the teacher who taught the lesson (Schön 1987). The three phases are intertwined: 'three component skills, but also an integrated set' (Jacobs et al. 2010, pp. 173, 174).

Noticing can be influenced by several factors, including mathematical knowledge for teaching (Kazemi et al. 2011; Schoenfeld 2011; Seidel and Stürmer 2014), beliefs (Erickson 2011; Goldsmith et al. 2014; Schoenfeld 2011), prior experience (Erickson 2011), context (Mitchell and Marin 2015; Seidel and Stürmer 2014) and pedagogical commitment of the teacher (Erickson 2011). It is therefore important that noticing be regarded 'within the wider context of the teachers' growing knowledge (resources), goals, and orientations' (Schoenfeld 2011, p. 237).

41.2.2 Frameworks for Analysing Professional Noticing

With the increase in research studies on teacher noticing, different frameworks have been developed to analyse and assess teachers' noticing. Examples are the Mathematical Quality of Instruction (MQI) analysis framework of Hill et al. (2008) as used by Mitchell and Marin (2015) to support noticing, the Van Es (2011) framework for learning to notice, the Santagata (2011) lesson analysis framework and the Choy (2015) FOCUS framework for task development. Other frameworks are focused on specific mathematical content, for example, the Fernández et al. (2013) framework for the development of teachers' noticing of students' mathematical thinking in the context of proportionality.

The Van Es (2011) framework was chosen for the research study in this paper as its focus on the content and levels of noticing and the trajectory for the development of noticing was useful to analyse growth and depth in pre-service teachers' reflections over time. The Van Es framework (Table 41.1) first focuses on two

Table 41.1 Framework for learning to notice student mathematical thinking (Van Es 2011, p. 139)

	What teachers notice	How teachers notice
Level 1 Baseline	Attend to whole class environment, behaviour, learning and teacher pedagogy	Form general impressions of what occurred
		Provide descriptive and evaluative comments
		Provide little or no evidence to support analysis
Level 2 Mixed	Primarily attend to teacher pedagogy Begin to attend to particular students' mathematical thinking and behaviours	Form general impressions and highlight noteworthy events
		Provide primarily evaluative with some interpretive comments
		Begin to refer to specific events and interactions as evidence
Level 3 Focused	Attend to particular students' mathematical thinking	Provide interpretive comments
		Refer to specific events and interactions as evidence
		Elaborate on events and interactions
Level 4 Extended	Attend to the relationship between particular students' mathematical thinking and between teaching strategies and student mathematical thinking	Provide interpretive comments
		Refer to specific events and interactions as evidence
		Elaborate on events and interactions
		Make connections between events and principles of teaching and learning
		On the basis of interpretations, propose alternative pedagogical solutions

central dimensions of noticing: *what* teachers notice and *how* teachers analyse what they notice. What teachers notice includes whom they notice (learners as class or group, individual learners, teacher, self) as well as the topic of their noticing (learner behaviour, mathematical reasoning, teacher pedagogy). How they analyse refers to an analytic, evaluative, interpretive or deep analysis of what they notice. Second, the framework provides a trajectory of four levels of development in the two dimensions: baseline, mixed, focused and extended noticing, as described in Table 41.1.

The practice of noticing is more likely to develop in contexts of collaborative sense-making of learner reasoning where teachers plan, observe and reflect on lessons together (Takahashi et al. 2013), such as lesson study.

41.2.3   Lesson Study as a Context for the Development of Noticing

Reflection and noticing are essential components of Japanese lesson study, a form of professional development aimed at the improvement of instruction (Lewis and Tsuchida 1998; Takahashi et al. 2013). Lesson study is the systematic and collaborative planning and reviewing of a research lesson by a community of teachers and a 'knowledgeable other' to bring together theoretical and practical learning in an authentic way. The lesson study cycle entails the collaborative planning of a lesson, the observation of the research lesson, and an in-depth post-lesson discussion (Lewis 2002; Pothen and Murata 2006). In some cases, the lesson is then revised and taught again, with another post-lesson reflection. The focus of a research lesson is on sense-making, building of mathematics concepts and problem solving. A lesson study group goes through multiple cycles of such inquiry over time, which leads to an improvement of their mathematical knowledge for teaching (Takahashi et al. 2013). This process fosters high-quality learning through high-quality teaching (Nishimura 2016). The improvement of teachers' professional knowledge and skills enables them to become expert teachers (high-quality teaching) who can progressively provide learners with opportunities for deep learning of content and in the process become independent learners and problem solvers (high-quality learning) (Sugiyama 2008; Takahashi et al. 2013). Groves et al. (2016) suggest that lesson study be adapted for other cultures and situations rather than adopting the Japanese model as is.

Noticing plays an important role in a lesson study cycle from the planning of the research lesson to observing and reflecting on the lesson. Mason's (2002) description of noticing as a set of practices that includes reflecting systematically, recognising choices and alternatives, preparing and noticing possibilities, and validating with others (p. 87) bears similarity to lesson study as deliberate practice (Miller 2011), as well as to the Gibbs' (1988) reflection cycle often used as a guide for pre-service teachers' reflection.

Professional noticing resonates strongly with the goals of lesson study—planning for student learning, noticing how the tasks and activities foster or hamper learning during the lesson, interpreting what happened during the lesson and deciding how to plan and teach the lesson more efficiently in the future.

41.3 Method

In a longitudinal project, a cohort of pre-service teachers' reflections in a problem-centred mathematics teaching and learning context over a period of two years were investigated. The purpose of the project was to explore pre-service teachers' reflections on their own and observed research lessons in order to improve the structure and development of reflective mathematical practice in an undergraduate teacher education program. The research questions driving the larger investigation dealt with aspects of the problem-centred approach that pre-service teachers reflect on and their noticing during mathematics lessons. This paper only focuses on the latter: the development of prospective teachers' noticing over a two-year period.

41.3.1 Context and Participants

At Stellenbosch University, mathematics teacher education for prospective foundation phase teachers specialising to teach in Grades R to 3 (6- to 9-year-olds) is nested in a problem-centred approach to the teaching of mathematics as described in the work of Murray et al. (1998), Hiebert et al. (1997) and Stein et al. (2008). To facilitate and support pre-service teachers' in adopting a problem-centred approach and develop the relevant knowledge, skills and dispositions, an adapted form of Japanese lesson study is implemented as one of the contexts for their professional development. Structured reflection opportunities during lesson study cycles help bridge the theory-practice gap in this mostly unfamiliar context for the prospective teachers.

Pre-service teachers conduct school observations for two weeks in the beginning of the second and third years and nine weeks of school practicum in their second, third and fourth years. These practicum experiences are supplemented by two cycles of micro lessons (Fishbowl[1]) in the first semester of the third year and a service-learning project in the first semester of the fourth year.

---

[1]The Fishbowl micro lessons comprise the teaching of a lesson by a pre-service teacher to a small group of learners while the lecturer and other pre-service teachers are observing behind one-way glass.

Lesson study is used in the mathematics education module for foundation phase (Grades R to 3) during Fishbowl in the third year and school practice in the fourth year. In the third year, students participate in groups of six or seven of their own choice and complete two lesson study cycles. The Fishbowl lesson study cycles are the first exposure of pre-service teachers to lesson study. In the fourth year, three to seven pre-service teachers are grouped together based on the locations of their school practice schools for one lesson study cycle. The lecturer acts as 'knowledgeable other' during the lesson study processes. Pre-service teachers in groups study the curriculum and mathematics content area of their chosen topic and formulate an over-arching goal as well as specific goals for their research lesson. Lesson planning includes considering a possible learning trajectory for the topic and where the lesson would fit into this trajectory, developing a real-world task for the lesson, anticipating possible learner strategies and misconceptions, considering connections between the possible strategies, planning questions to elicit learner reasoning and planning how to summarise the lesson at the end. The research lesson is then presented by each of the group members in their own classes and observed by some of the other group members. The lesson subsequently is refined and finally presented by one of the group members while the lecturer as 'knowledgeable other' and all other group members observe. A post-lesson discussion with thorough reflection-on-action by all observers follows. The mentor teachers at the schools where the pre-service teachers are doing school practice are not involved in the lesson study process. After a lesson study cycle, pre-service teachers are required to submit individual written reflections on the research lessons as a consolidation of learning (Lewis et al. 2009). These reflections have to be submitted a week after the research lesson and are guided by questions based on the Gibbs reflection cycle (Gibbs 1988). This reflection framework, used by all lecturers in the Department of Curriculum Studies, focuses on what transpired during the lesson, an analysis of what went well or wrong and why in order to make sense of the situation and the exploration of alternatives and culminates in an action plan.

Participants were undergraduate prospective teachers preparing to teach mathematics at the foundation phase level (6- to 9-year-olds in Grades R to 3). They were all female, as most of the population of foundation phase teachers in the region are. Participants' experience of noticing before the Fishbowl lesson study cycles in the third year was limited to doing video analyses of a number of lessons in class. Although the terminology and framework of noticing were not used, the lesson analysis was focused on noticing learner reasoning and strategies and using inferences from what they noticed in order to plan subsequent lessons. In the fourth year, a unit of lectures on noticing precedes the school practicum. In the research and lesson planning phases in the lesson study cycles in the third and fourth years, specific attention is given to anticipated learner strategies and planning for noticing of children's mathematical reasoning, and examples are discussed. Due to logistical arrangements, lesson study groups for the two lesson study cycles in the third year are the same, but change for the cycle in the fourth year. For the same reason, post-lesson discussions are shorter in the case of the two Fishbowl research lessons than they are for the research lesson in the fourth year. The post-lesson reflection colloquia of the different lesson study groups in the fourth year are videoed to capture student teachers' reflections, but they are not in the third year. All ethical requirements of Stellenbosch University were adhered to in the research study.

Fifty-six out of the cohort of 59 fourth-year pre-service teachers volunteered their third and fourth year reflections for analysis in the study. Participants were from different language groups (Afrikaans, English and Xhosa), and research lessons were taught in either Afrikaans or English. Post-lesson colloquia were conducted in Afrikaans, English or both, depending on the mother tongues of the participants; pre-service teachers often switched between languages, as groups consisted of students from different language groups. Written reflections were in the language of their choice.

Students' reflections did not yield many rich descriptions of learner reasoning. After scrutinising all reflections, eight students whose reflections included descriptions of learner reasoning in at least one of the three data sets were selected and their reflections further analysed using the Van Es (2011) framework. Due to space constraints, three prospective teachers' noticing as evidenced in their reflections on three research lessons have been selected to be developed as case studies for this paper. These three pre-service teachers were not members of the same group in any of the three lesson study cycles.

41.3.2 Data Generation

Data were generated at the end of three lesson study cycles, two in the third year and one in the fourth year, and comprised videoed post-lesson discussions and written reflections after each of these lesson study cycles. During Fishbowl lesson study cycles, only written reflections were generated, whereas written and verbal reflections were collected during the school practicum lesson study cycles. The reflections of pre-service teachers were on research lessons they either observed or taught themselves and were on entire lessons, as in the research of Santagata (2011), and not only on video clips of lessons. During post-lesson reflections, students' individual accounts of what they noticed during lessons were recorded for analysis. Further elaboration from the group on individuals' noticing was not included in the analysis covered in this paper.

41.3.3 Analysis

Video and document analysis of verbal and written reflections were conducted using the Van Es (2011) framework for learning to notice student reasoning to determine what and how pre-service teachers noticed during research lessons and to determine how their noticing developed over the two-year period. Pre-service teachers' written and videoed reflections were coded according to the four levels of the Van Es (2011) framework and an independent researcher checked the researcher's interpretations of the data.

41.4   Results

To determine growth and depth in noticing over a period of two years, pre-service teachers' written reflections on two research lessons for Fishbowl in the third year and one verbal or written reflection on a research lesson in the fourth year were analysed. Reflections were in some cases on lessons the pre-service teachers themselves taught and in some cases on observed lessons.

41.4.1   Case Studies

In the three case studies below, unless stated that the student teacher taught the lesson being discussed, all reflections were on observed research lessons.

Case Study 1: Myra (participants' real names are not used)

Myra's written reflections on observed Fishbowl research lessons in the third year were analysed, as were her reflections on her own lesson during the fourth-year lesson study.

Fishbowl research lesson 1: Myra's written reflection on the Grade 2 lesson about money focused on the whole class and the teacher (whom), her noticing comprised teacher pedagogy and learner behaviour (what) and she gave general impressions and evaluative comments (how). Examples: 'Learners were interested in the theme of the lesson' and 'The teacher did not tell the learners how to solve the problem, she gave them space to decide on their own strategies'. Myra's noticing for this research lesson has been categorized as Level 1 (baseline; Table 41.2).

Fishbowl research lesson 2: In her written reflection, Myra's noticing during this Grade 2 lesson on capacity measurement again was general and evaluative, focusing on learner behaviour and teacher pedagogy. Examples: 'The activities were focused and well explained' and 'The learners were so carried away with the water and activity that they did not focus on the math'. Her noticing still was on Level 1 (baseline; Table 41.2).

School practicum research lesson: Myra taught this Grade 2 lesson on capacity measurement, and her reflections (during the videoed post-lesson discussion and written reflection) showed noticing about the learners as group but also shifted towards particular learners' mathematical behaviour and reasoning with comments such as 'When they were estimating, I asked "What do you think?" One learner

Table 41.2 Noticing levels of pre-service teachers in the sample

	Baseline	Mixed	Focused	Extended
Research lesson 1 Fishbowl (3rd year)	Myra[3] Ruda[3]	Charlie[3]		
Research lesson 2 Fishbowl (3rd year)	Myra[3]	Charlie[3]	Ruda[1,3]	
Research lesson 3 School practicum (4th year)		Charlie[2, 3]	Myra[1,2,3] Ruda[1,2,3]	

[1]Taught research lesson
[2]Verbal reflection
[3]Written reflection

guessed 84, the next one said another number and the third thought deeply and then said "I will then say 82"' and 'Christian said that the group tried to get their measurement close to their estimation'. She evidently was starting to notice particular learners' mathematical behaviour and reasoning, albeit primarily evaluating comments. Myra's noticing therefore shows progress from Level 1 in the second year to Level 3 (focused) in the fourth year (Table 41.2).

Case Study 2: Ruda

Fishbowl research lesson 1: Ruda's noticing in her written reflection for this 3-D geometry research lesson for Grade 2 was on Level 1 (baseline; Table 41.2). Her comments were of a general and evaluative nature and concerned learner behaviour, teacher pedagogy and classroom control as the following three examples show: 'The lesson showed me how important classroom control is', 'The teacher used normal household objects; they were easy to relate to and when learners go home, they will be able to apply the knowledge they had learnt' and 'Learners were excited; the teacher did not tell them how to sort the objects, they were able to use their own ideas'.

Fishbowl research lesson 2: The Grade 2 fraction research lesson that Ruda taught elicited the following comments from her in a videoed post-lesson discussion: 'There were slight differences in how learners shared the last chocolate bar. Some continued to share 1–1 until they were finished, some just drew 2 lines to the same chocolate and others shared it in pieces with a line' and 'Even though they didn't know the terminology, they knew it wasn't a whole chocolate that each person received'. She noticed and interpreted the mathematical behaviour and reasoning of both the learners as a group and specific learners but did not suggest any teacher moves based on these responses; therefore, her reflections can be categorised as Level 3 (focused; Table 41.2).

School practicum research lesson: Ruda's noticing in this Grade 2 lesson on multiplication that she taught again was on Level 3 (focused; Table 41.2). In a videoed post-lesson discussion, she referred to and elaborated on individual learners' mathematical reasoning, interpreting their reasoning. Examples: 'When you ask them to explain, they don't explain what they did. Like Max, he counted 1, 2, 3, 4, 5, 6, 7, 8. He's not counting in 4s, he's counting in ones. But when you ask him, he says "I'm counting in 4s". I am sure if I asked him, he would be able to count 4, 8, 12...' and 'And like Melissa, she drew groups first, then she made lines in them, and then I asked "What did you do" and she said "I am counting in 4s". They don't always explain how they did it; they know they are supposed to be counting in 4s, so they say "I counted in 4s"'.

Ruda's noticing therefore progressed over the period of two years, from the Level 1 in the lesson study cycle one to the Level 3 in cycles two and three.

Case Study 3: Charlie

Fishbowl research lesson 1: This written reflection was on a Grade 2 lesson on mass measurement. Charlie highlighted a noteworthy event during the lesson and drew inferences from learner mathematical reasoning, but generalized her comments to the group of learners rather than commenting on individual learners' reasoning: 'Six out of 10 learners approached the problem incorrectly or miscalculated. It could be that they are not used to coming up with their own strategies or that the problem was too difficult for them to make sense of it'. Charlie tended to generalise individuals' reasoning to collective reasoning of the group. Her noticing for this lesson study cycle was categorised as Level 2 (mixed), bordering on Level 3 (focused; Table 41.2).

Fishbowl research lesson 2: This Grade 2 lesson also was on mass measurement. Charlie noticed and provided evaluative comments about an individual learner's mathematical behaviour and reasoning in her written reflection on the lesson: 'One boy's measurement units (he used nuts) were not quite enough to measure—he was quite confused when his first strategy didn't work. It was fantastic to see how he came up with another plan to solve the problem'. Her comments were descriptive and not interpretive; therefore, her comment is categorised as Level 2 (mixed; Table 41.2).

School practicum research lesson: This Grade 2 lesson on length measurement prompted comments about teacher pedagogy and the mathematical understanding of groups of learners as well as of individuals in a videoed post-lesson discussion: 'The focus shifted and a valuable discussion was lost. Learners did not have a discussion about how the different ways of measuring can influence their measurements and what they should be doing to get similar answers' and 'The teacher over-emphasised the way they should hold their hands when they are measuring. This became a requirement instead of the class understanding the concepts of measurement and the necessity to measure accurately'. Reference to and interpretation of specific events and actions show Level 2 noticing (mixed) bordering on Level 3 noticing (focused; Table 41.2).

41.4.2 Case Studies Summary

From Table 41.2 it is evident that verbal reflections during post-lesson discussions were on Level 2 or 3, while written reflections were spread over the first three levels. Furthermore, student teachers' reflections (Ruda: Lessons 2 and 3; Myra: Lesson 3) on their own lessons were all on the level of focused noticing (Level 3). Five of the six reflections in the third year (the first two lesson study cycles) were on the first two levels, with a shift to a higher level of noticing during the school practicum (third lesson study cycle) in the fourth year for two of the three pre-service teachers.

41.5 Discussion

In this study, which has been framed by research on reflection, noticing and lesson study, the Van Es (2011) framework for learning to notice has been used to track the development of pre-service teachers' noticing over the last two years of their studies. The backdrop of a problem-centred approach to the teaching of mathematics and a focus on the development of mathematical reasoning was especially important in the prospective teachers' four-year course as well as in the lesson study cycles during which the data were generated.

The findings indicate that two of the pre-service teachers made a positive shift towards higher levels of noticing from the third to the fourth year, although not to the highest level of extended noticing. This finding corroborates Korthagen et al. (2006) statement that pre-service teachers do not always adopt what they learn during their preparation programs. The third student teacher stayed on the same level for all three lesson study cycles. This is in line with Jacobs et al. (2010) and Santagata (2011) research, which indicated that a considerable number of pre-service teachers have trouble with noticing and attending to specific learner strategies. Post-lesson discussions elicited more productive noticing than written reflections, with more explicit references to individual learners' reasoning and more attempts at interpreting what they noticed. In written reflections, pre-service teachers tended to generalise individual learners' reasoning or comment on individuals' reasoning as being the collective reasoning of a group or the class. However, in written reflections, pre-service teachers made stronger connections between learner reasoning and behaviour and principles of teaching and learning, suggesting possible alternative teacher questions and moves. This might be a result of the nature of the four Gibbs (1988) reflection questions, which encourage more of a summary of what has been noticed, resulting in more general than specific comments.

Reflections on lessons that pre-service teachers themselves taught also showed more productive noticing than reflections on observed lessons. This could be due to the fact that the pre-service teacher teaching the lesson was more familiar with the learners in the classroom (Sherin et al. 2011b). Furthermore, although the lessons have been planned jointly, pre-service teachers may have taken more ownership of the lessons they taught (Takahashi et al. 2013) and may therefore have been more attuned to learners' reasoning during the lesson than the rest of the observing group.

During post-lesson discussions, the lecturer as 'knowledgeable other' first facilitated pre-service teachers reflections, where she focused their attention on the reasoning of individual learners, pointing out other alternatives and asking them to make sense of learner reasoning and its implications for teaching, therefore introducing her own higher level noticing and modelling the practice of noticing. This modus operandi is consistent with the practice of lesson study where the role of the 'knowledgeable other' includes bringing new knowledge about research and the curriculum, pointing out connections between theory and practice and helping teachers' 'to learn how to reflect on teaching and learning' (Takahashi 2013, p. 12). In the research of Van Es (2011), the facilitators of the video club also offered alternative perspectives and modelled to participants in the video club how to engage in the practice of noticing and reasoning about learner reasoning. This 'helped the group recognize multiple valid interpretations of a student idea and the value of further inquiry as ways to clarify the issue under discussion' (p. 148) and resulted in 'more substantive analyses of student thinking' (p. 148).

Kassim (2016), in a study of a cohort of pre-service foundation phase teachers at the end of their third year who were all following the same course at the same university, found that pre-service teachers perceived an improvement in their mathematical knowledge for teaching during their third year, enabling them to create effective mathematical instruction using learners' understanding and reasoning and to address their misconceptions. Participants in Kassim's study also perceived a marked change in their beliefs towards implementing a problem-centred approach 'where learners express their own understanding of the problem, engage learners' interactive discussions to enhance their reasoning and understanding' and 'assist learners to solve problems using their own strategies' (p. 377). It can therefore be speculated that a change in pre-service teachers' beliefs and mathematical knowledge during the third year might have contributed to their higher level of noticing in the fourth year.

The fact that participants were from different language groups and switched between languages during discussions could have had an influence on their expression of what and how they noticed. Although all prospective teachers are bilingual, they sometimes struggle to express themselves clearly in their second language (Setati et al. 2002; Webb and Webb 2008).

Research by Choy (2015), Posthuma (2012) and Takahashi et al. (2013) points to lesson study as a productive environment for reflection and the development of noticing in a teacher preparation program. These authors highlight aspects of lesson study that contribute to making it a productive context for fostering reflection and noticing. These aspects include focus on the development of mathematical thinking (Takahashi and McDougal 2016), a very thorough process of planning for student learning (Choy 2016), noticing of how tasks and activities foster learning during a lesson, interpretation of learner thinking during the lesson, and decision making about ways to plan and teach the lesson more efficiently in future (Lewis et al. 2009).

Pre-service teachers' noticing in the current study did not all progress as hoped, and more research is needed to determine the reasons behind these findings. Gaps in the pre-service teachers' preparation and the influence of different aspects in the process of lesson study such as group size, format of the reflections, limited time reflection after Fishbowl and the knowledge and beliefs of the pre-service teachers' could all have an influence and will have to be investigated.

41.6 Limitations and Future Research

The findings in this small-scale study cannot be generalised. Larger-scale longitudinal studies are needed to track the development of prospective teachers' noticing during their preparation programs and even into their first years of practice (Barnhart and Van Es 2015). The lesson study process fostered the development of more productive noticing, but pre-service teachers' noticing may also have been influenced by lecturers in other subjects encouraging them to reflect or the use of specific protocols to structure reflection. The noticing framework does not make provision for noticing of task characteristics or how learners responded to the task and can be extended to include such aspects of lessons.

More research is needed on the difference between prospective teachers' verbal and written reflections and the influence of their mathematical knowledge for teaching on their noticing and development of reflection skills. The role of lesson topics in noticing is also not clear.

41.7 Conclusion

Pre-service teachers need purposeful, structured opportunities and appropriate experiences to reflect on and make sense of learner mathematical reasoning: 'This expertise can be learned and … both teaching experience and professional development support this endeavor' (Jacobs et al. 2010, p. 191). Lesson study as situated in a problem-centred context where learners' mathematical reasoning takes centre stage has proven a productive context for such professional development and experiences. Noticing frameworks are useful for not only assessing growth and depth in noticing but should also be used as guides for scaffolding pre-service teachers' noticing skills when observing live as well as videoed lessons. Teaching for understanding is one of the hallmarks of a problem-centred approach to the teaching of mathematics, and noticing in teacher preparation programs needs to be framed in the bigger picture of learning to teach mathematics for understanding.

References

Anderson, L., & Krathwohl, D. (2001). *A taxonomy for learning, teaching, and assessing: A revision of bloom's taxonomy of educational objectives.* New York: Longman.

Ball, D., Lubienski, S., & Mewborn, D. (2001). Research on teaching mathematics: The unsolved problem of teachers' mathematical knowledge. In V. Richardson (Ed.), *Handbook of research on teaching* (4th ed., pp. 433–456). New York: Macmillan.

Ball, D., Thames, M., & Phelps, G. (2008). Content knowledge for teaching: What makes it special? *Journal of Teacher Education, 59*(5), 389–407.

Barnhart, T., & Van Es, E. (2015). Studying teacher noticing: Examining the relationship among pre-service science teachers' ability to attend, analyze and respond to student thinking. *Teaching and Teacher Education, 45,* 83–93.

Choy, B. (2014). Teachers' productive mathematical noticing during lesson preparation. In *38th Conference of the International Group for the Psychology of Mathematics Education (PME38) and the 36th Conference of the North American Chapter for the Psychology of Mathematics Education (PME-NA36)* (pp. 297–304). Vancouver, Canada.

Choy, B. (2015). *The FOCUS framework: Snapshots of mathematics teacher noticing* (Doctoral thesis in Philosophy in Mathematics Education). University of Auckland.

Choy, B. (2016). Snapshots of mathematics teacher noticing during task design. *Mathematics Education Research Journal, 28,* 421–440. https://doi.org/10.1007/s13394-016-0173-3.

Dewey, J. (1933). *How we think: A re-statement of the relation of reflective thinking in the educative process.* Chicago: Henry Regnery.

Erickson, F. (2011). On noticing teacher noticing. In M. Sherin, V. Jacobs, & R. Philipp (Eds.), *Mathematics teacher noticing: Seeing through teachers' eyes* (pp. 17–34). New York, NY: Routledge.

Fennema, E., Carpenter, T., Francke, M., Levi, L., Jacobs, V., & Empson, S. (1996). Mathematics instruction and teachers' beliefs: A longitudinal study of using children's thinking. *Journal for Research in Mathematics Education, 27,* 403–434.

Fernández, C., Llinares, S., & Valls, J. (2012). Learning to notice students' mathematical thinking through online discussions. *ZDM Mathematics Education, 44,* 747–759.

Fernández, C., Llinares, S., & Valls, J. (2013). Primary teacher's professional noticing of students' mathematical thinking. *The Mathematics Enthusiast, 10*(1&2), 441–468.

Galbraith, P. (2015). 'Noticing' in the practice of modelling as real world problem solving. In G. Kaiser, & H.-W. Henn (Eds.), *Werner Blum und seine Beiträge zum Modellieren im Mathematikunterricht: Festschrift zum 70. Geburtstag von Werner Blum* (pp. 151–166). Wiesbaden: Springer. https://doi.org/10.1007/978-3-658-09532-1_11.

Gibbs, G. (1988). *Learning by doing: A guide to teaching and learning methods.* Oxford: Further Education Unit, Oxford Brookes University.

Goldsmith, L., Doerr, H., & Lewis, C. (2014). Mathematics teachers' learning: A conceptual framework and synthesis of research. *Journal of Mathematics Teacher Education, 17*(1), 5–36.

Grossman, P., Hammerness, K., & McDonald, M. (2009). Redefining teaching, re-imagining teacher education. *Teachers and Teaching, 15*(2), 273–289. https://doi.org/10.1080/13540600902875340.

Groves, S., Doig, B., Vale, C., & Widjaja, W. (2016). Critical factors in the adaptation and implementation of Japanese Lesson Study in the Australian context. *ZDM Mathematics Education, 48,* 501–512. https://doi.org/10.1007/s11858-016-0786-8.

Hiebert, J., Carpenter, T., Fennema, E., Fuson, K., Wearne, D., Murray, H., et al. (1997). *Making sense: Teaching and learning mathematics with understanding.* Portsmouth: Heinemann.

Hiebert, J., Morris, A., & Glass, B. (2003). Learning to learn to teach: An 'experiment' model for teaching and teacher preparation in mathematics. *Journal of Mathematics Teacher Education, 6*, 201–222.

Hill, H., Blunk, M., Charalambous, C., Lewis, J., Phelps, G., Sleep, L., et al. (2008). Mathematical knowledge for teaching and the mathematical quality of instruction: An exploratory study. *Cognition and Instruction, 26*, 430–511.

Jacobs, V. R., Lamb, L. C., & Philipp, R. (2010). Professional noticing of children's mathematical thinking. *Journal for Research in Mathematics Education, 41*(2), 169–202.

Kassim, A. (2016). *Perceptions of pre-service teachers in Foundation Phase mathematics about their professional development* (Unpublished doctoral dissertation). Stellenbosch University.

Kazemi, E., Elliot, R., Mumme, J., Carroll, C., Lesseig, K., & Kelly Peterson, M. (2011). Noticing leaders' thinking about videocases of teachers engaged in mathematics tasks in professional development. In M. Sherin, V. Jacobs, & R. Philipp (Eds.), *Mathematics teacher noticing: Seeing through teachers' eyes* (pp. 188–203). New York, NY: Routledge.

Korthagen, F. (2007). The gap between research and practice revisited. *Educational Research and Evaluation, 13*(3), 303–310. https://doi.org/10.1080/13803610701640235.

Korthagen, F., Loughran, J., & Russell, T. (2006). Developing fundamental principles for teacher education programs and practices. *Teacher and Teacher Education, 22*(8), 1020–1041.

Lampert, M. (2001). *Teaching problems and the problems of teaching*. New Haven, CT: Yale University Press.

Lewis, C. (2002). *A handbook of teacher-led instructional change*. Philadelphia: Research for Better Schools.

Lewis, C., Perry, R., & Hurd, J. (2009). Improving mathematics instruction through lesson study: A theoretical model and North American case. *Journal of Mathemaitcs Teacher Education, 12*, 285–304. https://doi.org/10.1007/s10857-009-9102-7.

Lewis, C., & Takahashi, A. (2013). Facilitating curriculum reforms through lesson study. *International Journal for Lesson and Learning Studies, 2*(3), 207–217. https://doi.org/10.1108/IJLLS-01-2013-0006.

Lewis, C., & Tsuchida, I. (1998). A lesson is like a swiftly flowing river: Research lessons and the improvement of Japanese education. *American Educator, 14–17*, 50–52.

Llinares, S. (2013). Professional noticing: A component of the mathematics teacher's professional practice. *SISYPHUS Journal of Education, 1*(3), 76–93.

Mason, J. (2002). *Researching your own practice: The discipline of noticing*. London: Routledge Falmer.

Mason, J. (2011). Noticing: Roots and branches. In M. Sherin, V. Jacobs, & R. Philipp (Eds.), *Mathematics teacher noticing: Seeing through teachers' eyes* (pp. 35–50). New York, NY: Routledge.

Miller, K. (2011). Situation awareness in teaching. In M. Sherin, V. Jacobs, & R. Philipp (Eds.), *Mathematics teacher noticing: Seeing through teachers' eyes* (pp. 51–65). New York, NY: Routledge.

Mitchell, R., & Marin, K. (2015). Examining the use of a structured analysis framework to support prospective teacher noticing. *Journal of Mathematics Teacher Education, 18*, 551–575. https://doi.org/10.1007/s10857-014-9294-3.

Murray, H., Olivier, A., & Human, P. (1998). Learning through problem solving. In Olivier, A, & Newstead, K (Ed.), *Proceedings of the Twenty-second International Conference for the Psychology of Mathematics Education* (Vol. 1, No. 1, pp. 169–185). Stellenbosch.

Nishimura, K. (2016). Lesson study at the upper secondary level in Japan. In *Presentation, Lesson Study Mini Conference*. Nottingham, UK.

Paolucci, C., & Wessels, H. (2017). An examination of pre-service teachers' capacity to create mathematical modeling problems for children. *Journal of Teacher Education, 68(3)*, 330–344. https://doi.org/10.1177/002248711769763.

Posthuma, A. (2012). Mathematics teachers' reflective practice within the context of adapted lesson study. *Pythagoras, 33*(3). https://doi.org/10.4102/pythagoras.v33i3.140.

Pothen, B., & Murata, A. (2006). Developing reflective practitioners: A case study of pre-service elementary mathematics teachers' lesson study. In J. V. S. Alatore (Ed.), *Proceedings of the 28th Annual Meeting of the North American Chapter of the International Group for the Psychology of Mathematics Education* (Vol. 2, pp. 824–826). Merida, Mexico: Universidad Pedagogica Nacional.

Santagata, R. (2011). From teacher noticing to a framework for analyzing and improving classroom lessons. In M. Sherin, V. Jacobs, & R. Philipp (Eds.), *Mathematics teacher noticing: Seeing through teachers' eyes* (pp. 152–168). New York, NY: Routledge.

Santagata, R., Zannoni, C., & Stigler, J. (2007). The role of lesson analysis in pre-service teacher education: An empirical investigation of teacher learning from a virtual video-based field experience. *Journal of Mathematics Teacher Education, 10*(2), 123–140.

Schoenfeld, A. (2011). Noticing matters. A lot. Now what? In M. Sherin, V. Jacobs, & R. Philipp (Eds.), *Mathematics teacher noticing: Seeing through teachers' eyes* (pp. 223–238). New York: Routledge.

Schön, D. (1987). *Educating the reflective practitioner*. San Francisco: Jossey-Bass.

Seidel, T., & Stürmer, K. (2014). Modeling and measuring the structure of professional vision in pre-service teachers. *American Educational Research Journal, 51*(4). https://doi.org/10.3102/0002831214531321.

Setati, M., Adler, J., Reed, Y., & Bapoo, A. (2002). Incomplete journeys: Code-switching and other language practices in mathematics, science and english language classrooms in South Africa. *Language and Education, 16*(2), 128–149.

Sherin, M. G, Jacobs, V. R, & Philipp, R. A. (2011a). Situating the study of teacher noticing. In M. Sherin, V. Jacobs, & R. Philipp (Eds.), *Mathematics teacher noticing: Seeing through teachers' eyes* (pp. 3–13). New York, NY: Routledge.

Sherin, M. G., Jacobs, V. R., & Philipp, R. A. (2011b). *Mathematics teacher noticing: Seeing through teachers' eyes*. New York, NY: Routledge.

Sherin, M., Russ, R., & Colestock, A. (2011c). Accessing mathematics teachers' in-the-moment noticing Chapter 6. In M. Sherin, V. Jacobs, & R. Philipp (Eds.), *Mathematics teacher noticing: Seeing through teachers' eyes* (pp. 79–94). New York, NY: Routledge.

Stein, E., Engle, R., Smith, M., & Hughes, E. (2008). Orchestrating productive mathematical discussions: Five practices for helping teachers move beyond show and tell. *Mathematical Thinking and Learning, 10*, 313–340.

Stigler, J., & Hiebert, J. (1999). *The teaching gap. Best ideas from the world's teachers for improving education in the classroom*. New York: Free Press.

Sugiyama, Y. (2008). *Introduction to elementary mathematics education*. Tokyo: Toyokan Publishing Co. (In Japanese.)

Takahashi, A. (2013). *The role of the knowledgeable other in lesson study: Examining the final comments of experienced lesson study practitioners*. Mathematics Education Research Group of Australasia.

Takahashi, A., Lewis, C., & Perry, R. (2013). A US lesson study network to spread teaching through problem solving. *International Journal for Lesson and Learning Studies, 2*(3), 237–255.

Takahashi, A., & McDougal, T. (2016). Collaborative lesson research: Maximizing the impact of lesson study. *ZDM Mathematics Education*, 513–526. https://doi.org/10.1007/s11858-015-0752-x.

Van Es, E. (2011). A framework for learning to notice student thinking. In M. Sherin, V. Jacobs, & R. Philipp (Eds.), *Mathematics teacher noticing: Seeing through teachers' eyes* (pp. 134–151). New York: Routledge.

Webb, L., & Webb, P. (2008). Introducing discussion into multilingual mathematics classrooms: An issue of code switching. *Pythagoras, 67,* 26–32.

Chapter 42
Dialogues on Numbers: Script-Writing as Approximation of Practice

Rina Zazkis

Abstract Script-writing is a novel pedagogical approach and research tool in mathematics education. The goal of this chapter is to introduce the approach and exemplify its implementation. A script-writing task presents a prompt, which usually includes an incomplete argument or erroneous claim of a student. Prospective teachers address the prompt by creating a script for a dialogue—presenting an imaginary interaction between a teacher and her students, or among different students. In this chapter I exemplify several results of implementing script-writing tasks and discuss advantages of this approach. In particular, I focus on the concepts related to elementary number theory, prime numbers and factors of a number, and demonstrate how the understanding of these concepts can be explored and refined, as script-writers create characters who discuss particular claims. I suggest that engaging prospective teachers in script-writing is one possible way to support and improve preparation of mathematics teachers.

Keywords Script-writing · Prime numbers · Role-play · Approximation of practice · Lesson play

42.1 Introduction

How can we support and improve teacher development? Mathematics educators and researchers are investigating the issue and developing a variety of approaches by which the preparation of future teachers can be enhanced. In this chapter I present one such approach—script-writing—describe how it emerged and illustrate its implementation in several cases.

Watson and Mason (2005) suggested that "the fundamental issue in working with teachers is to resonate with their experience so that they can *imagine* [my italics] themselves 'doing something' in their own situation" (p. 208). I attempt to

R. Zazkis (✉)
Simon Fraser University, Burnaby, Canada
e-mail: zazkis@sfu.ca

G. Kaiser et al. (eds.), *Invited Lectures from the 13th International Congress on Mathematical Education*, ICME-13 Monographs,
https://doi.org/10.1007/978-3-319-72170-5_42

access this imagination by inviting prospective teachers to write scripts that describe instructional interactions between a teacher and students.

42.2 Script-Writing as Approximation of Practice

Script-writing (or scripting) is a valuable pedagogical strategy and an innovative research tool which I adopted and developed within the context of mathematics teacher education. In its initial implementations, this method was referred to as "lesson play" (Zazkis et al. 2013); however, the term script-writing extends to account for interactions that are not necessarily part of a lesson. Script-writing is a tool related to "approximations of practice" (Grossman et al. 2009), which "include opportunities to rehearse and enact discrete components of complex practice in settings of reduced complexity" (p. 283), and advocated as an essential part of teacher preparation.

I consider scripting as a form of role-playing in one's imagination. As such, I briefly describe role-playing in 2.1, focusing explicitly on role-playing on teacher education in 2.2, and then turn in 2.3 to script-writing and the use of script-writing in mathematics education. As teachers' scripts provide a lens to explore their knowledge, in 2.4 I attend briefly to the notion of knowledge of mathematics for teaching.

42.2.1 On Role-Playing

Role-playing is an unscripted "dramatic technique that encourages participants to improvise behaviors that illustrate expected actions of persons involved in defined situations" (Lowenstein 2007, p. 173). In other words, role-playing is "an 'as-if' experiment in which the subject is asked to behave as if he [or she] were a particular person in a particular situation" (Aronson and Carlsmith 1968, p. 26). Role-playing is used as an effective pedagogical strategy in a variety of fields (e.g., Blatner 2009), a few of which are mentioned here.

Traditionally role-playing is used in social studies classrooms in order to provide participants with more authentic experiences of historic events and people who experienced them (e.g., Cruz and Murthy 2006). It is used to explore the complexities of social situations, such as prejudice, and ethical issues (e.g., Lawson et al. 2010; McGregor 1993; Plous 2000). Participants, after engaging in role-playing, reported being better prepared to deal constructively with everyday instances of prejudice (Plous 2000) and generated more effective responses to prejudiced comments (Lawson et al. 2010). Additionally, role-playing was used with English language learners, where teachers used role-playing in an attempt to move from a prescribed dialogue to an improvisational one. In this context, Shapiro and Leopold (2012) suggested that implementing role-playing in a classroom provides a "space between practice and play [which] is a fertile ground for cognitive and linguistic growth" (p. 128).

Role-playing is used in the education of various groups of professionals in organizational research, where, for example, participants assume roles of performance evaluators or interviewers of job applicants (Greenberg and Eskew 1993). It is also prevalent in the training of health professionals, where the participants play the roles of a care-giver and a patient, practicing their clinical, diagnostic and patient management skills, and as such developing empathy and tolerance in a low-risk environment (e.g., Joyner and Young 2006). However, among various uses in developing professionals, the use of role-playing in teacher education is rare.

42.2.2 On Role-Playing in Teacher Education

In considering role-play in teacher education, Van Ments (1983) described it as experiencing a problem under unfamiliar constraints, as a result of which one's own ideas emerge and one's understanding increases. In this sense, role-playing can also be seen as role-training. It is aimed at increasing teachers' awareness of various aspects of their actual work. Yet, despite the known advantages, role-playing in teacher education is underdeveloped. While some authors advocate for this method and report on its implementation, this is most often done in the form of self-reports and anecdotal evidence of participants' experiences.

Kenworthy (1973) described a method in which one participant takes on a teacher-role while others take on the roles of various students (e.g., a slow student, a gifted student, a disruptive student). He considered this type of role-playing to be "one of the most profitable, provocative and productive methods in the education of social studies teachers" (p. 243). He claimed that engagement in role-playing activities helped participants anticipate difficulties they encounter in their classrooms and, as such, gain security from their successful experiences should they face similar situations on the job. Assigning participants teacher and student roles was also used in a skill training workshops to deal with disturbing behaviour (Jones and Eimers 1975). Teacher training via role-playing reduced disruptive student behaviour and demonstrated gains in productivity for most students.

More recently, in Palmer (2006) study, pre-service teachers took on the roles of children as their professor modelled science teaching. It was reported that teachers' self-efficacy increased and they were more open to the idea of implementing role-playing in their teaching. In Howes and Cruz (2009) research, students in an elementary science methods class were invited to assume roles of scientists and take part in an "Oprah Show" interview. In addition to learning about contributions of different scientists, this activity sharpened the prospective teachers' understanding of what science is and what image of science they wished to convey to their students.

In mathematics teacher education, role-playing tasks were used to provide an opportunity to imagine personal responses to a variety of situations (e.g., Maheux and Lajoie 2011; Lajoie and Maheux 2013). To extend participation, Lajoie and Maheux engaged groups of prospective teachers in planning the roles of different players in a given instructional situation, and then called upon representatives from the groups to enact the role-play.

42.2.3 On Script-Writing

Despite the recognized advantages, time and participation logistics cause significant limitations in role-playing. If we engage students in role-playing during class, only a few will be active players, while the majority will serve as the audience. To afford all students the opportunity to participate in a player's role, I introduced imagined role-playing, i.e., writing a script for a dialogue between characters: teacher and student(s). In this approach, participants are presented with a prompt that describes a problematic situation, a disagreement, a student error, or inappropriate reasoning. The script-writing task is to devise a dialogue between the characters that leads to a resolution.

Script-writing is novel in mathematics education research. Its roots, however, trace to the Socratic dialogue, a genre of prose in which a wise man leads a discussion, often pointing to flaws in the thinking of his interlocutor. The method echoes the style of Lakatos (1976) evocative *Proofs and Refutations* in which a fictional interaction between a teacher and students clarifies mathematical concepts and claims.

Initially, script-writing was introduced in mathematics teacher education as a "lesson play" (Zazkis et al. 2009, 2013). Juxtaposed to a classical lesson plan describing merely content and activities, the lesson play reveals how a teaching-learning interaction unfolds. Using the theatrical meaning of the word 'play', lesson play refers to a task in which teachers are asked to write a script for a lesson, or part of a lesson, presented as a dialogue between a teacher and students. The task was developed as a result of dissatisfaction with the traditional approach of creating "lesson plans", often used in teacher education. The lesson play was advocated as a tool for preparing to teach, as a diagnostic tool for teacher educators, and as a tool for researchers studying issues of didactics and pedagogy (Zazkis et al. 2013).

The evolution of lesson play tasks from a general request to create a 'play' to carefully designed prompts on which plays are based is described in Zazkis et al. (2013). In short, in the initial implementation of lesson plays, it was noted that prospective teachers attempted to avoid any 'problematic' situations, such as dealing with explicit student mistakes or misconceptions. We then refined the general instructions, asking teachers to address in their plays a student mistake or some "problematic issue". The resulting plays exemplified some variation of the following template: the teacher asks a question, student-A provides a wrong answer, a teacher asks whether someone has a different answer, student-B provides a correct answer, the teacher reiterates the correct answer, praises student-B and moves to a new question.

The explicit instruction that the plays should address some problematic situation made us realize that prospective teachers had a very limited repertoire of potential problems that students experience. Therefore, in later implementations, prospective teachers were presented with the beginning of a dialogue, referred to as a 'prompt', and asked to continue the conversation. The prompts usually presented a student's erroneous answer; the teachers were asked to identify what could have led to this error and how a conversation could guide a student towards a resolution. Furthermore, the teachers were explicitly asked to describe the setting, as they imagined it, in which the presented prompt took place.

Script-writing is both an instructional tool and a research tool for data collection. For example, in recent studies it was implemented to investigate prospective teachers' understanding of particular proofs (Koichu and Zazkis 2013; Zazkis 2014; Zazkis and Zazkis 2016), central concepts in geometry and number theory (Kontorovich and Zazkis 2016; Zazkis and Zazkis 2014), and the use of particular symbols (Zazkis and Kontorovich 2016). The participants had to identify problematic issues in the presented proofs or topics, and then clarify these by designing a scripted dialogue. The scripts revealed participants' personal understandings of the mathematical concepts involved, as well as what they foresaw as potential difficulties for their imagined students, and how they may address these difficulties by building on their knowledge of mathematics for teaching.

42.2.4 *Scripts as a Window on Knowledge of Mathematics for Teaching*

Knowledge of mathematics in and for teaching has received significant attention in mathematics education research (e.g., Rowland and Ruthven 2011). What appears to unify different and seemingly opposing perspectives is a view that teachers' mathematical knowledge is complex and has distinctive features that deserve research attention.

In this chapter, I illustrate how teachers' scripts provide a window to explore, and subsequently an opportunity to enhance, their knowledge of mathematics, and of teaching mathematics. Sections 42.3 and 42.4 exemplify research-informed design of prompts and provide excerpts from scripts composed by prospective elementary school teachers in response to these prompts. The excerpts are chosen to illustrate particular repeating approaches in imagining instructional interaction. At the time of data collection, the participants completed a unit on number theory (focusing on concepts of divisibility, factors and multiples, prime and composite numbers) in a "Principles of Mathematics" course designed explicitly for elementary school teachers.

42.3 Identifying Prime Numbers

What exactly is a prime number and how can it be identified? To investigate how prospective elementary school teachers address this question, they were invited to develop scripts (lesson plays) starting with the following prompts:

(A) Teacher   Why do you say 143 is prime?
      Johnny   Because 2, 3, 4, 5, 6, 7, 8 and 9 don't go into it.
(B) Teacher   Why do you say 37 is prime?
      Johnny   Because 2, 3, 4, 5, 6, 7, 8 and 9 don't go into it.

These prompts were developed based on the experience of teaching prospective elementary school teachers and on research conducted on their understanding of number theory related concepts. In particular, Zazkis and Liljedahl (2004) demonstrated that frequently the primality of a number is determined by checking the divisibility of a given number by small numbers only, at times focusing only on numbers for which divisibility rules were known. This served as a basis for prompt A.

In prompt B there is no apparent mistake; unlike 143, 37 is indeed a prime number. This prompt aimed at focusing prospective teachers' attention on the strategy the student employed to determine primality, rather than on working towards a correct answer.

The teachers were asked to develop an instructional situation in which they address Johnny's response. They could include other students in the conversation, as they found appropriate.

42.3.1 Following Prompt A

The main approaches[1] used in scripting a conversation were: attending to the size of possible factors and inviting students to consider factors larger than 9 (see Sect. 42.3.1.1), and using the divisibility rules (see Sect. 42.3.1.2).

42.3.1.1 On the Size of Factors

The following excerpt invites students' reflection on the existence of factors larger than 9. The conversation presented below continues after it has been confirmed that the numbers 2–9 were not factors of 143.

---

[1] My goal in this chapter is to exemplify several ideas attended to in the scripts, rather than to enumerate the occurrences.

T: Can a number that is bigger than 9 be a factor for a number?
S: I do not know, maybe.

[...]

T: Well now let's take the number 100 for a moment because it is nice and simple. Can 2 go into 100?
S: Yes.
T: How do you know?
S: Because $2 \times 50 = 100$.
T: And what is the divisibility rule that we just learned that can also help us?
S: 100 is an even number that ends in zero, so it is divisible by 2.
T: Good. So we know that 2 is a factor of 100, but is 50 not also a factor of 100?
S: Yes, 50 is a factor.
T: But 50 is a lot bigger than the number one and it still counts?
S: I guess so.
T: Okay, so knowing that a number can have factors that are bigger than nine, I want everyone to get out their calculators and see if they can find other factors for 143. It's fine to use trial and error for this question.

The play proceeds with the expected discovery of the factors of 143 and the expected conclusion. However, while this excerpt clearly shows a number (100) with a factor greater than 9 (50), it does not necessarily address the source of Johnny's difficulty, because 100 also has small number factors, like 2 and 5. However, the claim in the prompt that a number that is not divisible by 2–9 is prime is not based on the belief that a number cannot have a 'large' factor; rather, it is based on the belief that a small factor is always present. Such a belief was explored in Zazkis and Campbell (1996b) and in Zazkis and Liljedahl (2004). In particular, participants in these studies expected divisibility of composite numbers by "small primes," and this expectation co-existed with their awareness of infinitely many primes. This expectation is explained by a student in Zazkis and Campbell (1996b) study, who reasons that "when you factor a number into its primes [...] just the whole idea of factoring things down into their smallest parts [...] gives me an idea that those parts are themselves going to be small" (p. 216). As such, finding factors of 143 (with the help of a calculator), as the teacher directs students to do at the end of this excerpt, changes the students' ideas with respect to the primality of 143, but does not address the source of the presented confusion.

42.3.1.2 Focus on Divisibility Rules

While most scripts made use of divisibility rules (to determine or to confirm that numbers from 2 to 9 are not factors of 143), the following excerpt features the introduction of the divisibility rule for 11. But before getting to this divisibility rule, the class is invited to revisit what a prime is and what a factor is; then the play continues.

T: Okay class, did you wonder if 143 is divisible by numbers bigger than 9?
Johnny: No, because if 2–9 won't divide into 143 then any number made up of those numbers won't divide into it.
T: I see, well, what if we look at numbers higher than 9... what about 10?
Johnny: No, because the number would have to end in a zero to be divided by 10, like 100.
T: I see. Well do we know the rules for numbers higher than 10?
Johnny: No, I do not remember learning them.

[...]

Sue: Well, I tried 11.
T   How did you try 11?

(Sue shows a calculator)

T: Does anyone know the divisibility rule for 11?
Bobby: Yeah. For 11 you take the number *(T writes 143 on the board)* subtract the last digit from the first two, which equals 11 *(writes $14 - 3 = 11$)*. The answer is 11 so yes, 143 is divisible by 11.

(Johnny writes down the rule and tries to solve the problem himself)

Johnny: Well I see now that 11 goes into 143... so what is the other number?
T: Well, why don't we do the long division as a class. 11/143... How many times? Please work it out on your paper and put your hand up when you have the answer.

As is often the case in the plays, there is at least one student who recalls the desired rule. In fact, only a particular case of divisibility by 11 is demonstrated, but this does not appear to bother the teacher.

Sue's discovery of 11 being a factor 143, with the help of a calculator, is pushed aside by the teacher, who prefers to focus on divisibility rules. Further, the suggested strategy to find "the other number," that is, the other factor of 143, is to use long division. The developed script leaves the impression that performing division with a calculator is insufficient to reach a conclusion related to divisibility, that preference is given to rules or algorithms.

When it is confirmed that the 'rule' applies to 143, Johnny asks about the other factor of 143. The teacher invites the class to work on long division "as a class" (and it is interesting to consider why the teacher wants everyone to do it) in order to answer Johnny's question. But another way of addressing Johnny would be to ask whether the other factor is indeed needed. The teacher could use this opportunity to help students realize that only one factor is required to conclude that 143 is a composite number.

It is rather surprising that the script-writer assumed that elementary school students would be familiar with divisibility rules for 11 or 13. Those rules, which can be developed for any prime number, reveal some fascinating relationships among numbers (Eisenberg 2000). However, divisibility tests in the "calculator era" have little utility. The divisibility rules for 7 and beyond 10 are rarely discussed in current curricula for elementary school students or for prospective teachers of mathematics. Instead of using the rule as a method for determining divisibility, we see its role today more as an opportunity to engage in mathematical reasoning: either to develop, test and refine conjectured rules or to try to understand how and why the rules work.

Overall, teachers' dependence on divisibility rules presents a concern. It appears that, for them, the notion of divisibility is connected to a specific rule rather than to the multiplicative relationship of numbers. The rule is valued much more than other means of determining divisibility, such as long division or a calculator.

42.3.2   Following Prompt B

Even though all the scripts either acknowledged or checked that indeed 143 is not divisible by numbers smaller than 10, no script-writer seemed to question this strategy. That is, having acknowledged that the number is not divisible by 2 and 3, why was there a need to check for divisibility for 4, 6 or 9? It was in order to focus on the strategy, rather than the correctness of the decision, that script-writers were presented with prompt B.

In this case, the number under investigation is indeed prime, but the suggested student's answer includes unnecessary information and may again hint at an inappropriate strategy for checking for primes. However, the student strategy, refined in Sect. 42.3.2.1, was not the focus of most plays, rather, the focus was on the definition of primes (see Sect. 42.3.2.2) and on divisibility rules (Sect. 42.3.2.3).

42.3.2.1   Focus on Prime Factors (Only)

In the following excerpt, the strategy for determining primality of a number was explicitly acknowledged by a student-character. The conversation between two students in the excerpt below takes place after the teacher asked the class whether they agreed with the conclusion presented in the prompts and attributed to Student 1.

Student 2:   We only need to divide 37 by other primes. That is, for 37 you could try 2, 3, 5 and 7. You could stop at 7, as $7 \times 7 = 49$ which is bigger than 37. All the other numbers are composite numbers using these primes. So if the primes don't divide the number the composite cannot either.

Student 1:   So I do not need to try and divide all the numbers into 37 to see if it is prime?

Student 2:   All you have to do to find out if a number is prime is divide it by other prime numbers that if multiplied by themselves would be less or equal to the number you are looking at.
(the lesson continues in a different direction)

A correct strategy, which is an alternative to "try and divide all the numbers into 37," is attributed to Student 2. However, while the strategy is correctly summarized, the reason for it is not mentioned by the student and is not sought by the teacher. Further, the implementation of the strategy exemplified in this excerpt is different from the description of the strategy. That is, according to the description, one can stop at 5, while Student 2 explicitly claims "You could stop at 7, as $7 \times 7 = 49$ which is bigger than 37." Of course, there is no harm in checking divisibility by an additional prime number, even if it is inconsistent with the cited approach. It often happens that teachers can correctly describe the strategy for determining primality of a number, likely reciting what was learned, but they seem to lack trust in implementing it (Zazkis and Liljedahl 2004). That is, having checked divisibility by all the primes whose square is less than the number in question, they continue to check divisibility by other numbers, both prime and composite, "just in case" or "to be on the safe side." These actions are often connected to an inability to explain why only particular primes are to be considered. The lack of confidence in implementing the described strategy points to the potential lack of understanding of the strategy.

42.3.2.2   Focus on the Definition of Primes

The following excerpt begins with a discussion about the definition of primality, which exemplifies a common move (return-to-definition) in the lesson plays. There is no challenge to the approach used by Johnny.

T: What you say is true, but that's not how a prime number is defined. Do you remember the definition of a prime number?

S: Yea, sort of. A number that you can only divide by itself and one. Like 2, right?

T: Exactly. It can divide itself evenly. So what do we know about 37, if it is prime?

S: That it can be evenly divided by 1 and 37.

T: Great. So let's talk about the numbers you picked to try to divide 37. From 2 to 9. Can you please explain why you concluded that these numbers don't work?

S: I guess and checked with numbers 2–9 and I proved that it was prime. No matter what the divisor, unless it is 1 and 37, will not work.

T: Excellent. I think what we talked about is really useful; after all, the idea of prime numbers is quite difficult. I think I will get the class to stop their work and get them in their groups talking about prime numbers.

Let us focus on the teacher's response to a student's reasoning about why 37 is a prime number. Initially, the teacher appears satisfied with the student's conclusion, but not the reason for it because "that's not how a prime number is defined." It appears that there is an implicit expectation that a definition for a prime will be cited, rather than the results of checking for primality. A student recalls a definition of a prime as "A number that you can only divide by itself and one. Like 2, right?". Here, while the language attributed to a student can be improved, the teacher's rephrasing—"it can divide itself evenly"—does not communicate the idea of a number being prime. An extended discussion on the language related to number theory concepts is found in Zazkis et al. (2013). I bring this issue here to demonstrate that the scripts provide an opportunity to address the language of mathematics used by prospective teachers, both as a venue for analysis, and as a discussion in class.

Further, when the student offers the example of 2, the teacher does not take the opportunity to explore other examples, especially ones that are not even (and that might push on the student's idea of what dividing evenly might mean). Yet again I highlight this point to demonstrate that focusing on the examples found in scripts provides an additional layer of analysis, as well as the basis for a discussion with teachers on the choice of instructional examples.

42.3.2.3   Searching for a Number with Prime Factors Larger Than 9

In the next excerpt, the script-writer attempts to work on the problematic strategy proposed by Johnny. The following dialogue takes place after the definition of prime numbers was revisited.

Ms. L: Turn to your group of 3 and I'll give you your challenge. Your challenge is to find a number that is not a prime number *and* is also not divisible by 2, 3, 4, 5, 6, 7, 8 or 9. I will tell you that you don't need to look higher than 150.

Student 1: We have to test every number between 100 and 150 to see if it's divisible by 11 and 13.

Student 2: We don't need to test any even number since we know it's already divisible by 2.

Student 3: Oh yeah! And we don't have to test the numbers that end in 5 or 0 because they can be divided by 5!

Student 2: Here, let's use the whiteboard and write al the numbers it could be. *(Students write 101, 103, 107, 109, 111, etc. to 150.)*

Student 1: Oh, and remember Ms. L said that a number that has all its digits add up to something that can be divided by 3 means the whole number can be divided by 3. Yeah, that's one of the rules we wrote.

Student 2: Guys! Why don't we just use that list of divisibility tricks we made up in the first part of the class to cross off the other numbers? Then we won't have to do so much division. Our rules are still up on the board. Is that right, Ms. L.?
            (Students cross off numbers and are left with this list: 101, 103, 107, 109, 113, 121, 127, 131, 137, 139, 143, 149.)

Student 2: Now all we have to do is to test if those 12 numbers are divisible by 11 and 13

(Sharing the long division work, students find 121 and 143.)

Student 3: Ms. L., Look, we figured it out. It's 121 and 143! All the rest of the numbers up there are prime.

Ms. L: And 121 and 143 are what then?

Student 3: 121 and 143 are numbers that are not prime but they also don't divide by 2, 3, 4, 5, 6, 7, 8 or 9.

Ms. L: So are the divisibility rules from 2 to 9 always going to work to discover non-prime numbers?

Student 1: Nope!

Here, the attention has been diverted from the specific case of 37, and the students are working explicitly to produce an example where testing divisibility by the numbers from 2 to 9 is insufficient to conclude primality of a number. The students first sieve out multiples of numbers smaller than 10 from the list of numbers from 100 to 150, relying on divisibility rules that "are still up on the board", then check divisibility of the remaining numbers by 11 and 13, which results in identifying 121 and 143.

Students discover that checking divisibility by numbers from 2 to 9 is insufficient, or, in the words of the teacher not "always going to work to discover non-prime numbers." It is interesting to note that while the intention in this prompt was to focus on unnecessary steps in checking for primality of 37, this playwright focused on the strategy as being insufficient in some cases.

Despite the strength of this approach, there are further opportunities for the teacher to focus students' attention more squarely on the notion of primality than on numerical operations (division, especially). For example, while the students used their knowledge of divisibility rules and long division to find 121 and 143, the teacher might have intervened to show how these numbers can be constructed multiplicatively as 11×11 and 11×13. While it is possible that this particular script-writer will direct students' attention to the matter in the next lesson, research has shown that the connection between multiplication and division is frequently unaddressed in dealing with number theory related tasks. For example, prospective elementary teachers, when asked to find a 'large' 5-digit number divisible by 17 prefer to check for divisibility with calculator, rather than construct such a number by multiplying 17 by a 3 or 4-digit number (Hazzan and Zazkis 1999). Further, when asked to find a number with exactly 4 factors, participants' preferred strategy was to guess and check, rather than construct such a number as product of 2 primes (Zazkis and Campbell 1996a). As such, the strategy of sieving multiples of 2–9 on the list of numbers from 100 to 150 might have been the main approach that was available to the script-writer. As mentioned previously, presenting this approach for the scrutiny of prospective teachers can enrich their understanding of mathematics, which, in turn, shapes their pedagogical approaches.

42.4   Numbers and Their Factors

Is there a relationship between the size of a number and the number of its factors? To discuss this relationship, prospective teachers were presented with the following scenario:

> Bonnie and Clyde are discussing numbers and their factors. Bonnie claims that the larger a number gets, the more factors it will have. Clyde disagrees.

This prompt was developed as a result of having faced the repeated belief of students that the larger a number gets, the more factors it is expected to have (Zazkis 1999). This belief is consistent with the family of "intuitive rules" of the kind "more of A, more of B" discussed by Stavy and Tirosh (1996, 2000). For example, the erroneous idea that rectangles with a larger perimeter have larger areas is a frequent result of the reliance on this rule.

Prospective teachers were asked to develop a script for a conversation between these two characters that included their exchange of arguments. In particular, they were asked to consider examples and experiences that could have led Bonnie to this conclusion, the arguments that both sides use to convince each other, and what each one of them finds convincing. They were further asked to annotate the script analyzing the arguments of the characters and their examples.

The common tendency related to true or false decisions is exemplified in Sect. 42.4.1, a tendency to consider disconfirming evidence as particular counterexamples is discussed in Sects. 42.4.2 and 42.4.3. Furthermore, the particular role of the choice of examples to build a convincing, or a more convincing, argument is attended to in Sect. 42.4.4.

42.4.1   Who Is Right?

While no participant agreed with Bonnie (that larger numbers have more factors), there were various degrees of disagreement. A few scripts demonstrated a view in accord with mathematical convention by clearly rejecting Bonnie's claim. This is exemplified in the following:

Clyde: The answer to "True or False": As a number gets bigger the more factors it will have" is False. It may sometimes have more factors, but to say that it always does would be incorrect.

However, the verdict of 'false' to a statement that is 'sometimes true', or true in a large number of cases, is inconsistent with everyday reasoning. As such, even when a mathematically correct conclusion was drawn, some participants attempted to amend the theory by referring to a limited scope of applicability. Theory amendment is in accord with a mathematical/scientific norm in response to disconfirming evidence (Chinn and Brewer 1993), however, the amendment itself was usually incorrect. For example, consider the following views included in the commentary: *"Large numbers do not always have more factors. [...] Her statement could be true for even numbers that are increasing but it is not true for all numbers as a collective"*.

This statement exemplifies that a common tendency was to consider that Bonnie was "not totally wrong", "not completely right", or "only partly correct". This is consistent with intuitive application of fuzzy logic (e.g., Zazkis 1995) to a mathematical situation. Further, prime numbers were the most notable exceptions from the perceived 'rule', which is in accord with Chinn and Brewer (1993) category of "excluding data from the currently held theory", as is exemplified below.

42.4.2   Prime Numbers as Exceptions

Prime numbers immediately falsify Bonnie's initial claim. All script-writers attended to prime numbers, but this attention had different forms. Initial examples of small primes—such as 5 has fewer factors than 4, or that 7 has fewer factors than 6—were initially treated by Bonnie in many of the scripts as an anomaly. Consider the following reaction to disconfirming evidence:

Clyde: Exactly! Now haven't we just shown that larger numbers don't always have more factors?

Bonnie: Damn you and your tricks! No, I refuse to give in, maybe you have just selected the only two numbers that this general rule does not apply to. Maybe you chose an anomaly, the only exception to the rule; it's going to take more than just one counter example to persuade me!

Providing evidence that supports the claim was the usual reaction to the initial disconfirming evidence. However, as exemplified on the following commentary, *"Bonnie is selecting only composite numbers, and that is her mistake, she seems to be forgetting that there are more than just composite numbers."* This comment implicitly suggests that the script-writer may believe the statement to be correct for all composite numbers, that is, leaving out the primes. Other script-writers attributed this perception to their characters explicitly, as exemplified in the following excerpt:

Bonnie: The larger the number, the more factors it has.

Clyde: True, unless it's a PRIME NUMBER.

Bonnie: Why didn't you tell me this rule before, it could have helped save so much time!

Clyde: I wasn't sure myself either, I just didn't want you to think you were right so I denied it.

Students' tendency to reject evidence that is not in accord with their held beliefs was noted in several studies (e.g., Chinn and Brewer 1993; Edwards 1997). Given that script-writers were prospective teachers, this tendency of their script characters exemplifies their awareness of such behavior among students. However, in the assignment the participants were asked to acknowledge the erroneous claims of their characters in the accompanying commentary. When the characters' erroneous decisions were not noted, they are likely to be in accord with the script-writers' personal views. Excluding prime numbers from the generally accepted 'rule' was the most frequent conclusion attributed to Bonnie's character.

42.4.3 Powers of Primes as Exceptions

While prime numbers were the most frequently acknowledged exceptions, they were not seen as the only exceptions to the 'rule'. The other conclusion was to exclude powers of primes from the general assertion. The following dialogue was presented after several examples of prime numbers have been considered.

Bonnie: Prime numbers are the exception to the rule. They do not behave like other numbers. [...] The numbers that I am talking about when I say that the factors increase as the value of the numbers increase, are any number other than a prime.

Clyde: Okay, so what about the squares of prime numbers. For example, the square of 7 is 49, so its only factors are 1, 7 and 49. That means that a smaller number, like 12, actually has more factors than the larger number which is 49. I am confused.

Bonnie: Again, Clyde, we are looking at prime numbers in this situation. Any square of a prime will only have three factors just like you said. The same thing happens when you find the number of factors in the cube of a prime. But remember what I said before: prime numbers are the exception to the rule. [...] when leaving out numbers that can be factored into the base of a prime number, like you said, the rule does hold true.

Here Bonnie acknowledges prime numbers as exceptions, but later she is invited to look at squares of primes. As a result, the 'exceptions' to the rule are extended to include powers of primes. Though the expression "numbers that can be factored into the base of a prime number" used by both characters is colloquial, it is clear from the examples that this phrase refers to numbers whose prime factorization is a power of a single prime. This is yet another example of theory amendment, while the amendment itself is incorrect. As the script-writer does not comment on Bonnie's conclusion—that "the rule does hold true" once some numbers are excluded—it is likely that the teacher shared this belief. Other possible clusters of possible 'exceptions', such as the product of two large primes, were not discussed in any of the scripts.

42.4.4 On Counterexamples and Large Numbers

In all the scripts one counterexample was insufficient in convincing Bonnie to abandon her claim (recall Bonnie's claim in Sect. 42.4.1, "it's going to take more than just one counter example to persuade me!"). This shows the awareness of script-writers of the possible robust beliefs of their potential students, beliefs that they themselves may also possess. The following statement in a script-write's commentary summarizes this phenomenon:

> Bonnie insisted she was right until Clyde did more examples to prove her wrong. In order to thoroughly prove that this theory is a reliable one (without just taking someone's word for it), one must test the theory multiple times with various numbers. After picking a few strategic numbers, only a few examples are required before the trend can be seen that the size of the number does not influence the number of factors.

The tendency to treat counterexamples as exceptions, as in the case of 'large primes', was mentioned previously. However, counterexamples that included 'large' composite numbers that were close to each other had more convincing power than others. For example, comparing the number of factors of 512 (having 10 factors) and 513 (having 8 factors), or, in a different script, comparing the factors of 3800 and 3600 helped Bonnie reconsider her position. The script-writers demonstrated not only that several examples are essential, but also that examples with large numbers are—using a notion introduced by Mason (2006)—more 'exemplary', that is, are more likely to serve the intended purpose.

Attention to large numbers is an additional extension to the list of responses to "anomalous data" identified by Chinn and Brewer (1993): rechecking the evidence with additional and more convincing examples. Perhaps this extension is applicable to mathematics settings more than science settings.

42.5 Discussion

There are numerous advantages of the script-writing task for teachers, for teacher educators and for researchers. For teachers, advantages include the opportunity to examine their personal responses to students' erroneous perspectives, understand their origin and consider how students can be helped in overcoming their errors. Furthermore, scripting presents teachers with an opportunity to examine and extend their personal understanding of mathematics, without the need to "think on their feet", and to develop personal repertoires of general strategies to be used in future improvisations. As an "approximation of practice" (Grossman et al. 2009), expressing ideas in a form of a scripted dialogue affords a more thoughtful and pre-planned (rather than "real-time") response.

For mathematics teacher educators and researchers, there is an assumption that, when writing scripts, the writers present their personal views in the teacher's role. My experience suggests that even when a script involves a conversation among students, such as in the Bonnie and Clyde scenario, one of the characters acts in a "teacherly" mode. As such, advantages of scripting tasks for teacher educators include the opportunity to highlight a variety of appropriate pedagogical responses and direct teachers' further attention to learners. It is an opportunity to move beyond traditional "lesson planning", as a preparation for instruction. Script-writing introduces further variety to the tasks for prospective teachers in the design of "methods" courses.

For researchers, implementing scripting tasks provides a lens to examine teachers' ideas and discourse via their imagined actions and chosen words. It provides an opportunity to analyze both mathematical understanding and the chosen instructional approaches.

The goal of this chapter is to present an argument for, and exemplify the use of, script-writing tasks in mathematics teacher education and research. In the examples presented in Sect. 42.3, it is clearly demonstrated that prospective teachers are more comfortable to attend to student errors (reacting to the claim that 143 is prime) than to correct claims (that 37 is prime) followed by inefficient and incomplete justifications. This creates the need for further development of prompts in which correct answers are presented as a result of incorrect or insufficient reasoning and to extend a conversation with teachers that strengthens their attention to students' reasoning rather than to the correctness of the answer.

Furthermore, the examples presented in Sect. 42.4 highlight the teachers' tendency to implement fuzzy logic (e.g., Zazkis 1995), where the "middle ground" is sought between true and false. This accentuates the need for an explicit conversation on how the truth value of mathematical statements is determined, the issue that often remains implicit when mathematical encounters for prospective elementary teachers are considered.

Seeking ways to support and improve mathematics teacher education is a continuous challenge. I argue that engaging prospective teachers in script-writing is one possible way to address this challenge. Overall, the script-writing tasks, and the scripts written by prospective teachers, contribute to mathematics education on two arenas: they inform research on teachers' knowledge, and support instructional design in teacher education courses.

References

Aronson, E., & Carlsmith, J. M. (1968). Experimentation in social psychology. In G. Lindzey & E. Aronson (Eds.). *The handbook of social psychology* (Vol. 2, pp. 1–79). Reading, MA: Addison-Wesley.

Blatner, A. (2009). *Role playing in education*. Available: http://www.blatner.com/adam/pdntbk/rlplayedu.htm.

Chinn, C. A., & Brewer, W. F. (1993). The role of anomalous data in knowledge acquisition: A theoretical framework and implications for science instruction. *Review of Educational Research, 63*(1), 1–49.

Cruz, B., & Murthy, S. (2006). Breathing life into history: Using roleplaying to engage students. *Social Studies and the Young Learner, 18*(3), 4–8.

Edwards, L. D. (1997). Exploring the territory before proof: Students' generalizations in a computer microworld for transformation geometry. *International Journal of Computers for Mathematical Learning, 2*, 187–215.

Eisenberg, T. (2000). On divisibility by 7 and other low-valued primes. *International Journal for Mathematics Education in Science and Technology, 31*, 622–626.

Greenberg, J., & Eskew, D. E. (1993). The role of role-playing in organizational research. *Journal of Management, 19*(2), 221–241.

Grossman, P., Hammerness, K., & McDonald, M. (2009). Redefining teaching, re-imagining teacher education. *Teachers and Teaching: Theory and Practice, 15*(2), 273–289.

Hazzan, O., & Zazkis, R. (1999). A perspective on "give an example" tasks as opportunities to construct links among mathematical concepts. *Focus on Learning Problems in Mathematics, 21*(4), 1–14.

Howes, E. V., & Cruz, B. C. (2009). Role-playing in science education: An effective strategy for developing multiple perspectives. *Journal of Elementary Science Education, 21*(3), 33–46.

Jones, F. H., & Eimers, R. C. (1975). Role-playing to train elementary teachers to use a classroom management "skill package." *Journal of Applied Behavior Analysis, 8*, 421–433.

Joyner, B., & Young, L. (2006) Teaching medical students using role play: Twelve tips for successful role plays. *Medical Teacher, 28*(3), 225–229.

Kenworthy, L. S. (1973). Role-playing in teacher education. *Social Studies, 64*(6), 243–247.

Koichu, B., & Zazkis, R. (2013). Decoding a proof of Fermat's little theorem via script writing. *Journal of Mathematical Behavior, 32*, 364–376.

Kontorovich, I., & Zazkis, R. (2016). Turn versus shape: Teachers cope with incompatible perspectives on angle. *Educational Studies in Mathematics, 93*(2), 223–243.

Lajoie, C., & Maheux, J.-F. (2013). Richness and complexity of teaching division: Prospective elementary teachers' roleplaying on a division with remainder. In B. Ubuz (Ed.), *Proceedings of the Eighth Congress of European Research in Mathematics Education* (CERME 8), Manavgat-Side, Antalya, Turkey.

Lakatos, I. (1976). *Proofs and refutations*. Cambridge: Cambridge University Press.

Lawson, T. J., McDonough, T. A., & Bodle, J. H. (2010). Confronting prejudiced comments: Effectiveness of a role-playing exercise. *Teaching of Psychology, 37*, 256–261.

Lowenstein, A. J. (2007). Role play. In M. J. Bradshaw & A. J. Lowenstein (Eds.), *Innovative teaching strategies in nursing* (4th ed., pp. 173–182). Boston, MA: Jones and Bartlett.

Maheux, J.-F., & Lajoie, C. (2011). On improvisation in teaching and teacher education. *Complicity: An International Journal of Complexity and Education, 8*(2), pp. 86–92.

Mason, J. (2006). What makes an example exemplary: Pedagogical and didactical issues in appreciating multiplicative structures. In R. Zazkis & S. R. Campbell (Eds.), *Number theory in mathematics education: Perspectives and prospects* (pp. 41–68). Hillsdale, NJ: Lawrence Erlbaum Press.

McGregor, J. (1993). Effectiveness of role playing and antiracist teaching in reducing student prejudice. *Journal of Educational Research, 86*(4), 215–226.

Palmer, D. H. (2006). Sources of self-efficacy in a science methods course for primary teacher education students. *Research in Science Education, 36*, 337–353.

Plous, S. (2000). Responding to overt displays of prejudice: A role playing exercise. *Teaching of Psychology, 27*(3), 198–201.

Rowland, T., & Ruthven, K. (Eds.) (2011). *Mathematical knowledge in teaching*. Springer.

Shapiro, S., & Leopold, L. (2012). A critical role for role-playing pedagogy. *TESL Canada Journal, 29*(2), 120–130.

Stavy, R., & Tirosh, D. (1996). Intuitive rules in science and mathematics: The case of "more of A —more of B". *International Journal of Science Education, 18*(6), 653–667.

Stavy, R., & Tirosh, D. (2000). *How students (mis-)understand science and mathematics: Intuitive rules*. New York: Teachers College Press.

Van Ments, M. (1983). *The effective uses of role-play: A handbook for teachers and trainers.* London: Kogan Page.

Watson, A., & Mason, J. (2005). *Mathematics as a constructive activity: Learners generating examples.* Mahwah, NJ: Lawrence Erlbaum.

Zazkis, D. (2014). Proof-scripts as a lens for exploring students' understanding of odd/even functions. *Journal of Mathematical Behavior, 35,* 31–43.

Zazkis, D., & Zazkis, R. (2016). Prospective teachers' conceptions of proof comprehension: Revisiting a proof of the Pythagorean theorem. *International Journal of Mathematics and Science Education, 14,* 777–803.

Zazkis, R. (1995). Fuzzy thinking on non-fuzzy situations: Understanding students' perspective. *For the Learning of Mathematics, 15*(3), 39–42.

Zazkis, R. (1999). Intuitive rules in number theory: Example of "the more of A, the more of B" rule implementation. *Educational Studies in Mathematics, 40*(2), 197–209.

Zazkis, R., & Campbell. S. R. (1996a). Divisibility and multiplicative structure of natural numbers: Preservice teachers' understanding. *Journal for Research in Mathematics Education, 27*(5), 540–563.

Zazkis, R., & Campbell. S. R. (1996b). Prime decomposition: Understanding uniqueness. *Journal of Mathematical Behavior, 15*(2), 207–218.

Zazkis, R., & Liljedahl, P. (2004). Understanding primes: The role of representation. *Journal for Research in Mathematics Education, 35*(3), 164–186.

Zazkis, R., & Kontorovich, I. (2016). A curious case of superscript (-1): Prospective secondary mathematics teachers explain. *Journal of Mathematical Behavior, 43,* 98–110.

Zazkis, R., Liljedahl, P., & Sinclair, N. (2009). Lesson plays: Planning teaching versus teaching planning. *For the Learning of Mathematics, 29*(1), 40–47.

Zazkis, R., Sinclair, N., & Liljedahl. P. (2013). *Lesson play in mathematics education: A tool for research and professional development.* Dordrecht, The Netherlands: Springer.

Zazkis, R., & Zazkis, D. (2014). Script writing in the mathematics classroom: Imaginary conversations on the structure of numbers. *Research in Mathematics Education, 16*(1), 54–70.

Chapter 43
Equity in Mathematics Education: What Did TIMSS and PISA Tell Us in the Last Two Decades?

Yan Zhu

Abstract Equity in education has been a concern of almost all countries, whether developed, transitional, or in the progress of developing. It is believed that unequal education implies that human potential is being wasted. The present study focused on students with different characteristics as aggregate groups in an examination of similarities and differences in mathematics learning. The information analyzed here was mainly based on data from TIMSS and PISA databases. This investigation aims to paint an overall picture about gender equity, socioeconomic status, and indigenity equity in mathematics education over the last twenty years. It is hoped that the study can provide useful insights to individual education systems and further help them to identify more promising practices to narrow or even eliminate the existing between-system as well as within-system gaps.

Keywords Gender equity · Socioeconomic status equity · Immigrant background equity · PISA · TIMSS

43.1 Equity in Mathematics Education

Equity has been on the agenda of mathematics education research for at least four decades. For instance, Fennema's (1974) seminal work was about male-female differences in mathematics achievement and Fennema and Sherman (1977) approached the topic via affective perspectives. In the first *Handbook of Research on Mathematics Teaching and Learning* (Grouws 1992), there are two chapters dealing with this issue. Year 1995 witnessed the publication of two books that concerned research on equity within mathematics education (Rogers and Kaiser 1995; Secada et al. 1995). Pais (2012) noted that this interest in equity has proliferated theories in mathematics education research that progressively deemphasized cognitive psychology as an interpretative framework for mathematics learning

Y. Zhu (✉)
East China Normal University, Shanghai, People's Republic of China
e-mail: yzhu@kcx.ecnu.edu.cn

© The Author(s) 2018
G. Kaiser et al. (eds.), *Invited Lectures from the 13th International Congress on Mathematical Education*, ICME-13 Monographs,
https://doi.org/10.1007/978-3-319-72170-5_43

in favor of more socio-cultural oriented frameworks. Consistently, Gutiérrez (2010) has claimed that "sociocultural theories, once seen on the fringe of a mainly cognitive field, now take their place squarely within mainstream mathematics education journals like *JRME*" (p. 2). The National Council of Teachers of Mathematics, in *Principles and Standards for School Mathematics* (NCTM 2000), identifies as its first guiding Principle that "Excellence in mathematics education requires equity—high expectations and strong support for all students" (p. 11).

While the growing attention to equity in mathematics education promotes 'talk' of equity becoming more mainstream in the mathematics education community (Gutiérrez and Dixon-Roman 2011), the term often has different meanings to different people. In fact, when referring to educational equity, the term *equality* is often used interchangeably. It can be seen that the two terms have close similarities but with important distinctions. The Oxford English Dictionary defines *equity* as "the quality of being fair and impartial" and *equality* as "the state of being equal, especially in status, rights, or opportunities". Based on these two definitions, the former appears to be more about being impartial with the latter more about being the same. In this sense, differences related to individual needs and requirements are recognized and treated in the notion of equity; in contrast, everyone will be regarded and treated in the same manner without focusing on specific needs and requirements in the notion of equality.

Regarding the subject of mathematics, NCTM (2000) has highlighted that equity does not mean that every student should receive identical instruction; instead, reasonable and appropriate accommodations should be made to promote access and attainment for all students. Similarly, Secada (1989) argued that the two notions are not synonymous, remarking that rather than striving for equality, people should work towards equitable inequalities that reflect the needs and strengths of individuals. In fact, as Volmink (1994) described mathematics as a field that is the "sole creation of a few, singularly brilliant … individuals" (p. 51), it is unrealistic to expect all individuals to achieve equally in this elitist field. Correspondingly, it would become more important and meaningful to investigate the source of the differences as well as identify the reasons underlying so as to pursue *equitable inequalities* that reflect individuals' specific strengths and needs.

Arnaud (2001) and Arnesson (2001), respectively, did some concise analyses on equity and equality in a more general sense, and their work problematized the conceptualization of the two notions. Arnaud commented that, due to being associated with fairness and impartiality, the notion of equity could possibly be regarded as a means to bring harmony into progressive societies and/or solve conflicts in some legal cultures. Arnesson, on the other hand, proposed that, as equality has been linked to the basic idea of being the same, it could bring up with two issues: one is about who should be the same and the other is about how important to be the same. Although there exist ambiguities and disagreements with respect to the two notions, Arnaud (2001) pointed out that the relationship between them appears to be a newly contemporary and significant notion (also see Herrera 2007). To a certain extent, this idea supports Hutmacher's (2001) call for the need of a clear conceptual framework with the potential to be a starting point in studying equity and equality.

Not only are there similar concepts to equity, but also the notion itself can be understood in many different ways. For instance, the Organization for Economic Cooperation and Development (OECD) defines equity in education through two dimensions: fairness and inclusion (Field et al. 2007). This suggests that, by fairness, equity implies ensuring that personal and socio-economic circumstances (e.g., gender, ethnic origin, family background) should not be obstacles to achieving education potential, while by inclusion, it implies ensuring all students to reach at least a basic minimum level of skills. It shall be noted that the two dimensions are closely intertwined. More specifically, equitable education is expected to support students to reach their learning potential without either formally or informally pre-setting barriers or lowering expectations (OECD 2012). In this sense, tackling school failure may help to overcome the effects of social deprivation, which often causes school failure (Field et al. 2007).

The NCTM research committee also remarks that there are multiple concepts encompassed in the notion of equity and they can be classified into either *conditions* of learning or the *outcomes*. According to Lipman (2004), the former can be described as "equitable distribution of material and human resources, intellectually challenging curricula, educational experiences that build on students' cultures, languages, home experiences, and identities, and pedagogies that prepare students to engage in critical thought and democratic participation in society" (p. 3). From the perspective of *outcomes*, Gutstein (2000) defined equity as "obliterating the differential and socially unjust outcomes in mathematics education" (p. 26). The mentioned outcomes could include students' achievement and participation in mathematics, their powers of analyzing and reasoning, and their abilities to "critique knowledge or events" (Gutstein et al. 2005).

Brown (2006) differentiated equity into horizontal equity and vertical equity. Correspondingly, horizontal equity refers to equal treatment of those who are similar to each other and vertical equity refers to unequal but equitable treatment of those with different needs, which is designed to reduce inequality. It is suggested that horizontal equity is a starting point that can be used to help achieve vertical equity. In this sense, vertical equity will look into whose situation can be improved, and then how to make the improvement.

Gutiérrez (2009) proposed a four-dimension model for the notion of equity in mathematics learning including: *access*, *achievement*, *identity*, and *power*. In particular, *access* relates to the tangible resources that students have available to them to participate in mathematics, and *achievement* refers to observable results for students at all level of mathematics. The two were further characterized by Gutiérrez as the dominant axis of equity, which measure how well students "play the game" of mathematics as it currently stands. Moreover, *access* is suggested to be a precursor to *achievement*. The dimension *identity* concerns not only students' pasts (e.g., the contributions of their ancestors), but more about a balance between themselves and others. In other words, students need to have opportunities to see themselves reflected in the curriculum while having a view onto a broader world. The issue of *power* is raised because equity is more than having students "be themselves and better themselves" via doing mathematics. According to Gutiérrez

(2007), it cannot be called equity if mathematics as a filed and/or people's relationship has no changes. Gutiérrez further proposed *identity* and *power* to make up the critical axis of equity, as they challenge the static formalism embedded in traditions. This axis relates to students' ability to "change the game". On this axis, *identity* can be regarded as a precursor to *power*.

Although there are different understandings and interpretations about the notion of equity, it is suggested that the relevant research in mathematics education can be used to help in understanding the causes for the inequalities and identifying strategies to reduce the disparities and the effects of these inequalities (Rohn 2013). Furthermore, some researchers have even claimed that only focusing on equity and equality is not enough, and additional attention should be further given to liberation. Liberation refers to working to challenge and reverse the effects of structural oppression in society. No matter what standpoints people take, it has generally been believed that high performing education systems are those that combine equity and quality, where all students are given opportunities for a good quality education.

43.2   Factors Contributing to Inequity in Education

There are many factors that may influence inequitable opportunities and outcomes in education, such as gender, income and socio-economic status, ethnicity, indigeneity, culture, religion, language, geographical location, etc. (Wood et al. 2011). Different countries usually would use different sub-sets from this set of categories to define diversity, and assess how equitable their education systems are. Among these factors, gender, typically, appeared to be the most widely used category (Clancy and Goasstellec 2007).

Within the field of mathematics education research, gender was historically the initial dimension of equity researched widely, and later served as the springboard for emphases on, or in combination with, the other dimensions of equity (Forgasz and Rivera 2012). In earlier times, a wide range of international research studies reported gender inequities, with most favoring males. For instance, Maccoby and Jacklin's (1974) review of close to 1600 studies of gender differences concluded that boys were better in mathematics and physical sciences, whereas girls were better in reading and writing. Later reviews, adopting more sophisticated meta-analytical techniques, consistently reported similar patterns of gender difference, although the magnitudes of the differences were smaller (e.g., Wider and Powell 1989; Willingham and Cole 1997). In fact, mathematics was traditionally stereotyped as a male domain and societal influences tended to suggest that mathematical learning was not particularly appropriate for girls (e.g., Damarin 1995; Fennema 2000; Leder 1992).

In order to explain the potential gender gap in mathematics learning, various theories have been explored (see Wider and Powell 1989). One strand of such explorations looks into biological differences to support innate differences in spatial ability, higher order thinking, or brain development. Nevertheless, some researchers have pointed out that such differences are small and their relationships with mathematics test performance are tenuous (e.g., Guiso et al. 2008). Kane and Mertz (2011) further argued that if gender differences are primarily a consequence of innate, biologically determined differences, they should be expected to be similar across countries, regardless of culture, and should remain constant across time. Another strand of research emphasizes societal factors, highlighting how girls are socialized into believing that mathematics is not important, useful, doable, or part of a girl's identity. For instance, West and Zimmeran (1987) remarked that a person's gender is not simply an aspect of what one is, but more fundamentally it is something that one does recurrently in interaction with others; they called this "doing gender". According to Kaiser (2003), the social construction of gender forms the theoretical base of many empirical studies on the relationship between gender and mathematics.

In more recent studies, however, researchers have observed that the gender differences in mathematics have not only narrowed substantially over time, but sometimes have even been eliminated (e.g., Halpern et al. 2005; Hyde and Mertz 2009; Spielman 2008). In fact, there are some researchers who have started including boys' educational needs into their work on gender (e.g., Forgasz and Leder 2001; Lingard et al. 2002; Weiner et al. 1997). Regarding the complexity of the gender gap in mathematics achievement, Ellison and Swanson (2010) attributed it to the differences that exist between tests and systems.

Compared to gender, students' socioeconomic status (SES) was not enunciated as a problem in the field of mathematics education until the 1980s, when the "social turn" was advanced (Lerman 2000, 2006). Since the 1990s, the number of studies investigating the relation between SES and students' mathematics achievements has increased, with growing importance given to periodic, international, standardized, comparative studies as TIMSS and PISA (Valero and Meaney 2014). On the other hand, studies investigating the connection between people's social and economic position and school achievements emerged much earlier, at the beginning of the 20th century (see Valero et al. 2015). The *Coleman Report* (Coleman et al. 1966) was one of the first large scale national surveys that acknowledged socioeconomic status as a major predictor of educational achievement (Knapp and Woolverton 2004).

No matter whether it is education in general, or specific to mathematics education, the existing research consistently demonstrates a positive correlation between students' socioeconomic status and their academic achievement level. Such findings have been reported in both international large-scale assessments, and school level assessments. Sirin's (2005) meta-analysis on SES and academic achievement in journal articles published between 1990 and 2000 revealed a medium to strong association. Rothman and McMillan (2003) further identified that the relationship within schools was relatively small, although significant; while that between schools was much larger and significant.

Following the "age of migration" (Castles and Miller 2003), many countries now host a substantial and growing population of immigrants, a considerable number of whom are children. Consequently, immigrant children's educational performance has become one of policymakers' core concerns. The observations that the educational performance of children with an immigration background often differs from that of their host countries, and is also different from their countries of origin; these two important macro-level factors invite explanations. At the individual level, the relevance of classic background attributes for explaining the educational achievement of immigrant children has been well documented (Kao and Thompson 2003). Some frequently mentioned attributes include socioeconomic status, parental income, and cultural capital (e.g., number of books at home), language spoken at home, and age of arrival in the host country.

Focusing on the learning of mathematics, it is suggested that there is a lot to learn for a newcomer, and the least problematic may be mathematics (Bishop 2006). Besides the subject, a learner's mathematics practices would be shaped and negotiated by classroom participants with various levels of shaping power. In particular, classroom teachers have power of the formal and institutional kind, classmates or peers play a fundamental role at the level of being near equals, that is, equality between the one who chooses to exercise influence and the one who is chosen to be influenced. In some sense, the learner has the most power over his/her own learning, such as choosing how much effort to expand, whom to listen to, and whose views to respect. Moreover, the learning is also a product of the learner's cultural and social history, shaped in large by his/her family life and outside-school life experiences. In this regard, parents are particularly influential.

While a variety of factors at different levels (i.e., macro, mezzo, micro) could cause inequality in education, it should be noted that the inequalities and injustices often do not work in isolation, but rather a combination of two or more of this diversity of dimensions. For instance, the status of being poor and living in a rural area could increase disadvantage several times over (Morely et al. 2009; UNESCO 2008). Moreover, in many countries, some of these factors have clear historical roots and trajectories. Consequently, co-occurrences of low achievement among some ethnic minority groups and those of low socio-economic status are then not unusual. In fact, in some countries that were regarded as 'highly developed', structural inequalities also existed (United Nations Development Program 2009). For instance, in the UK and the US, ethnicity and low socioeconomic status appeared to be two of the main risk factors for students' underperformance in schools. While a list of such contributory attributes is far from exhaustive, identifying causal relations between the sources and consequences of educational disadvantage is important for making effective policy recommendations.

43.3 What Do TIMSS and PISA Tell About Equity in Mathematics Education?

43.3.1 TIMSS Versus PISA

The Trends in International Mathematics and Science Study (TIMSS) is one of the studies established by the International Association for the Evaluation of Educational Achievement (IEA). It aims to measure the extent to which students have mastered the topics and skills as appeared in school curricula. A *Pilot Twelve-Country Study*, conducted in 1959 to 1962, was the very first IEA study, and increasingly more education systems participated in its later cycles. The term *TIMSS* first appeared in 1995, known as *the Third International Mathematics and Science Study,* and was renamed *the Trends in International Mathematics and Science Study* in 1999 and onwards. Meanwhile, the series of studies were conducted in regular four-year cycles from 1995. The most recent study was TIMSS 2015 with more than 60 systems participating. In most of the cycles, 4th and 8th graders' achievement in mathematics and science were assessed. TIMSS uses the curriculum as its major organizational aspect. Three curriculum layers are envisaged: intended curriculum (i.e., the subject intended for students to learn, and how the education system should be organized to facilitate this learning); implemented curriculum (i.e., what is actually taught in the classroom, who teaches it, and how is it taught); and attained curriculum (i.e., what it is that students have learned, and what they think about the subject).

The Program for International Student Assessment (PISA) is another worldwide large-scale study, which is under the auspices of the Organization for Economic Co-operation and Development (OECD). It aims to look "at young people's ability to use their knowledge and skills in order to meet real-life challenges rather than how well they had mastered a specific school curriculum" (OECD 2005, p. 9). The PISA study series was first implemented in 2000 and then repeated every three years. The most recent was PISA 2015, with 71 countries/economies participating. Due to its focus on the practicalities of students' skills, PISA uses the term *literacy* referring to "the capacity of students to apply knowledge and skills in key subject areas and to analyse, reason and communicate effectively as they pose, solve and interpret problems in a variety of situations".

Coessens et al. (2014) differentiated the two international large-scale studies from four perspectives. The first is that TIMSS focuses on curriculum-related tasks, while PISA is literacy based. This links to the second difference, that is, TIMSS items are more knowledge oriented, while PISA items are aimed at life skills. Third, TIMSS focuses on the extent to which students have mastered mathematics and science as they appear in school curricula, while PISA aims to capture the ability to use mathematical and scientific knowledge and skills to meet real-life challenges. In short, TIMSS focuses more on pure mathematical performance, while PISA focuses more on the practicalities of mathematical skills. Fourth, TIMSS is explicitly organized around two frameworks, a curriculum framework and an assessment framework, while PISA focuses on skills for future life rather than on the grasp of the school curriculum. There are also researchers differentiating the two studies from the perspective of targeted populations. For instance, Harlen (2001) highlighted that while TIMSS assessed the progress of students at particular grade levels, and so at different ages for countries, PISA was concerned with 15-year-old students' performance as an indicator of the outcomes of compulsory education. Similarly, Lester (2007) summarized the difference as TIMSS holding a grade-specific structure versus PISA holding an age-specific structure.

Regarding the issue of equity, it appears that PISA gives it more explicit attention. In particular, PISA defines the notion of equity as "to provide all students, regardless of gender, family background or socio-economic status (SES), with opportunities to benefit from education" (OECD 2013a, p. 13). In this sense, equity implies more than everyone having the same results, but everyone, regardless of his/her background, should be offered access to quality educational resources and opportunities to learn. As a result, one's gender, SES, or immigrant background should then have little or no impact on his/her performance. The following sections of this chapter will focus on what TIMSS and PISA have found regarding equity in mathematics education in the last two decades, from the perspectives of three important personal background aspects: gender, socioeconomic status, and immigrant background.

43.3.2 Gender Equity

Gender equity has become one of the most prominent issues in education reform efforts worldwide, with international organizations and governments having increasingly recognized that gender equity strengthens democracy, and serves as a hallmark of an inclusive society that values and capitalizes on the contributions of all its members.

Related to the subject of mathematics, gender has always been an issue of concern that is investigated in the IEA study series. As early as in the *Pilot Twelve-Country Study* (13-year-olds), girls were, in general, observed to be outperformed by boys. In the *First International Mathematics Study* (FIMS), Keeves (1973) found that boys performed better than girls in all ten original FIMS countries in terms of overall mathematics achievement, with some variations in the magnitudes of the differences at the 13-year-old level (Population I) across the countries. Based on the data from all the twelve FIMS countries, Steinkamp et al. (1985) again reported that boys outperformed girls in 10 countries in overall mathematics achievement, and in eight countries the differences reached statistical significance. Husén (1967) further claimed that while the gender differences in favor of boys appeared to be a global phenomenon, the differences in favor of girls were observed within some countries, although overall, the differences were insignificant. Besides test performance, Steinkamp et al. further identified three important contextual variables for gender differences in mathematics learning including: student attitudes, the opportunity to learn, and the amount of homework. Compared to the differences at the secondary school level, those at the pre-university level (Population II) were greater (Keeves 1973). Harnisch et al. (1986) suggested that the gender differences were pervasive across cultures, and that non-biological factors played a role in determining the magnitudes.

Interestingly, the terms used in the discussion of differences between the sexes changed from FIMS to SIMS (*Second International Mathematics Study*), with "gender differences" gaining prominence over "sex differences". Hanna (2000) argued that such a change may imply that "gender" could be a term more appropriate for describing psychological, social, attitudinal, and cultural characteristics, while "sex" could be one reserved for immutable biological characteristics. The SIMS revealed that boys outperformed girls significantly in seven out of the 19 countries, girls outperformed boys significantly in four, and no significant gender differences were found in the remaining eight countries (Baker and Jones 1993). Based on the results, Baker and Jones pointed out that the gender differences in SIMS varied in both size and direction among countries. Furthermore, Hanna (2000) claimed that while the gender differences varied widely from country to country, between-country differences were smaller than within-country ones.

In *the Third International Mathematics and Science Study* (TIMSS) at the fourth grade, the mathematics achievements in most countries were approximately the same for boys and girls, although in three countries statistically significance differences were found in favor of males. Similarly, the differences at the eighth grade level were also small or negligible overall. However, all the statistically significant differences were consistently found to favor male students (in eight countries). A rather different finding was revealed at the twelfth grade, where males in most countries had significantly higher average achievement than females in both mathematics literacy and in advanced mathematics.

Based on the review of gender differences in mathematics achievement from FIMS to SIMS to TIMSS, Hanna (2003) proposed "the end of gender differences" (p. 209). According to Baker and Wiseman (2005), among the countries participating in only the 1960s and 1990s assessments, and those in just the 1980s and 1990s assessments, the proportion of countries with statistically significant male-dominated gender differences in mathematics scores declined from 33 to 9% from the 1960s to 1990s, and from 35 to 18% from the 1980s to 1990s.

Such a diminishing tendency further continued in the later TIMSS studies. In particular, most of the gender differences found in TIMSS 1999 were negligible, and no country showed a significant increase in difference over time. In TIMSS 2003, gender differences in favor of girls matched gender differences in favor of boys in terms of number and magnitude. TIMSS 2007 revealed negligible gender differences at the fourth grade in roughly half of the participating countries; in the remaining countries, girls had higher achievement in about half and boys had higher achievement in the other half. Interestingly, at the eighth grade, TIMSS 2007 found that, on average, girls had higher achievement than boys. In TIMSS 2011, female fourth graders' average mathematics score was only 1 point lower than that of male students, while female eighth graders' average mathematics score was 4 points higher than that of male students. Moreover, at the fourth grade, of the 50 participating countries, about 24 had significant gender differences with all but four in favor of boys; at the eighth grade, of the 42 countries, about 20 had significant gender differences, with seven favoring boys and 13 favoring girls.

While TIMSS is more concerned about school curriculum related mathematics achievement, PISA focuses on everyday skills-related mathematics achievement. Based on the PISA data, Forgasz and Hill (2013) argued that the gap between boys and girls on mathematics literacy had widened. It can be seen that in the first PISA study, boys tended to perform at somewhat higher levels in most countries, with an average gap of 11 points, and 17 out of the 42 participating countries (40.5%) revealed statistically significant differences in favor of boys. The PISA 2003 mathematics assessment consistently revealed an overall gender difference of 11 score points in favor of boys. Of the 41 participating countries, 27 (65.9%) revealed a gender difference in favor of boys, and one with a difference in favor of girls. PISA 2006 similarly revealed more than 60% of the participating countries (35 out of 57) having boys outperforming girls in mathematical literacy at a significant level. In both PISA 2009 and PISA 2012, the proportions of countries with an advantage in favour of boys in mathematics literacy went down to 61.4% and 56.9%, respectively. Researchers claimed that a wider gender gap among the least and most able students was actually masked (OECD 2013b). In fact, in most countries, the most able girls lagged behind the most able boys. For instance, in PISA 2009 it was found that only 3.4% of girls compared to 6.6% of boys were at the top performance in mathematics literacy.

Based on both the TIMSS and PISA databases, Baye and Monseur (2016) analyzed gender differences, from an international perspective, from 1995 to 2015 via the use of effect sizes and variance ratios. The results showed that the sizes of the gender differences varied according to student proficiency levels. In particular, at the lower tail of the distribution, effect sizes were close to zero or in favor of girls, while systematically at the upper tail, boys were more proficient. The largest gender difference in mathematics literacy was observed on PISA 2003 for the most proficient students (i.e., percentile 95). Baye and Monseur claimed that such a tendency was more obvious at the secondary level of education, and in PISA rather than in TIMSS. In fact, at the eighth grade level, the TIMSS data revealed that the tendency for boys to outperform girls at the upper end of the distribution had decreased over time, which is consistent with the overall pattern.

43.3.3   Socioeconomic Status Equity

Regarding students' socioeconomic status, PISA constructs a composite index, *Economic, Social and Cultural Status* (ESCS), which is derived from three variables related to family background: highest level of parental education, highest parental occupation, and the number of possession in the home. While information about parental education and home possessions were also collected in TIMSS, the study constructed a composite index in a less consistent way. In particular, TIMSS 1999 and TIMSS 2011 created an index of *Home Educational Resources* (BSDGHER), while in other years the emphasis in the investigation was more focused on the item base.

In general, no matter whether the influence of students' socioeconomic status on their mathematics achievement was investigated via individual items or via a composite index, all the TIMSS studies consistently revealed that students having more books in the home and parents with more education achieved better scores in mathematics. This pattern was observed in all of the participating countries and in all of the years. On the other hand, the data also revealed that students at the high level of *Home Educational Resources* were relatively rare in most countries. In TIMSS 1999, there were just 9% of eighth graders in this category on average, and their mathematics achievement was 109 score points higher than those in the low category (19%). Similarly, in TIMSS 2011, about 17% of fourth graders were in the high category (i.e., *many resources*), and their mathematics achievement was 109 score points higher than those in the low category (i.e., *few resources*: 9%); about 12% of eighth graders were in the high category and their mathematics achievement was 107 score points higher than those in the low category (20.5%).

PISA 2003 shows that although poor performance in school does not automatically follow from a disadvantaged home background, home background remains one of the most powerful factors influencing performance. The average performance gap in mathematics between students in the top quarter of the PISA index of occupational status and those in the bottom quarter amounts to an average 93 score points, which is more than one-and-a-half proficiency levels in mathematical literacy. PISA 2012 confirms that in all countries, students from socioeconomically disadvantaged backgrounds show lower levels of mathematics achievement than their better-off peers. In particular, the performance difference between the advantaged (the top quarter of socio-economic status) and the disadvantaged (the bottom quarter of socio-economic status) students is 90 score points, which is equivalent of more than two years of schooling and more than one PISA proficiency level.

In order to investigate the influence of students' socioeconomic status on their mathematics achievement in both TIMSS and PISA, Adamson (2010) constructed the SES index for TIMSS which mirrored the PISA index, although only including the variables for home possessions and parental education, followed by assigning students to SES quintiles. Taking national economic conditions into account, Adamson found that when income per capita increases, students' mathematics achievement also increases for both low and high SES students. However, the achievement differences between the two groups of students remained large. In particular, the difference is about one SD on PISA 2003 and nearly three-quarters of a SD on TIMSS 2003. Furthermore, the study revealed that country-level income inequity interacted with SES in a way that partially negated the SES significance on PISA. Comparatively, a straightforward relationship between income inequality and achievement was shown on TIMSS, with increasing income inequality correlating significantly with lower mathematics achievement in all models. In addition, as income inequality increased, high SES students achieved even higher mathematics scores than their lower SES peers.

43.3.4 Immigrant Background Equity

While both TIMSS and PISA ask students about their birthplaces, as well as their parents', the two study series look into the issue in different ways. In particular, PISA identified three immigrant backgrounds including: native students (who have at least one parent born in the country of assessment), second-generation students (who were born in the country of assessment but whose parents were foreign-born), and first-generation students (who were foreign-born and whose parents were also foreign-born). TIMSS also classified three immigrant backgrounds but with parents' birthplaces being the main indicators: native students (both parents born in the country of assessment), half-and-half immigrant students (one parent born in the country of assessment), and immigrant students (neither parent born in the country of assessment).

Hastedt (2016) adopted PISA's immigrant definitions to analyze trends in the percentages of immigrant students in both fourth and eighth grades, as well as achievement differences between immigrant and native students using TIMSS data from the 1995 to 2007 cycles. The results showed that for both grade levels, the percentage of immigrant students increased incrementally over the years. In particular, the first-generation immigrant population increased between 1995 and 2007 in a large number of countries. Regarding students' mathematical literacy performance, the data overall revealed that immigrant students were outperformed by native students. For instance, in TIMSS 1995, native eighth grader students significantly outperformed first-generation immigrant students in 17 out of 37 countries (46%), as well as second-generation immigrant students in 10 countries (27%). The corresponding percentages of countries in TIMSS 1999, TIMSS 2003, and TIMSS 2007 were 32 and 16%, 76 and 31%, and 76 and 38%, respectively.

PISA 2012 reported that across OECD countries, 11% of the students had an immigrant background and they tended to be socioeconomically disadvantaged in comparison to their native peers (OECD 2015). In the mathematical literacy assessment, the students with immigrant background scored an average of 34 points lower than native students, and an average of 21 points lower after accounting for socioeconomic differences. In fact, immigrant students are 1.70 times more likely than native students to perform in the bottom quarter of the performance distribution. The achievement differences were even larger in PISA 2003, although the percentage of students across OECD countries who had an immigrant background was slightly lower (9%). In that year, immigrant students scored 47 points lower in mathematical literacy than their native peers, and 33 points lower when controlling for socioeconomic status. There is one promising finding from the comparison between PISA 2003 and PISA 2012: immigrant students' socioeconomic status profile was slightly more advantaged than that of immigrant students in 2003. This appears to suggest that, on average across OECD countries, immigrant students face less socioeconomic and performance disadvantage. On the other hand, the significant disadvantage in mathematical literacy performance was still evident among immigrant students in PISA 2012.

Andon et al. (2014) did a quantitative synthesis of the immigrant achievement gap across OECD countries using data from both TIMSS and PISA between 2000 and 2009. The study found a significant mean effect size for mathematics ($d = 0.38$). Moreover, the analysis revealed a larger gap in TIMSS than PISA. Andon et al. argued that this may be due in part to the type of content assessed, that is, TIMSS evaluated formal mathematics knowledge, and PISA items are more applied in nature and posed within real-world scenarios which require mathematics. They suggested that immigrant students fared better on items that tell a story, provide more context, and allow them to apply their experience and knowledge, as in the PISA. Consequently, immigrant students performed less poorly on PISA than on TIMSS relative to native students.

43.3.5 Concentration of Disadvantages Related to Inequity

It is suggested that underperformance of students with particular personal characteristics can be partly linked to the fact that these students tend to be concentrated in groups. For instance, immigrant students may settle in neighborhoods with other immigrants when they move to a new country. Similarly, students with low socioeconomic status may also more often group together. The potential result of such a concentration is that large differences in student performance are likely to exist at both the school and national levels.

Focusing on mathematics, two cross-national comparisons in students' achievement outcomes have been conducted on the TIMSS and PISA, respectively. With the analysis of the data from TIMSS 1995, TIMSS 1999, TIMSS 2003 and TIMSS 2007, Zopluoglu (2012) found that the proportions of differences that occurred at the student level were, in general, decreasing at both the fourth and eighth grade levels. For instance, there was about 58% of the difference in fourth graders' mathematics achievement related to students' individual differences in TIMSS 1995, and the percentage decreased to 25% in TIMSS 2007. Although the magnitude of the change was smaller, the shrinking pattern was also clearly observed at the eighth grade (TIMSS 1995: 50% vs. TIMSS 2007: 36%). Comparatively, the change in the proportions of school-level differences across the years was smaller at the fourth grade level (TIMSS 1995: 16% vs. TIMSS 2007: 21%) and nearly maintained at the same level at the eighth grade level (TIMSS 1995: 25% vs. TIMSS 2007: 25%). In fact, a greater change was revealed at the national level. The proportion of national level differences at the fourth grade in TIMSS 2007 (54%) is more than twice that in TIMSS 1995 (26%) and about 1.5 times that at the eighth grade (TIMSS 1995: 25% vs. TIMSS 2007: 39%).

Similar to Zopluoglu's (2012) findings, Uno's (2013) analysis with PISA mathematical literacy data also revealed that the differences of students' performances largely occurred at the individual student level. However, some inconsistences were also revealed in the PISA-based investigation. For instance, the proportion of student-level differences nearly remained at the similar level from PISA 2003 (56%) to PISA 2012 (54%). While Zopluoglu's TIMSS-based study found that the proportion of national level differences generally increased, Uno reported that the differences in students' mathematics performance in PISA occurred at the national level decreased between 2003 (16%) to 2012 (10%). Two differences in the research design between TIMSS and PISA may be related to these results: content focus and grade level. More investigations are needed for clarification. Furthermore, another common pattern can be found from Zopluoglu's study of TIMSS data and Uno's study of PISA data. Both found that the differences at the school level revealed a tendency to widen.

43.4 Final Remarks

It is clear that inequity in mathematics learning could be associated with individual students' personal characteristics (e.g., gender, family socioeconomic status, immigrant background) and their combinations. Inequity is also related to the countries, the schools, and the communities where the students are studying and living, which, to a certain extent, can be regarded as a concentration of effects. Although general large-scale international student assessments such as PISA and TIMSS may not be the best means to pinpoint the origin of the differences in students' mathematics learning outcomes, they may be the best means to observe changes in the differences in students' mathematics achievement (Forgasz 2010). Flores (2007) made a comment that efforts to document and eliminate the achievement gap are appropriate and necessary, as achievement gap is more of a symptom than a root issue.

On the other hand, while most equity-based mathematics education research appears to focus on "gap gazing", Lubienski and Gutiérrez (2008) have called for a broader focus to address equity issues such as identity and experience. Referring to Gutiérrez's (2009) four-dimension model of equity, equity-based research in mathematics education should move beyond the "dominant axis (i.e., access and achievement) and include the "critical axis" (i.e., identity and power) so as to allow students not only to *play* with mathematics but also to *change* mathematics.

References

Adamson, F. M. (2010). *How does context matter? Comparing achievement scores, opportunities to learn, and teacher preparation across socio-economic quintiles in TIMSS and PISA* (Unpublished doctoral dissertation). Stanford University, Stanford.

Andon, A., Thompson, C. G., & Becker, B. J. (2014). A quantitative synthesis of the immigrant achievement gap across OECD countries. *Large-scale Assessments in Education, 2*(7), 1–20.

Arnaud, A. J. (2001). Equity. In N. J. Smelser & P. B. Paultes (Eds.), *International encyclopedia of social and behavioral sciences* (Vol. 7, pp. 4729–4734). Amsterdam: Elsevier.

Arnesson, R. J. (2001). Equality: Philosophical aspects. In N. J. Smelser & P. B. Paultes (Eds.), *International encyclopedia of social and behavioral sciences* (Vol. 7, pp. 4724–4729). Amsterdam: Elsevier.

Baker, D. P., & Jones, D. P. (1993). Creating gender equality: Cross-national gender stratification and mathematical performance. *Sociology and Education, 66,* 91–103.

Baker, D. P., & Wiseman, A. W. (2005). The declining significance of gender and the rise of egalitarian mathematics education. In D. P. Baker & G. K. LeTendre (Eds.), *National differences, global similarities: World culture and the future of schooling* (pp. 16–33). Stanford, CA: Stanford University Press.

Baye, A., & Monseur, C. (2016). Gender differences in variability and extreme scores in an international context. *Large-scale Assessments in Education, 4*(1), 1–16.

Bishop, A. (2006). The transition experience of immigrant secondary school students: Dilemmas and decisions. In G. de Abreu, A. J. Bishop & N. C. Presmeg (Eds.), *Transitions between contexts of mathematics practices* (pp. 53–79). Dordrecht: Kluwer Academic Publishers.

Brown, K. (2006). "New" educational injustices in the "new" South Africa: A call for justice in the form of vertical equity. *Journal of Educational Administration, 44*(5), 509–519.

Castles, M. J., & Miller, S. (2003). *The age of migration* (3rd ed.). Basingstoke: Palgrave Macmillan.

Clancy, P., & Goastellec, G. (2007). Exploring access and equity in higher education: Policy and performance in a comparative perspective. *Higher Education Quarterly, 61*(2), 136–154.

Coessens, K., Francois, K., & Van Bendegem, J. P. (2014). Olympification versus aesthetization: The appeal of mathematics outside the classroom. In P. Smeyers & M. Depaepe (Eds.), *Educational research: Material culture and its representation* (Vol. 8, pp. 163–178). Switzerland: Springer International Publishing.

Coleman, J. S., Campbell, E. Q., Hobson, C. J., McPartland, J., Mood, A. M., Weinfeld, F. D., et al. (1966). *Equality of educational opportunity.* Washington, DC: Government Printing Office.

Damarin, S. (1995). Gender and mathematics from a feminist standpoint. In W. G. Secada, E. Fennema, & L. B. Adajian (Eds.), *New directions for equity in mathematics education* (pp. 242–257). New York: Cambridge University Press.

Ellison, G., & Swanson, A. (2010). The gender gap in secondary school mathematics at high achievement levels: Evidence from the American mathematics competition. *The Journal of Economic Perspectives, 24*(2), 109–128.

Fennema, E. (1974). Mathematics learning and the sexes: A review. *Journal for Research in Mathematics Education, 5,* 126–139.

Fennema, E. (2000, May). *Gender and mathematics: What is known and what do I wish was known?* Paper presented at the 5th annual forum of the National Institute for Science Education. Detroit, MI.

Fennema, E., & Sherman, J. (1977). Sex-related differences in mathematics achievement, spatial visualization and affective factors. *American Educational Research Journal, 14,* 51–71.

Field, S., Kuczera, M., & Pont, B. (2007). *No more failures: Ten steps to equity in education.* Paris: OECD.

Flores, A. (2007). Examining disparities in mathematics education: Achievement gap or opportunity gap? *The High School Journal, 91*(1), 1–15.

Forgasz, H. J., Becker, J. R., Lee, K., & Steinthorsdottir, O. (Eds.). (2010). *International perspectives on gender and mathematics education.* Charlotte, NC: Information Age Publishing.

Forgasz, H. J., & Hill, J. C. (2013). Factors implicated in high mathematics achievement. *International Journal of Science and Mathematics Education, 11*, 481–499.

Forgasz, H., & Leder, G. (2001). "A+ for girls, B for boys": Changing perspectives on gender equity in mathematics. In B. Atweh, H. Forgasz, & B. Nebres (Eds.), *Sociocultural research on mathematics education* (pp. 347–366). Mahwah: Erlbaum.

Forgasz, H., & Rivera, F. (Eds.). (2012). *Towards equity in mathematics education: Gender, culture and diversity.* Heidelberg: Springer.

Grouws, D. A. (Ed.). (1992). *Handbook of research on mathematics teaching and learning.* New York: Macmillan.

Guiso, L., Monte, F., Sapienza, P., & Zingales, L. (2008). Culture, gender, and math. *Science, 320,* 1164–1165.

Gutiérrez, R. (2007). Context matters: Equity, success, and the future of mathematics education. In T. S. Lamberg & L. R. Wiest (Eds.), *Proceedings of the 29th Annual Conference of the North American Chapter of the International Group for the Psychology of Mathematics Education* (pp. 1–18). Lake Tahoe, NV: University of Nevada.

Gutiérrez, R. (2009). Framing equity: Helping students "play the game" and "change the game". *Teaching for Excellence and Equity in Mathematics, 1*(1), 5–8.

Gutiérrez, R. (2010). The sociopolitical turn in mathematics education. *Journal for Research in Mathematics Special Equity Issue,* 1–32.

Gutiérrez, R., & Dixon-Román, E. (2011). Beyond gap gazing: How can thinking about education comprehensively help us (re)envision mathematics education? In B. Atweh, M. Graven, W. Secada, & P. Valero (Eds.), *Mapping equity and quality in mathematics education* (pp. 21–34). New York, NY: Springer.

Gutstein, E. (2000). Increasing equity: Challenges and lessons from a state systemic initiative. In W. G. Secada (Ed.), *Changing the faces of mathematics: Perspectives on multiculturalism and gender equity* (pp. 25–36). Reston, VA: NCTM Publishing.

Gutstein, E., Middleton, J. A., Fey, J. T., Larson, M., Heid, M. K., Dougherty, B., et al. (2005). Equity in school mathematics education: How can research contribute? *Journal for Research in Mathematics Education, 36*(2), 92–100.

Halpern, D. F., Wai, J., & Saw, A. (2005). A psychobiosocial model: Why females are sometimes greater than and sometimes less than males in math achievement. In A. M. Gallagher & J. C. Kaufman (Eds.), *Gender differences in mathematics: An integrative psychological approach* (pp. 48–72). Cambridge: Cambridge University Press.

Hanna, G. (2000). Declining gender differences from FIMS to TIMSS. *International Reviews on Mathematical Education, 32*(1), 11–17.

Hanna, G. (2003). Reaching gender equity in mathematics education. *The Educational Forum, 67,* 204–214.

Harlen, W. (2001). The assessment of scientific literacy in the OECD/PISA project. *Studies in Science Education, 36,* 79–104.

Harnisch, D. L., Steinkamp, M. W., Tsai, S., & Walberg, H. J. (1986). Cross-national differences in mathematics attitude and achievement among seventeen-year-olds. *International Journal of Educational Development, 6*(4), 233–244.

Hastedt, D. (2016). *Mathematics achievement of immigrant students.* Amsterdam, The Netherlands: Springer.

Herrera, L. M. (2007). Equity, equality and equivalence—A contribution in search for conceptual definitions and a comparative methodology. *Revista Espanola de Education Comparada, 13,* 319–340.

Husén, T. (Ed.). (1967). *International study of achievement in mathematics: A comparison of twelve countries* (Vols. I & II). New York: Wiley.

Hutmacher, W. (2001). Introduction. In W. Hutmacher, D. Cochrance, & N. Bottani (Eds.), *In pursuit of equity in education* (pp. 1–22). Dordrecht: Kluwer Academic Publishers.

Hyde, J. S., & Mertz, J. E. (2009). Gender, culture, and mathematics performance. *Proceedings of the National Academy of Sciences, 106,* 8801–8807.

Kaiser, G. (2003). Feminist frameworks for researching mathematics. In N. Pateman, B. Dougherty, & J. Zilliox (Eds.), *Proceedings of the 2003 Joint Meeting of PME and PMENA* (Vol. 1, pp. 157–160). Honolulu: University of Hawaii.

Kane, J. M., & Mertz, J. E. (2011). Debunking myths about gender and mathematics performance. *Notices of the American Mathematical Society, 59*(1), 10–21.

Kao, G., & Thompson, J. S. (2003). Racial and ethnic stratification in educational achievement and attainment. *Annual Review of Sociology, 29*, 417–442.

Keeves, J. (1973). Differences between the sexes in mathematics and science courses. *International Review of Education, 19*, 47–62.

Knapp, M. S., & Woolverton, S. (2004). Social class and schooling. In J. A. Banks & C. A. M. Banks (Eds.), *Handbook of research on multicultural education* (2nd ed., pp. 50–69). San Francisco, CA: Jossey-Bass.

Leder, G. C. (1992). Mathematics and gender: Changing perspectives. In D. A. Grouws (Ed.), *Handbook of research on mathematics teaching and learning* (pp. 597–622). New York: Macmillan.

Lerman, S. (2000). The social turn in mathematics education research. In J. Boaler (Ed.), *Multiple perspectives on mathematics teaching and learning* (pp. 19–44). Westport: Ablex Publishing.

Lerman, S. (2006). Cultural psychology, anthropology and sociology: The developing 'strong' social turn. In J. Maasz & W. Schloeglmann (Eds.), *New mathematics education research and practice* (pp. 171–188). Rotterdam: Sense.

Lester, F. K. (Ed.). (2007). *Second handbook of research on mathematics teaching and learning*. Greenwich, CT: NCTM/Information Age.

Lingard, B., Martino, W., Mills, M., & Bahr, M. (2002). *Addressing the educational needs of boys*. Canberra, Australia: Commonwealth Department of Education, Science and Training.

Lipman, P. (2004, April). *Regionalization of urban education: The political economy and racial politics of Chicago-metro region schools*. Paper presented at the annual meeting of the American Educational Research Association, San Diego, CA.

Lubienski, S. T., & Gutiérrez, R. (2008). Bridging the gaps in perspectives on equity in mathematics education. *Journal for Research in Mathematics Education, 39*(4), 365–371.

Maccoby, E. E., & Jacklin, C. N. (1974). Gender segregation in childhood. In H. W. Reese (Ed.), *Advances in child development and behavior* (Vol. 20, pp. 239–288). New York: Academic Press.

Morely, L., Leach, F., & Lugg, R. (2009). Democratising higher education in Ghana and Tanzania: Opportunity structures and social inequalities. *International Journal of Educational Development, 29*, 56–64.

National Council of Teachers of Mathematics (NCTM). (2000). *Principles and standards for school mathematics*. Reston, VA: Author.

Organization for Economic Cooperation and Development (OECD). (2005). *PISA 2003 technical report*. Paris: Author.

Organization for Economic Cooperation and Development (OECD). (2012). *Equity and quality in education: Supporting disadvantaged students and schools*. OECD Publishing. http://dx.doi.org/10.1787/9789264130852-en.

Organization for Economic Cooperation and Development (OECD). (2013a). *PISA 2012 results: Excellence through equity* (Vol. II). Paris: OECD Publishing.

Organization for Economic Cooperation and Development (OECD). (2013b). *PISA 2012 results: Students' engagement, drive and self-beliefs* (Vol. III). Paris: OECD Publishing.

Organization for Economic Cooperation and Development (OECD). (2015). *PISA in focus*. Paris: OECD Publishing.

Pais, A. (2012). A critical approach to equity. In O. Skovsmose & B. Greer (Eds.), *Opening the cage: Critique and politics of mathematics education* (pp. 49–92). Sense Publishers.

Rogers, P., & Kaiser, G. (Eds.). (1995). *Equity in mathematics education: Influences of feminism and culture*. London: The Falmer Press.

Rohn, D. (2013). Equity in education: The relationship between race, class, and gender in mathematics for diverse learners. *Urban Education Research and Policy Annuals, 1*(1), 12–22.

Rothman, S., & McMillan, J. (2003). *Influences on academic in literacy and numeracy*. Research Report 36. Camberwell, VIC: ACER.

Secada, W. G. (1989). Educational equity versus equality of education: An alternative conception. In W. G. Secada (Ed.), *Equity and education* (pp. 68–88). New York: Falmer.

Secada, W. G., Fennema, E., & Adajian, L. (Eds.). (1995). *New directions for equity in mathematics education*. Cambridge, USA: Cambridge University Press.

Sirin, S. R. (2005). Socioeconomic status and academic achievement: A meta-analysis. *Review of Educational Research, 75*(3), 417–453.

Spielman, L. J. (2008). Equity in mathematics education: Unions and intersections of feminist and social justice literature. *ZDM Mathematics Education, 40*, 647–657.

Steinkamp, M. W., Harnisch, D. L., Walberg, H. L., & Tsai, S. (1985). Cross-national gender differences in mathematics attitude and achievement among 13-year-olds. *The Journal of Mathematical Behavior, 4*, 259–277.

United Nations Development Programme. (2009). *Human development report 2009 overcoming barriers: Human mobility and development*. New York: Author.

United Nations Educational, Scientific and Cultural Organization (UNESCO). (2008). *Education for all global monitoring report 2009*. Oxford, UK: Oxford University Press.

Uno, M. (2013). *National institutional context and educational inequality: A multilevel analysis of variance in family SES effects on academic achievement across OECD countries* (Unpublished doctoral dissertation). University of Minnesota, Minneapolis.

Valero, P., Graven, M., Jurdak, M., Martin, D., Meaney, T., & Penteado, M. (2015). Socioeconomic influence on mathematical achievement: What is visible and what is neglected. In Cho S. J. (Ed.), *The Proceedings of the 12th International Congress on Mathematical Education: Intellectual and Attitudinal Challenges* (pp. 285–301). New York: Springer.

Valero, P., & Meaney, T. (2014). Trends in researching the socioeconomic influences on mathematics achievement. *ZDM Mathematics Education, 46*(7), 1–10.

Volmink, J. (1994). Mathematics by all. In S. Lerman (Ed.), *Cultural perspectives on the mathematics classroom* (pp. 51–67). Norwell, MA: Kluwer.

Weiner, G., Arnot, M., & David, M. (1997). Is the future female? Female success, male disadvantage, and changing gender patterns in education. In A. H. Halsey, P. Brown, H. Lauder, & A. S. Wells (Eds.), *Education, economy, culture, and society* (pp. 620–630). Oxford: Oxford University Press.

West, C., & Zimmeran, D. H. (1987). Doing gender. *Gender and Society, 1*, 125–151.

Wider, G. Z., & Powell, K. (1989). *Sex differences in test performance: A survey of the literature* (College Board Report No. 89–3). New York: College Entrance Examination Broad.

Willingham, W. W., & Cole, N. S. (1997). *Gender and fair assessment*. Hillsdale, NJ: Erlbaum.

Wood, E., Levinson, M., Postlethwaite, K., & Black, A. (2011). *Equity matters*. Brussels: EI. EN/FR.

Zopluoglu, C. (2012). A cross-national comparison of intra-class correlation coefficient in educational achievement outcomes. *Egitimde ve Psikolojide Olcme ve Degerlendirme Dergisi, 3*(1), 242–278.